HISTORY, PHILOSOPHY AND SOCIOLOGY OF SCIENCE

Classics, Staples and Precursors

HISTORY, PHILOSOPHY AND SOCIOLOGY OF SCIENCE

Classics, Staples and Precursors

Selected By

YEHUDA ELKANA
ROBERT K. MERTON
ARNOLD THACKRAY
HARRIET ZUCKERMAN

STUDIES
IN THE
HISTORY AND METHOD OF SCIENCE

EDITED BY

CHARLES SINGER

Volumes I and II

ARNO PRESS
A New York Times Company
New York – 1975

Reprint Edition 1975 by Arno Press Inc.

Reprinted from copies in
The Princeton University Library

Reprinted by permission of Nancy W. Underwood

HISTORY, PHILOSOPHY AND SOCIOLOGY OF SCIENCE:
Classics, Staples and Precursors
ISBN for complete set: 0-405-06575-2
See last pages of this volume for titles.

Publisher's Note : The plates in this book
have been reproduced in black and white
for this edition.

Manufactured in the United States of America

Library of Congress Cataloging in Publication Data

Singer, Charles Joseph, 1876-1960, ed.
Studies in the history and method of science.

(History, philosophy, and sociology of science)
Reprint of the 1917-1921 ed. published by the
Clarendon Press, Oxford.
1. Science--History. 2. Medicine--History. I. Title. II. Series.
Q125.S6 1975 509 74-26291
ISBN 0-405-06617-1

STUDIES

IN THE

HISTORY AND METHOD OF SCIENCE

WIESBADEN CODEX B. fo. 1r

Plate I. HILDEGARD RECEIVING THE LIGHT FROM HEAVEN

STUDIES
IN THE
HISTORY AND METHOD OF SCIENCE

EDITED BY

CHARLES SINGER

OXFORD
AT THE CLARENDON PRESS
1917

OXFORD UNIVERSITY PRESS
LONDON EDINBURGH GLASGOW NEW YORK
TORONTO MELBOURNE CAPE TOWN BOMBAY
HUMPHREY MILFORD
PUBLISHER TO THE UNIVERSITY

PRINTED IN ENGLAND
AT THE OXFORD UNIVERSITY PRESS

INTRODUCTION

The record of men and of movements, History teaches us the growth and development of ideas. Our civilization is the final expression of the two great master-thoughts of the race. Seeking an explanation of the pressing phenomena of life, man has peopled the world with spiritual beings to whom he has assigned benign or malign influences, to be invoked or propitiated. To the great 'uncharted region' (Gilbert Murray) with its mysteries, his religions offer a guide; and through 'a belief in spiritual beings' (Tylor's definition of religion) he has built an altar of righteousness in his heart. The birth of the other dominant idea, long delayed, is comparatively recent. 'The discovery of things as they really are' (Plato) by a study of nature was the great gift of the Greeks. Knowledge, *scientia*, knowledge of things we see, patiently acquired by searching out the secrets of nature, is the basis of our material civilization. The true and lawful goal of the sciences, seen dimly and so expressed by Bacon, is the acquisition of new powers by new discoveries—that goal has been reached. Niagara has been harnessed, and man's dominion has extended from earth and sea to the air. The progress of physics and of chemistry has revolutionized man's ways and works, while the new biology has changed his mental outlook.

The greater part of this progress has taken place within the memory of those living, and the mass of scientific work has accumulated at such a rate that specialism has become inevitable. While this has the obvious advantage resulting from a division of labour, there is the penalty of a narrowed horizon, and groups of men work side by side whose language is unintelligible to each other.

Here is where the historian comes in, with two definite objects, teaching the method by which the knowledge has been gained, the evolution of the subject, and correlating the innumerable subdivisions in a philosophy at once, in Plato's words, a science in itself as well as of other sciences. For example, the student of physics may know Crookes's tubes and their relation to Röntgen, but he cannot have a true conception of the atomic theory without a knowledge of Democritus; and the exponent of Madame Curie and of Sir J. J. Thomson will find his happiest illustrations from the writings of Lucretius. It is unfortunate that the progress of science makes useless the very works that made progress possible; and the student is too apt to think that because useless now they have never been of value.

The need of a comprehensive study of the methods of science is now widely recognized, and to recognize this need important Journals have been started, notably *Isis*, published by our Belgian colleague George Sarton, interrupted, temporarily we hope, by the war; and *Scientia*, an International Review of Scientific Synthesis published by our Italian Allies. The numerous good histories of science issued within the past few years bear witness to a real demand for a wider knowledge of the methods by which the present status has been reached. Among works from which the student may get a proper outlook on the whole question may be mentioned Dannemann's *Die Naturwissenschaften in ihrer Entwicklung und in ihrem Zusammenhange*, Bd. IV; *De la Méthode dans les Sciences*, edited by Félix Thomas (Paris: Alcan); Marvin's *Living Past*, 3rd ed. (Clarendon Press, 1917); and Libby's *Introduction to the History of Science* (Houghton Mifflin & Co., 1917).

This volume of Essays is the outcome of a quiet movement on the part of a few Oxford students to stimulate a study of the history of science. Shortly after his appointment to the Philip Walker Studentship, Dr. Charles Singer (of Magdalen College) obtained leave from Bodley's Librarian and the Curators to have a bay in the Radcliffe Camera set apart for research work in the history of science and a safe installed to hold manuscripts; and (with Mrs. Singer) offered £100 a year for five years to provide the necessary fittings, and special books not already in the Library. The works relating to the subject have been collected in the room, the objects of which are:

First, to place at the disposal of the general student a collection that will enable him to acquire a knowledge of the development of science and scientific conceptions.

Secondly, to assist the special student in research: (*a*) by placing him in relationship with investigations already undertaken; (*b*) by collecting information on the sources and accessibility of his material; and (*c*) by providing him with facilities to work up his material.

In spite of the absence of Dr. Singer on military duty for the greater part of the time, the work has been carried on with conspicuous success, to use the words of Bodley's Librarian. Ten special students have used the room. Professor Ramsay Wright has made a study of an interesting Persian medical manuscript. Professor William Libby, of Pittsburg, during the session of 1915–16, used the room in the preparation of his admirable *History of Science* just issued. Dr. E. T. Withington, the well-known medical historian, is making a special study of the old Greek writers for the new edition of Liddell and Scott's *Dictionary*. Miss Mildred Westland has helped Dr. Singer with the Italian medical manuscripts. Mr. Reuben Levy has worked at the Arabic medical manuscripts of Moses Maimonides. Mrs. Jenkinson is engaged on a study of early medicine and magic. Dr. J. L. E. Dreyer, the distinguished historian of Astronomy, has used the room in connexion with the preparation of the *Opera Omnia* of Tycho Brahe. Miss Joan Evans is engaged upon a research on mediaeval lapidaries. Mrs. Singer has begun a study of the English medical manuscripts, with a view to a complete catalogue. How important this is may be judged from the first instalment of her work dealing with the plague manuscripts in the British Museum. With rare enthusiasm and energy Dr. Singer has himself done a great deal of valuable work, and has proved an intellectual ferment working far beyond the confines of Oxford.

I have myself found the science history room of the greatest convenience, and it is most helpful to have easy access on the shelves to a large collection of works on the subject. Had the war not interfered, we had hoped to start a *Journal of the History and Method of Science* and to organize a summer school for special students—hopes we may perhaps see realized in happier days.

Meanwhile, this volume of essays (most of which were in course of preparation when war was declared) is issued as a *ballon d'essai.*

WILLIAM OSLER.

CONTENTS

LIST OF PLATES

ILLUSTRATIONS IN TEXT

SCIENTIFIC VIEWS AND VISIONS OF SAINT HILDEGARD

A STUDY IN EARLY RENAISSANCE ANATOMY

DR. JOHN WEYER AND THE WITCH MANIA

THE SCIENTIFIC VIEWS AND VISIONS OF SAINT HILDEGARD (1098-1180)

By Charles Singer

I. Introduction

In attempting to interpret the views of Hildegard on scientific subjects, certain special difficulties present themselves. First is the confusion arising from the writings to which her name has been erroneously attached. To obtain a true view of the scope of her work, it is necessary to discuss the authenticity of some of the material before us. A second difficulty is due to the receptivity of her mind, so that views and theories that she accepts in her earlier works become modified, altered, and developed in her later writings. A third difficulty, perhaps less real than the others, is the visionary and involved form in which her thoughts are cast.

But a fourth and more vital difficulty is the attitude that she adopts towards phenomena in general. To her mind there is no distinction between physical events, moral truths, and spiritual experiences. This view, which our children share with their mediaeval ancestors, was developed but not transformed by the virile power of her intellect. Her fusion of internal and external universe links Hildegard indeed to a whole series of mediaeval visionaries, culminating with Dante. In Hildegard, as in her fellow mystics, we find that ideas on Nature and Man, the Moral World and the Material Universe, the Spheres, the Winds, and the Humours, Birth and Death, and even on the Soul, the Resurrection of the Dead, and the Nature of God, are not only interdependent, but closely interwoven. Nowadays we are well accustomed to separate our ideas into categories, scientific, ethical, theological, philosophical, and so forth, and we even esteem it a virtue to retain and restrain our thoughts within limits that we deliberately set for them. To Hildegard such classification would have been impossible and probably incomprehensible. Nor do such terms as *parallelism* or *allegory* adequately cover her view of the relation of the material and spiritual. In her mind they are really interfused, or rather they have not yet been separated.

Therefore, although in the following pages an attempt is made to estimate her scientific views, yet the writer is conscious that such a method must needs interpret her thought in a partial manner. Hildegard, indeed, presents to us scientific thought as an undifferentiated factor, and an attempt is here made to separate it by the artificial but not unscientific process of dissection from the organic matrix in which it is embedded.

The extensive literature that has risen around the life and works of Hildegard has come from the hands of writers who have shown no interest in natural knowledge, while those who have occupied themselves with the history of science have, on their side, largely neglected the period to which Hildegard belongs, allured by the richer harvest of the full scholastic age which followed. This essay is an attempt to fill in a small part of the lacuna.

II. Life and Works

Hildegard of Bingen was born in 1098, of noble parentage, at Böckelheim, on the river Nahe, near Sponheim. Destined from an early age to a religious life, she passed nearly all her days within the walls of Benedictine houses. She was educated and commenced her career in the isolated convent of Disibodenberg, at the junction of the Nahe and the Glan, where she rose to be abbess. In 1147 she and some of her nuns migrated to a new convent on the Rupertsberg, a finely placed site, where the smoky railway junction of Bingerbrück now mars the landscape. Between the little settlement and the important mediaeval town of Bingen flowed the river Nahe, spanned by a bridge to which still clung the name of the pagan Drusus (see Fig. 1). At this spot, a place of ancient memories, secluded and yet linked to the world, our abbess passed the main portion of her life, and here she closed her eyes in the eighty-second year of her age on September 17, 1180.

Hildegard was a woman of extraordinarily active and independent mind. She was not only gifted with a thoroughly efficient intellect, but was possessed of great energy and considerable literary power, and her writings cover a wide range, betraying the most varied activities and remarkable imaginative faculty. The best known, and in a literary sense the most valuable of her works, are the books of visions. She was before all things an ecstatic, and both her *Scivias* (1141–50) and her *Liber divinorum operum simplicis hominis* (1163–70) contain passages of real power and beauty. Less valuable, perhaps, is her third long

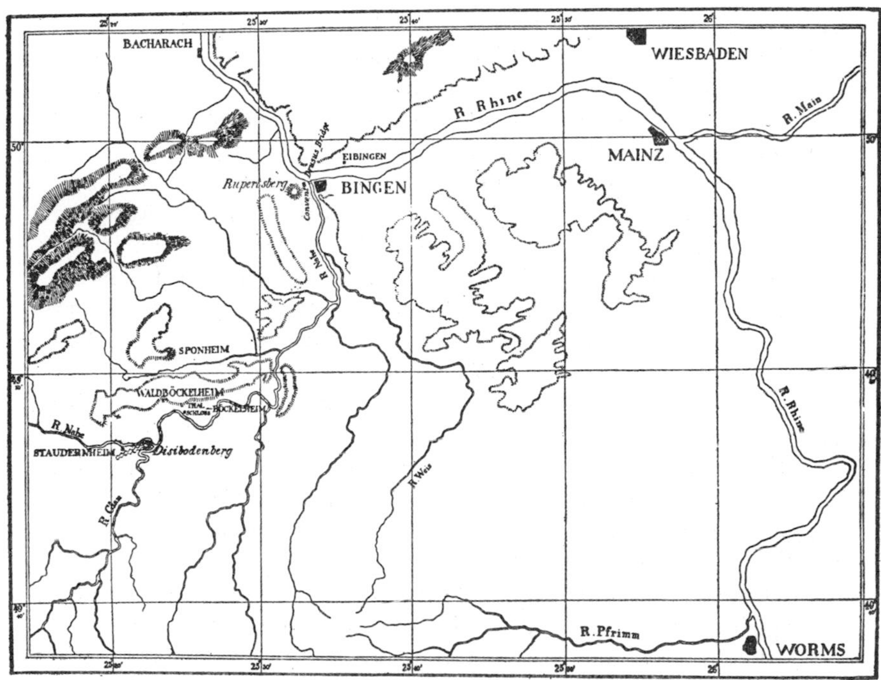

FIG. 1. THE HILDEGARD COUNTRY

mystical work (the second in point of time), the *Liber vitae meritorum* (1158–62). She is credited with the authorship of an interesting mystery-play and of a collection of musical compositions, while her life of St. Disibode, the Irish missionary (594–674) to whom her part of the Rhineland owes its Christianity, and her account of St. Rupert, a local saint commemorated in the name ‘ Rupertsberg ’, both bear witness alike to her narrative powers, her capacity for systematic arrangement, and her historical interests. Her extensive correspondence demonstrates the influence that she wielded in her own day and country, while her *Quaestionum solutiones triginta octo*, her *Explanatio regulae sancti Benedicti*, and her *Explanatio symboli sancti Athanasii ad congregationem sororum suorum* give us glimpses of her activities as head of a religious house.

Her biographer, the monk Theodoric, records that she also busied herself with the treatment of the sick, and credits her with miraculous powers of healing.[1] Some of the cited instances of this faculty, as the curing of a love-sick maid,[2] are, however, but manifestations of personal ascendancy over weaker minds; notwithstanding her undoubted acquaintance with the science of her day, and the claims made for her as a pioneer of the hospital system, there is no serious evidence that her treatment extended beyond exorcism and prayer.

For her time and circumstance Hildegard had seen a fair amount of the world. Living on the Rhine, the highway of Western Germany, she was well placed for observing the traffic and activities of men. She had journeyed at least as far north as Cologne, and had traversed the eastern tributary of the great river to Frankfort on the Main and to Rothenburg on Taube.[3] Her own country, the basin of the Nahe and the Glan, she knew intimately. She was, moreover, in constant communication with Mayence, the seat of the archbishopric in which Bingen was situated, and there has survived an extensive correspondence with the ecclesiastics of Cologne, Speyer, Hildesheim, Trèves, Bamberg, Prague, Nürnberg, Utrecht, and numerous other towns of Germany, the Low Countries, and Central Europe.

Hildegard's journeys, undertaken with the object of stimulating spiritual revival, were of the nature of religious progresses, but, like those of her contemporary, Bernard of Clairvaux, they were in fact largely directed against the heretical and most cruelly persecuted Cathari, an Albigensian sect widely spread in the Rhine country of the twelfth century, whom Hildegard regarded as 'worse than the Jews'.[4] In justice to her memory it is to be

[1] *Vita Sanctae Hildegardis auctoribus Godefrido et Theodorico monachis*, lib. iii, cap. 1. The work has been frequently reprinted and is in Migne, *Patrologia Latina*, vol. 197, col. 91 ff. This volume will be quoted here simply as 'Migne'.

[2] Migne, col. 119.

[3] The erroneous statement in some of her biographies that she journeyed to Paris is based on a misunderstanding.

[4] Cardinal J. B. Pitra, *Analecta sacra*, vol. viii, p. 350, Paris, 1882. This volume will here be quoted simply as 'Pitra'.

Folio 17 r col. b

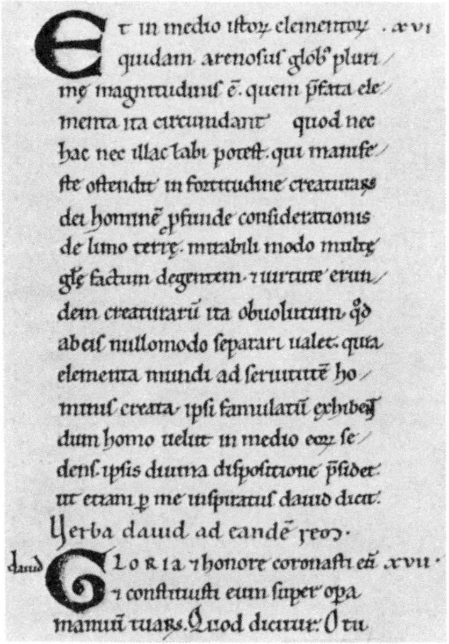

Folio 32 v col. b

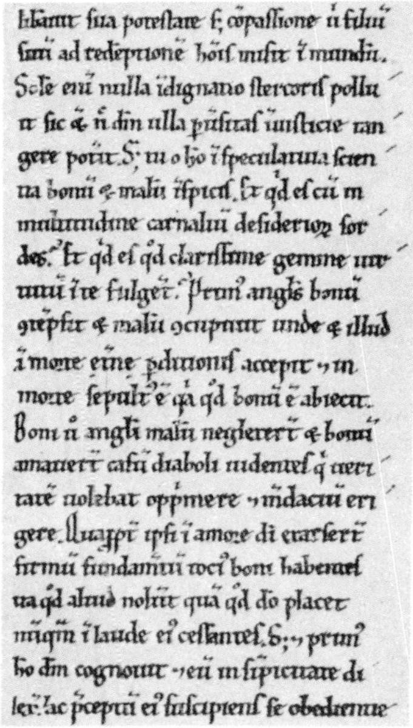

Folio 205 r col. b

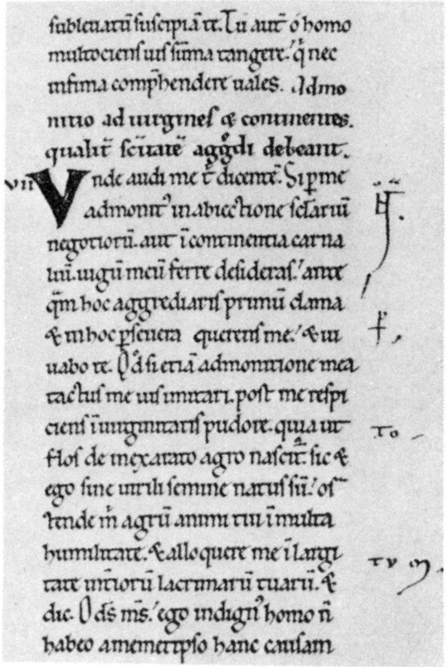

PLATE II. THE THREE SCRIPTS OF THE WIESBADEN CODEX B

Plate III. TITLE-PAGE OF THE HEIDELBERG CODEX OF THE *SCIVIAS*

recalled that she herself was ever against the shedding of blood, and had her less ferocious views prevailed, some more substantial relic than the groans and tears of this people had reached our time, while the annals of the Church had been spared the defilement of an inexpiable stain.

Hildegard's correspondence with St. Bernard, then preaching his crusade, with four popes, Eugenius III, Anastasius IV, Adrian IV, and Alexander III, and with the emperors Conrad and Frederic Barbarossa, brings her into the current of general European history, while she comes into some slight contact with the story of our own country by her hortatory letters to Henry II and to his consort Eleanor, the divorced wife of Louis VII.[1]

To complete a sketch of her literary activities, mention should perhaps be made of a secret script and language, the *lingua ignota*, attributed to her. It is a transparent and to modern eyes a foolishly empty device that hardly merits the dignity of the term 'mystical'. It has, however, exercised the ingenuity of several writers, and has been honoured by analysis at the hands of Wilhelm Grimm.[2]

Ample material exists for a full biography of Hildegard, and a number of accounts of her have appeared in the vulgar tongue. Nearly all are marred by a lack of critical judgement that makes their perusal a weary task, and indeed it would need considerable skill to interest a detached reader in the minutiae of monastic disputes that undoubtedly absorbed a considerable part of her activities. Perhaps the best life of her is the earliest; it is certainly neither the least critical nor the most credulous, and is by her contemporaries, the monks Godefrid and Theodoric.[3]

[1] Pitra, p. 556.

[2] Wilhelm Grimm, 'Wiesbader Glossen', in Moriz Haupt's *Zeitschrift für deutsches Alterthum*, Leipzig, 1848, vol. vi, p. 321. Thescript is reproduced in the ill-arranged and irritating work of J. P. Schmelzeis, *Das Leben und Wirken der heiligen Hildegardis*, Freiburg im Breisgau, 1879; and in Pitra, p. 497. The subject has been summarized by F. W. E. Roth in his *Lieder und unbekannte Sprache der h. Hildegardis*, Wiesbaden, 1880.

[3] A short sketch of her life of yet earlier date has survived. It is from the hand of the monk Guibert and was probably written in 1180: Pitra, p. 407. The best modern account of her is by F. W. E. Roth in the *Zeitschrift für kirchliche Wissenschaft und kirchliches Leben*, vol. ix, p. 453, Leipzig, 1888. Less critical but more readable is the essay by Albert Battandier, 'Sainte Hildegarde, sa vie et ses œuvres', in the *Revue des questions historiques*, vol. xxxiii, pp. 395–425, Paris, 1883.

The title of 'saint' is usually given to Hildegard, but she was not in fact canonized. Attempts towards that end were made under Gregory IX (1237), Innocent IV (1243), and John XXII (1317). Miraculous cures and other works of wonder were claimed for her, but either they were insufficiently miraculous or insufficiently attested.[1] Those who have impartially traced her life in her documents will agree with the verdict of the Church. Hers was a fiery, a prophetic, in many ways a singularly noble spirit, but she was not a saint in any intelligible sense of the word.

III. Bibliographical Note

There is no complete edition of the works of Hildegard. For the majority of readers the most convenient collection will doubtless be vol. 197 of Migne, *Patrologia Latina*. This can be supplemented from Cardinal J. B. Pitra's well-edited *Analecta sacra*, the eighth volume of which contains certain otherwise inaccessible works of Hildegard,[2] and is the only available edition of the *Liber vitae meritorum per simplicem hominem a vivente luce revelatorum*.

Manuscripts of the writings of our abbess are numerous and are widely scattered over Europe. Four of them are of special importance for our purpose, and are here briefly described.

(A) is a vast parchment of 480 folios in the Nassauische Landesbibliothek at Wiesbaden. This much-thumbed volume, still bearing the chain that once tethered it to some monastic desk, is written in a thirteenth-century script. There is evidence that it was prepared in the neighbourhood of Hildegard's convent, if not in that convent itself. It is interesting as a collection of those works that the immediate local tradition attributed to her, and is thus useful as a standard of genuineness.[3] Reference will be made to it in the following pages as the *Wiesbaden Codex A*. Its contents are as follows:

1. Liber Scivias.
2. Liber vitae meritorum.

[1] The 'Acta inquisitionis de virtutibus et miraculis sanctae Hildegardis' are reprinted in Migne, col. 131.

[2] This volume is supplemented by 'Annotationes ad Nova S. Hildegardis Opera' in *Analecta Bollandiana*, vol. i, p. 597, Brussels, 1882.

[3] This Wiesbaden MS. has been fully described by Antonius van der Linde, *Die Handschriften der Königlichen Landesbibliothek in Wiesbaden*, Wiesbaden, 1877.

3. Liber divinorum operum.
4. Ad praelatos moguntienses.
5. Vita sanctae Hildegardis. By Godefrid and Theodoric.
6. Liber epistolarum et orationum. This collection contains 292 items, and includes the Explanatio symboli Athanasii, the Exposition of the Rule of St. Benedict, and the Lives of St. Disibode and St. Rupert.
7. Expositiones evangeliorum.
8. Ignota lingua and Ignotae litterae.
9. Litterae villarenses.
10. Symphonia harmoniae celestum revelationum.

(B) is also at Wiesbaden, and will be cited here as the *Wiesbaden Codex B*. It contains the *Scivias* only, and is a truly noble volume of 235 folios, beautifully illuminated, in excellent preservation, and of the highest value for the history of mediaeval art. It has been thoroughly investigated by the late Dom Louis Baillet,[1] who concluded that it was written in or near Bingen between the dates 1160 and 1180. Its miniatures help greatly in the interpretation of the visions, illustrating them often in the minutest and most unexpected details. In view of the great difficulty of visualizing much of her narrative, these miniatures afford to our mind strong evidence that the MS. was supervised by the prophetess herself, or was at least prepared under her immediate tradition. This view is confirmed by comparing the miniatures with those of the somewhat similar but inferior Heidelberg MS. (C).

Both the miniatures and the script of the Wiesbaden Codex B are the work of several hands. There are three distinct handwritings discernible (Plate II). The earliest is attributed by Baillet in his careful work to the twelfth century, while the later writing is in thirteenth-century hands.[2] It thus appears to us that while Hildegard herself probably supervised the earlier stages of the preparation of this volume, its completion took place subsequent to her death. This view is sustained by the fact

[1] Louis Baillet, 'Les Miniatures du Scivias de sainte Hildegarde', in the *Monuments et Mémoires publiés par l'Académie des Inscriptions et Belles-Lettres*, Paris, 1912, especially pp. 139 and 145.

[2] We are inclined to place the preparation of this remarkable MS. at a slightly later date than that attributed to it by Baillet. As Wiesbaden is at present inaccessible we have reproduced the facsimiles in Plate II from Baillet's monograph.

that some of the later miniatures are far less successful than the earlier figures in aiding the interpretation of her text.

The two Wiesbaden MSS. appear to have remained at the convent on the Rupertsberg opposite Bingen until the seventeenth century. They were studied there by Trithemius in the fifteenth century, and one of them at least was seen by the Mayence Commission of 1489. Later they were noted by the theologians Osiander (1527) and Wicelius (Weitzel, 1554), and by the antiquary Nicolaus Serarius (1604). In 1632, during the Thirty Years' War, the Rupertsberg buildings were destroyed, the MSS. being removed to a place of safety in the neighbouring settlement at Eibingen, where they were again recorded in 1660 by the Jesuits Papenbroch and Henschen.[1] At some unknown date they were transferred to Wiesbaden, where they were examined in 1814 by Goethe,[2] and a few years later by Wilhelm Grimm,[3] and where they have since remained.

(C) This MS. is at the University Library at Heidelberg. It also contains only the *Scivias*, and it is the only known illuminated MS. of that work except the Wiesbaden Codex B. The Heidelberg MS. was prepared with great care in the early thirteenth century, only a little later than its fellow, but its figures afford little aid in the interpretation of the text. Thus, for instance, the Heidelberg diagram of the universe (Plate IV) is of a fairly conventional type which quite fails to illustrate the difficult description. The obscurities of the text are, however, at once explained by a figure in the Wiesbaden Codex B (Fig. 2): we thus obtain further indirect evidence of the personal influence of Hildegard in the preparation of that MS. The representation of Hildegard in the Heidelberg MS. (Plate III) shows no resemblance to those in the Wiesbaden Codex B (Plate I) or in the Lucca MS. (Plates VI to IX), which will now be described.

(D) is an illustrated codex of the *Liber divinorum operum simplicis hominis* at the Municipal Library at Lucca. It contains ten beautiful miniatures, some of which are here reproduced (Plates VI to IX and XI), as they are of special value for the

[1] For the history of these MSS. see A. van der Linde, loc. cit., pp. 30–6.

[2] Goethe, 'Am Rhein, Main und Neckar', *Cotta's Jubiläums-Ausgabe,* vol. xxix, p. 258.

[3] Wilhelm Grimm in M. Haupt's *Zeitschrift für deutsches Alterthum,* vi, p. 321, Leipzig, 1847.

interpretation of Hildegard's theories on the relation of macrocosm and microcosm.

This Lucca MS. was described and its text printed in 1761 by Giovanni Domenico Mansi,[1] a careful scholar, who was himself

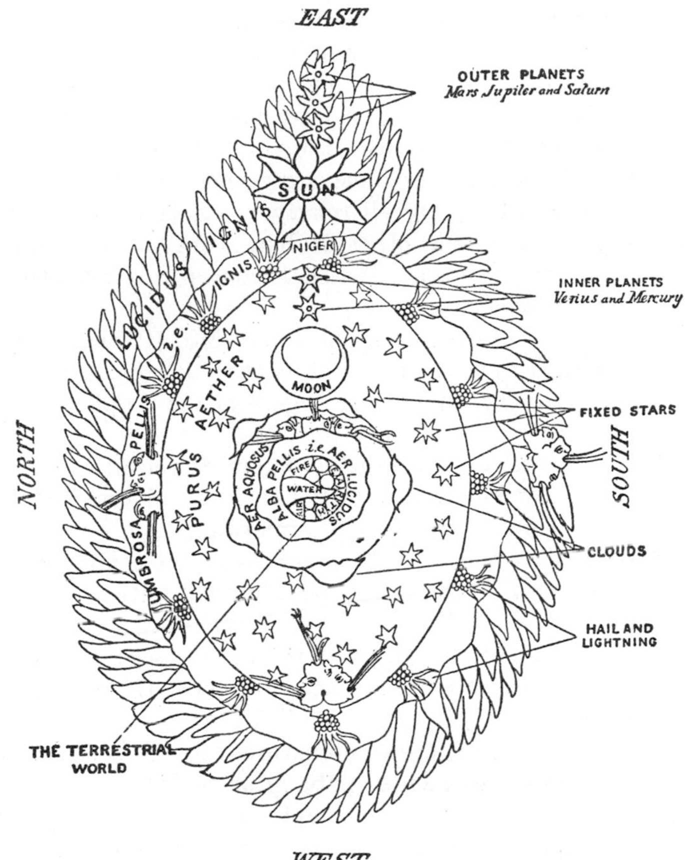

FIG. 2. HILDEGARD'S FIRST SCHEME OF THE UNIVERSE
Slightly simplified from the Wiesbaden Codex B, folio 14 r.

sometime Archbishop of Lucca. Mansi concluded that it was written at the end of the twelfth or the beginning of the

[1] In Étienne Baluze, *Miscellanea novo ordine digesta et non paucis ineditis monumentis opportunisque animadversionibus aucta opera ac studio J. D. Mansi*, 4 vols., Lucca, 1761–6; see vol. ii, p. 377.

thirteenth century. On palaeographical grounds a slightly later date would nowadays probably be preferred (Plate V *b*).

The work consists of ten visions, each illustrated by a figure. The date, character, and meaning of these miniatures raise special problems to which only very superficial reference can here be made. Unfortunately but little work has been done on early Italian schools of miniaturists, and it is not a subject on which any exact knowledge can yet be said to exist.[1]

Of these ten miniatures we may dismiss the last five in a few words. The sixth to the tenth visions are of purely theological interest, and the miniatures illustrating them are by a different hand to the rest. They are all relatively crude products, which appear to us to resemble other Italian work of the period at which the MS. was written. We shall concentrate our attention on the first five miniatures.

The first three miniatures of the Lucca MS. (Plates VI to VIII) may be attributed to the same hand on the following grounds:

1. All have a very similar inset figure of the prophetess below the main picture.

2. The character of the principal figure of the first miniature (Plate VI) is almost identical with the curious universe-embracing double-headed figure of the second miniature (Plate VII).

3. The features and draughtsmanship of the central figure of the second miniature (Plate VII) are identical with those of the third (Plate VIII).

4. The beasts' heads arranged round the second miniature (Plate VII) are exactly reproduced in the third miniature (Plate VIII).

Now although these three miniatures are in some respects unique, they contain elements enabling us to date them with an approach to accuracy. These elements are to be found especially in the central figure of the second and third miniatures (Plates VII and VIII).

About the middle of the thirteenth century, as Venturi has shown,[2] there was a well-marked change in Northern Italy in the traditional representation of the form on the Cross. This change was followed with almost slavish accuracy, and the new form is well represented by a painting in the Uffizi Gallery (Plate X).

[1] Cf. J. A. Herbert, *Illuminated Manuscripts*, London, 1911, p. 160.

[2] A. Venturi, *Storia dell' arte italiana*, Milan, *in progress*, vol. V, p. 16.

It is this figure of Christ which is reproduced by our miniaturist. The central figure of Plates VII and VIII resembles that of the Uffizi crucifix, for instance, in the general pose of the body, in the position of the legs and of the arms, in the treatment of the abdominal musculature, in the method of outlining the muscles of the legs and of the arms, and in a minute and very constant detail by which the outline of the left side is continued with the fold of the groin, thus giving an impression of the left thigh being advanced on the right. Furthermore, the somewhat Byzantine cast of countenance of the figure can be closely paralleled from Northern Italian work of the same period. We therefore regard these first three miniatures of the Lucca MS. as dating from about the middle of the thirteenth century.

The remaining two miniatures (Plates IX and XI) offer special difficulties. Plate XI (illustrating the fifth vision) presents us with no complete human figures, except the small and probably copied inset of the prophetess below the miniature. The faces bear some resemblance to those of the last five miniatures; the wings, on the other hand, to those of the first miniature (Plate VI). It is perhaps possible that this miniature was the work of an early thirteenth-century artist, and that the wings and some other details were added by a later hand. The abnormal orientation, east to the left and south above, suggests that we have here to do with some special influence.

The most anomalous of all is, however, the beautiful fourth miniature (Plate IX). This picture has a general feeling of the early Renaissance, though it is hard to find in it any definite humanistic element. The nude female figure in the upper left quadrant is especially striking. No parallel to it is to be found in the thirteenth-century Italian miniatures that have so far been reproduced, and it appears to us difficult to date the miniature anterior to the fourteenth century at the very earliest. It is, in any event, by a different hand to the others. The rashes on the patients in the two upper and the right lower quadrants are perhaps an attempt to render the fatal 'God's tokens' of those waves of pestilence that devastated the Italian peninsula in the fourteenth century.

Whatever the date of these miniatures, however, they reproduce the meaning of the text of the *Liber divinorum operum* with a convincing certainty and sureness of touch. This work is the most difficult of all Hildegard's mystical writings. Without the clues provided by the miniatures, many passages in it are wholly incomprehensible. It appears to us therefore by no means improbable that the traditional interpretation of Hildegard's works, thus preserved to our time by these miniatures and by them alone, may have had its origin from the mouth of the prophetess herself, perhaps through another set of miniatures that has disappeared or has not yet come to light.[1]

IV. The Spurious Scientific Works of Hildegard

The scientific views of Hildegard are embedded in a theological setting, and are mainly encountered in the *Scivias* and the *Liber divinorum operum simplicis hominis.* To a less extent they appear occasionally in her *Epistolae* and in the *Liber vitae meritorum.*

Two works of non-theological tone and definitely scientific character have been printed in her name. One of these was recently edited under the title *Beatae Hildegardis causae et curae.*[2] A single MS. only of this work is known to exist, and is now deposited in the Royal Library of Copenhagen.[3] It is an ill-written document of the thirteenth century, and the original work probably dates from this period. It has none of the characteristics of the acknowledged work of Hildegard, and indeed the only link with her name is the title, which is written in a hand different from that of the text (Plate v *a*). Nothing could be more unlike the ecstatic but well-ordered and systematic work of the prophetess of Bingen than the prosy disorder of the *Causae et curae.* Linguistically, also, it differs entirely from the typical writings of Hildegard, for it is full of Germanisms, which never interrupt the eloquence of her authentic works. Again, Hildegard's tendency to theoretical speculation, as for instance

[1] We are unable to concur with Baillet, however, that there is enough evidence to suggest that the miniaturists of the Lucca MS. had consulted the Wiesbaden illuminations. Baillet, loc. cit., p. 147.

[2] *Hildegardis causae et curae edidit Paulus Kaiser*, Leipzig, B. G. Teubner, 1903. The MS. was brought to light by C. Jessen in the *Sitzungsberichte der kaiserl. Akademie der Wissenschaften, Mathematisch-naturwissenschaftliche Klasse,* Band xlv, Heft 1, p. 97, Vienna, 1862. See also the same author in *Botanik in kulturhistorischer Entwickelung,* pp. 124–6, Leipzig, 1862, and in the *Anzeiger für Kunde der deutschen Vorzeit,* 1875, p. 175. An imperfect edition appeared in 1882 in Pitra, p. 468, under the title *Liber compositae medicinae de aegritudinum causis signis atque curis.*

[3] Royal Library of Copenhagen, MS. Ny. Kgl. Saml., No. 90 b.

From the HEIDELBERG CODEX OF THE *SCIVIAS*

PLATE IV. THE UNIVERSE

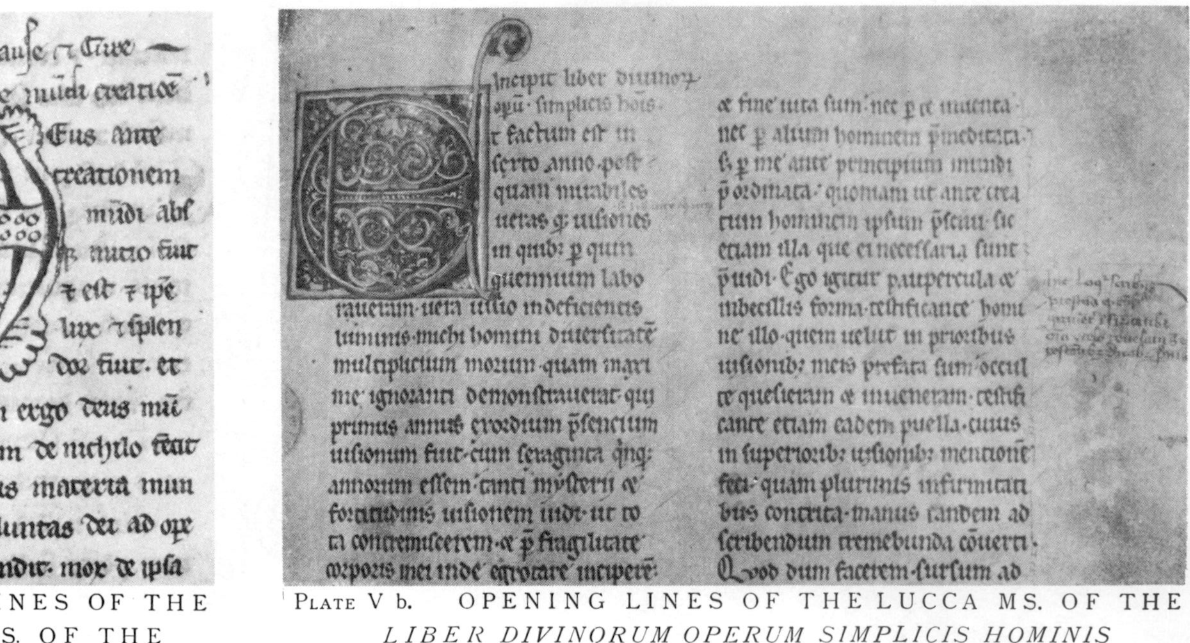

Plate V a. OPENING LINES OF THE COPENHAGEN MS. OF THE *CAUSAE ET CURAE*

Plate V b. OPENING LINES OF THE LUCCA MS. OF THE *LIBER DIVINORUM OPERUM SIMPLICIS HOMINIS* on the nature of the elements or on the form of the Universe, finds no place in the scrappy paragraphs of this apocryphal compilation.

A second work, of somewhat similar character, is entitled *Subtilitatum diversarumque creaturarum libri novem.* This is clearly a compilation, and numerous passages in it can be traced to such sources as Pliny, Walafrid Strabus, Marbod, Macer, the Physiologus, Isidore Hispalensis, Constantine the African, and the *Regimen Sanitatis Salerni,* only the last three of which exerted a traceable influence on the genuine works of our authoress. Nevertheless this *Liber subtilitatum* was early printed as Hildegard's work, along with a treatise attributed with as little justification to another woman writer, Trotula, one of the ladies of Salerno, whose name was also a household word in the Middle Ages, and was freely attached to medical writings with which she had little or nothing to do.[1] It is true that Hildegard's contemporary biographer, the monk Theodoric, assures us that she had written *De natura hominis et elementorum, diversarumque creaturarum,*[2] but there is nothing to suggest that the *Liber subtilitatum* is intended thereby.

The modern scholars Daremberg and Reuss have edited the *Liber subtilitatum* as Hildegard's composition,[3] and the work attracted the attention of Virchow,[4] but notwithstanding the

[1] *Experimentarius medicinae continens Trotulae curandarum Aegritudinum muliebrium . . . item quatuor Hildegardis de elementorum, fluminum aliquot Germaniae, metallorum, . . . herbarum, piscium & animantium terrae, naturis et operationibus.* Edited by G. Kraut, Strasbourg, J. Schott, 1544. The work often ascribed to Trotula is somewhat similar to the spurious medical works of Hildegard. Like them, it was probably written early in the thirteenth century. Trotula herself lived in the eleventh century, a generation or two before Hildegard. On Trotula see Salvatore de Renzi, *Collectio Salernitana,* vol. i, p. 149, Naples, 1852.

[2] In the *Vita,* lib. ii, cap. 1; Migne, col. 101.

[3] Migne, col. 1125. See also F. A. Reuss, *De Libris physicis S. Hildegardis commentatio historico-medica,* Würzburg, 1835, and 'Der heiligen Hildegard Subtilitatum diversarum naturarum creaturarum libri novem, die werthvollste Urkunde deutscher Natur- und Heilkunde aus dem Mittelalter' in the *Annalen des Vereins für Nassauische Alterthumskunde und Geschichtsforschung,* Band vi, Heft 1, Wiesbaden, 1859.

[4] Rudolf Virchow, 'Zur Geschichte des Aussatzes und der Spitäler, besonders in Deutschland', in Virchow's *Archiv für Pathologie,* vol. xviii, p. 285, &c., Berlin, 1860.

authority of these names, the objections which apply to the genuineness of the *Causae et curae* are also valid here :

(*a*) The *Liber subtilitatum* is not included in the Wiesbaden Codex A.

(*b*) The phrase *De natura hominis et elementorum diversarumque creaturarum*, used by Theodoric as a description and by Reuss as a title,[1] would lead one to expect great emphasis on the nature of the elements and their entry into the human frame. Such emphasis is not, in fact, discoverable in the *Liber subtilitatum*, which, moreover, does not treat of human anatomy or physiology.

(*c*) On the other hand, the genuine *Liber divinorum operum simplicis hominis* does lay stress on these points. This is possibly therefore the work to which Theodoric refers, and to it his description certainly applies well.

(*d*) As in the *Causae et curae*, there are linguistic difficulties that prevent us attributing the *Liber subtilitatum* to Hildegard. Such, for instance, is the number of Germanisms as well as the marked difference from the style and method of her acknowledged work.

(*e*) There are statements in the *Liber subtilitatum* that can scarcely be attributed to our authoress. Having largely explored the Rhine basin, and corresponding constantly with writers beyond the Alps, how could she possibly derive all rivers, Rhine and Danube, Meuse and Moselle, Nahe and Glan, from the same lake (of Constance) as does the author of the *Liber subtilitatum* ? [2]

(*f*) Furthermore, although that spurious work has a chapter *De elementis*, it reveals none of Hildegard's most peculiar and definite views as to their nature, origin, and fate,[3] nor does it refer to the sphericity of the earth, to the vascular system of man, to

[1] Reuss, in Migne, cols. 1121 and 1122, states on Theodoric's authority that Hildegard had written a *book* on this subject : 'Exstat inter libros virginis fatidicae superstites opus argumenti partim physici partim medici, "De natura hominis, elementorum diversarumque creaturarum" in quo, ut Theodoricus idem fusius exponit, secreta naturae prophetico spiritu manifestavit.' But Theodoric does not in fact anywhere speak of a *special work* with this title or of this character. What he does write is as follows (*Vita*, lib. ii, cap. i, Migne, col. 101) : 'Igitur beata virgo . . . librum visionum . . . consummavit et *quaedam* de natura hominis et elementorum, diversarumque creaturarum, et quomodo homini ex his succurrendum sit, aliaque multa secreta prophetico spiritu manifestavit.'

[2] Migne, cols. 1212 and 1213.

[3] As detailed in the *Liber vitae meritorum*, Pitra, p. 228, and in many places in the *Liber divinorum operum* and *Scivias*.

the humours and their relation to the winds and the elements, or to a dozen other points on which, as we shall see, Hildegard had views of her own.

Before leaving the subject of Hildegard's apocryphal works, brief reference may be made to the *Speculum futurorum temporum*, a spurious production to which her name is often attached. It exists in innumerable MSS., and has been frequently edited and translated. It is the work of Gebeno, prior of Eberbach, who wrote it in 1220, claiming that he extracted it from Hildegard's writings. Another work erroneously attributed to Hildegard is entitled *Revelatio de fratribus quatuor mendicantium ordinum*, and is directed against the four mendicant orders—Franciscans, Dominicans, Carmelites, and Augustinians. It also has been printed, but is wholly spurious, and was probably composed towards the latter part of the thirteenth century.

V. Sources of Hildegard's Scientific Knowledge

In the works of Hildegard we are dealing with the products of a peculiarly original intellect, and her imaginative power and mystical tendency make an exhaustive search into the origin of her ideas by no means an easy task. With her theological standpoint, as such, we are not here concerned, and unfortunately she does not herself refer to any of her sources other than the Biblical books; to have cited profane writers would indeed have involved the abandonment of her claim that her knowledge was derived by immediate inspiration from on high. Nevertheless it is possible to form some idea, on internal evidence, of the origin of many of her scientific conceptions.

The most striking point concerning the sources of Hildegard is negative. There is no German linguistic element distinguishable in her writings, and they show little or no trace of native German folk-lore.[1] It is true that Trithemius of Sponheim (1462–1516), who is often a very inaccurate chronicler, tells us that Hildegard 'composed works in German as well as in Latin, although she had neither learned nor used the latter tongue except for

[1] An exception must be made for the *lingua ignota*, which is presumably hers. The absence of Germanisms in her other writings may be partly due to the work of an editor. See the *Vita* by Theodoric, Migne, col. 101. Also the birth scene (see chapter ix below) is perhaps adapted from a German folk-tale.

simple psalmody '.[1] But with the testimony before us of the writings themselves and of her skilful use of Latin, the statement of Trithemius and even the hints of Hildegard[2] may be safely discounted and set down to the wish to magnify the element of inspiration.[3] So far from her having been illiterate, we shall show that the structure and details of her works betray a considerable degree of learning and much painstaking study of the works of others. Thus, for instance, she skilfully manipulates the Hippocratic doctrines of miasma and the humours, and elaborates a theory of the interrelation of the two which, though developed on a plan of her own, is yet clearly borrowed in its broad outline from such a writer as Isidore of Seville. Again, as we shall see, some of her ideas on anatomy seem to have been derived from Constantine the African, who belonged to the Benedictine monastery of Monte Cassino.[4]

[1] Johannes Trithemius, *Chronicon insigne Monasterii Hirsaugensis, Ordinis St. Benedicti*, Basel, 1559, p. 174.

[2] Migne, col. 384.

[3] It is not enough to suppose with some of her biographers that the visions were dictated by Hildegard and were latinized by a secretary. The visions imply a good deal of study and considerable book-learning. Among many reasons for believing that she had a very serviceable knowledge of Latin are the following :

(*a*) She was well acquainted with the Biblical writings and quotes them aptly and frequently.

(*b*) She was regarded by her contemporaries as an authority on scriptural interpretation and on Church discipline, and was frequently consulted by them on these subjects.

(*c*) She pleaded in person before clerical tribunals.

(*d*) One of the least remarkable and most credible of her ' miracles ', the expounding of certain letters found upon an altar-cloth (Migne, col. 121), depends entirely on a knowledge of Latin.

(*e*) In the *Liber divinorum operum* (Migne, col. 922) she writes 'firmamentum *celum* nominavit quoniam omnia *excellit* ', a derivation taken from Isidore and incomprehensible to one ignorant of Latin. There are many other passages in her works in which the sense depends on the Latin usage of a word.

(*f*) No mention of this ignorance is made by Guibert in the short sketch of her life that he wrote almost immediately after her death (1180 ; see Pitra, p. 407). On the contrary, he suggests that she had been an industrious student.

(*g*) The *Liber divinorum operum* may especially be pointed out among her works as betraying a very considerable degree of learning. Notably her elaborate doctrine of the macrocosm and microcosm must have involved extensive reading.

The general question of Hildegard's knowledge of Latin has also been discussed by Pitra and by Albert Battandier in the *Revue des questions historiques*, vol. xxxiii, p. 395, Paris, 1883.

[4] See chapter viii.

Hildegard lived at rather too early a date to drink from the broad stream of new knowledge that was soon to flow into Europe through Paris from its reservoir in Moslem Spain. Such drops from that source as may have reached her must have trickled in either from the earlier Italian translators or from the Jews who had settled in the Upper Rhineland, for it is very unlikely that she was influenced by the earlier twelfth-century translations of Averroes, Avicenna, Avicebron, and Avempace, that passed into France from the Jews of Marseilles, Montpellier, and Andalusia.[1] Her intellectual field was thus far more patristic than would have been the case had her life-course been even a quarter of a century later.

Her science is primarily of the usual degenerate Greek type, disintegrated fragments of Aristotle and Galen coloured and altered by the customary mediaeval attempts to bring theory into line with scriptural phraseology, though a high degree of independence is obtained by the visionary form in which her views are set. She exhibits, like all mediaeval writers on science, the Aristotelian theory of the elements, but her statement of the doctrine is illuminated by flashes of her own thoughts and is coloured by suggestions from St. Augustine, Isidore Hispalensis, Bernard Sylvestris of Tours, and perhaps from writings attributed to Boethius.

The translator Gerard of Cremona (1114–87) was her contemporary, and his labours made available for western readers a number of scientific works which had previously circulated only among Arabic-speaking peoples.[2] Several of these works, notably Ptolemy's *Almagest*, Messahalah's *De Orbe*, and the Aristotelian *De Caelo et Mundo*, contain material on the form of the universe and on the nature of the elements, and some of them probably reached the Rhineland in time to be used by Hildegard. The

[1] It is, however, just possible that she had consulted the astrological work that had been translated from the Arabic by Hermann the Dalmatian for Bernard Sylvestris, and is represented in the Bodleian MSS. Digby 46 and Ashmole 304.

[2] See Baldassare Boncompagni, *Della vita e delle opere di Gherardo Cremonese, Traduttore del secolo duodecimo, e di Gherardo di Sabbionetta, Astronomo del secolo decimoterzo*, Rome, 1851 ; also K. Sudhoff, 'Die kurze "Vita" und das Verzeichnis der Arbeiten Gerhards von Cremona, von seinen Schülern und Studiengenossen kurz nach dem Tode des Meisters (1187) zu Toledo verabfasst', in *Archiv für Geschichte der Medizin*, Bd. viii, p. 73, November 1914.

Almagest, however, was not translated until 1175, and was thus inaccessible to Hildegard.[1] Moreover, as she never uses an Arabic medical term, it is reasonably certain that she did not consult Gerard's translation of Avicenna, which is crowded with Arabisms.

On the other hand, the influence of the Salernitan school may be discerned in several of her scientific ideas. The *Regimen Sanitatis* of Salerno, written about 1101, was rapidly diffused throughout Europe, and must have reached the Rhineland at least a generation before the *Liber Divinorum Operum* was composed. This cycle of verses may well have reinforced some of her microcosmic ideas,[2] and suggested also her views on the generation of man,[3] on the effects of wind on health,[4] and on the influence of the stars.[5]

On the subject of the form of the earth Hildegard expressed herself definitely as a spherist,[6] a point of view more widely accepted in the earlier Middle Ages than is perhaps generally supposed. She considers in the usual mediaeval fashion that this globe is surrounded by celestial spheres that influence terrestrial events.[7] But while she claims that human affairs, and especially human diseases, are controlled, under God, by the heavenly cosmos, she yet commits herself to none of that more detailed astrological doctrine that was developing in her time, and came to efflorescence in the following centuries. In this respect she follows the earlier and somewhat more scientific spirit of such writers as Messahalah, rather than the wilder theories of her own age. The shortness and simplicity of Messahalah's tract on the sphere made it very popular. It was probably one of the earliest to be translated into Latin ; and its contents would account for

[1] Another translation of the *Almagest* was made in Sicily in 1160, direct from the Greek. See C. H. Haskins and D. P. Lockwood, 'The Sicilian Translators of the Twelfth Century and the First Latin Version of Ptolemy's *Almagest*', in *Harvard Studies in Classical Philology*, xi. 75, Cambridge, Mass., 1910. It is wholly improbable that Hildegard had access to this rendering, which is only known from a single MS. of the fourteenth century.

[2] De Renzi, *Collectio Salernitana*, vol. i, p. 485, and vol. v, p. 50.

[3] De Renzi, i. 486 and 495 ; v. 51 and 70.

[4] De Renzi, i. 446 ; v. 3.

[5] De Renzi, i. 485–6 ; v. 50–2.

[6] *Scivias*, Migne, col. 403, and *Liber Divinorum Operum*, Migne, col. 868 and elsewhere.

[7] *Scivias*, Migne, col. 404, and throughout the *Liber Divinorum Operum*.

the change which, as we shall see, came over Hildegard's scientific views in her later years.

The general conception of the universe as a series of concentric elemental spheres had certainly penetrated to Western Europe centuries before Hildegard's time. Nevertheless the prophetess presents it to her audience as a new and striking revelation. We may thus suppose that translations of Messahalah, or of whatever other work she drew upon for the purpose, did not reach the Upper Rhineland, or rather did not become accepted by the circles in which Hildegard moved, until about the decade 1141–50, during which she was occupied in the composition of her *Scivias*.

There is another cosmic theory, the advent of which to her country, or at least to her circle, can be approximately dated from her work. Hildegard exhibits in a pronounced but peculiar and original form the doctrine of the macrocosm and microcosm. Hardly distinguishable in the *Scivias* (1141–50), it appears definitely in the *Liber Vitae Meritorum* (1158–62),[1] in which work, however, it takes no very prominent place, and is largely overlaid and concealed by other lines of thought. But in the *Liber Divinorum Operum* (1163–70) this belief is the main theme. The book is indeed an elaborate attempt to demonstrate a similarity and relationship between the nature of the Godhead, the constitution of the universe, and the structure of man, and it thus forms a valuable compendium of the science of the day viewed from the standpoint of this theory.

From whence did she derive the theory of macrocosm and microcosm? In outline its elements were easily accessible to her in Isidore's *De Rerum Natura* as well as in the Salernitan poems. But the work of Bernard Sylvestris of Tours, *De mundi universitate sive megacosmus et microcosmus*,[2] corresponds so closely both in form, in spirit, and sometimes even in phraseology, to the *Liber Divinorum Operum* that it appears to us certain that Hildegard must have had access to it also. Bernard's work can be dated

[1] Pitra, pp. 8, 114–16, 156, and 216.

[2] The work of Bernard Sylvestris has been printed by C. S. Barach and J. Wrobel, Innsbruck, 1876. His identity, his sources, and his views are discussed by Charles Jourdain, *Dissertation sur l'état de la philosophie naturelle . . . pendant la première moitié du XIIe siècle*; by A. Clerval, *Les Écoles de Chartres au Moyen Âge*, Paris, 1895, p. 259, &c.; by R. L. Poole, *Illustrations of the History of Mediaeval Thought*, London, 1884, p. 116, &c.; and by J. E. Sandys, *History of Classical Scholarship*, Cambridge, 1903, vol. i, p. 513, &c.

between the years 1145–53 from his reference to the papacy of Eugenius III. This would correspond well with the appearance of his doctrines in the *Liber Vitae Meritorum* (1158–62) and their full development in the *Liber Divinorum Operum* (1163–70).

Another contemporary writer with whom Hildegard presents points of contact is Hugh of St. Victor (1095–1141).[1] In his writings the doctrine of the relation of macrocosm and microcosm is more veiled than with Bernard Sylvestris. Nevertheless, his symbolic universe is on the lines of Hildegard's belief, and the plan of his *De arca Noe mystica* presents many parallels both to the *Scivias* and to the *Liber Divinorum Operum.* If these do not owe anything directly to Hugh, they are at least products of the same mystical movement as were his works.

We may also recall that at Hildegard's date very complex cabalistic systems involving the doctrine of macrocosm and microcosm were being elaborated by the Jews, and that she lived in a district where Rabbinic mysticism specially flourished.[2] Benjamin of Tudela, who visited Bingen during Hildegard's lifetime, tells us that he found there a congregation of his people. Since we know, moreover, that she was familiar with the Jews,[3] it is possible that she may have derived some of the very complex macrocosmic conceptions with which her last work is crowded from local Jewish students.

The Alsatian Herrade de Landsberg (died 1195), a contemporary of Hildegard, developed the microcosm theory along lines similar to those of our abbess, and it is probable that the theory, in the form in which these writers present it, reached the Upper Rhineland somewhere about the middle or latter half of the twelfth century.

Apart from the Biblical books, the work which made the deepest impression on Hildegard was probably Augustine's *De Civitate Dei,* which seems to form the background of a large part of the *Scivias.* The books of Ezekiel and of Daniel, the Gospel of Nicodemus, the Shepherd of Hermas, and the Apocalypse, all contain a lurid type of vision which her own spiritual experiences

[1] The works of Hugh of St. Victor are published in Migne, *Patrologia Latina,* clxxv–clxxvii.

[2] The Kalonymos family furnished prominent examples.

[3] Charles Singer, 'Allegorical Representation of the Synagogue, in a Twelfth-Century Illuminated MS. of Hildegard of Bingen', *Jewish Quarterly Review,* new series, vol. v, p. 268, Philadelphia, 1915. For further evidence of Hildegard's acquaintance with the Jews see Pitra, p. 216 ; and Migne, cols. 967 and 1020–36.

From the LUCCA MS. fo. 1v

PLATE VI. NOUS PERVADED BY THE GODHEAD AND CONTROLLING HYLE

From the LUCCA MS. fo. 9 r

PLATE VII. NOUS PERVADED BY THE GODHEAD EMBRACING THE MACROCOSM WITH THE MICROCOSM would enable her to utilize, and which fit in well with her microcosmic doctrines. Ideas on the harmony and disharmony of the elements she may have picked up from such works as the Wisdom of Solomon and the Pauline writings, though it is obvious that Isidore of Seville and the *Regimen Sanitatis Salerni* were also drawn upon by her.

Her figure of the Church in the *Scivias* reminds us irresistibly of Boethius' vision of the gracious feminine form of Philosophy. Again, the visions of the punishments of Hell which Hildegard recounts in the *Liber Vitae Meritorum* [1] bear resemblance to the work of her contemporary Benedictine, the monk Alberic the younger of Monte Cassino, to whom Dante also became indebted.[2]

Hildegard repeatedly assures us that most of her knowledge was revealed to her in waking visions. Some of these we shall seek to show had a pathological basis, probably of a migrainous character, and she was a sufferer from a condition that would nowadays probably be classified as hystero-epilepsy. Too much stress, however, can easily be laid on the ecstatic presentment of her scientific views. Visions, it must be remembered, were 'the fashion' at the period, and were a common literary device. Her contemporary Benedictine sister, Elizabeth of Schönau, as well as numerous successors, as for example Gertrude of Robersdorf, adopted the same mechanism. The use of the vision for this purpose remained popular for centuries, and we may say of these writers, as Ampère says of Dante, that 'the visions gave not the genius nor the poetic inspiration, but the form merely in which they were realized'.

The contemporaries of Hildegard who provide the closest analogy to her are Elizabeth of Schönau (died 1165), whose visions are recounted in her life by Eckbertus; [3] and Herrade de Landsberg, Abbess of Hohenburg in Alsace, the priceless MS. of whose *Hortus Deliciarum* was destroyed by the Germans in the siege of Strasbourg in 1870.[4] With Elizabeth of Schönau,

[1] Pitra, p. 51 et seq.

[2] Catello de Vivo, *La Visione di Alberico, ristampata, tradotta e comparata con la Divina Commedia*, Ariano, 1899. For a comparison of Dante's visions and those of Hildegard see Albert Battandier in the *Revue des questions historiques*, vol. xxxiii, p. 422, Paris, 1883.

[3] Reprinted in Migne, vol. 195.

[4] Herrade de Landsberg, *Hortus Deliciarum*, by A. Straub and G. Keller, Strasbourg, 1901, with two supplements.

who lived in her neighbourhood, Hildegard was in frequent correspondence. With Herrade she had, so far as is known, no direct communication; but the two were contemporary, lived not very far apart, and under similar political and cultural conditions. Elizabeth's visions present some striking analogies to those of Hildegard, while the figures of Herrade, of which copies have fortunately survived, often suggest the illustrations of the Wiesbaden or of the Lucca MSS.

VI. The Structure of the Material Universe

To the student of the history of science, Hildegard's beliefs as to the nature and structure of the universe are among the most interesting that she has to impart. Her earlier theories are in some respects unique among mediaeval writers, and we possess in the Wiesbaden Codex B a diagram enabling us to interpret her views with a definiteness and certainty that would otherwise be impossible.

Hildegard's universe is geocentric, and consists of a spherical earth,[1] around which are arranged a number of concentric shells or zones. The inner zones are spherical, the outer oval, and the outermost of all egg-shaped, with one end prolonged and more pointed than the other (Fig. 2). The concentric structure is a commonplace of mediaeval science, and is encountered, for instance, in the works of Bede, Isidore, Alexander of Neckam, Roger Bacon, Albertus Magnus, and Dante. To all these writers, however, the universe is spherical. The egg-shape is peculiar to Hildegard. Many of the *Mappaemundi* of the Beatus and other types exhibit the *surface* of the habitable earth itself as oval, and it was from such charts that Hildegard probably gained her conception of an oval universe. In her method of orientation also she follows these maps, placing the east at the top of the page where we are accustomed to place the north.[2]

It is unfortunate that she does not deal with geography in the restricted sense, and so we are not in full possession of her views on the antipodes, a subject of frequent derision to patristic and of misconception to scholastic writers. She does, however, vaguely

[1] For sphericity of earth see especially Migne, cols. 868 and 903.

[2] In her later *Liber Divinorum Simplicis Hominis* this method of orientation is varied both in the text and also in the Lucca illustrations.

refer to the inversion of seasons and climates in the opposite hemisphere,[1] though she confuses the issue by the adoption of a theory widespread in the Middle Ages and reproduced in the *Divina Commedia,* that the antipodean surface of the earth is uninhabitable, since it is either beneath the ocean or in the mouth of the Dragon[2] (Plate XI, cp. Fig. 4). The nature of the antipodean inversion of climates was clearly grasped by her contemporary, Herrade de Landsberg (Fig. 5).

Hildegard's views as to the internal structure of the terrestrial sphere are also somewhat difficult to follow. Her obscure and confused doctrine of Purgatory and Hell has puzzled other writers besides ourselves,[3] nor need we consider it here, but she held that the interior of the earth contained two vast spaces shaped like truncated cones, where punishment was meted out and whence many evil things had issue.[4] Her whole scheme presents analogies as well as contrasts to that of her kindred spirit Dante.[5] Hildegard, however, who died before the thirteenth century had dawned, presents us with a scheme far less definite and elaborated than that of her great successor, who had all the stores of the golden age of scholasticism on which to draw.

In Hildegard's first diagram of the universe, which is of the nature of an 'optical section', the world, the *sphaera elementorum* of Johannes Sacro Bosco and other mediaeval writers, is diagrammatically represented as compounded of earth, air, fire, and water confusedly mixed in what her younger contemporary, Alexander of Neckam (1157–1217), calls 'a certain concordant discord of the elements'. In the illustrations to the Wiesbaden Codex B the four elements have each a conventional method of representation, which appears again and again in the different miniatures (Fig. 2 and Plates XII and XIII).

Around this world with its four elements is spread the atmosphere, the *aer lucidus* or *alba pellis,* diagrammatically represented, like the earth which it enwraps, as circular. Through this *alba pellis* no creature of earth can penetrate. Beyond are

[1] Migne, col. 906.
[2] Migne, cols. 903–4.
[3] See H. Osborn Taylor, *The Mediaeval Mind,* vol. i, p. 472, London, 1911.
[4] Migne, cols. 904–6.
[5] H. Osborn Taylor, *The Mediaeval Mind,* i. 468, 471 ; ii. 569. See also A. Battandier, *Revue des questions historiques,* vol. xxxiii, p. 422, Paris, 1883.

ranged in order four further shells or zones. Each zone contains one of the cardinal winds, and each cardinal wind is accompanied by two accessory winds, represented in the traditional fashion by the breath of supernatural beings.

Of the four outer zones the first is the *aer aquosus,* also round, from which blows the east wind. In the outer part of the *aer aquosus* float the clouds, and according as they contract or expand or are blown aside, the heavenly bodies above are revealed or concealed.

Enwrapping the *aer aquosus* is the *purus aether,* the widest of all the zones. The long axis of this, as of the remaining outer shells, is in the direction from east to west, thus determining the path of movement of the heavenly bodies. Scattered through the *purus aether* are the constellations of the fixed stars, and arranged along the long axis are the moon and the two inner planets. From this zone blows the west wind. The position and constitution of this *purus aether* is evidently the result of some misinterpretation of Aristotelian writings.

The next zone, the *umbrosa pellis* or *ignis niger,* is a narrow dark shell, whence proceed the more dramatic meteorological events. Here, following on the hints of the Wisdom of Solomon (chap. v) and the Book of Job (chap. xxxviii), are situated the diagrammatically portrayed treasuries of lightning and of hail. From here the tempestuous north wind bursts forth. This *ignis niger* is clearly comparable to the *dry earthy exhalation* that works of the Peripatetic school regard as given off by the outer fiery zone. The presence of the *ignis niger* thus suggests some contact on the part of the authoress with the teaching of the *Meteorologica* of Aristotle.[1]

The outermost layer of all is a mass of flames, the *lucidus ignis.* Here are the sun and the three outer planets, and from here the south wind pours its scorching breath (Fig. 2).

The movements of the four outer zones around each other, carrying the heavenly bodies with them, are attributed to the

[1] The *Meteorologica* had been translated about 1150 by Aristippus, the minister of William the Bad of Sicily. The version of Aristippus passed quickly into circulation (Valentine Rose, ' Die Lücke im Diogenes Laërtus und der alte Übersetzer ' in *Hermes,* i. 376, Berlin, 1866), but hardly soon enough for Hildegard's *Scivias,* which was completed about 1150. It is, of course, possible that the references to the *ignis niger* are later interpolations, but this is very unlikely in view of the way in which she speaks of this vision in the *Liber Divinorum Operum.*

winds in each zone. The seasonal variations in the movements of the heavenly bodies, along with the recurring seasons themselves, are also determined by the prevalent winds, which, acting as the motive power upon the various zones, form a celestial parallelogram of forces. In this way is ingeniously explained also why in spring the days lengthen and in autumn they shorten until in either case an equinox is reached (Fig. 2).

'I looked and behold the east and the south wind with their collaterals, moving the firmament by the power of their breath, caused it to revolve over the earth from east to west; and in the same way the west and north winds and their collaterals, receiving the impulse and projecting their blast, thrust it back again from west to east. . . .

'I saw also that as the days began to lengthen, the south wind and his collaterals gradually raised the firmament in the southern zone upwards towards the north, until the days ceased to grow longer. Then when the days began to shorten, the north wind with his collaterals, shrinking from the brightness of the sun, drove the firmament back gradually southward until by reason of the lengthening days the south wind began yet again to raise it up'[1] (Plates VII and VIII).

Intimately bound up not only with her theory of the nature and structure of the universe but also with her eschatological beliefs is Hildegard's doctrine of the elements. Before the fall of man these were arranged in a harmony,[2] which was disturbed by that catastrophe (Plate XII *a*),[3] so that they have since remained in the state of mingled confusion in which we always encounter them on the terrestrial globe. This *mistio*, to use the mediaeval Aristotelian term, is symbolized by the irregular manner in which the elements are represented in the central sphere of the diagram of the universe (Fig. 2). Thus mingled they will remain until subjected to the melting-pot of the Last

[1] Migne, cols. 789–91.

[2] Migne, col. 389.

[3] Plate XII *a*. The elements are represented in their original order undisturbed by the Fall. Uppermost is the *purus aether* or *aer lucidus* containing the stars and representing the element *air* in Hildegard's cosmic system. Next comes *water*. Below, and to the left, is a dark mass separating into tongues, one of which is formed into a serpent's head. These tongues are flames of *fire*. Below, and to the right, are plants and flowers emblematical of *earth*. The serpent, the enemy, vomits over a cloud of stars (signifying the fallen angels) that are borne downward by the falling Adam. In the four corners of the miniature the symbols of the elements are again displayed.

Judgement (Plate XIII),[1] when they will emerge in a new and eternal harmony, no longer mixed as matter, but separate and pure, parts of the new heaven and the new earth (Plate XII *b*).[2]

'But the heavens and the earth, which are now, . . . are kept in store and reserved unto fire against the day of judgment and perdition of ungodly men. . . . But the day of the Lord will come . . . in the which the heavens shall pass away with a great noise, and the elements shall melt with fervent heat, the earth also and the works that are therein shall be burned up. . . . Nevertheless we, according to his promise, look for new heavens and a new earth, wherein dwelleth righteousness' (2 Peter iii. 7, 10, and 13).

So Hildegard, acting on a scriptural hint, is enabled to dematerialize her doctrine of the after-things.

But although since man's fall the elements have lost their order and their harmony on this terrestrial orb, yet is that harmony still in part preserved in the celestial spheres that encircle and surround our globe; and water, air, earth, and fire have each their respective representatives in the four concentric zones, the *aer aquosus*, the *purus aether*, the *umbrosa pellis*, and the *lucidus ignis* (Fig. 2). These are the 'superior elements' which still retain some at least of their individuality and primal purity. From each of their spheres blows, as we have seen, one of the cardinal winds, and each wind partakes of the elemental character of the zone whence it issues, and has a corresponding influence

[1] Plate XIII. Above, in a circle, sits the Heavenly Judge. He is flanked on either side by groups of angels bearing the cross and other symbols. The lower circle exhibits the final destruction of the elemental Universe. The four winds and their collaterals are here subjecting the elements to the crucible heat of their combined blasts. Strewn among the elements can be seen men, plants, and animals. Between the circles is an angel sounding the last trump, and holding the recording roll of good and evil deeds. He faces the throng of the righteous who are rising from their bones, while he turns his back on the weeping crowd of those doomed to torment. Below these latter crouches Satan, now enchained.

[2] Plate XII *b*. In the highest circle is the Trinity flanked to the left by the Virgin and to the right by the Baptist, with Cherubim below. In the middle circle are two groups, the Saints above and the Prophets and Apostles below. In the lowest circle are the elements, now rearranged in their eternal harmony; uppermost of these is the *purus aether* now separated from the *aer lucidus* and containing the stars; on either side are light-coloured flame-like processes representing the *air*; below the aether is *water*, indicated by a zone of undulating lines; then comes the *earth* symbolized, as usual, by a group of plants. Below and to the side of *earth* are dark-coloured flames of fire, now controlled and confined to this lowest rung.

on man's body, since each of the four humours is specifically affected by the element to which it corresponds.

'Then I saw that by the diverse quality of the winds, and of the atmosphere as they in turn sweep through it, the humours in man are agitated and altered. For in each of the superior elements there is a breath of corresponding quality by which, through the power of the winds, the corresponding element [below] is forced to revolve in the atmosphere, and in no other way is it moved. And by one of those winds, with the agency of sun, moon, and stars, the atmosphere which tempers the world is breathed forth'[1] (Plate VII).

This doctrine of the relation of the various winds to the four elements and through them to the four humours is found in the *De Rerum Natura* of Isidore of Seville, and is occasionally illustrated in European MSS. from the ninth century onward,[2] but we meet it set forth with special definiteness in the twelfth century in the translations from Messahalah. It is encountered also in the work of Herrade de Landsberg. In and after the thirteenth century it had become a commonplace.

The description we have given of the universe was in the main set forth by Hildegard in her first work, the *Scivias* (1141–50).[3] Subsequently she became dissatisfied with the account she had given, and while not withdrawing it, she sought in the *Liber Divinorum Operum* (1163–70) so to modify the original presentment as to bring it more into line with accepted views. Thus she writes: 'There appeared to me in vision a *disk* very like that object which I saw twenty-eight years ago of the form of an *egg*, in the third vision of my book *Scivias*. In the outer part of the disk there was as it were the *lucidus ignis*, and beneath it the circle of the *ignis niger* was portrayed . . . and these two circles were so joined as to be one circle.' There was thus one outer zone representing the fire. 'Under the circle of the *ignis niger* there was another circle in the likeness of the *purus aether* which was of the same width as the two conjoined [outer] fiery circles. And below this circle again was the circle of the *aer aquosus* as wide as the *lucidus ignis*. And below this circle was yet another

[1] Migne, col. 791.

[2] See Ernest Wickersheimer, 'Figures médico-astrologiques des neuvième, dixième et onzième siècles', in the *Transactions of the Seventeenth International Congress of Medicine, Section XXIII, History of Medicine*', p. 313, London, 1913.

[3] Migne, cols. 403–14.

circle, the *fortis et albus lucidusque aer* . . . the width whereof was as the width of the *ignis niger*, and these circles were joined to make one circle which was thus again of width equal to the outer two. Again, under this last circle yet another circle, the *aer tenuis*, was distinguishable, which could be seen to raise itself as a cloud, sometimes high and light, sometimes depressed and dark, and to diffuse itself as it were throughout the whole disk. . . . The outermost fiery circle perfuses the other circles with its fire, while the watery circle saturates them with its moisture. [cp. Wisdom of Solomon, xix. 18–20]. And from the extreme eastern part of the disk to the extreme west a line is stretched out [i. e. the equator] which separates the northern zones from the others '[1] (see Fig. 3 and Plates VII and VIII).

The earth lies concentrically with the *aer tenuis*, and its measurements are given thus: 'In the midst of the *aer tenuis* a globe was indicated, the circumference of which was everywhere equidistant from the *fortis et albus lucidusque aer*, and it was as far across as the depth of the space from the top of the highest circle to the extremity of the clouds, or from the extremity of the clouds to the circumference of the inner globe'[1] (Fig. 3).

In her earlier work, the *Scivias*, Hildegard had not apparently realized the need of accounting for the independent movements of the planets other than the sun and moon. She had thus placed the moon and two of the moving stars in the *purus aether*, and the sun and the three remaining moving stars in the *lucidus ignis*. Since these spheres were moved by the winds, their contained planets would be subject to the same influences. In the *Liber Divinorum Operum*, however, she has come to realize how independent the movements of the planets really are, and she invokes a special cause for their vagaries. 'I looked and behold in the outer fire (*lucidus ignis*) there appeared a circle which girt about the whole firmament from the east westward. From it a blast produced a movement from west to east in the opposite direction to the movement of the firmament. But this blast did not give forth his breath earthward as did the other winds, but instead thereof it governed the course of the planets.'[2] The source of the blast is represented in the Lucca MS. as the head of a supernatural being with a human face (Plate VIII).

These curious passages were written at some date after 1163, when Hildegard was at least 65 years old. They reveal our pro-

[1] Migne, col. 751. [2] Migne, col. 791.

From the LUCCA MS. fo. 27 v

PLATE VIII. THE MACROCOSM THE MICROCOSM AND THE WINDS

Plate IX. From THE LUCCA MS. fo. 37r

CELESTIAL INFLUENCES ON MEN ANIMALS AND PLANTS

phetess attempting to revise much of her earlier theory of the universe, and while seeking to justify her earlier views, endeavouring also to bring them into line with the new science that was now just beginning to reach her world. Note that (*a*) the universe has become round; (*b*) there is an attempt to arrange the zones according to their density, i. e. from without inwards, fire, air (ether), water, earth; (*c*) exact measurements are given; (*d*) the watery zone is continued earthward so as to mingle with the

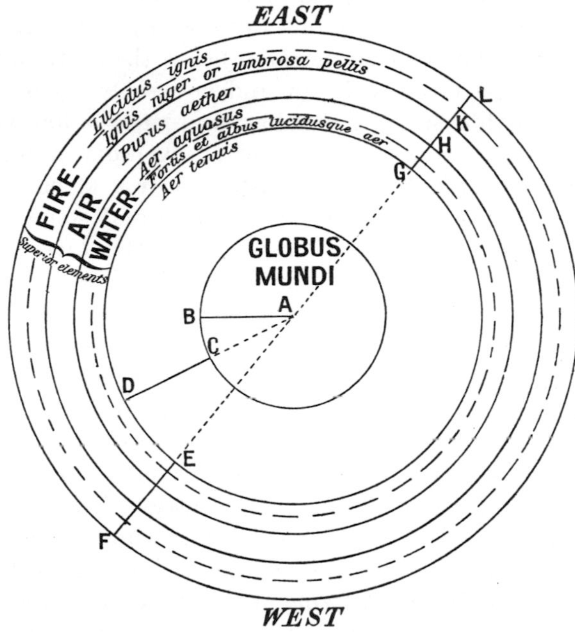

Fig. 3. HILDEGARD'S SECOND SCHEME OF THE UNIVERSE
Reconstructed from her measurements. AB, CD, and EF are all equal to each other, as are also GH, HK, and KL. The clouds are situated in the outer part of the *aer tenuis*, and form a prolongation downwards from the *aer aquosus* towards the earth.

central circle. In all these and other respects she is joining the general current of mediaeval science then beginning to be moulded by works translated from the Arabic. Her knowledge of the movements of the heavenly bodies is entirely innocent of the doctrine of epicycles, but in other respects her views have come to resemble those, for instance, of Messahalah, one of the simplest and easiest writers on the sphere available in her day. Furthermore, her conceptions have developed so as to fit in with the macrocosm-microcosm scheme which she grasped about the year 1158. Even in her latest work, however, her theory of the universe exhibits differences from that adopted by the schoolmen, as may be seen by comparing her diagram with, for example, the scheme of Dante (Fig. 4).

Like many mediaeval writers, Hildegard would have liked to imagine an ideal state of the elemental spheres in which the rarest, fire, was uppermost, and the densest, earth, undermost. Such a scheme was, in fact, purveyed by Bernard Sylvestris and by Messahalah. Her conceptions were however disturbed by the awkward facts that water penetrated below the earth, and indeed sought the lowest level, while air and not water lay immediately above the earth's surface. Mediaeval writers adopted various devices and expended a great amount of ingenuity in dealing with this discrepancy, which was a constant source of obscurity and confusion. Hildegard devotes much space and some highly involved allegory both in the *Scivias* and in the *Liber Divinorum Operum* to the explanation of the difficulty, while Dante himself wrote a treatise in high scholastic style on this very subject.[1]

VII. Macrocosm and Microcosm

The winds and elements of the outer universe, the macrocosm, become in Hildegard's later schemes intimately related to structures and events within the body of man himself, the microcosm, the being around whom the universe centres. The terms *macrocosm* and *microcosm* are not employed by her, but in her last great work, the *Liber Divinorum Operum*, she succeeds in most eloquent and able fashion in synthesizing into one great whole, centred around this doctrine, her theological beliefs and her physiological knowledge, together with her conceptions of the working of the human mind and of the structure of the universe. The work is thus an epitome of the science of the time viewed through the distorting medium of this theory. In studying it the modern reader is necessarily hampered by the bizarre and visionary form into which the whole subject is cast. Nevertheless the scheme, though complex and difficult, is neither incoherent nor insane, as at first sight it may seem. On the contrary, it is a highly

[1] The *Quaestio de Aqua et Terra* is doubtless a genuine, albeit the least pleasing, production of the great poet. The genuineness is established by Vincenzo Balgi in his edition, Modena, 1907.

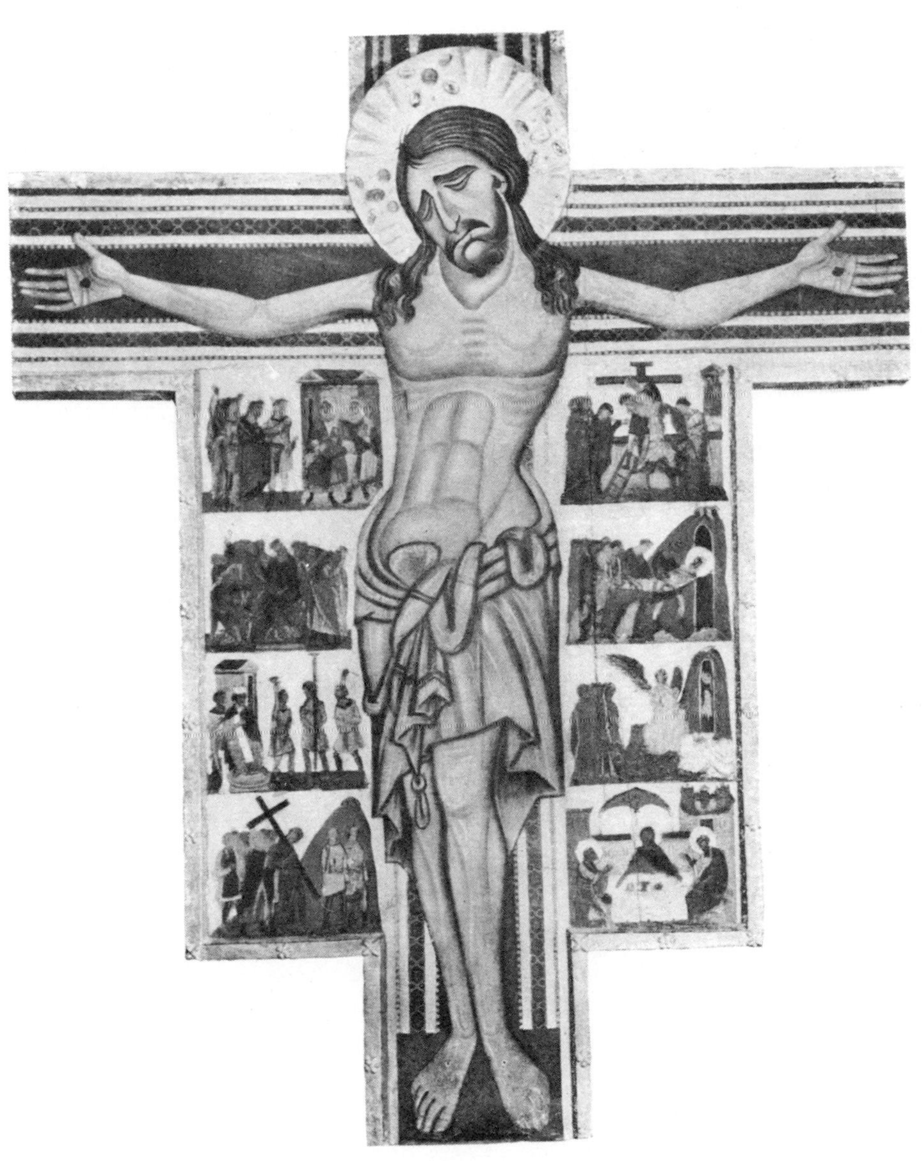

PLATE X. A CRUCIFIX IN THE UFFIZI GALLERY
About the middle of the XIIIth Century

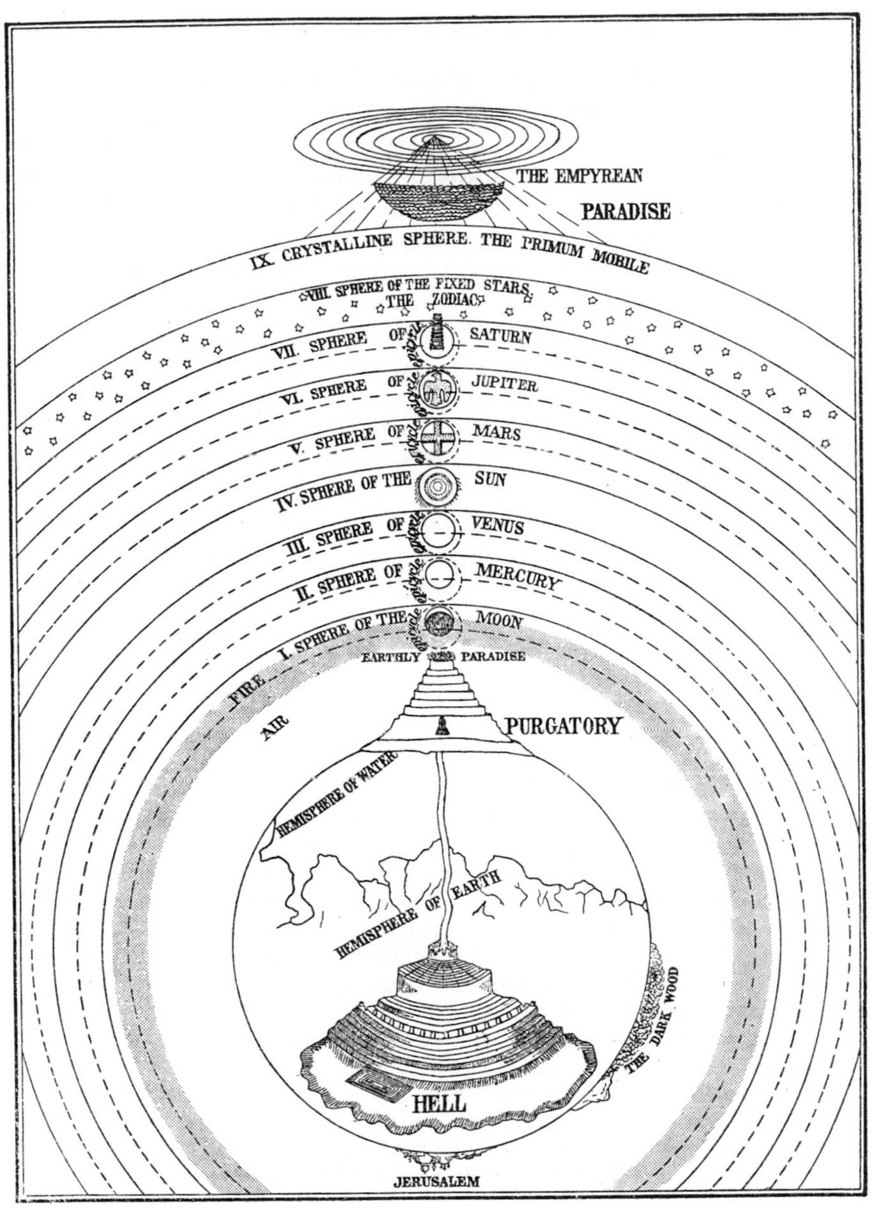

FIG. 4. DANTE'S SCHEME OF THE UNIVERSE

Slightly modified from Michelangelo Caetani, duca di Sermoneta, *La materia della Divina Commedia di Dante Allighieri dichiarata in VI tavole,* Monte Cassino, 1855.

systematic and skilful presentment of a cosmic theory which for centuries dominated scientific thought.

As an explanation of the complexity of existence which thinkers of all ages have sought to bring within the range of some simple formula, this theory of the essential similarity of macrocosm and microcosm held in the Middle Ages, during the Renaissance, and even into quite modern times, a position comparable to that of the theory of evolution in our own age. If at times it passed into folly and fantasy, it should be remembered that it also fulfilled a high purpose. It gave a meaning to the facts of nature and a formula to the naturalist, it unified philosophic systems, it exercised the ingenuity of theologians, and gave a convenient framework to prophecy, while it seemed to illumine history and to provide a key and meaning to life itself. Even now it is not perhaps wholly devoid of message, but as a phenomenon in the history of human thought, a theory which appealed to such diverse scientific writers as Seneca, Albertus Magnus, Paracelsus, Gilbert, Harvey, Boyle, and Leibnitz, is surely worthy of attention.

In essaying to interpret the views of our authoress on this difficult subject, we rely mainly on the text of the *Liber Divinorum Operum*, supplemented by the beautiful illuminations of that work which adorn the Lucca MS. The book opens with a truly remarkable vision (Plate VI):

'I saw a fair human form and the countenance thereof was of such beauty and brightness that it had been easier to gaze upon the sun. The head thereof was girt with a golden circlet through which appeared another face as of an aged man. From the neck of the figure on either side sprang a pinion which swept upward above the circlet and joined its fellow on high. And where on the right the wing turned upward, was portrayed an eagle's head with eyes of flame, wherein appeared as in a mirror the lightning of the angels, while from a man's head in the other wing the lightning of the stars did radiate. From either shoulder another wing reached to the knees. The figure was robed in brightness as of the sun, while the hands held a lamb shining with light. Beneath, the feet trampled a horrible black monster of revolting shape, upon the right ear of which a writhing serpent fixed itself.' [1]

The image declares its identity in words reminiscent of the Wisdom literature or of passages in the hermetic writings, but which seem in fact to be partly borrowed from Bernard Sylvestris.

[1] Migne, col. 741.

From the LUCCA MS. fo. 86 v

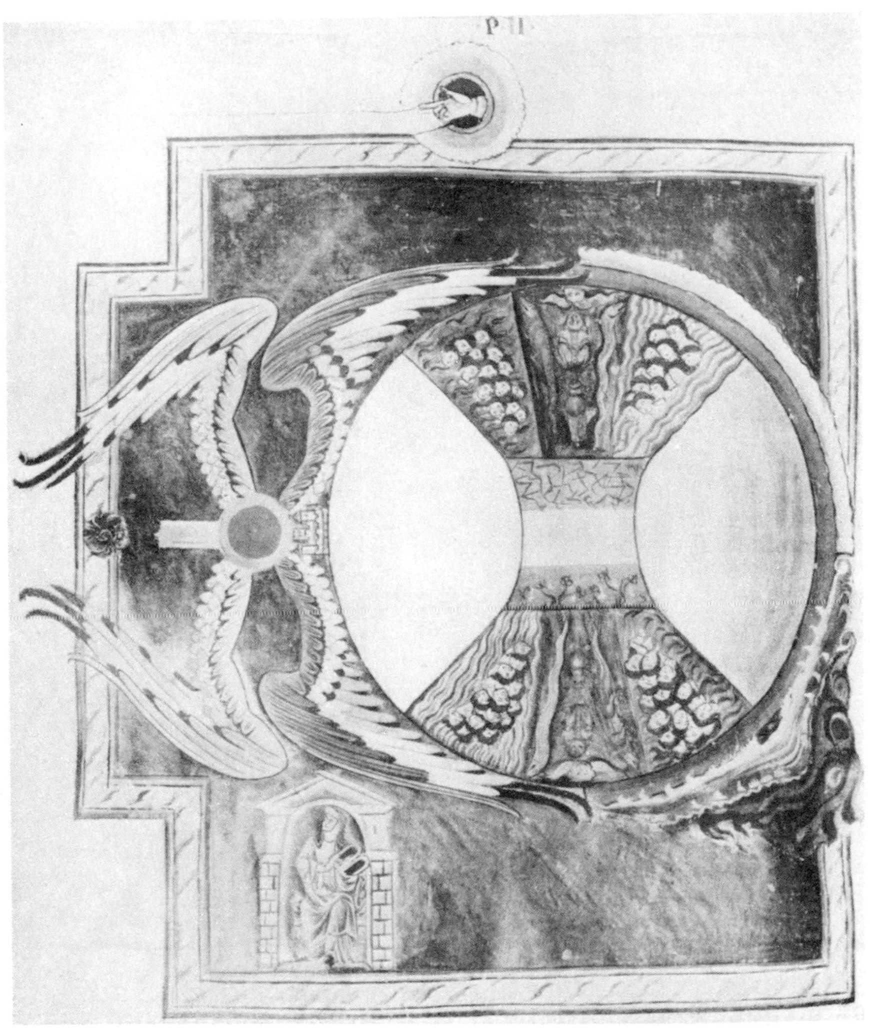

PLATE XI. THE STRUCTURE OF THE MUNDANE SPHERE

WIESB. COD. B fo. 4 r

PLATE XII a. MAN'S FALL AND THE DISTURBANCE OF THE ELEMENTAL HARMONY

WIESB. COD. B fo. 224 v

PLATE XII b. THE NEW HEAVEN AND THE NEW EARTH

'I am that supreme and fiery force that sends forth all the sparks of life. Death hath no part in me, yet do I allot it, wherefore I am girt about with wisdom as with wings. I am that living and fiery essence of the divine substance that glows in the beauty of the fields. I shine in the water, I burn in the sun and the moon and the stars. Mine is that mysterious force of the invisible wind. I sustain the breath of all living. I breathe in the verdure and in the flowers, and when the waters flow like living things, it is I. I formed those columns that support the whole earth. . . . I am the force that lies hid in the winds, from me they take their source, and as a man may move because he breathes so doth a fire burn but by my blast. All these live because I am in them and am of their life. I am wisdom. Mine is the blast of the thundered word by which all things were made. I permeate all things that they may not die. I am life.' [1]

Hildegard thus supposes that the whole universe is permeated by a single living spirit, the figure of the vision. This spirit of the macrocosm, the *Nous* or 'world spirit' of the hermetic and Neoplatonic literature, the impersonated *Nature*, as we may perhaps render it, is in its turn controlled by the Godhead that pervades the form and is represented rising from its vertex as a second human face. Nature, the spirit of the cosmic order, controls and holds in subjection the hideous monster, the principle of death and dissolution, the *Hyle* or primordial matter of the Neoplatonists, whose chaotic and anarchic force would shatter and destroy this fair world unless fettered by a higher power.

With the details of the visionary figure we need not delay,[2] but we pass to the description of the structure of the macrocosm itself, to which the second vision is devoted (Plate VII). Here appears the same figure of the macrocosmic spirit. But now the head and feet only are visible, and the arms are outstretched to enclose the disk of the universe which conceals the body. Although the macrocosm now described is considerably altered from Hildegard's original scheme of the universe, she yet declares, 'I saw in the bosom of the form the appearance of a disk of like sort to that which twenty-eight years before I had seen in the third vision, set forth in my book of *Scivias*'.[3] The zones of this disk are

[1] Migne, col. 743.

[2] It is outside our purpose to attempt a full elucidation of Hildegard's allegory. The eagle in the right wing signifies the power of divine grace, while the human head in the left wing indicates the powers of the natural man. To the bosom of the figure is clasped the Lamb of God.

[3] Migne, col. 751.

then described (Plates VII, VIII, and XI and Fig. 2). They are from without inwards:

(*a*) The *lucidus ignis*, containing the three outer planets, the sixteen principal fixed stars, and the south wind.

(*b*) The *ignis niger*, containing the sun, the north wind, and the materials of thunder, lightning, and hail.

(*c*) The *purus aether*, containing the west wind, the moon, the two inner planets, and certain fixed stars.

(*d*) The *aer aquosus*, containing the east wind.

(*e*) The *fortis et albus lucidusque aer*, where certain other fixed stars are placed.

(*f*) The *aer tenuis*, or atmosphere, in the outer part of which is the zone of the clouds.

From all these objects, from the spheres of the elements, from the sun, moon, and other planets, from the four winds each with their two collaterals, from the fixed stars, and from the clouds, descend influences, indicated by lines, towards the figure of the macrocosm.

The microcosm is then introduced.

'And again I heard the voice from heaven saying, "God, who created all things, wrought also man in his own image and similitude, and in him he traced [*signavit*] all created things, and he held him in such love that he destined him for the place from which the fallen angel had been cast."'[1]

The various characters of the winds are expounded in a set of curious passages in which the doctrine of the macrocosm and microcosm is further mystically elaborated. An endeavour is made to attribute to the winds derived from the different quarters of heaven qualities associated with a number of animals.[2] The conception is illustrated and made comprehensible by the miniatures in the Lucca MS. (Plates VII and VIII).

'In the middle of the disk [of the universe] there appeared the form of a man, the crown of whose head and the soles of whose feet extended to the *fortis et albus lucidusque aer*, and his hands were outstretched right and left to the same circle. . . . Towards these parts was an appearance as of four heads; a leopard, a wolf, a lion, and a bear. Above the head of the figure in the zone of the *purus aether*, I saw the head of the leopard emitting a blast from its mouth, and on the right side of the mouth the blast, curving itself somewhat backwards, was formed into a crab's

[1] Migne, col. 744. [2] *Liber Divinorum Operum*, part i, visions 2 and 3.

head . . . with two chelae; while on the left side of the mouth a blast similarly curved ended in a stag's head. From the mouth of the crab's head, another blast went to the middle of the space between the leopard and the lion; and from the stag's head a similar blast to the middle of the space between the leopard and the bear . . . and all the heads were breathing towards the figure of the man. Under his feet in the *aer aquosus* there appeared as it were the head of a wolf, sending forth to the right a blast extending to the middle of the half space between its head and that of the bear, where it assumed the form of the stag's head; and from the stag's mouth there came, as it were, another breath which ended in the middle line. From the left of the wolf's mouth arose a breath which went to the midst of the half space between the wolf and the lion, where was depicted another crab's head . . . from whose mouth another breath ended in the same middle line. . . . And the breath of all the heads extended sideways from one to another. . . . Moreover on the right hand of the figure in the *lucidus ignis,* from the head of the lion, issued a breath which passed laterally on the right into a serpent's head and on the left into a lamb's head . . . similarly on the figure's left in the *ignis niger* there issued a breath from the bear's head ending on its right in the head of [another] lamb, and on its left in another serpent's head. . . . And above the head of the figure the seven planets were ranged in order, three in the *lucidus ignis,* one projecting into the *ignis niger* and three into the *purus aether.* . . . And in the circumference of the circle of the *lucidus ignis* there appeared the sixteen principal stars, four in each quadrant between the heads. . . . Also the *purus aether* and the *fortis et albus lucidusque aer* seemed to be full of stars which sent forth their rays towards the clouds, whence . . . tongues like rivers descended to the disk and towards the figure, which was thus surrounded and influenced by these signs.' [1]

The third vision is devoted to an account of the human body, the microcosm (Plate VIII), with a comparison of its organs to the parts of the macrocosmic scheme, together with a detailed account of the effects of the heavenly bodies on the humours in man, the whole brought into a strongly theological setting. Some of these views are set forth below in the chapter on anatomy and physiology.

The fourth vision explains the influence of the heavenly bodies and of the superior elements on the power of nature as exhibited on the surface of the earth. It is illustrated by a charming miniature in the Lucca MS. (Plate IX).

[1] Migne, cols. 752–5.

'I saw that the upper fiery firmament was stirred, so that as it were ashes were cast therefrom to earth, and they produced rashes and ulcers in men and animals and fruits.' These effects are shown in the left upper quadrant of Plate IX, where the ashes are seen proceeding from the *lucidus ignis*, the 'upper fiery firmament'. Two figures are seen, a female semi-recumbent, who lifts a fruit to her mouth, and a male figure fully recumbent, on whose legs a rash is displayed. The trees also in this quadrant show the effects of the ashes, two of them being denuded of fruit and foliage.

'Then I saw that from the *ignis niger* certain vapours (*nebulae*) descended, which withered the verdure and dried up the moisture of the fields. The *purus aether*, however, resisted these ashes and vapours, seeking to hold back these plagues.' These vapours may be seen in the right upper quadrant of Plate IX. They descend from the *ignis niger*, attenuate for a space in the *purus aether*, and then descend through the other zones on to an arid and parched land. Here are two husbandmen; one sits forlornly clasping his axe, while the other leans disconsolately upon his hoe. On the legs of the latter a rash may be distinguished.

'And looking again I saw that from the *fortis et albus lucidusque aer* certain other clouds reached the earth and infected men and beasts with sore pestilence, so that they were subjected to many ills even to the death, but the *aer aquosus* opposed that influence so that they were not hurt beyond measure.' This scene is portrayed in the right lower quadrant of Plate IX. Here is a husbandman in mortal anguish. He has gathered his basket of fruit and now lies stricken with the pestilence. His left hand is laid on his heart, while his right hangs listless on his thigh, pointing to tokens of plague upon his legs. Beyond lies the dead body of a beast on which a carrion bird has settled.

'Again I saw that the moisture in the *aer tenuis* was as it were boiling above the surface of the earth, awakening the force of the earth and making fruits to grow.'[1] This happier scene is represented in the left lower quadrant of Plate IX. Here the beneficent fertilizing influence is falling on trees and herbs and the happy husbandmen are reaping its results.

The main outline of the *Liber Divinorum Operum* is, we believe, borrowed from the work of Bernard Sylvestris of Tours, *De mundi*

[1] Migne, col. 807.

From WIESBADEN CODEX B fo. 224 r

PLATE XIII. THE LAST JUDGEMENT AND FATE OF THE ELEMENTS

From BIBL. NAT. MS. LAT. 5543 fo. 136 r

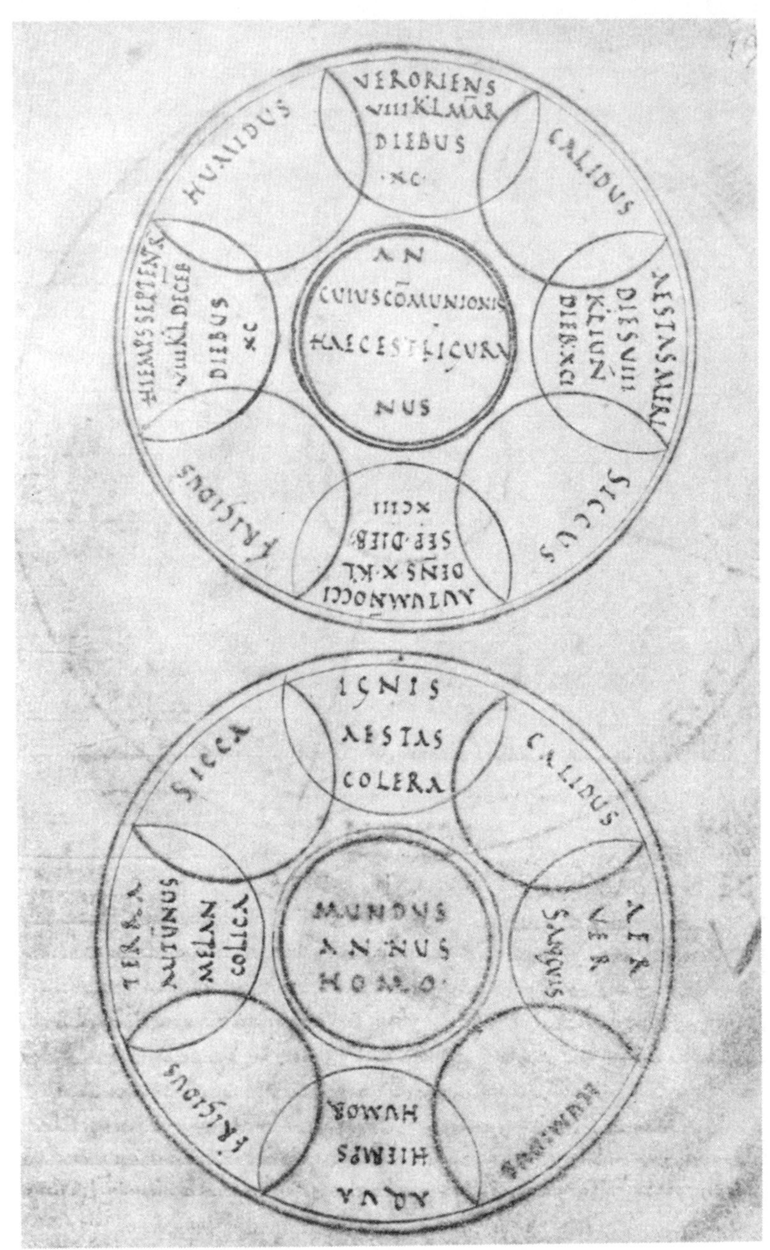

PLATE XIV. DIAGRAM OF THE RELATION OF HUMAN AND COSMIC PHENOMENA
IXth Century

universitate libri duo sive megacosmus et microcosmus.[1] In this composition by a teacher at the cathedral school of Chartres,[2] the gods and goddesses of the classical pantheon flit across the stage, for all the world as though the writer were a pagan, and the work might be thought to be the last one from which our pious authoress would borrow. The *De mundi universitate* is alternately in prose and verse and betrays an acquaintance with the classics very rare at its date. 'The rhythm of the hexameters is clearly that of Lucan, while the vocabulary is mainly of Ovid.'[3] The mythology is founded mainly on the *Timaeus.* The eternal *seminaria* of created things are mentioned, and it has been conjectured that the work exhibits traces of the influence of Lucretius,[4] but the general line of thought is clearly related to Neoplatonic literature. Thus the *anima universalis* of Neoplatonic writings can be identified with the *Nous* or *Noys* of Bernard. This principle is contrasted with primordial matter or *Hyle.* The parallel character of the *Liber Divinorum Operum* and the *De mundi universitate* can be illustrated by a few extracts from the latter. It will be seen that although the general setting is changed, yet Hildegard's figure of the spirit of the macrocosm is to be identified with Bernard's *Noys. Hyle,* on the other hand, becomes in Hildegard's plan the monstrous form, the emblem of brute matter, on which the spirit of the universe tramples.

'In huius operis primo libro qui Megacosmus dicitur, id est maior mundus, Natura ad Noym, id est Dei providentiam, de primae materiae, id est hyles, confusione querimoniam quasi cum lacrimis agit et ut mundus pulchrius petit. Noys igitur eius mota precibus petitioni libenter annuit et ita quatuor elementa ab invicem seiungit. Novem ierarchias angelorum in coelo ponit. stellas in firmamento figit. signa disponit. sub signis orbes septem planetarum currere facit. quatuor ventos cardinales sibi invicem opponit. Sequitur genesis animantium et terrae situs medius. . . .

'In secundo libro qui Microcosmus dicitur, id est minor mundus, Noys ad Naturam loquitur et de mundi expolitione gloriatur et in operis sui completione se hominem plasmaturam pollicetur. Iubet

[1] The work is printed by C. S. Barach and J. Wrobel, Innsbruck, 1876. The writers, however, confuse Bernard Sylvestris of Tours with his somewhat older contemporary, Bernard of Chartres.

[2] A. Clerval, *Les Écoles de Chartres au Moyen Âge*, Paris, 1895.

[3] J. E. Sandys, *History of Classical Scholarship*, Cambridge, 1903, vol. i, p. 515.

[4] R. Lane Poole, *Illustrations of the History of Mediaeval Thought in the Departments of Theology and Ecclesiastical Politics*, Oxford, 1884, pp. 118, 219.

igitur Uraniam, quae siderum regina est, et Physin, quae rerum omnium est peritissima, sollicite perquirat. Natura protinus iubenti obsequitur et per caelestes circulos Uraniam quaeritans eam sideribus inhiantem reperit. eiusque itineris causa praecognita se operis et itineris comitem Urania pollicetur. . . . Subitoque ibi Noys affuit suoque velle eis ostenso trinas speculationes tribus assignando tribuit & ad hominis plasmationem eas impellit. Physis igitur de quatuor elementorum reliquiis hominem format et a capite incipiens membratim operando opus suum in pedibus consummat. . . .

'Noys ego scientia et divinae voluntatis arbitraria ad dispositionem rerum, quem ad modum de consensu eius accipio, sic meae administrationis officia circumduco. . . .

'(Noys) erat fons luminis, seminarium vitae, bonum bonitatis divinae, plenitudo scientiae quae mens altissimi nominatur. Ea igitur noys summi & exsuperantissimi Dei est intellectus et ex eius divinitate nata natura. . . . Erat igitur videre velut in speculo tersiore quicquid generationi quicquid operi Dei secretior destinarat affectus.' [1]

Hildegard's conception of macrocosm and microcosm, which was thus probably borrowed from Bernard Sylvestris, has analogies also to those well-known figures illustrating the supposed influence of the signs of the zodiac on the different parts of the body.[2] Such figures, with the zodiacal symbols arranged around a figure of Christ, may be seen in certain MSS. anterior to Hildegard,[3] while the influence of the 'Melothesia', to give it the name assigned by Porphyry, has been traced through its period of efflorescence at the Renaissance (Plates XV,[4] XVI,[5] and

[1] Barach and Wrobel, *loc. cit.*, pp. 5–6, 9 and 13.

[2] For a general consideration of these figures see K. Sudhoff, *Archiv für Geschichte der Medizin*, i. 157, 219 ; ii. 84.

[3] E. Wickersheimer, 'Figures médico-astrologiques des neuvième, dixième et onzième siècles', *Transactions of the Seventeenth International Congress of Medicine, Section XXIII, History of Medicine*, p. 313, London, 1913.

[4] The MS. from which Plate XV is taken (*Paris, Bibl. nat., Latin* 7028) is entitled *Scholium de duodecim zodiaci signis et de ventis.* It was once the property of St. Hilaire the Great of Poitiers. The legend above our figure reads, 'Secundum philosophorum deliramenta notantur duodecim signa ita ab ariete incipiamus'. The relation of the signs to the parts of the body is different in this eleventh-century MS. from that which was widely accepted in the astrology of the thirteenth and fourteenth centuries as illustrated in Plate XVI.

[5] The MS. from which Plate XVI is taken (*Paris, Bibl. nat., Latin* 11229) was written about the end of the fourteenth century. It has been described by K. Sudhoff, *Arch. f. Gesch. d. Med.*, ii. 84, Leipzig, 1910. The relation of the central figure to the signs of the zodiac in this plate bears a manifest resemblance

XVII,[1] compare with Plates VII and VIII) right down to our own age and country, where it still appeals to the ignorant and foolish.[2]

Hildegard often interprets natural events by means of a peculiarly crude form of the doctrine, as when she describes how 'if the excess of waters below are drawn up to the clouds (by the just judgment of God in the requital of sinners), then the moisture from the *aer aquosus* transudes through the *fortis et albus lucidusque aer* as a draught drunk into the urinary bladder ; and the same waters descend in an inundation'.[3]

Again, events in the body of man are most naively explained on the basis of the nature of the external world as she has pictured it.

'The humours at times rage fiercely as a leopard and again they are softened, going backwards as a crab ;[4] or they may show their diversity by leaping and goring as a stag, or they may be as a wolf in their ravening, and yet again they may invade the body of man after the manner of both wolf and crab. Or else they may show forth their strength unceasingly as a lion, or as a serpent they may go now softly, now violently, and at times they may be gentle as a lamb and at times again they may growl as an angered bear, and at times they may partake of the nature of the lamb and of the serpent.' [5]

to the relation of the central figure to the beasts' heads in Plate VII. The lines which cross and recross the figure in Plate VII are analogous also to the lines of influence of Plate XVI. The verse above the figure in Plate XVI is taken from the *Flos medicinae scholae Salerni*; cp. de Renzi, loc. cit., i. 486. This Melothesia and that of the next figure is identical with that propounded in Manilius, ii. 453 (edition of H. W. Garrod, Oxford, 1911).

1 Plate XVII is from an early German block book. It exhibits a scheme closely parallel to Plate VII. The universe in Plate XVII is represented as a series of concentric spheres, *earth* innermost, followed by *water*, *air*, and *fire*. In the outermost zone hover the angels who have replaced the beast's head of Hildegard's scheme. The whole world is embraced by the figure of the Almighty, much as in Plate VII.

2 See E. Wickersheimer, ' La médecine astrologique dans les almanachs populaires du XXe siècle', *Bulletin de la Société française d'histoire de la médecine*, x (1911), pp. 26–39.

3 Migne, col. 757. This phrase is reproduced in a mediaeval Irish version of the work of Messahalah. See Maura Power, *An Irish Astronomical Text*, Irish Text Society, London, 1912.

4 The word *cancer* is here used, but the crab goes sideways, not backwards. By *cancer* Hildegard, who had never seen the sea, probably means the crayfish, an animal fairly common in the Rhine basin. It is the head of a crayfish or lobster that is figured in the miniatures of the vision of the macrocosm in the Lucca MS., and a similar organism frequently serves for the sign Cancer in the mediaeval zodiacal medical figures, as in Plate XV of this essay.

5 Migne, cols. 3, 791–2.

Having completed her general survey of the macrocosm (Vision II), and having investigated in detail the structure of man's body, the microcosm, in terms of the greater universe (Vision III), and discussed the influence of the heavenly bodies on terrestrial events (Vision IV), Hildegard turns to the internal structure of the terrestrial sphere (Vision V). This vision is illustrated by the figure in the Lucca MS. reproduced in Plate XI.

Upon the surface of the earth towards the east stands the building which symbolizes the *aedificium* of the church, a favourite conception of our authoress. This church is surmounted by a halo, whence proceed a pair of pinions which extend their shelter over a full half of the earth's circumference. As for the rest of the earth's surface, part is within the wide-opened jaws of a monster, the Destroyer, and the remainder is beneath the surface of the ocean. Within the earth are five parts analogous, as she would have us believe, to the five senses. An eastern clear arc and a western clouded one signify respectively the excellence of the orient where Zion is situated, and the Cimmerian darkness of the occidental regions over which the shadow of the dragon is cast. Centrally is a quadrate area divided into three zones where the qualities of heat and cold and of a third intermediate 'temperateness' (*temperies*) are stored. North and south of this are two areas where purgatory is situate. Each is shaped like a truncated cone and composed also of three sectors. Souls are seen suffering in one sector the torment of flame, in another the torment of water, while in the third or intermediate sector lurk monsters

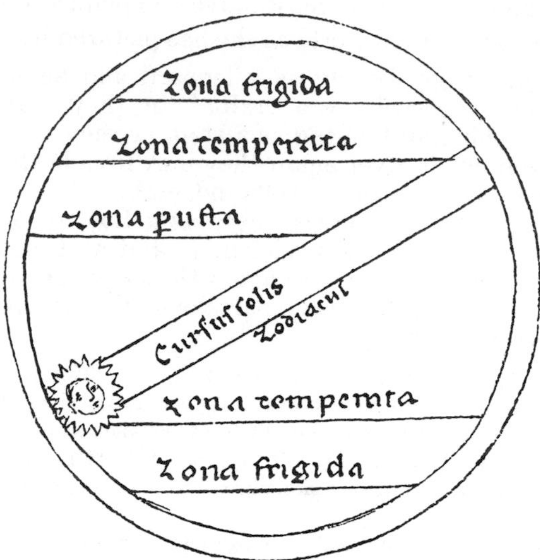

FIG. 5. From Herrade de Landsberg's *Hortus deliciarum*, after Straub and Keller.

From BIBL. NAT. MS. LAT. 7028 fo. 154 r

PLATE XV. AN XIth CENTURY FRENCH MELOTHESIA

From BIBL. NAT. MS. LAT. 11229 fo. 45 v

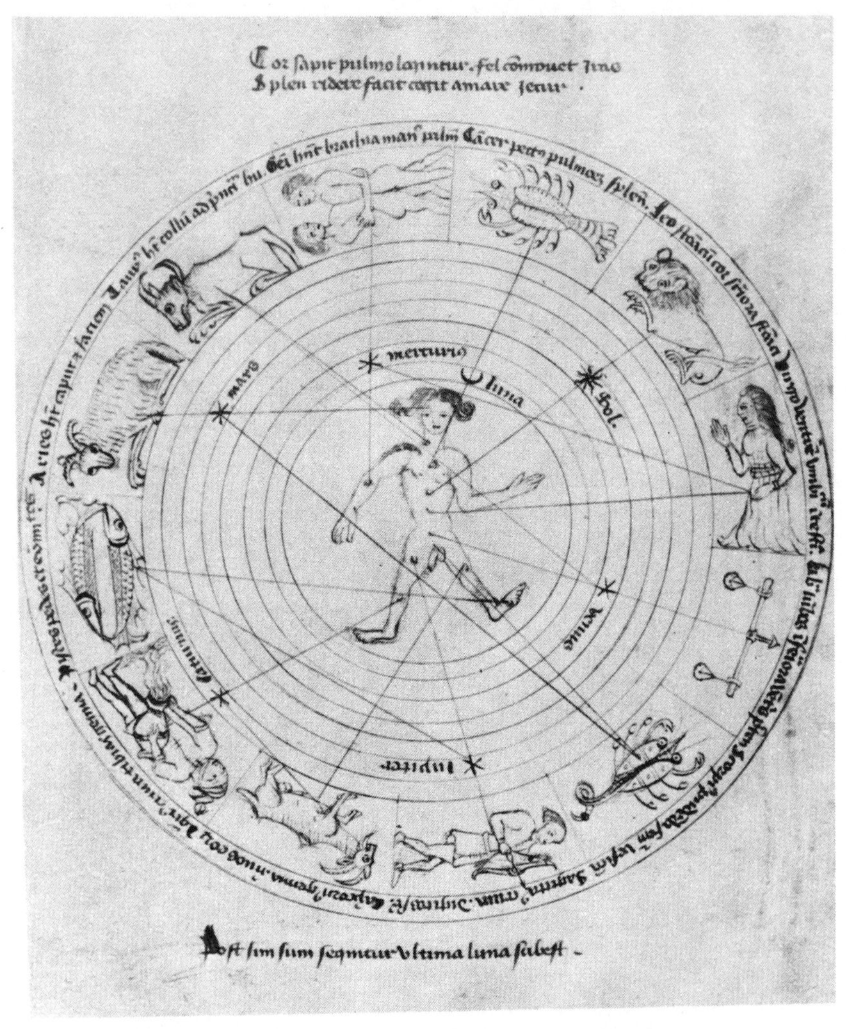

PLATE XVI. A MELOTHESIA OF ABOUT 1400

From the *SYMBOLUM APOSTOLICORUM*

Plate XVII. A GERMAN BLOCK BOOK
First Half of XVth Century. Heidelberg University Library

From BODLEIAN MS. ASHMOLE 399 fo. 18r

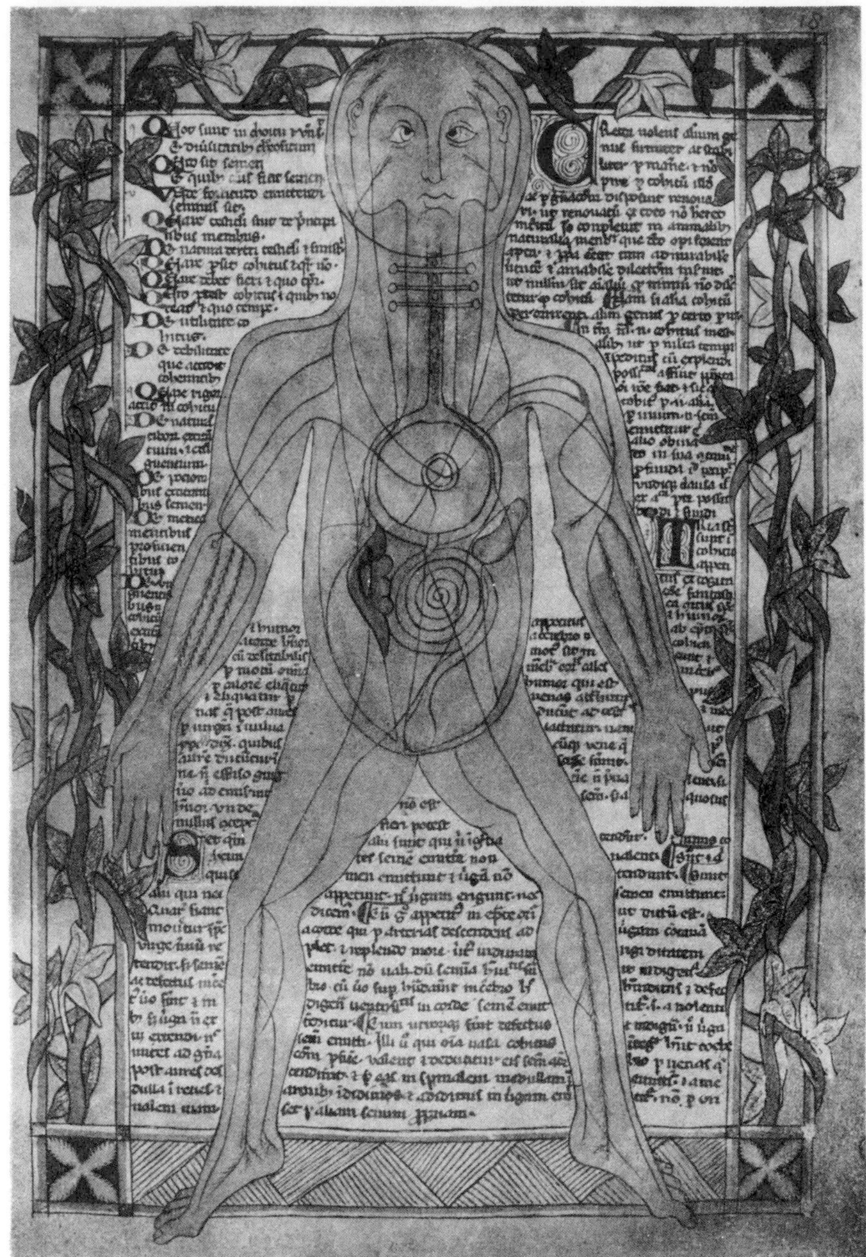

PLATE XVIII. AN ANATOMICAL DIAGRAM OF ABOUT 1298
From the Five-Figure Series. Cp. Plate XXXIII

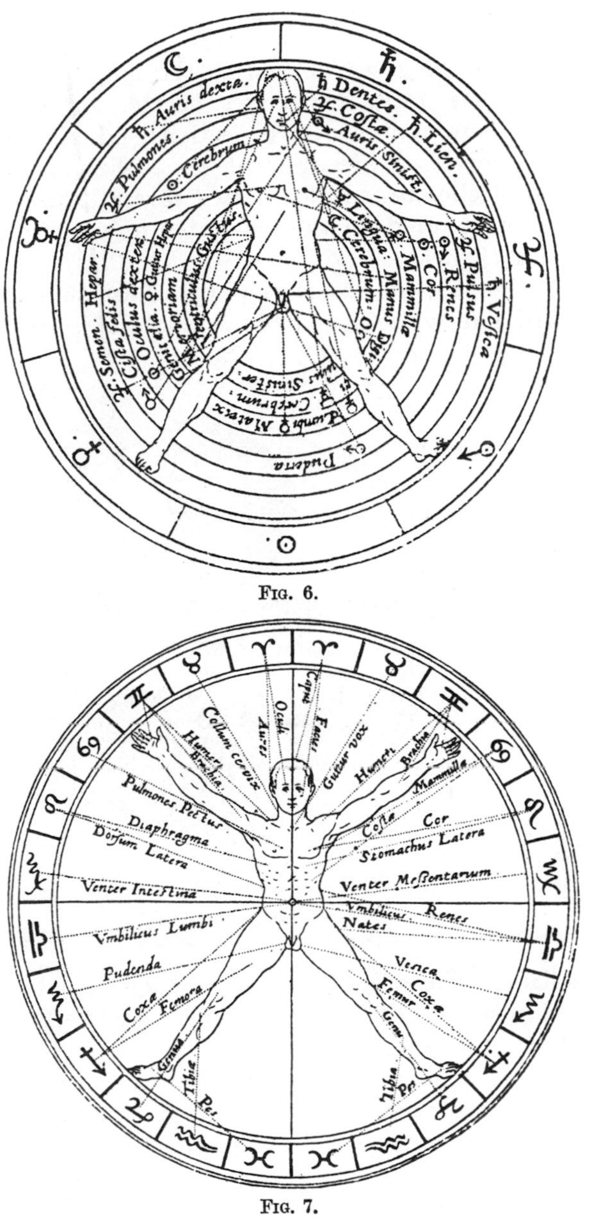

Fig. 6.

Fig. 7.

MELOTHESIAE

From R. Fludd, *Historia utriusque cosmi*, Oppenheim, 1619, pp. 112 and 113.

and creeping things which add to the miseries of purgatory or at times come forth to earth's surface to plague mankind. These northern and southern sections exhibit dimly by their identically reversed arrangement the belief in the antipodean inversion of climate, an idea hinted several times in Hildegard's writings, but more definitely illustrated by a figure of Herrade de Landsberg (Fig. 5).

Macrocosmic schemes of the type illustrated by the text of Hildegard and by the figures of the Lucca MS. had a great vogue

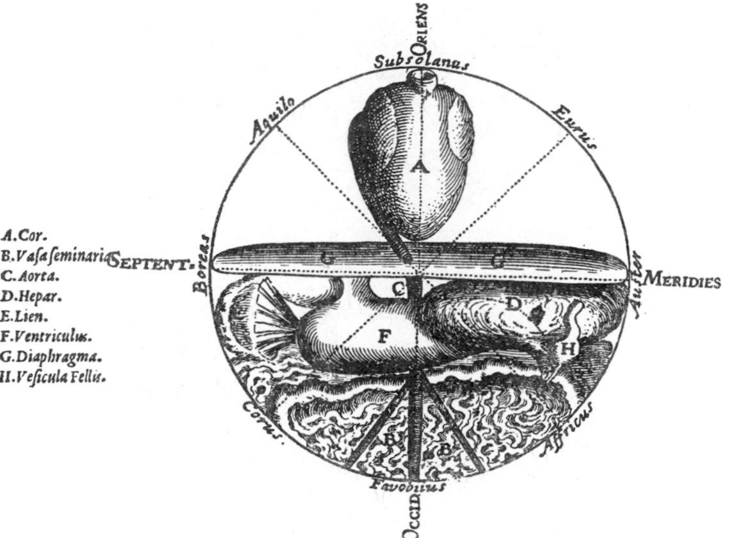

Fig. 8. THE MICROCOSM
From R. Fludd, *Philosophia sacra seu astrologia cosmica*, Frankfurt, 1628, p. 52.

in mediaeval times, and were passed on to later ages. Some passages in Hildegard's work read curiously like Paracelsus (1491–1541),[1] and it is not hard to find a link between these two difficult and mystical writers. Trithemius, the teacher of Paracelsus, was abbot of Sponheim, an important settlement almost within sight of Hildegard's convents on the Rupertsberg and Disibodenberg. Trithemius studied Hildegard's writings with great care and attached

[1] An illustration of this parallelism between Paracelsus and Hildegard is afforded by certain passages in the *Labyrinthus medicorum errantium* and the *Scivias*, lib. i, vis. 4. Especially compare p. 279 et seq. of Huser's edition of the *Opera*, Strasbourg, 1603, with Migne, col. 428.

much importance to them, so that they may well have influenced his pupil. The influence of mediaeval theories of the relation of macrocosm and microcosm is encountered among numerous Renaissance writers besides Paracelsus, and is presented to us, for instance, by such a cautious, balanced, and scientifically-minded humanist as Fracastor. But as the years went on, the difficulty in applying the details of the theory became ever greater and greater. Facts were strained and mutilated more and more to make them fit the Procrustean bed of an outworn theory, which at length became untenable when the heliocentric system of Copernicus and Galileo replaced the geocentric and anthropocentric systems of an earlier age. The idea of a close parallelism between the structure of man and of the wider universe was gradually abandoned by the scientific, while among the unscientific it degenerated and becâme little better than an insane obsession. As such it appears in the ingenious ravings of the English follower of Paracelsus, the Rosicrucian, Robert Fludd, who reproduced, often with fidelity, the systems which had some novelty five centuries before his time (Figs. 6, 7, and 8). As a similar fantastic obsession this once fruitful hypothesis still occasionally appears even in modern works of learning and industry.[1]

VIII. Anatomy and Physiology

Hildegard's ideas on these subjects are set out in the fourth vision of the *Liber Divinorum Operum*, which is devoted to a description of man's body according to the macrocosmic scheme. This setting makes her account by no means easy to read, while it increases the difficulty of tracing the origin of her views.

The list of works containing anatomical descriptions available to a German writer in the early Middle Ages is not long. Avicenna was hardly yet accessible, and only such scraps of Galen as appear in Constantine and the Salernitans. The available works may be enumerated thus:

(*a*) The short *Anatomia porci* of Copho of Salerno, dating from about 1085.[2]

[1] A good example is furnished by a work of Isaac Myer, *Qabbalah. The philosophical writings of Solomon ben Yehudah ibn Gebirol or Avicebron and their connection with the Hebrew Qabbalah and Sepher ha-Zohar*, Philadelphia, 1888.

[2] The most accessible edition is in S. de Renzi's *Collectio Salernitana*, vol. ii, p. 388.

(*b*) An anonymous Salernitan anatomy,[1] written about 1100 and largely based on Copho and Constantine.

(*c*) The *Liber de humana natura* of Constantine the African, written probably between 1070 and 1085 at Monte Cassino.[2]

(*d*) Constantine's *De communibus medico cognitu necessariis locis*, written about the same time as the above.[3] This work is in four books, of which the second, third, and fourth are devoted to anatomy and physiology.

(*e*) Here may be placed also Constantine's translation of the *Viaticum* of Isaac Judeus. Both these latter works of Constantine are long and technical, and designed for the use of the trained physician.

In addition to these there was in the Middle Ages a definite anatomic tradition, which expressed itself constantly in:

(*f*) A series of five anatomical diagrams representing respectively the arteries, veins, bones, nerves, and muscles[4] (see Plate XXXIII, opposite page 92 of the present volume). These diagrams were copied in the most servile fashion for centuries, and something very like them has remained in use to this day in Tibet.[5] The versions, whether in Persia or England, in Germany or Italy, were remarkably uniform.

(*g*) In several MSS. there has been found attached to these remarkable diagrams a short text describing the five systems, arteries, veins, nerves, bones, and muscles. This text, however, purporting to be from Galen, has little

[1] Printed in de Renzi, vol. ii, p. 391.

[2] Printed in *Methodus medendi certa clara et brevis*, Basel, Henricus Petrus, 1541, p. 313.

[3] Printed in *Summi in omni philosophia viri constantini africani medici operum reliqua*, Basel, Henricus Petrus, 1539, p. 24.

[4] Karl Sudhoff, *Tradition und Naturbeobachtung*, Leipzig, 1907; *Ein Beitrag zur Geschichte der Anatomie im Mittelalter*, Leipzig, 1908; 'Drei weitere anatomische Fünfbilderserien aus Abendland und Morgenland' (with Ernst Seidel) and 'Abermals eine neue Handschrift der anatomischen Fünfbilderserie' in *Archiv für Geschichte der Medizin*, Leipzig, 1910 and 1914.

[5] E. H. C. Walsh, 'The Tibetan Anatomical System', in the *Journal of the Royal Asiatic Society*, London, October 1910, p. 1215; Berthold Laufer, *Beiträge zur Kenntnis der Tibetanischen Medizin*, Berlin, 1900; and K. Sudhoff, 'Weitere Beiträge zur Geschichte der Anatomie im Mittelalter', in the *Archiv für Geschichte der Medizin*, vol. viii, p. 143, Leipzig, 1914.

From WIESBADEN CODEX B fo. 22r

PLATE XIX. BIRTH. THE ARRIVAL AND TRIALS OF THE SOUL

From WIESBADEN CODEX B fo. 25r

Plate XX. DEATH. THE DEPARTURE AND FATE OF THE SOUL

relation to the figures, which it does not really explain, and it should therefore be regarded as a separate work.[1]

Of these seven sources it appears to us that (*c*) and (*f*)—the short *De humana natura* of Constantine, and the five-figure series—are those on which Hildegard drew. The absence of Arabisms and the scarcity of technical anatomical terms in her writings, her failure to distinguish between veins and arteries, the absence of anything of the nature of myology or osteology, together with the neglect of the spinal marrow as an important organ, make it very unlikely that she consulted Constantine's longer works or the Salernitan authorities or the text of the five-figure series. Her anatomical descriptions resemble those of Constantine's shorter work, on the other hand, in the description of the three vesicles of the brain and their relations to the faculties of the mind, in the treatment of the five senses, in the view of the influence of the planets on the child and the emphasis laid on epilepsy, as well as in the absence of any distinction between arteries and veins, and in the loose doctrines of the humours and of the causes of deformities and monstrosities. In some of these respects also her account of the human body presents points of resemblance to the *De hominis membris ac partibus* of Hugh of St. Victor,[2] with whom, however, her contact appears to be less close than with Constantine.

We may infer that Hildegard had consulted anatomical diagrams and was accustomed to this method of representing the organs from a passage descriptive of the microcosm, in which she says that 'in the mouth of the figure in whose body was the disk, I saw a light brighter than the light of day, in the form of threads, some circular, some in other geometrical forms, and some shaped like human members belonging to the figure, which was clearly portrayed on the disk upright and accurately limned'.[3] These 'circles and geometrical figures' fairly describe the highly diagrammatic manner in which the five-figure series represents the internal organs, and several points suggest that she does indeed refer to this series. Her description of the abdominal muscles

[1] This text, critically treated, has been printed by K. Sudhoff, who, however, regards it as related to the figures: *Archiv für Geschichte der Medizin*, vol. iii, p. 361, Leipzig, 1910.

[2] Hugh of St. Victor, *De bestiis et aliis rebus*, iii. 60.

[3] Migne, col. 755.

(*umbilicus*) 'covering the viscera like a cap', her general descriptions of the vessels (*venae*) and the muscles, and especially her account of the vessels of the leg and of the intimate relations of the main *venae* to the organ of hearing, fits in perfectly with the form of these remarkable diagrams (Plate XVIII).

We here render some of the most important of her general anatomical descriptions:

'The humours may pass to the liver, where wisdom is tested, having been already tempered in the brain by the strength of the spirit, and having absorbed its moisture so that now it is plump, strong, and healthy.

'In the right of man is the liver and its great heat, so that the right is swift to act and to work;[1] but towards the left are heart and lung, which fortify the body for its task and receive their heat from the liver as from a furnace. But the vessels of the liver, affected by the agitation of the humours, trouble the venules of the ear of man and sometimes confound the organ of hearing. . . .

'I saw also that sometimes the humours seek the navel, which covers the viscera as a cap, and holds them in, lest they be dissipated, and maintains their course and preserves the heat both of them and of the veins. . . . But sometimes the humours seek the loins (*lumbos*),[2] which mock, deceive, and endanger the virile powers and which are held in place by nerves and other vessels; in which, nevertheless, reason flourishes so that man may know what to do and what to avoid. . . .

'And the same humours go to the vessels of the reins and of other members, and pass in their turn to the vessels of the spleen, and then to the lungs and to the heart; and they meet the viscera on the left where they are warmed by the lungs, but the liver warms the right-hand side of the body. And the vessels of the brain, heart, lung, liver, and other parts carry strength to the reins, whose vessels descend to the legs, strengthening them; and returning along with the leg vessels, they unite with the virile organ or with the womb as the case may be.

'And as the stomach absorbs food, or as iron is sharpened on a stone, so do they bring the reproductive power to those parts.

[1] An idea that occurs in Aristotle, *Parts of Animals*, ii, c. 2, but is rejected by Galen.

[2] Early mediaeval writers held that the *lumbus*, which we have rendered *loin*, was intimately connected with the sexual faculties. Thus Hugh of St. Victor (1095–1141), *De bestiis et aliis rebus*, iii. 60 'Lumbi a libidinis lascivia dicti, quia in viris causa corporeae voluptatis in ipsis est, sicut in umbilico feminis. Unde et ab Iob in exordio sermonis dictum est, *accinge sicut vir lumbos tuos*, ut in his esset resistendi praeparatio, in quibus est libidinis usitata dominandi occasio.'

Prima visio tertie partis.

T ego
homo
sumpta
ab aliis
homini
bus que
nō sū dig
na nomi
nari homo ppt transgressionē legis
dī. cū deberem ēē iusta & sum iniusta.
n qd di creatura sum ipsi grā. que
me etiā saluabit. uidi ad orientem.
& ecce illic conspexi uelut lapidem
unū totū integrū imm̄se latitudi
nis atq; altitudinis. habentē ferreū
colorē. & sup ipsum candidā nubē.
ac sup eā positū regalem tronum ro
tundū. in quo sedebat quidā uiuens
lucidus mirabilis glē. tanteq; clarita
tis ut nullatenus eū pspicue possem
intueri. habens qsi in pectore suo li
mū nigrū & luculentū. tante lati
tudinis ut alicuius magni hominis
pectus ē. circūdatū lapidib; p̄ciosis
atq; margaritis. Et de ipso lucido
sedente in trono ptendebat magnus
circulus auri coloris ut aurora. cui
amplitudinē nullom̄ cōphendere
potui. girans ab oriente ad septentrio
nē & ad occidentē atq; ad meridiem.

Plate XXI. THE FALL OF THE ANGELS

'Again, the muscles of the arms, legs, and thighs contain vessels full of humours; and just as the belly has within it viscera containing nourishment, so the muscles of arms, legs, and thighs have both vessels and the [contained] humours which preserve man's strength. . . . But when a man runs or walks quickly, the nerves about the knees and the venules in the knees become distended. And since they are united with the vessels of the legs, which are numerous and intercommunicate in a net-like manner, they conduct the fatigue to the vessels of the liver, and thus they reach the vessels of the brain, and so send the fatigue throughout the body. But the vessels from the reins pass rather to the left leg than to the right, because the right leg gets its strength more from the heat of the liver. And the vessels of the right leg ascend as far as the renal and kindred vessels, and these latter vessels unite with those of the kidney. And the liver warms the reins which lie in the fatness derived from the humours. . . .

'The humours in man are distributed in just measure. But when they affect the veins of the liver, his humidity is decreased and also the humidity of the chest is attenuated; so that thus dried, he falls into disease of such a nature that the phlegm is dry and toxic and ascends to the brain. There it produces headache and pain in the eyes and wasting of the marrow, and thus if the moon is in default he may develop the falling evil [epilepsy].

'The humidity also which is in the umbilicus is dispersed by the same humours, and turned into dryness and hardness, so that the flesh becomes ulcerated and scabby as though he were leprous, if indeed he do not actually become so. And the vessels of his testicles, being adversely affected by these humours, similarly disturb the other vessels, so that the proper humidity is dried up within them; and thus, the humours being withdrawn, impetigos may arise . . . and the marrow of the bones and the vessels of the flesh are dried up, and so the man becomes chronically ill, dragging out his days in languor.

'But sometimes the humours affect breast and liver . . . so that various foolish thoughts arise . . . and they ascend to the brain and infect it and again descend to the stomach and generate fevers there, so that the man is long sick. Yet again they vex the minor vessels of the ear with superfluity of phlegm; or with the same phlegm they infect the vessels of the lung, so that he coughs and can scarce breathe; and the phlegm may pass thence into the vessels of the heart and give him pain there, or the pain may pass into the side, exciting pleurisy; under such circumstances also, the moon being in defect, the man may lapse into the falling sickness.' [1]

[1] Migne, cols. 792–3.

Sometimes Hildegard's anatomical ideas can be paralleled among her contemporaries. Thus the following passage on the relationship of the planets to the brain is well illustrated by a diagram of Herrade de Landsberg.

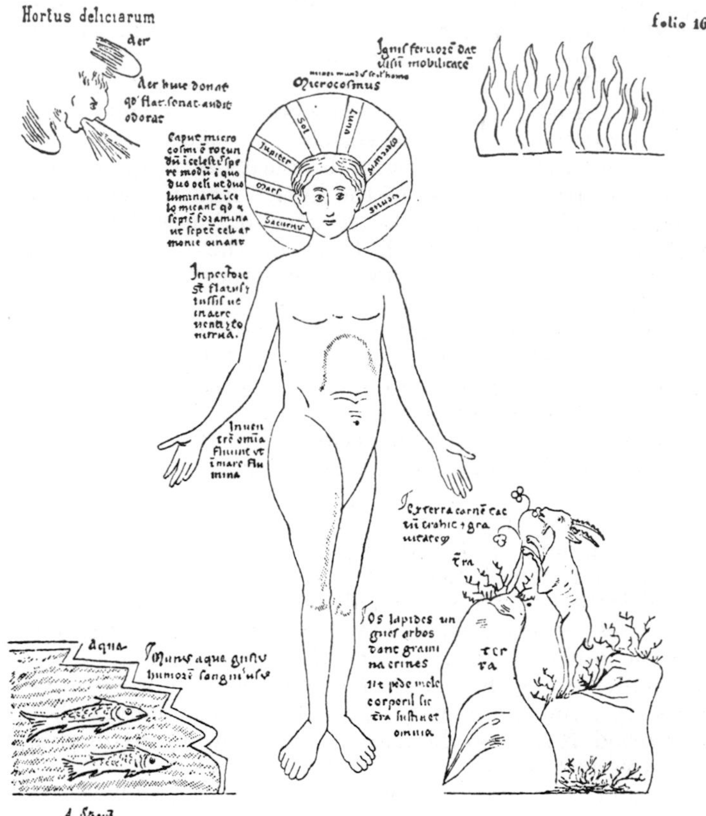

Fig. 9. From Herrade de Landsberg's *Hortus deliciarum*, after Straub and Keller's reproduction.[1]

[1] The legend reads as follows: 'Minor mundus scilicet homo. *Microcosmus.* [Then on the head the names of the seven planets.] Caput microcosmi est rotundum in celestis spere modum in quo duo oculi ut duo luminaria in celo micant quod & septem foramina ut septem celi armonie ornant. In pectore sunt flatus & tussis ut in aere uenti & tonitrua. In uentrem omnia fluunt ut in mare flumina. Os lapides ungues arbos dant gramina crines Ut pede mole[m] corporis sic terra sustinet omnia. [At the four corners the following legends:] Aer huic donat quod flat. sonat. audit. odorat. Ignis feruorem dat uisum mobilitatem. Aqua. Munus aque gustus humorem sanguinis usus. Ex terra carnem tactum trahit & gravitatem.'

From WIESBADEN CODEX B fo. 41 v

PLATE XXII. THE DAYS OF CREATION AND THE FALL OF MAN

'From the summit of the vessel of the brain to the extremity of the forehead seven equal spaces can be distinguished. Here the seven planets are designated, the uppermost planet in the highest part, the moon in front, the sun in the middle and the other planets distributed among the other spaces' (Fig. 9).

IX. Birth and Death and the Nature of the Soul

The method by which the soul enters the body is set forth in a very striking vision in the *Scivias* and is illustrated in the Wiesbaden Codex B by a no less remarkable miniature (Plate XIX). The soul, which contains the element of wisdom, passes into the infant's body while yet within the mother's womb. The *Wisdom of God* is represented as a four-square object, with its angles set to the four quarters of the earth, this form being the symbol of stability. From it a long tube-like process descends into the mother's womb. Down this there passes into the child a bright object, described variously as 'spherical' and as 'shapeless', which 'illumines the whole body' and becomes or develops into the soul.

The birth scene is strikingly portrayed. In the foreground lies the mother with the head and shoulders supported and the right arm raised. In her womb is the infant in the position known to obstetricians as a 'transverse presentation'. Around the child may be distinguished clear traces of the uterine membranes. Near the couch are ranged a group of ten figures who carry vessels containing the various qualities of the child. Above and to the left the Evil One may be seen pouring some noxious substance into one of these vessels, or perhaps abstracting some element of good. The whole scene suggests the familiar fairy tale in which, while all bring pleasant gifts to the child's birth, there comes at last the old witch or the ill-used relative who adds a quota of spitefulness.

The scene is described and expounded as follows:

'Behold, I saw upon earth men carrying milk in earthen vessels and making cheeses therefrom. Some was of the thick kind from which firm cheese is made, some of the thinner sort from which more porous [*tenuis*] cheese is made, and some was mixed with corruption [*tabes*] and of the sort from which bitter cheese is made. And I saw the likeness of a woman having a complete human form within her womb. And then, by a secret disposition of the Most High Craftsman, a fiery sphere having none of the lineaments of a human body possessed the heart of the form, and reached the brain and transfused itself through all the members. . . .

And I saw that many circling eddies possessed the sphere and brought it earthward, but with ever renewed force it returned upward and with wailing asked, "I, wanderer that I am, where am I?" "In death's shadow." "And where go I?" "In the way of sinners." "And what is my hope?" "That of all wanderers."'[1] The vision is explained as follows: 'Those whom thou seest carrying milk in earthen vessels are in the world, men and women alike, having in their bodies the seed of mankind from which are procreated the various kinds of human beings. Part is thickened because the seed in its strength is well and truly concocted, and this produces forceful men to whom are allotted gifts both spiritual and carnal. . . . And some had cheeses less firmly curdled, for they in their feebleness have seed imperfectly tempered, and they raise offspring mostly stupid, feeble, and useless. . . . And some was mixed with corruption . . . for the seed in that brew cannot be rightly raised, it is invalid and makes misshapen men who are bitter, distressed, and oppressed of heart, so that they may not lift their gaze to higher things.[2] . . . And often in forgetfulness of God and by the mocking devil, a *mistio* is made of the man and of the woman and the thing born therefrom is deformed, for parents who have sinned against me return to me crucified in their children.'[3] (Compare Constantine *De humana natura,* sections 'De perfectione' and 'De impeditione'.)

Hildegard thus supposes that the qualities and form of a child are inherited from its parents, but that two factors, the formless soul from the Almighty and the corrupt fluid instilled by the devil, also contribute to the character of offspring. This is the usual mediaeval view and is broadly portrayed in the figure.

The strange conception of the body being formed from the seed, as cheese is precipitated and curdled from milk, is doubtless derived from a passage in the Book of Job:

> 'Hast thou not poured me out as milk,
> And curdled me like cheese?
> Thou hast clothed me with skin and flesh,
> And knit me together with bones and sinews' (Job x. 10, 11).[4]

When the body has thus taken shape there enters into it the soul which, though at first shapeless, gradually assumes the form of its host, the earthly tabernacle; and at death the soul departs

[1] Migne, col. 415. [2] Migne, col. 421. [3] Migne, col. 424.

[4] The Aristotelian writings also compare the transformation of the material humours into the child's body with the solidification of milk in the formation of cheese.

From WIESBADEN CODEX B. fo 47r

Plate XXIII. THE VISION OF THE TRINITY

through the mouth with the last breath, as a fully developed naked human shape, to be received by devils or angels as the case may be (Plate xx).

During its residence in the body the soul plays the part usually assigned to it in the earlier mediaeval psychology, before the ideas of Nemesius and Ibn Ghazali had been elaborated and systematized by Albert and Aquinas. Hildegard regards the brain as having three chambers or divisions, corresponding to the three parts of man's nature, an idea encountered in the writings of St. Augustine. Parallel to these there are, she tells us,

'three elements in man by which he shows life; to wit, soul (*anima*), body (*corpus*), and sense (*sensus*). The soul vivifies the body and inspires the senses; the body attracts the soul and reveals the senses; the senses affect the soul and allure the body. For the soul rules the body as a flame throws light into darkness, and it has two principal powers or limbs, the intellect (*intellectus*) and the will (*voluntas*); not indeed that the soul has limbs to move itself, but that it manifests itself thereby as the sun declares himself by his brightness. . . . For the intellect is attached to the soul as the arms to the body: for as the body is prolonged into arms with fingers and hands attached, so the intellect is produced from the soul by the operation of its various powers.' [1]

We need follow Hildegard no further into her maze of micro-cosmology, in which an essential similarity and relationship is discovered between the qualities of the soul, the constitution of the external cosmos, and the structure of the body, a thought which appears as the culmination of her entire system and provides the clue to the otherwise incomprehensible whole.[2]

X. The Visions and their Pathological Basis

For the physical accompaniments and phenomena of Hildegard's visions we have three separate lines of evidence: her own account; the statements of her contemporary biographers, Theodoric and Godefrid; and the miniatures of the Wiesbaden Codex B, probably prepared under her supervision.

It is clear that despite the length and activity of her life, Hildegard did not enjoy normal health. From a very early age she was the subject of trances and visions, and from time to time she was prostrated with protracted illness.

[1] Migne, col. 425.

[2] Especially in the *Liber Divinorum Operum*, pars 1, vis. iv.

'God punished me for a time by laying me on a bed of sickness so that the blood was dried in my veins, the moisture in my flesh and the marrow in my bones, as though the spirit were about to depart from my body. In this affliction I lay thirty days while my body burned as with fever, and it was thought that this sickness was laid upon me for a punishment. And my spirit also was ailing, and yet was pinned to my flesh, so that while I did not die, yet did I not altogether live. And throughout those days I watched a procession of angels innumerable who fought with Michael and against the dragon and won the victory. . . . And one of them called out to me, "Eagle! Eagle![1] why sleepest thou? . . . All the eagles are watching thee. . . . Arise! for it is dawn, and eat and drink." And then the whole troop cried out with a mighty voice, . . . "Is not the time for passing come? Arise, maiden, arise!" Instantly my body and my senses came back into the world; and seeing this, my daughters who were weeping around me lifted me from the ground and placed me on my bed, and thus I began to get back my strength.

'But the affliction laid upon me did not fully cease; yet was my spirit daily strengthened. . . . I was yet weak of flesh, timid of mind, and fearful of pain . . . but in my soul I said, "Lord! Lord! all that Thou puttest upon me I know to be good . . . for have I not earned these things from my youth up?" Yet was I assured He would not permit my soul to be thus tortured in the future life. . . .[2] Thus was my body seethed as in a pot . . . yet gave I thanks to God, for if this affliction had not been from Him I had surely not lived so long. But although I was thus tortured, yet did I, in supernal vision, often repeat, cry aloud, and write those things which the Holy Spirit willed to put before me.

'Three years were thus passed during which the Cherubim pursued me with a flaming sword . . . and at length my spirit revived within me and my body was restored again as to its veins and marrows, and thus I was healed.'[3]

This illness of Hildegard was the longest and the most typical, but by no means the only one through which she passed. She describes her affliction as continuing for long periods, but there can be little doubt, from her history, that during much of the time she was able to carry on some at least of her functions as head of a religious house.

The condition from which she was suffering was clearly a functional nervous disorder; this is sufficiently demonstrated by her

[1] The eagle is frequently in mediaeval writings a symbol of the power of divine grace.

[2] Migne, col. 110.

[3] Migne, col. 111.

WIESB. COD. B. fo. 213v

SEDENS LUCIDUS

From WIESBADEN CODEX B fo 153r

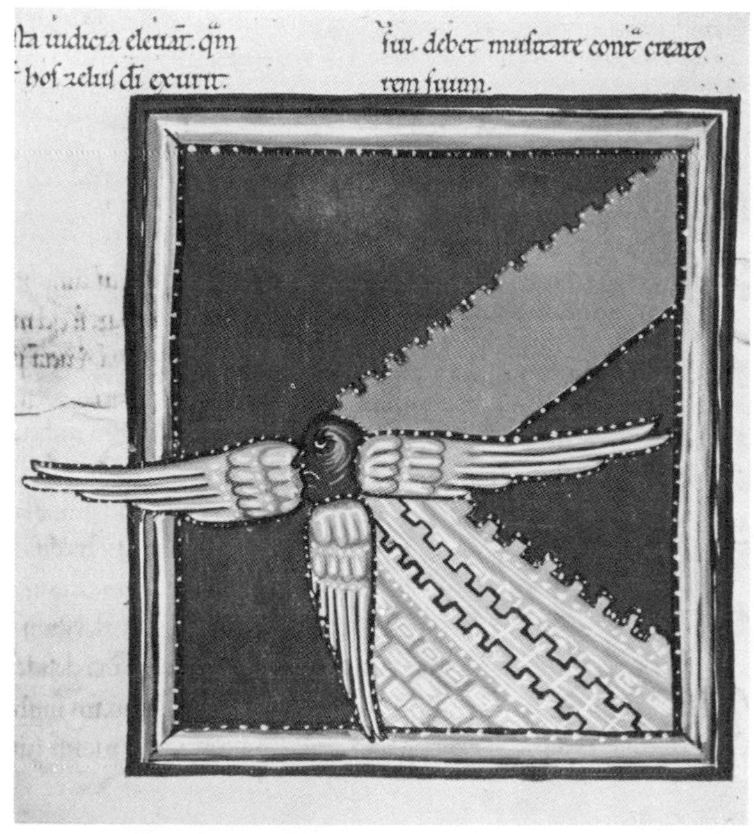

Plate XXIV. ZELUS DEI

repeated complete recoveries, her activity between the attacks, and the great age to which she lived. At first sight, the long procession of figures and visions suggests that she might have been the victim of a condition similar to that of which Jerome Cardan has left us so complete a personal record. But on reading the books of visions, the reader will easily convince himself that we are not here dealing with a dream-state. The visions are indeed essentially vivid. 'These visions which I saw', she repeatedly assures us, 'I beheld neither in sleep, nor in dream, nor in madness, nor with my carnal eyes, nor with the ears of the flesh, nor in hidden places; but wakeful, alert, with the eyes of the spirit and with the inward ears I perceived them in open view and according to the will of God. And how this was compassed is hard indeed for human flesh to search out.' [1]

Nevertheless, though the visions exhibit great originality and creative power—the reader will often be reminded of William Blake—all or nearly all present certain characters in common. In all a prominent feature is a point or a group of points of light, which shimmer and move, usually in a wavelike manner, and are most often interpreted as stars or flaming eyes. In quite a number of cases one light, larger than the rest, exhibits a series of concentric circular figures of wavering form; and often definite fortification figures are described, radiating in some cases from a coloured area. Often the lights gave that impression of *working*, boiling or fermenting, described by so many visionaries, from Ezekiel onwards.

This outline of the visions the saint herself variously interpreted. We give examples from the more typical of these visions, in which the medical reader or the sufferer from migraine will, we think, easily recognize the symptoms of scintillating scotoma. Some of the illuminations, here reproduced in their original colours, will confirm this interpretation.

'I saw a great star most splendid and beautiful, and with it an exceeding multitude of falling sparks which with the star followed southward. And they examined Him upon His throne almost as something hostile, and turning from Him, they sought rather the north. And suddenly they were all annihilated, being turned into black coals . . . and cast into the abyss that I could see them no more' [2] (Plate XXI).

[1] Migne, col. 384.

[2] *Scivias*, lib. iii, vis. 1; Migne, col. 565.

This vision, illustrated by the beautiful figure of stars falling into the waves, is interpreted by her as signifying the *Fall of the Angels.*

The concentric circles appear in numerous visions, and notably in that of the *Days of the Creation of the World and the Fall of Man*, illustrated by what is perhaps the most beautiful of all the miniatures of the Wiesbaden Codex B (lib. ii, vis. 1, Plate XXII). It is in this concentric form that Hildegard most frequently pictures the Almighty, and the idea again appears in the eleventh miniature, here reproduced in its original colours, which she describes as 'a most shining light and within it the appearance of a human form of a sapphire colour which glittered with a gentle but sparkling glow' (lib. ii, vis. 2, Plate XXIII). Appearances of this type are recorded again and again.

The type with fortification figures is encountered in a whole series of visions, of which we reproduce the account and illumination of the *Zelus Dei* (lib. iii, vis. 5, Plate XXIV, lower section).

> 'I looked and behold a head of marvellous form . . . of the colour of flame and red as fire, and it had a terrible human face gazing northward in great wrath. From the neck downward I could see no further form, for the body was altogether concealed . . . but the head itself I saw, like the bare form of a human head. Nor was it hairy like a man, nor indeed after the manner of a woman, but it was more like to a man than a woman, and very awful to look upon.
>
> 'It had three wings of marvellous length and breadth, white as a dazzling cloud. They were not raised erect but spread apart one from the other and the head rose slightly above them . . . and at times they would beat terribly and again would be still. No word uttered the head, but remained altogether still, yet now and again beating with its extended wings.'

From the head extended a series of fortification lines, and this peculiar form of vision is reproduced on several occasions and variously interpreted (Plate XXIV, upper section). It is united with similar visions in what we regard as a reconstructed conception of exceedingly complex structure. This she claims to see separately, and she interprets it as the *aedificium* of the city of God (Plate XXV). Such reconstructed visions are clearly of a different type and origin to the simple group in which a shining light or group of lights is encountered and interpreted as a speaking figure.

From THE WIESBADEN CODEX B. fo. 30r

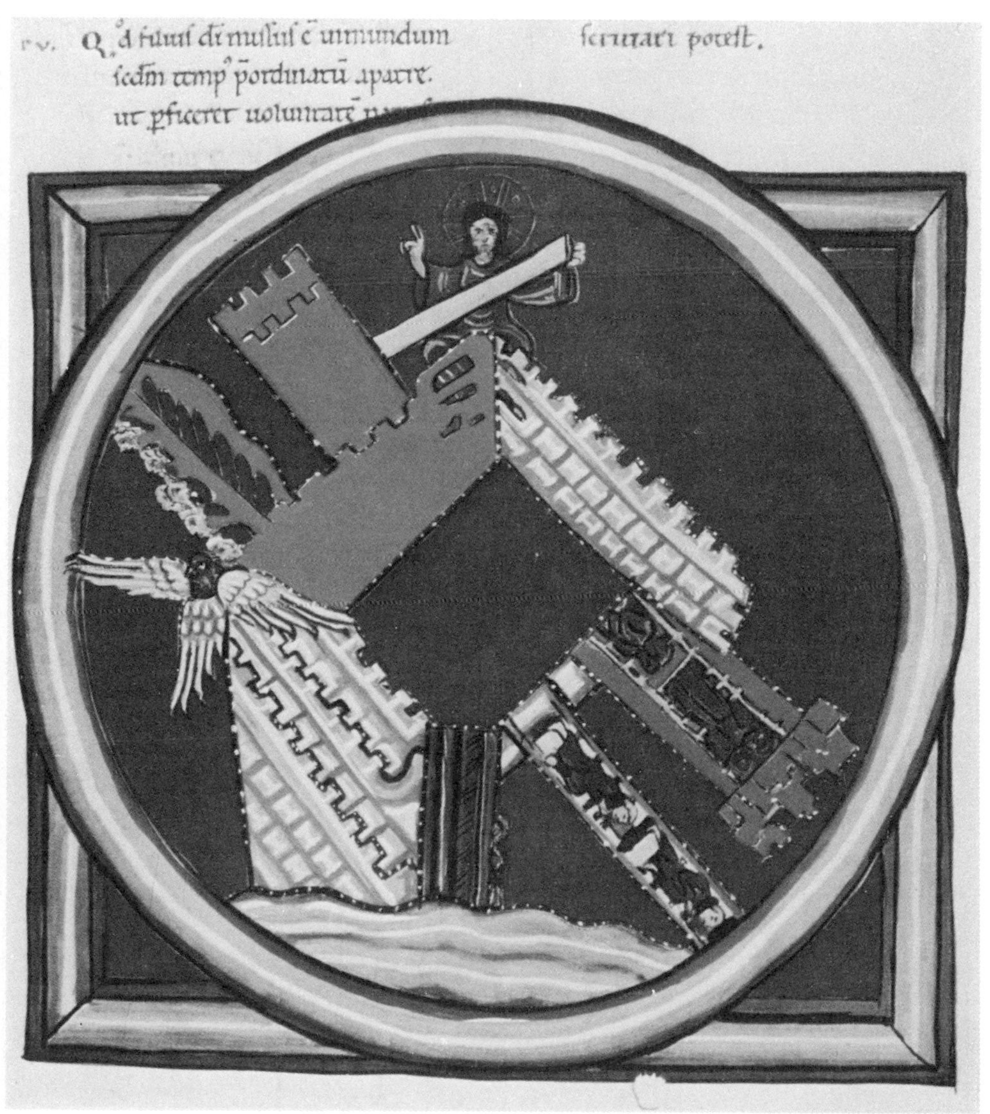

Plate XXV. THE HEAVENLY CITY

Hildegard's visions, perhaps without exception, contain this element of a blinding or glittering light, which she interprets in a more or less spiritual manner. We terminate our account with the passage in which she sums up her experiences of it.

'From my infancy', she says, 'up to the present time, I being now more than seventy years of age, I have always seen this light in my spirit and not with external eyes, nor with any thoughts of my heart nor with help from the senses. But my outward eyes remain open and the other corporeal senses retain their activity. The light which I see is not located but yet is more brilliant than the sun, nor can I examine its height, length, or breadth, and I name it the "cloud of the living light". And as sun, moon, and stars are reflected in water, so the writings, sayings, virtues, and works of men shine in it before me. And whatever I thus see in vision the memory thereof remains long with me. Likewise I see, hear, and understand almost in a moment and I set down what I thus learn....

'But sometimes I behold within this light another light which I name "the Living Light itself".... And when I look upon it every sadness and pain vanishes from my memory, so that I am again as a simple maid and not as an old woman.[1] . . .

'And now that I am over seventy years old my spirit according to the will of God soars upward in vision to the highest heaven and to the farthest stretch of the air and spreads itself among different peoples to regions exceeding far from me here, and thence I can behold the changing clouds and the mutations of all created things; for all these I see not with the outward eye or ear, nor do I create them from the cogitations of my heart . . . but within my spirit, my eyes being open, so that I have never suffered any terror when they left me.'[2]

[1] Migne, col. 18.

[2] Migne, col. 18.

NOTE.—The author's thanks are due to the Rev. H. A. Wilson, Mr. C. C. J. Webb, and Mr. R. R. Steele, who have read the proofs of this article and have made valuable suggestions; to Mr. J. A. Herbert of the MS. Department of the British Museum, who drew his attention to the work of Herrade de Landsberg; and to Mr. M. H. Spielmann, who brought to his uotice the crucifix figured in Plate X. He owes a special debt of gratitude to the late Dom Louis Baillet of Oosterhoot for his courtesy and generosity in lending him reproductions of the illuminations of the Weisbaden Codex. Baillet was a young scholar of great promise, whose early death is a severe loss to the knowledge of mediaeval science.

The author has also to thank Professor Henrici of the Nassauische Landesbibliothek at Weisbaden, Professor Wille and Professor Sillib of the Universitätsbibliothek at Heidelberg, and Signor Boselli of the R. Bibleotica Governativa at Lucca, who have all given him exceptional facilities for the study of the treasures under their charge.

JOHN WILFRED JENKINSON

John Wilfred Jenkinson was born in 1871, and came from Bradfield to Exeter College, Oxford, with a classical scholarship in 1890. After taking his degree in *Literae Humaniores* he came, in 1894, to University College, London, where he devoted himself with extraordinary and never-flagging energy to biological studies.

Without having had the usual preliminary scientific teaching, he brought, on the other hand, a well-trained mind to bear on his new work, and the rapidity and completeness with which he acquired his scientific equipment was one of the most striking and interesting points in his career. Jenkinson very soon turned to original investigation, and from the first he showed a predilection for Embryology.

For a short time he held a post at one of the great London hospitals, but he soon returned to Oxford to join the teaching staff of the Department of Comparative Anatomy. He used the opportunity of University vacations to work in the laboratory of the late Professor A. A. W. Hubrecht at Utrecht, where part of his first published research was written. During the fifteen years of life that remained to him, he established himself as the foremost English writer on Embryology, devoting himself especially to its experimental aspect, a line of work in which he will rank as one of the pioneers.

Jenkinson became Doctor of Science in 1905, and in the same year he married Constance Stephenson. In 1906 he was appointed University Lecturer in Embryology, and in 1909 he was elected to a Research Fellowship at Exeter College.

Jenkinson's mind was not of the type that matures early, but one felt in him a power of solid intellect that gained in force from year to year. The gap in the ranks of British Science caused by his death has been generally recognized, but his loss seems greatest to those personally acquainted with him, who know that he had by no means reached the zenith of his powers.

Jenkinson led a single-minded and unselfish life, wholly free from worldly and ignoble ambitions. Of simple and winning humour, happy in his domestic life and absorbed in his studies, he represented the very best type of scientific worker.

He was gifted with a powerful physique, and on the outbreak of war he became an ardent member of the Oxford Volunteer

Training Corps. His qualities of calm courage and high sense of duty marked him out as a valuable officer. Although forty-three years of age, he took a commission in the 12th Worcester Regiment in January, 1915, and was promoted Captain in the following April. On May 10 he left for the Dardanelles, having been selected for service with the 2nd Royal Fusiliers. He was killed in action on June 4, only ten days after his arrival at the Gallipoli peninsula.

BOOKS AND PAPERS BY J. W. JENKINSON

1. 'A Re-investigation of the Early Stages of the Development of the Mouse.' *Quart. Jour. Micr. Science*, xliii. 1900.

2. 'Observations on the Histology and Physiology of the Placenta of the Mouse.' *Tijdschr. Nederland. Dierkund. Vereen.*, vii (2). 1902.

3. 'Observations on the Maturation and Fertilization of the Egg of the Axolotl.' *Quart. Jour. Micr. Science*, xlviii. 1905.

4. 'Remarks on the Germinal Layers of Vertebrates and on the Significance of Germinal Layers in general.' *Mem. and Proc. Manchester Lit. and Phil. Soc.* 1906.

5. 'Notes on the Histology and Physiology of the Placenta in Ungulata.' *Proc. Zool. Soc.* 1906.

6. 'On the Effects of certain Solutions upon the Development of the Frog's Egg.' *Arch. Ent.-Mech.*, xxi. 1906.

7. 'On the Relation between the Symmetry of the Egg and the Symmetry of the Embryo in the Frog (*Rana temporaria*).' *Biometrika*, v. 1906.

8. *Experimental Embryology*. Oxford, 1909.

9. 'On the Relation between the Symmetry of the Egg, the Symmetry of Segmentation, and the Symmetry of the Embryo in the Frog.' *Biometrika*, vii. 1909.

10. 'The Effects of Sodium Chloride on the Growth and Variability of the Tadpole of the Frog.' *Arch. Ent.-Mech.*, xxx (2). 1910.

11. 'Vitalism.' *Hibbert Journal*. 1911.

12. 'On the Development of Isolated Pieces of the Gastrulae of the Sea-Urchin *Strongylocentrotus lividus*.' *Arch. Ent.-Mech.*, xxxii. 1911.

13. 'On the Effect of certain Isotonic Solutions on the Development of the Frog.' *Arch. Ent.-Mech.*, xxxii. 1911.

14. 'On the Origin of the Polar and Bilateral Structure of the Egg of the Sea-Urchin.' *Arch. Ent.-Mech.*, xxxii. 1911.

15. 'The Development of the Ear-Bones in the Mouse.' *Jour. Anat. and Phys.* vi (3). 1911.

16. 'Growth, Variability, and Correlation in Young Trout.' *Biometrika*, viii. 1912.

17. *Vertebrate Embryology*. Oxford, 1913.

18. 'The Effect of Centrifugal Force on the Structure and Development of the Egg of the Frog.' *Quart. Jour. Micr. Science*. 1914.

19. 'The Placenta of a Lemur.' *Quart. Jour. Micr. Science*. July 1915.

20. *Three Lectures on Experimental Embryology*. Oxford, 1917.

VITALISM

By J. W. Jenkinson

In one of the oldest biological treatises in the world, the soul or life of an organism is defined in the most general way as an activity of a natural organic living body—ἐντελέχεια σώματος φυσικοῦ ὀργανικοῦ δυνάμει ζωὴν ἔχοντος—life being autonomous nutrition and growth and decay. The activity may, however, be latent or patent, passive or active, sleeping or waking, without losing its peculiar characters. It is substance (οὐσία), but substance as 'form' as opposed to the material substance of the body, and the living body is therefore also a substance in a double sense.

It is not identical with the body; but as form, proportion (λόγος), activity (ἐνέργεια), essence (τὸ τί ἦν εἶναι), it is related to the body, mere matter (ὕλη), and potentiality (δύναμις) in just the same way as the seal is related to the wax; and the body is the instrument whereby it effects its purposes; though subsequent in time, it is prior in thought to the body, as all activities are to the materials with which they operate.

At the same time neither it nor its parts are separable from the body, with the exception, possibly, of mind (νοῦς); it is indeed the actual or possible functioning of the body, like the seeing of the eye or the cutting of the axe, and with the disappearance of the capacity of this functioning the soul itself also perishes. Lastly, it is a cause (ἀρχὴ καὶ αἰτία) in a triple sense: first, as the source of motion; secondly, as that for the sake of which the body exists; and thirdly, as its essence (οὐσία) or formal cause. The soul or life is of several kinds, which form together an ascending series each member of which is necessarily involved in those above it.

The lowest is the nutritive soul (θρεπτική), found in all living things, and the only soul possessed by plants. It is defined as motion in respect of nutrition, decay, and growth, processes which involve alteration (ἀλλοίωσις) in the body; and its functions (ἔργα) are to utilize food for the maintenance and reproduction of the form of the body, and to control and limit growth.

The second is the perceptive soul (αἰσθητική), the possession of which distinguishes animals from plants. This also is a kind of alteration (ἀλλοίωσις τίς) and consists in being moved and affected. The fundamental and indispensable perception is touch (ἁφή), for it is concerned in the acquisition of the food. It is invariably present: the others may or may not, some or all, coexist with it.

Thirdly, some animals are possessed of a capacity for locomotion, and the performance of this function requires again a special kind of soul.

Lastly, there is the reasoning soul (διανοητικά) or mind (νοῦς). This is found in man alone, unless there be other beings similar to him, or even nobler than he. Mind alone is eternal and separable from the body.

Though the observation and experiment of modern science would doubtless find much to alter in the details of these simple definitions, yet it must be conceded that, by what is certainly a most fortunate guess if it is not the most wonderful insight, Aristotle has laid his finger on the cardinal point of modern physiological doctrine. For, putting aside for the moment the mental faculties, it is here laid down in the clearest manner that not only the functions of growth and decay, nutrition, and reproduction, but also the capacity of responding to stimuli are to be ultimately resolved into some kind of movement of the particles of which the body is composed. Life, in short, as we might say with Virchow, is a mode of motion.

The biology of to-day distinguishes living from inanimate bodies by the possession and exercise of the three principal properties or functions of metabolism, irritability, and reproduction; and further, the body which performs these functions is not only composed of chemically complex substances—proteids—which are not found in things that are not alive, but possesses a structure. In no case, even the simplest, is the organism a mere homogeneous lump of protoplasm, but it has parts or organs, visibly different from one another, and obviously correlated with the activities appropriated to each; and it is the preservation of that structure, in the individual and in the race, which is the end towards which the collective performance of all these functions, or the life of the organism, is apparently directed.

Some of these peculiarities are shared by certain things that are not commonly regarded as alive. Crystals have of course a definite structure; they can divide, and when broken they can make good the missing part, but they do not assimilate to the substance of their own bodies a food-material which is less complex than it, and they are not irritable.

The differences, indeed, between the living and the lifeless are so profound, that it is not to be wondered at that there should have been in all ages natural philosophers who have held that living activities are phenomena *sui generis*, differing *toto caelo* from the properties exhibited by lifeless bodies, and never by any conceivability to be expressed in terms of these.

This doctrine is vitalism.

It exists in several varieties, but one at least is of very ancient lineage and can be traced back through mediaeval times to the biological speculations of the Greeks.

Whether Aristotle really held the vitalistic views which have since been attributed to him is a matter we shall have to discuss later on, but it is certain that in the writings of Galen there is to be found a theory of life which bears the stamp of Aristotelian influence, and was destined to hand that influence on to future generations. Galen admits the sensitive soul of Aristotle as the peculiarity of animals, and the rational soul for man, but substitutes for the nutritive soul certain works of nature attraction, repulsion, retention, alteration. And further, the rational soul is no longer immortal, but perishable, and is dependent on the body, where its seat is in the brain; it is material or quasi-material, a πνεῦμα, most efficient when dry.

After a long interval this doctrine reappears in the sixteenth century in the writings of Vesalius, who tells us that the heart has a vital soul, the liver a natural soul, while there is elaborated in the ventricles of the brain an animal spirit or principal soul.

Meanwhile, however, the conception of life as something material had been discarded by Paracelsus for the belief that the soul, or as he called it, the 'Archaeus', by which the chemical processes of the body are governed, is not a material but a spiritual force, a view restated by Stahl more than a hundred years afterwards. 'The events of the body', says this author, 'may be rough-hewn by chemical and physical forces, but the soul will shape them to its own ends, and will do that by its instrument, motion.'

This, of course, is vitalism, and vitalism in its extreme or 'animistic' form. The idea recurs later on in the biology of

Treviranus. To be living is to have a soul, he tells us, and the conscious *Lebenskraft* employs the forces of the material world to form the organism. 'Das Weitzenkorn hat allerdings Bewusstsein dessen, was in ihm ist und aus ihm werden kann, und träumt wirklich davon.' Though he adds quaintly enough, 'Sein Bewusstsein und seine Träume mögen dunkel genug sein'. It is curious to observe the revival, at the beginning of the twentieth century, of this mediaeval mysticism in the speculative writings of so accomplished an experimentalist as Hans Driesch.

Driesch is an embryologist who in his earlier days had enunciated an invaluable analytical theory of development, a theory which suggests that while the formation of the first or elementary organs that appear in the embryo or larva—such structures as the larval gut or sense-organ, or the germ-layers—depends upon the presence in the germ of certain specific organ-forming substances (and this is a fact which has since been abundantly demonstrated by experiment), the origin of parts that appear later in development may be accounted for by the action of the first-formed structures upon one another, these actions being in the nature of physiological responses to stimuli; and for this also some evidence has been produced. On this view differentiation is a mechanical process, set in motion by fertilization or some other cause, and, given a certain initial structure of the germ or ovum, given the presence in it of a certain number of parts or substances capable of acting upon one another with a fixed co-ordination or harmony of the stimuli and the responses, given further a proper constitution of the external environment, then a definite result must follow, the production of an organism which is like the parents that gave it birth.

But in his later treatises this hypothesis has been repudiated, and, by a remarkable *volte-face*, replaced by a dogma of a wholly different kind. For now it is urged that no merely material factors can possibly account either for the harmony of development—the due co-ordination of mutually reacting parts; or for the secondary harmony of composition—the formation of complex organs by the union of tissues; or for the functional harmony seen in the activities of the adult.

For example, it is asserted that any fragment of an egg of a sea-urchin, if not too small (not less than $\frac{1}{32}$ of the egg), can give rise to a whole and normal larva. We are told that the cells of the segmented ovum may be disarranged to any extent by various means, such as raising the temperature, diluting the sea-water, removing the calcium from the sea-water, or by shaking, without prejudice to the ultimate normality of development. Each part of the ovum can therefore, according to the needs of the case, give rise to any part of the resulting organism. 'Jeder Teil kann nach Bedürfniss jedes.'

And thirdly, when the gastrula of a sea-urchin is transversely divided into two, each half, it is stated, develops into a diminished whole larva in which the gut becomes divided into the characteristic three regions, and all the other organs are formed in correct proportion.

For each of these acts of development in the whole uninjured larva an explanation may conceivably be given in terms of formative stimuli exerted by the originally distinct parts of the egg and calling forth responses in other parts. A mechanism may be thought of which, when set in motion, will achieve a certain end in accordance with its own pre-established harmony; but a mechanism which can be subdivided *ad libitum*, or almost *ad libitum*, and the parts of which will still achieve the same end, will still behave as wholes with their parts co-ordinated in the same ratio, temporally and spatially! Such a mechanism is inconceivable; for to ensure the uniform result, the relative amounts and positions of the necessary substances must be imagined as identical in every possible fragment of the egg that is not too small. Something is therefore required to superintend, to co-ordinate the causes of development in the case not only of the part but of the whole egg as well; and this something is not material. A corroborative proof of the inadequacy of the purely material explanation—the causal explanation in the ordinary sense of the phrase—may be derived from a consideration of certain other vital processes. The facts of acclimatization and immunity betray an extraordinary adaptability of the organism to a change in its environment; an organ will adapt itself structurally to an alteration, quantitative or qualitative, of function [Roux's 'Functional Adaptation']; lost parts can be regenerated; and then there is the physiology of the nervous system.

In all these cases of 'regulation'—and indeed in all other responses to stimuli—the same element, inexplicable in chemical and physical terms, exists and must exist in development. This entity is not a form of energy, but a vital constant, analogous to the constants or ultimate conceptions of mechanics and physics and chemistry and crystallography, but not reducible to these, just as these cannot be translated into one another.

Driesch describes it as a rudimentary feeling and willing, a 'psychoid', 'morphaesthetic' or perceptive of that form which is the desired end towards which it controls and directs all the material elements of differentiation, like the grain of wheat of Treviranus, dreaming dimly of its destiny. It is thus a *vera causa*—an unconditional and invariable antecedent—a psychical factor which can intervene in the purely physical series of causes and effects, and for it he revives the Aristotelian term 'Entelechy'.

Such is the 'vitalism' introduced by Hans Driesch, a teleological theory clearly, but no mere metaphysical doctrine of final causes: rather a dynamic teleology which not only sees an end in every organic process, but postulates an immaterial entity to guide the merely mechanical forces towards the realization of that end.

Such a theory is open to very serious criticism from both the scientific and the philosophical side. But before we pass to that criticism let us turn aside to examine some of the other aspects under which the Proteus of Vitalism presents himself.

Thus the modern physiologist Bunge, while owning that it would be a lack of intelligence to expect to make with our senses discoveries in living nature of a different order to those revealed to us in inorganic nature, yet insists that we must transfer to the objects of our sensory perception, to the organs, to the tissue elements, and to every minute cell, something which we have acquired from our own consciousness, something, that is to say, which is not motion, and is not in space, but is in time only.

The essence of vitalism, so Bunge would have it, lies in starting from what we know, the internal world, to explain what we do not know, the external world. We can only remark that this position appears to rest upon an epistemological confusion, for Bunge has evidently failed to distinguish between the idealism which teaches that the world of nature, including our own bodies, only exists in so far as it is an object of knowledge, that reality is ultimately ideal, and the 'animism' which, as we have seen, gives every object, at least every living object, in nature a directive consciousness of its own. The former does not lie immediately within the scope of the present inquiry; the latter we shall have occasion to discuss again.

How far the tenets of animism are to be attributed to Johannes Müller is not very clear. For while Müller maintained that an organism is due to an idea which regulates its structure, is the cause of its harmony, and is in action in the organism itself, exerting on it a formative power, yet he held that the process was unconscious. Müller indeed distinguished explicitly between the vital and the mental or conscious principle, for in the operations of the former the manifestation of design is the result of necessity, not of choice. At the same time the two resemble one another in being homogeneous, in existing throughout the mass of the organism which they animate, and in being divided together with the organism (as in regeneration) without suffering any diminution or change of their powers.

In this conception of the unconscious idea there may possibly be some confusion between the formal and the final cause, between the idea of the end to be realized, present at the beginning in the mind of the artificer, and the end itself. The former is animism; the latter is sound enough as metaphysics, but is not science at all.

There is still another school of vitalists which, while not going so far as to commit itself to a belief in a 'psychoid', yet proclaims in no uncertain voice the autonomy of the organism, and not content with the assertion that at present we have not succeeded in reducing the activities of the organism to chemical, physical, and mechanical processes, maintains the utter futility of such endeavour, and pronounces over the hidden mysteries of life an eternal *Ignorabimus*.

Some such view as this we must, I think, attribute to Dr. Haldane. 'In biology', he says, 'the phenomena which are or ought to be observed from the very beginning are not physical and chemical phenomena as self-existent events, but these phenomena as expressions of the activity of living organisms. It is the living organism, and not the physical phenomenon, which is the reality for biology.' His belief in organic autonomy is based on the physiology of metabolism, secretion and absorption, the circulation of the blood, and the nervous system. Thus in discussing the blood, after pointing to the constancy in its volume and composition, he proceeds: 'Neither starvation nor ingestion of food and drink materially affect it: liquid injected into it is got rid of with remarkable rapidity; and any loss of blood by bleeding is soon replaced. This vital metabolism of the circulatory system is doubtless due chiefly to the activity of its lining endothelium, which most certainly does not play the mere mechanical part which has often been attributed to it. The other so-called "mechanisms" can likewise be shown to have all the characteristics of the living body, inasmuch as they actively maintain their structure, just as the organism as a whole does so. There is thus no warrant for calling them mechanisms, and thus ignoring what is one of their essential characteristics.' In passages such as these we seem to catch an echo of Müller's unconscious idea, and again we ask ourselves, Are we dealing with a final or a formal cause ? Indeed, Dr. Haldane insists that his ground conception is teleological.

There is still one other vitalistic theory to which we must allude, although its interest is now merely historical. This is the belief in a special vital material, unlike the material of which lifeless bodies are composed, and endowed with a special vital force, different from but co-ordinate with the forces of mechanics and physics.

In his *Histoire Générale des Animaux* Buffon, after referring to the obvious peculiarities of animals and vegetables—that their actions are directed to an end, the conservation of a durable species—proceeds to elaborate a thesis in which it is held that they are composed of organic germs, and that germs of the same kind are distributed throughout nature, lifeless as well as living. When an animal or plant dies, its body is dissolved into these germs, which are then scattered abroad ; when it assimilates, it is by separating these ubiquitous particles from the brute inorganic portion of the food. The former is utilized for its own growth, the latter it gets rid of by evacuation and excretion. Lifeless matter is therefore never converted into living material.

Another advocate of the doctrine of a vital force, a property of the tissues of the body, and at perpetual war with those inorganic forms which tend to their destruction, was the physiologist Bichat. Such a conception as this could not of course survive the rise of modern chemistry. Its death-knell was sounded when Lavoisier and Laplace showed that the bodies of organisms were composed of the same elements as are found in inanimate nature, and it has long since passed into the limbo of discredited speculations.

Apart from this, vitalistic theories would appear to be in the main of two kinds.

First, there is the metaphysical vitalism which tells us we can never explain the living in terms of the lifeless, insists on the permanent separation of the sciences of biology on the one hand from chemistry and physics on the other, and preaches the autonomy of the organism without venturing to tell us in what that autonomy consists.

Secondly, there is the psychological theory of animism which posits an autonomous psychical entity to preside over the chemical and mechanical operations of the body, whether already formed or in process of development, and to direct them towards its own ends, the conservation and reproduction of that body's specific form.

A third party, halting between two opinions, suggests an unconscious idea, without, however, clearly explaining whether this is to be taken in a metaphysical or a psychological sense. Frankly opposed to vitalism in all its forms is the conception of the living body as a mechanism. This has also an honourable ancestry behind it. How far the biology of Aristotle is to be looked upon as mechanistic we shall presently have to inquire, but in Galen the soul is certainly material, or quasi-material, as we have already observed. It is, however, in the physiology of Descartes that mechanism first appears unmistakably in its modern guise.

For Descartes the body is simply an earthly machine. The nerves are tubes up which—in sensation—the animal spirits flow to the brain only to be reflected (whence our term reflex action) down other tubes to the muscles.

'All the functions of the body', he tells us, 'follow naturally from the sole disposition of its organs, just in the same way that the movements of a clock or other self-acting machine or automaton follow from the arrangement of its weights and wheels. So that there is no reason on account of its functions to conceive that there exists in the body any soul, whether vegetative or sensitive, or any principle of movement other than the blood and its animal spirits agitated by the heat of the fire which burns continually in the heart and does not differ in nature from any of the other fires which are met with in inanimate bodies.'

The rational soul, the soul which thinks, that is, understands, wishes, imagines, remembers, and feels, is not material. Yet it always acts through the machine, though that machine can go on perfectly well without the soul. 'When the body has all its organs properly arranged for a particular movement it has no need of the soul to carry them out. All movements, even those which we call voluntary, depend principally on the same disposition of the organs.

F 2

One and the same cause renders the dead body unfit to produce the movements and leads the soul to quit the body.'

The biology of Descartes appears to have been accepted by contemporary physiologists like van Helmont and Borelli, and certainly commended itself to another philosopher of eminence, Leibnitz. Like Descartes, Leibnitz also affirms that the body is a machine or natural automaton; unlike Descartes, however, he refuses to believe that the mind directs the machine in any way. Rather there is a complete series of psychical parallel to a complete series of physical events, and between the two a pre-established harmony.

Although the details of Cartesian physiology have long since been exploded, yet the mechanical principle which that philosophy enunciated so clearly has persisted and has indeed proved to be the rock on which modern physiological science has been built. For, when once the chemists had discovered animal and plant structure to be composed of elements found in lifeless bodies, and had proved that compounds found only in the organism could yet be synthesized *in vitro*, there was no longer any reason why the properties of the compounds should be considered as of a different order to the properties of their component elements. A method applicable to one was applicable to the other, and as Claude Bernard has put it, mechanical, physical, and chemical forces are the only effective agents in the living body, and they are the only agencies of which the physiologist has to take account.

The substances of which the living body is made up are no doubt extremely complex, yet none the less—to quote a more recent writer, Verworn—'physiology is in the last resort the chemistry of the proteids'. This is the principle that has now for nearly a century guided and stimulated research into the functions of the organism: to this principle physiologists, too numerous to name, have not been ashamed to subscribe: under its banner some of the proudest triumphs of the science have been won. Yet it is precisely this which modern or neo-vitalism has challenged and asks us to relinquish in favour of a theory of psychoids or a pseudo-metaphysical view of life.

The vitalistic position may be assailed from two points, the scientific and the philosophical.

In the first place the vitalist asserts that mechanism is inadequate to explain the phenomena of metabolism, of transmission of nervous stimuli, or of development. It is upon the last of these that Driesch lays special stress.

He has urged, as we have seen, that although a mechanical explanation might be given (such an explanation has indeed been put forward by himself) of the specific differentiation of the organism by supposing the first-formed elementary organs, developed out of the substances given in the initial structure of the germ, to act and react upon one another in accordance with a certain harmony, provided for by the same structure; yet a mechanism which can be subdivided *ad libitum* or almost *ad libitum*, and each part of which will still give rise to a complete organism, is not to be conceived. The answer to this objection has, however, been supplied by the experiments of Driesch himself and of many others. For though it is true that each of the first two, four, eight, or even in some cases each of the first sixteen cells into which the fertilized ovum becomes segmented, can, when separated from its fellows, give rise to a complete organism, yet in all cases there comes a time when the parts cease to be totipotent and produce not whole but partial structures.

This invariable restriction of potentialities, which occurs earlier in some cases than in others, and is not due to mere deficiency of substance, is not hard to account for.

Those substances on the presence of which in the ovum, as experiment has taught us, the formation of the elementary organs of the embryo or larva depends, are arranged in different cases in different ways: and they certainly may be, and very frequently are, so distributed that while each of the first four cells contains a like quantity of each of these specific substances, arranged in it exactly as they were in the whole ovum, the next division will sunder these materials in such a way that of the resulting eight blastomeres four will have more of one of the primary egg-substances, less of another; the amounts apportioned to the other four being in just the inverse ratio of this: and the result will be a difference in the fate of the cells when they are isolated from one another. In those of the one group the proportions of the organs developed out of these substances will not be the same as they are in the other. This is precisely the result which experiment has revealed; it is exactly this result which Driesch has ignored, or rather attempted to explain away.

It is evident, then, that to some extent the parts of this

mechanism are interchangeable, that it can be subdivided, and that each part, brought now under new conditions, will still possess the potentialities of the whole, just as such a mechanism as a rocket, out of which, under the appropriate stimulus, a certain pattern of stars is developed, might be subdivided into two or more rockets of half size or less. There is, however, a limit to this interchangeability, while if the subdivision be carried beyond a certain point the totipotence of the parts is lost.

If the number of these organ-forming substances given in the germ were very large, as large, let us suppose, as the total number of separately inheritable characters, it might indeed be difficult to imagine a mechanism divisible into even two totipotent parts. But from the need for this assumption we are saved by the second part of Driesch's own *Analytische Theorie*, which accounts for subsequent processes of differentiation by attributing the production of new parts to the mutual interactions of those that are the first to appear. For this also experimental evidence, though meagre, is not lacking, while a close parallel is found in the dependence of certain bodily functions upon substances—the hormones of Professor Starling—secreted by other organs.

In the second place the vitalist maintains that the processes of metabolism defy, nay more, always will defy, chemical and physical analysis. The first part of this statement may be a true description of the knowledge of to-day, but the existence in the living body of the same elements as are met with elsewhere, the synthesis of complex organic substances, the establishment of the equivalence of the energy which leaves the body as mechanical work or heat to that which enters it in chemical form in the food, should surely make us hesitate before abandoning all hope of attaining to a chemistry of life.

And thirdly, there are physiologists who believe that the complex phenomena presented to us in the activities of the nervous system are susceptible of a purely mechanical explanation.

'A feature', says Gotch, 'which more particularly suggests spontaneous cellular activity is the well-known fact that centrifugal discharges may continue after the obvious centripetal ones have ceased. This is pre-eminently the case when the central mass is rendered extremely unstable by certain chemical compounds, such as strychnine, &c. There are, however, suggestive indications in connexion with such persistent discharges. The more completely all the centripetal paths are blocked by severance and other means, the less perceptible is such persistent discharge, and since nervous impulses are continually streaming into the central mass from all parts, even from those in apparent repose, it would seem that could we completely isolate nerve-cells, their discharge would probably altogether cease.' Even in the hyper-excitable condition produced by strychnine the spinal motor nerve cells do not discharge centrifugal impulses when cut off from the centripetal connexions. The physiologist, therefore, has 'definite grounds for believing that, as far as present knowledge goes, both the production and cessation of central nervous discharges are the expression of propagated changes and that these changes reveal themselves as physico-chemical alterations of an electrolytic character. The nervous process, which rightly seems to us so recondite, does not, in the light of this conception, owe its physiological mystery to a new form of energy, but to the circumstance that a mode of energy displayed in the non-living world occurs in colloidal electrolytic structures of great chemical complexity.'

To all these considerations we must add the fact that life did once originate upon this planet from matter which was not alive, and that even now some inorganic phenomena present at least remote analogies with certain vital processes. Such are the structure, the spontaneous division, and the regeneration of crystals.

We turn now to the philosophical objections that may be raised to vitalistic speculations ; and here we must be careful to distinguish what we may term the psychological from the metaphysical form of the theory.

Driesch has maintained that the belief in a morphaesthetic psychoid finds support in the philosophies of Kant and Aristotle. Let us examine the merits of this claim.

Like the scientists of to-day, Kant, in his *Critique of the Teleological Judgement*, lays it down as a rule that the mechanical method, by which natural phenomena are brought under general laws of causation and so explained, should in all cases be pushed as far as it will go, for this is a principle of the determinant judgement. There are cases, however, in which this alone does not suffice. The possibility of the growth and nutrition, above all of the reproduction and regeneration of organisms, is only fully intelligible through another quite distinct kind of causality, their purposiveness. Organisms are not mere machines, for these have simply moving power. Organisms possess in themselves formative power of a self-propagating kind, which they communicate to their materials. They are, in fact, natural purposes, both cause and effect of themselves, in which the parts so combine that they are reciprocally both end and means, existing not only by means of one another but for the sake of one another and the whole. The whole is thus an end which determines the process, a final cause which brings together the required matter, modifies it, forms it, and puts it in its appropriate place. Such purposiveness is internal, for the organism is at once its own cause and an end to itself, not merely a means to other ends, like a machine whose purposiveness is relative and whose cause is external.

Such is the principle of the teleological judgement. It is a heuristic principle rightly brought to bear, at least problematically, upon the investigation of organic nature, by a distant analogy with our own causality according to purposes generally, and indispensable to us, as anatomists, as a guiding thread if we wish to learn how to cognize the constitution of organisms without aspiring to an investigation into their first origin.

Could our cognitive faculties rest content with this maxim of the reflective judgement it would be impossible for them to conceive of the production of these things in any other fashion than by attributing them to a cause working by design, to a Being which would be productive in a way analogous to the causality of intelligence. Natural science, however, needs not merely reflective but determinant principles which alone can inform us of the possibility of finding the ultimate explanation of the world of organisms in a causal combination for which an understanding is not explicitly assumed, since the principle of purposes does not make the mode of origination of organic beings any more comprehensible. And then, in a passage remarkable for its prophetic insight, Kant proceeds to show how this might be. This 'analogy of forms', he says, 'which with all their difference seem to have been produced according to a common original type, strengthens our suspicion of an actual relationship between them in their production from a common parent, through the gradual approximation of one genus to another—from those in which the principle of purposes seems to be best authenticated, that is from man down to the polype, and again from this down to mosses and lichens, and finally to the lowest stage of nature noticeable by us, namely, crude matter'.

And so the whole technic of nature, which is so incomprehensible to us in organized beings that we believe ourselves compelled to think a different principle for it, seems to be derived from matter and its powers according to mechanical laws like those by which it operates in the formation of crystals. A purposiveness must, however, be attributed even to the crude matter, otherwise it would not be possible to think the purposive form of animals and plants.

Although there are doubtless in the *Critique* many obscurities and inconsistencies, to which we cannot allude now, the general meaning of Kant's reflections upon organisms is perfectly clear. He who would 'complete the perfect round' of his knowledge must think not only in beginnings but in ends. The end in the case of a living being is apparently plain—it is the maintenance and reproduction of its form; the end in the case of the cosmic process is to be sought in the ethical, or, in Kantian phraseology, the 'practical' concept of the freedom of the moral consciousness of man.

Such a position is quite intelligible, philosophically, but the testimony it brings to the theory of the psychoid is of very doubtful value, as Driesch is well aware. He complains indeed that Kant's teleology is descriptive or 'static', rather than 'dynamic', as is perfectly true, except in the case of man, a point of which Driesch naturally makes the most. There are, no doubt, passages where Kant speaks of 'a cause which brings together the required matter, modifies it, forms it and puts it in its appropriate place'; but against these must be set the explicit statement 'that if the body has an alien principle (the soul) in communion with it, the body must either be the instrument of the soul—which does not make the soul a whit more comprehensible—or be made by the soul, in which case it would not be corporeal at all.' Vitalism can glean small comfort from this. Let us turn, then, to the second authority.

As we have seen already, the souls or functions of nutrition and perception are, in the Aristotelian biology, ultimately to be expressed as alterations or movements of the particles of the body; mind alone is separable from body and eternal.

In the development of the individual organism the mind comes in from outside, but the two souls of lower order are present in the *σπέρμα*, or *κύημα*, as Aristotle calls it, which results from the commingling of the male and female elements, or, as we should say, the fertilized ovum. The material and efficient causes of development are not, however, both contributed by each of the parents.

The teaching of Aristotle is that the matter is provided by the female and the female alone. The egg (or catamenia in mammals) is described as being mere matter (ὕλη), body (σῶμα), potentiality (δύναμις), passive (παθητικόν) and merely quantitative, although it is true that a sort of soul, the nutritive, is somewhat grudgingly conceded to it, since unfertilized eggs appear in some sense to be alive. The male element, on the other hand, provides the principle of motion (ἀρχὴ τῆς κινήσεως) and the form (εἶδος); it is qualitative, it is activity, it produces the perceptive soul, if it is not itself that soul, and it is responsible for the 'correct proportionality' (λόγος) of the organization. The male element contributes only motion, but no matter; it acts upon the female element as rennet acts when it coagulates milk, except that the analogy is incomplete, since the γονή brings about a qualitative and not merely a quantitative change in the material on which it operates. To this it imparts the same kind of motion which itself possesses, the motion which was present in the particles of the food in its final form from which it was itself derived. The communication of this motion is enough to set going the machinery (αὐτόματον); the rest then follows of itself in proper order.

Lastly, the sperm of the male acts like a cunning workman who makes a work of art, using heat and cold as the workman uses his tools: for this heat and this cold could never of themselves—by coagulations and condensations—produce the form of the body as the older naturalists had supposed, regarding only the efficient and ignoring the formal and the final cause: for the organic body is not what it is because it is produced in such and such a fashion, rather it is because it is to be such and such that it must be developed as it is.

And here lies the kernel of the whole matter. For while Aristotle has made it perfectly plain that, according to his idea, the soul, at least its nutritive and perceptive faculties, is to be regarded as a function of matter and that this function may be ultimately expressed in terms of movement, and further that development is a mechanism which is set going by the communication of motion proceeding from the 'soul' of the male element and derivable eventually from the motions into which the 'functions' or 'soul' of the parent can be resolved to the mere matter which the female provides, it is equally evident that he does not regard this mechanical explanation—in terms of material and efficient causes—as satisfactory or complete. But when we inquire why, he gives us no certain nor consistent answer. On the one hand, there are passages in which he tells us that there must be something which controls the material forces and imposes on them a limit and proportionality of growth ; that the soul makes use of them as the artist makes use of his implements, and such passages are naturally interpreted by Driesch in the sense of a 'dynamic' teleology ; it is the ψυχή which superintends and controls, and the ψυχή is 'entelechy'.

Elsewhere, however, we are informed that even the proportionality of the developing parts is simply the outcome of the motion imparted by the male, which is *actu* what the female element only is *potentia*.

Moreover, it may be questioned whether Aristotle ever intended to imply more than an 'analogy with the causality of purpose' when he uses the figure of the workman and his implements to illustrate his meaning of the formal cause. The formal cause of a work of art is an intelligible *vera causa*; it is the idea in the mind of the artist antecedent to the execution of the work; but the formal or final cause of an organism, the end which it apparently strives to attain, can only be said by a metaphor to be prior in time to the existence of the organism itself. Prior in thought, however, it certainly is, for it is only the performance of its functions (ἐντελέχεια) by the organism complete in all its parts that makes the mere mechanism of development comprehensible to us ; the process, therefore, exists for the sake of the end. Only as efficient cause is the soul prior in time ; only so far as it is prior in thought can it be said to be a final cause.

Such a teleology is, it is obvious, indistinguishable in principle from the position in which Kant leaves us. It is the position adopted by Driesch himself in his earlier *Analytische Theorie*, but abandoned in the *Vitalismus* in favour of a theory of 'psychoids'.

Now quite apart from the meaning which Aristotle may or may not have intended to convey, there are grave objections to this belief. This 'psychoid', to which the name 'entelechy' is surely misapplied, this rudimentary feeling and willing, which is aware of the form it desires to produce, must be psychically at least as complex as the phenomena it is designed to account for, and stands, therefore, as much in need of explanation as they ; as Kant has observed, this will involve us at once in an infinite series of such entities. In fact it is only a photograph of the problem, and not a solution at all.

Again, when we ask what the *modus operandi* of this cause is, we get no reply either from Driesch or from any other neovitalist. The objection that the intervention of a psychical cause in a physical process is unintelligible, an objection which would probably appeal to many, may be waived, for in the last resort the connexion between any—even simple mechanical—causes and effects is equally hard to understand.

It may, however, be doubted whether these entities are not being multiplied beyond necessity, and whether the progress of science would not be better served by an adherence to a simpler philosophy. But even when it has discarded the psychoid we find vitalism still denying the possibility of mechanical explanation, still preaching the autonomy of the organism. The 'dynamic' teleology of Driesch has only disappeared to be replaced by the metaphysical doctrine of the final cause.

We may point out, perhaps, in passing, that the organism is by no means as autonomous as might be desired. The end towards which the creature strives, the maintenance and reproduction of its own specific form, is not a constant *terminus ad quem*, for species are as mortal as individuals: nor is it always achieved; the autonomy of a worm, which, bisected in a certain way, regenerates a tail instead of a head, or of a frog, which, after a particular injury, develops six legs instead of two, has surely renounced its rights. But, setting this aside, it must be seriously questioned whether any good purpose is served in biological discussion by decrying the value of mechanical conceptions or by confounding two distinct orders of thought. The questions are grave ones: for the issue at stake is no less than the existence of physiology as the science of the causes of living activities.

'Recte ponitur', said Francis Bacon, 'vere scire esse per causas scire.' The maxim of the great founder of modern inductive science has been the lode-star of biology in the past, and is still its watchword to-day. By exact observation and crucial experiment, utilizing every canon of induction, the activities of the living organism are to be brought under wide general laws of causation, which will be, in the first instance, physiological laws—of response to stimuli, of metabolism, and of growth: by means of these laws predictions can be made, and verified as often as we please. But no bar can legitimately be set to the scope of human inquiry; the thought process will not rest here, and ultimately it may be possible to state the widest generalizations of biology in chemical and physical, and these again in purely mechanical terms. The maintenance and evolution of form in the individual, as well as the larger evolution of form in the race, become but the final terms in a far vaster cosmic process, from 'homogeneity to heterogeneity'.

The idea is, of course, perfectly familiar: it is the analysis of purely physical causes, carried to its extreme limit. Phenomena are thought out in terms not of origins merely, but of one origin, and that one origin is the only mystery that remains. This unification of the sciences has always been and must still remain the dream and the faith and the inspiration of the scientific man, and could such an edifice of the intellect ever be realized, the task of science would have been completed. Only when this purely deterministic method has been pushed as far as it will go does science leave off; only where science leaves off does philosophy begin.

There is an order of time, and there is an order of thought. Science works in the order of time, and necessarily so: for although science can never say what constitutes the invariable link between antecedent and consequent which it terms causal, yet it rightly speaks of the first as cause, determining the second as effect, since it is its function to predict from the past which is known to the future which is not.

But the outlook of philosophy is different. Dissatisfied with the endless regress of cause and effect, sceptical of first causes and original homogeneities, out of which by no conceivability could any heterogeneity have ever been developed, philosophy looks to the end.

The activities of living organisms at least appear to be directed to an end; they are apparently purposive, and it is this purposiveness which lends to biology, though built on the fundamental conceptions of chemistry and physics, peculiar features of its own, and is, of course, answerable for the teleological language which biologists so frequently employ. And by a knowledge of the end, the view of science, to which *qua* science it cannot too rigidly confine itself, will doubtless be supplemented and enlarged.

But, plain and definite though the end of an individual life may be, the end of the race—of the human or any other race—the end of the universe, are things only to be guessed at, and all we are left with is an indefinite series of evolving systems emerging out of an infinite past and fading into an infinite future.

In the final issue, indeed, the last effect is as delusive an *ignis fatuus* as the first cause. The philosophy which has rejected one must divest itself of the other, and seek its end, if anywhere, in the logical *prius* of the mind, which, though last in time, is yet first in thought, since through it alone can that ordered knowledge of nature which we call science be born and brought to perfection.

From the Italian translation of 'KETHAM', VENICE 1493

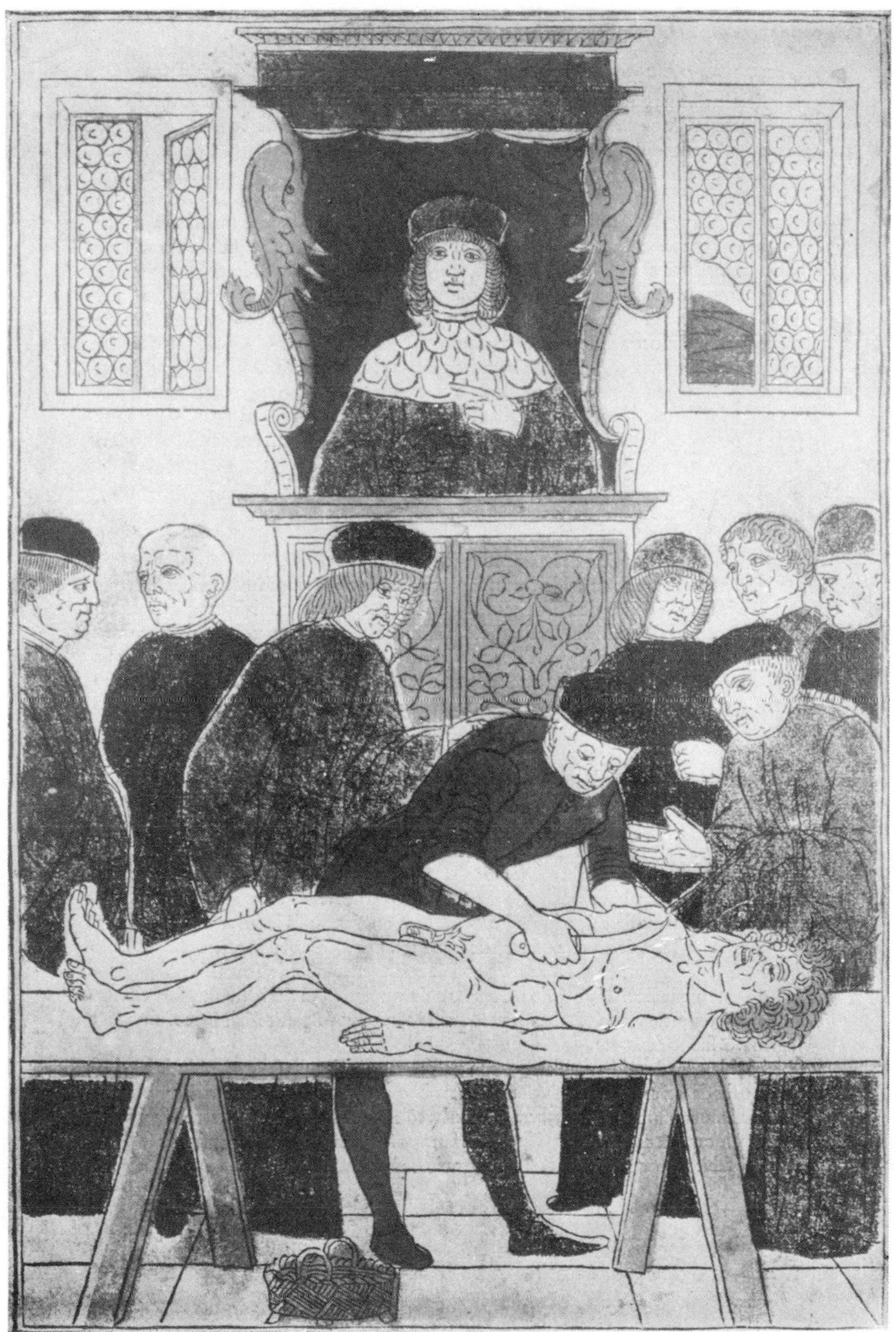

PLATE XXVII. MUNDINUS (?) LECTURING ON ANATOMY

BIBLIOTHÈQUE NATIONALE MS. fr. 2030
Written in 1314

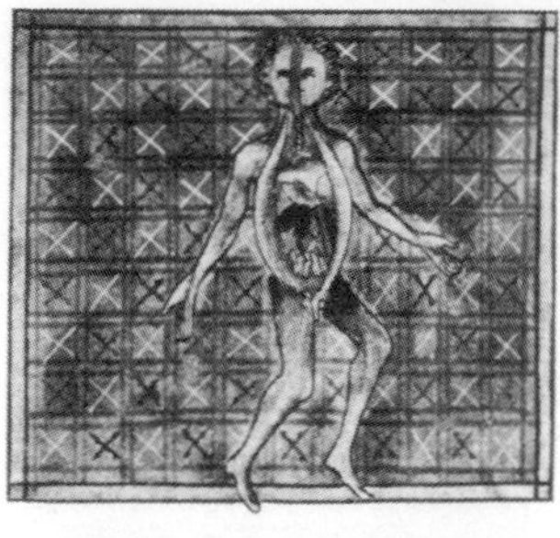
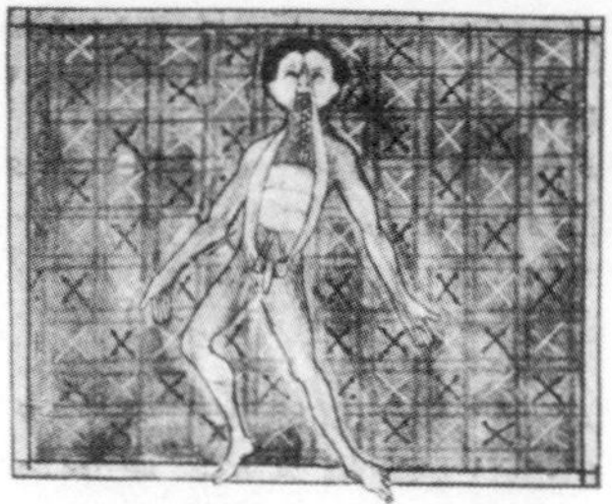
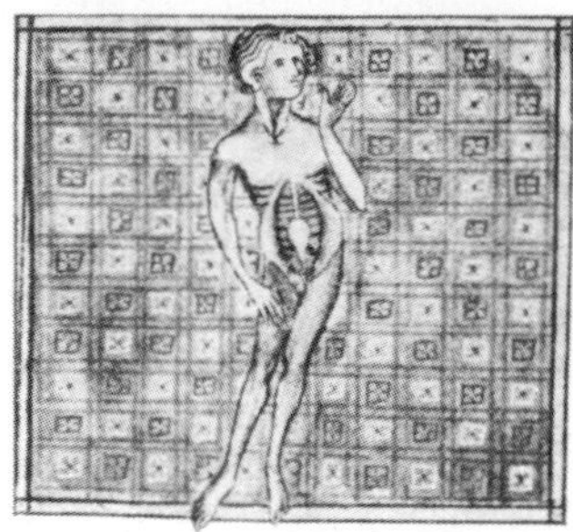
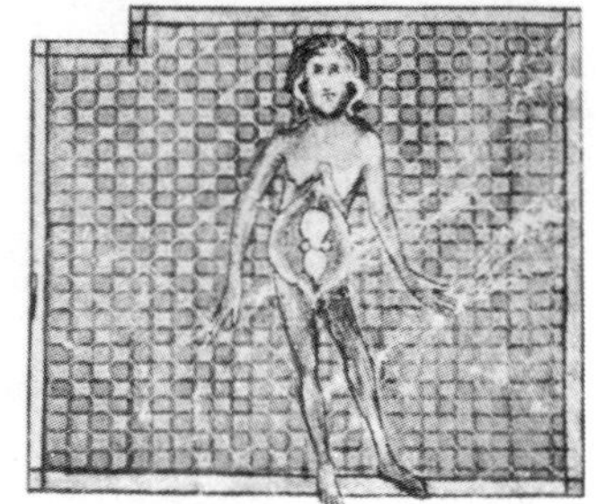

TO ILLUSTRATE THE ANATOMY OF HENRI DE MONDEVILLE

BODLEIAN MS. ASHMOLE 399 fo. 34 r

PLATE XXVIII. A DISSECTION SCENE
circa 1298

A STUDY IN EARLY RENAISSANCE ANATOMY,

WITH A NEW TEXT:

THE ANOTHOMIA OF HIERONYMO MANFREDI (1490)

BY CHARLES SINGER

TEXT TRANSCRIBED AND TRANSLATED BY A. MILDRED WESTLAND

I. ANATOMY IN THE FOURTEENTH AND FIFTEENTH CENTURIES

THERE was little or no progress in the knowledge of anatomy between the death of Mondino in 1327 and the sixteenth century. This appears the more remarkable when we recall how widespread was the practice of dissection during the period. In France, at the University of Montpellier, public dissections were decreed in the year 1377,[1] and Catalonian Lerida followed suit in 1391.[2] At Bologna, where dissection had long been customary, it received official recognition in the University Statutes in 1405,[3] and the same event took place at Padua in 1429. Public anatomies were instituted at the University of Prague in 1460, of Paris in 1478, and of Tübingen in 1485.[4] For these 'Anatomies' the bodies of executed criminals were usually employed, and therefore the number of subjects available varied greatly in different localities.[5] In addition to

[1] *Cartulaire de l'Université de Montpellier (1180–1518)*, Montpellier, 1894, p. 21.

[2] Dates of the institution of dissection at this and other Universities are given by F. Baker in *Bulletin of the Johns Hopkins Hospital*, vol. xx, p. 331, Baltimore, 1909.

[3] Statuti dell' Università di Medicina e di Arti del 1405, Rubr. lxxxxvi ('De anothomia quolibet anno fienda') in the *Statuti delle Università e dei collegi dello Studio bolognese*, edited by Carlo Malagola, Bologna, 1888, p. 289.

[4] J. Säxinger, *Ueber die Entwickelung des medizinischen Unterrichts an der Tübinger Hochschule*, Tübingen, 1884, pp. 5 and 10.

[5] How rarely dissections were conducted in some of the Universities may be gathered from the first statutes of the medical faculty of Tübingen, dated 1497. These ordain a dissection *every three or four years*. Not till 1601 was an anatomy

Fig. 1. From the French translation of Bartholomaeus Anglicus, Lyons, 1482. The first printed picture of dissection.

MONTPELLIER BIBL. DE LA FACULTE DE MEDECINE
MS. fr. 184 fo. 14r

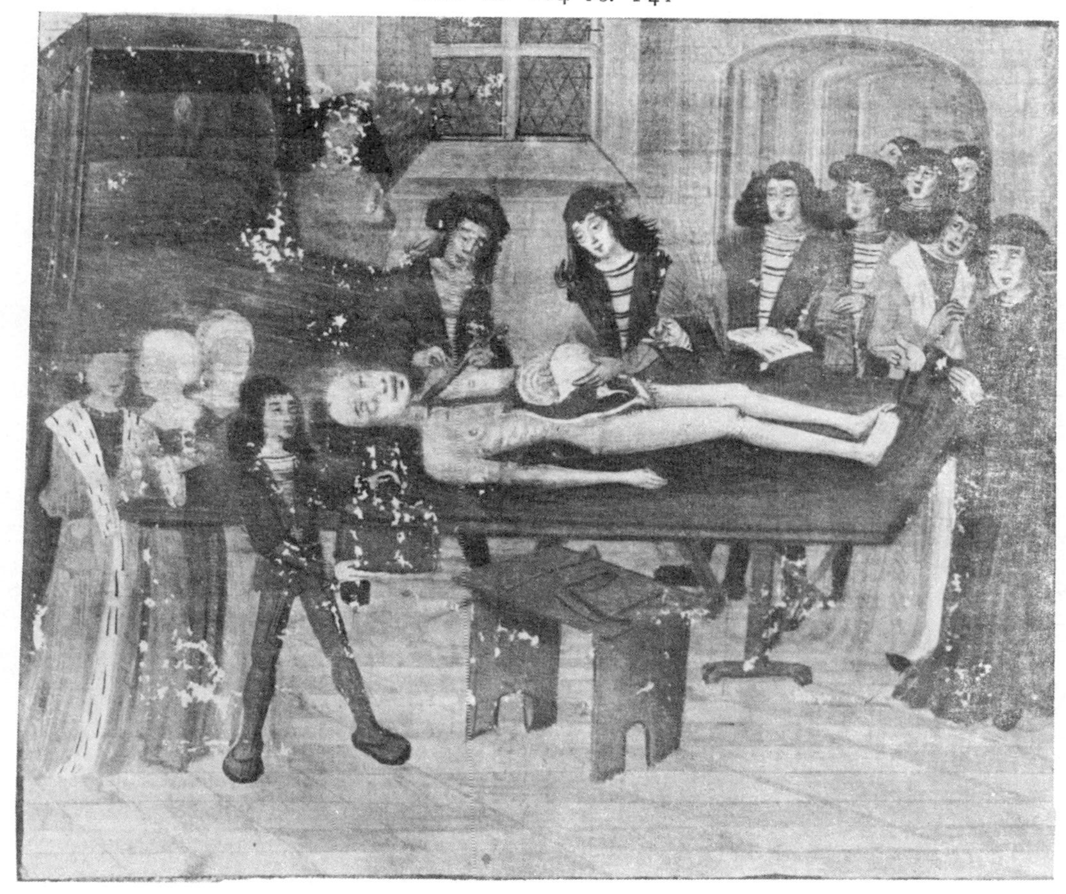

PLATE XXIX. A POST-MORTEM EXAMINATION. Late XIVth Century

VATICAN MS. HISPANICE 4804 fo. 8r

PLATE XXX a. A DEMONSTRATION OF SURFACE MARKINGS

Second half of XVth Century

BRISTOL REFERENCE LIBRARY MS. fo. 25 r

PLATE XXX b. A DEMONSTRATION OF THE BONES TO ILLUSTRATE GUY DE CHAULIAC

First half of XVth Century

these regular dissections, there was certainly a considerable amount of post-mortem examination, surreptitious (Plate XXVIII *b*[1]), or even open (Plate XXIX[2]), long before Benivieni published his memorable list of cases.[3]

That so much industry was rewarded by so small an increase in knowledge may probably be attributed to the method adopted.

held at Tübingen even once a year (see Säxinger, loc. cit.). Even at Montpellier in the sixteenth century the scarcity was so great that Rondelet (1507–66) was on one occasion reduced to dissect the body of his son. For this terrible incident see A. Portal, *Histoire de l'Anatomie et Chirurgie*, Paris, 1770, vol. i, p. 522; A. Haller, *Bibliotheca anatomica*, Lib. iv, § clxxxiv, Leyden, 1774, vol. i, p. 205; and A. O. Goelicke, *Introductio in historiam litterariam anatomes*, Frankfurt, 1738, p. 136. There was, however, a relatively plentiful supply of subjects in the Italian Universities and especially at Bologna and Padua in the fourteenth, fifteenth, and sixteenth centuries (cp. A. Haller, *Bibliotheca anatomica*, introduction to Lib. v, p. 218). This was perhaps due to the utterly depraved state of public and private morals to which the peoples of the peninsula had been reduced by the excesses of the tyrants and the condottieri.

[1] Plate XXVIII *b* is perhaps the earliest representation of the practice of dissection yet brought to light. It is described in Charles Singer, 'Thirteenth-Century Miniatures illustrating Medical Practice', *Proceedings of the Royal Society of Medicine, Section of the History of Medicine*, 1916, vol. ix, pp. 29–42.

[2] Plate XXIX: a post-mortem scene in the late fourteenth century, from a French MS. of the *Grande Chirurgie* of Guy de Chauliac, Bibliothèque de la Faculté de Médecine de Montpellier, MS. 184 français, folio 14 recto. The scene is laid in the bedroom of the deceased. In the left-hand top corner is the bed, by the side of which a female figure, partly obliterated, is praying. Below and to the left are two other female figures, and a man richly dressed in an ermine-trimmed robe. These are presumably the relatives of the dead. The corpse, that of a woman, has been placed on a bare table and is opened from the larynx to the symphysis pubis. In front stands a lad holding a round wooden vessel for the reception of the viscera, and farther to the right is a stool on which are placed two or three instruments. The physician, in full canonicals, is at the extreme right of the picture. The actual process of examination is being made by three of his assistants. To the left the first of these deepens, with a knife, the incision that has already been made over the sternum, the second is grasping with his two hands and rolling up the great omentum so as to display the viscera beneath, and the third holds a wand in his right hand, with which he points to the abdomen, while in his left he carries a book. Five others throng into the room from a passage which opens into it.

[3] Antonio Benivieni, *De abditis nonnullis ac mirandis morborum et sanationum causis*, Florence, 1506. In the description of Case 32, Benivieni expresses surprise at having been refused permission to perform a post-mortem examination, as though it were unusual for him to meet rebuffs of the kind. 'Experimento comprobare volentes, corpus incidere tentavimus sed nescio qua superstitione negantibus cognatis, voti compotes fieri nequivimus.'

The so-called 'anatomies' were conducted in the most formal manner. Bertuccio, for example, who succeeded Mondino as professor of Surgery at Bologna, was accustomed, as we learn from his pupil Guy de Chauliac, to give short systematic anatomical demonstrations on a fixed and rigid method.[1] The occupant of the chair at this period was indeed no professor in the modern sense of the word. To expound the tradition of anatomy as it had reached him was regarded as the limit of his duty. Of any attempt to extend the bounds of knowledge, of any systematic endeavour to correct or improve the anatomical views of his predecessors, we find little or no trace. Indeed, at Padua it was expressly laid down in the statutes that the exposition of anatomy should follow the very words of Mondino.[2]

Fig. 2. Title-page of Mellerstadt's edition of the *Anatomy* of Mondino, Leipzig, 1493. The scene is laid in the open air.[3]

Early figures portraying the teaching of anatomy (Plate XXVII and Figs. 1–3, 5) usually show us a medical doctor sitting at a desk, well removed from the subject of dissection, and reading from his text-book the description of the part. Meanwhile an assistant, who is usually also a doctor, performs the actual work of dissection. The professor of

[1] See E. Nicaise, *La Grande Chirurgie de Guy de Chauliac*, p. 30, Paris, 1890.

[2] 'Ut Anatomici explicationem ipsius Mundini sequantur', Francesco Maria Colle, *Storia scientifico-letteraria dello Studio di Padova*, 4 vols., Padua, 1824–5, vol. iii, p. 108.

[3] Martin von Mellerstadt, also called Pollich or Polich.

Surgery, to whom the teaching of anatomy was entrusted, stands by with a pointer to indicate the different organs.

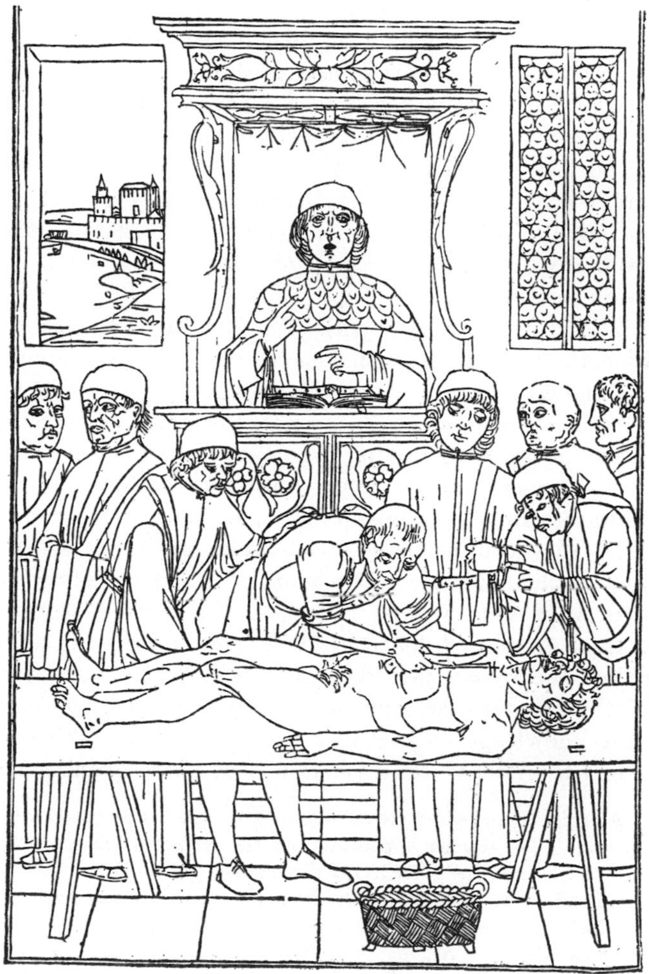

Fig. 3. A DISSECTION SCENE
From the Venice 1495 edition of 'Ketham' (compare Plate xxvii).

Sometimes the professor changes places with the reader at the desk. In some later MSS. the teacher is figured as himself handling the body and demonstrating to his pupil (Plate xxx *a*[1]

[1] Plate xxx *a*, from a late fifteenth-century Provençal translation of the *Grande Chirurgie* of Guy de Chauliac. Vatican Library, MS. hispanice 4804, folio

and *b*[1]), but there is evidence that the miniatures portraying this are the work of artists unfamiliar with dissection and with the teaching of anatomy.

The study of anatomy had to contend with two great difficulties, want of subjects for dissection, and faith in the written word.

Thus, at Bologna, where it was arranged that every medical student of over two years' standing should attend an *Anatomy* once a year, no less than twenty students were admitted to see the anatomy of each man, and thirty to the anatomy of each woman.[2] This was all the practical instruction received. Some other Universities had to be content with the cadaver of a single criminal per annum for the whole body of students.

In the *first* period during which the human body was dissected in Europe, the thirteenth century, a certain amount of progress was certainly made, despite the rarity of subjects. The rebirth of learning in the thirteenth century was not, however, as favourable to anatomical progress as might have been hoped. Galen, indeed, ceased to be a mere name, and the Latin translations of his text, or of its adumbra in the writings of the Arabians, became ever more familiar. On the other hand, with more authoritative texts in their hands, men were but the more inclined to follow the evil scholastic way, and to trust rather to the written words of the master than to the evidence of their own senses. Thus it came about that the *second* period, which covers the fourteenth and most of the fifteenth century, was really stationary so far as the first-hand knowledge of anatomy was concerned. With the last decade of the fifteenth century, however, there opens a new and

8 recto. A professor and pupil are examining a wasted corpse placed on a trestle in the open air. The teacher is pointing out the surface markings.

[1] Plate xxx *b*, from the French Guy de Chauliac MS. in the Bristol Reference Library, folio 25 recto. The MS. dates from between the years 1420 and 1435; cp. Norris Mathews, *Early Printed Books and MSS. in the Bristol Reference Library*, Bristol, 1899, p. 70; J. A. Nixon, 'A New Guy de Chauliac MS.', in *Transactions of the XVIIth Internat. Cong. of Med., Sect. of Hist. of Med.*, London, 1914, p. 419; and Charles Singer, 'The Figures of the Bristol Guy de Chauliac MS. *circa* 1430', *Proceedings of the Royal Society of Medicine, Section of the History of Medicine*, 1917, vol. x, pp. 71–91. The figure shows a professor and pupil. The former is demonstrating the bones of a skeleton.

[2] The number of female criminals being less than the number of male criminals, Ludovico Frati states (*La vita privata di Bologna dal secolo XIII al XVII*, Bologna, 1900, pp. 116–18) that only two anatomies *in all* were held each year, and thirty students admitted to the female and twenty to the male dissection. This would mean far less than *two dissections a year for each student* of over two years' standing.

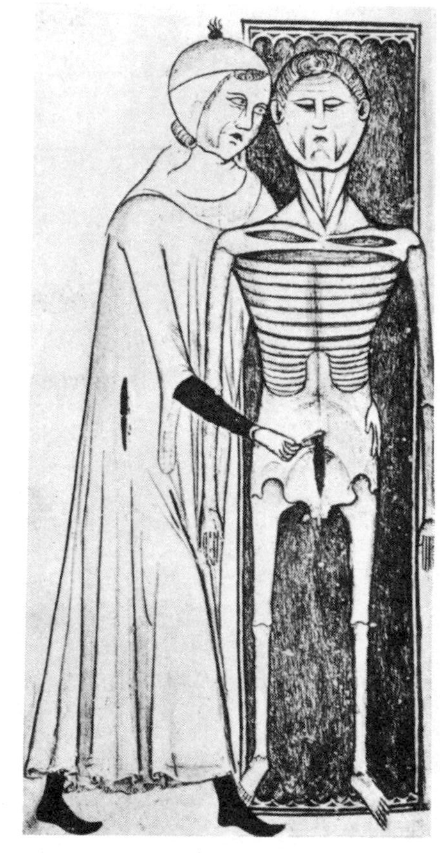

PLATE XXXI. From the MS. of GUY DE VIGEVANO of 1345 at CHANTILLY

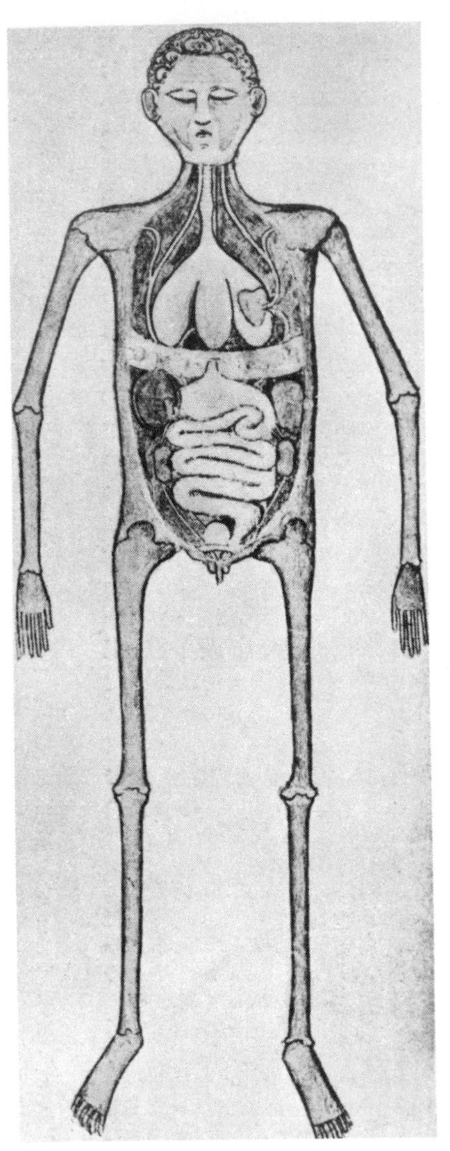

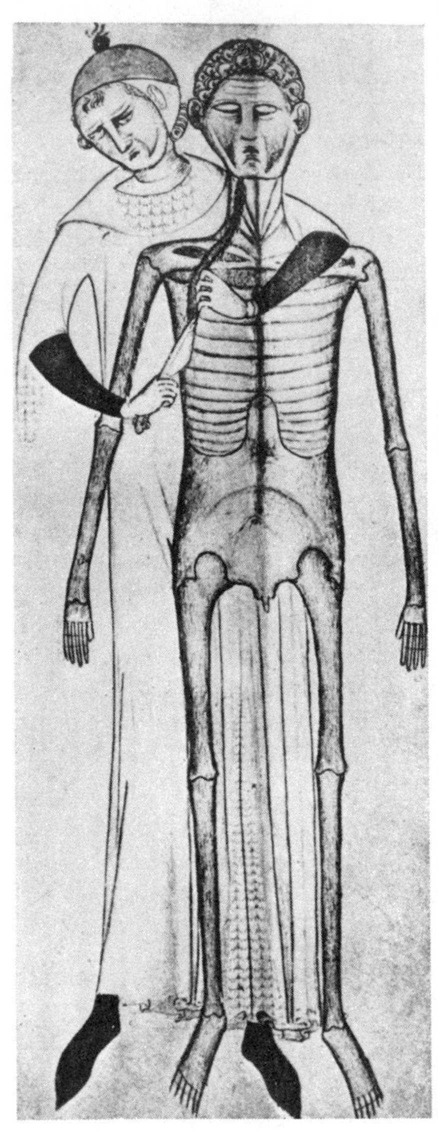

PLATE XXXII. From the MS. of GUY DE VIGEVANO of 1345 at CHANTILLY

FIG. 4. From the English translation of Bartholomaeus Anglicus, printed by Wynkyn de Worde, 1495. The first picture of dissection in an English-printed book.

VENETIIS ANNO .D. M.CCCCC.XXXV.

FIG. 5. A LECTURE ON ANATOMY
From the 1535 Venice edition of Berengar of Carpi's Commentary on Mondino.

third period in the history of our subject. From that time dates the true era of anatomical renaissance, which may be regarded as continuing until the commencement of modern anatomy with the great work of Vesalius in 1543.

We have said that throughout the second period, the formal demonstrations based on the declaimed text of Galen or Avicenna or Mondino were practically the sole opportunities afforded to either teacher or pupil for the investigation of the minuter details of the human frame. But in making this statement concerning the arrest of anatomical progress, we must expressly exclude the products of the mighty genius of Leonardo da Vinci (1452–1519), whose anatomical researches were without influence, and remained long unnoticed.[1] We must also omit evidence gathered from the work of such early Renaissance painters as Antonio Pollaiuolo (1429–98) or Andrea del Verrocchio (1435–88), for these pursued the study of anatomy in a special field and with a special object.[2] Furthermore, there are a number of artists of similar date of whose anatomical studies we have no direct evidence, but who yet outlined the muscles of the nude human figure in such a way as leads us to suppose that they had investigated the superficial structures at least of flayed parts. Such is the suggestion of some of the work of Luca Signorelli (*c.* 1442–*c.* 1524), and of Andrea Mantegna (died 1506). With such reservations, however, it is probably true that no evidence is forthcoming until the last decade of the fifteenth century of any advance from the standpoint of Mondino.[3]

But if descriptive anatomy developed slowly in the hands

[1] The anatomical works of Leonardo have now been rendered accessible in *Tredici Foglie della Royal Library di Windsor. Leonardo da Vinci, Quaderni d'anatomia . . . Pubblicati da O. C. L. Vangensten, A. Fonahm, H. Hopstock*, Christiania, 1911, &c.

[2] Pollaiuolo and Verrocchio only studied surface anatomy, so far as is known. For a summary of the anatomical work of these painters see M. Duval and E. Cuyer, *Histoire de l'Anatomie plastique*, p. 20, Paris, 1898.

[3] It has been suggested that Giammatteo Ferrari da Grado (Matthaeus de Gradibus), who was professor of Medicine at Pavia 1432–72, made original contributions to anatomy. He wrote no separate work on anatomy, but his observations on the ovaries (which he was perhaps the first to call by that name) appear in his *Practica*, Milan, 1471, and in his *Expositiones super vigesimam secundam Fen tertii canonis Avicennae*, Milan, 1494. An interesting account of Ferrari's life and work is given by his descendant, H. M. Ferrari, in *Une Chaire de Médecine au XVe siècle; Un professeur à l'université de Pavie de 1432 à 1472*, Paris, 1899. In this work the claim that De Gradibus was an original and independent observer is effectively disposed of.

of the physicians, the art of graphic representation of anatomical structures was still more backward. Several groups of anatomical drawings of mediaeval date have come down to our time, but examination of them shows that they have been drawn without direct reference to the human frame. Some of these figures are of the crude type known as the 'five-figure series' (Plate XXXIII), mere traditional diagrammatic sketches.[1] Hardly better or more instructive are the series of dissections which illustrate certain MS. works of Henri de Mondeville (Plate XXVIII *a*)[2] and Guido de Vigevano (Plates XXXI and XXXII), 1345.[3] A few sketches representing the separate organs have also survived (Fig. 6),[4] but these never suggest that the draughtsman had before him the structure which he seeks to depict, and the drawings appear to have been made in order to illustrate contemporary physiological theory rather than observed anatomical fact. Even the magnificent illuminated Dresden Codex of Galen, prepared in France or Flanders as late as the second half of the fifteenth century, betrays not the slightest first-hand knowledge of anatomy.[5] Although the

[1] At least six Western copies of this series, besides three or more of oriental origin, have now been detected. The Western MSS. and their dates are as follows:

(*a*) Munich, Hof- und Staatsbibliothek, Cod. lat. monacensis 13002, before 1158.
(*b*) Munich, Hof- und Staatsbibliothek, Cod. lat. monacensis 17403, *circa* 1250.
(*c*) Bodleian Library, MS. Ashmole 399, *circa* 1290.
(*d*) Dresden, Kgl. Öffentl. Bibliothek, Codex 310, before 1323.
(*e*) Bodleian Library, MS. e Museo 19, before 1344.
(*f*) Library of Count F. Zdenho von Lobkowicz in Raudnitz, of 1399.

See E. Seidel and K. Sudhoff, especially 'Drei weitere anatomische Fünfbilderserien aus Abendland und Morgenland', in *Archiv für Gesch. der Med.*, iii, p. 165, Leipzig, 1910.

[2] Cp. K. Sudhoff in *Ein Beitrag zur Gesch. der Anatomie im Mittelalter*, Leipzig, 1908.

[3] E. Wickersheimer, 'L'Anatomie de Guido de Vigevano, médecin de la reine Jeanne de Bourgogne (1345)', in *Archiv für Geschichte der Med.*, vii. 1, Leipzig, 1914. M. Wickersheimer has kindly given permission for the reproduction of the figures in Plates XXXI and XXXII.

[4] Notably the MS. Roncioni 99, dating from the first half of the twelfth century, in the University Library of Pisa, reproduced by K. Sudhoff in the *Archiv für Gesch. der Med.*, vii, Tafel xiv, 1914. Also separate organs are depicted in the Bodleian MS. Ashmole 399, dating from the end of the thirteenth century, reproduced in Fig. 6.

[5] The miniatures of the Dresden Codex have been studied by L. Choulant, *Geschichte und Bibliographie der anatomischen Abbildung nach ihrer Beziehung auf anatomische Wissenschaft und bildende Kunst*, Leipzig, 1852, and in the *Archiv*

illustrations of this MS. are prepared with the utmost technical skill, they yet show us a teacher exhibiting to his pupils a heart of the form found on playing-cards, and other anatomical figures scarcely more faithful to the facts (Plate XXXIV).

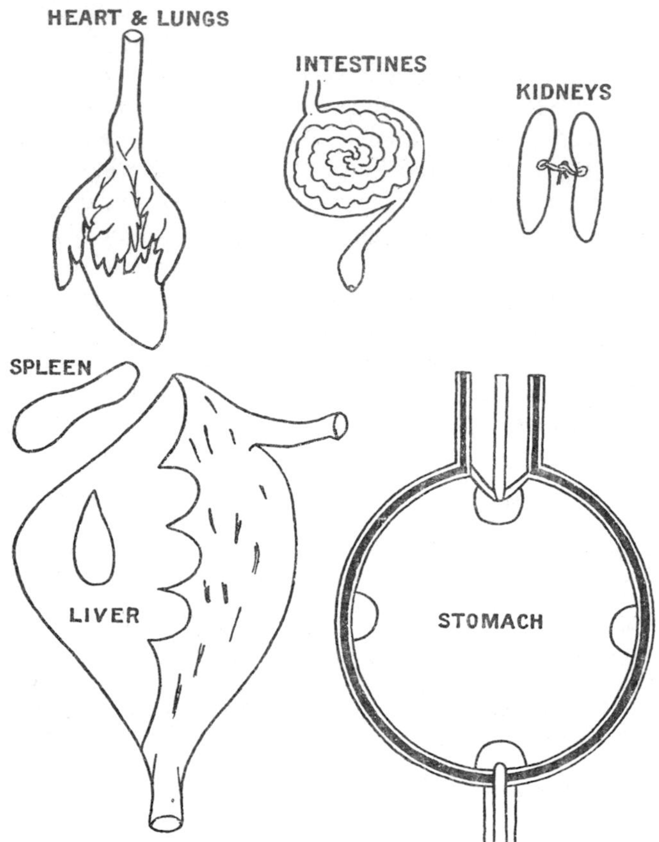

FIG. 6. DIAGRAMS OF THE INTERNAL ORGANS
After Bodleian Library MS. Ashmole 399 of about 1298, fos. 23 recto–24 recto.

The spirit of investigation of the artist who perforce went direct to nature, dissecting with his own hands and observing

für die zeichnenden Künste, II. Jahrgang, Leipzig, 1856, p. 264. More recently the MS. has been most carefully described and its miniatures reproduced by E. C. van Leersum and W. Martin, *Miniaturen der lateinischen Galenos-Handschrift der kgl. öffentl. Bibliothek in Dresden, in phototypischer Reproduktion*, Leyden, 1910. We have to thank Dr. Van Leersum of Leyden for kind permission to reproduce the figures of Plate XXXIV.

with his own eyes (Plate XXXVI), showed itself indeed far more fruitful than the tedious *ex cathedra* methodization of the professor.[1] Yet the system of the schools needed to be combined with the freedom of the artist for the production of an effective anatomical work. What the projected treatise of Marcantonio della Torre (1473–1506) might have been we may guess from the anatomical sketches of Leonardo da Vinci (Plate XXXV), who was to have been associated with him in the work.[2] In the event, however, the medical schools had to wait yet another generation before the subject was placed on a sound basis by André Vesale.

The Mondino pamphlet—for it is little more—used since its author's death in 1327 as a text-book in the schools of northern Italy, was first printed in 1478. Not until the last decade of the fifteenth century did there appear another work bearing evidence of the hand of a practical anatomist. This was an Italian translation of Ketham's *Fasciculus medicinae*, impressed at Venice in the year 1493.[3] The volume comprises Mondino's pamphlet and a collection of other medical tracts that were probably put together by Giorgio di Monteferrato from the work of a writer of the previous century, for their contents are traceable to a fourteenth-century MS.[4] The text is neither original nor remarkable, but the Venice volume derives its importance from certain figures which appear in it for the first time.

Two of these plates are of great interest both intrinsically and also in relation to the history of anatomy. One of them is the magnificent representation of a dissection scene, which is regarded as perhaps the finest example of book illustration produced during the first century of typography[5] (Plate XXVII). This work of the

[1] Cp. P. Triaire, *Les leçons d'anatomie et les peintres hollandais aux XVIe et XVIIe siècles*, Paris, 1887.

[2] For della Torre and his projected work on anatomy, see G. Cervetto, *Di alcuni illustri anatomici italiani del decimoquinto secolo*, p. 46, Verona, 1842; also L. Choulant, *Geschichte der anatomischen Abbildung*, p. 5, Leipzig, 1852.

[3] The first edition appeared in Venice in 1491 and is in Latin. It is of less typographical interest.

[4] K. Sudhoff, 'Eine Pariser "Ketham" Handschrift aus der Zeit König Karls VI (1380–1422)', in *Archiv für Geschichte der Medizin*, vol. ii, p. 84, Leipzig, 1909; 'Neue Beiträge zur Vorgeschichte des Ketham', in *Archiv für Geschichte der Medizin*, vol. v, p. 280, Leipzig, 1912.

[5] Prince d'Essling, *Les livres à figures vénitiens de la fin du XVe siècle et du commencement du XVIe*, part i, vol. ii, p. 56, Florence and Paris, 1908.

'maître aux dauphins', as the unknown artist is called by critics,[1] is doubly interesting, for it is the subject of an experiment in colour printing, no less than four pigments being laid on by means of stencils. As early as 1457 the method of stencilling was employed for colouring the initials of a Psalter, and in 1485 Erhard Ratdolt in an astronomical work added yellow to the earlier red and black. The figure from which our plate is taken represents, however, the first attempt at a complex colour scheme and leads up to the work of Hugo da Carpi.[2]

In this picture the professor, a youthful figure perhaps intended to represent Mondino himself, is shown standing at a desk which hides his book. Around a corpse, laid on a trestle table before him, there cluster a number of men in doctor's robes. Their valid faces are sufficient to convince us that the artist is here presenting us with portraits. One of the listeners has removed his robe and stands with upturned sleeves and knife in hand, ready to make the first incision on the direction of the doctor, who points to the part with a wand held in the left hand. In the impression of 1495 and in those of later date, the book appears above the desk, the attitudes of the students are somewhat changed, and many other details are altered. In all these, however, the blocks have been recut and the result is artistically inferior [3] (Fig. 3).

The second plate from the 1493 Ketham with which we are here concerned is the outline of a female body, in a traditional pose,[4] laid open to exhibit some of the internal organs (Fig. 7). These had clearly been sketched from the object, and therefore this drawing, the first printed figure of its kind, may be said to introduce the new era for the investigation of the human frame. The anatomical renaissance had begun. Into a discussion of the full development of that age we cannot now enter. But the MS. of Manfredi, with which we have here to deal, was written at the very dawn of the new era and is itself one of its earliest documents.

[1] Eugène Piot, *Le Cabinet de l'amateur*, nouv. série, Paris, 1861, 'Le maître aux dauphins', p. 354 et seq. The dolphins are seen on either side of the chair in Plate XXVII.

[2] Duc de Rivoli, *Bibliographie des livres à figures vénétiens*, p. 110, Paris, 1893.

[3] Cp. G. Albertotti, *Nuove osservazioni sul 'Fasciculus medicinae' del Ketham*, Padua, 1910.

[4] See K. Sudhoff, 'Weibliche Situsbilder von ca. 1400–1543', in *Tradition und Naturbeobachtung*, p. 79, Leipzig, 1907. The number and character of the indication lines attached to this figure suggest that the block from which the impression has been taken had previously been used for some other publication. This work, however, if it exists, has not yet come to light.

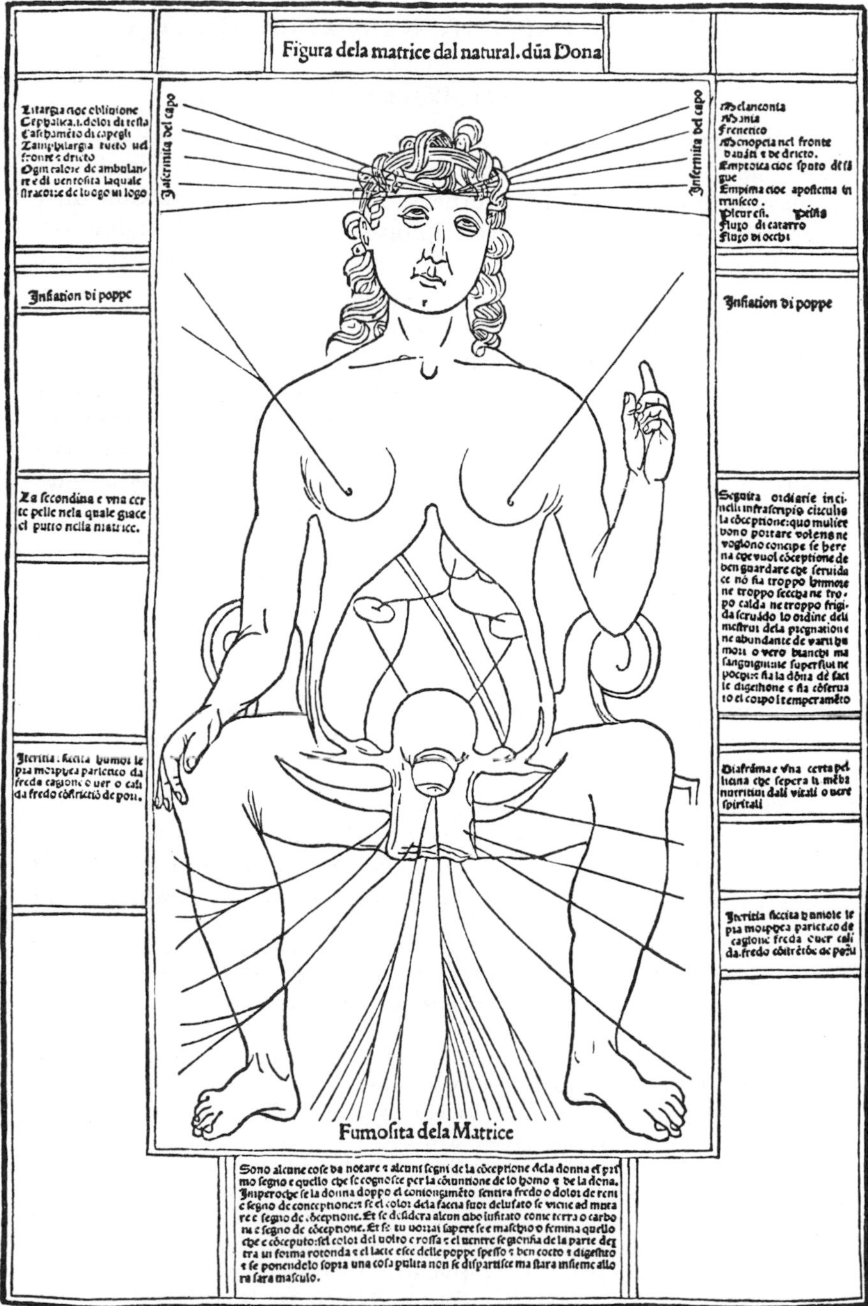

Fig. 7. A FEMALE FIGURE LAID OPEN TO SHOW THE WOMB AND OTHER ORGANS
From the 1493 Venice edition of 'Ketham' translated into Italian. This is the first printed anatomical figure drawn from the object.

II. Bolognese Works on Anatomy

An organized Medical Faculty existed at Bologna at least as early as 1156,[1] though the first record of dissection there is of considerably later date. In February 1302 a certain Azzolino died under suspicious circumstances. Poison was suspected, an inquest was held and a post-mortem examination ordered. The investigation was conducted by two physicians and three surgeons, who unanimously agreed 'that the said Azzolino assuredly met his death by no poison, but on the contrary, we assert that the quantity of blood collected in the great vein known as the *vena chilis* [vena cava][2] and in the veins of the liver adjacent thereunto, has prevented the due movement of the *spiritus* throughout the body, and has thus produced the diminution or rather extinction of the innate heat and thereby induced a rapid post-mortem discoloration. Of this condition we have assured ourselves *by the evidence of our own senses and by the anatomization of the parts.*'[3]

The first anatomical document emanating from the University of Bologna is, however, of still earlier date, and is the work of William of Saliceto (1210 ?–80). This writer was educated at Bologna, and it is claimed that he was the first to dissect the human body there.[4] His *Cyrurgia*, which was completed in 1275 (*editio princeps*, Piacenza, 1476), is divided into five books, of which the fourth and shortest is devoted to anatomy. Its descriptions are brief and concise. They are often clearly the result of actual observation, and they show hardly any trace of the absurd and irritating teleology that the influence of the Arabians and

[1] Michele Medici, *Della vita e degli scritti degli anatomici e medici fioriti in Bologna dal comincio del secolo XIII*, Bologna, 1853; *Compendio storico della scuola anatomica di Bologna dal Rinascimento delle Scienze e delle Lettere a tutto il Secolo XVIII*, Bologna, 1857.

[2] The mediaeval term, 'vena chilis', lasted in anatomy until the end of the sixteenth century and probably later. '*Chilis*' is a corruption of the Greek κοίλη. This hybrid name was abandoned by Vesalius (*Fabrica*, 1543 Basle edition, p. 376) in favour of the title 'vena cava'.

[3] The passage is translated from Michele Medici, *Compendio storico*, pp. 10–11.

[4] See A. Laboulbène, 'Les anatomistes anciens', in *Revue scientifique pour la France et pour l'Étranger*, vol. xxxviii, p. 641, Paris, 1886; Robert Ritter von Töply in Puschmann, Pagel, and Neuburger, *Handbuch der Geschichte der Medizin*, vol. ii, p. 197, Jena, 1903; G. Martinotti, 'L'insegnamento dell' Anatomia in Bologna prima del secolo XIX', in *Studi e Memorie per la storia dell' università di Bologna*, vol. ii, p. 51, Bologna, 1911.

PLATE XXXIII. The FIVE-FIGURE SERIES BODLEIAN MS. ASHMOLE 399, about 1292 Fos. 18 r–22 r

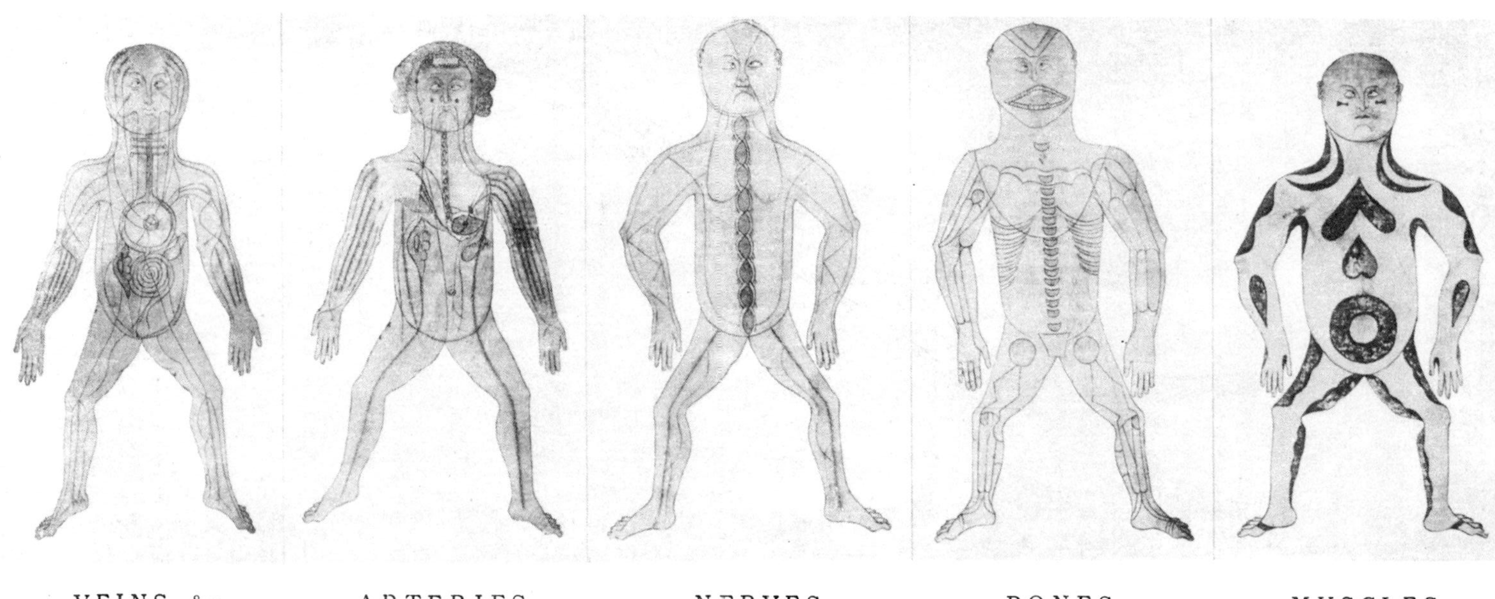

VEINS, &c. ARTERIES NERVES BONES MUSCLES

From the DRESDEN GALEN MS.

PLATE XXXIV. DEMONSTRATIONS OF ANATOMY

Second half of XVth Century

of Galen made customary in early anatomical literature. The anatomy of Saliceto appears to us very sensible and so far as it goes practical. It betrays the method rather of the Salernitan than of the Arabian anatomical writings, and is on the whole the best European work of the kind before the Renaissance. It was, however, soon replaced by the text-book of Mondino di Luzzi (1285–1327),[1] which, though inferior to that of Saliceto, held the field until the subject was revolutionized by Vesalius.

Mondino was professor at Bologna till his death in 1327. His work, easily accessible in one of its many editions, ' is corrupted by the barbarous leaven of the Arabian schools, and his Latin defaced by the exotic nomenclature of Avicenna and Rhazes '.[2] But it is not the language alone that has suffered. The schoolman's attitude, well fitted for the classification of ideas, is an ill instrument for the investigation of Nature, and in the scholastic Mondino the very basis of scientific judgement is undermined, so that he readily accepts the views of the ancients against what must often have been the evidence of his own senses. The work, however useful to the contemporary student, was thus essentially reactionary as against the efforts of the earlier Salernitan anatomists and of William of Saliceto. This is the more remarkable because it is quite clear that he was accustomed to demonstrate on the actual body—a privilege denied to the early Salernitan school,—and he was, moreover, a popular and successful teacher. His work is a manual of dissection rather than a treatise on anatomy. This, added to its conciseness and brevity, strengthened its appeal to the ' practical ' man—an epithet claimed then, as now, by the majority of stupid and unpractical people. The personal influence and enthusiasm of its author no doubt helped also towards the phenomenal success of this work, which for two hundred years held a position without rival as the text-book of the medical schools of Italy, where even as late as the sixteenth century Mondino ' was still worshipped by all the students as a very god '.[3]

[1] An intermediate anatomist was Gulielmo Varignana, who was professor of Medicine in Bologna, and is recorded as having opened for judicial purposes, on February 15, 1302, the corpse of one alleged to have been poisoned. See Michele Medici, op. cit. The investigation is referred to above.

[2] Dr. Craigie in his excellent account of the History of Anatomy, in the ninth and subsequent editions of the *Encyclopaedia Britannica*.

[3] ' Mundinus quem omnis studentium universitas colit ut deum ', J. Adelphus in his edition of Mondino, Strassburg, 1513.

Mondino was succeeded in the chair of Surgery at Bologna by his pupil, the Lombard Bertuccio, who died in the Black Death of 1347. Bertuccio's surviving work is unnoteworthy, but he was the anatomical teacher of Guy de Chauliac, whose *Surgery*[1] is of great value and was very influential in standardizing practice, especially in the north and west of Europe. Nevertheless it appears to us that the anatomical section is the weakest part of Guy's great work. The teleology that is a blot in Mondino has here become a perfect plague, and Guy's anatomy consists of one-third description and two-thirds wearisomely reiterated reasons for the existence of imperfectly described structures. Through Guy de Chauliac the anatomical tradition of Mondino passed over into the University of Montpellier.

A later fourteenth-century Bolognese writer was Tommaso di Garbo (died 1370), who did little but comment on Avicenna. A surgeon of the next generation, however, Pietro d'Argellata, deserves to be remembered for his description of the examination of the body of Pope Alexander V, who died suddenly at Bologna on May 4, 1410. His account throws light on the customary procedure and may be rendered here.[2]

'I ordered the attendants', he says, 'first to cut the abdomen from the *pomegranate* [i. e. the Adam's apple or laryngeal cartilage[3]] to the os pectinis [i. e. the symphysis pubis]. Then, so that they should not rupture the intestines, I myself sought the rectum and ligatured it in two places and then cut it between. Next I removed all the intestines as far as the duodenum and dealt with them as with the rectum, and so I had the intestines clean and without fetor. After this I extracted the liver, seizing its ligaments; then the spleen and then the kidneys, and these were all placed together in a jar. I now passed to the spiritual members [i. e. the thorax] and removed lung and heart and all their ligaments. Then I ligatured the *meri* [the Arabian term for oesophagus] and removed the stomach. When this had been done there were some who wished to remove the tongue but knew not how. I however cut under the chin and extracted the tongue

[1] *Editio princeps*, Lyons, 1478.

[2] Pietro de Argellata, *Cirurgia*, 'Incipit liber primus cirurgie magistri Petri de la Cerlata' (!), Venice, 1492. Quotation from lib. v, tract. 12, chap. 3. An earlier edition which we have not seen was printed in Venice in 1480.

[3] The 'pomegranate' sometimes also means the xiphisternum. It is not clear which is implied here.

through that hole, together with *trachea arteria* [trachea] and *meri*. Then I passed to the *arteria adorti* [aorta] and *vena chilis* [vena cava]. Lastly I removed the ligatured remnant of the intestines as far as the anal margin.'

Giovanni da Concoreggio (died 1438), who was lector in Surgery at Bologna in the early part of the fifteenth century, left a few anatomical observations of little note,[1] and not very much more can be said for his successors and Manfredi's contemporaries Gabriele Gerbi (de Zerbis, died 1505) and Alessandro Achillini (1463–1512). Gerbi [2] does little but repeat in the most verbose fashion the work of Mondino and of Avicenna, some of whose errors, however—e. g. the three ventricles of the heart—he omits. He wrote also an anatomy of the infant, or rather of the foetus,[3] and a treatise taken mainly from Avicenna's *De generatione embryonis*. Like all his work, these are in the full scholastic style of a professor of Logic, a position to which, in fact, he ultimately attained.

Achillini's work [4] is but a slight advance on that of Gerbi. It is really little else than a note-book for students, and gives the baldest directions for dissection, accompanied by a few comments taken from Avicenna. Achillini occasionally ventures to criticize Mondino, and his work has at least the advantage of brevity. He has a claim to be remembered in that he was the first to describe the duct of Wharton and is said to have been the first to describe the ear ossicles, malleus and incus. Achillini, like Gerbi, was a windy and very 'scholastic' disputator. He was best known to his contemporaries as a supporter of the philosophy of Averroes. In 1506, when driven from Bologna with the other supporters of Bentivoglio, he became professor of Philosophy at Padua.

With Giacomo Berengario da Carpi we come at length to one who definitely advanced the science, and who may be regarded as the first modern anatomist, so far as printed works are concerned. He was professor of Surgery from 1502 to 1527, and during

1 Giovanni da Concoreggio, *Lucidarium et Flos Medicinae*, Giunta, Florence, 1521. It contains a few scattered anatomical points.

2 De Zerbis, *Liber Anatomiae corporis humani et singulorum membrorum illius*, Venice, 1502.

3 Reprinted in the *Anatomia* of Johannes Dryander, Marburg, 1537.

4 Alessandro Achillini, *Annotationes anatomiae*, Bologna, 1520. This work is also included in the 1502 edition of De Zerbis' *Liber Anatomiae*.

that period published his great anatomical work.[1] This volume, though modestly put forward as a commentary on Mondino, is in reality an original contribution of great value. It is the earliest anatomical treatise that can properly be described as having figures illustrating the text (Fig. 8).[2] Carpi does not hesitate to criticize the work on which he comments—as for instance when he denies

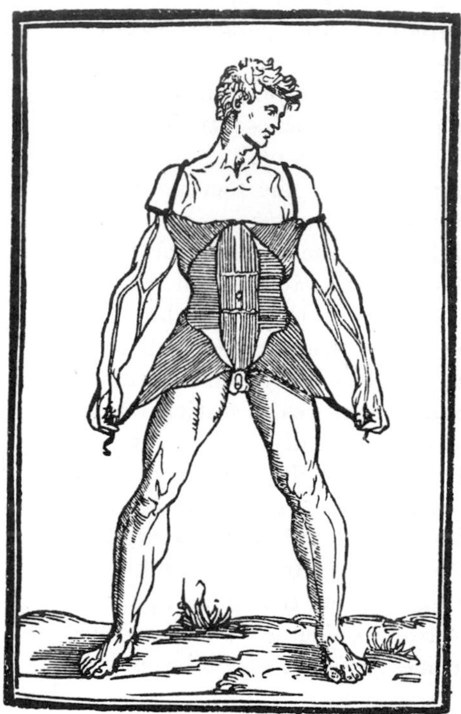

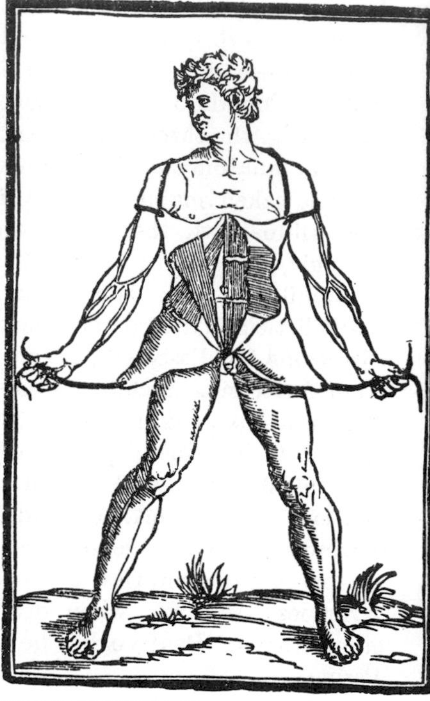

Fig. 8. THE ABDOMINAL MUSCLES
From Berengar of Carpi's Commentary on Mondino, Bologna, 1521.

the existence of the 'rete mirabile' below the brain, though descriptions of the 'rete mirabile' had been based on the statement

[1] *Carpi commentaria cum amplissimis additionibus super anatomia mundini una cum textu eiusdem in pristinum et verum nitorem redacto*, Bologna, 1521. An earlier and less important edition of Carpi was the *Anathomia Mundini noviter impressa ac per Carpum castigata* that appeared at Bologna in 1514.

[2] The figures in Ketham and in the wretched productions of Johannes Adelphus (J. A. Muelich), of Hundt, and of Peyligk can hardly be said to illustrate the text of anatomical treatises.

From a drawing in the Library, WINDSOR CASTLE

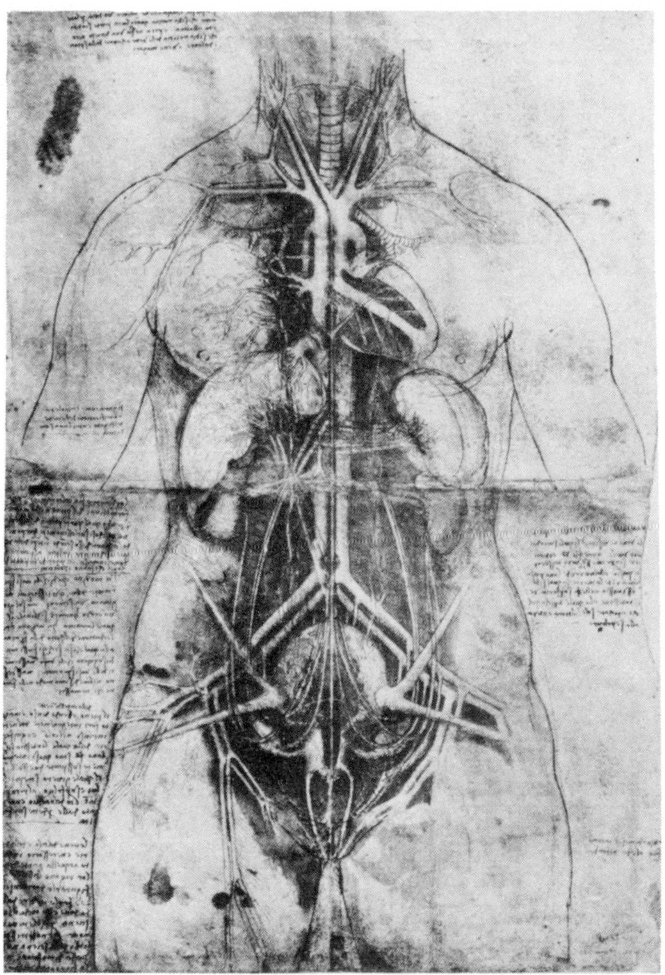

PLATE XXXV. VIEW OF THE INTERNAL ORGANS
LEONARDO DA VINCI

From a Drawing in the ASHMOLEAN MUSEUM, OXFORD, attributed to BARTOLOMEO MANFREDI (1574?–1602)

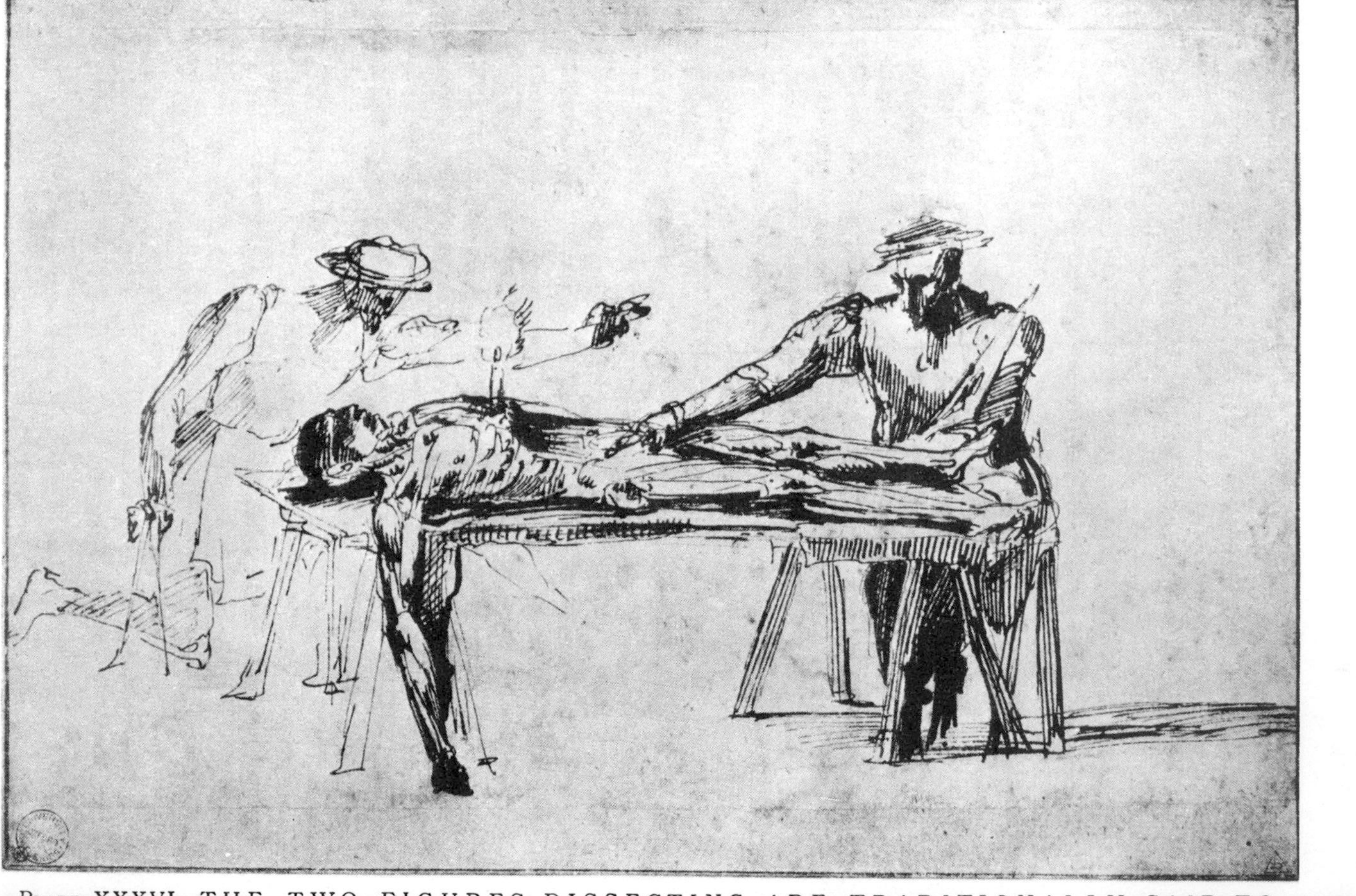

PLATE XXXVI. THE TWO FIGURES DISSECTING ARE TRADITIONALLY SAID TO of no less an authority than Galen. Furthermore he was the first to describe the vermiform appendix, and he gave the earliest correct account of several other organs, e. g. the choroid plexus and the olfactory nerves. He was an industrious dissector, and he tells us that he had examined more than a hundred bodies.

With Carpi we close our series of Bolognese anatomists. Into that group we now proceed to fit the writer with whom we are here specially concerned, Hieronymo Manfredi.

III. Hieronymo Manfredi

Hieronymo Manfredi was a member of a family that had already for more than two centuries provided distinguished citizens, and especially physicians, to the city of Bologna.[1] He was born about the year 1430 and was educated at the University of Bologna. Here in 1455 he was *laureatus* in Philosophy and Medicine, and here he became professor of the latter subject in 1463.[2]

During the second half of the fifteenth century, a perfect mania for the study of astrology infected Italy and penetrated equally into the Court, the Church, and the Academy. The profession of Medicine was far from immune, and at the University of Bologna, where a chair of Astrology had long been established,[3] the study was pursued with ardour and enthusiasm. Here Manfredi early devoted himself to that will-o'-the-wisp, the pursuit of which absorbed and sterilized many of the best intellects of his day. By the year 1469 he was already regarded as an authority on the vainest of studies,[4] and as the years went on he seems to have devoted himself to it ever more and more. The generally credulous character of Manfredi's astrological ideas may be

[1] Albano Sorbelli, *Le Croniche Bolognesi del Secolo XIV*, Bologna, 1900; *La Signoria di Giovanni Visconti a Bologna*, Bologna, 1901; Michele Medici, loc. cit., p. 4.

[2] Giovanni Fantuzzi, *Notizie degli scrittori bolognesi*, Tom. v, p. 196, Bologna, 1786.

[3] Hastings Rashdall, *The Universities of Europe in the Middle Ages*, 3 vols., Oxford, 1895, vol. i, p. 244.

[4] He is mentioned in this capacity by Niccolò Burzio, *Bononia illustrata*, Bologna, 1494. We have been unable to consult this work, which is quoted by Fantuzzi, loc. cit. See also Ferdinando Gabotto, *Bartolomeo Manfredi e l'Astrologia alla Corte di Mantova*, Torino, 1891, p. 19.

gathered from the page of his *Prognosticon ad annum 1479* which we here reproduce (Fig. 10).

The history of Manfredi's connexion with the University of Bologna may be briefly told. He appears for the first time on the professorial roll in 1462, when we find him giving the 'extraordinary' lectures on Philosophy, a subject then regarded as under especial charge of the physicians. In 1465 he was conducting the 'ordinary' course in Philosophy, and at the same time giving occasional lectures on Medicine. In the following year he was called to the chair of Theoretical Medicine, and in 1469 he helped the Faculty out of a difficulty by giving lectures on 'Astronomia' in place of the aged professor Giovanni de Fundis. The latter died in 1474, and from that date onward Manfredi assumed responsibility for the course on 'Astronomia'. Among the colleagues who joined him were Gabriele de Gerbi, who became lecturer on Logic in 1476, Filippo Beroaldo, who became lecturer on Rhetoric and Poetry in 1479, and Alessandro Achillini, who became lecturer on Logic in 1484.[1]

Such was the regard for Manfredi's powers of astrological prediction that to all the University announcements of his course of lectures on Astronomy is added 'cum hoc quod faciat iudicium et tachuinum'.[2] In spite of his proficiency in the science, however, he was unable to foretell his own death. Giovanni Pico della Mirandola writes of him thus derisively:

'quo anno [1493] obiit omnimoda[m] uite incolumitate[m] fuerat pollicitus Hieronymus manfredus astrologus nostra aetate singularis: a quo tamen nihil mirandum minus praeuisam aliorum mortem: qui nec suam ipse praeuiderit: nam cum proxima estate uita sit functus: in istius tame[n] anni publico uaticinio qui s[cilicet] ei fuit fatalis: multa & mira sequenti anno dicturum se non semel pollicebatur. Qui nescio oppignoratam fidem quomodo reluet: nisi forte de caelo uerius nunc terrena despiciat q[uam] de terra oli[m] caelestia suspiciebat.'[3]

Manfredi died in 1493 and was buried in the church of Santa

[1] Manfredi's University career is extracted from Umberto Dallari, *I rotuli dei lettori legisti e artisti dello studio bolognese dal 1384 al 1799*, Bologna, vol. i, 1888, and Luigi Nardi and Emilio Orioli, *Chartularium Studii Bononiensis*, Imola, vol. i, 1907. [2] See also P. A. Orlandi, *Notizie degli scrittori bolognesi*, Bologna, 1714.

[3] Johannes Franciscus Picus Mirandula, *Disputationes adversus astrologos*, Lib. ii, cap. 9, Bologna, 1495. Our quotation is from the original 1495 edition, not from the slightly variant *édition contrefaite*.

Margarita in Bologna. This church no longer exists, but it contained in the eighteenth century a tomb bearing the inscription:

HIERON. MANFREDO BONON. PHILOSOPHO AC MEDICO
SVAE AETATIS NEMINI SECVNDO ASTRONOMORVMQVE
CITRA INVIDIAM FACILE PRIMARIO.
POSVIT SVPERSTES IOAN. FILIVS
SVISQVE POSTERIS.
VALE ATQVE ILLVM
VALERE OPTA.[1]

Manfredi left a widow, Anna, who was still living in 1496 with a household of ten persons in the Via S. Margarita.[2] The houses on one side of this street backed on the very walls of the buildings belonging to the 'University of Medicine',[3] and we may suppose that Hieronymo Manfredi had resided here on that account. His surviving son, Giovanni, lived hard by in the Via S. Antonio di Padoa.

It cannot be said that Manfredi's printed works suggest great scientific attainments. All are permeated by the same astrological obsession. They comprise the following:

(*a*) The *editio princeps* of Ptolemy's *Cosmographia* and *Tabulae Cosmographiae*, the best-known printed work to which Manfredi's name is attached. He was associated in its production with the famous scholar Filippo Beroaldo, and the finely produced volume was published at Bologna in 1472 (?),[4] and dedicated to the memory of Pope Alexander V (died 1410). It is interesting as containing the first printed map of England (Fig. 9). At the end of the work we read:

'Accedit mirifica imprimendi tales tabulas ratio. Cuius inuentoris laus nihil illorum laude inferior. Qui primi l[itte]rarum

1 G. Fantuzzi, loc. cit., p. 197.

2 U. Santini, 'Cenni statistici sulla Popolazione del Quartiere di S. Proclo in Bologna', in *Atti e Memorie della R. Deputazione di Storia Patria per le Provincie di Romagna*, series 3, vol. xxxiv, pp. 366 and 367, Bologna, 1906.

3 See map of the old University buildings of Bologna prefixed to Francesco Cavazza, *Le Scuole dell' antico studio bolognese*, Milan, 1896.

4 The date 1462, clearly printed on this edition, is certainly erroneous, since there was no printing-press at Bologna till 1471. A. E. Nordenskiöld (*Facsimile Atlas till Kartografiens äldesta Historia*, Stockholm, 1889, p. 12) consider that 1472 is the true date, but the point is not yet finally settled. See J. A. J. de Villiers, 'Famous Maps in the British Museum', in *Geographical Journal*, vol. liv, London, August 1914, p. 173. Albano Sorbelli, in his authoritative *I Primordi della Stampa in Bologna*, Bologna, 1908, does not mention Manfredi's edition of Ptolemy among the earliest printed Bolognese works (1471–5).

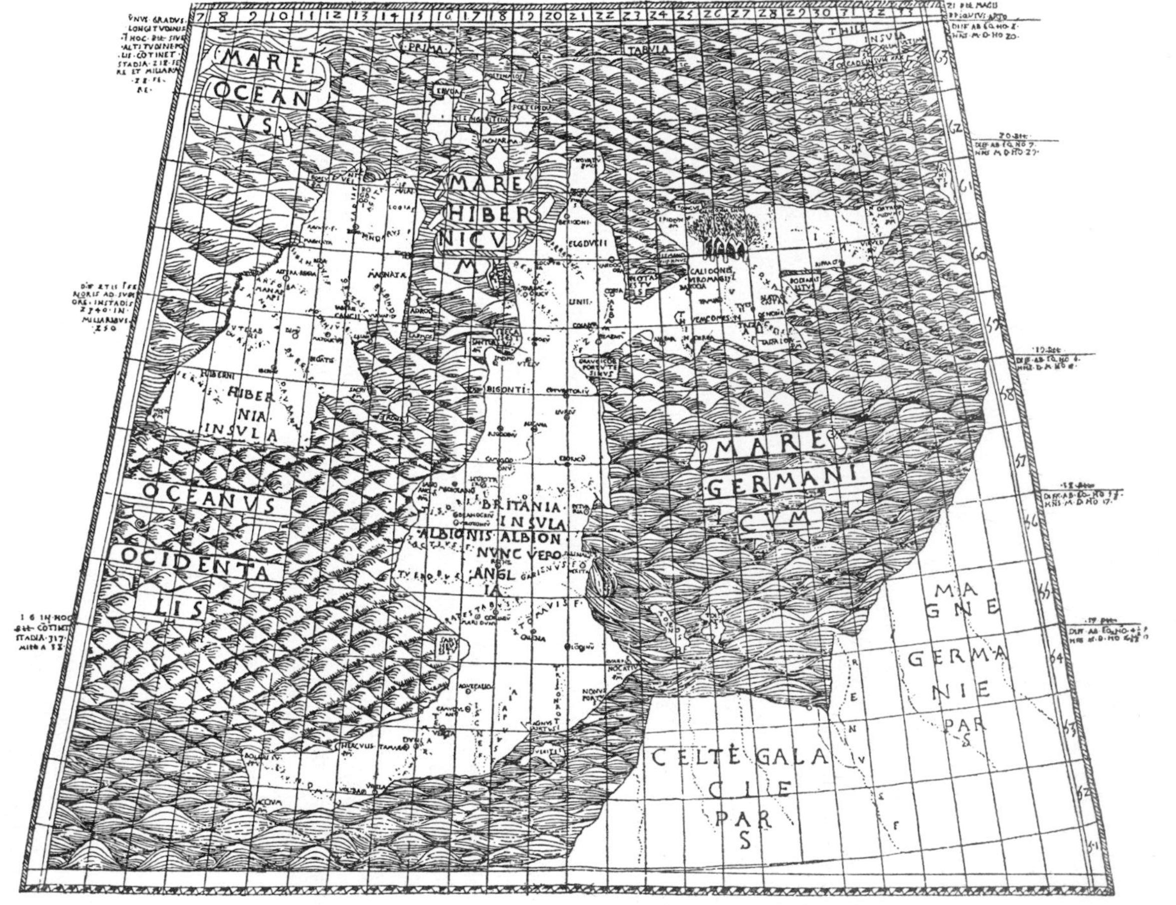

FIG. 9. THE FIRST PRINTED MAP OF ENGLAND. From the 1472 (?) Bologna Ptolemy, edited by Manfredi and others.

imprimendarum artem pepererunt in admirationem sui studiosissimum quemque facillime conuertere potest. Opus utrumque summa adhibita diligentia duo Astrologiae peritissimi castigaueru[n]t Hieronimus Mamfredus & Petrus bonus. Nec minus curiose correxerunt summa eruditione prediti Galeottus Martius & Colla montanus. Extremam emendationis manum imposuit philippus b[e]roaldus.'

(*b*) *Liber de homine: cuius su[n]t libri duo. Primus liber de conservatione sanitatis. . . .* [Liber secundus de causis in homine circa compositione[m] eius], Bologna, 1474. The work is in Italian, and consists of a number of paragraphs, each beginning with the word 'perchè'. There is a servile dedicatory epistle in Latin addressed to Giovanni Bentivoglio. The first book is concerned with diet, and occupies two-thirds of the volume. The second book answers questions on the subject of physiognomy and bears resemblance in many passages to the *Anatomy.* It is taken in the main from the pseudo-Aristotelian *Problemata.* The book is without pagination or figures. It is well printed, and illuminated examples are not infrequently encountered.

This work was very popular. In 1478, during the lifetime of its author, it was audaciously pirated at Naples with the following *incipit*: 'Incomenza el Libro chiamato della uita costumi natura & om[n]e altra cosa pertine[n]te tanto alla conservatione della sanita dellomo quanto alle cause et cose humane. Co[m]posto per *Alberto Magno* filosofo excellentissimo.'

In 1497, after Manfredi's death, the work appeared in black-letter folio at Bologna, with its author's original dedication slightly altered. The text in this edition commences, 'Perchel sophio nele cose che noi viuemo: & lo indebito modo del viuere nostro: induce in noi egritudine'.

In 1507 it appeared at Venice in small black-letter quarto as *Opera noua intitulata Il perche utilissima ad intendere la cagione de molte cose.* By this title, *Il Perchè*, the work, which ran through numerous editions, has usually been known. It continued to be reprinted as late as 1668.

(*c*) A treatise on the Plague: *Tractato degno & utile de la pestile[n]tia co[m]posto p[er] el famosissimo philosopho medico & astrologo maestro Hieronymo di manfredi da Bologna,* Bologna, 1478. This was translated into Latin by the author himself in the same year. The work owes much to Avicenna, but contains some original clinical observations, and shows a certain independence of the prevailing spirit of the age by quoting opinions of contemporary as well as of ancient physicians. The remedies

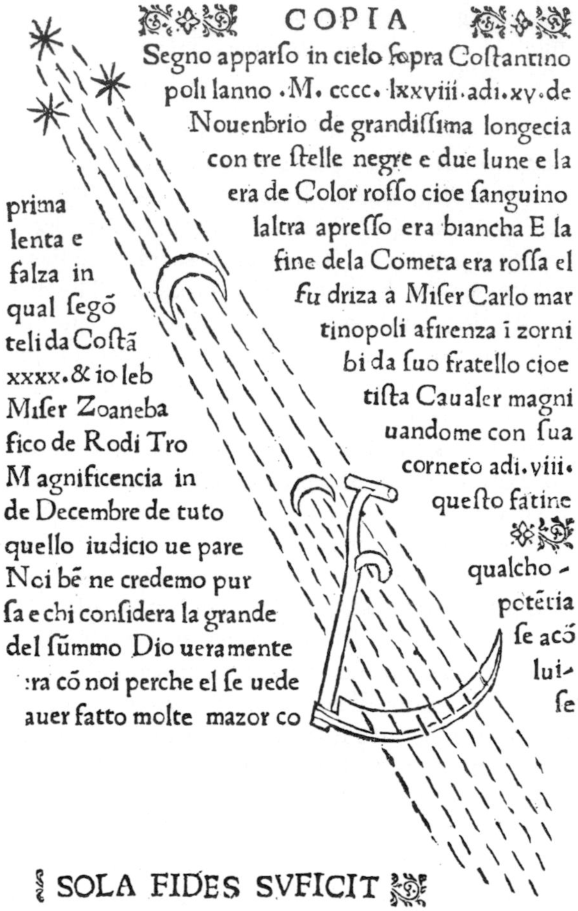

COPIA

Segno apparſo in cielo ſopra Coſtantino
poli lanno .M. cccc. lxxviii. adi. xv. de
Nouenbrio de grandiſſima longecia
con tre ſtelle negre e due lune e la
era de Color roſſo cioe ſanguino
laltra apreſſo era biancha E la
fine dela Cometa era roſſa el

prima
lenta e
falza in
qual ſegō
teli da Coſtā
xxxx. & io leb
Miſer Zoaneba
fico de Rodi Tro
M agnificencia in
de Decembre de tuto
quello iudicio ue pare
Noi bē ne credemo pur
ſa e chi conſidera la grande
del ſūmmo Dio ueramente
:ra cō noi perche el ſe uede
auer fatto molte mazor co

fu driza a Miſer Carlo mar
tinopoli afirenza ī zorni
bi da ſuo fratello cioe
tiſta Caualer magni
uandome con ſua
corneto adi. viii.
queſto fatine

qualcho-
poteria
ſe acō
lui-
ſe

SOLA FIDES SVFICIT

FIG. 10. The last page of Manfredi's *Prognosticon ad annum 1479*, Bologna, 1478.

are similar to those recommended by John of Bourdeaux in his widely distributed tract on the plague, and are probably derived ultimately from the *Regimen Sanitatis Salerni.*

(*d*) *Prognosticon ad annum 1479*, Bologna, 1478. We reproduce the terminal page of this work (Fig. 10).

(*e*) *Prognosticon anni 1481,* in which is embodied *Oratio contra turcos & hostes Christianorum,* s. l. Jan. 1481.

From his tomb in the Church of S. Giacomo Maggiore at Bologna

Plate XXXVII. GIOVANNI BENTIVOGLIO II

BRIT. MUS. MS. ROY. 7 F VIII, fo. 50 v

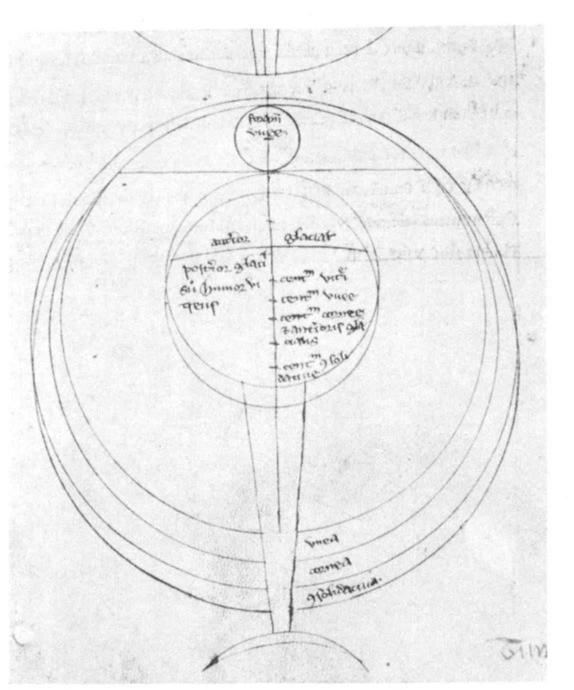

PLATE XXXVIII a. ROGER BACON'S DIAGRAM OF THE EYE.
XIIIth Century

From a drawing in WINDSOR CASTLE

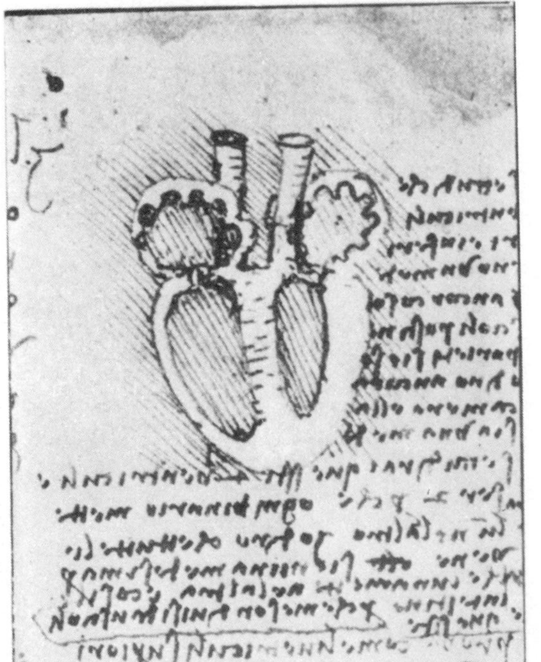

PLATE XXXVIII b. LEONARDO DA VINCI'S DIAGRAM OF THE HEART
Early XVIth Century

(*f*) *Centilogium de medicis et infirmis*, Bologna, 1488. With a dedication to Bentivoglio. This short work is wholly astrological, and consists of one hundred precepts concerning the relationship of the stars to various diseases and conditions. Reprinted Venice, 1500, and Nuremberg, 1530.

The following three works are attributed to Manfredi, but are not mentioned in Hain, Copinger, or Reichling's lists of Incunabula; we have not seen any of them and their existence is doubtful.

(*g*) *Ephemerides astrologicae operationes medicas spectantes*, mentioned in the *Biographisches Lexikon der hervorragenden Aerzte* of E. Gurlt and A. Hirsch. Possibly it represents another edition of (*e*).

(*h*) *Quaestiones subtilissimae super librum aphorismorum*, Bologna, 1480 (?), mentioned by Haller.[1] Possibly it represents another edition of (*b*).

(*i*) *Chiromantia secundum naturae vires ad extra*, Padua, 1484, mentioned by Haller.[1]

IV. The Manuscript Anatomy of Manfredi

The MS. of Manfredi's *Anatomy* is in the Bodleian Library at Oxford (Canon. Ital. 237, Western 20287). It is a fairly preserved small quarto parchment, originally of forty-nine folios, of which the third and fourth are missing. The writing is in the fine Italian hand that the printed type of the period was accustomed to imitate. There are no figures or illuminations, but the titles are rubricated in burnished gold or in colours.

There is no reference to this work in any account of Manfredi, and the volume itself appears to be quite unknown. Neither the man nor his work is mentioned in Medici's detailed history of the anatomical school at Bologna[2] nor in Martinotti's recent study on the same topic,[3] nor is any MS. of Manfredi included in Mazzatinti's monumental catalogue of the MSS. in the Italian libraries.[4]

[1] Albrecht von Haller, *Bibliotheca anatomica*, Zürich, 1774–7, vol. ii, p. 738.

[2] Michele Medici, *Compendio storico della Scuola anatomica di Bologna dal Rinascimento delle Scienze e delle Lettere a tutto il Secolo XVIII*, Bologna, 1857, folio.

[3] G. Martinotti, 'L'insegnamento dell' anatomia in Bologna prima del secolo XIX', in *Studi e Memorie per la Storia dell' Università di Bologna*, vol. ii, Bologna, 1911.

[4] Mazzatinti, *Inventari dei Manoscritti delle Biblioteche d'Italia*, Forli & Firenze, 1890–1915, vols. i to xxiii, in progress.

Manfredi's MS. is written in the involved Italian of the day, with sentences of inordinate length. These general characters of style are encountered also in his published works. The dedication is in Latin, of the same unpleasing quality, and is couched in the usual subservient manner. It is addressed to Giovanni Bentivoglio, and in it Manfredi relates that

'Your illustrious lordship Johannes Bentivolus in this present year 1490 with your usual humanity condescended on one occasion to watch the dissection of a corpse. . . . It was then that you saw the wonderful works of Nature in the anatomy . . . and you parentally urged me, Hieronymo Manfredi, to inscribe to your most noble name this work on anatomy. . . . I therefore extracted this work as best I might from various works of antiquity and abbreviated it. I have not followed their order, but I have so composed it that the work should be pleasing to your lordship.

'Accept then, O great and powerful lord, this work on the anatomy of the human body inscribed to your noble name! Accept it with your customary benevolence and humanity and in a kindly and gracious spirit, for it will be pleasing to you and will delight you greatly, for it is a worthy work!'

The Giovanni Bentivoglio (Plate XXXVII), with adulation of whom Manfredi was thus accustomed to plaster his works, was the second of the name and was the son of Annibale Bentivoglio. In the year 1462 he became head of the republic of Bologna, and played there much the same rôle as did Lorenzo de' Medici at Florence. He adorned Bologna with numerous buildings,[1] and acted as patron of the arts and the sciences. The Palazzo dei Bentivogli still stands as a memorial to him and his family. A stern and high-handed tyrant, he held his position until 1506, when he was expelled and the city reverted to the papacy. He died two years later.

It is remarkable to find a man of Bentivoglio's eminence and position taking an interest in the practical study of anatomy. Other Italian rulers, Lorenzo de' Medici among them, encouraged and legalized the practice of dissection, but probably Bentivoglio is the only one recorded as having patronized an 'anatomy' in person. The interest taken in the subject by the heads of states must have been of great value to the artists whose patrons they were.

[1] Lino Sighinolfi, *L'Architettura Bentivolesca in Bologna e il Palazzo del Podestà*, Bologna, 1909.

The MS. is a unique copy, and was doubtless written for presentation to Bentivoglio. That it was never printed is perhaps due to the fact that Manfredi died within a comparatively short time of its composition. It represents the most satisfactory post-mediaeval account of the human frame until the appearance of the work of Berengario da Carpi in 1521. It is more complete than the work of William of Saliceto or of Mondino or the anatomy erroneously attributed to Richardus Anglicus; it is more natural than the book of Gabriele de Gerbi, and is far superior to the crude contemporary sketches of Hundt, Peyligk, and Achillini, while it wastes less space than Guy de Chauliac on teleology, though it has none of the charm of the work of that great surgeon. In one respect at least, viz. the spirit in which it is written, Manfredi's *Anatomy* is original and probably unique for its age. There is no reason to doubt the assurance of the dedication that it was composed for the edification of the tyrant of Bologna, and for the simple purpose of setting forth the wonderful structure of man's body without thought of any medical application.

The sources of the MS. are obvious. It is in the main a re-arranged and on the whole improved Mondino, but amplified by reference to translations from Galen, Rhazes, Haly Abbas, and Avicenna. Guy de Chauliac has perhaps also been used. The work gives a general impression of being the product of a practical dissector, and it provides us with a good example of early Renaissance anatomy as taught in the Italian schools before the reforms of Vesalius. It is perhaps the first complete treatise on its subject written originally in the vernacular.[1] It exhibits, however, no other original features nor any considerable departures from its sources, and it may be taken to represent, with but little modification, the tradition of Mondino as developed at his own University of Bologna at the end of the fifteenth century.

Manfredi's work, however, if not original is at least eclectic, and the variety of its sources indicates a dawning consciousness of the unwisdom of trusting to the infallibility of any one writer. The work is thus in a sense intermediate between the early printed versions of Mondino, such as that of 1478, and the edition

[1] Several short sketches or tractates on anatomy in the vernacular are however known. Thus a Provençal anatomical tractate of the thirteenth century has been published by K. Sudhoff in his *Beitrag zur Gesch. der Anatomie im Mittelalter*, Leipzig, 1908.

published in 1528 by Berengario da Carpi with its frank commentary of the master. All represent stages towards the freedom of the later Renaissance investigators.

We reproduce the text in full, and the passages on the head, on the eye, and on the heart, are rendered into English. All are similar to the accounts of Mondino. We are able to illustrate them by figures from contemporary works, and thus to give an idea of the limits of the anatomical knowledge of the day.

V. Translation of Selected Passages from the Anatomy, with Commentary

(a) THE HEAD

Tractate i, Chapter 2

(folio 5 verso) There are ten layers of the head.

The *first* is the *hair* made by nature for the better protection of the head from external things, and also for beauty.

The *second* part is the *skin*, which has here to be very thick, so that the hair may be firmly embedded, having its roots thick and long; and also to be a better shield and covering for the bone and brain, since there is no muscular part here.

The *third* part is the *flesh*, developed only on the face, the temples, and about the jaws, not on the other parts.

The *fourth* part is an external membrane called *almochatim* [Arabian term for cranial periosteum] which, when the skin is raised, appears to be continuous and covers the whole cranium. And nature made this membrane firstly so that the skin which is soft should not come into contact with the hard bone, secondly that the bone of the head should have sensation through it, and thirdly that the internal membrane of the head, called *dura mater*, should, by means of this membrane, be attached to the bone of the cranium by certain nerves and ligaments. These, issuing through the commissures of the bones, have thus their origin in the aforesaid internal membrane, while on emerging through the bone, they weave themselves into or rather compose the external membrane called *almochatim*.

The *fifth* part is the *skull*. This is a bone like a cap, inside the cavity of which is located the brain. In the skull are four bones sutured together. Nature made the skull not of one but of many pieces, firstly, so that if harm should fall on one part it might not spread to the others; secondly, so that by their joints or rather sutures [Italian *cusiture*=sewings], the humours of the brain might be the better exhaled; and thirdly, so that when there is need of applying medicines, these might the better penetrate to the parts within.

Hence it is that four pieces of bone are sutured and joined together by nature in a denticulate fashion, so that they might be the firmer and stronger. Nor are they bound with ligaments as are the joints, for these would not have been so strong, and furthermore the bones of the head do not need to move.

These sutures are five in number, three being true and two false. The true sutures are those which pass right through the bone, while the false do not. Of the true sutures one is in the anterior part and is called *coronal*; it is made like the letter C, and stretches from right to left of the head, the two wings of the C being directed towards the forehead. The second true suture extends along the length of the head, beginning from the coronal and reaching the back part of the head. It is like a shaft or rather arrow that goes backwards from the brow, wherefore it is called *sagittal* ——(. The third true suture is in the posterior part and is called laudal, for it is made like a Λ, the letter called by the Greeks *lauda*. The sagittal suture extends from the coronal to the lauda)——(.

The false sutures are two, one on each side. They are called *cortical* because they do not penetrate.

Now if we consider these five sutures we shall see that there are four bones articulated together. One is the forehead bone [frontal] which begins at the coronal and ends below at another suture, which itself begins as a branch of the coronal suture and proceeds by way of the eyebrow to the corresponding branch [of the other side] ɑ.

A second bone is behind and terminates at the laudal suture. There are two other bones which form the temples. These terminate at the false sutures which themselves begin at the laudal and end at the coronal suture.

The *sixth* part [of the head] consists of two membranes. One of these is called *dura mater*, and lies in contact with the cranium. The other is called *pia mater* and is in contact with and covers the brain. And nature contrived it thus, having great solicitude for this latter member, that while close to the bone, it should yet not be touched by it. Wherefore, taking due precautions, she made the one [membrane] harder than the other. Furthermore she made two membranes, so that if harm befell one of them, it might not be communicated to the underlying brain.

In the *pia mater* are woven certain veins by which the brain is nourished. [The brain is] everywhere covered by it except on the posterior part; because this part being dry, it has no need of this membrane, as have the anterior and middle parts. The two membranes in many places penetrate the substance of the brain, dividing it into a right and a left, a front and a back section. By this division, divers cells or rather small chambers are made therein, in which the soul (*anima*) performs its divers operations, for which reason it is necessary that these parts should be of different structure.

When the two membranes are raised, the *seventh* part of the head, namely the brain itself, appears. The brain is wrought by nature so that the *vital spirit* from the torrid heart should be tempered by its cold, for here it is converted into *animal spirit*, which is the beginning of the perceptive (*cognoscitiue*) and motive processes.

The brain is of a substance like marrow, white, soft, and viscous, and from it the nerves arise. The anterior part is moister, softer, and less cold than the posterior because the senses [*sentimenti* = senses + mental processes], which are themselves moist and soft, have here their origin. In the posterior part the motor nerves arise, and it is therefore drier and firmer.

The brain is divided into three parts or ventricles. The first ventricle or anterior part is itself divided into two, right and left, and is moreover larger than any of the other ventricles, for in this first ventricle nature has placed the two faculties subservient to perception (*al cognoscere*). One of these is called *common sensation* (*senso comune*); in it the external senses terminate as at a centre and deliver the *images* or rather *species* of sensible things, so that this faculty may perceive and distinguish between one sensible thing and another, and also comprehend the operations of particular senses; which two things none of these [senses of themselves can do]. The other faculty of the first ventricle is called *fantasia* and by some *imagination*; it retains and preserves the *species* of sensible things in the absence of the material objects themselves.

When thou examinest the first ventricle thou wilt see three things before thou comest to the second ventricle.

[*a*] The first is itself double, and is formed of the very substance of the brain, so that it forms the base of the anterior ventricle both right and left [= corpora striata].

[*b*] To the side of this is another thing like a subterranean worm, red as blood, yet tethered by certain ligaments and nervelets [=choroid plexus and taenia semicircularis]. And this worm when it lengthens itself closes these passages, and thus blocks the path between the first ventricle and the second. Nature has wrought it thus, so that when a man wills he may cease from cogitation and thought; and similarly when, on the other hand, he would think and contemplate, this worm contracts itself again and opens these passages and thus frees the way between one ventricle and another.

[*c*] The third structure is a little lower and is a *lacuna* or rounded concavity [= infundibulum]. In the middle of this is a hole which passes down towards the palate, and this lacuna provides also a direct passage which descends from the middle ventricle to its *colature* [=sieve-like structure, i. e. certain parts of the sphenoid bone]. And this lacuna has around it certain large round eminences which support the veins and arteries that ascend to the ventricle. This passage is wide above and narrow below, and by it the first and second ventricles purge themselves of their superfluities, but the anterior part [of the first ventricle] purges itself more by the colature of the nose [=cribriform plate]. Thus nature has made two passages to cleanse the superfluities of the brain.

When thou hast seen these three structures there will appear the second or middle ventricle which is as a passage and transit from the anterior to the posterior ventricle. Here are two faculties. One, the *estimative*, deduces [Italian *elicere*] the insensible from the sensible. The other, called the *cognitive*, comprehends both things sensible and things insensible, synthesizing and analysing them (*componendo e dividendo*). These [two] faculties in the middle ventricle minister to the intellect. Now all the other faculties described, and even the power of memory, are found in brute animals, but this [intellectual power] is encountered in man alone.

Now will appear the third ventricle in the posterior part; and it is hard, for it gives rise to the greater part of the motive nerves which are of a strong and firm nature. This ventricle is pyramidal in shape, and culminates in an apex directed upwards where images of visible things (*spetie*) are conserved, for these are better stored in a strait than in an ample space; but the part below is wide to receive these images, which are better received in an ample than in a strait place. This ventricle has two functions: it gives rise to the spinal cord [*nucha*, an Arabian term] and motor nerves; and it is also the storehouse of the *memorative* faculties.

From what [has been said] it will be apparent that when the back of the head is injured, the memory immediately suffers; when the middle part is injured, the estimative and cognitive faculties suffer; and when the anterior part is injured, the faculties of common sensation and of imagination (*fantasia*) suffer. And thus it is that the doctors have become aware of the location of these powers.

This being disposed of, thou wilt next raise the brain carefully so as not to break the nerves. Commencing now with the part in front, there will first appear two small fleshy protuberances like two nipples, of like substance to the brain in which they originate, and covered by a thin membrane, the pia mater. These are the olfactory organs, wherein is the sense of smell.

From the brain arise seven pairs of nerves. Proceed therefore farther with the anterior part, and thou wilt see the first pair of these nerves, which are large, and called the *nervi optici*. These have their origin in the front ventricle of the brain and proceed towards the eyes. But before they pass through the pia mater, they join together, and at their place of union there is a perforated spot. Galen maintains that these nerves only join or rather unite, but do not intersect, so that the nerve that comes from the right after union returns again towards the right, and similarly with the nerve coming from the left, which after the union returns towards the left eye.[1] But Rhazes maintains the contrary,[2] although the opinion of Galen is the more common. These nerves are subservient to sight, and they are united so that the images of the things received by the two eyes and conveyed by the two nerves should return in unity; so that one thing should not appear as two.

[1] Cf. Galen, *De usu partium corporis humani*, Lib. x, chap. 12.

[2] Cf. Rhazes, *Almansur*, i. 4.

After these two nerves, raise the brain towards its middle and thou wilt see another pair of nerves, thin and firm, which also go to the eyes, to give them voluntary movement, controlling certain muscles.

Farther on thou wilt see the third pair of nerves, one part of which goes to the face to give it sensation and voluntary movement, while another part goes to give taste to the tongue. Yet a third part of these nerves mingles with the fourth [1] pair of nerves, and together they descend to give sensation to the diaphragm, stomach, and other viscera. A certain part also of the fourth [1] pair of nerves goes to give sensation to the palate.

Then there is the fifth pair of nerves [which] go to the *petrous* bone around the ear; and of these nerves there are framed in the ear-holes certain membranes, which are the organs of hearing.

Next there is the sixth pair of nerves, which divides into three parts. One part goes to the muscles of the throat, the second to the muscles of the shoulders, and the third and largest descends to the epiglottis and to the diaphragm, and spreads into the chest, the heart, and the lungs, accompanying the nerves of the third pair. From the nerves of this sixth pair which go to the epiglottis arise the nerves of the voice, called *reversive.*

The seventh pair of nerves arise at the back of the brain and give voluntary movement to the tongue.

Of these seven pairs of nerves, the first two pairs originate in the anterior part of the brain, the third pair originates between the anterior and posterior parts, while the remaining four pairs originate in the posterior part.

Proceeding still farther, the brain may be completely raised, and the eighth part of the head will appear, that is, the two membranes situated below the brain. When these in turn are raised there will appear the ninth part, which is a certain net called *rethe mirabile,* because it is composed of exceedingly strong and marvellous texture, augmented by certain very fine arteries which are branches of arteries that ascend from the heart, and are called the *apoplectic arteries.* In these arteries of this net is contained the vital spirit, sent from the heart to be changed to animal spirit. That the spirit may be the better modified and distributed, nature made these arteries very fine, and separated

[1] Manfredi here follows Mondino, who confuses Galen's fourth pair with Galen's sixth pair of nerves.

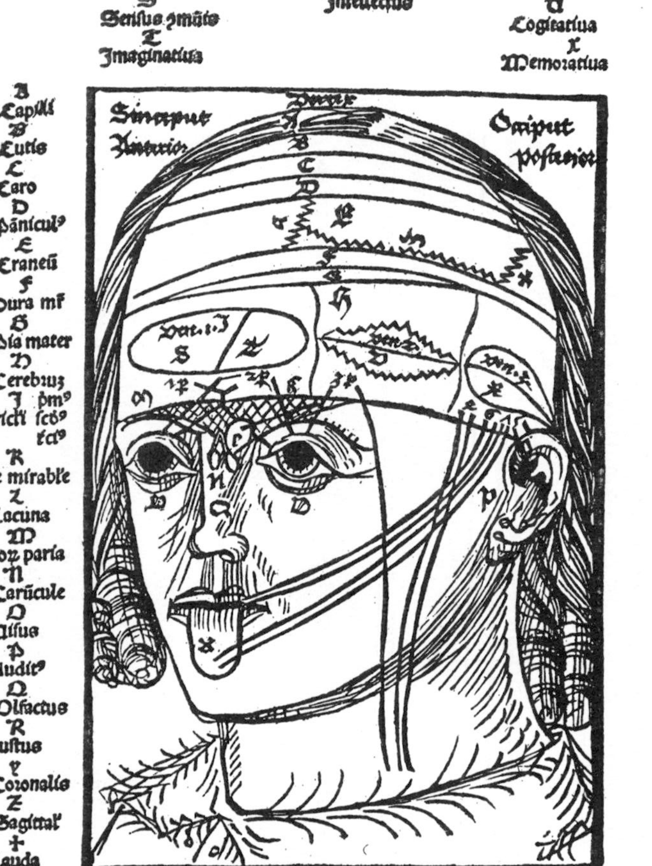

Tactus per totum corpus expādit per nervos spondilium.

FIG. 11. From M. Hundt, *Antropologium, de hominis dignitate natura et proprietatibus*, Leipzig, 1501. The figure shows the ten layers of the head, the cerebral ventricles and cranial nerves, and the relation of the nerves to the

FIG. 12. THE LAYERS OF THE HEAD
From the *Anatomia* of Johannes Dryander, Marburg, 1537.

them into very small branches so that the spirit should be minutely divided. Nature placed the rethe mirabile under the brain because it was necessary to guard its site carefully, and also that the moist vapours of the brain which fall upon the net, obstructing it, should induce natural sleep.

After all these things thou wilt see the basal bone which is the tenth and last part of the head, and called *basilar*, because it is the base and foundation of the whole head; and it was made hard so that the superfluities which descend to it should not putrefy it. This bone can be seen to be formed of many other bones articulated together. It is divisible into the petrous bones and the bones of the nose and eyes and two other lateral bones which can only be seen by means of disarticulation. [Folio 10 verso, line 22.]

The ten parts or layers of the head are a commonplace of the anatomy of the period, taken from Avicenna. We may illustrate the division by the crude contemporary diagram of Fig. 11, which is improved in the later drawing reproduced in Fig. 12.

Manfredi's account of the brain itself is amplified from Mondino. The division of this organ into three ventricles, each associated with a corresponding division of the mental functions, was very familiar to medical writers of the fifteenth century. The idea is found among Western writers as early as St. Augustine (354–430), and is encountered in the writings of Roger Bacon (1214–94). It had long been popularized in mediaeval psychology by the writings of Albertus Magnus (1206–80). The anatomical distinction is found in Haly Abbas, Avicenna, and Rhazes, and in some of the best MSS. of the latter writer a rough diagram of the ventricles is given.[1] These writers are all clearly indebted to the anatomy of Galen,[2] but on the psychological side Albertus Magnus probably drew mainly either from Ghazali[3] (1059–1111), who in turn derived his inspiration from Nemesius (fourth century) and Johannes Damascenus (died 756), or else from

[1] See P. de Koning, *Trois Traités d'Anatomie arabes*, Leyden, 1903, p. 47.

[2] See J. Wiberg, 'The Anatomy of the Brain in the Works of Galen and 'Ali 'Abbas; a comparative historical-anatomical study', *Janus*, vol. xix, p. 17 and p. 84, Leyden, January and March, 1914.

[3] See A. Schneider, 'Die Psychologie Alberts des Grossen', p. 160, in *Beiträge zur Geschichte der Philosophie des Mittelalters*, Band iv, Heft 5, Munich, 1903.

De anima.

FIG. 13. From *Illustrissimi philosophi et theologi domini Alberti magni compendiosum insigne ac perutile opus Philosophiae naturalis*, Venice, 1496, showing the ventricles of the brain.

early writers of the Salernitan tradition, such as Constantine[1] (eleventh century), or Petrocello[2] (twelfth century), who drew largely on Theophilus (seventh century).[3]

This outline of a tripartite division of the brain and its cavities was closely followed throughout the Middle Ages, as was also the curiously naïve and excessively 'materialistic' psychology to which it gave rise, and which Manfredi adopts. We illustrate his views of the relationship of the different parts of the brain and their parallelism in mental processes, from a series of diagrams extracted from contemporary works (Figs. 13–18).

The brain was

[1] Constantine Africanus, *De communibus medico cognitu necessariis locis*, Lib. iii, cap. 11, Edition Henricus Petrus, Basel, 1541.

[2] *Practica Petrocelli Salernitani. Epistola. Quot annis latuit medicina.* S. de Renzi, *Collectio Salernitana*, Naples, 1852–9, vol. iv, p. 189.

[3] A very elaborate study of the doctrine of the three vesicles of the brain has recently been made by Walther Sudhoff, 'Die Lehre von den Hirnventrikeln', in the *Archiv für Gesch. der Med.*, Leipzig, 1914, vol. vii, p. 149.

regarded by mediaeval and early Renaissance anatomists as having two channels of discharge through which the *phlegm*, the especial product of this organ, could be evacuated when in excess. One of these channels communicated with the anterior ventricle of the brain and poured its secretion into the nose. It may be identified with the *anterior colature* or cribriform plate. The second, the *lacuna*, led down from the second ventricle and poured its secretion into the pharynx. It may be identified with the infundibulum, pituitary body, and 'cella turcica'. The term 'pituitary' which we still use is derived from its supposed association with the 'pituita' or phlegm. At an early date this process was connected with the four humours (Fig. 14). The rest of the description of the brain can be easily followed. The comparison of the choroid plexus to a worm is very common. The suggestion originated with Galen and was developed by the Arabians.

COMPARATIVE TABLE OF ANCIENT AND MODERN NOMENCLATURE OF CRANIAL NERVES.

Mondino and Manfredi following Galen, especially in the περὶ χρείας τῶν ἐν ἀνθρώπου σώματι μορίων. *De usu partium corporis humani.*	*Modern usage.*
Not regarded as separate nerves.	I. Olfactory nerves.
I. τὰ μαλακὰ νεῦρα τῶν ὀφθαλμῶν.	II. Optic nerves.
II. τὰ κινητικὰ τῶν ἀμφ' αὐτοὺς μυῶν.	III. Oculomotor nerves.
Not mentioned.	IV. Trochlear nerves.
III. τρίτη συζυγία. IV. τετάρτη συζυγία. Mondino and Manfredi confuse Galen's fourth pair and Galen's sixth pair.	V. Trigeminal nerves.
Not mentioned by Manfredi. By Galen probably united with II.	VI. Abducent nerves.
V. πέμπτη συζυγία.	VII. Facial nerves. VIII. Auditory nerves.
VI. ἕκτη συζυγία.	IX. Glossopharyngeal nerves. X. Vagi. XI. Accessory nerves of Willis.
VII. ἑβδόμη συζυγία.	XII. Hypoglossal nerves.

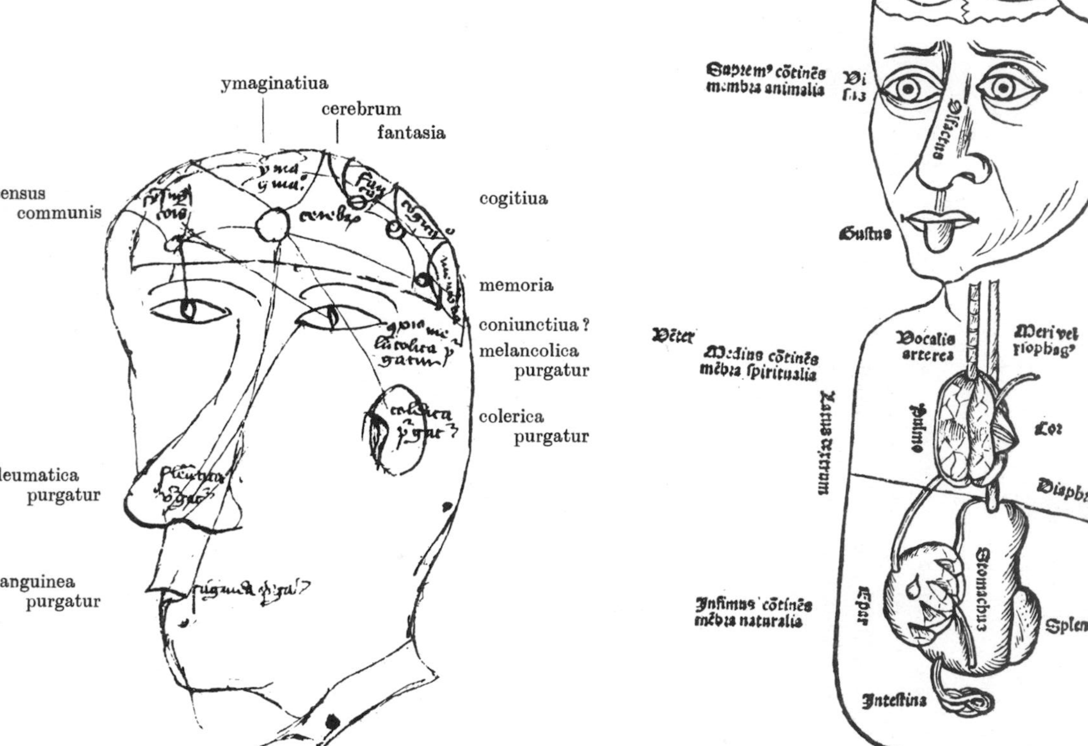

Fig. 14. Diagram of the senses, the humours, the cerebral ventricles, and the intellectual faculties. MS. Sloane 2156, folio 11 recto, in the British Museum, being a copy written in 1428 of the *De Scientia Perspectiva* of Roger Bacon.

Fig. 15. From K. Peyligk's *Philosophiae naturalis compendium*, Leipzig, 1489. Illustrating the general ideas on anatomy current at the Renaissance.

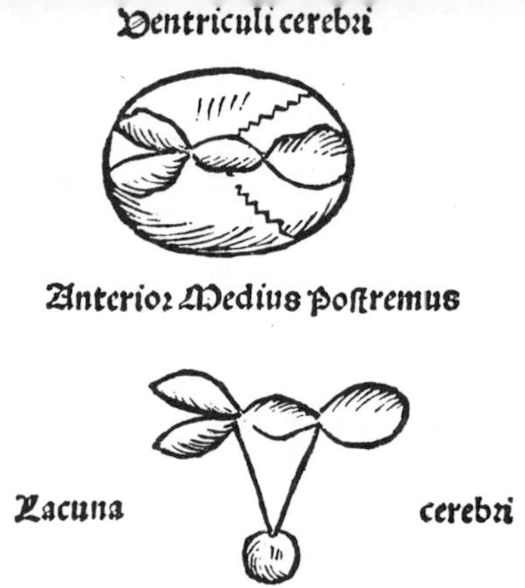

Ventriculi cerebri

Anterior Medius Postremus

Lacuna cerebri

FIG. 16. The cerebral ventricles from above and from the side. According to K. Peyligk, *Philosophiae naturalis compendium*, Leipzig, 1489.

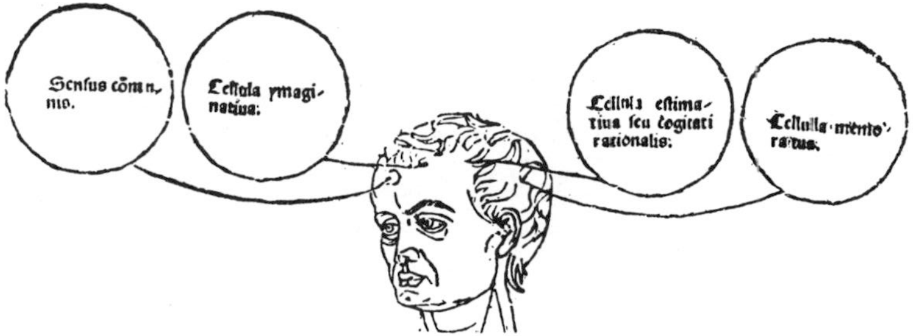

Sensus commu-nis.

Cellula ymagi-natiua:

Cellula estima-tiua seu cogitati rationalis:

Cellula memo-ratiua.

FIG. 17. The localization of cerebral functions. From the Italian edition of 'Ketham', *Fasciculus Medicinae*, Venice, 1493.

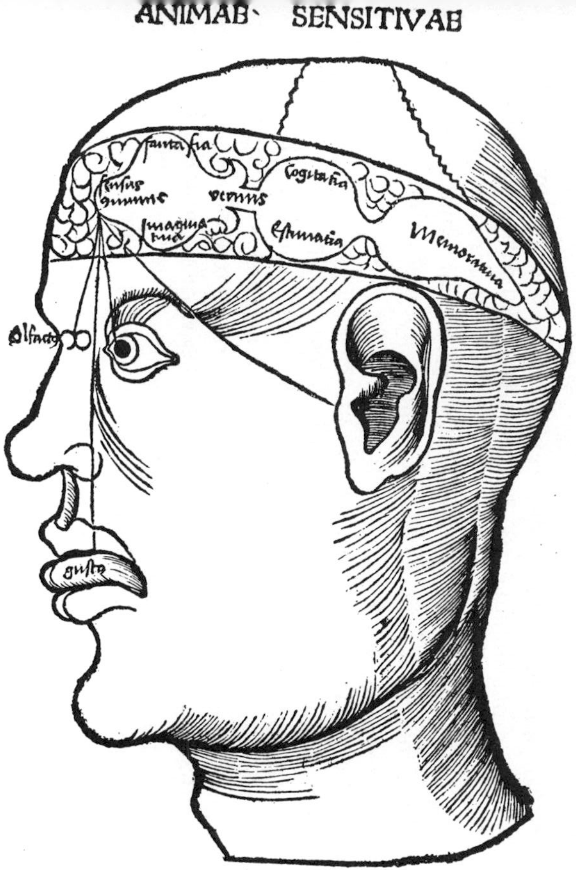

ANIMAE SENSITIVAE

FIG. 18. From G. Reisch, *Margarita philosophiae*, Leipzig, ? 1503. Diagram of the ventricles and the senses with their relation to the intellectual processes according to the doctrine of the Renaissance anatomists.

The nomenclature of the cranial nerves adopted by Manfredi is taken from Mondino and is almost identical with that of Galen, whose classification is summarized above.[1] Manfredi's description of Galen's fourth pair is confused and inadequate, but his account of Galen's sixth pair is an improvement upon Mondino.

The 'rete mirabile' is an interesting survival of Galenic anatomy. This structure is hardly present in man, but is developed in the lower animals, and especially in calves, upon whose bodies Galen worked. The father of physiology regarded the 'rete mirabile' as the place where the psychic pneuma was elaborated.[2] Galen's findings in the lower animals were assiduously transferred to the human body, to which his descriptions are much less applicable, while his views on the pneuma lasted in more or less misunderstood form well into the seventeenth century.

(*b*) THE EYE

Tractate i, Chapter 3

(folio 11 recto) The socket of the eye is not over-depressed, for it has to receive the images (*spetie*) of visible things. Nor does it project greatly, lest it should be liable to injury from exterior violence. For the eyes of man being very soft and susceptible, nature provided eyebrows as a shield above, and eyelids as protectors in front, and made moreover the projections of the maxillae and the nose, so that the eyes should be guarded on every side. So great was the solicitude of nature for these members.

Seven are the tunics of the eye and three its humours. Three front coatings join with three coatings at the back like six shields, the edges of every pair joining each to each, the outer being larger and containing the others. The seventh tunic is largest of all, and encloses the whole eye, and therefore it is called *conjunctiva* because it joins and surrounds the whole eye except the place where the pupil is, and that small part [is covered] by the

[1] See F. G. A. Stumpff, *Historia nervorum cerebralium ab antiquissimis temporibus usque ad Willisium nec non Vieussensium. Dissertatio inauguralis*, Berlin, 1841; C. Daremberg, *Œuvres anatomiques, physiologiques et médicales de Galien*, Paris, 1854, p. 583, &c.; G. Helmreich, ΓΑΛΗΝΟΥ, *περὶ χρείας μορίων*, Leipzig, 1909; and Theodor Beck, 'Die Galenischen Hirnnerven in moderner Beleuchtung', in *Arch. für Gesch. der Med.*, vol. iii, p. 110, Leipzig, 1910.

[2] Galen, *De usu partium*, ix. 4; *De Hippocratis et Platonis decretis*, vii. 3.

cornea. Now this first tunic where it covers the outside part is seen to be white.

The second tunic in its front part is called *cornea* because it resembles horn in its substance and colour ; and this covering is transparent, so that the images of visible things may penetrate through it. And it is also solid and large and composed of four membranes, so that being near external things it should not receive hurt. With this [corneal tunic] is united posteriorly another tunic [the third] called *sclerotic,* i. e. hard. These two coverings have their origin in the membrane about the brain, that is in the dura mater, just as the first tunic arises from the membrane over the skull, called *almochatim.*

The fourth tunic as to its front part is called *uvea* [because] it is like a seed of a black grape, and in its midst is a hole called the pupil. Nature made this tunic opaque so that the visual spirit should be conserved and not dissipated by the light outside. Moreover nature made the opening in the tunic that the image might penetrate freely ; while it is narrow, so that the visual spirit should be concentrated. Thus when the said pupil, or rather hole, dilates more than usual, either naturally or accidentally, the sight becomes imperfect. [The uveal tunic] joins posteriorly the fifth tunic, called *secundina* because it is made like the after-birth, i. e. the membrane in which the child is enveloped in its mother's womb, and it arises from the pia mater.

The sixth coating in front is called *arachnoid* because it is formed after the manner of a spider's web, and posteriorly it joins the seventh coating, called *retina,* because it is made like a net.

Between the uvea and the arachnoid anteriorly there is a humour called *albugineus,* like the white of an egg, to moisten the eye and to preserve the convexity of the cornea. In a dead man this humour dries up, and the cornea falls and is flattened, and then the vulgar say that there appears a curtain before the eyes which is an infallible sign of death. Also this humour holds the pupil open ; therefore when it dries up the pupil contracts.

Between the two last tunics, i. e. the arachnoid and the retina, which have their origin from the optic nerve, there are two humours. These are the *vitreous* humour, so called from its likeness to liquified glass, and the *crystalline* humour, from its likeness to a crystal. This is also called the *grandid,* because it is like a hailstone ; and it is somewhat hard and round, but flattened anteriorly where it receives the images of visible things, and posteriorly pyramidal shape and pointed. And here is completed the act of seeing. In the posterior part it is surrounded by the vitreous humours by which it is nourished. The crystalline

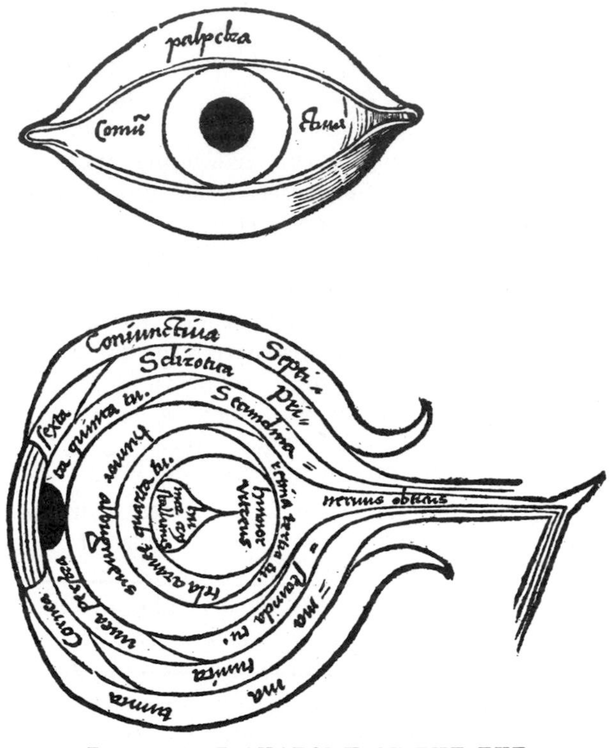

FIG. 19. THE ANATOMY OF THE EYE
From G. Reisch, *Margarita philosophiae*, Leipzig, ? 1503. Showing the seven tunics and three humours of the eye according to the doctrines of Renaissance anatomists.[1]

humour is convex anteriorly and the vitreous posteriorly. And the optic nerves come to the eyes and convey the images seen by the eyes to [the seat of] common sensation and to the other internal faculties. [Folio 12 verso, line 7.]

A great deal of attention was paid by the Arabians to the diseases and the structure of the eye, and the essentials of Manfredi's description are to be found in Rhazes, Hunain ben Ishak,

[1] The first edition of the work appeared in 1496.

and Haly Abbas. The tradition presented by these writers passed early into Western science, and is reproduced, for example, in the works of Constantine Africanus and in the well-known anatomy to which the name of Richardus Anglicus (Richard of Wendover) has become attached[1] (cp. Fig. 19). Avicenna's description of the eye is somewhat different, and gave rise to the tradition reproduced in the works of John of Peckham and of Roger Bacon (Plate XXXVIII *a*), and it influenced the views of Leonardo and even perhaps of Vesalius (Fig. 20). The views on the anatomy of the eye expressed by Rhazes, Hunain ben Ishak, and Haly Abbas were, on the whole, more widely accepted than those of Avicenna.

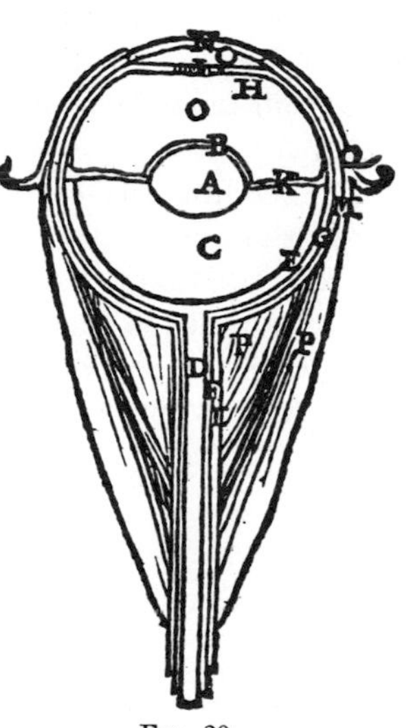

FIG. 20.
THE ANATOMY OF THE EYE
From Vesalius, *De humani corporis fabrica*, Basel, 1543, p. 643. A, Crystalline humour; O, Albugineous humour; C, Vitreous humour; N, Cornea; Q, Conjunctiva; M, Sclerotica; G, Secundina; H, Uvea; K, Arachnoidea; E, Retina.

The treatment of the eye was always felt to be hardly within the range of the ordinary practitioner of surgery, and its structure, as we learn from Guy de Chauliac,[2] was not usually treated in the general course of anatomy. The custom was rather to refer the student to special works such as those of Jesu Aly or of Alcoatim.

Manfredi's description of the anatomy of the eye is that generally accepted at the end of the fifteenth and the beginning of the sixteenth centuries, and is unusually clear for its date. It represents a considerable advance on such writers as Henri de Mondeville (1260–1320)[3] or the pseudo Richardus Anglicus, and is far superior to the descriptions of the eye dating from the fourteenth and fifteenth centuries recently brought to light

[1] The so-called *Anatomia Richardi Anglici*, which has been printed by Robert Ritter von Töply (Vienna, 1902), is really the same as the pseudo-Galenic *Anatomia vivorum*, to which Richard's name was not attached until the fourteenth century. See Christoph Ferckel, *Archiv für die Gesch. der Naturwissenschaften und der Technik*, vol. vi, p. 78, Leipzig, 1912, and K. Sudhoff, *Archiv für Gesch. der Medizin*, vol. viii, p. 71, Leipzig, 1915.

[2] E. Nicaise, *La Grande Chirurgie de Guy de Chauliac*, p. 45, Paris, 1890.

[3] J. Pagel, *Die Anatomie des Heinrich von Mondeville*, Berlin, 1889, p. 37.

by Sudhoff.[1] We reproduce as illustrating Manfredi a diagram taken from the *Margarita philosophica* of Gregorius Reisch (died 1525). This represents the earliest printed figure of any value of the anatomy of the eye (Fig. 19).[2] We give for comparison the figure from a thirteenth-century MS. of Roger Bacon (Plate XXXVIII *a*), representing the rival tradition of Avicenna and Alhazen that influenced Leonardo da Vinci and other contemporaries of Manfredi. These figures may be compared with that of Vesalius (1543, Fig. 20), whose description of the eye is less free from traditional bias than are most parts of his epoch-making work.

In reading any early description of the eye, it is to be remembered that until the nineteenth century the 'emanation theory' prevailed. Light was regarded as of the nature of a stream of particles emitted from the object seen, and the act of vision was considered as a collision of this emanation with an emission of something from the eye itself, called in mediaeval writings the 'visual spirit'.

(*c*) THE HEART

Tractate ii, Chapter 3

(folio 19 verso) Then you will see in the midst of the lung the heart, covered by its membranes. [It is thus situated] that the air attracted by this lung should cool it, and that thus the heat and spirit of the heart be tempered. This member is the most important of the four [principal members], because it is the first to live and the last to die. It is of medium size compared with the other members of man, but compared with the hearts of other animals it is very large, because man, in a quantitative and not an intensive sense, has more natural heat than other animals. It is pyramidal, that is in the form of a flame; because it is of excellent warmth,

[1] For the whole question of early figures of the eye consult K. Sudhoff, 'Augenanatomiebilder im 15. und 16. Jahrhundert' in his *Illustrationen medizinischer Handschriften und Frühdrucke*, Leipzig, 1907; and the same writer's recent article on 'Augendurchschnittsbilder aus Abendland und Morgenland' in *Archiv für Gesch. der Medizin*, vol. viii, p. 1, Leipzig, 1915.

[2] Our figure from the *Margarita philosophiae* has been taken from the 1503 edition, the earliest to which we have had access. A figure in the *Philosophiae naturalis compendium* of K. Peyligk, dated Leipzig, 1489, is so inferior as to be negligible in this connexion.

therefore it is necessary that it should be of a shape resembling a flame. Its figure is also called 'pine-shaped', because it is wide below and narrow above, being thus formed that distinction could better be made between its cavities or ventricles; moreover, had it been made of a shape all uniform as is the lower part, it would be too heavy and ponderous.

This member is situated in the middle of the entire body, measured in every direction; that is, in the middle between the upper and lower parts: in the middle also between front and back and right and left, like a king standing in the midst of his kingdom, and this was done that it might give the strength of life equally to all the members; and although the heart as regards its foundation and base be in the middle, yet its point declines to the left below the left breast, so that it warms the left side as the liver warms the right.

This member is sustained and strengthened by a certain cartilaginous bone. For since it is continually moving, it needs some point of purchase to support it in its movements. Moreover, it has a certain fatty layer on the outside which prevents the heart from drying and keeps it moist: and there are certain veins and arteries dispersed through its substance: and it is formed also of a kind of hard flesh so that it may sustain many and forceful movements; also it is formed of longitudinal, latitudinal, and transverse fibres, so that it may have the power to attract, retain, and expel.

This member has three ventricles or chambers, like the brain. One ventricle is on the right side, the second on the left, and the third in between. The right ventricle towards the liver has two orifices. One is towards the liver and is very large. Into this there enters a vein called *vena chilis*, which arises in the convexity of the liver and brings the blood from the liver to the heart. In that right ventricle the blood is purified, and then sent by the heart to all the other members.

Now since the heart attracts by this orifice of the *vena chilis* more than it expels, therefore nature ordains that in the moment of contraction when the blood is expelled this orifice closes, and when the heart dilates it opens.

Moreover there are three little valves (*hostiolitti*) or doors opening from without inward, and these valves are not very depressed; so that by this same orifice only part of the purified blood is expelled to the other members, because part goes to the lungs and the remainder forms the vital spirit; therefore nature ordains that these valves do not entirely close. From the *vena chilis*, before it enters the cavity of the heart, there arises another vein, which surrounds the root of the heart; and from it are given off branches which disperse themselves through the substance of the heart, and from the blood of that vein the heart nourishes itself.

The right ventricle towards the lung has another orifice into which opens the *arterial vein*, bringing the blood from the heart to nourish the lung: in this orifice also are three valves (*hostioli*) opening from within outward and closing from without inward, in the opposite way to the valves of the other orifice; and this is so that they should entirely close. Hence by this orifice the heart during the period of contraction can expel, and yet during the period of its dilatation cannot attract anything through it as was done in the first orifice.

The left ventricle of the heart has its sides denser and thicker than the sides or walls of the right ventricle; and this for three reasons: Firstly, because in the right ventricle is contained the blood, which is heavy, while in the left ventricle there is spirit, which is very light; therefore in order that the heart should not be heavier and more ponderous on one side than on the other, it was necessary to compensate in this manner, that is, that the left ventricle should be thicker in its walls than the right. In the second place, the spirit being more subtil and more volatile (*resolubile*) than blood, it needs a stronger habitation and better supports. Thirdly, the left ventricle is much warmer than the right, because in it is generated the spirit from the blood, by a great heat which makes that blood more subtil; and heat is better preserved in a substance that is dense and thick.

In the cavity of this ventricle near its root are two orifices: one is the orifice of an artery called *artharia adorti* [= aorta], because it has immediate origin in the heart and because it is the source of all the others: by this artery the heart sends the generated spirit to all the members; and the very subtil blood is mixed with the spirit when the heart contracts. For which reason there are at the entrance of this orifice three valves, which close entirely from the outside inwards; and they open from the inside outwards, and this orifice is very deep.

The other orifice is that of the *venal artery* which conveys the air from the lung to cool the heart and transports warm vapours from the heart to the lung as has been said above; and in this orifice are two valves which do not entirely close: and they are well raised so that they can better apply themselves to the sides [edges] of the heart when it sends out the spirit: these are marvellous works of nature, as is also the central ventricle of the heart, for this ventricle has not one cavity but many; these are small but wide, and more numerous on the right than on the left; and nature contrived thus, so that the blood which goes from the right ventricle to the left to be converted continually into spirit becomes thin in these cavities.

And by this thou canst see that four things have birth in the heart. The first is the artery called *adorti*, the second is the *vena chilis*, the third is the *arterial vein*, and the fourth the *venal artery*.

Also thou wilt see in the heart certain membranous parts like *auricles*, or rather like small ears, able to dilate or contract; these are contrived by nature in order that when overmuch blood or spirit is generated the heart can dilate so as to contain it; and also that the heart may contract when there is no such abundance.

And it is here that Galen asks, Why did not nature make the heart so large that it could contain every increase of blood or spirit without the addition of these membranes? Galen replies that this was first because the heart would have been too large and therefore too heavy; secondly, because as it is not always generating a great quantity of blood and spirit, if the heart had been too large, its cavity would usually have been empty: but these auricles dilate with the accumulation of blood or spirit, and contract with its decrease.

The heart is surrounded by a firm and nervous membrane, like a little house in which it is placed as in a tabernacle to defend it from accidents. This capsule is very dilated, that the heart in its dilations and movement may not be impeded thereby, and therefore nature made this capsule so that it should contain a certain dewy moisture with which the heart is bathed and moistened so that in its continual movement it should not become dry. For when this water be dried up, then the heart itself is desiccated, and emaciates and dries up all the body.

The description of the heart follows Mondino closely. Occasionally a phrase or two is reminiscent of Mondeville. The trite conception of the heart as a king in its necessarily central position was very frequently repeated by writers in the Middle Ages. To Harvey, who had a certain mediaeval element in his mentality, it seems to have appealed, and he used it in his *Prelectiones Anatomiae*,[1] and chose it to introduce his great work on the circulation of the blood.[2] The heart was similarly described as 'flame-shaped', because it was regarded as the source of animal heat. The idea that it is the first to live and the last to die comes from Aristotle.[3] The *bone* in the heart also comes from Aristotle.[4] The idea was quite familiar to mediaeval anatomists, who frequently endeavoured to identify the *bone* with the firm tissue around the orifices of the aorta and pulmonary artery. The reader may be reminded that a true 'os cordis' is in fact to be found in some mammalia.

Mondino, followed by Manfredi, describes the action of the heart and blood-vessels mainly according to the views of Galen, but without any very clear or connected statement. The 'third ventricle' especially has its origin in a misunderstanding.

This mythical structure is an attempt to combine the views of Aristotle and of Galen. Aristotle, who probably never dissected a human body, derived his anatomical conceptions largely from cold-blooded animals, in some of which the heart is provided with three cavities. He considered that the heart had three chambers, the largest being on the right, the smallest on the left, and one of intermediate size between the two. As far as they can be identified, the largest was the right ventricle plus the right auricle, the smallest or left chamber was the left auricle, while the intermediate cavity appears to have been the left ventricle.[5]

[1] W. Harvey, *Prelectiones anatomiae universalis*, reproduced in facsimile from the author's MS. notes, London, 1886, folio 72 recto.

[2] W. Harvey, *Exercitatio anatomica de motu cordis et sanguinis*, Frankfort, 1628. The opening passage of the dedication to Charles I may be translated as follows: 'Most serene king, the heart of animals is the basis of their life, the sun of their microcosm, that from which all strength proceeds. The king is in like manner the basis of his kingdom, the sun of his world, the heart of the commonwealth, whence all power derives, all grace appears.'

[3] *Historia animalium*, vi. 3.

[4] *Historia animalium*, ii. 11 ; *De Partibus animalium*, iii. 4.

[5] *Historia animalium*, i. 14 and iii. 3 ; *De Partibus animalium*, iii. 4. The question of the identi y of these chambers is a difficult one. We have followed T. E. Lones, *Aristotle's Researches in Natural Science*, London, 1912, p. 137, where the conflicting views are summarized.

Galen's description differed altogether from that of Aristotle. He tells us expressly and somewhat contemptuously that 'it is no marvel if Aristotle erred in many anatomical matters, a man who thought forsooth that the heart in the larger animals had three chambers'.[1] Galen always describes the heart as having but two chambers, the right and left ventricles, a wholly subordinate part being assigned to the auricles. These latter were regarded as safety-valves, expanding to hold superfluous blood when the chambers of the heart to which they correspond become overfilled.

No third ventricle is described by Rhazes or Haly Abbas,[2] but Avicenna, in his *Canon*, makes an effort to combine the views of Aristotle and Galen. Speaking of the anatomy of the heart (lib. iii, fen. xi, chap. 1) he describes the ventricular portion as follows: 'In the heart are three cavities, two large, and a third as it were central in position. So that the heart has [*a*] a receptacle [the right ventricle] for the nutriment with which it nourishes itself—this nutriment is thick and firm like the substance of the heart; [*b*] a place where the pneuma is formed [the left ventricle], being engendered of the subtil blood; and [*c*], thirdly, a canal between the two.'[3] A somewhat similar account is given in Constantine's translation of Isaac.[4] The idea soon crept into European medicine, for in a Pisan MS. dating from the first half of the thirteenth century[5] a crude figure of a three-chambered heart is to be found (Fig. 21).

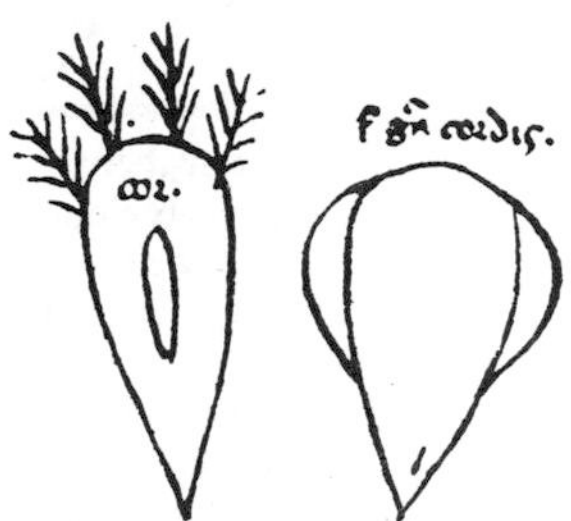

FIG. 21. THE HEART
From the Roncioni MS. (Pisa 99) after Sudhoff.

[1] Galen, Περὶ ἀνατομικῶν ἐγχειρήσεων, Book 7 (157); *καὶ θαυμαστὸν οὐδέν, ἄλλα τε πολλὰ κατὰ τὰς ἀνατομὰς Ἀριστοτέλη διαμαρτεῖν, καὶ ἡγεῖσθαι τρεῖς ἔχειν κοιλίας ἐπὶ τῶν μεγάλων ζώων τὴν καρδίαν*, Kühn, ii. 62.

[2] Haly Abbas expressly denies its existence, chap. 21.

[3] P. Koning, *Trois traités d'anatomie arabes*, Leyden, 1903, 687, renders the passage as follows: 'Dans le cœur il y a trois cavités, deux grandes et une autre qui se trouve pour ainsi dire au milieu, afin que le cœur ait un dépôt pour la nourriture avec laquelle il se nourrit, nourriture épaisse et forte, semblable à la substance du cœur, ensuite un endroit où se forme un pneuma qui y est engendré d'un sang subtil et enfin un canal entre ces deux.'

[4] *Pantechni. Theorice*, lib. iii, cap. 22. Here, however, only two *concauitates* are described and between them a *foramen: quod a quibusdam vocatur tertia concauitas: sed non est ita.*

[5] The MS. Roncioni 99, reproduced by K. Sudhoff in *Archiv für Gesch. der Med.*, vol. vii, Tafel XIV, Leipzig, 1914.

The first translator of the *Canon of Avicenna*, Gerard of Cremona, whose work appeared towards the end of the twelfth century, improved on his original. 'In it [the heart] are three ventricles; two are large, and the third as it were between, which Galen called the fovea or non-ventricular meatus, so that there may be

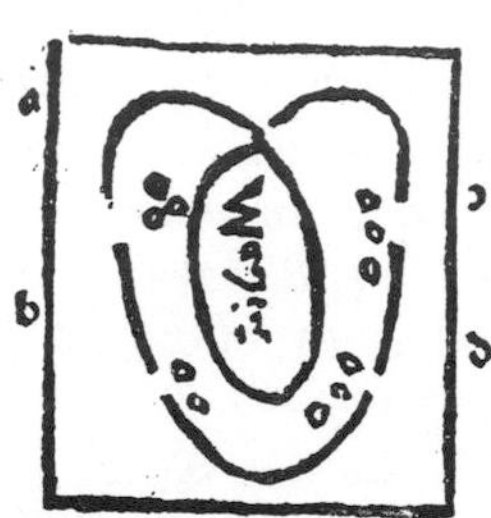

Argumentū rei cum interpre tatione Jo. Adelphi.
A ¶ Artarie adorti ꝑ quā mittit cor ſpm̄ ad oīa corꝑis mēbra qñ ꝯſtringit. Ei⁹ hoſtiola claudunt ꝑfecta clauſiōe ab extra ad intus/⁊ aperiuntur econuerſo.
B ¶ Artarie venalis portātis vaporē a corde ad pulmonē et attrahentis aerem a pulmone ad cor. Cuius hoſtiola imperfecte claudunt/hñs tunicā vnicā/qꝛ natura paꝝ ſollicita eſt de eo quod per ipſum tranſit.
C ¶ Vene chilis/per eius orificiū trahit cor ſanguinē ab epate/⁊ mādat ad oīa mēbra. Claudit hora expulſionis ⁊ aperit hora dilatatōis. Eius hoſtiola aperiunt ab extra ad intus/⁊ imperfecte clauduntur.
D ¶ Vene arterialis que portat ſanguinē ad pulmonem a corde: arterialē ſcꝫ. Duaꝝ tunicaꝝ ꝓpter acceſſuꝫ ei⁹ ad membꝝ ꝯtinui motus. Et qꝛ portat ſanguinē colericū valde ſubtilē: eius hoſtiola aperiunt ab intus ad extra/⁊ claudunt ecōuerſo perfecte. Per hoc orificiū cor tm̄ a ſe expellit hora ꝯſtrictōis/⁊ nihil retinet hora dilatatōis. Hoꝝ duoꝝ arterie venalis ⁊ vene arterialis/ꝯtrariū hꝫ Galie. viij. de vtilitate ꝑticulariū. ix. ca. Et de iuuamētis mēbroꝝ. vij. c. vt & Gentilis li. xxxv. cap. primo.

Fig. 22. From Johannes Adelphus, *Mundini de omnibus humani corporis interioribus menbris Anathomia*, Strassburg, 1513. The diagram shows the two lateral ventricles and the 'central' ventricle. By a printer's error the letters *c* and *d* are transposed. The *arteria adorti* is the aorta, the *arteria venalis* the pulmonary vein, the *vena chilis* the vena cava, and the *vena arterialis* the pulmonary artery. The auricles are ignored, as is frequently the case in works of the period, and the pulmonary veins are represented as opening directly into the ventricles.

a receptaculum for the thick and strong nourishment, like to the substance of the heart, with which it is nourished, and also a storehouse for the pneuma (spiritus) generated in it from the subtil blood. And between the two are channels or meatuses.' [1] Henri

[1] The passage in the *Editio princeps* of Gerard of Cremona's translation runs as follows (folio 96 recto): 'Et in ipso sunt tres ventres, scilicet duo ventres magni et venter quasi medius quem Galienus nominavit foveam aut meatum

de Mondeville (died about 1320), by going direct to Galen, avoided some of the errors of Avicenna, with whom, however, he still describes three ventricles.[1] Mondino does little but copy the Arabian, whom Manfredi also follows.

We may terminate our description of the mythical third ventricle by quoting from Bartholomew the Englishman. His encyclopaedia written about 1260 was translated into English in 1397, and printed by Thomas Berthelet [2] in the 27th year of the reign of Henry VIII (1535), when Bartholomew's work was still extremely popular. Berthelet's rendering runs as follows :

'And the hert hath ij holownesses, one in the left syde, that cometh sharpe : and one in the ryght side, that is within : And

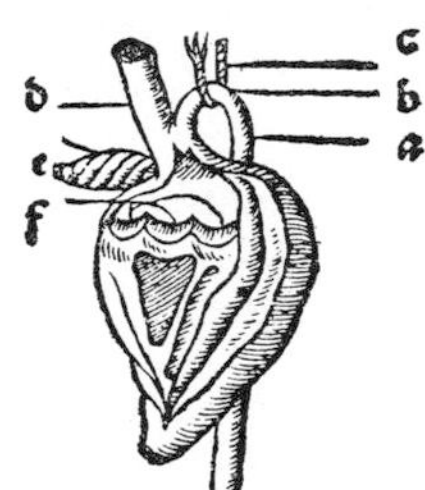
c
d
b
e
a
f
A. b. bedeut den gantzen stamm der grossen hertzaderen / so sich herab durch alle glider erstreckt. c. der obern hauptadern theyl. d. ist die groß lufftröre / Trachea genant. e. bedeut das öhrlin des hertzen. f. zeygt das ort des hertzgeblüts / so da ist ein bekrefftigung anderer lebhaffter glider.

FIG. 23. From Hans von Gersdorff, *Feldt und Stattbüch bewerter Wundartznei*, Frankfurt, 1556. The trachea (*d*) is represented as opening directly into the heart.

these two holownesses ben called the wombes of the hart. And betwene these two wombes is one hole, that some men call a veyne, other an holowe way. And this hole is brode afore the ryghte syde, and streyte afore the left syde. And that is nedefulle to make the bloode subtyll, that commeth from the ryght wombe to the lefte, and so the spirite of lyfe may be bredde the easelyer in the lefte wombe.'

In order to understand why all these authors invoked the existence of the third ventricle, regarded by some of them as a passage between the other two, we must turn to the physiological beliefs of the age. It must be recalled that before the demonstration of the circulatory movement of the blood a

non ventrem, ut sit ei receptaculum nutrimenti quo nutriatur spissum forte simile substantiae ipsius & minera spiritus generati in ipso a sanguine subtili. Et inter ambos sunt viae ut meatus.'

[1] J. L. Pagel, *Die Chiru gie des Heinrich von Mondeville*, Berlin, 1892, p. 45.

[2] Bartholomaeus Anglicus, *De Proprietatibus Rerum*, London, 1535. Our quotation is from p. liiii.

certain amount of communication was believed to exist between right and left ventricles. The complicated nature of the ventricular cavities and the intricacy of the columnae carneae promoted the idea of the presence of minute passages in the interventricular septum. Even so astute an observer as Leonardo da Vinci considered that 'the ventricles are separated by a *porous* wall, through which the blood of the right ventricle penetrates into the left ventricle, and when the right ventricle shuts, the left opens and draws in the blood which the right one gives forth' (Plate XXXVIII *b*).[1]

Although the third ventricle is described in all the twenty-five editions of Mondino, many of which are illustrated, they present no drawing of it except the wretched little diagram of J. A. Muelich (Johannes Adelphus) in 1513, which we here reproduce (Fig. 22). The confusion, however, to which the idea of a third ventricle gave rise influenced anatomy almost as late as the seventeenth century, and is illustrated in the anatomical figures of a late edition of Hans von Gersdorff (1556),[2] where the trachea is actually shown opening into the left ventricle (Fig. 23). It was Vesalius who took the first great step towards the discovery of the circulation of the blood, by firmly maintaining that the interventricular septum was solid and contained neither passages nor intermediate ventricle.[3]

VI. Italian Text

MS. Canonici Ital. 237

Hyeronimi manfredi ad Magnificum & potentem dominum ac militem Iohannem Bentiuolum insequens opus de corporis humani anothomia exordium.

[folio 1 verso] Opportet de sapientia admirari creatoris ut XV° de utilitate particularum scribitur a Galieno. Cum enim membrorum nostri corporis admirabilem Galienus aspiceret Armoniam predictum sermonem explicauit: ut nos ad dei sublimis et gloriosi admiranda opera commoueret: Quamuis nostra cognitio a dei compraehensione deficiat: unde et Seneca XL[a] epistola

[1] Leonardo da Vinci, *Quaderni d'anatomia . . . Pubblicati da O. C. L. Vangensten, A. Fonahn, H. Hopstock*, Christiania, 1911.

[2] Hans von Gersdorff, *Feldt und Stattbüch bewerter Wundartznei*, edition Frankfurt, 1556.

[3] Ancient views on the cardiac system, including those of Mondino, are admirably reviewed by J. C. Dalton in his *Doctrines of the Circulation*, Philadelphia, 1884.

ad Lucillum ait quid deus sit incertum est habitat in nobis: Sed deum mouemur inuocare eius sapientiam mirabiliter contemplantes. Quanta enim fuerit summi opificis in producendo res sapientia quanta eius solicitudo et prudentia opera profecto nature declarant: unde et psalmista mirabilia sunt opera tua deus, et alibi celi enarrant gloriam dei et opera manuum eius annuntiat firmamentum. Quis enim talia et tanta inspitiens creatorem suum abneget et eius potentiam? Inscipiens quidem erit hic iuxta illud psalmiste dixit inscipiens in corde suo non est deus. Sublimis autem dei multiplitia et diuersa fuere opera. Creauit enim duplitia entium genera scilicet corruptibilia et incorruptibilia; et in utrisque suam admirabilem sapientiam, suamque [folio 2 recto] infinitam potentiam ostendit. Totam enim entis latitudinem nihil prorsus de spetiebus, quas ab aeterno in mente sua retinuit obmittens perfulciuit, et eas quas ab aeterno in sua habebat essentia ad aliud esse procreauit, ut in indiuiduis esse haberent: quae in suae maiestatis lumine existebant: et uniuscuiusque spetiei modo perfecit ac uarietates per esse quod in singularibus habent (natura mediante & cum lege) imposuit. Admirantur angelorum caetus obstupent hominum intellectus tantae maiestatis opera mirabilia: ut hoc summo bono: hoc perfectissimo ente nihil melius excogitari possit. O admirabilem maiestatem, O deitatem incompraehensibilem, O inefabilem potentiam: Quis te negliget? Quis te non insequetur? Quis in operibus tuis non delectabitur?

Omnis igitur qui in operum dei gloriosi intuitu delectatur, hic prudens et non inscipiens est: hic dignus homo: hic intellectu non caret. Cum igitur tua illustris Dominatio Iohannes bentiuole magnanimis praesenti anno ex sui qua solet humanitate ad cuiusdam hominis defuncti anothomiam uno semel uidere non fuerit dedignata ob sui intellectus dignitatem qui semper alta intelligere concupiscit, cumque tu opera tam naturae miranda in anothomizato incaepisti uidere corpore tunc haec intelligendi creuit animus, tua digna [folio 2 verso] creuit uoluntas: Et me hyeronimum Manfredum ad hoc opus de anothomia intitulatum materno sermone tuo dignissimo nomini inscribere concitasti: (ut omnino sicut debeor) rem gratam tuae faciam dominationi: In hoc enim tui agnoui dignitatem intellectus, tui ingenii solertiam quod in rebus naturae mirandis tuum peruoluas intellectum. Hoc enim opusculum quantum melius potui ex uariis antiquorum uoluminibus exserpsi ac id abreuiaui: nec eumdem forte tenui ordinem ut illi: et ipsum materno composui sermone ut opus hoc delectabilius tuae sit magnificentie.

Accipe igitur magnifice et potens domine hoc opus de corporis humani anothomia tuo dignisimo nomini intitulatum, ea benignitate et humanitate, qua soles: et animo illari ac gratioso id accepta: qui satis tibi erit delectabile et perplacebit quia dignum est opus: Vale miles magnanimis, et solito ama.

Finis prohemii.

[Here a folio is missing.]

[folio 3 recto] a li nerui lequale hano origine da le extremita di musculi: Unde e da sapere che li musculi sono compositi de nerui, corde, e ligamenti e carne facti da la natura a dare el moto uoluntario, Impero da le soe extrimita escono queste tale corde e uadono a membri che se debano mouere: e quando se retraheno li dicti musculi consequenter se se retraheno le lor corde: & finaliter i membri: et similiter quando se dilatano i musculi se dilatano etiam le corde & consequenter i membri.

Li ligamenti sono etiam simili a nerui facti a ligare le iuncture de le osse e non li dette la natura sentimento como fece a li nerui & a le corde acio che per el molto mouimento e fricatione de le iuncture non doleseno.

Le Artarie sono de substantia neruosa & ligamentale in longo extense e concaue: ne le quale se contene el sangue sutilissimo & depurato et el spirito uitale el quale e mandato dal core a dare uita a tuti i membri: et hano origine da esso core: & impero hebeno doe tuniche acio chel sangue sutile & el spirito uitale non usisseno fuora.

Le vene sono simile a lartarie ma sono quiete e non se moueno, ma hano origine dal figato et in esse se contene el sangue grosso cum li altri humuri che non e cusi depurato ne [folio 3 verso] cusi sutile como e el sangue de le artarie: impero non li fece senon una tunicha: per che quelo sangue non era cosi sutile chel potesse penetrare fuora ne anche non bisognandose mouere non era suspitione de rompersi como ne le artarie che era neccessario a mouerse per refrigerare el core atrahendo laiere frigido & expellendo fuora li fumi caldi da esso.

Li panniculi sono composti e texuti de fili neruosi sutilissimi che non se posseno uedere e sono questi paniculi spissi e sutili e sono de molte manerie: Alcuni forno facti a continere e coprire a Alcuni membri e custodirli ne la sua figura e substantia como sono li paniculi che copreno el cerebro e molti altri di li quali poi diremo: Alcuni altri panniculi sono facti a suspendere uno membro a laltro como li rognoni sono aligati a laschina mediante uno certo paniculo: Alcuni altri paniculi sono facti acio che alcuni membri che non hano sentimento recceuano qualche sentimento per el panniculo: nel quale sono inuolti como sono el pulmone el figato la milza & i rognoni li quali sono priuati de sentimento impero la natura aciascuno di loro li fece uno panniculo doue fusseno inuolti per la casone dicta.

Da poi tuti questi membra hauendo la natura ordito el corpo de lhomo de [folio 4 recto] li predicti bisogno reimpire le uacuita e reimpille de carne: Fece aduncha la natura la carne per reimpire le uacuita che rimangono da lorditura de nerui uene & altri membri dicti.

Praeterea e da sapere che la natura ha dato aciascuno di li predicti membri quatro uirtu. Una e uirtu atratiua per laquale ha ad atrahere el nutrimento suo a se del quale el membro se ha a nutricare: La seconda uirtu e digestiua per laquale el nutrimento atrato se digerisse & conuertese ne la sustantia del membro: La terza uirtu si e retentiua per laquale el nutrimento atrato se retiene debito tempo acio che la uirtu digestiua possa perficere la sua operatione circha quello: La quarta uirtu e expulsiua laquale ha expellere le superfluita che se generano dal nutrimento ne la digestione.

Anche e da sapere che la natura nel corpo de lhuomo ha facto quatro membri principali como quatro signori et aciascuno di loro li ha dato una casa o uero uno palazo a sua custodia doue habite cum certe camare o uero stantie che hano aseruirli al suo bisogno: El primo membro principale e signore e el cerebro al quale li fece la natura el capo cum le sue circumstantie per suo habitaculo e dette a questo membro che lui fusse principio e radice de tuto el sentimento e moto de tuto el [folio 4 verso] corpo: dal quale tuti li altri membri recceueno el sentire: e el mouere, & a questo membri li dette etiam cinque uirtu cognoscitiue exteriore cio e li cinque sentimenti e cinque altre uirtu cognoscitiue interiore che deserueno a lo intellecto.

El secondo membro principale e signore si e el Core alquale la natura ha dato la sua casa cio e el pecto cum le sue adiacentie: et aquesto membro li ha dato la uirtu de la uita dal quale proceda la uita in tuti li altri membri como da uno primo principio.

El terzo membro principale e signore e el Figato alquale dette la natura per suo domicilio el uentre inferiore cum li altri membri circumstanti che sono neccessarii a la sua operatione e dette a questo membro la uirtu nutritiua chel fusse principio e radice del nutricare de tuti li membri.

El quarto membro principale fu li testiculi e la sua casa e la bursa laquale li contene et aquilli deseruено piu altri membri como poi se uedera et a questi testiculi ha dato la natura la uirtu generatiua cio e de generare el sperma o uero seme el quale habia una uirtu generatiua che possa produre una cosa simile a colui dal quale se decide tale sperma : et questo fu facto per conseruare lhuomo in spetie non se possendo conseruare in indiuiduo.

Ultra questi quatro membri principali e suoi domicilii [folio 5 recto] ha facto la natura alcuni altri membri cio e el collo cum la gola che fusse uia e transito dal primo membro principale cio e cerebro ali altri membri principali et etiam a tute laltre parte & per altre utilita quale noi da poi diremo.

Item ha facto la natura le braza e le mane che hauesseno a pigliare el cibo e mandarlo al luoco conueniente et etiam per che lhuomo solo uiue per arte lequale non se possono perficere senza le braza e mano.

Item fece le cosse, gambe e piedi acio se potesse mouere da luocho a luocho secondo li soi bisogni.

Noi aduncha poneremo la Anothomia de tuti li membri e parte dicte : Comenciando per ordine dal cerebro e da la sua casa et consequenter descendendo per insino apiedi.

Capitulum secundum tractatus primi de anothomia capitis et omnium contentorum in eo.

Fece la natura el capo ossuoso per magiore tutela del cerebro : el quale essendo inmobile non li bisogno hauere musculi : Et per che el cerebro ne lhuomo e magiore che ne li altri animali secondo la sua grandeza impero bisogno chel capo de lhuomo fusse etiamdio grande per rispecto de li altri animali : Et etiam bisogno li meati del capo ne lhuomo essere piu distincti essendo piu dedito al cognoscere.

La figura [folio 5 verso] del capo naturale e rotonda compressa da dui canti como sel fusse una cera rotonda compressa cum le mano da la parte drita e da la stancha faria doe eminentie una dinanzi e laltra de drieto e la parte drita e stancha rimaneriano piane : Bisogno fusse rotondo acio fusse piu capace et etiam che fusse piu securo e risguardato da nocumenti exteriori a li quali e molto exposito : Bisogno etiam essere facto cum quelle eminentie acio che li meati del cerebro hauesseno megliore distinctione et acio che li cinque sentimenti exteriori hauesseno origine da la eminentia anteriore.

Diece sono le parte del capo : La prima e li capilli quasi capitis pili facti da la natura a magiore tutela del capo da le cose exteriore et etiam per belleza : La seconda parte del capo e la cute la quale bisogno essere molto grossa acio che li capilli fusseno ben firmi hauendo le radice sue molte grosse e longhe et etiam che fusse megliore scuto e cooperimento de losso et del cerebro non li essendo parte musculose : La terza parte si e la carne laquale solo e ne la fronte e ne le tempie e circha le masselle e non in le altre parte : La quarta parte e uno panniculo exteriore chiamato almochatim elquale appare in continenti como e liuata su la cute e copre tuto losso del craneo de fuora : Et fece la natura questo panniculo [folio 6 recto] acio che lacute che e molle non tochasse incontinenti losso che e duro : Et etiam acio che losso del capo hauesse sentimento per questo panniculo : Et tertio anche acio che el paniculo interiore del capo chiamato Duramater mediante questo panniculo stesse suspeso a losso del craneo cum certi nerui e ligamenti che escono per le comissure del dicto osso et hano origine dal dicto panniculo interiore & uscendo fuora de losso texono o uero componeno quello panniculo exteriore dicto Almochatim : La quinta parte e el craneo cioe osso facto como uno capello nela concauita del quale glie locato el cerebro: & in questo craneo furno quatro ossa cusite insieme e la natura non fece questo osso uno ma de piu pezi acio che achadando nocumento in una parte non comunicasse a laltre parte: Et etiam acio che per quelle comissure o uero cusiture potesseno meglio exhalare fuora le fumusitade dal cerebro: Et tertio acio che bisognando la uirtu de le medicine applicate potesseno meglio penetrare ale parte dentro quisti aduncha quatro pezi de osso furno da la natura cusiti et insieme ionti in modo de denti acio fusse piu fermi e forti et non furno facti in modo che se potesseno uincare como fano le iunture per che non seriano state cusi forte: et etiam [folio 6 verso] per che non bisognaua a losso del capo mouerse: Et queste comissure sono cinque cio e tre uere e doe mendose: Le comissure uere sono quelle che passano tuto losso et le mendose non passano: De le uere comissure una si e ne la parte anteriore chiamata coronale et e facta a modo de uno C e protende da la parte drita a la stancha del capo et ha li branchi uerso la fronte. La secunda comissura uera si protende per la longheza del capo comencianda da la comissura coronale ala parte posteriore como una friza o uero sagitta che uene da larcho, impero e chiamata sagitale ——C. La terza comissura e ne la parte posteriore chiamata laudale facta a modo de uno Λ, per abacho chiamato dal greco lauda: e la comissura sagittale protende da la coronale a la laudale)——C.

Le comissure mendose sono due da ciascaduno lato una cio e dal drito e dal stanco e sono dicte corticale per che non passano.

Et se noi consideremo per queste cinque comissure hauemo quatro ossi cusiti insieme: Uno si e losso de la fronte che comenza dala comissura coronale e termina uerso la parte inferiore a una altra comissura la quale comenza da uno brancho de la comissura coronale e procede a presso le ciglie de li ochii a laltro brancho Ɑ. Laltro osso si e de drieto el [folio 7 recto] quale se termina a la comissura laudale e dui altri ossi da le tempie che se terminano da le comissure mendose le quale comenzano da la comissura laudale a la comissura coronale.

La sexta parte sono doi paniculi uno chiamato Dura mater el quale e in continenti de poi el craneo: e laltro se chiama pia mater el quale incontinente copre el cerebro e questo fece la natura hauendo grande solicitudine di questo membro acio che in continenti non fusse tocho da losso ma processe per piu mezi che uno fusse piu duro che laltro: Et anche fece dui panniculi acio che se la cadesse nocumento in uno de loro non comunicasse al cerebro in continente. Ne la pia matre sono texute certe uene per le quale se nutrisse el cerebro e si lo copre per tuto excepto la parte posteriore per che essendo quella parte sicca non bisogno di questo paniculo como la parte anteriore e meza. Questi dui panniculi in piu luochi penetrano la sustantia del cerebro et se lo diuide in parte drita e parte sinistra et in parte anteriore & parte posteriore: et per queste tale diuisione furno fabrichate nel capo diuerse celule o uero camerette ne le quale produce lanima diuerse operatione per che bisognaua che queste tale parte fuseno de diuerse complexione.

E leuati adoncha questi dui panniculi apparera La [folio 7 verso] Septima parte del capo: et e esso cerebro facta da la natura acio che el spirito uitale mandato dal core calidissimo sia contemperato da la frigidita de esso cerebro: et iue douenti spirito animale elquale e principio de le operatione cognoscitiue & motiue: e questo cerebro e una sustantia medulare biancha molle e uiscosa a cio che da essa hauesseno origine li nerui: ma la parte dinanci fu generata piu humida e molle & mancho frigida che la parte posteriore per che da la parte anteriore hano origine li sentimenti li quali sono molli & humidi ma da la parte posteriore hano origine li nerui motiui li quali bisognano essere piu sicci e forti : Questo cerebro aduncha se diuide in tri uintriculi ouero tre parte : El primo uentriculo o parte anteriore e diuisa in doe, cio e dextra e sinistra : et e magiore che nesuno de li altri uentriculi : et in questo primo uentriculo li pose la natura doe uirtu deseruente al cognoscere una se chiama senso comune doue se terminano li altri sensi exteriori como al suo centro et deferiscono le imagine o uero spetie de le cose sensiue a quello luocho acio che quella uirtu cognosca e distingua tra una cosa sensibile e laltra et etiam cognosca le operatione di li sentimenti particulari lequale doe cose non puo fare nesuno de quilli.

Laltra uirtu de questo primo uentriculo se [folio 8 recto] chiama fantasia et apresso alcuni se chiama imaginatiua laquale ha a retinere et conseruare le spetie de le cose sensibile ne la absentia de le cose sensibile. Quando tu harai ueduto el uentriculo primo tu uederai tre cose inanzi che uegni al uentriculo secondo. La prima si e doe anche cio e una cosa facta de la sustantia del cerebro in modo de doe anche che sono fundamento del uentriculo anteriore cusi da la dextra como da la sinistra parte : et dal lato di ciascuna ancha glie una altra cosa facta a modo de uno uerme subterraneo rosa se sanguinea ligata de certi ligamenti e neruitti el quale uerme quando se alonga chiude quelle anche et consequenter chiude la uia tra el primo uentriculo et el secondo et questo fece la natura acio che lhuomo quando uole posse cessare da le cogitatione e dal considerare et similiter quando uole considerare e pensare questo uerme se contrahe et contrahendosi apre quelle anche et consequenter apre la uia che e tra uno uentriculo e laltro : La terza cosa che tu uederai un poco piu de sotta e una lacuna cio e una certa conchauita rotonda che tra allongo nel mezio de laquale glie uno bucho che ua gioso al palato et a questo bucho li occorre una uia drita laquale descende dal uentriculo di mezo al colatorio e questa lacuna ha circumquaque eminentie grande rotonde facte a sustentare [folio 8 verso] le uene et artharie che ascendeno a dicti uentriculi : e quello bucho e lato di sopra e stretto in fonde e per questa lacuna el primo e secondo uentriculo purgano le sue superfluitade benche la parte anteriore piu se purghi per li colatorii del naso : Unde queste doe uie fece la natura ad expurgare le superfluita del cerebro.

Quando adoncha tu hauerai ueduto queste tre cose incontinente te apparera el secondo uentriculo del mezo el quale e como una uia et uno transito dal primo uentriculo al posteriore : in questo uentriculo sono doe uirtu una chiamata extimatiua laquale ha elicere cose insensate da le cose sensate. Laltra uirtu se chiama cogitatiua laquale cognosce cusi le cose sensate como le cose insensate componendo e diuidendo : e questa uirtu in mediate deserue a lo intellecto : et tute le altre uirtu dicte et anche la uirtu memoratiua se ritrouano ne li animali bruti, ma questa solo se retruoua ne lhuomo.

Dapoi te occorrera el terzo uentriculo situato ne la parte posteriore duro per che e principio de la piu parte di nerui motiui liquali bisogno essere piu forti e duri : Questo uentriculo e de figura pyramidale cio e facto in ponta e la ponta si e ne la parte superiore doue ha aconseruare le spetie per che meglio se riserua la cosa [folio 9 recto] in stretto luocho che in amplo : e la parte di sotto e lata per che ha a receuere le spetie e meglio se receue in luocho amplo che stretto : Due adoncha utilita se ha da questo uentriculo una che e principio de la nucha e di li nerui mottiui. Laltra si e che e camera de la uirtu memoratiua.

E per questo appare che quando e offesa la parte posteriore del capo in continenti se offende la memoria e quando se offende la parte de mezo se offende la uirtu extimatiua & cogitatiua & offesa la parte dinanzi se offende el senso comune e la fantasia et in questo modo ueneno in cognitione li medici de li luochi de le dicte uirtu.

Facto questo tu leuarai el cerebro ligieramente chel non si rompa alcuno neruo e comezarai da la parte di nanzi & incontinenti te apparerano doe carne picole in modo de doi capi de mamille simile ala sustantia de cerebro per che nascono da quello et sono coperte dal paniculo subtile cio e da la pia matre e queste sono lorgano de lo oderato doue e la uirtu olphatiua.

Dal cerebro nascono septe para de nerui: procedi adoncha piu oltra ne la parte dinanci e uederai el primo paro de dicti nerui liquali sono grandi chiamasi nerui obtitii de li quali la origine e dal cerebro ne li uentriculi anteriori e procedeno uerso li ochii ma nanci che escano la pia matre se coniongeno [folio 9 verso] et in luocho de la sua unione sono perforati: Uolse Galieno che dicti nerui solo se coniongeseno o uero se unisseno e non se incrutiasseno ma quello neruo che uiene dala parte drita da poi la unione ritorna pure dala parte drita et similiter quello che uiene da la parte sinestra da poi la unione ritorna uerso lochio sinistro: Ma Rasis uolse el contrario benche la opinione de Galieno sia piu comune: questi nerui deserueno al uedere e fu necessario che se uniseno acio che le spetie de la cosa che se uede receuuta in doi ochii e portata per doi nerui ritorni a unita acio che una cosa non appara doe.

Dapoi li dicti nerui leua el cerebro secondo la sua medieta e uederai uno altro pare de nerui subtili et duri li quali uengono similiter a li ochii a darli el mouimento uoluntario componendo certi musculi.

Da poi tu uederai el terzo pare de nerui di quali una parte se ne ua ala faza a darli el sentire e el mouere uoluntario et anche una parte de quisti ua a dare el gusto a la lengua: Un altra parte de dicti nerui se mescola insieme cum el quarto [1] pare de nerui et descendeno insieme gioso a dare sentimento al Diafragma et al stomaco et alaltre uiscera: Una certa parte de li nerui del quarto [1] pare se ne ua a dare el sentimento al palato.

Da poi e el quinto pare de nerui se ne ua a li ossi petrosi liquali sono apresso [folio 10 recto] le orechie e de questi nerui ne li buchi de lorechie se componeno certi panniculi liquali sono organo de lo audire.

Da poi e el sexto pare de nerui che se diuide in tre parte una parte ua ali musculi de la gola: Laltra parte ua ali musculi de le spalle la terza parte che e magiore de le altre descende gio a lo epyglotto e nel diafragma se sparge nel pecto nel core e nel polmone a compagnandosi insieme cum li nerui del terzo pare dicti: Et anche da li nerui di questo sexto pare quali uadeno gio a lo epyglotto se generano li nerui de la uoce chiamati reuersiui dili quali piu disotto se uedera.

Dapoi e el septimo pare de nerui ha origine da la parte posteriore del cerebro e uadeno a dare el mouimento a la lingua uoluntario: De questi septe para de nerui li primi doi pari hano origine da la anteriore parte del cerebro: el terzo pare ha origine dal mezo de lanteriore e posteriore parte: li altri quatro para de nerui hano origine da la parte posteriore.

E dapoi quisti procedendo piu oltre leua tuto el cerebro & apparera la octaua parte del capo cioe doi panniculi posti sotto el cerebro li quali leuati apparerati la nona parte che e una certa rethe laquale se chiama rethe mirabile per che e contexta de una tessetura fortissima et miraculosa multiplicata de certe artharie sutilissime: lequale sono [folio 10 verso] rami de alcune artharie che ascendeno dal core chiamate artharie apopletice: & in queste artharie di questa rethe se contiene el spirito uitale mandato dal core acio che douenti animale: et acio che questo spirito meglio se alterasse e disponesse fece la natura quelle artharie sutilissime diuise per minime

[1] Manfredi here follows Mondino, who confuses Galen's fourth pair with Galen's sixth pair of nerves.

parte acio che questo spirito fusse diuiso anche in minime parte : et pose la natura questa rethe mirabile sotto el cerebro perche bisogno hauere de molta custodia onde lo situo in luoco tutissimo et etiam acio che le humidita uaporese del cerebro che cadeno sopra questa rethe opilandola inducesse el somno naturale.

Da poi tute queste cose uederai losso basilare che e la decima et ultima parte del capo e chiamasi basilare per che e base e fondamento de tuto el capo e fu facto duro acio che le superfluita che descendono a lui non lo putrefesse : e questo osso e diuiso in molti altri ossi como se puo uedere cociandelo. Onde se diuide ne le osse petrose e ne li ossi del naso e ne le ossi de li ochii & in doi altri ossi laterali li quali non se possono uedere se non per uia de decocione.

Capitulum tertium de anothomia oculorum et membrorum deseruientium uisui.

Le ossa del naso forno cauernose e porrose acio che le superfluita del cerebro possano meglio de[folio 11 recto]scendere e lo odore ascendere.

Dapoi scinde tuti doi li ossi de gliochii e uederai la colligantia loro cum li nerui obtitii e cum li nerui motiui : e el loco de li ochii non fu molto in profondo per che douea receuere le spetie de le cose uisibile : ne anche fu tropo eminente acio non receuesse lesione da le cose exteriore : Et essendo li ochii molto molli e passibili ne lhuomo fece la natura li supercilii acio fusseno custoditi da le cose che descendeno de su in gioso e fece le palpebre che fuseno custoditi da le cose che uengono da fuora dentro : e fece le eminentie de le maxille et anche el naso in mezo che da ogne lato e per ogne uerso fusseno custoditi : tanto fu la solicitudine che hebe la natura di questo membro.

Septe sono le tuniche e tri humori di liquali e composto lochio tre tuniche anteriore se coiongeno cum tre altre posteriore como se fusseno sei scutelle che cum la bocha ogne doe se coniongesseno e che doe fussene magiore che contineseno le altre doe e poi li e la septima tunica che e magiore de tute e contene tuto lochio : e pero se chiama coniontiua per che congionge e circunda tuto lochio excepto el luocho de la pupilla e quello pocho de la cornea che appare e questa e la prima tunicha comenzando da le parte de fuora et e biancha.

La seconda tunicha ne la parte dinanci se si chiama cornea [folio 11 verso] per che se asomiglia al corno quanto ala substantia e quanto al colore : e fu questa tunicha transparente acio che le spetie de le cose uisibile potesseno penetrare per essa e fu etiam sollida e grossa composita de quatro pellicule e questo fu per che e propinqua a le cose exteriore non receuesse nocumento da esse e cum questa tunicha ne la parte posteriore se conionge un altra tunica dicta scliroticha cio e dura e queste doe tuniche hano origine dal paniculo di sotto el craneo cio e da la dura matre cusi como la prima tunicha ha origine dal panniculo disopra el craneo dicto almochatim.

La quarta tunicha ne la parte dinanzi se chiama uuea a similitudine de uno grano de uua negra et in el mezo di quella glie uno buco che se chiama la pupilla : fece la natura questa tunica obscura acio chel spirito uisiuo se confortasse e che non si resoluesse dal lume exteriore : e fece quello buco in questa tunica acio che le spetie potesseno penetrare senza impedimento e fecelo stretto acio chel spirito uisiue fusse unito : Onde quando dicta pupilla o uero buco se alargha oltra el debito o per natura o per accidente se impedisse el uedere : e ne la parte de drieto se li coniongne la quinta tunica dicta secundina per che e facta a similitudine de la secondina cio e paniculo nel quale se inuoltano li putti nel uentre [folio 12 recto] de la matre et hano origine de la pia matre.

La sexta tunicha se chiama ne la parte dinanzi aranea per che e facta in modo de una tela de ragno a la quale ne la parte posteriore se li coniongne la septima tunicha chiamata arethina per che e facta in modo de una rethe : et in mezo de la tunicha uuea et de la aranea da la parte dinanzi glie uno humore dicto albugineo facto a similitudi^ne de uno albumo de ouo facto per humettare lochio et acio che la tunicha cornea stia suleuata impero in li homini che moreno quando questo humore se desicca cade la cornea e se si spiana et a lhora dice el uulgo che appare una tela dinanzi da gliochii et e signo infalibile de la morte : Et anche questo humore tiene la pupilla apperta impero quando se sicca se stringe la pupilla : Nel meze de le due ultime tuniche cio e aranea et arethina lequale hano origine da nerui obtitii li sono dui humuri cio e uno humore uitreo a similitudine de uno uetro liquefacto : Laltro humore e dicto cristallino a similitudine del cristallo : dicto etiam grandineo a similitudine de una grandine et e alquanto duro e rotondo cum una certa planitie ne la parte anteriore doue se receueno le spetie de le cose uisibile : e ne la parte posteriore e de figura pyramidale cio e che e facta in ponta : et iue se conpisse [folio 12 verso] lacto del uedere : e ne la parte posteriore e circumdato da lhumore uitreo dal quale se nutrisse : e questo humore cristalino declina piu uerso la parte anteriore e lhumore uitreo uerso la parte posteriore. Et a li ochii uengono li nerui obtitii per li quali se de portano le spetie uisibile da gliochii al senso comune et ali altri sensi interiori.

Capitulum quartum de anothomia aurium et membrorum deseruientium auditui.

Expedito questo tu uederai le orechie poste da doi lati del capo in mezo de lanteriore parte e posteriore acio che la uoce o uero sono se potesse audire da ogne canto cio e da la parte drita e stancha dinanzi e de drieto de sopra e disotto : non furno situate da la parte dinanzi per che iue li sono gliochii el gusto e lolphato : non furno poste de drieto per che seriano state tropo distante dal senso comune : forno poste sotto la tonsura di capilli per che se piu sopra fusene stato poste seriano state uelate da cepilli e da quelle co^se che se portano in capo.

Furno le orechie rotonde acio fusseno piu capace de laere sonoro : non furno ossuose acio che per qualche percussione o caso non se rompeseno : forno adoncha carthilaginose acio che fusseno piu sonore : non furno etiamdio carnose ne paniculare per che non hauerebeno seruata la figura e composi-tione [folio 13 recto] debita.

Hebbe uno buco ritorto e non dritto como quello de le limache acio che se facesse megliore reuerberatione de laiere sonoro in esse : et anche ne aiere disproporcionato ne sono si tropo forte senza misura peruenisse a lorgano de laudito : e questo buco e uelato de uno paniculo duro texuto de fili neruosi che hano origine dal quinto pare de nerui del cerebro et de fili ligamentali che hano origine da losso petroso al quale se termina el dicto buco : ne la concauita del quale li e el neruo auditiuo cio e nel quale se compisse laudito et e texuto in modo de uno panniculo : et e continuo a la dura matre nel quale se contiene uno certo spirito auditiuo dal principio de la generatione iue complantato : et apresso di quello li e una certa uisichetta ne laquale e posto un certo aiere connaturale el quale deserue a laudito.

Capitulum quintum de anothomia nasi et aliorum membrorum deseruientium olphatui.

Le osse de le maxille comenzano da la comissura che e tra el craneo e losso basilare in luocho che e ne la fine del sopracilio e de la fronte et procede uerso la parte posteriore a presso losso petroso doue se termina lorechia e terminano ne la parte di sotto a li denti : de liquali poi uederemo la nothomia.

El naso e composito de doi ossi figurati [folio 13 verso] secondo la forma de doi trianguli che hano le ponte in su uerso el collatorio : et sono lati ne la parte de sotto. Onde el naso e piu largo di sotto che di sopra e queste ossa furno sutile acio che fusseno ligiere e non graue : ne anche furno tropo dure per che non li bisognaua in quello luocho grande forteza.

Fu etiam el naso composto de tre carthilagine cio e doe ne lextremita de doi ossi acio che le parte molle cio e la cute e li musculi inmediate non fusseno tochi da le osse dure e che le nare stesseno aperte e se potesseno dilatare e constringere secondo la neccessita de laiere atrato & expulso e questo non se harebe potuto fare se solo fusse stato ossuoso.

La terza carthilagine diuide el naso per mezo per el longo et e piu dura ne la parte superiore che ne la inferiore : Onde furno facti doi meati e buchi acio che uscendo le superfluita per uno laltro deseruisse a laiere atrahato et expulso : Onde essendo uno meato solo ne lexito de le superfluita harebbe impedito el transito de laiere : questi doe meati peruengono al collatorio cio [e] uno buco che e ne losso basilare et similiter iue sono perforati li dui panniculi che copriuano el cerebro per insino a le caronchole mamillare : lequale sono ne lextremita de le due parte del uentriculo anteriore del cerebro como e stato dicto.

El naso etiam fu composto de doi musculi [folio 14 recto] picoli acio che essendo grandi non impedisseno glialtri musculi de la faza cio e quilli che sono ne le maxille che moueno i labri : et similiter glialtri musculi.

El naso fu composto per molte rasone : prima per euentare el cerebro : Secundo ad atrahere laiere : nel quale sono le spetie de le cose odorabile : e cusi deserue a lolphato : Tertio acio che le littere prolate meglio se distinguano come el buco grande de la fistola o uero zalamella deserue ala distinctione di soni : Quarto acio che per questo meato se expurgaseno le superfluita del cerebro.

Capitulum sextum de anothomia oris palati dentium uuulae faucum et linguae.

Ne la bocha sono doi labri uno disotto e laltro si e disopra composti de nerui carne cute e panniculo de una mirabile comixtione in modo che la cute e la carne e li nerui et el panniculo non se posseno seperare insieme : e questo fu facto acio che hauendo bisogno quisti labri di mouerse per ognie uerso bisogno che fusseno cusi composti per che non se posse fare in quello luocho musculi per la graueza grande che seria stata : el paniculo che copre i labri nasce da la tunicha intrinsecha del meri cio e de la uia che ua a lo stomaco : et consequenter se continua per questo modo cum la tunicha interiore del stomaco cusi como etiam dio tute le altre parte de la bocha se [folio 14 verso] continuano acio chel sentimento del stomaco se conformi al sentimento de la bocha et per questo appare che quando el de uenire uomito a qualche uno trema lo labro inferiore.

Da poi li labri sono trentadoi denti sedeci superiori et sedici inferiori: de li inferiori doi sono dicti duali: doi altri incisiui: doi altri canini: quatro maxillari: et sei molari che sono in tuto sedici: & altratanti superiori. Forno facti li denti: prima per masticare el cibo acio che meglio si digesta: secundo per la uoce et distinctione de la eloquela cusi como furno facti li labri. Onde quilli che manchano de denti o de labri non proferiscono bene.

Da poi tu uederai el palato el quale ha una certa concauita ne la sumittade acio che la uoce habbia el suo tono: et etiam chel cibo quando se masticha meglio si possa reuolgere per bocha:

Ne la fine del palato tu uederai una carne pendente in modo de uno grano duua: impero si chiama uuula: et e de substantia rara e spongiosa per che fu facta principalmente a receuere la humidita che descende dal capo acio non descenda a membri inferiori impero spesso se tumefa dicta uuula: fu facta etiamdio acio che temperasse et modulasse la uoce refrangendo laiere che uiene dal polmone: et etiam che lo aiere atrahato al polmone lo ritenga al quanto repercutiendolo acio che cusi frigido non peruenga al polmone [folio 15 recto] ma alquanto alterato: e per questa rasone appare che quilli che hano tagliata la uuula sono molto catarosi impero comandano i medici che non se taglie quando e apostomata: ma che se cauterige cum fuoco.

Dapoi la uuula sono le fauce: e sono li luochi ampli glandosi disposti a receuere le superfluitade del cerebro impero facilmente se apostemano.

Dapoi e la lingua laquale e fabricata et ligata a losso posteriore del capo dicto lauda facto a modo de uno Λ per abacho e fu composta di carne, panniculo, uene, artharie, et noue musculi: e forno facti tanti musculi in essa per che se douea molto mouere per ogne uerso secondo el bisogno de la loquela: Et fu in essa piu uene artharie e nerui che in qualoncha altro membro rispetto de la sua grandeza: et fu facta la lingua acio che fusse organo del gusto per nerui che uengono dal terzo pare di nerui gia dicto circa la sua radice: et sono de due facta nerui che uengono a la lingua cio e uno paro di nerui motiui a darli el moto: et uno altro paro di nerui sensitiui a darli el gusto: et tu uederai che li nerui motiui piu se profondano ne la lingua per darli el mouere: et li nerui sensitiui sono piu expansi ne la superficie: et nel suo panniculo a darli el gusto e el tacto: Fu etiam facto la lingua che deseruisse al proferire de le parole: et etiam a reuolgere el cibo per bocha quando se masticha.

Circa [folio 15 verso] la radice de la lingua da ciascuno lato sono carne glandose facte acio che generasseno la humidita saliuale che hauesse a humetare la lengua acio che non se siccase per tanti mouimenti che ha in se: et in queste carne glandose sono dui buchi che poria intrare uno stile e per quilli buchi se distilla la humidita saliuale. Sotto la lingua sono doe uene grande uiride da le quale poi procedeno piu altre uene.

Et nota che la megliore lingua quanto al deseruire al parlare e la lingua che e mediocre ne la longitudine e sua latitudine cio e che non sia tropo longa ne tropo larga: e che apresso de la ponta et extremita sua exteriore sia sutile per che la lingua che e longa larga e grossa o uero tropo picola non e conueniente al parlare.

Nota etiam che la lingua ha colligantia cum el cerebro mediante li nerui che uengono ad essa et cum el figato mediante le uene: et cum el core mediante le artharie et cum el stomaco mediante el meri: et cum el polmono mediante la cana de esso polmone: impero in ciascuna infirmita i signi de la lengua sono molti efficaci a iudicare di tale infirmita: e quiue se finisse la anothomia del primo membro principale cio e [el] cerebro e del suo habitaculo.

Tractatus secundus de anothomia membrorum spiritualium et secundi membri principalis: capitulum primum de anothomia gule et colli.

[folio 16 recto] Finito el primo membro principale e ueduta la anothomia del suo habitaculo e de le altre camare deseruente a quello resta a uedere la anothomia di gli altri membri principali : E prima uederemo la nothomia del collo e de la gola che e condutto e meato dal primo membro principale a glialtri. Diciamo adoncha che la gola si e uno certo spatio nel quale sono doe uie una che mena el cibo al stomaco : e questa se chiama meri : Laltra uia mena laiere al pulmone a rifrigerare el core : & etiam mena fuora laiere e uapori caldi da esso core : Onde se tu scarni el collo e la gola tu uederai certi musculi longitudinali sopra liquali nota le uene da tuti doi li canti : et eleuati quilli musculi tu uederai doe carne ala forma de doe mandole ne la radice de la lingua : una da ciascuno lato : de le quale habiamo dicto parte disopra : et anche noi dicemo che sono como doe orechiette picole, e sono neruose acio [che] siano forte et aiuteno a fare penetrare laiere a la canna del polmone : et etiam queste tale amigdale hano a congregare una certa humidita per humettare la lingua como e stato dicto et per humetare etiam la canna del polmone acio [che] non se dessiccasse : et anche acio che reimpisseno i luochi uacui de la gola : et anche acio che fusseno scuto e tutella de le uene & artharie che ascendeno al capo : Onde per questo collo e gola passano le uene dal figato ascendendo al cerebro a darli el nutrimento [folio 16 verso] per esso anche passano le artharie che ascendeno dal core al cerebro a darli la uita : et acio chel spirito uitale per esse uada al rethe mirabile delquale e stato dicto douenti animale e chiamase queste artharie apopletice per che quando se opillano generano la poplesia cio e el male de la gozola prohibendo el transito del spirito. Per questo etiam collo passano i nerui che descendeno dal capo ai membri inferiore a darli el sentire et el mouere : e tute queste parte potrai uedere escarnando e tagliando el collo e la gola per lo longo.

Capitulum secundum de anothomia pulmonis et tracheae artharie; id est cane pulmonis.

Vediamo hora la anothomia del core el quale e laltro membro principale : e del suo domicilio nel quale e anche collocato el polmone como quello che serue ad esso core.

Volse Aristotile chel core fusse el primo principio e cagione de tute le operatione del corpo : e che fusse principio del sentire e del mouere e del nutrire e del uiuere e che li era solo uno membro principale : e che el cerebro e el figato erano suoi ministri : ma questo non piaque a Galieno ne a li altri medici liquali per hora noi seguitemo.

El domicilio adoncha del core si e el luocho del pecto circundato da le coste dala parte dinanzi e da la parte de drieto [folio 17 recto] da uno certo panniculo chiamato mediastino e da la parte di sopra el comenza dal principio de la canna del polmone et terminase ale parte di sotto a uno paniculo chiamato diafragma. Comentiamo aduncha ala parte disopra cio e dal principio de la canna del polmone e diciamo che el meri cio e la uia del cibo et la trachea artharia cioe la canna del polmone che e uia de lo hanelito comentiano in uno medesimo luocho : Et impero fece la natura uno coopertorio al principio de la canna del polmone de una carne carthilaginosa e panniculosa anexa al palato sotto luuula e questa carne copre lorificio de essa canna del polmone el quale orificio si chiama epiglotto : acio che ne lhora del transglutire niente del cibo e del poto descendesce a la uia del polmone per che indurebbe suffocatione : Impero aduiene che se uno ridendo transglutisse qualche cosa ua al polmone et appare che lhuomo se soffochi per che ne lhora del ridere se apre lo epiglotto : Lieua adoncha el meri da la trachea artharia acio che tu uidi la compositione sua : ma sapii che el meri e la trachea facilmente se seperano per insino al epyglotto cio e al orificio de essa trachea ma circa lo epyglotto cum dificulta se seperano per che la tunica del meri si e dispersa ne lo epyglotto : e questo fece la natura sagacemente acio che ne lhora del transglutire del cibo quando [folio 17 verso] el meri se lieua uerso la bocha ad atrahere el cibo anche lo epyglotto se lieua acio che remanendo gioso per la sua dureza non impedisse el transito del cibo.

La trachea artharia o uero canna del polmone e composita de anuli carthilaginosi e panniculosi e de ligamenti che continuano quilli anuli insieme facta da la natura a transportare laiere al polmone per auentare el core : & a transportare fuora i uapori caldi da esso et etiam fu facta a formare la uoce ne la sua extremita cio e ne lo epyglotto : Questa canna bisogno che fusse carthilaginosa et alquanto dura et non pelliculare e molle perche bisognaua stare aperta essendo uia de laiere : e non fu etiam ossuosa per che douea essere flexibile per la formatione de la uoce : et anche se fusse ossuosa impediria el transito del cibo per el meri quando fusse tropo : Et per questa ragione la carthilagine di questa canna non fu una ma furno piu continuate per certe pellicole insieme : e queste sono facte como certi semicirculi in modo de uno C per che se fusse una carthilagine seria dura e comprimirebbe el meri et impediria el transito del cibo. Onde questa cana ne la parte anteriore e carthilaginosa per che uerso quella parte non tocha el meri et anche acio che sia piu difesa da le cose exteriore ma uerso la parte posteriore e pelliculare per insino a lo epyglotto : La quale poi tuta e carthilaginosa [folio 18 recto] per la ragione dicta : e questa canna del pulmone non descende se non insino a la furcula sotto laquale e incontinenti situato el pulmone : et el sito de essa e ne la parte dinanzi : et dritamente procede e non storta acio che laiere habbia piu libero ingresso : et lo epyglotto che e principio di questa canna si e tuto carthilaginoso acio che sia piu sonoro : et e apresso la bocha acio che sia instrumento dela uoce : laquale poi ne la bocha douenta locutione per che la uoce finalmente ne lhuomo se ordina al parlare. Questo epyglotto e composto de tre carthilagine e uinti musculi : Una carthilagine si e ne la parte anteriore e chiamasi clipeale a modo de uno capello : Laltra si e ne la parte posteriore uerso el meri e questa non ha nome : La terza si e in mezo di queste doe et in essa e una lenguetta in modo de una lingua de zalamella e chiamasi questa carthilagine fistula de lo epyglotto per che como la fistula se ordina nel sono cusi questa carthilagine si e ordinata al canto e la melodia : Questo epyglotto etiam e composto de uinti musculi a dare el moto uoluntario secondo el bisogno de formare la uoce : e dodeci di quisti sono da la parte di dentro e octo dala parte de fuori et a quisti musculi uengono dui nerui che hano origine dal sexto pare de nerui del cerebro dicti : di quali una parte descende per insino al core e poi comenza a reascendere per insino a lo epyglotto impero [folio 18 verso] sono dicti nerui reuersiui li quali sono nerui de la uoce e quando sono alo epyglotto se spargeno inquisti uinti musculi a darli el sentire e el mouere. Questi nerui forno reuersiui e non directi per molte cagione : prima acio [che] fusseno piu forti per che quanto el neruo e piu remoto dal cerebro tanto e piu sicco e forte : La seconda acio [che] fuseno facti a modo de uno freno da cauallo acio chel cerebro meglio mouesse lo epyglotto secondo lo imperio de la sua uolunta mediante questi nerui como lhuomo moue el cauallo al suo libito mediante el freno: La terza cagione e per che la uoce non solo depende dal cerebro como dal principio del moto uoluntario ma etiam depende dal core como da quello nel quale se formano i concepti del cerebro et consequenter i concepti de la uoce: bisogno adoncha che dicti nerui comunicasseno al core: La quarta cagione e per che quisti nerui douendo uegnire ali musculi predicti bisogno che uigniseno al principio de dicti musculi e non a la fine: et el principio di quisti musculi de lo epyglotto e ne la parte inferiore.

Da poi la trachea artharia tu uederai el pulmone ala compositione del quale concorreno piu parte ramificate como fili sutili ad ordire la sua substantia: La prima parte che entra ne la substantia del polmone si e la trachea artharia laquale [folio 19 recto] como gionge a la furcula del pecto se diuide in doe parte: una ua al dritto e laltra al sinistro del pulmone e ciascuna di quelle se diuide in doe altre parte cio e superiore et inferiore: e ciascuno de quilli rami: se diuide etiam in rami minori e cusi diuidendosi peruengono a rami minimi como fili e circundano tuta la substantia del pulmone. Una altra parte che ordisse la substantia del pulmone si e una certa uena che ha origine dal uentriculo dritto del core laquale porta el sangue sutile dal core a nutrire el pulmone: e chiamasi uena arthariale Vena per che non pichia arthariale per che e composta de doe tuniche como sono le artharie: e questa uena se ramifica ne la substantia del pulmone como la trachea artaria.

La terza parte che compone el pulmone si e una certa artharia che nasce dal sinistro uentriculo del core dicta artharia uenale: Artharia per che pichia Venale per che e composta de una tunica como le altre uene et per questa artharia se transporta dal pulmone al core laiere che uiene da la trachea artharia a refrigerare esso core: Et perquesta artharia etiam se manda dal core al pulmone laiere e uapori caldi e dal polmone poi escono fuori per essa trachea e questa artharia similiter se ramifica como le altre doe parte predicte: Onde li rami de la trachea [folio 19 verso] e de lartharia uenale e uena arthariale compongono tuto el pulmone in modo de una rethe: et i buchi de questa rethe reimpisse una certe carne molle spongiosa laquale proprio e substantia de esso pulmone: Et tute queste quatro parte predicte sono inuolute da uno certo panniculo che ha origine da uno panniculo che e sotto le coste chiamato pleura del quale poi se dira per questo panniculo ha el pulmone el sentimento per che el pulmone non sente secondo la sua substantia.

Et nota che li rami de la trachea artharia sono magiori che li rami de la uena arthariale, et de la artharia uenale per che nascono da magiore troncho et etiam nota che el pulmone e magiore ne la parte dritta che ne la stancha per che dal lato stancho glie el core che occupa quello luocho: Similiter e magiore ne la parte posteriore che ne la parte anteriore: Questo membro sie como flabello del core a refrigerarlo et etiam a mondificarlo da li uapuri che continue se generano in esso: impero e seruo e ministro del core.

Capitulum tertium de anothomia cordis quod est secundum membrum principale.

Dapoi te apparera el core nel mezo del pulmone cooperto da le sue penole acio che laiere atrahatto da esso pulmone lo refrigere, e del suo caldo e spirito se tempri: Questo membro tra lialtri quatro e principalissimo per che e el primo che ne la generatione [folio 20 recto] uiue et e lultimo che more. Questo membro e de mediocre quantita per rispecto di li altri membri de lhuomo: ma per rispecti di li cori de lialtri animali e molto grande perche lhuomo ha piu del caldo naturale che glialtri animali quantitatiue et non intensiue : Et e di figura pyramidale cio e de la forma del fuocho per che esso e de excellente calidita impero bisogno che fusse de una figura che asomigliasse a la figura del fuocho : e questa tale figura se chiama pigneale cio e simile ala figura de una pigna laquale e lata disotto e strecta di sopra et di tale figura fu facto acio che meglio se facessono distinctione de le sue cellule o uentriculi : et etiam se fusse stato de una figura tuta uniforme como e la parte disotto seria stato tropo graue e ponderoso. Questo membro e situato nel mezo de tuto el corpo tolti uia glie extremi cio e nel mezo de le parte superiore et inferiore : nel mezo de le parte dinanzi e de drieto, e nel mezo de la parte dritta e sinistra como uno re che sta nel mezo del suo regname e questo fu facto acio [che] potesse equalmente dare la uirtu de la uita a tuti membri : E benche el core sia quanto al suo fondamento et ala sua base nel mezo tamen secondo la sua ponta declina al lato stancho sotto la mamilla sinistra acio che riscaldasse la parte sinistra como el figato riscalda la parte dritta : e questo [folio 20 verso] membro se sustenta e ferma de uno certo osso cartilaginoso per che e in continuo mouimento : bisogno aduncha che hauesse uno apogiamento alquale se fermasse nel suo mouimento : Et e etiam composto de una certa pinguedine ne la parte exteriore acio che prohibisca chel core non se desichi tenendolo humectato : Et e composto di certe uene et artharie disperse per la sua substantia : et e composto etiamdio de una certa carne dura per che haueua a sustignire de molti e forti mouimenti : Et etiam fu composto de uili longitudinali latitudinali e transuersali per che bisognaua che hauesse uirtu de atrahere retignire et expellere : E questo membro ha tri uentriculi o uero tre cellule como ha el cerebro. Uno uentriculo e dal lato dritto e laltro dal lato stancho e el terzo e in mezo : el uentriculo dritto uerso el figato : el quale ha doi orificii : uno e uerso el figato et e molto grande nel quale entra una uena chiamata uena chilis laquale nasce dal gibbo del figato e porta el sangue dal figato al core : Et in questo uentriculo dextro del core se puriffica quello sangue e cusi purificato poi lo manda el core a tuti li altri membri : e per che per questo orificio ha el core piu ad atrahere che ad expellere impero ordino la natura che ne lhora de la constrictione quando de expellere che questo orificio se chiudesse : e che [folio 21 recto] quando el core se dilatta se aprisse : Et iui sono tre hostiolitti o uero usitti liquali se apreno da fuora adentro : e questi hostioli non sono molto depressi e per che per questo medesimo orificio se expelle el sangue depurato aglialtri membri ma non tuto per che una parte ua al polmone e de laltra parte se ne fa spirito uitale : impero ordino la natura che quisti hostioli non se chiudesseno in tuto : E da questa uena chilis inanzi che entri la concauita del core nasce un altra uena laquale circunda la radice del core e da quella nascono alcuni rami che se disparghono per la substantia del core : E del sangue de questa uena se nutrisse esso core.

Uno altro orificio ha questo uentriculo destro uerso el pulmone nel quale entra la uena arthariale che porta el sangue dal core a nutrire el pulmone : Et in questo orificio li sono etiam tri hostioli liquali se apreno de la parte dentro a la parte difuori e se chiudeno da la parte difuori a la parte di dentro per el contrario di li hostioli de laltro orificio : e questo e per che in tuto se chiudeno : Onde per questo orificio el core ne lhora de la constrictione solo ha ad expellere : e ne lhora de la sua dilatatione non ha ad atrahere alcuna cosa como faceua nel primo orificio.

El uentriculo sinistro del core ha i lati piu densi e piu spissi che li lati o uero parieti del uentriculo dextro : e questo [folio 21 verso] fu per tre ragione : La prima per che nel uentriculo dextro se de contenere el sangue el quale e graue E nel uentriculo sinistro se de continere el spirito el quale e molto ligiero : acio aduncha chel core non fusse piu graue e ponderoso da una parte che da laltra bisogno recompensare in questo modo cio e che lo uentriculo stancho hauesse piu groseza ne li suoi parieti che el dextro : La seconda cagione e che essendo el spirito piu suttile e piu resolubile chal sangue bisogno adoncha che el suo habitaculo hauesse piu grosso e de megliore sponde : La terza cagione si e per che el uentriculo sinistro e molto piu caldo cha el dextro per che iui se genera el spirito dal sangue per una grande calidita che suttiglia quello sangue e la calidita meglio se conserua nel subiecto denso e grosso :

Ne la concauita di questo uentriculo circa la sua radice li sono dui orificii : uno si e lorificio de una artharia chiamata artharia adorti per che inmediate ha origine dal core e per che e principio de la origine de tute le altre : per laquale artharia manda el core el spirito generato a tuti i membri : et etiam el sangue molto suttile insieme cum el spirito e questo fa quando el core se constringe : Onde nel principio di questo orificio li sono tri hostioli liquali in tuto se chiudeno da la parte difuori a quella dentro : e se se apreno da la parte dentro a la parte difuori e questo [folio 22 recto] orificio e molto profundo.

Laltro orificio si e de lartharia uenale laquale transporta laiere dal polmone a refrigerare el core e transporta i uapori caldi dal core al polmone como e stato dicto disopra : Et in questo orificio li sono doi hostioli che non se chiudeno altuto : Et sono molto eleuati acio che se apogiono melglio a la sponda del core quando el manda el spirito : Queste sono mirabile opere de la natura como anche mirabile opera fu nel uentriculo mezo del core per che questo uentriculo non ha una concauita ma piu lequale sono picole ma larghe e piu nela drita parte che la sinistra : E questo fece la natura acio chel sangue che ua dal drito uentriculo al sinistro per conuertersi in spirito continuamente se uegna suttigliando per quelle concauita.

Et per questo tu poi uedere che dal core nascono quatro cose cio e lartharia chiamata adhorti : Laltra si e la uena chilis : la terza si e la uena arthariale : e la quarta si e artharia uenale.

Anche uederai nel core certe parte pelliculare & in modo de auricule o uero orechiette apte a dillatarsi e constringersi facte da la natura acio che quando nel core se genera molto sangue o molto spirito se potesse el core dilatare a contenire quello sangue o quello spirito multiplicato et anche se constrinza quando non glie tanta habundantia di sangue o de spirito.

E qui adimanda Galieno [folio 22 verso] per che non fece la natura el core si grande che potesse continere ogne multitudine di sangue e de spirito senza quilli adittamenti di quelle pellicule. Risponde Galieno che questo fu : prima perche el core seria stato tropo grande : et consequenter tropo ponderoso : Secundario per che non se generando sempre molta quantita de sangue o de spirito sel core fusse stato tropo grande per la piu parte de le uolte la concauita del core seria stata uacua : ma queste tale auricule se dillatano ne lo aduenimento del sangue o del spirito e cusi se stringono ne la paucita soa.

Questo core e circumdato da uno panniculo duro neruoso o uero pelliculare facto in modo de una cassetta nel quale e posto el core como in uno suo tabernaculo a diffensarlo da le cose occurrente : Et e questa capsula molto dilatata acio chel core ne la sua dillatatione e mouimento non fusse agrauato da essa : Et etiam fece la natura questa capsula acio che continesse una certa aquosita rorida de laquale se bagnasse et humetasse el core acio che per el suo continuo mouimento non se sichasse : Onde quando questa aqua che e ne la capsula del core sie desiccata etiam se desicca esso cuore et consequenter se demacra e desicca tuto el corpo.

Capitulum quartum de anothomia trium panniculorum interiorum scilicet mediastine, pleure, & diafragmatis.

[folio 23 recto] Tri sono li panniculi interiori diquesto domicilio del core : Uno che se chiama mediastino che diuide la concauita del pecto per mezo cio e la parte dinanzi da la parte de drieto et consequenter diuide el polmone per mezo : e questo panniculo non e neruoso ne anche e ueramente uno continuo como li altri paniculi : e questo ha facto la natura per alcune utilita : prima acio che se una parte del polmone receuesse nocumento di qualchi superflui humuri che se agregasseno in quella non peruegnisse el nocumento e non regurgitasse quella materia a laltra parte; Secundario acio che tenesse suspeso e ligato el polmone al pecto.

El secondo panniculo chiamato pleura e uno panniculo duro e neruoso e molto grande : el quale copre tute le coste da la parte dentro : impero ha colligantia cum tuti li membri liquali se contengono ne la concauita del pecto e questo panniculo fece la natura acio che cuprisse tuti quilli membri a sua tutela; et acio che li paniculi dili membri tuti del pecto hauesseno principio et origine da quello.

El terzo panniculo se chiama Diafragma e da Aristotile e chiamato diazona per che e como una cintura che cinge per mezo: Questo panniculo e musculoso cio e carnoso e neruoso et e situato ne la fine del pecto e de le coste e ne la parte dinanzi quanto a la parte sua [folio 23 verso] carnosa e continuato cum le carthilagine de le coste mendose, e ne la parte posteriore e continuato cum la duodecima spondile doue sono le rene : De le coste e di li spondili poi noi diremo.

La utilita de questo panniculo prima fu acio chel seperasse li membri spirituali da li membri naturali cio e el secondo domicilio dal terzo acio che li fumi leuati da le feze non peruegniseno a li membri spirituali : Secundario per che ha a mouere el pulmone al mouimento de lo hanelito : e questo panniculo benche cingha per mezo oblique tamen et non ex directo: e la cagione di questa obliquita sie che da questo panniculo insieme cum el myrach del quale poi noi diremo se comprimino le feze che sono ne lintestini ne lhora de la egestione como se fusseno tra doe asse de uno torchio : E quanto a la parte meza di questo panniculo laquale e neruosa e panniculosa e colligato cum el pulmone per darli el mouimento como e stato dicto mediante i nerui quali uengono ad esso dal cerebro e da la nucha e per questo appare la cagione de la diuersita de el Diafragma e de li altri musculi per che li altri musculi nel luocho doue se congiongeno cum el membro quale debeno mouere sono como corde e ne li altri luochi sono carnosi per che sono facti principaliter a mouere le osse : ma nel diafragma e tuto el contrario per che fu instituito principalmente [folio 24 recto] a mouere el pulmone e non le ossa, e per questo appare chel diafragma sie rotondo cum una certa longitudine e che la sua substantia e musculosa e cordosa e che le utilitade sue sono tre : Prima acio che sia principio del moto de lo hanelito : Secundo acio che diuida tra membri spirituali e naturali: Tertio acio che aiuti el mirach ad expellere le superfluita quale sono ne lintestini.

Capitulum quintum de anothomia pectoris seu toracis continentis membra spiritualia.

Dicto di li membri che sono contenuti dentro dal pecto : poniamo adesso la anothomia de esso pecto : e disopra habiamo dicto che glie uno paniculo chiamato pleura quale copre tute le coste da la parte di dentro : Da poi quello panniculo tu uederai le ossa le quale sono di doe maniere cio e le coste e li spondili che sono como sponde doue se apogiano le coste lequale sono dodece da ciascuno lato cio e septe uere e cinque mendose : Le coste uere sono continuate cum li spondili a coprire et perficere el pecto : ma le mendose non : et una costa non attinge laltra ne la extremita acio che meglio se possa dilatare e constringere el pecto : Li spondili sono septe che se coniungono cum le septe coste uere mediante certe cartilagine lequale sono tra luno e laltro : e da queste carthilagine cum le sue ossa [folio 24 verso] se compone uno membro chiamato la furcula del pecto facta a modo de una furcula bifurchata ; e ne la extremita sua li e una certa carthilagine facta a modo de uno scuto a custodire la bocha del stomaco e chiamasi pomo granato : Da li lati de le coste mendose sono certe carthilagine.

Da poi uenendo a le parte de fuora : sono alcuni musculi di li quali alcuni sono a dillatare el pecto e sono dui musculi del Diafragma posti ne le parte inferiore del pecto : et hano a dillatare el Diafragma et consequenter el pecto ne la parte inferiore doue e una grande spaciosita : Item li sono dui altri musculi liquali sono nel collo et hano a dillatare la concauita superiore del pecto la quale e picola. Item sono altri musculi ne la schina doue e la origine de le coste, e comenzano apresso la origine de la prima costa : Item sono molti altri musculi picoli liquali cum difficulta se possono uedere ne la anothomia : e tuti quisti musculi predicti sono solo a dillatare.

Alcuni altri musculi sono a dilatare e constringere e sono situati tra le coste perche tra ciascune doe coste li sono doi musculi di liquali uno ha li uili latitudinali a dillatare, e laltro ha li uili transuersali a constringere.

Oltra questi musculi appare la pinguedine le mamille e la cute : La cute e la pinguedine e asai manifesta [folio 25 recto] impero solo noi direme de la anothomia de le mamille e haueremo fornito la anothomia del secondo domicilio e del secondo membro principale.

Capitulum sextum de anothomia mamillarum et de utilitatibus earum.

La figura de le mamille si e in modo de una çucha rotonda per che bisognaua essere capace del sangue che se ha a conuertere in lacte e la figura rotonda e piu capace cha le altre : et etiam per che le mamille sono como scuto del core impero doueano hauere una figura piu secura da li nocumenti : e questa tale figura e la rotonda.

Le mamille hebbeno doi capi picoli acio che la creatura potesse suciare el lacte : E la substantia sua si e certe carne glandose le quale de sua natura sono frigide acio che el sangue douenti biancho in esse e questo non se fa senon per infrigidatione del dicto sangue.

La quantita de le mamille ne la dona e magiore che nel maschio per che bisognaua generare el lacte ne la dona e non nel maschio. Et etiam essendo la femina piu frigida chel maschio bisogno essere magiore le mamille in esse acio che facesseno magiore reuerberatione del caldo al core et per questa reuerberatione lo fortifficaseno.

Le mamille ne lhuomo forno facte due como in tuti li altri animali che generano una o doe creature : ma ne [folio 25 verso] glialtri animali che generano piu figlioli sono facte piu mamille.

Ne lhuomo forno situate nel pecto e ne li altri animali nel uentre : e questo fu per molte casone : La prima secondo Galieno e chel sangue del quale se genera el lacte deba essere ben digesto impero bisogno essere propinque al core ne lhuomo per la cui calidita quello sangue fusse meglio digesto : ma ne li altri animali molta quantita de tale sangue superfluo ua a conuertirse in corni o in altri membri.

La seconda cagione asegna Aristotile che li altri animali hano le gambe dinanzi molto strette et impero hano el pecto molto stretto : ma ne lhuomo el pecto e amplo : onde non potete la natura situare le mamille ne glialtri animali como ne lhuomo.

La terza cagione si e chel core de lhuomo hebbe bisogno de essere piu deffensato che el core de li altri animali li quali li hano pili disopra impero fece la natura le mamille como defensaculo ne lhuomo che non ha pili inquelle parte.

Le mamille hano colligantia cum el core e cum el figato per una certa uena che ascende dal figato ad esse mamille : ha etiam dio colligantia cum la matrice mediante certe uene che uengono da la matrice ad esse e procedeno quelle uene tortuose acio che continuamente se asuttiglie el sangue e meglio se digesta a conuertirse in lacte.

[folio 26 recto] *Tractatus tertius de anothomia tertii membri principalis scilicet epatis et eidem deseruientibus : capitulum primum de anothomia stomaci.*

Veduto de doi membri principali et di li suoi ministri et etiam de li suoi domicilii vediamo mo la anothomia de doi altri membri principali cio e figato e testiculi et di li membri che sono suoi ministri et etiam de li suoi domicilii : E noi determinaremo de tuti dui quisti inquesto tractato per che li membri che deserueno a la generatione non hanno distincto domicilio da li membri nutritiui : E questo domicilio comenza dal pomo granato che copre la bocha del stomaco del quale habiamo dicto e dura per insino al petenechio inclusiue includendoli la uirga e li testiculi : et questo e quanto per lo longo, ma quanto per el largo dura da uno fiancho a laltro e per el profondo dura da la cute de lombelico che copre el corpo dinanzi dale coste ingioso per insino a laschina de drieto :

Inquesto domicilio li sono contenuti di molti membri cio e stomaco, intestini, figato, fele, milza misinterii, girbo, rognoni, vesica, testiculi, vasi spermatici, matrice ne la femina, e la uirga ne lhuomo de liquali membri solo dui sono principali cio e el figato et li testiculi secondo Galieno, o uasi spermatici secondo Aristotile.

Noi adoncha sequitaremo secondo el nostro ordine consueto comentiando a li membri superiori e descendendo a linferiori. Comentiaremo [folio 26 verso] adoncha dal stomaco e dal meri che e uia del cibo ad esso stomaco, E noi habiamo dicto di sopra che como la cana del pulmone era conducto de laiere cusi el meri era conducto del cibo e del poto : E che la bocha de la cana del pulmone e la bocha del meri erano congionte insieme per la rasone iue dicta.

La sustantia di questo meri sie pelliculare e molle como la cana del pulmone e pelliculare e carthilaginosa e bisogno chel meri fusse molle acio potesse dilatarsi quando lhuomo piglia tropo cibo, et anche questo meri non sta aperta como fa la cana del pulmone ma per la sua mollitie una parte cade sopra laltra.

La substantia del meri e composta de doe tuniche una intrinsecha che ha certi uili o neruetti longitudinale che sono facti ad atrahere el cibo : e laltra sie exteriore ne laquale sono uili latitudinali facti ad expellere quello che e stato atratto da la tunicha interiore : benche la prima tunicha sie piu principale che la seconda.

La quantita del meri e magiore che non e la quantita de la cana del pulmone per che el meri ua piu longo che non fa essa cana : Onde el meri ua per insino al diafragma e desotto da esso se continua cum la bocha del stomaco onde el stomaco e incontinenti sotto el diafragma : Et anche el meri e magiore in largheza per che hauea a passare per esso cosa piu grosa che non e [folio 27 recto] laiere.

Questo etiam Meri e posto piu nel profondo uerso le parte posteriore cio e uerso la schina doue ua a ritrouare la bocha del stomaco laquale bocha e uerso le parte posteriore : per che la bocha del stomaco e ligata ala schina ex directo in el principio de la sua ligatura cio e a la decima terza spondile sotto el diafragma : el quele se termina ala duodecima spondile e poi consequenter procede el stomaco aligandosi ali spondili de le rene.

Questo stomaco sie cella del cibo et e quasi in mezo de tuto el corpo como e stato dicto del core : per che essendo como lauezo doue se ha a cocere el cibo bisogno essere in mezo acio chel receuesse calore da tute le parte e da tuti li membri circumstanti : et non fu posto el stomaco apresso de la bocha per la rasone dicta : Tu uederai adoncha el stomaco hauere sopra si el core e el diafragma e desotto el misinterio e lintestini : da la parte dritta el figato el quale lo abraza cum cinque sue penole : da la parte sinistra la milza laquale li rende calore mediante le sue artharie : da la parte dinanzi ha una rethe chiamata el Girbo : da la parte de drieto li musculi de la schina e una uena grande e una artharia che passa per la schina como poi se uedera : da tuti quisti membri receue calore el stomaco acio che coza bene el cibo.

E ben chel stomaco sia situato sopra de la Schina niente di meno la parte sua superiore declina al [folio 27 verso] lato stancho, e la parte inferiore al lato drito : e questo fu per che ne la parte dritta li e el figato molto eleuato ne le parte superiore, e la milza ne la parte stancha e piu de pressa : impero la parte superiore del stomaco non se potete locare ne la parte dritta per che el figato occupaua quello luocho ma ben se potete locare ne la sinistra cio e disopra dala milza doue li era uacuita Item per che disotto dal figato li sono glintestini suttili e gracili liquali occupano pocho luocho et iue remane una grande concauita impero fu locata la parte inferiore del stomaco iue a reimpire quella concauita : Et per che etiam ne la parte stancha disotto da la milza apresse de le rene glie uno intestino molto grosso chiamato colon el quale occupa uno grande luocho impero non se potete locare dicta parte inferiore nel lato stancho.

Una altra cagione per laquale el stomaco non fu posto a presso de la bocha e perche apresso de la bocha bisognorno essere i membri de lo hanelito ad atrahere laiere : Et anche per che el bisognaua che glintestini fusseno continuati cum el stomaco, e bisognaua che glintestini fusseno disotto dal diafragma.

Et per questo appare che per molte cagione el stomac non fu locato per el dritto ma per lo storto e per lo obliquo : la prima si e gia dicta acio reimpisse la uacuita de la parte dritta e stancha : La seconda per che essendo lhuomo de statura dritta non retigniria bene el cibo [folio 28 recto] ma subito uscirebe fuori per la bocha disotto : La terza cagione per che bisognaua chel stomaco receuesse da la milza quanto a la bocha superiore lhumore melenconico a darli lapetito : et quanto a la bocha disotto bisono che receuesse lhumore collerico dal figato : et impero bisogno che la bocha superiore del stomaco fusse dal lato stanco doue e la milza e la bocha inferiore fusse dal lato dritto doue e el figato.

E per questo appare chel stomaco ha colligantia cum la milza per certe nene che portano lhumore melenconico ad esso : et ha similiter colligantia cum el figato per molte altre uene che li portano el nutrimento dal figato : et ha colligantia cum el core mediante una grande artharia che e posta sotto esso : et ha colligantia cum el cerebro mediante uno certo neruo el quale ua ala bocha del stomaco et iue se sparge e diuidese circa la superiore parte de esso stomaco.

La figura del stomaco fu rotonda acio che fusse piu tuta da li nocumenti extrinseci et acio anche che fusse piu capace per che bisognaua continere di molto cibo : Ma non fu perfectamente rotonda per la rasone dicta per che bisognaua che una parte declinasse al lato dritto e laltra al lato stanco impero e di figura arcuale in modo de una cucha ritorta e fu molto grande el stomaco acio potesse receuere grande quantita de cibo.

El stomaco e composto de due tuniche : Una interiore laquale e neruosa e laltra [folio 28 verso] exteriore e carnosa : Et la prima tunica neruosa e piu grossa e spessa che la seconda per che hauea a tochare el cibo acio che non receuesse nocumento da esso e per che se potesse dilatare e constringere secondo el bisogno de la quantita del cibo : ma la tunicha exteriore fu piu suttile onde e da notare che la tunicha interiore bisogno essere neruosa per molte rasone: prima per che in essa & de essere lapetito e el sentimento e non e dubio che meglio se sente la cosa quando senza mezo ocorre al sentimento : ma la exteriore fu carnosa facta a digerire et alterare el cibo : la alteratione e digestione se puo ben fare per mezo e non occorrendo in mediate a la cosa : Questa tunicha adoncha exteriore e piu suttile che la interiore per che e aiutata dai membri circumstanti a digerere : non bisogno essere adoncha tropo grossa.

La tunicha interiore e deputata ad atrahere el cibo et a retignirlo debito tempo per insino che se digestisse : impero ha alcuni uili longitudinali ne la superficie interiore mediante li quali atrahe a se el cibo : e ne la superficie exteriore ha alcuni uili transuersali per liquali ritiene el cibo Et la tunicha exteriore ha a digerire el cibo et consequenter ha ad expelerlo quando e digesto : impero in essa certi uili latitudinali sono posti per liquali ha ad expellere el cibo digesto :

La bocha del stomaco superiore e piu lata che non e la inferiore per che [folio 29 recto] per la bocha disopra hauea intrare el cibo grosso indigesto e per la bocha disotto hauea uscire el cibo suttile e digesto : E quisti doi orificii non sono facti molto eminenti ma la parte inferiore del stomaco e piu disotto che la bocha inferiore acio chel cibo se retegna et similiter la parte superiore del stomaco e piu eminente e piu insuso che non e la bocha superiore acio che essendo el stomaco pieno de cibo inclinandosi lhuomo cum la bocha in giu non ritornasse el cibo fuora.

Doe adoncha sono le utilita del stomaco : Una ad appetere el cibo necessario per tuto el corpo : e questo fa per la tunicha neruosa interiore e laltra e a digerere el cibo e questo fa per la tunica exteriore carnosa.

Capitulum secundum tractatus tertii de anothomia intestinorum et misinterii.

Dapoi il stomaco li sequitano glintestini li quali sono sei reuoluti cioe tri suttili e tri grossi et non fu ne lhuomo uno solo intestino recto ma furno piu e circumuoluti acio chel cibo longo tempo se continesse nel stomaco et intestini per che se cusi non fusse bisogneria che lhuomo fusse in continua asumptione de cibo, et in continua egestione e seria stato lhuomo molto occupato in tale uile operatione Et anche sel fusse stato uno solo intestino recto non seria stato tuto el cibo da ciascuna parte de lo [folio 29 verso] intestino toco et consequenter non seria stato exsiccata tuta lhumiditade del cibo : Acio adoncha tuta la humidita del cibo sia desiccata et atratta al figato e che niente o pocha non rimanga ne le feze : Furno facti piu intestini circumuoluti : El primo adoncha intestino e chiamato duodeno et e suttile e chiamasi duodeno perche e longo quanto e dodice uolte el dito grosso di quello tale : Et in questo intestino li entra el cibo como e digesto nel stomaco per la bocha de sotto de esso stomaco chiamata portonaria o uero pylerum cum la quale se continua questo intestino duodeno. Digesto andoncha el cibo nel stomaco se apre questo portonario e manda la uirtu expulsiua del stomaco questo tale cibo ne lo intestino duodeno : A questo intestino ua uno canale o uero condutto dal fele per el quale se porta la collera ad esso intestino. Da poi questo intestino li sequitan uno altro intestino suttile chiamato ieiuno perche e la piu parte del tempo uacuo per doe ragione : Prima per che e dritto e non inuoluto : La seconda per che una grande multitudine de collera pura uiene ad esso per quello medesimo condutto che ua al duodeno : e questa collera mordica lo intestino e fa descendere gioso el cibo.

Dapoi sequita el terzo intestino suttile chiamato ileon per che e situato circa gli ilii id est li fianchi : Onde in questo intestino glie uiene el dolore iliaco cio e dolore de fianco e questo [folio 30 recto] intestino hebbe molte inuolutione, et anche ad esso peruengono de molte uene picole dal figato chiamate mesaraiche : E questo fece la natura acio che el figato atrahesse la humorosita dal cibo per quelle uene, onde a questo intestino li peruengono piu uene mesaraiche che nesuno di li altri.

Dapoi questi tri intestini sutili sucedeno li grossi : E questo fu facto per che quanto el cibo uiene piu descendendo tanto piu douentano dure le feze e piu grosse impero bisogno che glintestini inferiori fusseno piu ampli che li superiori.

El primo adoncha intestino grosso che sequita ali suttili si e chiamato monoculo, non per che habia solo uno oreficio per che questo seria impossibile anzi ne ha doi como li altri uno per elquale atrahe el cibo e laltro per elquale expelle : ma per che quisti doi oreficii inquesto intestino sono uno a presso de laltro como coiuncti e non dispartiti como ne glialtri impere appare hauere solo uno oreficio, onde per questo monoculo e chiamato : Et anche chiamato sacco per che pende la sua concauita como un sacco stando li suoi orificii de sopra : Questo intestino e situato ne la parte dritta apresso lancha e disotto dal rognone dritto. E fu facto acio che retinesse el cibo ançi lo reuerberase a li intestini superiori e prohibisse che non descendesse acio che in quilli intestini se esuccasse dal figato la sua humidita como e stato dicto.

Da poi questo intestino sequita laltro grosso [folio 30 verso] chiamato colon per che ha piu colli o uero cellule ne lequale el stercho recceue la sua forma.

Questo intestino ha de molte inuolutione circa el rognon stancho e poi ascende e copre la milza e poi se declina a la parte dritta uendo piu uerso le parte exteriore e copre el stomaco.

E per questo appare la cagione per che fu locato sopra del stomaco e de sopra tuti li altri intestini : questo fu per che era piu ignobile de lialtri, e como membro piu ignobile fu posto uerso le parte exteriore et anche per che le feze se indurano in esso acio che hauesse qualche humidita dal girbo del quale poi noi uederemo. Laltra cagione de cio e che essendo questo intestino facto a continere et expellere le feze ma piu ad expelere impero bisognaua ad esso uenire piu collera che hauesse a stimulare la uirtu expulsiua piu che ne glialtri : impero sopra di quello ne la parte dritta una penula del figato doue e alligata la cesta del fele como appare al sentimento : e questo fu che de sopra de questo intestini li peruenisse la collera oltra quella che ua a la sua concauita como etiam ua a le concauita de glialtri intestini.

La substantia di questo intestino e grossa e sollida facta cusi per la uentosita grande che se genera in esso laquale fa dolore fortissimo chiamato dolore collico : Et in questo intestino se generano certi uermi longi [folio 31 recto] et altre manerie de uermi chiamati lombrici.

Da poi e lultimo intestino chiamato intestino dritto de el quale la extremita et oreficio inferiore se chiama ano o uero culo : e uasene uerso el fiancho stancho doue poi comenza lo intestino colon predicto. In questo intestino recto li sono una grande moltitudine de uene meserayce che uengono a sugare se qualche humidita fusse rimasta ne le feze.

Quisti sono adoncha li sei intestini liquali sono alligati a la schina mediante uno certo membro chiamato misinterio o uero intriglio quasi interiora tenens che non solo glintestini ma tute le uiscere sono alligate per questo interiglio ala schina et impero questo membro fu composto de uene, corde, panniculi, e ligamenti acio potesse ligare li predicti membri : Et e etiam e composto de una sustantia seposa e pingue acio che li membri duri como sono li spondili non se congiongesseno senza mezo cum li membri molli cio e cum li intestini e le altre uiscere acio che el molle non receuesse nocumento dal duro. Le altre uacuita di questo membro sono reimpite de certe sustantie glandose, facte etiam acio che sustentino le uene meseraiche che sono disperse in questo membro : et forse che sono facte etiam a generare la humidita che humetti la feze de glintestini acio che piu tosto lubrichi : et impero uedemo che mangiando cibi duri [folio 31 verso] niente dimeno quello che nesce per egestione e liquido.

Capitulum tertium de anothomia epatis quod est tertium membrum principale : et de uenis orientibus ab eo.

Vediamo mo del terzo membro principale situato in questo palazo et e el figato alquale deseueno tuti li altri membri che sono posti quiue. El figato naturalmente e situato sotto el diafragma et non sotto le coste uere, ma una parte de esso sta sotto le parte mendose : benche ne lhuomo morto appara essere locato tuto sotto le coste, e questo e per che li membri spirituali ne lhumo morto sono molto anihilati et el figato ua a reimpire le uacuita derelicte : impero quando tu fai la anothomia tu dei eleuare el corpo morte e tirare in gioso el figato acio chel uada al suo luocho naturale.

La quantita del figato fu molto granda ne lhuomo per che e molto sanguineo e de natura calda e humida.

El figato sie composto de certe uene diuise e disperse in modo de una rethe et le uacuita sue reimpisse una certa carne rossa che e como sangue coagulato : Et per queste uene se si sparze el cibo digesto nel stomaco chiamato chile cio facto in modo de suco dorzo che cusi douenta nel stomaco, e questo fu facto acio che se diuidesse in parte picole che tuto el figato potesse tochare tuto quello chilo acio che meglio lo conuertisse in sangue : Ma nel stomaco non sono tal uene doue se hauesse a receuere el cibo ma solo [folio 32 recto] li fece una concauita per che li cibi che se pigliano sono molto grossi che non harebono potuto penetrare per dicte uene. Questa decocione che se fa nel figato a conuertere el chile in sangue piu se compisse ne la parte superiore : et impero quella parte e piu solida e dura : Hebbe el figato cinque penule benche ne lhuomo non siano sempre diuise che se possano uedere.

Questo figato ha doe parte cio e la parte gibosa e la parte concaua, et ha colligantia cum el core per una certa uena che nasce dal suo gibo e uasene al core, et e chiamata uena chilis : Et etiam ha colligantia cum el diafragma alquale sta suspenso Et similiter a li spondili de la schina ala quale e alligato mediante un certo paniculo : Onde ha dui panniculi uno chel suspende e liga al diafragma e ala schina e laltre chel copre e sel circunda. Dala gibosita sua nasce la uena chilis laquale porta el sangue al core de laquale habiamo gia dicto. E da la parte sua concaua ne nasce unaltra chiamata porta o uero uena concaua e questa uena ha cinque rami : cusi como sono cinque penule del figato ne le quale entrano quisti cinque rami E poi quando escono fuora del figato sono da poi otto de lequale doe sono molte picole che male se possono discernere ma si le altre sei : De lequale una ua ala dextra parte del stomaco a nutrire la tunicha sua exteriore et maxime la parte inferiore : Laltra uena ua a la milza et e asai grande de laquale nel mezo del [folio 32 verso] suo transito nasce un ramo che descende gioso a nutrire lintriglio e portali el sangue piu aquoso : Da poi quando questa uena sa proxima a la milza nasce un altro ramo el quale ua a nutrire la parte sinistra inferiore del stomaco da poi sucede piu oltra e uasene ala concauita de la milza et iue se diuide in doi rami cio e inferiore e superiore : linferiore ramo descende gioso a nutrire el girbo quanto ala parte sua sinistra : el ramo superiore passa per le concauita de la milza e diuidise in doi altri rami di liquali uno ua a nutrire la parte superiore sinistra del stomaco, Laltro ua circa la bocha superiore del stomaco a portarli lhumore melenconicho per incitare lo apetito : Laltro ramo che rimane ua a la milza anutricarla.

La terza uena di queste sei sene ua al lato stancho e uaseno alo intestino recto a sucare se qualche humidita uiuatiua fusse rimasta ne le feze.

La quarta uena se ne ua a la superiore parte dritta del stomaco per nutrirla : La quinta uæna ha doe parte una ua a nutrire la dritta parte del girbo, laltra parte se ne ua alo intestino colon a sucare quello che e rimasto ne le feze de humidita et anche a nutrirlo : et impero el girbo molto se congionge cum lo intestino colon ne la parte dritta : La sexta uena sene ua alo intestino ieiuno et a lialtri intestini suttili a sucarli e nutrirli.

La figura del figato debba [folio 33 recto] essere lunare in modo de una luna quando e piu che meza. Questo membro ha quatro uirtu una atratiua per la quale atrahe el chilo a se : La seconda retentiua per laquale lo ritiene debito tempo acio che la terza uirtu che e digestiua lo conuerta in sangue : La quarta uirtu e expulsiua per laquale manda el sangue a tuti i membri a nutricarli : et cum esso sangue manda anche el spirito nutritiuo el quale se genera in esso figato.

Capitulum quartum. Tractatus tertii de anothomia chistis fellis.

El fele si ha uno uase como una cista doue se contiene lhumore collerico et e apicata a la meza penula del figato acio che depuri el sangue da lhumore collerico : e fu situato nel concauo e non nel gibo acio che piu facilmente potesse mandare la colera aglintestini a incitare la uirtu expulsiua che mandi fuora le feze.

Et ha doe parte cio e el collo che porta la collera e la uesica chela contiene : El collo a certa distantia rimane uno : E dapoi se diuide in doi rami uno ua amezo del figato ad attrahere la collera da esso Laltro ramo descende alo intestino duodeno et questo se diuide anche in doi altri rami uno ua al fondo del stomaco a confortare la digestione e questo ramo e picolo per che non bisognaua [folio 33 verso] andare tropo collera al stomaco per non incitare tropo la uirtu expulsiua del stomaco ad expellere, ma solo a confortare como e stato dicto : Et impero quilli che hano questo rame molto grande sono chiamati da medici infelici impero che sempre & al continuo regurgita su al stomaco la collera :

E per questo appare che questo membro ha colligantia cum el stomaco, intestini e figato e chel se nutrisse per certe uene et artharie che uadeno ad esso cio e a la sua concauita : et anche peruengono a lui alcuni nerui a darli el sentimento : Onde ha anche colligantia cum el core e cum el cerebro.

Questo membro si e di figura oblonga cum una certa rotondita e la sua substantia e pelliculare cio e in modo de una pellicula facta per le utilita sopradicte.

Capitulum quintum de anothomia splenis et de eius uiuamentis.

Dal lato stanco sotto le coste mendose li e la milza laquale cum el suo concauo al lato del stomaco stanco se glie apozia : E quanto ala parte sua gibosa e alligata ala Schina et al panniculo dicto siphac mediante alcuni panniculi sutili

Et non fu posta cusi insu o uero in luocho alto como el figato ma piu ingioso: Et e di figura quadrangulare per che ha areimpire la concauita sinistra circumstante del stomaco che e di tale figura : ma e piu grossa ne la parte disopra et e piu sutile ne la parte inferiore a modo de una lingua.

E questo [folio 34 recto] membro e composto de una certe carne spongiosa acio che meglio receua lhumore grosso melenconico alquale finalmente e ordinata, Et anche e composta di uene et artharie molte, & de uno paniculo che linuolge Onde appare che la milza ha colligantia cum el figato, lintriglio, girbo, & cum el stomaco, cum le coste e cum el diafragma, et ha anche colligantia cum el core mediante certe artharie che uengono ad esse acio chel sangue grosso melenconico per el calore di queste artharie se suttigliasse e digerisse : Et anche acio che riscaldasse la sinistra parte del stomaco a laquale lui se apogia.

Fu facto questo membro per molte utilita : Prima acio chel mondificasse el sangue da lhumore melenconico el quale atrahe asi : Secondo fu facto a contra operare ala calidita del core e del figato : Tertio acio che excitasse lo appetito transmitendo lhumore melenconico a la bocha de esso stomaco.

Capitulum sextum de anothomia girbi siue rethis cooperientis stomacum & intestina.

Appare uno certo panniculo chiamato el Girbo o uero la rethe el quale copre el stomaco da la parte dinanzi : e ne lhuomo tuti glintestini : e non ne lialtri animali : E questo fu facto ne lhuomo per che tra glialtri animali de equale quantita la uirtu digestiua piu debile ne lhuomo : et etiam per che glintestini suoi per la suttilita de la cute sono piu dispositi a [folio 34 verso] receuere li nocumenti exteriori : Et impero appare la utilita di questo membro per la quale fu principalmente facto : et e acio chel confortasse la uirtu digestiua nel stomaco e de glintestini reuerberando el caldo naturale ad essi : Onde narro Galieno de uno che fu uulnerato e cauato li fu el girbo e da poi che fu guarito non potete mai ben padire.

Et impero bisogno che fusse composto di tre sustantie cio e prima de doi panniculi subtili acio che continesse glialtri membri et etiam per che douea essere ligiero e che se potesse dilatare : et anche fu spesso acio che reuerberasse piu la calidita ali membri predicti : Secundo e composto de una assungia seposa la quale hauesse ariscaldare essendo la natura de lasongia molto propinqua al caldo : Tertio e composto di certe artharie e uene lequale molto riscaldano.

Et per questo appare chel girbo ha colligantia cum el stomaco cum la milza e cum glintestini, Et maxime cum lo intestino colon cum liquali lui si termina cooperendoli : Et etia ha colligantia cum li membri da liquali ha origine : onde nasce da uno certo panniculo carnoso da la schina tra el diafragma per che a questo panniculo seglie terminano do extremita del panniculo chiamato siphac del quale poi noi diremo : Lequale extremita compogono el girbo : Et etiam per che iue glie una uena grande et etiam artharia [folio 35 recto] de lequale apresso el stomaco nascono certe uene et artharie picole lequale componeno el girbo : Ha etiam colligantia cum lintriglio dal quale nasce la sua songia seposa laquale reimpie le sue uacuita.

Per insino adoncha qui habiamo ueduto la anothomia del girbo, del stomaco, de glintestini, de lintriglio, del figato, del fele, e de la milza andiamo mo a glialtri membri di questa terza casa.

Capitulum septimum de anothomia membrorum urine scilicet renum & uesice et aliorum membrorum deseruentium eis.

Vediamo la anothomia dele rene. Onde tu uederai che da la uena chilis che nasce dal gibo del figato se fa uno ramo grande che descende gioso a le parte inferiore, e quando questo ramo e indritto de lerene se diuide in doi altri rami di liquali uno ua al rognone dritto e laltro al rognone stanco cio e a le sue concauita e chiamase uene emulgente : E gliorificii di queste doe uene non sono indritto uno dilaltro ma uno piu elto et e quello del rognone dritto e laltro piu basso cio e quello che ua al rognone stanco : Et questo fu perche el rognone dritto si e piu de sopra per che el rognone dritto e piu caldo cha el stanco, e de natura del caldo e distare disopra benche a le uolte acada chel rognone stanco sia disopra al dritto et alhora el rognone stanco uira essere piu caldo che el dritto : ben che questo sia [folio 35 verso] rare uolte.

Queste uene deportano la aquosita del sangue che e inutile al nutrimento del corpo a le rene et consequenter ala uesica : laquale esce poi fuora per urina : E per che cum questa aquosita e mescolato anche del sangue impero bisogno fare a la natura che el se colasse ne le rene in modo chel sangue mescolato cum questa aquosita rimanesse, e laquosita sola pasasse ala uesica : et impero se tu scindi el rogne ne la parte gibosa per lo longo per insino ala concauita tu uederai uno panniculo como uno panno raro per el quale puo passare la aquosita ma el sangue non impero quilli che hano aperto questo panniculo o uero colatorio orinano sangue. E questo panniculo si genera da la uena emulgente dicta laquale intrando ne la concauita del rognone se rariffica in modo de uno colatorio.

E bisognorno essere dui rognoni e non uno per che era molta quantita daquosita laquale uno solo rognone non haueria potuto atrahere sel non fusse stato molto grande e non se seria posuto debitamente situare sel non hauesse facto qualche eminentia in quello luoco che seria stato molte deforme.

Quisti rognoni sono picoli in comparatione de li altri membri interiori e sono de una figura alquanto rotonda acio che fusseno capaci di magiore quantita, et etiam che fusse piu tuto da li nocumenti extrinseci ; E furno etiam alquanto longhi acio che li suoi oreficii cio e el superiore doue entra [folio 36 recto] laquosita e loreficio inferiore doue esce haueseno megliore distintione : a loreficio di sotto segli continua uno porro chiamato Uritides cioe che porta la urina da le rene a la uesica : Onde sono dui porri uritides como sono doi rognoni : Et in quisti rognoni ale uolte se genera la preda de molte harenule per la calidita de le rene la quale desicca certa humidita fleumaticha laquale se genera nel stomaco per indigestione, e poi sene ua al figato, et tandem se ne uiene ale rene, et iue per la calidita de esse rene se conuerte in harenule et tandem se conuerte in preda : laquale poi si discerne dala preda generata ne la uesica per che la preda de le rene e rossa e quella de la uesica e biancha. Li homini adoncha che hano fredo el stomaco e calde le rene sono disposti ala generatione de la preda et maxime hauendo li meati de lurina stricti.

Leuate adoncha le rene e ueduti i porri uritides tu uederai che terminano al mezo de la uesicha e non forano la uesicha ex directo cum uno bucho grande ma cum piu busitti picoli et obliqui facti tra una tunica e laltra de la uesica o uero tra el cooptorio e la tunica e non uno indritto de laltro, e questo fu acio che quando la uesica fusse piena de urina ritornasse la urina indrieto ale rene, anzi quanto la uesica e piu piena de urina tanto piu se chiudeno dicti buchi.

La uesica e composta de doe tuniche quanto al suo fundo ma quanto al [folio 36 verso] suo collo e composta de carne e musculo Item e composta de nerui e de uene e de artharie ad atrahere laquosita dale rene et consequenter ad expellerla fuora per la uena.

E per questo appare che tuti quisti membri dicti cioe uene emulgente rognoni porri uritides e la uesica sono facti de la natura a mondificare el sangue che de nutrire el corpo de la predicta aquosita e mandarla fuora per urina.

Et impero li rognoni furno de sustantia e carne dura acio non fusse mordicata et corrosa de lacuita de lurina e da alcuni humori acuti che molte uolte se mescola cum essa urina.

Questi rognoni hebbeno dui paniculi uno che li copre e questo li da el sentimento, e laltro chel liga e suspende a la schina et anche questo li da el sentire : e ciascuno di questi doi panniculi e composto de uno certo neruo che nasce da la nucha de li spondili de la schina in luocho chiamato alchatim che e luocho a lo indritto de le rene et etiam e composto de uno certo ligamento che nasce da quilli medesimi spondili.

E per questo appare che hano colligantia cum el cerebro e la nucha et cum la schina mediante li nerui di li predicti panniculi, et hanno colligantia cum el core mediante certe artharie che nascono da lartharia adorthi e cum el figato mediante le uene emulgente, e [folio 37 recto] cum la uesica mediante li porri uritide liquali sono certi canili stricti per liquali passa la aquosita urinale da le rene a la uesica como e stato dicto, E questa uesica ha una grande concauita laquale e neruosa et el suo collo e carnoso e musculoso acio che quando bisogna lhuomo expella la urina e quando bisogna lui la ritengha et congiongese el collo de la uesica cum la uirga ne li maschii, nel quale collo insieme cum la uirga e uno bucho per loquale se urina: ma ne le femine lextrimita del collo de lauesica se termina apresso a dua dita al oreficio de uulua : et el collo de la uesica ne li maschii e piu longo che ne le done.

E per questo appare che sel se incide la uesica nel collo se puo consolidare ma se si taglia nel fondo non si puo saldare, per che el collo e musculoso e carnoso, et el fonde da la uesica e neruoso.

Et el collo de la uesica ne li homini ha tre tortuosita, ne le quale se ritiene lurina acio che facilmente non esca fuori senza uolunta de lhuomo ma ne le femine non ha sino una tortuosita, et el collo ne le femine e piu largo che ne li maschii : Et el fondo de la uesica e composto de doe tuniche como e stato dicto, e la tunica interiore e doe uolte piu grossa che la exteriore per che inmediate tocha la urina.

A la uesica peruengono nerui da la nucha et anche le uene da la uena chilis et etiam certe artharie da la artharia adhorthi. Et nel collo suo e solo uno [folio 37 verso] musculo che circunda esso collo del quale la utilita e a retinere la urina secondo el bisogno e la uolunta de lhuomo, E quando lhuomo uole urinare se relassa quello musculo : et alhora li musculi del uentre de liquali diremo constringeno la uesica e mediante la uirtu expulsiua mandano fuora lurina.

Tractatus quartus de anothomia membrorum generationis capitulum primum, de anothomia matricis et uasorum spermaticorum in mulieribus.

Veduto la anothomia de tri membri principali e signori li quali cum li soi ministri sono producti da la natura a conseruare lo indiuiduo poniamo adesso la anothomia del quarto membro principale el quale e facto a conseruare la spetie. Et benche anche noi non habiamo fornito la anothomia del domicilio del terzo membro principale per che in uno medesimo domicilio quasi sono locati dicti membri cum li suoi ministri Diciamo adoncha che i membri de la generatione in alcune cose conuenene ne li maschii e ne le femine: prima quanto a la origine per nascono circa le rene in questo modo che li uasi che sono ne la parte sinistra e li uasi che sono ne la parte dritta nascono desopra de le rene, cio e le loro uene da la uena chilis e le lore artharie da lartharia adorthi Onde appare per questo che li uasi spermatici ne li maschii e ne le femine sono decusi da el core e da el figato e questa e la seconda conuenientia.

Ma etiam sono differenti per che ne le femine questi uasi se terminano a la matrice [fol. 38 recto] nel luocho exteriore doue sono li loro te testiculo anzi propriamente parlando non sono ueramente testiculi como ne gli maschii anzi sono como testiculi de lepore Onde fuora de la matrice se riuolgono e se contexeno e le concauita diquella texitura se reimpiseno di certe carne minute glandose : E sono facti ne le femine acio che generino una certa humidita saliuale laquale e cagione de la delectatione de cohito ne la femina.

Da poi quisti uasi spermatici penetrano la matrice per insino a la concauita e li suoi oreficii di quisti uasi ne la concauita de la matrice se chiamano cotilidoni cio e legamenti per che mediante quilli sta ligata la creatura ala matrice : e per questi oreficii uene el sangue mestruo ala femina : Et alcuni di questi uasi peruengono a la bocha de la matrice a portar li la humidita saliuale gia dicto : Et da queste uene ramificate nascono doe uene da ciascun lato cio e una che penetra nel panniculo chiamato mirach et ascendeno per insina che peruengono ale mamille a deportare el sangue a quelle : Et nota che quanto piu ascendeno tanto piu se acostano a la cute di fuora : et sono piu manifeste : ma nel mirach sono piu oculte e questo e contrario ne la porcha o altri animali che hano le mamille nel mirach : Queste uene nascono da la matrice e se manifestano nel mirach doue sono poste le mamille.

E dapoi queste uene ascende dal [folio 38 verso] profondo del pecto indrito al pomo granato una certe uena laquale uene ale mamille a cuocere el sangue che se de conuertire in lacto e non appare senon una uena.

El luocho de la matrice e che le situata ne la concauita del luocho chiamato alchatim, laquale concauita e circundata da certi spondili dela schina per insino a la cauda da la parte de drieto, ma da la parte dinanzi e circundata da la parte che se chiama petenechio : onde la matrice e locata inmediate tra lo intestino recto el quale e como colcitra sua da la parte posteriore e fra la uesica da la parte dinanzi et el collo de la uesica e piu eminente cha el collo de la matrice benche la concauita de la matrice sia piu profonda che la concauita de la uesica : et la matrice e postà nel mezo preciso tra el lato dritto e el stanco.

Questa matrice ha colligantia quasi cum tuti li membri superi cio cum el core mediante certe artharie e cum el figato mediante certe uene, e cum el cerebro mediante molti nerui, e cum el stomaco mediante nerui e uene : et ha colligantia cum li membri di mezo cio e cum el diafragma le rene, et mirach: per che mediante quisti e alligata ali predicti ha maxime colligantia cum le mamille como e stato dicto : Ha etiam colligantia cum li membri inferiori cio e cum la uesica mediante el suo collo : et e similiter cum lo intestino colon.

Et e alligata a le [folio 39 recto] anche mediante alcuni ligamenti grossi e forti li quali apresso de la matrice sono larghi e grossi et apresso le anche sono suttili como corne che sono nel capo de glianimali et impero sono chiamati corni de la matrice.

La figura sua e quadrangulare cum certa rotondita : et ha el collo inferiore longo et hebbe questa figura acio che meglio se potesseno distinguere le cellule o uero camerette che sono ne la sua concauita e sono septe tre ne la parte dritta e tre ne la parte stancha e una ne la sumita o uero mezo e queste celule sono certe concauita ne la matrice ne lequale el sperma cum el sangue mestruo se possano continere et coagulare et consequenter alligarsi a li oreficii de le uene.

La quantita de la matrice fu mediocre secondo la quantita de la uesica, ma e magiore in una femina che in laltra per che la femina che fa figlioli ha magiore matrice che la sterile et similiter la femina che e usa al cohito lha magiore che la uergene et similiter la matrice de la giouene e magiore che quella de la puta e de la uechia e per altre cagione narrate da medici puo essere questa diuersita.

La sua sustantia e neruosa e pelliculosa acio che se possa dillatare a continere la creatura : et e molto spessa e grossa.

Le parte exteriore de la matrice sono queste cio e li lati difuori aliquali sono alligati li testiculi e anche sono li uasi seminarii e [folio 39 verso] le sue corne di liquali tuti habiamo dicto : et el suo collo del quale lextremita se chiama uulua : e questo collo e longo quanto e uno palmo como e la uirga de lhuomo et e lato e dillatabile : et impero pelliculoso : Et ha le rughe o uero crespe in modo de sangue sughe acio che la uirga de lhuomo nela confricatione del cohito se le induca tintalatione e consequenter dolceza : Et ne lextremita di questa uulua sono doe pellicole che se lieuano e deprimeno sopra el dicto oreficio acio che prohibiscano lo introito de laiere o di qualche cosa extrinsecha nel collo de la matrice o uero uesica como la uirga de lhuomo e custodita da la pellicula del preputio.

E la bocha de la matrice e molto neruosa facta in modo de una bocha de uno cagnolo nouamente nato o uero meglio a modo de una tench uechia : et e ualata de uno uele suttile ne le uergene e ne le uiolate se rompe et impero se sanguina.

Facto e adoncha questo membro da la natura per la conceptione : et ne lhomo fu facto anche acio che mondificasse tuto el corpo de la femina dal superfluo sangue indigesto el quale se genera in essa per la sua frigidita, e nel maschio non e cusi : ma li altri animali non hano questo fluxo mestruale per che tale superfluita che se genera in loro se conuerte in pelle in pili in unghie in rostri e penne e simili membri di quali lhuomo e priuato.

[folio 40 recto] *Capitulum secundum de anothomia uasorum spermaticorum et testiculorum in viris seu masculis*

Dicto di uasi spermatici e testiculi de le femine diciamo di quilli di li maschii : Onde e da sapere che li uasi spermatici sono de doe manerie, alcuni sono uasi che preparano el sperma e quisti descendeno da luochi predicti ali testiculi & circa la parte superiore de essi se inuolgeno intanto che fano in modo de uno sacho o uero de una bursa e questi non intrano la sustantia dili testiculi e questi sono uenosi e neruosi

Alcuni altri uasi sono dilatorii liquali portano el sperma preparato ne li altri uasi dicti a li testiculi e questi se continuano cum li predicti et sono piu neruosi : e quanto uano piu ascendendo da li testiculi sono tanto piu neruosi et ascendeno per insino a losso del petenechio : et alhora se profondano dentro apresso el collo de la uesica e finaliter procedeno al meato de la uirga nel luocho che e nel bucho de losso del petenechio e per doe meati che sono iue mandano el sperma fora da li testiculi el quale fu preparato prima negli altri uasi e mandano quello sperma nel canale de la uirga e poi la uirga el manda fuori.

Et li testiculi ne lhuomo maschio sono di fuora e non detro como e ne le femine onde li uasi spermatici del maschio non sono terminati dentro dal mirach o uero dentro dal corpo ma escono fuora e se copulano a li [folio 40 verso] testiculi como a doi suspensorii o uero contrapeso, Et quisti uasi sono cooperti & uelati de uno panniculo chiamato didimo el quale nasce del paniculo siphach del quale poi noi diremo, e questo didimo se ha uno oreficio chiuso ne la fine de dicti uasi et in processo se dillata e tanto procede dilatandosi che infine di quello se dillatta ala quantita de li testiculi et iue fa una bursa la quale se chiama borsa di testiculi : onde appare che questo didimo fu facto a continere e custodire li testiculi : et li uasi spermatici che peruengono ad essi.

Et in questa borsa glie sono posti doi testiculi facti de sustantia glandosa rotondi facti secondo li medici a generare e produre el sperma per che benche el sia preparato ne li uasi spermatici tamen non recceue in essi la debita forma specifica ma da li testiculi. Et secondo el philosopho Aristotile el sperma perfectamente se produce ne li uasi spermatici e che li testiculi furno facti como doi contrapesi a retinere i uasi aper ne la proiectione del sperma.

Capitulum tertium de anothomia uirgae et de musculis ani : & de quinque uenis emoroydalibus.

Ultimo e la uirga continuata cum lo collo de la uesica carnoso e e continuata cum esso cum molti ligamenti e corde lequale nascono da losso del petenechio insieme cum certi nerui [folio 41 recto] che nascono da la nucha : et impero questo membro e molto sensibile et extensible ; Et anche e continuata la uirga cum gran uene che nascono dal ramo de la uena che descende ale parte inferiore et similiter e continuata cum grande artharie lequale nascono da quella artharia laquale se bifurcha ale doe anche : onde a la lingua et ala uirga uengono magiore uene et arthariae che a nesuno altro membro a tanto pertanto : Et impero queste uene & artharie nel luocho chiamato peritoneon cio e tra loreficio del culo et el luocho di testiculi sono inuolute e sono molto grande : et iue e el principio de la uirga : Et per questo la uirga e tuta cauernosa e le sue cauernosita se reimpino de uentuosita laquale se genera in quelle artharie et alhora se driza la uirga : Onde se tu scindi per lo longo la uirga insino al suo canale et apparerano dui buchi predicti et etiam le sue cauernosita.

La quantita de la uirga o uero longheza sie duno palmo como e quello del collo de la matrice.

La sustantia de la uirga sie neruosa excepto la extremita sua che se chiama preputio.

Da poi a lextremita delo intestino recto chiamato anus tu trouerai certi musculi che apreno & asera o quello oreficio et similiter ne lextremita del dicto oreficio li sono cinque uene terminato ad esso chiamate uene [folio 41 verso] emoroydale per lequale in alcuni homini a certi tempi esce di molto sangue.

Capitulum quartum de anothomia mirach : quod est domicilium predictorum duorum membrorum principalium.

Dapoi che noi habiamo ueduto de doi membri principali uno che serue al nutrimento di li membri a conseruare el corpo e laltro a conseruare la spetie : et anche de li suoi ministri resta a uedere del suo domicilio el quale e comune a tuti quilli el quale se chiama mirach.

Questo mirach o uero questo domicilio si e composto de cinque parte cio e cute pinguedine uno certo panniculo carnoso e certi musculi cum le sue corde et el siphac : de tute queste cinque parte se constituisse uno cooperculo et una casa ne laquale se contengono li membri predicti.

E questo tale domicilio fu posto di sotto da li altri per la ignobilita di membri che se contengono in esso : Onde contiene alcuni membri deputati a purgare le feçe e le superfluita lequale essende graue descendeno a le parte inferiore.

Questo domicilio non potette essere ossuoso ma fu carnoso et pelliculoso acio che secondo li bisogni se potesse dillatare et intumescere como ne la femina pregnante o uero in colui che ha pigliato troppo cibo o uero ne lo ydropico o per qualche altra cagione bisognasse infiare el uentre, sel fusse ossuoso non se potria fare questo.

[folio 42 recto] La prima parte di questo mirach si e la cute de fuora circa laquale sono da considerare piu luochi : Uno si e corespondente ala bocha del stomaco che una cartiligine che copre quello e chiamasi pomo granato como e stato dicto.

Laltro luocho si e la parte che e sopra el stomaco sopra de lombelico circa a quatro dita.

El terzo luocho si e la parte umbelicale cio e doue e lombellico cum el quale sta alligata la creatura nela matrice cum le uene de essa matrice : et impero ne le parte interiore de lombelico appare una certa uena che se continua cum esso, et passa per el gibo del figato e per questa uena se porta el sangue da le uene de la matrice al figato de la creatura et inquesto modo se nutrisse nel uentre de la matre : Ma questa uena quando lhuomo e nato se priua di sangue per che mancha la sua operatione quale facea alhora : Et impero continuamente se ua diminuendo quella uena, onde ne li uechii appare molte minore che ne li gioueni : Et similiter cum questa uena descende una certa artharia a lombelico de la creatura laquale quando e ne lombelico descende gioso e uasene a lartharia adorthi apresso li spondili de le rene et di li fianchi e questa artharia simelmente se ua deleguando e continue appare minore como e stato dicto de la predicta uena, E questa artharia tu uederai exscarnando apresso lombelico et apparerati in forma de uno neruo o de una corda [folio 42 verso] El quarto luocho se chiama sumen, di sotto da lo imbilico quatro dita et e una parte ne laquale se terminano alcune uene ala cute per le quale la creatura nel uentre de la matre manda fuora le sue aquosita : e queste uene e questa tale parte si e piu manifesta ne li puti che non sono nati che ne li perfecti perche essendo queste uene frustrate da la sua operatione se uadeno anullande.

El quinto luocho si e el petenechio doue sono li membri genitali.

Da poi anche tu hai a considerare le parte laterale cioe li li fianchi e li ypocondrii uno da la parte dritto sotto el quale sta el figato e laltro da la parte mancho doue e locata la milza.

Dapoi la cute apparerati incontinenti la pinguedine la quale e molto piu grande nel porcho che ne lhuomo.

Dapoi et tertio te apparera uno panniculo el quale e composto de carne e nerui.

Quarto di sotto a questo panniculo li sono etiam octo musculi di liquali doi sono longitudinali che protendeno per el longo dal clipeo de la bocha del stomaco insino a lossa del petenechio, e quisti musculi non hano gran corde senon ligamentale, Quatro altri sono transuersali dui superiori e dui inferiori: Li superiori nascono da le parte di sopra a presso le coste et terminano a certe corde circa le ossa del petenechio inquesto modo che la corda [folio 43 recto] dritta ua alingioso al musculo che uiene da la parte sinistra, et la corda stancha ua gioso al musculo che uiene da la parte dritta : Onde le corde se incrociano ne la parte inferiore. Li altri dui musculi transuersali sono inferiori per che comentiano da le ossa del petenechio et de le anche e se terminano a certe corde in questo medesimo modo che la corda dritta ua al musculo sinistro e la sinistra ua al musculo dritto, e le corde se incrociano como e stato dicto.

Doi altri sono latitudinali cusi dicti per che li fili di liquali se componeno protendendeno secondo el lato : Et uno di quisti musculi e dal lato dritto e laltro del lato stancho, e sono piu manifesti et anche la sua origine apresso de la schina uerso la parte superiore : e quisti musculi latitudinali insieme cum li longitudinali se intersecano ne li anguli dritti.

La utilita di quisti musculi sie prima acio che deffendeseno li membri interiori da li nocumenti extrinseci, et anche che li riscaldaseno reuerberando la loro calidita a le parte dentro La seconda e acio che aiutino ad expellere le superfluita dal pecto e le superfluita dele feze et etiam ad expellere la creatura fuora, e queste sono utilita comune a quisti octo musculi : Ma piu particularmente parlando Li musculi [folio 43 verso] longitudinali sono facti primo ad atrahere, secondo ad expellere, onde expelleno contrahendo li suoi uili liquali contratti comprimeno glintestini uerso el diafragma como se fosseno tra doe mane che li comprimeseno e per questo modo expelleno fuora le feze : Et per che glintestini hano bisogno maxime di queste doe operatione cio e de atrahere et expellere impero quisti musculi furno grandi.

Ma li musculi latitudinali sono solo facti ad expellere : & impero sono piu apresso glintestini et fano questa expulsione comprimendo la parte da laquale deno expellere: Et per che la expulsione se fa da suso ingioso impero furno locati piu tosto ne le parte superiore che inferiore.

Li transuersali furno facti a retinere e questo fano mediante li suoi uili transuersali, E questo bisogno fare la natura acio che le superfluita gio descese non reascendeseno impero fece li dui transuersali superiori et anche hebbe intentione che le feçe non descendeseno molto ueloce mente anzi se retignisseno tanto che el figato le potesse bene esuccarle como e stato dicto impero fece altri du imusculi transuersali inferiori: liquali sono minori che li superiori per che magiore fu intentione de la natura a fare che le feze non reascendeseno cha che uelocemente non descendeseno.

La quinta parte de questo mirach [folio 44 recto] si e uno panniculo suttilissimo e molto duro chiamato siphac et fu facto acio chel prohibisse che li musculi dicti non comprimeseno i membri naturali e per questo fu neruoso acio che se possa dillatere e constringere quando quilli membri se dillateno e se constringeno E fu sutile acio che quello non li agrauasse. Et fu duro acio che facilmente el non se rompesse per che quando se rompe accade quella passione che si chiama crepatura.

E fu facto etiam questo panniculo acio che el lighe glintestini a la schina et acio che tuti li panniculi de li altri membri interiori che se contengono in esso habiano origine da quello: Et etiam acio chel prohibisca che glintestini non se rompano quando se infiano de uentosita, e perquesto appare la anothomia de tuto el mirach el quale e domicilio de tuti li altri membri gia dicti.

Tractatus quintus de anothomia partium extremarum & ossium Capitulum primum de anothomia ossium et neruorum quae sunt a collo usque ad caudam.

Expediti li quatro membri principali cum li loro ministri e cum li loro domicilii. Vediamo mo la anothomia de le parte extreme cio e braza cum le mano, et de le cosse cum li piedi ma prima uederemo de le ossa nerui e nucha comentiando dal collo per infino a la cauda.

Diciamo adonca che el collo fu facto per el pulmone e per la sua cana ne li animali che respirano: et inquesto collo sono septe [folio 44 verso] ossa chiamati spondili, et sono piu suttili de glialtri inferiori per che sono sustentati da quilli: Et benche siano suttili pur sono molto duri e firmamente congionti acio che non si dislacaseno, et anche che non receuesseno nocumento da le cose extrinsece Et quisti spondili benche siano piu suttili de li altri pur hano el bucho magiore per che la nucha e piu grossa nel collo che in alcuna pare di li altri spondili e questo fu per che iue ha la sua origine.

Dapoi quisti septe spondili li sono altri spondili che se chiamano spondili de le coste e sono dodece secondo el numero de le coste de lequale septe sono uere e cinque mendose.

Da poi sono li spondili de le rene liquali sono cinque, e sono molto grossi e grandi per che sono fondamento e sustentaculo de li altri spondili.

Da poi sono alcuni altri spondili liquali sono ne plichatura che e da la schina a la cauda e sono tri minori di li predicti per che se doueano congiongere cum li spondili de la cauda liquali sono picoli.

Ultimo sono li spondili de la cauda et quiui sono molte differentie de buchi per liquali passano li nerui, e queste tale diuersitade se uedeno meglio nel corpo cotto o uero perfectamente esiccato.

Et in ciascuno spondile e posta la nucha la quale e una medula simile ala substantia del cerebro senon che e piu uiscosa e piu salda et ha [folio 45 recto] origine da esso cerebro, el quale essendo diuise in doe parte cio e ne la parte dritta e ne la parte mancha impero ne la superficie di questa nucha appare uno filo che la diuide per mezo cio e la parte dritta da la parte stancha: E fu facta da la natura acio desse el sentire e el mouere a tuto el corpo dal capo ingio onde la nucha e ditta uicaria del cerebro.

Da la nucha in ciascaduno spondile nasce uno pare de nerui che uanno a dare el sentire e el mouere a certi e uarii membri. E per che li spondili sono intuto trenta impero sono trenta para de nerui secondo el numero di li spondili: et poi dala cauda ne nasce un altro pare de nerui onde sono intuto trentauno pare de nerui oltra quelli sei para ditti disopra che nascono dal cerebro.

Capitulum secundum de anothomia brachiorum et manuum.

Le braze e le mano sono composti de cute pinguedine, carne, uene, corde, ligamenti, ossa.

Tu uederai una uena che penetra per sotto la lasina del brazo e procede per la parte domestica e uasene ala curuatura del brazo et appare ne la parte inferiore de gubito e chiamasi basilica e poi protende piu oltra descendendo gioso a la mano ne la parte siluestra e uasene tra doi digiti cio e el digito picolo chiamato auriculare et el suo proximo digito [folio 45 verso] chiamato annulare, e chiamasi questa uena iue sylen e coresponde ala basilica como suo ramo.

Vederai simelmente un altra uena che uiene per la parte domestica del brazo ne la parte superiore de gubito e chiamasi cephalica per che e uacua del capo et nasce da una uena che ascende al capo e questa uena piu oltra procede uerso la mano e uassene ne la siluestra parte tra e il dito grosso e lindice e chiamasi saluatella e corresponde a la cephalica.

Un altra uena uederai ne la curuatiua del brazo in mezo de le predicte como uno ramo continuato cum tute doe e chiamase uena media o uero uena comune.

Da poi le uene tu uederai di molti musculi e molte corde grande e grosse. Li musculi furno facti a dare el moto uoluntario al quale deserueno etiam esse corde.

Dapoi tu uederai le ossa et comentiando ala spala tu uederai prima uno osso chiamato spatula de simile figura como e una spatula de legno el quale e largo disotto acio che non impedischa el pecto e le coste, et e strecto di sopra acio che cum laltro osso che tu uederai chiamato aiutorio meglio se firme : et impero ne la extremita superiore di questa spatula glie una concauita superficiale rotonda acio che in esse sia situata la extrimita de lo aiutorio rotonda del quale el capo primo e rotondo locato ne la extremita de losso de la spatula poi nel mezo se obliqua uerso la domestica parte acio che nel plicare et [folio 46 recto] amplexare de le cose sia piu habile : Et lo extremo di questo aiutorio ha quasi doe eminentie per che se congionge cum dui ossi chiamati focilli, et in mezo de quelle parte eminente ha piu disopra una certa concauita ne laquale entra lextremita del focille inferiore laquale e facta a modo de uno instrumento da trare laqua acio che sia piu ferma la sua coniunctione et el focille inferiore e piu longo che el superiore per che linferiore sustenta el superiore : Ma tuti dui conuegono in questo che ne li extremi sono piu grossi che nel mezo per che da li extremi loro nascono ligamenti e iuncture, et nel mezo glie sono musculi che supliseno a la loro sutilita : Et el focille superiore non procede dritamente como linferiore acio che sia cagione de plicare la mano et el brazo.

Dapoi questi do focilli glie la resetta de la mano ne la quale sono octo ossi in doe schiere cioe quatro per schiera : Dapoi sono le ossa del pectine de la mano perche e facta ala forma de uno pectine e sono quatro correspondenti a quatro digiti per che al dito grosso non corresponde alcuno osso di questo pectine per che non e in schiera cum li altri digiti.

Dapoi sono le ossa de li cinque digiti, et hano tre ossi per digito che sono intuto quindice : Da poi sono le corde che uano ale iuncture et ultimo la carne laquale e molto piu ne la parte domestica, et da li lati ma pocha ne la parte siluestra per che plicandosi [folio 46 verso] ne la parte domestica non recceuesse lesione per la dureza de li ossi e che non accadesse uacuita alcuna dai lati e da poi li sono le onghie a coprire la cornosita che e ne lextremita de dicti digiti.

Capitulum tertium de anothomia cossarum tibiarum et pedum.

Vediamo mo ultimamente de la anothomia de le cosse, gambe e piedi. Diciamo adoncha che scorticando le cosse tu trouerai doe uene grande che sono ramificate dal troncho de la uena chilis che descende gioso el quale quando e nel fine de li spondili dele rene se diuide in doi rami uno ua ala gamba dritta e laltro ala stancha et similiter se ramifica el troncho de larthoria adorthi che descende: e ciascuno di quisti doi rami in ciascuna gamba se diuide in doi altri rami uno descende per el dritto e per la domestica parte de la gamba et chiamas Saphena per che flobotomata e uacua dai membri naturali e genitali et appare questa uena sopra del genochio e sopra la cauichia del piede, e desotto nel calchagno et appare anche nel pectine del pede.

Un altro ramo se obliqua et intra apresso la iunctura de la scia[1] o del galbue[1] impero e chiamata siatica: ondo per la obliquatione che fa circa queste iuncture flobotomata uale ne le sue passione: Et appare questa uena intuti i predicti [folio 47 recto] luochi como e dicto de la saphena.

Ne la parte siluestra ua escarnando e lieua su li musculi e le corde e uederai prima losso del petenechio sopra del quale sono fabricati li spondili de la schina et consequenter tuto el corpo ne la parte inferiore ha una concauita ne laquale e locata la extremita rotonda laquale extremita se chiama uertebro e nel mezo di quisti doi da la parte dentro lie uno certo ligamento e questa iunctura di questi doi ossi se chiama scia: et impero el dolore che uiene iui se chiama dolore sciatico.

Dapoi tu uederai losso grande de la cossa el quale e magiore de tute le ossa che sono nel corpo per che sustentaculo de tuto el corpo: Et hebbe una grande concauita acio che fusse piu ligiero e che hauesse molta merolla. Et per che potesse meglio sustentare non lo fece dritto la natura ma ne la extremita fecelo pighato uerso la domestica parte: e nel mezo sie plicato e conuexo.

Dapoi questo osso nela iunctura del genochio sono dui ossi dicti focilli de la gamba ma uerso la parte dinanzi di quella iunctura glie uno osso chiamato patella facto in modo de una patella acio che la iunctura fusse piu forte: e questa iunctura e facta de ligamenti como fusse ligata per uno groppo Et el focille che e ne la parte domestica e magiore e piu grosso per che ha piu a sustentar el peso [folio 47 verso] del corpo e quello de la parte siluestra e piu sutile e curto facto solo chel sia apogio del magiore.

Da poi lie losso de la caichia cum el quale se congiongeno li dicti focilli, et e losso del calcagno grosso quadrangulato sotto del quale e una cute grossa e callosa molto.

Da poi e uno osso facto in modo de una u nauicula quadrangulare alquanto longo.

Dapoi e la rasetta del pede composto de tri ossi e non de octo como fu la resetta de la mano perche el pede douea stare firmo e non mouersi a retinire qualche cosa como la mano.

Da poi li e el pectine composto da cinque ossi per che el dito grosso e inschiera cum li altri.

Da poi sono le ossa dili digiti che sono quatuordice cioe dui al dito grosso e tri per ciascaduno de li altri. Da poi sono certi musculi e molte corde a mouere contrahendo e dillatando i digiti et ultimo li sono le onghie che copreno la carnosita de li cime de li digiti como e stato dicto di digiti de le mane. E cussi a laude de dio habiamo compiuto quello che era nostra intentione e quello che dal principio noi prometessimo di narrare.

[1] These four words are very indistinct. The last is half erased and *scia* is written *s^c^ia*.

THE BLESSING OF CRAMP-RINGS

A CHAPTER IN THE HISTORY OF THE TREATMENT OF EPILEPSY

By Raymond Crawfurd

The origin of this ceremony of blessing rings, by the kings and queens of England, for the cure of epilepsy and other spasmodic disorders, appears to be well attested by the evidence of many contemporary records. All alike refer it back to Edward the Confessor, or, to be more exact, to the ring which was one of the sacred relics in the shrine of the Confessor in his abbey of Westminster. Caxton, in the *Golden Legend*,[1] tells the tale of this wonderful ring, as follows:

'When the blessed King Edward had lived many years, and was fallen into great age, it happened he came riding by a church in Essex called Havering, which was at that time in hallowing, and should be dedicated in the honour of our Lord and S. John the Evangelist; wherefore the king for great devotion lighted down and tarried, while the church was in hallowing. And in the time of procession, a fair old man came to the king and demanded of him alms in the worship of God and S. John the Evangelist. Then the king found nothing ready to give, ne his almoner was not present, but he took off the ring from his finger and gave it to the poor man, whom the poor man thanked and departed. And within certain years after, two pilgrims of England went into the holy land to visit holy places there, and as they had lost their way and were gone from their fellowship, and the night approached, and they sorrowed greatly as they that wist not whither to go, and dreaded sore to be perished among wild beasts; at the last they saw a fair company of men arrayed in white clothing, with two lights borne afore them, and behind them there came a fair ancient man with white hair for age. Then these pilgrims thought to follow the light and drew nigh. Then the old man asked them what they were, and of what region, and they answered that they were pilgrims of England, and had lost their fellowship and way also. Then this old man comforted them goodly, and brought them into a fair city where was a fair

[1] Life of St. Edward.

cenacle honestly arrayed with all manner of dainties, and when they had well refreshed them and rested there all night, on the morn this fair old man went with them, and brought them in the right way again. And he was glad to hear them talk of the welfare and holiness of their king S. Edward. And when he should depart from them, then he told them what he was and said: I am John the Evangelist, and say ye unto Edward your king that I greet him right well, by the token that he gave to me this ring with his own hands at the hallowing of my church, which ring ye shall deliver to him again. And say ye to him that he dispose his goods, for within six months he shall be in the joy of heaven with me, where he shall have his reward for his chastity and for his good living. . . . And when he had delivered to them the ring he departed from them suddenly. And soon after they came home and did their message to the king, and delivered to him the ring, and said that S. John the Evangelist sent it to him.'

Shortly after this Edward departed this life, and was laid in his abbey of Westminster, where the usual abundant harvest of miraculous cures was enacted at his shrine. In the above story we have also the explanation of one synonym of epilepsy, the 'morbus sancti Iohannis'.

The further history of the ring may be gleaned from several sources, but notably from a MS. by one Richard Sporley, a monk of the abbey, entitled, 'De fundacione ecclesie Westm', dated A.D. 1450, and now in the British Museum.[1]

St. Edward's ring was deposited with his corpse in the tomb in A.D. 1066. He was translated at midnight of October 13, 1163, when his body was found to be incorrupt. Abbot Lawrence took the robes from the body and made them into three copes, and gave the ring as a sacred relic to the Abbey:

'Dompnus Laurentius quondam abbas huius loci . . . sed et annulo eiusdem (Sancti Edwardi) quem Sancto Iohanni quondam tradidit, quem et ipse de paradiso remisit, elapsis annis duobus et dimidio, postea in nocte translationis de digito regis tulit, et pro miraculo in loco isto custodiri iussit.'

The story of the ring is also depicted in the miniatures of a beautiful illuminated Norman-French MS. Life of St. Edward the King, dating from the thirteenth century, and now in the University Library at Cambridge.[2] The single miniature reproduced here (Plate XXXIX) shows seven blind men, restored to

[1] British Museum MS. Cotton. Claud. A. viii, ff. 32, 33, and *Archaeol. Journal*, London, June, 1864.

[2] MS. Ee. iii. 59.

From CAMBRIDGE UNIVERSITY LIBRARY MS. Ee. iii. 59

PLATE XXXIX. MIRACLES AT THE TOMB OF EDWARD THE CONFESSOR

XIIIth Century

sight, kneeling at the shrine, while a priest reads the *Te Deum*. At the sides of the shrine are figures on pillars of St. John as the palmer (left), and St. Edward with his ring (right). No cure of epilepsy, so-called cramp, is depicted among the many miraculous cures recorded in the MS. The earliest extant records of the use of the ring for this purpose date from the reign of Edward II.

Anstis[1] cites the following entry from the last chapter of the *Constitutions of the Household of Edward II*: 'Item le Roi doit offrer de certein le jour de grant vendredi a crouce. v *s*. queux il est acustumez receivre devers lui a la mene le chapelein afair ent anulx a donner pur medicine az divers gentz': the language, however, of the entry leaves little room to doubt that the custom was already an established one. At his coronation, too, Edward II offered a pound of gold wrought into a figure representing St. Edward holding a ring, and a mark of gold, or eight ounces, worked into the figure of a pilgrim putting forth his hand to receive the ring: and the presumption is that this gold was to be converted into cramp-rings.

We have detailed accounts of the manner of this ceremony of hallowing cramp-rings dating from early Tudor times, and there is sufficient evidence in the brief notices of earlier date to show that the ceremonial observed by the Plantagenet kings was essentially similar. On Good Friday, when the king went to adore the cross, he used to make an offering of money, which was redeemed by a sum of equivalent value: the money so received was converted into rings, which were subsequently hallowed by the king. In Tudor times the hallowing of the rings took place on Good Friday, so that the offering of the money must have been made at some previous time, or this part of the ritual may have actually become obsolete. The change of custom was effected some time between 9 Edward IV (1470–1) and 13 Henry VIII (1521–2), and was probably therefore the work of Henry VII, who, as we know, materially altered the kindred ceremonial of Touching for the Evil.

A MS. copy of the Orders of the King of England's Household, 13 Henry VIII, preserved in the Bibliothèque Nationale at Paris,[2] contains, 'The Order of the Kynge, on Good Friday, touching the cominge to Service, Hallowinge of the Crampe Rings, and Offeringe and Creepinge to the Crosse'. It is quoted

[1] *History of the Garter*, vol. i. [2] MS. 9986.

in extenso in the Northumberland Household Book,[1] and also by Mansell in his *Monumenta Ritualia.*[2] It runs as follows:

'First the king to come to the closett or to the chappell with the lords and noblemen wayting on him, without any sword to bee borne before him on that day, and there to tarry in his travers till the bishop and deane have brought forth the crucifix out of the vestry (the almoner reading the service of the cramp rings) layd upon a cushion before the high altar, and then the huishers shall lay a carpet before y^t^ for ye king to creepe to the crosse upon: and y^t^ done, there shall be a fourme set upon the carpet before the crucifix, and a cushion layd before it for the king to kneele on; and the Master of the jewell house shal be ther ready with the crampe rings in a basin or basins of silver: the king shall kneele upon the sayd cushion before the fourme, and then must the clerke of the closett bee ready with the booke conteyninge ye service of the hallowing of the said rings, and the almoner must kneel upon the right hand of the king, holding of the sayd booke, and when y^t^ is done the king shall rise and go to the high altar, where an huisher must be ready with a cushion to lay for his grace to kneele upon, and the greatest Lord or Lords being then present shall take the basin or basins with the rings and bear them after the king, and then deliver them to the king to offer; and this done the queen shall come down out of her closett or travers into the chappell with ladies and gentlewomen wayters on her, and creepe to the crosse; and that done she shall returne againe into her closett or travers, and then the ladies shall come downe and creepe to the crosse, and when they have done, the Lords and noblemen shall in likewise.'

Creeping to the Cross seems to have been practised in noble households as well as in that of the king. The following entry is found in the Northumberland Household Book[3] (*temp.* Henry VIII):

'Item my Lord useth and accustometh yerely when his Lordship is at home to cause to be delyveride for the Offerings of my Lordis Sone and Heire the Lord Percy upon the said Good Friday When he crepith the Crosse ij*d.* Ande for every of my Yonge Maisters my Lords Yonger Sonnes after j*d.* to every of them for their Offerings when they Crepe the Cross the said Good Friday iiij*d.*'

Many of the entries in the accounts of the Plantagenet kings show that the homage was paid to the Gneyth Cross. This cross was held in great veneration, and, according to tradition, was made of wood from the true Cross presented by a pilgrim to Richard Cœur de Lion: no satisfactory explanation of its name is forth-

[1] p. 36. [2] Vol. iii. [3] p. 334.

coming. It seems to have been transferred from place to place. Under Edward I we find it in the royal chapel of the Priory of Plympton: under Edward II in the royal chapel within the Tower: under Edward III in the private chapel of the royal Manor of Clipstone, and later in the same reign in St. George's Chapel at Windsor, where it was in the time of Henry VII. The purpose of the ceremony is set forth in a Proclamation of February 26, 30 Henry VIII, now in the possession of the Society of Antiquaries: 'On Good Friday it shall be declared howe creepyng of the Crosse signifyeth an humblynge of ourselfe to Christe before the Crosse, and the kyssynge of it a memorie of our redemption made upon the Crosse.' When Convocation, in A.D. 1536, abolished some of the old ceremonies, on the ground that they were superstitious, this of Creeping to the Cross was retained as a laudable and edifying custom.

The following records, taken from the Household Books and Account Rolls of the times, serve to establish the continuity of the ceremonial subsequent to its first mention in the time of Edward II.

In the Eleemosyna Roll of 9 Edward III [1] occurs the following entry:

'In oblacione domini Regis ad crucem de Gneythe die parasceues in capella sua, infra mannerium suum de Clipstone, in precium duorum florencium de Florencia, xiiii die Aprilis, vi*s*. viij*d*., et in denariis quos posuit pro dictis florenciis reassumptis pro annulis medicinalibus inde faciendis, eodem die, vi*s*.: summa xii*s*. viii*d*.' [For the offering of the lord King to the Gneythe Cross on Good Friday in his chapel, in his manor of Clipstone, to the value of two florins, on the 14th day of April, vi*s*. viii*d*., and for the pence bestowed in redemption of the said florins for the making of medicinal rings, on the same day, vi*s*.: total xii*s*. viii*d*.]

Again, in the Eleemosyna Roll of the following year, 10 Edward III: [2]

'In oblacione domini Regis ad crucem de Gneyth in die parasceues apud Eltham, xxix die Marcii, v*s*., et pro iisdem denariis reassumptis pro annulis inde faciendis per manus Iohannis de Crokeford eodem die, x*s*.'

In this entry the name of the almoner is introduced, and the form of the account is abbreviated by omitting repetition of the substituted v*s*.

1 *Gent's. Mag.*, N.S., vol. i; British Museum MS. Cotton Nero C. viii, f. 209.

2 MS. Cotton Nero C. viii, ff. 212, 213b, and *Gent's. Mag.*, N.S., vol. i.

And in 11 Edward III:[1]

'In oblacione domini regis ad crucem de Gneyth in capella sua in pcħo de Wyndesore die parasceues, v*s*., et pro totidem denariis reassumptis pro annulis inde faciendis v*s*.'

Here the sum total is omitted: the three entries, though mutually explanatory, show how puzzling becomes a too strict economy of words.

Entries substantially the same as these may be seen in the Wardrobe Accounts of 12–14 Edward III.[2]

One more entry from the Account Books of John de Ypres, 44 Edward III, is perhaps worth quoting, as it seems to point definitely to the rings being made in this instance of both gold and silver:

'In oblacionibus Regis factis adorando crucem in capella sua infra castrum suum de Wyndesore die parasceues in pretio trium nobilium auri et quinque solidorum sterling xxv*s*.—In denariis solutis pro iisdem oblacionibus reassumptis pro annulis medicinalibus inde faciendis, eodem die xxv.'

The offering of both gold and silver money would seem to bear out the suggestion as to the material of the rings, as we know that in later times both metals were used. It is, of course, arguable that the larger sum of money indicates only a greater demand for the rings.

Richard II's Account Books[3] show that he maintained the practice of his grandfather. The following is from an account of the Controller of the Wardrobe in his reign:

'in denar̄ solut̄ decano capelle Regis pro eisdem oblacionibus reassumpt̄ pro anulis medicinal̄ inde faciendis, xxv*s*.'

The substituted money seems to have been actually laid on the altar, and removed thence to be made into rings: this will explain payment being made in this case to the Dean of the Chapel Royal.

Henry IV could ill afford to dispense with any of the prerogatives of royalty, and we find him offering 25 shillings in the chapel of the palace of Eltham for the making of medicinal rings.[4]

It is no matter for surprise that no mention should be forthcoming of cramp-rings in the reign of Henry V, most of which was

[1] loc. cit. [2] Record Office, Exchqr. Tr. of R., Mis. Book 203, pp. 150–3.

[3] Record Office, Exchqr. Q. R., Accounts 403/10.

[4] British Museum Harleian MS. 319: Household Accounts of Henry IV, 1405–6.

spent beyond the shores of England, and in the propagation rather than in the relief of disease. A passage, however, in the literary remains of Sir John Fortescue [1] taken from a tract entitled *Defensio Iuris Domus Lancastriae,* now to be seen in the Cotton Collection at the British Museum, and referable to the year A. D. 1462, seems to show that the practice had not been allowed to lapse during his memory, which ranged over the reigns of Henry IV, V, and VI. The translated passage runs thus:

'Many duties likewise are incumbent on the Kings of England in virtue of the kingly office, which are inconsistent with a woman's nature, and Kings of England are endowed with certain powers by special grace from heaven, wherewith Queens in the same country are not endowed. The Kings of England at their very anointing receive such an infusion of grace from heaven, that by touch of their anointed hands they cleanse and cure those infected with a certain disease, that is commonly called the King's Evil, though they be pronounced otherwise incurable. Epileptics too, and persons subject to the falling sickness, are cured by means of gold and silver devoutly touched and offered by the sacred anointed hands of the kings of England upon Good Friday, during divine service (according to the ancient custom of the Kings of England); as has been proved by frequent trial of rings, made of the said gold and silver and placed on the fingers of sick persons in many parts of the world. The gift is not bestowed on Queens, as they are not anointed on the hands.'

The passage also brings out the fact that the use of both gold and silver rings had long been customary.

We have abundant evidence of the maintenance of the ceremony under Edward IV in a number of separate entries. Thus in an Eleemosyna Roll of 8 Edward IV is the following: 'Pro eleemosyna in die parasceves c. marc. et pro annulis de auro et argento pro eleemosyna Regis eodem die.' And in a Liber Niger Domus Regis Edwardi IV: 'Item, to the Kynge's offerings to the crosse on Good Friday, out from the counting-house for medycinable rings of gold and silver, delyvered to the jewell house xxv*s*.' And again in a Privy Seal Account of 9 Edward IV: 'Item, paid for the King's Good Fryday rings of gold and silver xxxiii*l*. vi*s*. viii*d*.' Edward IV seems to have aimed at fortifying himself upon the throne by a liberal use of the Royal Gift of Healing, and I have elsewhere expressed my belief, in the absence of any written evidence, that

[1] See R. Crawfurd, *King's Evil,* p. 45, Oxford, 1911.

it was in his reign, and not in that of Henry VII, as commonly believed, that the dole of the angel to those touched by the King for the Evil was instituted. Cramp-rings are mentioned in the Comptroller's Accounts of 20 Henry VII, but the Tudors certainly devoted their healing powers chiefly to sufferers from the Evil.

There is a passage in the *Historia Anglicana* of Polydore Vergil,[1] the Italian, who came to live in England in A.D. 1502, and wrote his history during the reigns of Henry VII and VIII, which shows the nature of the patients for whom these sacred rings were used.

'Iste annulus in eodem templo (scil. Westmonasterii), multâ veneratione perdiu est servatus, quod salutaris esset membris stupentibus valeretque adversus comitialem morbum, cum tangeretur ab illis, qui eiusmodi tentarentur morbis. Hinc natum, ut reges postea Angliae consueverint in die Parasceues, multâ coeremoniâ sacrare annulos, quos qui induunt, hisce in morbis omnino nunquam sunt.'

Besides true epileptics, they were used for those who had palsied limbs: this is interesting as suggesting the inclusion of Jacksonian epilepsy, and perhaps hemiplegia, and the resulting contractures in these conditions may have contributed to the confusion with contractures from other causes, such as chronic rheumatism. We have to bridge over in some such way the gap between their conception of 'cramp' and ours.

In the will of John Baret of Bury St. Edmunds,[2] dated 1463, is a bequest to 'my lady Walgrave' of a 'rowund ryng of the Kynges silver'; and also to 'Thomais Brews, esquiyer, my crampe ryng with blak innamel, and a part silvir and gilt.' And in 1535 Edmund Lee bequeaths to 'my nece Thwartow my gold ryng w^t^ a turkes, and a crampe ryng of gold w^t^ all.'

There are even earlier bequests than this of healing-rings,[3] but not specifically termed cramp-rings: they are simply spoken of as 'vertuosi'. Thus Thomas de Hoton, rector of Kyrkebymisperton, in 1351, bequeathed to his chaplain 'j. zonam de serico, j. bonam bursam, j. firmaculum, et j. anulum vertuosum. Item, domino Thome de Bouthum j. par de bedes de corall, j. anulum vertuosum.' Talismanic rings, inscribed with the names of the three Magi, Caspar, Melchior, Balthazar, were used as preservatives from epilepsy in Plantagenet times.

[1] Lib. i, chap. 8. [2] *Bury Wills*, p. 35, Camden Soc., ed. Tymms.

[3] *Archaeol. Journal*, London, vol. iv, p. 78.

The royal cramp-rings enjoyed no monopoly in the cure of epilepsy, as is shown by an extract from a medical treatise written in the fourteenth century: [1]

'For the Crampe. Tak and ger gedine on Gude Friday, at fyfe parriche kirkes, fife of the first penyes that is offerd at the crosse, of ilk a kirk the first penye: than tak them al and ga before the crosse and say V. pater nosters in the worschip of fife wondes, and bere thaim on the V. dais, and say ilk a day als mekyl on the same wyse: and then gar mak a ryng thar of with owten alay of other metel, and writ with in Jasper, Batasar, Altrapa, and writ with outen Ih' c. nazarenus; and sithen tak it fra the goldsmyth upon a Fridai, and say V. pater nosters als thu did before and use it alway afterward.'

The 'fife wondes' are, of course, the five wounds of the crucified Jesus.

A silver ring, made of five sixpences contributed by five different bachelors, conveyed by a bachelor to the hand of a smith that was also a bachelor, was another reputed remedy for epilepsy; and its virtue was enhanced, if none of the bachelors knew for what purpose or to whom it was given.[2]

In Berkshire, rings made from a piece of silver collected at the Communion found favour, and they were more efficacious if collected on Easter Sunday. Devonshire preferred a ring made of three nails or screws that had been used to fasten a coffin, and that had been dug out of a churchyard.[3]

Cramp-rings hallowed by the King of England enjoyed repute beyond the shores of England.[4] Lord Berners, the translator of Froissart, when ambassador to Charles V, writing to 'my Lorde Cardinall's grace from Saragoza, the xxi daie of June, 1510', says: 'If your grace remember me with some crampe rynges ye shall do a thynge muche looked for, and I trust to bestow thaym well, with Godd's grace, who evermor preserve and encrease your moste reverent astate.' Among various charms that Charles V carried about with him were 'gold rings from England against cramp'.[5]

In A.D. 1518 we find the President of the College of Physicians lending his patronage to the royal cramp-rings. In a letter to the Parisian scholar, Guillaume Budé,[6] Thomas Linacre writes

[1] British Museum MS. Arundel, fol. 23b, and *Gent's. Mag.*, N.S., vol. i, p. 49.

[2] *Gent's. Mag.*, 1794.

[3] Brand, *Pop. Antiq.*, ii. 598.

[4] *Gent's. Mag.*, N.S., vol. i, p. 49.

[5] *Cloister Life of Charles V*, p. 109.

[6] Brewer, *State Papers; Budaei Epistolae*, June 10, 1518, 4223.

that he 'has sent him some rings consecrated by the King as a charm against Spasms': and on July 10, 1518, Budé replies to him from Paris that he has 'received his letter with the rings on July 6', and has distributed among the wives of his relatives and friends the eighteen rings of silver and one of gold he received from Linacre, telling them that they were amulets against slander and calumny.

Even the hard-headed Scot was not proof against the magnetism of the royal rings. A letter from Dr. Thomas Magnus, Warden of Sibthorpe College, Nottinghamshire, to Cardinal Wolsey,[1] written in A.D. 1526 says:

'Pleas it your Grace to write that M. Wiat of his goodnes sent unto me for a present certaine cramp ringges, which I distributed and gave to sondery myne acquaintaunce at Edinburghe, amonges other to M. Adame Otterbourne, who, with oone of thayme, releved a mann lying in the falling sekenes, in the sight of myche people: sethenne which tyme many requestes have been made unto me for cramp ringges, at my departing there, and also sethenne my comyng from thennes. May it pleas your Grace therefore to show your gracious pleasure to the said M. Wyat, that some ringges may be kept and sent into Scottelande; whiche after my poore oppyniyoun shulde be a good dede, remembering the power and operacion of thaym is knowne and proved in Edinburgh, and that they be gretly required for the same cause both by grete personnages and other.'

When Bishop Gardiner was in Rome in A.D. 1529, Anne Boleyn wrote him the following letter:[2]

Master Stephyns,

I thank you for my letter, wherein I perceive the willing and faithful mind that you have to do me pleasure, not doubting, but as much as is possible for man's wit to imagine, you will do. I pray God to send you well to speed in all your matters, so that you would put me to the study, how to reward your high service: I do trust in God you shall not repent it, and that the end of this journey shall be more pleasant to me than your first, for that was but a rejoicing hope, which causing the like of it, does put me to the more pain, and they that are partakers with me, as you do know: and therefore I do trust that this hard beginning shall make the better ending.

Master Stephyns, I send you here cramp-rings for you and Master Gregory, and Mr. Peter, praying you to distribute them as you think best. And have me kindly recommended to them both,

[1] *Gent's. Mag.*, loc. cit., and British Museum MS. Cotton Calig. B. ii, fol. 112.

[2] Burnet, *Hist. of Reformation*, part ii, book ii, record 24.

as she that you may assure them, will be glad to do them any pleasure, which shall be in my power. And thus I make an end, praying God send you good health.

Written at Grenwiche, the 4th day of April,

By your assured friend,

ANNE BOLEYN.

[To Master Stephyns this be delivered.]

Burnet [1] refers to this letter, as follows:

'When he [Gardiner] went to Rome, in the year 1529, Anne Boleyn writ a very kind letter to him, which I have put in the Collection (Records No. 24). By it, the reader will clearly perceive that he was then in the secret of the King's designing to marry her as soon as the divorce was obtained. There is another particular in that letter, which corrects a conjecture which I had set down in the beginning of the former book concerning the cramp rings that were blessed by King Henry, which I thought might have been done by him after he was declared head of the Church.[2] That part was printed before I saw this letter: but this letter shows they were used to be blessed before the separation from Rome: for Anne Boleyn sent them as great presents thither. This use of them had been (it seems) discontinued in King Edward's time: but now, under Queen Mary, it was designed to be revived, and the office for it was written out in a fair MS. yet extant, of which I have put a copy in the Collection (No. 25). But the silence in the writers of that time makes me think it was seldom if ever practised.'

Queen Mary's Manual, of which we shall have more to say later, seems to have been the source from which Burnet transcribed the Office. In his time it was in the library of R. Smith, titular Bishop of Chalcedon.

Numerous allusions in the records of the De Lisle family bear testimony to the popularity of cramp-rings in the reign of Henry VIII.[3] Edward Seymour, Earl of Hertford and afterwards Duke of Somerset, writes to Lady Lisle, in 1537:

'Hussey told me you were very desirous to have some cramp-rings against the time that you should be brought a bedd. . . . I send by the present messenger 18 cramp-rings, which you should have had long ago.' [4]

John Husee writes from London on April 17, 1535, to his mistress, Lady Lisle: 'I send you by Mr. Degory Gramefilld 59 cramp rings

[1] *Hist. of Reformation*, part ii, book ii.

[2] A.D. 1534.

[3] *Lisle Papers* and *Notes and Queries*, 5th series, vol. ix, p. 514.

[4] *Lisle Papers*, xi. 15.

of silver, that Christofer Morys giveth you, and one of gold';[1] and again, on May 2, 1538: 'Cramp-rings I can get none out of the jewel-house. Mr. Wyll^ms says the King had the most part of gold, but has promised me twelve silver.'[2]

In a letter of May 13, 1536, John Husee combines denunciation of Anne Boleyn with a promise of cramp-rings to Lady Lisle:

'Madam, I think verily that if all the books and cronycles were totally revolved and to the uttermost proscuted and tried, which against wymen hath been pennyd, contryvyd, and wryten, syns Adam and Eve, these same were I think verily nothing in comparison with that which hath been done and committed by Anne the Queen. . . . I think not the contrary but she and all they shall suffre. John Williams hath promised me some cramp rings for your Ladyship';[3]

and again, six days after:

'Your ladyship shall receive of this berer 9 cramp-rings of silver. John Williams says he never had so few of gold as this year. The king had the most part himself: but next year he will make you amends.'[4]

This day, May 19, 1536, was the day of Anne Boleyn's execution.

Margaret Mylynton, in 1516, bequeaths to 'my dame Croche my best gown and a kercheve, and my cramp-ring'.[5] There is nothing, however, to show that it had received the royal benediction.

Andrew Boorde, in his *Introduction of Knowledge*, says, 'the kynges of Englande doth halowe every yere crampe rynges, ye which rynges worne on ones fynger doth help them whych hath the crampe'; and again, in his *Breviarie of Health*, published in 1547, but written during the lifetime of Henry VIII: 'The kynges majesty hath a great helpe in this matter, in hallowing crampe rynges, and so given without money or petition.' Boorde was medical attendant to Thomas, eighth Duke of Norfolk, Lord President of the Council and uncle of Anne Boleyn, and by him was recommended to the notice of Henry VIII, who employed him much in State business, but not, so far as is known, in a medical capacity. His testimony therefore is peculiarly reliable, and shows

[1] *Lisle Papers*, xi. 111.
[2] *Ibid.*, xii. 43.
[3] *Ibid.*, xii. 58.
[4] *Ibid.*, xii. 60.
[5] Registry of Wills, Archdeaconry of Norwich.

that Henry VIII maintained the ceremony throughout his reign, as is borne out by the scattered references we have adduced from other contemporary sources.

In 1547, after the death of Henry VIII, Gardiner sent a letter to Ridley, which contains the following passage:

'The late king used to bless cramp rings both of gold and silver, which were much esteemed everywhere, and when he was abroad they were often desired from him. The gift he hoped the young king would not neglect. He believed the invocation of the name of God might give such a virtue to holy water as well as to the water of baptism,' and further he speaks of the rings as endued 'by the special gift of curation ministered to the king of this realm'.[1]

That Edward VI did not relinquish the practice of blessing cramp-rings, as has been supposed, and as Burnet submits, is conclusively proved by an entry in the Household Accounts of the year 1553, before his death. Under the heading 'Oblations' is 25 shillings for the redemption of rings commonly called medicine rings, to be made of gold and silver.[2]

It was little likely that Mary would allow a Catholic ceremonial to lapse for want of royal patronage. In the Appendix to *Illustrations of the manners and expences of antient times in England, in the 15th, 16th, and 17th centuries, deduced from the accompts of churchwardens and other authentic documents, London, 1797, 4to*, printed in the same year, is a list of the New Year's gifts presented by Queen Mary in 1556, among which we find:

'Item, deliuerid by the queins commandement—to the said Robert Raynes, in broken golde, to make crampe rings, etc. Item, more deliuerid the same time, to make cramp ringes, in broke plate of silu' theise parcelles,' &c.

But there is further the evidence of the actual existence of Queen Mary's Illuminated MS. Manual, in the Library of the Roman Catholic Cathedral of Westminster, giving the Office of the Blessing of Cramp-rings in Latin, with rubrics in English showing it to be the form made use of by herself. It also contains a miniature painting of Queen Mary performing the service of consecration. The whole office is transcribed below. A full description of the Manual will be found in the *Proceedings of the Society of Anti-*

[1] Burnet, *Hist. of Reformation*, part ii, book i, § 12.

[2] British Museum Additional MS. 35184, Household Account, 1553.

quaries,[1] at a meeting of whom it was shown and described by Sir Henry Ellis. Sparrow Simpson has also described it in the *Journal of the Archaeological Association*, 1871. It is a 'small quarto volume, eight and a half inches in height by six and three-eighths in width'. Cardinal Wiseman, to whom it formerly belonged, has written on the fly-leaf, 'Queen Mary's manual for blessing cramp-rings and touching for the Evil. Bound 1850.' The cover is spangled with roses and fleurs-de-lis, together with the Queen's monogram MR. 'The volume consists of nineteen leaves of vellum, each surrounded with a rich border, and filled either with miniatures or with the two offices which it comprises. Then follow four ruled leaves and fifteen plain leaves without manuscript. . . . On the recto of leaf 1 the royal arms of Philip and Mary are emblazoned, surrounded by a garter and surmounted by a crown. A rich border containing the rose, the fleur-de-lis, and the pomegranate, together with a shield bearing the cross of St. George, completes the decorations of the page.' The red and white roses represent Queen Mary's double title to the throne of England as the heiress of the houses of Lancaster and York, the fleur-de-lis her claim to the throne of France, and the pomegranate of Granada her descent from Ferdinand and Isabella. The Cross of St. George is derived from the shield of the Order of the Garter. 'On the verso of this leaf is an illumination (Plate XL) representing the interior of a chapel with an altar furnished with curtains, candle-sticks, and crucifix. At a prayer-desk before the altar kneels the Queen; before her is an open book, and on either side two golden basins containing cramp-rings.' Leaves 2 to 10 contain 'certayn prayo's to be vsed by the quenes heighnes in the consecration of the crampe ryngs'. A study of the rubrics, which are in English, suffices 'to show the essentials for the consecration of the rings: the prayers, the royal touch, the holy water. . . . The recto of leaf 11 is filled with an illumination of the Crucifixion with St. Mary and St. John. In the border are the instruments of the Passion—the spear, the reed and sponge, the hammer and pincers, three nails, two scourges, and (a very unusual addition) a centre-bit of the same form as that now in use. On the verso of this leaf is a very interesting full-page illumination. At a prayer-desk, on which is an open book, kneels the Queen, turning to the right (the dexter side of the picture), wearing the head-dress familiar to us in all her portraits.

[1] Series i, vol. ii, p. 292.

PLATE XL. QUEEN MARY TUDOR BLESSING CRAMP-RINGS
From QUEEN MARY'S MS. MANUAL fo. 1v
Library of the Roman Catholic Cathedral at WESTMINSTER

Before her kneels a sufferer, apparently a young man, whose bare and swollen neck the Queen holds between her two hands. Behind him, holding open the collar of the patient's coat, kneels the "clarke of the closett" in a cassock and gown, and with a tonsured head. On the left of the prayer-desk stands "the chaplen", a bald-headed, venerable man in a long cassock, a somewhat short surplice with full sleeves, and the "stole abowte his neck" ordered in the rubric, reading the appointed office. The Queen wears a brown dress cut square at the neck, white sleeves, and a lace ruff and waist-bands. The office for the healing follows, commencing on folio 12*a*, and ending on folio 19*a*.

'The rubrics are in red ink, bright and fresh; and each page has a rich border of scrolls, leaves, flowers, and fruit, with occasional figures of children, &c. I enumerate the most important subjects. Folio 1*b*, David with head of Goliath, St. George and the Dragon, and a child with a skull; folio 2*b*, arms of the city of London; folio 3*a*, VERITAS TEMPORIS FILIA (the Queen's favourite motto), with a sword and sceptre; folios 3*b* and 4*a*, large terminal figures with grapes; folio 4*a*, arms of France and England quarterly; folio 4*b*, DÑS MIHI ADIVTOR; folios 5*a* and *b*, portcullis and rose; folios 6*a* and *b*, PACIENTIA and PRVDĒTIA, with allegorical figures; folios 7*a* and *b*, CHARITAS and IVSTICIA; folios 8*a* and *b*, FIDES and SPES; folios 9*a* and *b*, FORTITVDO and TEMPERANCIA.'

With the death of Mary, the ceremonial seems finally to have fallen into disuse. There is, however, a passage in the *Historia Anglicana Ecclesiastica* of Nicholas Harpsfield,[1] which was written entirely in the reign of Elizabeth, which seems to throw some doubt on the point. The words are as follows:

'Quin et annulus ille, de quo diximus, magna in Westmonasteriensi Londini coenobio postea reverentia reservatus, adversus comitialem morbum multis profuit: indeque etiam ortum, ut ad sacram parasceuen Reges Angliae certos annulos statis quibusdam precibus et caerimoniis consecrare consueverint, adversus eundem morbum salutares. Quae consuetudo et ad nostra usque tempora perducta est, multique huiusmodi annulorum beneficium, nostra etiam aetate, senserunt.' [And further the above-mentioned ring was reverently preserved afterwards in the monastery of Westminster in London, and relieved many of epilepsy. That too was the origin of the custom of the Kings of England on Good Friday consecrating certain rings with set prayers and ceremonies, for the

[1] Ed. 1622, p. 219.

cure of the same disease. Which custom has persisted even down to our own times, and many even in our own lifetimes have derived benefit from rings of this kind.]

Nicholas Harpsfield, though he did not write till the reign of Elizabeth, was born as early as A.D. 1519, so that his words are consistent with discontinuance of the ceremony after the time of Queen Mary.

It remains to consider what diseased states were embraced by the term 'cramp'. Epilepsy, convulsions, and rheumatism certainly. All these terms have in common the idea of muscular contraction or spasm, and their relation in usage to one another may be represented graphically as under:

Convulsion ⟍
Epilepsy ⟋ Cramp = Rheumatism.

Confusion of these terms is far more marked in medical than in lay writers; but at the same time there is little doubt that the conservative sentiment inspired by the royal ceremonial kept the term 'cramp' alive in a sense that was all but obsolete in the common diction.

Chaucer applies 'crampe' to muscular spasm:

> But wel he felte about his herte crepe . . .
> The crampe of death, to streyne him by the herte.[1]

Linacre, as we have seen, speaks of cramp-rings, in 1518, as a charm against *spasms*, while about the same year Polydore Vergil speaks of the royal cramp-rings as a cure for the *morbus comitialis*. Each of these two writers clearly indicates epilepsy. In 1526 Magnus speaks definitely of cramp-rings as relieving a man lying in the *falling-sickness*, a term habitually applied to epilepsy. Nicholas Harpsfield too, writing in the middle of the reign of Elizabeth, speaks of cramp-rings blessed by the kings as remedies for the *morbus comitialis*. In all probability royal cramp-rings were used for epilepsy and epilepsy only, but it is quite possible, and I am inclined to think probable, that other cramp-rings had a less exclusive use.

Bacon's description of cramp in his *Natural and Experimental History* is fairly explicit and obviously does not embrace epilepsy: 'The cramp cometh of contraction of sinews, which is manifest, in that it comes either by cold or dryness.'

[1] *Troilus*, book iii, 1069.

Shakespeare recognizes both epilepsy and rheumatism as entities apart from cramp. Epilepsy he seems to associate more with falling than with convulsion: thus, of the fit that attacked Caesar when the crown was offered to him, he writes:

Casca. He fell down in the market-place, and foamed at mouth, and was speechless.
Brutus. 'Tis very like: he hath the falling-sickness.[1]

'Cramp' is used by Shakespeare for muscular spasms or contractures, and he links the term on the one side to rheumatism, and on the other to convulsions, in the following passages:

> For this, be sure, to-night thou shalt have cramps,
> Side-stitches that shall pen thy breath up.[2]

Parolles says:

'In a retreat he outruns any lackey: marry, in coming on he has the cramp.'[3]

Prospero says:

> Go, charge my goblins that they grind their joints
> With dry convulsions; shorten up their sinews
> With aged cramps.[4]

and

'Leander . . . went but forth to wash him in the Hellespont, and being taken with the cramp was drowned.'[5]

Robert Bayford, in his *Enchiridion Medicum* published in 1655, includes both wry-neck and convulsions under the heading cramp, but he treats epilepsy separately on the ground that, as we know to be the case, it is not always associated with convulsions. He has no word 'rheumatism' at all.

Pepys (1664) carried about with him a hare's foot as a charm against *colic*, i. e. against muscular spasm. Among Indians, Norwegians, and Central Africans, the foot of an elk was a charm against epilepsy. Pepys also recites a charm against cramp:

> Cramp, be thou faintless
> As our Lady was sinless
> When she bare Jesus.

In this charm the word cramp seems to refer to the painful *muscular spasms* of labour. Pepys, as we know, suffered from colic, but not

[1] *Julius Caesar*, I. ii.
[2] *Tempest*, I. ii.
[3] *All's Well that Ends Well*, IV. iii.
[4] *Tempest*, IV. i.
[5] *As You Like It*, IV. i.

from epilepsy, so in using a hare's foot as a charm against colic he was probably employing a charm against epilepsy. In like manner the 'rheumatic ring' of to-day seems to be the lineal descendant of the cramp-ring of aforetime, and the confusion of nomenclature has doubtless not affected its efficacy. Folk-medicine serves rather to confirm than to elucidate the confusion, for in Suffolk moles' feet are carried as a charm against rheumatism, but in Sussex against cramp. In Devonshire a dried frog is worn as a cure for fits.

Boswell, in his description of Johnson at the time of their tour to the Hebrides, uses the word 'cramp' in its earlier significance. 'His head,' he says, 'and sometimes also his body, shook with a kind of motion like the effect of a palsy: he appeared to be frequently disturbed by cramps, or convulsive contractions, of the nature of that distemper called St. Vitus's dance.'

It may be asked how it came about that rings were used in the first instance as a remedy for epilepsy. It has occurred to me that their use may have originated in the time-honoured belief that an epileptic seizure may be aborted by ligature of a limb or part above the situation in which the warning 'aura' commences. Galen, Alexander of Tralles, Rhazes, and Avicenna, among the earlier writers on medicine, all recommend the measure.

THE OFFICE OF CONSECRATING THE CRAMP-RINGS

Certain prayers to be used by the queen's highness, in the consecration of the cramp-rings.

Deus misereatur nostri et benedicat nos Deus, illuminet vultum suum super nos et misereatur nostri.

Ut cognoscamus in terra viam tuam, in omnibus gentibus salutare tuum.

Confiteantur tibi populi Deus, confiteantur tibi populi omnes.

Laetentur et exultent gentes, quoniam iudicas populos in aequitate, et gentes in terra dirigis.

Confiteantur tibi populi Deus, confiteantur tibi populi omnes, terra dedit fructum suum.

Benedicat nos Deus, Deus noster, benedicat nos Deus, et metuent eum omnes fines terrae.

Gloria Patri, et Filio, et Spiritui Sancto.

Sicut erat in principio, et nunc, et semper, et in saecula saeculorum, Amen.

Omnipotens sempiterne Deus, qui ad solatium humani generis, varia ac multiplicia miseriarum nostrarum levamenta uberrimis gratiae tuae donis ab inexhausto benignitatis tuae fonte manantibus incessanter tribuere dignatus es, et quos ad regalis sublimitatis fastigium extulisti, insignioribus gratiis ornatos, donorumque tuorum organa atque canales esse voluisti, ut sicut per te regnant aliisque praesunt, ita te authore reliquis prosint, et tua in populum beneficia conferant : preces nostras propitius respice, et quae tibi vota humillime fundimus, benignus admitte, ut quod a te maiores nostri de tua misericordia sperantes obtinuerunt, id nobis etiam pari fiducia postulantibus concedere digneris. Per Christum Dominum nostrum. Amen.

The rings lying in one bason, or more, this prayer to be said over them.

Deus coelestium terrestriumque conditor creaturarum, atque humani generis benignissime reparator, dator spiritualis gratiae, omniumque benedictionum largitor, immitte Spiritum Sanctum tuum Paracletum de coelis super hos annulos arte fabrili confectos, eosque magna tua potentia ita emundare digneris, ut omni nequitia lividi venenosique serpentis procul expulsa, metallum a te bono conditore creatum, a cunctis inimici sordibus maneat immune. Per Christum Dominum nostrum. Amen.

Benedictio annulorum.

Deus Abraham, Deus Isaac, Deus Iacob, exaudi misericors preces nostras, parce metuentibus, propitiare supplicibus, et mittere digneris sanctum Angelum tuum de coelis qui sanctificet ✠ et benedicat ✠ annulos istos, ut sint remedium salutare omnibus nomen tuum humiliter implorantibus, ac semetipsos pro conscientia delictorum suorum accusantibus, atque ante conspectum divinae clementiae tuae facinora sua deplorantibus, et serenissimam pietatem tuam humiliter obnixeque flagitantibus ; prosint denique per invocationem sancti tui nominis omnibus istos gestantibus, ad corporis et animae sanitatem. Per Christum Dominum nostrum. Amen.

Benedictio.

Deus qui in morbis curandis maxima semper potentiae tuae miracula declarasti, quique annulos in Iuda patriarcha fidei arrabonem, in Aarone sacerdotale ornamentum, in Dario fidelis custodiae symbolum, et in hoc regno variorum morborum remedia esse voluisti, hos annulos propitius ✠ benedicere et ✠ sanctificare digneris : ut omnes qui eos gestabunt sint immunes ab omnibus Satanae insidiis, sint armati virtute coelestis defensionis, nec eos infestet vel nervorum contractio, vel comitialis morbi pericula, sed sentiant te opitulante in omni morborum genere levamen. In nomine Patris ✠ et Filii ✠ et Spiritus Sancti ✠ . Amen.

Benedic anima mea Domino : et omnia quae intra me sunt nomini sancto eius. *Here follows the rest of that Psalm.*

Immensam clementiam tuam misericors Deus humiliter imploramus, ut qua animi fiducia et fidei sinceritate, ac certa mentis pietate, ad haec impetranda accedimus, pari etiam devotione gratiae tuae symbola fideles prosequantur : facessat omnis superstitio, procul absit diabolicae fraudis suspicio, et in gloria tui nominis omnia cedant : ut te largitorem bonorum omnium fideles tui intelligant, atque a te uno quicquid vel animis vel corporibus vere prosit, profectum sentiant et profiteantur. Per Christum Dominum nostrum. Amen.

These prayers being said, the queen's highness rubbeth the rings between her hands, saying :

Sanctifica Domine annulos istos, et rore tuae benedictionis benignus asperge, ac manuum nostrarum confricatione, quas, olei sacra infusione externa, sanctificare dignatus es pro ministerii nostri modo, consecra, ut quod natura metalli praestare non possit, gratiae tuae magnitudine efficiatur. Per Christum Dominum nostrum. Amen.

Then must holy water be cast on the rings, saying :

In nomine Patris, et Filii, et Spiritus Sancti. Amen. Domine Fili Dei unigenite, Dei et hominum Mediator, Iesu Christe, in cuius unius nomine salus recte quaeritur, quique in te sperantibus facilem ad Patrem accessum conciliasti, quem quicquid in nomine tuo peteretur, id omne daturum, cum certissimo veritatis oraculo ab ore tuo sancto, quum inter homines versabaris homo pronunciasti, precibus nostris aures tuae pietatis accommoda, ut ad thronum gratiae in tua fiducia accedentes, quod in nomine tuo humiliter postulavimus, id a nobis, te mediante, impetratum fuisse, collatis per te beneficiis, fideles intelligant. Qui vivis et regnas cum Deo patre in unitate Spiritus Sancti Deus, per omnia saecula saeculorum. Amen.

Vota nostra quaesumus Domine, Spiritus Sanctus qui a te procedit, aspirando praeveniat, et prosequatur, ut quod ad salutem fidelium confidenter petimus, gratiae tuae dono efficaciter consequamur. Per Christum Dominum nostrum. Amen.

Maiestatem tuam clementissime Deus, Pater, Filius, et Spiritus Sanctus, suppliciter exoramus, ut quod ad nominis tui sanctificationem piis hic ceremoniis peragitur, ad corporis simul et animae tutelam valeat in terris, et ad uberiorem felicitatis fructum proficiat in coelis.

Qui vivis et regnas Deus, per omnia saecula saeculorum. Amen.

THE CEREMONIES OF BLESSING CRAMP-RINGS, ON GOOD FRIDAY, USED BY THE CATHOLICK KINGS OF ENGLAND

The psalme 'Deus misereatur nostri, etc.', with the 'Gloria Patri'.

May God take pity upon us and blesse us : may he send forth the light of his face upon us, and take pity on us.

That we may know thy ways on earth : among all nations thy salvation.

May people acknowledge thee, O God : may all people acknowledge thee.

Let nations rejoice and be glad, because thou judgest people with equity : and doest guide nations on the earth.

May people acknowledge thee, O God, may all people acknowledge thee : the earth has sent forth her fruit.

May God blesse us, that God who is ours : may that God blesse us : and may all the bounds of the earth feare him.

Glory be to the Father, and to the Son : and to the holy Ghost.

As it was in the beginning, and now, and ever : and for ever and ever. Amen.

Then the king reades this prayer :

Almighty eternal God, who by the most copious gifts of thy grace flowing from the unexhausted fountain of thy bounty, hast been graciously pleased for the comfort of mankind, continually to grant us many and various means to relieve us in our miseries, and art willing to make those the instruments and channels of thy gifts, and to grace those persons with more excellent favours, whom thou hast raised to the royal dignity ; to the end that as by thee they reign and govern others, so by thee they may prove beneficial to them, and bestow thy favours on the people : graciously heare our prayers and favourably receive those vows we powre forth with humility, that thou mayest grant to us, who beg with the same confidence, the favour which our ancestors by their hopes in thy mercy have obtained, through Christ our Lord. Amen.

The rings lying in one bason, or more, this prayer is to be said over them :

O God, the maker of heavenly and earthly creatures, and the most gracious restorer of mankind, the dispenser of spiritual grace, and the origin of all blessings : send downe from heaven thy holy Spirit the Comforter upon these rings, artificially fram'd by the workman, and by thy greate power purify them so, that all the malice of the fowle and venomous serpent be driven out ; and so the metal, which by thee was created, may remaine pure and free from all the dregs of the enemy, through Christ our Lord. Amen.

The blessing of the rings.

O God of Abraham, God of Isaac, God of Jacob, heare mercifully our prayers. Spare those who feare thee. Be propitious to thy suppliants, and graciously be pleased to send downe from heaven thy holy angel : that he may sanctify ✠ and blesse ✠ these rings : to the end they may prove a healthy remedy to such as implore thy name with humility, and accuse themselves of the sins which ly upon their conscience : who deplore their crimes in the sight of thy divine clemency, and beseech with earnestnes and humility thy most serene pity. May they in fine by the invocation of thy holy name become profitable to all such as weare them, for the health of their soule and body, through Christ our Lord. Amen.

A blessing.

O God, who has manifested the greatest wonders of thy power by the cure of diseases, and who were pleased that rings should be a pledge of fidelity in the patriark Judah, a priestly ornament in Aaron, the mark of a faithful guardian in Darius, and in this kingdom a remedy for divers diseases : graciously be pleased to blesse ✠ and sanctify ✠ these rings, to the end that all such as weare them may be free from all snares of the devil, may be defended by the power of celestial armour, and that no contraction of the nerves or any danger of the falling-sickness may infest them, but that in all sort of diseases by thy help they may find relief. In the name of the Father, ✠ and of the Son, ✠ and of the Holy Ghost ✠ . Amen.

Blesse, O my soule, the Lord : and let all things which are within me praise his holy name.

Blesse, O my soule, the Lord : and do not forget all his favours.

He forgives all thy iniquities : he heales all thy infirmities.

He redeemes thy life from ruin : he crownes thee with mercy and commiseration.

He fils thy desires with what is good : thy youth like that of the eagle shal be renewed.

The Lord is he who does mercy : and does justice to those who suffer wrong.

The merciful and pitying Lord : the long sufferer and most mighty merciful.

He will not continue his anger for ever : neither wil he threaten for ever.

He has not dealt with us in proportion to our sins : nor has he rendred unto us according to our offences.

Because according to the distance of heaven from earth : so has he enforced his mercies upon those who feare him.

As far distant as the east is from the west : so far has he divided our offences from us.

After the manner that a father takes pity of his sons; so has the Lord taken pity of those who feare him : because he knows what we are made of.

He remembers that we are but dust : man like hey such are his days : like the flower in the field so wil he fade away.

Because his breath wil passe away through him, and he wil not be able to subsist : and it wil find no longer its owne place.

But the mercy of the Lord is from all eternity : and will be for ever upon those who feare him.

And his justice comes upon the children of their children : to those who keep his wil.

And are mindful of his commandements : to performe them.

The Lord in heaven has prepared himselfe a throne : and his kingdom shall reign over all.

Blesse yee the Lord all yee angels of his, yee who are powerful in strength : who execute his commands, at the hearing of his voice when he speakes.

Blesse yee the Lord all yee vertues of his : yee ministers who execute his wil.

Blesse yee the Lord all yee works of his throughout all places of his dominion : my soule praise thou the Lord.

Glory be to the Father, and to the Son : and to the holy Ghost.

As it was in the beginning, and now, and ever : and for ever and ever. Amen.

Wee humbly implore, O merciful God, thy infinit clemency : that as we come to thee with a confident soule, and sincere faith, and a pious assurance of mind : with the like devotion thy beleevers may follow on these tokens of thy grace. May all superstition be banished hence, far be all suspicion of any diabolical fraud, and to the glory of thy name let all things succeede : to the end thy beleevers may understand thee to be the dispenser of all good ; and may be sensible and publish, that whatsoever is profitable to soul or body, is derived from thee : through Christ our Lord. Amen.

These prayers being said, the king's highnes rubbeth the rings between his hands, saying :

Sanctify, O Lord, these rings, and graciously bedew them with the dew of thy benediction, and consecrate them by the rubbing of our hands, which thou hast been pleased according to our ministery to sanctify by an external effusion of holy oyle upon them ; to the end that what the nature of the mettal is not able to performe, may be wrought by the greatnes of thy grace : through Christ our Lord. Amen.

Then must holy water be cast on the rings, saying :

In the name of the Father, and of the Son, and of the holy Ghost. Amen.

O Lord, the only begotten Son of God, mediatour of God and men, Jesus Christ, in whose name alone salvation is sought for, and to such as hope in thee givest an easy access to thy Father ; who when conversing among men, thyself a man, didst promise by an assured oracle flowing from thy sacred mouth, that thy Father should grant whatever was asked in thy name ; lend a gracious eare of pity to these prayers of ours : to the end that approaching with confidence to the throne of thy grace, the beleevers may find by the benefits conferrd upon them, that by thy mediation we have obteined, what we have most humbly beg'd in thy Name ; who livest and reignest with God the Father, in the unity of the holy Ghost, one God, for ever and ever. Amen.

Wee beseech thee, O Lord, that the Spirit, which proceedes from thee, may prevent and follow on our desires ; to the end that what we beg with confidence for the good of the faithful, we may efficaciously obteine by thy gracious gift : through Christ our Lord. Amen.

O most clement God ; Father, Son, and holy Ghost : wee supplicate and beseech Thee, that what is here performed by pious ceremonies to the sanctifying of thy name, may be prevalent to the defense of our soule and body on earth ; and profitable to a more ample felicity in heaven. Who livest and reignest God, world without end. Amen.

DR. JOHN WEYER AND THE WITCH MANIA

By E. T. Withington

The value of every new truth or discovery is relative, and depends upon the state of ideas or knowledge prevalent at the time. Should it go greatly beyond this, it may lose much in practical effect, like good seed falling on unprepared soil; but the discoverer is no less worthy of praise though he be so far in advance of his fellows that they refuse to accept his teaching, and persecute instead of honouring him. Posterity, however, often ignores former conditions, especially in an era of rapid progress, for the quicker the advance the sooner will the early stages be forgotten, however important and difficult they may have been.

EFFIGIES IOANNIS WIERI ANNO ÆTATIS LX. SALVTIS M.D.LXXVI.

Among those who were so far beyond their age that the truths they proclaimed not only were rejected by the majority but brought them into danger was Dr. John Weyer, the first serious opponent of the witch mania. He stood almost alone. His attack on the witch-hunters, though it marks the turn of the tide, was followed by more than a century of cruelty, injustice, and superstition; yet our ideas on the subject are now so entirely altered that it is hard to imagine the value and danger of the service he performed, and his name was almost forgotten even by members of his own profession, when his biography was published by Dr. K. Binz in 1885.[1]

[1] *Dr. Johann Weyer, der erster Bekämpfer des Hexenwahns*, Bonn, 1885, 2nd ed., Berlin, 1896. Also J. Geffcken, 'Dr. Johann Weyer' in *Monatshefte der Comenius*

Let us try to get some idea of the nature of the witch mania, that we may better appreciate the courage and intelligence of this ancient physician.

In the second half of the fifteenth century a new age began in Western Europe. The revival of Greek, the invention of printing, and the discovery of America gave fresh ideas and new prospects to mankind. But, as the sun's rays were believed to breed serpents in fermenting matter, so amid this ferment of new life and light rose a hideous monster, more terrible than any fabled dragon of romance or superstition of the darkest ages, which for generations satiated itself on the tears and blood of the innocent and helpless. This was the witch mania. For two centuries the majority of theologians and jurists in Western Europe were convinced that vast numbers of their fellow creatures, especially women, were in league with the devil, that they had sexual intercourse with him or his imps, and that he bestowed on them in exchange for their souls the power of injuring their neighbours in person or property. They thought it their duty to search out these witches, to force from them, by the most terrible tortures they could devise, not only confessions of their own guilt, but also denunciations of their associates, and finally to put them to death, preferably by burning. In consequence, many thousands of innocent persons of all ages and ranks, but especially poor women, were judicially murdered, after being first compelled by unspeakable torments to commit moral suicide by declaring themselves guilty of unmentionable crimes, and to involve their dearest friends and relations in a similar fate. There is no sadder scene in the whole tragicomedy of human history.

There had been nothing like it in the darkest of the dark ages, there was nothing like it among the far more ignorant and superstitious adherents of the Eastern Church. The witch mania in its extreme form has been manifested only by the Catholics and Protestants of the sixteenth and seventeenth centuries and by some tribes of African savages.

In early Christian times, witchcraft was recognized as a relic of paganism, but it was not feared. Christ had overcome the powers of darkness, and His true followers need fear no harm from them. A canon of the Church, at least as early as the ninth

Gesellschaft 3, 1904; J. Janssen and L. Pastor, *Geschichte des deutschen Volkes*, 8 vols., Freiburg im Breisgau, 1898–1903, viii. 600 ff.

century, declared that women who thought they rode through the air with Diana or Herodias were only deluded by the devil, and that those who believed human beings could create anything, or change themselves or others into animal forms, were infidels and worse than heathens; and confessors were instructed to inquire into and inflict penance for the belief that witches could enter closed doors, make hail-storms, or kill persons without visible means.[1]

In the enlightened sixteenth century, any one who professed his disbelief that witches could ride through the air, change themselves into cats, or make caterpillars and thunder-storms, would have had an excellent chance of being burnt as a heretic or concealed sorcerer. St. Boniface (680–755) classed belief in witches and werewolves among the works of the devil, and St. Agobard of Lyons (779–840) declared the idea that witches caused hail and thunderstorms to be impious and absurd.[2] The laws of Charlemagne made it murder to put any one to death on charge of witchcraft, and in the eleventh century King Coloman of Hungary asserted briefly, 'Let no one speak of witches, seeing there are none'.[3] Few, indeed, were quite so sceptical as this; still witchcraft was in the Middle Ages looked upon by the educated in a half-contemptuous fashion, and even those who openly professed sorcery frequently escaped with no worse punishment than penance, banishment, or an ecclesiastical scourging.

This may be well illustrated by a story told in the life of the learned Dominican, St. Vincent of Beauvais. An old woman once (1190–1264) came to a priest in his church and demanded money from him, saying she had done him a great service, for that, when she and her companions, who were witches, had entered his bedroom the previous night, she had prevented them from injuring him. 'But how', asked the priest, 'could you enter my chamber, seeing that the door was locked?' 'Oh,' said the witch, 'that matters naught to us, for we go through keyholes as easily as through open doors.' 'If what you say is true,' replied the holy man, 'you shall not lack a reward, but I must first have proof of it.' With these words, he locked the church door, and began

[1] Jean Hardouin (Harduinus), *Collectio regia maxima conciliorum graecorum et latinorum*, 12 vols., Paris, 1715, i. 1506; H. C. Lea, *History of the Inquisition of the Middle Ages*, 2nd ed., 3 vols., London, 1906, iii. 494; W. G. Soldan and H. Heppe, *Geschichte der Hexenprocesse*, 2 vols., Stuttgart, 1880, i. 132.

[2] Lea, loc. cit., iii. 414.

[3] Soldan and Heppe, loc. cit., i. 128, 139.

vigorously to beat the old woman with the handle of the crucifix he carried, asking her, when she complained, why she did not escape through the keyhole.[1]

The great Pope Nicholas I (died 867) strongly condemned the use of torture to induce confessions, and Gregory VII (died 1085) forbade inquisition to be made for witches and sorcerers on occasions of plague or bad weather.[2] Later, the inquisitorial process, combined with torture to enforce denunciations, became the chief agent in spreading and maintaining the witch mania.

The Eastern Church remained in this mediaeval stage, and never developed a witch mania. In the West the change seems to have been brought about mainly by two causes, the development of heresies and the increasing prominence of the devil.

There is no doubt that the Albigensian and other heresies of the twelfth and thirteenth centuries contained Manichean elements. It was taught that there were two divinities—one perfectly good, the creator of the invisible spiritual world, the other the creator of the material world, the Demiurgus, a being capable of evil passions, wrath, jealousy, &c., who was identified with the Jehovah of the Old Testament.[3] It required very little to confound this Demiurgus with Satan, the Prince of this world ; after which it was easy to look upon Satan as a being not entirely evil, as Lucifer, son of the morning, the disinherited son or brother of God, a natural object of worship for the oppressed and discontented.[4]

The serfs, equally tyrannized over by bishop and noble, the relics of the persecuted sects Waldenses and Cathari,[5] sought refuge, like Saul of old, in forbidden arts, and thus sects of Luciferans, or devil-worshippers, arose (especially in Germany and France) whose numbers were exaggerated by the fear and horror of the orthodox.[6]

[1] See also Lea, loc. cit., iii. 434, on this mildness of the Church up to the fourteenth century.

[2] Soldan and Heppe, loc. cit., i. 136.

[3] Lea, loc. cit., i. 91.

[4] The Paulicians were accused of teaching that the devil created this world, but seem merely to have taken such texts as John xii. 31, xiv. 30 ; 2 Cor. iv. 4 ‘ in their plain and obvious sense ’. F. C. Conybeare, *Key of Truth, A Manual of the Paulician Church of Armenia*, Oxford, 1898, 46.

[5] The term ‘ Cathari ’ was said to come ‘ from their kissing Lucifer under the tail in the shape of a cat ’. Lea, loc. cit., iii. 495.

[6] Lea, loc. cit., i. 105, ii. 334, &c. The main evidence is Conrad of Marburg’s report to Pope Gregory XI, 1233 : ‘ A tissue of inventions ’, but ‘ apparently doubted by no one ’.

At the same time the devil acquired more importance in other ways. That fearful calamity, the Black Death, seemed to display his power over both the just and the unjust; while the Great Schism in which each pope excommunicated the other, handing him and his adherents over to Satan, put every one not absolutely certain of being on the right side in reasonable fear of the powers of darkness.

The belief in the great activity and power of the devil and his servants the sorcerers was further supported by the vast authority of St. Thomas Aquinas (1225–74), whose ingenuity enabled him to explain away those ancient canons which seemed opposed to the more extreme views. Thus the synod of Bracara (A. D. 563) had declared the doctrine that the devil can produce drought or thunder-storms to be heresy; to which the Doctor Angelicus replied that though it is doubtless heresy to believe the devil can make natural thunder-storms, it is by no means contrary to the Catholic faith to hold that he may, by the permission of God, make artificial ones.[1]

For these and other reasons, the devil assumed greater prominence during the fourteenth and fifteenth centuries than ever before. Men believed that he might appear to them from behind every hedge or ruin, that his action was to be seen in almost all pains and diseases, but that he was to be dreaded most of all when he entered into a league with some man or woman. Thus everything was ready for the outbreak of witch mania when, in 1484, Pope Innocent VIII by his bull *Summis desiderantes* gave the sanction of the Church to the popular beliefs concerning witches, such as sexual intercourse with devils, destruction of crops, and infliction of sterility and disease on man and beast.

The charge of sorcery had usually been employed in earlier times either to check learned men who seemed to be going too far, or tending to heresy in their researches, as in the case of the physicians Arnold of Villanova (1240–1312) and Peter of Abano (1250–1320), or to crush individuals and societies who were politically dangerous, as with Joan of Arc, the Duchess of Gloucester, and the Templars—the Church being called in to aid the civil power. Now it was the Church which called upon the civil power to assist in a crusade against witches and sorcerers as being the worst and most dangerous of heretics.

[1] Quodlibet, xi. 10; Soldan and Heppe, loc. cit., i. 143; Lea, loc. cit., iii. 415.

In the Middle Ages it was held that a man who called up the devil, knowing it to be wrong, was not a heretic but merely a sinner. But if he thought it was not wrong, or that the devil would tell him the truth, or that the devil could do anything without God's permission, he was also a heretic, since these beliefs are contrary to Church doctrine. In the fifteenth century it was taught that all sorcerers are heretics, *maleficus* being, according to the learned authors of the *Malleus Maleficarum*, a contraction of *male de fide sentiens* or heretic.[1]

Nor was the identification of heresy and witchcraft illogical, whatever we may think of the etymology. The Church is the kingdom of God, heretics form the kingdom of the devil, and just as the Church possesses saints who see visions, work miracles, and commune with Christ face to face, so there are specially eminent heretics, saints of the devil's church, who work miracles and have obscene intercourse with their master. All true Christians are potential saints, all heretics potential sorcerers, for all have committed treason against the divine Majesty, though only some may have entered into a definite compact with the enemy. The former, if they repent, may hope for perpetual imprisonment; the latter are to be put to death whether they repent or not.

This view was also of advantage to the Church, for it increased the horror of heresy and facilitated its suppression. The laity had never entirely reconciled themselves to the sight of their apparently harmless neighbours being tortured and burnt for differences in abstract belief, but almost every one was ready to torture and burn a sorcerer, and local outbreaks of witch-hunting were frequently started by mob violence. In 1555 it was declared by the Peace of Augsburg that no one should suffer in life and property for his religion; but to take a Lutheran, call him a sorcerer, confiscate his goods, and force him by torture to confess that he was led into his errors by the devil himself, seems to have been too great a temptation for the prince-bishops who headed the 'counter-reformation' in South Germany to resist. That this was partly the cause of the great witch-burnings in the bishoprics of Würzburg, Bamberg, Fulda, and Trèves is evidenced by the large proportion of male victims, and by the frequent and signi-

[1] H. Institoris and J. Sprenger, *Malleus Maleficarum*, editio princeps, Cologne, 1486, and frequently reprinted until the end of the seventeenth century. See especially pars 1, quaestio 2.

ficant appearance of the phrase 'is also Lutheran' in the official reports.

As soon as the Reformation was established, Protestants vied with Catholics as witch-hunters. Eager to show that they were in no way inferior to their opponents in zeal for the Lord and enmity against Satan and his servants, they had the advantage of being able to follow the scriptural injunction, 'Thou shalt not suffer a witch to live', without previously explaining away ancient canons and decrees of Church synods which seemed to throw doubt on the very existence of the more typical forms of witchcraft. Nor did they hesitate to attack their rivals with similar weapons. If Protestants were burnt as sorcerers at Würzburg, we find the first Danish Lutheran bishop, Peter Palladius, recommending the zealous members of his flock to seek out the so-called wise women of their neighbourhoods on pretence of having some disease. If then the latter use paternosters, holy water, or invocations of saints, they are probably not only Catholics but witches, and should be treated accordingly.[1]

Almost all the victims of the witch mania were executed on their own confession, extorted in the vast majority of instances by torture or the fear of torture. In England, where torture was theoretically illegal, confessions were comparatively rare, and nearly all died protesting their innocence. The few exceptions prove the rule; thus Elinor Shaw and Mary Philips, almost the last witches legally executed in England, 1705, confessed because they were threatened with death if they refused, and promised release if they pleaded guilty,[2] while others were induced to admit their guilt by being kept awake several nights, and forced to run up and down their cells till utterly exhausted, methods almost as effectual in producing 'a readiness to confess' as the rack or the thumbscrew.[3]

Nearly all the confessions were to a similar effect. From Lisbon to Liegnitz, from Calabria to Caithness, the central point of the story was the 'sabbat', an assembly of witches and sorcerers in some barren spot where they adored a visible devil, indulged in

[1] J. Diefenbach, *Der Hexenwahn*, Mainz, 1886, p. 299.

[2] The story of Elinor Shaw and Mary Philips, as well as many other accounts of witchcraft, may be read in two volumes entitled *Rare and Curious Tracts illustrative of the History of Northamptonshire*, Northampton, 1876 and 1881.

[3] F. Hutchinson, *Historical Essay*, London, 1718, cap. iv.

feasts, dances, and sexual orgies, reported what evil they had done and plotted more.

A few examples will therefore suffice, and they may be best taken from the *Daemonolatria*[1] of Nicholas Remy, Inquisitor of Lorraine, who burned nearly 900 witches and sorcerers in fifteen years, 1575–90.

He proves the reality of the witch dances as follows: A boy named John of Haimbach confessed that his mother took him to a sabbat to play the flute. He was told to climb up into a tree that he might be heard the better, and was so amazed by what he saw that he exclaimed: 'Good God! where did this crowd of fools and lunatics come from?' Thereupon he fell from the tree and found himself alone with a dislocated shoulder. Otillia Velvers, who was arrested soon after, confirmed the whole story, as did also Eysarty Augnel, who was burnt the following year. So too, Nicholas Langbernard, while going home in the early morning of July 21, 1590, saw in full daylight a number of men and women dancing back to back, some of them with cloven hoofs. He cried out 'Jesus' and crossed himself, upon which all vanished except a woman called Pelter, whose broomstick dropped, and who was then carried off by a whirlwind. The grass was afterwards found to be beaten down in a circle with marks of hoof-prints. Pelter and two other women were arrested and confessed they were present, as also did John Michael, who said he was playing the flute in a tree, and fell down when Nicholas crossed himself, but was carried off in a whirlwind, his broomstick not being at hand.

'What further evidence', asks the inquisitor, 'can any one require?' The only possible objection, viz. that they were phantoms or spirits of people whose bodies were asleep in their beds, is worthless, 'it being the pious and Christian belief that soul and body when once parted do not reunite till the day of judgement'.

The food at these sabbats usually included the flesh of unbaptized children, and was always abominable. A certain Morel said he was obliged to spit it out, at which the demon was much enraged. 'Dancing opens a large window to wickedness,' and is therefore specially encouraged by the devil, but the dances cause great exhaustion, just as his feasts cause loathing, and his money changes to dung or potsherds. 'Barberina Rahel, and nearly all others, declared they had to lie in bed two days after a witch dance,

[1] *Daemonolatriae libri tres,* Lyons, 1595.

but even the oldest cannot excuse themselves, and the devil beats them if they are lazy.' The music is horrible; every one sings or plays what he likes, a favourite method being to drum on horse skulls or trees. Sometimes the devil gives a concert of his own, at which all are required to applaud and show pleasure; those who do not are beaten so that they are sore for two days, as Joanna Gransandeau confessed.

All are compelled to attend and give an account of their evil deeds under heavy penalties. C. G. said 'he was beaten till he nearly died for failing to attend a sabbat, and for curing a girl whom he had been told to poison. The devil also carried him up into the air over the river Moselle, and threatened to drop him unless he swore to poison a certain person.' The witch Belhoria was attacked by dropsy because she refused to poison her husband. If they failed in their attempts on others, they were compelled to poison their own children, or destroy their own property.

Antonius Welch was asked to lend his garden for a witch dance. He refused, and found it full of snails and caterpillars. Men of little faith have objected that only God can create, for 'without Him nothing is made that was made'; but why should not demons collect vast numbers of insects in a moment? Look at the well-known rain of frogs, blood, &c. This is doubtless done by devils out of mere sport: how much more would they do for love of harm? The making of thunder-storms is harder to believe, but has been admitted by more than 200 condemned witches and sorcerers. Almost all confessed that they could creep into locked rooms and houses in the form of small animals, and resuming their natural shape commit all sorts of crimes, showing, says Remy, what a peril they are to mankind.

A worthy comrade of Remy was Peter Binsfeld, suffragan Bishop of Trèves and foremost opponent of John Weyer. He is said to have burnt no fewer than 6,500 persons and to have so desolated his diocese that in many villages round Trèves there was scarcely a woman left. His *Tractatus de confessionibus maleficorum* [1] begins with the following case, which with those mentioned above affords a complete view of the usual witch confessions. John Kuno Meisenbein, a youth about eighteen years old, was studying 'poetry and the humaner letters' at the High School in Trèves, when he confessed to the authorities that his mother,

[1] Trèves, 1595.

brother, sister, and self were all in league with the devil. He said that in his ninth year his mother had initiated him as a sorcerer, and had carried him up the chimney on a goat to a heath near Trèves, where he took part in the usual sabbat and had intercourse with a female demon named Capribarba. The mother, Anna Meisenbein, a woman of good position, had already escaped to Cologne, but a son and daughter were arrested, strangled, and burned. 'They died with much sorrow and penitence.' The eldest son, John Kuno, thereupon urged the judges to use all means to capture his mother, 'that by punishment and momentary death in this world she might escape eternal damnation'.

Moved by this most creditable and merciful petition (*honestissima et plenissima misericordiae petitione excitatus*), the prior wrote to his friends at Cologne, and the unhappy woman was arrested and taken back to Trèves. At first she protested her innocence, 'but when more severe tortures were employed' she made the usual admissions. Having lost a baby, she had, for a moment, doubted the goodness of God. Whereupon a man in black raiment appeared at the side of the bed, and promised if she would renounce God and serve him he would give her peace of mind. She did so, and he became her lover, and gave her money, which however vanished. He called himself Fedderhans, and had asses' feet. Then follows the usual story of the sabbat. 'This woman', concludes the bishop, 'was burnt alive October 20, 1590, and had a good end.' They offered to behead John Kuno as a reward for his filial piety and repentance, but he said he was unworthy of such a favour and was therefore strangled and burnt. 'He had a most edifying end,' says the bishop, who proceeds to comment upon sexual intercourse between witches, sorcerers, and demons, 'which is so certain that it is an impudence to deny it, as St. Augustine saith,[1] being supported by the confessions of learned and unlearned, and by all the doctors of the Church, though a few medical men, advocates of the devil's kingdom [an obvious reference to Weyer, whom he abuses in the preface], have dared to deny it'.[2]

It is not our purpose to try and discover what amount of truth is contained in the immense farrago of absurdities comprised in the

[1] *Civ. Dei*, xv. 23.

[2] Peter Binsfeld, *Tractatus de confessionibus maleficorum*, Trèves, 1595, pp. 37–44, 230, &c. Binsfeld often refers to this case as proving the reality of disputed forms of witchcraft and the soul-saving work of the witch-hunters.

witch confessions. Actual nocturnal meetings of peasants, either to celebrate heathen rites or to plot against their oppressors, or merely to enjoy rude dances and music, as the negro in the Southern States was supposed to play the banjo nightly after his labours on the plantation, may or may not have assisted in spreading and confirming the belief in the sabbat, but they were not necessary. The whole story of child murder, obscene worship of a demon, dances and sexual orgies, was ready to hand long before. It had been applied in classic times to the worshippers of Isis and Bacchus, by the pagans to the early Christians, by the orthodox to the first heretics, to the Jews, to the Templars, and in our own day we have seen very similar charges brought against the Freemasons. All these sets of people had known meeting-places—the witches had none; they must therefore meet on some barren moor or mountain and be carried there supernaturally. Once started, the belief spread rapidly. Indeed we know from contemporary writers that it was a common subject of village gossip, and if any wretched victim had any doubt as to what she was expected to confess, the gaoler and judges were always ready with hints or leading questions.

One learned German [1] has attributed the whole witch mania to the *Datura Stramonium*, or thorn-apple, a plant introduced into Europe about this time. Women dosed themselves with this drug, or applied it in ointments, and forthwith had hallucinations of broomstick rides and witch dances. Others look upon belladonna as the principal agent, and one ardent investigator took dangerous doses of it in the hope of experiencing the adventures of a mediaeval sorcerer, but without definite effect. A similar experiment has recently been made by Kiesewetter, the historian of 'Spiritualism'. He used the witch ointments described by Baptista Porta and others, but could produce nothing more diabolical than dreams of travelling in an express train.[2] Others, again, have supposed that the badly baked rye bread of the period must have produced an immense amount of nightmare among the poorer classes. The power of suggestion, doubtless, had a very real influence both on the victims and their judges, and with the aid of narcotics may not infrequently have produced vivid dreams of dancing and other intercourse with demons.

[1] L. Meyer, *Die Periode der Hexenprocesse*, Hannover, 1882.

[2] K. Kiesewetter, *Die Geheimwissenschaften*, Leipzig, 1895, p. 579 f.

No doubt many persons were quite ready to become witches or sorcerers, and some really believed they had acquired such powers. Cases are recorded in which formal agreements, duly signed in blood, and awaiting the devil's acceptance, were discovered, and resulted in the arrest and burning of the would-be wizard. Others took pleasure in the terror the reputed powers inspired, and may have sometimes caused or increased it by the use of actual poisons.

But these formed but a small minority of the vast army of victims ; and even when some real criminal was arrested or some half-insane person voluntarily ' confessed ', she was encouraged or compelled to denounce her supposed associates, and thus often involved scores of innocent acquaintances in her own awful fate.

The witch-hunters are not to be blamed for believing in witchcraft, or even for carrying out the scriptural injunction ' Thou shalt not suffer a witch to live '. It is the methods they employed, compared with which the procedure of a Jeffreys or a Caiaphas was just and merciful, which cannot be excused by any talk about the spirit of the age, which brought agony and death to many thousands of innocent men, women and little children, and which excited the fiery and righteous indignation of Dr. John Weyer.

According to Pascal, men never do wrong so thoroughly and so cheerfully as when they are obeying the promptings of a false principle of conscience. To which we may add that men are never more cruel and unjust than when they are in a fright. The witch-hunters, most of them at least, were pious and conscientious men. They appeal to God, the Church, and the Bible at every step. Nicholas Remy, for instance, after torturing and burning over 800 of his fellow creatures, retired from work thinking he had done God and man good service. But one thing troubled his conscience. He had spared the lives of certain young children, and merely ordered them to be scourged naked three times round the place where their parents were burning. He is convinced that this was wrong, and that they will all grow up into witches and sorcerers. Besides, if God sent two she-bears to slay the forty and two children who mocked Elisha, of how much greater punishment are those worthy who have done despite to God, His Mother, the saints, and the Catholic religion ?[1] He hopes his sinful clemency will not become a precedent—a fear which was quite unnecessary, for scores

[1] Op. cit., ii. 2 (p. 200).

of children under twelve were burnt for witchcraft ; and the one plea which even then respited the most atrocious murderess did not always avail a witch, since it was believed that her future child, if not the actual offspring of the devil, would infallibly belong to his kingdom.

But the witch-hunters were urged on by fear as well as by piety, for not only did they think themselves exposed to personal attacks from the devil and his allies, but they believed there was a vast and increasing society of men and women in league with the evil one, and that the fate of the world depended on its suppression.

All the machinery, therefore, which the Roman emperors had devised for their protection against treason and the Church for the suppression of heresy was brought into action against the witches, for witchcraft was the acme of treason and heresy, a *crimen laesae maiestatis divinae.*[1]

For a description of the methods employed we cannot do better than go to the *Malleus Maleficarum*,[2] the guide and handbook of the witch-hunters.

All proceedings in cases of witchcraft, say the reverend authors, must be on the plan recommended by Popes Clement V and Boniface VIII, ' summarie, simpliciter, et de plano, ac sine strepitu ac figura iudicii ', a harmless looking phrase which swept away at a stroke all the safeguards which the lawyers of pagan Rome and the ruder justice of ancient Gaul and Germany had placed around accused persons. There are, says the *Malleus*,[3] two forms of criminal procedure : (1) the old legal or *accusatorial* form where the prosecutor offers to prove his charge and to accept the consequences of failure, which must be carefully avoided as being dangerous and litigious ; and (2) the *inquisitorial*, where a man denounces another either from zeal for the faith, or because called upon to do so, but takes no further part nor offers to prove his charge, or where a man is suspected by common report and the judge makes inquiry, and this method must always be preferred. The inquisitors, on entering a new district, should issue a proclamation calling on all persons to give information against

[1] *Malleus*, pars i, quaestio 1, p. 6, edit. 1596.

[2] By H. Institoris and J. Sprenger. Between 1486 and 1596 several editions were printed in specially small form ' that inquisitors might carry it in their pockets and read it under the table '.

[3] iii. 1 (p. 337 f.).

suspected witches on pain of excommunication and temporal penalties. Any one may be compelled, by torture if necessary, to give evidence, and if he refuses must be punished as an obstinate heretic. Other sorcerers, or the man's wife and family, are lawful witnesses against, but not for, the accused. Criminals and perjured persons, if they show zeal for the faith, may be admitted to give evidence. Priests, nobles, graduates of universities, and others legally exempt from torture are not exempt in the case of witch trials.[1]

'Delation,' the scandal of imperial Rome, was not only encouraged but enforced, and in some places, as at Milan, boxes were put in the churches, into which any one might drop an anonymous denunciation of his neighbour.

Names of informers are not to be revealed under penalty of excommunication; the advocate, if there is one, need be told the charges only. This advocate must not be chosen by the accused but by the inquisitor, and he must refuse the case if it seems to him unjust or hopeless. He must not use legal quibbles or make delays or appeals, and is to be specially warned that if he be found a protector of heretics or a hinderer of the inquisition, he will incur the usual penalties for those heinous crimes. If he reply that he defends the person, not the error, this avails not, for he must make no defence which interferes with proceeding *summarie, simpliciter, et de plano*.[2] After this it is not surprising to find that those accused of witchcraft were rarely defended by an advocate.

Faith need be kept with heretics and sorcerers 'for a time only'.[3] Therefore an inquisitor may promise not to condemn a person if he confesses, and then pass sentence after a few days, or if of very tender conscience by the mouth of another. It is also lawful to introduce persons, *etiam mulieres honestae*, to the accused who promise to find means for their escape if they will teach them some form of witchcraft. This, say the authors, is a most successful method for getting convictions.[4]

Torture, though it may not be repeated on the same charge, may be continued as long as necessary, and any fresh evidence justifies a repetition. Finally the accused may be burnt without confession if the evidence is strong enough, or he may be kept

[1] *Malleus*, iii. 4, p. 344.
[2] iii. 10.
[3] iii. 14.
[4] iii. 16.

in prison for months or years, when the *squalor carceris* may induce him to confess his crimes.[1]

Such are the proceedings recommended against persons suspected of or denounced for witchcraft, and they conclude appropriately with the hideously hypocritical formula with which they were delivered over to be burnt: 'Relinquimus te potestati curiae secularis, deprecantes tamen illam ut erga te citra sanguinis effusionem et mortis periculum suam sententiam moderetur',[2] which means, according to the *Malleus*, that sorcerers are to be burned even though they repent, while repentant heretics may be imprisoned for life.

What was meant by the *squalor carceris* may be seen from the following description by an eye-witness, Pretorius:[3]

> 'Some [of the dungeons] are holes like cellars or wells, fifteen to thirty fathoms (?) deep with openings above, through which they let down the prisoners with ropes and draw them up when they will. Such prisons I have seen myself. Some sit in great cold, so that their feet are frost-bitten or frozen off, and afterwards, if they escape, they are crippled for life. Some lie in continual darkness, so that they never see a ray of sunlight, and know not whether it be night or day. All of them have their limbs confined so that they can hardly move, and are in continual unrest, and lie in their own refuse, far more filthy and wretched than cattle. They are badly fed, cannot sleep in peace, have much anxiety, heavy thoughts, bad dreams. And since they cannot move hands or feet, they are plagued and bitten by lice, rats, and other vermin, besides being daily abused and threatened by gaolers and executioners. And since all this sometimes lasts months or years, such persons, though at first they be courageous, rational, strong, and patient, at length become weak, timid, hopeless, and if not quite, at least half idiotic and desperate.'

Yet all this was not considered torture, and if some poor wretch, after a year of it, went mad, or preferred a quick death to a slow one, her confession was described as being 'entirely voluntary and without torture'.

As to the torture itself, it combined all that the ferocity of savages and the ingenuity of civilized man had till then invented. Besides the ordinary rack, thumb-screws, and leg-crushers or Spanish boots, there were spiked wheels over which the victims were drawn with weights on their feet; boiling oil was poured on

[1] iii. 14. [2] iii. 29–31, repeated with slight variations.
[3] *Von Zauberei und Zauberern*, p. 211; Soldan and Heppe, i. 347.

their legs, burning sulphur dropped on their bodies, and lighted candles held beneath their armpits. At Bamberg they were fed on salt fish and allowed no water, and then bathed in scalding water and quicklime. At Lindheim they were fixed to a revolving table and whirled round till they vomited and became unconscious, and on recovery remained in so dazed a state that they were ready to confess anything.[1] At Neisse they were fastened naked in a chair 'with 150 finger-long spikes in it' and kept there for hours. And so effective were these tortures that nine out of ten innocent persons preferred to die as confessed sorcerers rather than undergo a repetition of them.

The Jesuit Father Spee, a worthy successor of John Weyer, accompanied nearly two hundred victims to the stake at Würzburg in less than two years. At the end of this time his hair had turned grey and he seemed twenty years older, and on being questioned as to the cause, declared that he was convinced that all these persons were innocent. They had, he said, at first repeated the usual confession, but on being tenderly dealt with had one and all protested their innocence, adjuring him at the same time not to reveal this, for they would much rather die than be tortured again. He added that he had received similar reports from other father confessors.[2] A few years later, 1631, he plucked up courage to publish anonymously his *Cautio Criminalis*, in which he exclaims:

> 'Why do we search so diligently for sorcerers? I will show you at once where they are. Take the Capuchins, the Jesuits, all the religious orders, and torture them—they will confess. If some deny, repeat it a few times—they will confess. Should a few still be obstinate, exorcise them, shave them: they use sorcery, the devil hardens them, only keep on torturing—they will give in. If you want more, take the Canons, the Doctors, the Bishops of the Church—they will confess. How should the poor delicate creatures hold out? If you want still more, I will torture you and then you me. I will confess the crimes you will have confessed, and so we shall all be sorcerers together.'[3]

[1] The Lindheim cases are recorded by G. C. Horst, afterwards pastor of the place, in his *Dämonomagie*, 2 vols., Frankfort, 1818, and *Zauberbibliothek*, 6 vols., Mainz, 1821–6. See also O. Glaubrecht, *Die Schreckensjahre von Lindheim*, 1886.

[2] *Cautio Criminalis*, Rinteln, 1631, Dubium xix (p. 128). He calls himself 'Sacerdos quidam'.

[3] Dubium xx (p. 153).

In the most notorious of judicial murders, we read that the judges had some difficulty owing to a disagreement between the witnesses. This rarely troubled the witch-hunters. At Lindheim a woman was accused of having dug up and carried off the body of an infant, which, under torture, she admitted, denouncing four others as her accomplices. But on the grave being opened, the body was found uninjured. The inquisitors at once decided that this must be a delusion of the devil, and all five women were burned. A man confessed, under torture, that he was a were-wolf, and in that form had killed a calf belonging to a neighbour; the latter, however, said he had never lost a calf, though two or three years ago two hens had disappeared, he believed through witchcraft. The accused was burnt, for what need had they of witnesses? Had they not heard his confession?[1]

It was even laid down as a principle that doubtful points must be decided 'in favour of the faith'—in other words, against the accused. 'If a sorcerer retracts his denunciations at the stake, it is not void, for he may have been corrupted by friends of the accused. Also when witnesses vary, as they often do, the positive assertion is always to be believed,' says Bishop Covarivias, a prominent member of the Council of Trent. In which he is supported by the jurist Menochius of Padua, 'ne tam horrendum crimen occultum sit'.

Anything might start a witch-hunting, and once started it increased like an avalanche. If an old woman happened to be out of doors in a thunder-storm; if the winter was prolonged; if there was a more than usual number of flies and caterpillars; if a woman had a spite against her neighbour, some one might be denounced and forced in turn to denounce others. The prolonged winter of 1586 in Savoy, for instance, resulted in the burning of 113 women and two men, who confessed, after torture, that it was due to their incantations.

It is thus not difficult to understand how, in the diocese of Como, witches were burnt for many years at an average rate of 100 per annum; how in that of Strassburg 5,000 were burnt in twenty years, 1615–35; how in the small diocese of Neisse 1,000 suffered between 1640–50, insomuch that they gave up the stake and pile as being too costly, and roasted them in a specially prepared oven; and how the Protestant jurist Benedict Carpzov could boast not

[1] Horst, *Zauberbibliothek*, ii. 374, and *Dämonomagie*, ii. 412.

only of having read the Bible through fifty-three times, but also of having passed 20,000 death sentences, chiefly on witches and sorcerers.[1]

One of Carpzov's victims is specially interesting to medical men, the Saxon physician, Dr. Veit Pratzel, who on one occasion (1660) produced twenty mice by sleight of hand in a public-house, probably for the sake of advertisement. He was denounced as a sorcerer, tortured and burnt, while his children were bled to death in a warm bath by the executioner, lest they should acquire similar diabolical powers.[2]

A like fate befell the servant of a travelling dentist at Schwersenz in Poland. The dentist, John Plan, left his assistant in the town to attract attention by conjuring tricks, while he went to sell his infallible toothache tinctures in the neighbouring villages. On his return next evening, he was horrified to see the body of the unfortunate man hanging on the town gallows, and was told on inquiry that he was an evident sorcerer who had made eggs, birds, and plants before everybody in the market-place. He had therefore been arrested, scourged, put on the rack, and otherwise tortured till he confessed he was in league with the devil. Whereupon the town council, 'out of special grace and to save expense', had, instead of burning him, mercifully condemned him to be hanged. The dentist fled in terror to Breslau.[3]

But it was by no means necessary to be so foolhardy as this to fall into the hands of the witch-hunters. A woman at Lindheim was noticed to run into her barn as the inquisitorial officials came down the street. She had never been accused or even suspected of witchcraft, but was nevertheless immediately arrested, and brought more dead than alive to the chief inquisitor, Geiss,[4] who declared her flight justified the strongest suspicion. Exposed to the most extreme torture, she confessed nothing, but at length, at the question whether she had made a compact with the devil, one of the inquisitors declared he saw her nod her head. This was enough; she was burnt; probably a happy fate under the circumstances, for she thus escaped being forced by further tortures to give details of her imaginary crime and to denounce her neighbours.

[1] Soldan and Heppe, ii. 209.
[2] Soldan and Heppe, ii. 130.
[3] J. H. Böhmer, *Ius ecclesiasticum*, 5 vols., Halle, 1738–43, v. 35.
[4] Horst, *Dämonomagie*, ii. 377.

Once in the clutches of the witch-hunters, the unfortunate victim was confronted by a series of dilemmas from which few escaped. A favourite beginning was to ask whether he believed in witchcraft. If he said 'Yes', he evidently knew more of the subject; if 'No', he was *ipso facto* a heretic and slanderer of the inquisition; if in confusion he tried to distinguish, he was *varius in confessionibus*,[1] and a fit subject for immediate torture. If he confessed under torture, the matter was, of course, settled; if he endured manfully, it was evident that the devil must be aiding him. If a mark could be found on his body which was insensible and did not bleed when pricked, it was the devil's seal and a sure sign of guilt; but if there was none, his case was no better, for it was held that the devil only marked those whose fidelity he doubted, so that a suspected person who had no such mark was in all probability a specially eminent sorcerer.[2]

Then came the water test, of which there is no better account than the report sent by W. A. Scribonius, Professor of Philosophy at Marburg, to the town council of Lemgo in 1583:

'When I came to you, most prudent and learned consules, 26th September, there were, two days later on St. Michael's eve, three witches burnt alive for divers and horrible crimes. The same day three others, denounced by those aforesaid, were arrested, and on the following day about 2 p.m. for further proving of the truth were thrown into water to see whether they would swim or not. Their clothes were removed and they were bound by the right thumb to the left big toe and vice versa, so that they could not move in the least. They were then cast three times into the water in the presence of some thousands of spectators, and floated like logs of wood, nor did one of them sink. And it is also remarkable that almost at the moment they touched the water a shower of rain then falling ceased, and the sun shone, but when they were taken out it started raining as before.'

On request of the burgomaster, he investigated 'the philosophy' of this, and, though he could find nothing definite, had no doubt of its value as a test of witchcraft. 'The physician Weyer rejects it as absurd and fallacious, but he can produce no good arguments or examples against it, and may therefore be ignored.' Perhaps witches are made lighter because possessed by demons who are 'powers of the air' and often carry them

[1] *Malleus*, iii. 14 (p. 370).

[2] Father Spee gives a long list of these dilemmas, *Cautio Criminalis*, Dubium li.

through the air. All who float have afterwards confessed, therefore though not scriptural nor of itself sufficient to convict, the swimming test is not to be despised.[1]

With regard to the number of victims, even sober historians, such as Soldan, speak of millions, but if we take three-quarters of a million for the two centuries 1500–1700, it will give a rate of ten executions daily, at least eight of which were judicial murders.

Even more pathetic than the notice of 800 condemned in one body by the senate of Savoy[2] are the long lists of yearly executions preserved in the fragmentary records of small towns and villages. Thus at Meiningen, between 1610–31 and 1656–85, 106 suffered—in 1610 three, 1611 twenty-two, 1612 four, &c. &c., the intervening records being omitted owing to war. Similar notices have survived at Waldsee, Thun in Alsace, and many other hamlets, where through a long series of years we read of one to twenty persons burnt annually, some of them being previously 'torn with red-hot pincers'.[3]

At Würzburg the Prince-bishop, Philip of Ehrenberg, is said to have burnt 900 in five years (1627–31), and we have terrible lists of twenty-nine of the burnings, almost all of which include young children. Here are two of them:

'In the thirteenth burning, four persons: the old court smith, an old woman, a little girl of nine or ten years, a younger girl her sister.'

'In the twentieth burning, six persons: Babelin Goebel, the prettiest girl in Würzburg; a student in the fifth form who knew many languages and was an excellent musician, instrumental and vocal; two boys from the new minster, twelve years old; Babel Stepper's daughter; the caretaker on the bridge.'[4]

At Bamberg the Prince-bishop, John George, 1625–30, burnt at least 600 persons, and his predecessors had been hardly less vigorous witch-hunters. He was ably seconded by his suffragan, Bishop Förner, and two doctors of law, Braun and Kötzendörffer, who besides the ordinary torture implements, salt fish and quicklime baths, found a so-called prayer stool or bench covered with

[1] *De sagarum natura et potestate, deque his recte cognoscendis et puniendis deque purgatione earum per aquam frigidam epistola*, Lemgo, 1583. Also in Sawr, *Theatrum de Veneficiis*, 1856.

[2] Lea, iii. 549.

[3] Haas, *Die Hexenprocesse*, Tübingen, 1865.

[4] Soldan and Heppe, ii. 46, and elsewhere.

spikes, on which the victim was forced to kneel, and a cage with a sharp ridged floor on which he could not stand, sit, or lie without torment, of great value in extorting confessions. The record of their deeds has been published by Dr. F. Leitschuh,[1] librarian of Bamberg, and contains, among other cases, that of the Burgomaster, John Junius, which throws more light on the nature of the witch trials than do volumes of second-hand history.[2]

John Junius, a man universally respected, had been five times Burgomaster of Bamberg, and held that office in June 1628, when he was arrested on a charge of sorcery. He protested his innocence though six witnesses declared, under torture, that they had seen him at the witch dances. On June 30 he endured the torment of the thumb-screws and leg-crushers (Spanish boots) without confession. Then they stuck pins in him and found a 'devil's mark', and finally drew him up with his arms twisted backwards, but he would admit nothing. Next day, however, when threatened with a repetition of the torture, he broke down, made the usual confession (including intercourse with a female demon who turned into a he-goat), and denounced twenty-seven persons whose names and addresses are given.[3] He was condemned to be beheaded and burnt, but before his death wrote the following letter to his daughter:

'Many hundred thousand good-nights, my dearest daughter Veronica! Guiltless was I taken to prison, guiltless have I been tortured, guiltless I must die. For whoever comes here must either be a sorcerer, or is tortured until (God pity him) he makes up a confession of sorcery out of his head. I'll tell you how I fared. When I was questioned the first time, there were present Dr. Braun, Dr. Kötzendörffer, and two strangers. Dr. Braun asked me, "Friend, how came you hither?" I answered, "Through lies and misfortune." "Hear you," said he, "you're a sorcerer. Confess it willingly or we'll bring witnesses and the executioner to you." I said, "I am no sorcerer. I have a clear conscience on this matter, and care not for a thousand witnesses, but am ready to hear them." Then the chancellor's son, Dr. Haan, was brought out. I asked, "Herr Doctor, what do you know of me? I never had anything to do with you, good or bad." He answered, "Sir, it is a judgement matter, excuse me for witnessing against you. I saw you at the dances." "Yes, but how?" He did not know. Then I asked the commissioners to put him on oath, and examine him properly.

[1] *Beiträge zur Geschichte des Hexenwesens in Franken*, Bamberg, 1883.

[2] 48 ff.

[3] Official report, given by Leitschuh in appendix.

"The thing is not to be arranged as you want it," said Dr. Braun; "it is enough that he saw you." I said, "What sort of witness is that? If things are so managed, you are as little safe as I or any other honourable person." Next came the chancellor and said the same as his son. He had seen me, but had not looked carefully to see who I was. Then Elsa Hopffen. She had seen me dancing on Haupt's moor. Then came the executioner and put on the thumbscrews, my hands being tied together, so that the blood spurted from under the nails, and I cannot use my hands these four weeks, as you may see by this writing. Then they tied my hands behind and drew me up. I thought heaven and earth were disappearing. Eight times they drew me up and let me fall so that I suffered horrible agony. All which time I was stark naked, for they had me stripped.

'But our Lord God helped me, and I said to them, "God forgive you for treating an innocent man like this; you want not only to destroy body and soul, but also to get the goods and chattels." [At Bamberg, two-thirds of the property of convicted sorcerers went to the bishop, and the rest to the inquisitors.] "You're a rascal," said Dr. Braun. I replied, "I am no rascal, but as respectable as any of you; but if things go on like this, no respectable man in Bamberg will be safe, you as little as I or another." The doctor said he had no dealings with the devil. I said, "Nor have I. Your false witnesses are the devils, your horrible tortures. You let no one go, even though he has endured all your torments."

'It was Friday, 30th June, that, with God's help, I endured these tortures. I have ever since been unable to put my clothes on or use my hands, besides the other pains I had to suffer innocently.

'When the executioner took me back to prison, he said to me, "Sir, for God's sake confess something, whether true or not. Think a little. You can't stand the tortures they'll inflict on you, and even if you could you wouldn't escape, though you were a count, but they'll go through them again and again and never leave you till you say you are a sorcerer, as may be seen by all their judgements, for all end alike." Another came and said the bishop had determined to make an example of me which would astonish people, and begged me for God's sake to make up something, for I should not escape even though I were innocent, and so said Neudecker and others.

'Then I asked to see a priest, but could not get one. . . . And then this is my confession as follows, but all of it lies.

'Here follows, dearest child, what I confessed that I might escape the great torments and agonies, for I could not have endured them any longer. This is my confession, nothing but lies, that I had to make on threat of still greater tortures, and for which I must die.

'"I went into my field, and sat down there in great melancholy, when a peasant girl came to me and said, 'Sir, what is the matter? Why are you so sorrowful?' I said I did not know, and then she sat down close to me, and suddenly changed into a he-goat and said, 'Now you know with whom you have to do.' He took me by the throat and said, 'You must be mine, or I'll kill you.' Then I said, 'God forbid.' Then he vanished and came back with two women and three men; bade me deny God, and I did so, denied God and the heavenly host. Then he baptized me and the two women were sponsors; gave me a ducat, which turned into a potsherd."

'Now I thought I had got it over, but they brought in the executioner, and asked where I went to the witch dances. I did not know what to say, but remembered that the chancellor and his son and Elsa Hopffen had mentioned Haupt's moor and other places, so I said the same. Then I was asked whom I had seen there. Replied I did not recognize any. "You old rascal, I must get the executioner to you. Was the chancellor there?" Said "Yes." "Who else?" "I recognized none." Then he said, "Take street by street, beginning from the market." Then I had to name some persons. Then Long Street. I knew nobody; had to name eight persons. . . . Did I know any one in the castle? I must speak out boldly whoever it was. So they took me through all the streets till I could and would say no more. Then they gave me to the executioner to strip, shave off my hair, and torture me again. "The rascal knows a man in the market-place, goes about with him daily, and won't name him." They meant Dietmeyer, so I had to name him.

'Next they asked what evil I had done. I replied, "None." The devil bade me to, and beat me when I refused. "Put the rascal on the rack." So I said I was told to murder my children but killed a horse instead. That wasn't enough for them. I had also taken a sacramental wafer and buried it. When I said this they left me in peace.

'There, dearest child, you have all my confession, for which I must die, and it is nothing but lies and made-up things, so God help me. For I had to say all this for fear of the tortures threatened me, besides all those I had gone through. For they go on torturing till one confesses something; be he as pious as he will, he must be a sorcerer. No one escapes, though he were a count. And if God does not interfere, all our friends and relations will be burnt, for each has to confess as I had.

'Dearest child, I know you are pious as I, but you have already had some trouble, and if I may advise, you had better take what money there is and go on a pilgrimage for six months, or somewhere where you can stay for a time outside the diocese till one sees what will happen. Many honourable men and women in Bamberg go to church and about their business, do no evil, and have clear consciences as I hitherto, as you know, yet they come to the witch prison, and if they have a tongue to confess, confess they must, true or not.

'Neudecker, the chancellor, his son, Candelgiesser, Hofmeister's daughter, and Elsa Hopffen all denounced me at once. I had no chance. Many are in the same case, and many more will be, unless God intervenes.

'Dear child, keep this letter secret so that nobody sees it, or I shall be horribly tortured and the gaoler will lose his head, so strict is the rule against it. You may let Cousin Stamer read it quickly in private. He will keep it secret. Dear child, give this man a thaler.

'I have taken some days to write this. Both my hands are lamed. I am in a sad state altogether. I entreat you by the last judgement, keep this letter secret, and pray for me after my death as for your martyred father . . . but take care no one hears of this letter. Tell Anna Maria to pray for me too. You may take oath for me that I am no sorcerer, but a martyr.

'Good-night, for your father, John Junius, will see you never more.

24th July, 1628.'

On the margin is written :

'Dear child, six denounced me : the chancellor, his son, Neudecker, Zaner, Ursula Hoffmaister, and Elsa Hopffen, all falsely and on compulsion as they all confessed. They begged my pardon for God's sake before they were executed. They said they knew nothing of me but what was good and loving. They were obliged to name me, as I should find out myself. I cannot have a priest, so take heed of what I have written, and keep this letter secret.'

The letter is still preserved, with its crippled handwriting, in the library at Bamberg. This case is beyond comment. It is like the trial of Faithful at Vanity Fair, but with rack and thumb-screw in place of a jury. Yet it is but a moderate sample of those outrages on justice and humanity called witch trials. Men rarely held out long, but, did space permit, we might tell stories of many heroic women who endured ten, twenty, even fifty repetitions of torture, till they died on the rack or in the dungeon rather than falsely accuse themselves or their neighbours.[1]

[1] Maria Hollin at Nördlingen (1593) withstood fifty-six repetitions of torture, and was finally 'dismissed' on the terms mentioned (Janssen, op. cit., viii. 719).

For when once arrested, the victim had small hope of acquittal, and in the most favourable cases, when there was no external evidence, and no amount of torture could induce a 'confession', the accused was sent back friendless and crippled to her home, which she was forbidden to leave, having first sworn to have no more dealings with the devil, and to take no proceedings against her accusers. To acquit her would imply that an innocent person had been tortured, a thing naturally repugnant to the tender consciences of the inquisitors.

Nor was the mania confined to any special class. Protestants vied with Catholics, and town councils with bishops in cruelty and injustice. At Nördlingen they had a special set of torture instruments which the Protestant town council lent to neighbouring district authorities, with the pious observation that 'by these means, and more especially by the thumb-screw, God has often been graciously pleased to reveal the truth, if not at first, at any rate at the last'.[1]

It is obvious from the above cases that the main cause of the continuance of the witch-burnings, and of the number of the victims, was the use of torture to obtain denunciations. The instances in which insane persons accused themselves or others seem to have been fewer than we might have expected.

Then, as now, there were melancholics who thought they had committed the unpardonable sin, and in those days the unpardonable sin might be represented by an imaginary compact with the devil. Then, as now, the 'mania of persecution' was a prominent symptom in some forms of insanity, and the idea of being bewitched by some old woman corresponded to the modern dread of detectives, electric batteries, or telephones.

Some of the supposed signs of witchcraft resemble those of mania and melancholia. Thus maniacs sometimes collect dirt for money, and witches often confessed that the devil's money changed to dirt. Melancholics mutter to themselves, look on the ground, and avoid society, all of which were considered signs of witchcraft. But then red hair and left-handedness were no less infallible indications.

Insanity and crime were indeed present at the witch trials, but they were at least as obvious in the accusers and judges as in the

[1] The Nördlingen authorities acquired an evil eminence in this frightfulness, which they termed 'eine heilsame Tortur' (Soldan, ii. 470).

victims, and the first man who was bold enough to say so was Dr. John Weyer. Though a few feeble protests may have been made by others, it was from the medical profession that the first determined opposition came. Mystics like Paracelsus and Cardan might encourage the superstition ; pious and able members of the profession like Ambroise Paré and Sir Thomas Brown might give it their sanction, but it was the physician Cornelius Agrippa who first successfully defended a witch at the risk of his own life,[1] and it was his pupil John Weyer who first declared open war against the witch-hunters and invoked the vengeance of heaven upon their atrocities.

'The feareful abounding at this time in this countrie of those detestable slaves of the divell, the witches or enchanters hath moved me (beloved reader) to dispatch in post the following treatise of mine, not in any wise (as I protest) to serve for a shewe of my learning and ingine, but only (moved of conscience to preasse thereby) so far as I can, to resolve the doubting hearts of manie both that such assaults of Satan are most certainly practised, and that the instruments thereof merit most severely to be punished, against the damnable opinions of two principally in our age, whereof the one called Scot, an Englishman, is not ashamed in public print to denie that there can be such a thing as witchcraft and so maintains the old error of the Sadduces in denying of spirits, the other called Wierus, a German physition sets out a publike apologie for all these crafts-folks, whereby procuring for their impunity, he plainly bewrayes himself to have been of that profession.'

Thus did our 'British Solomon', James I, commence his *Daemonologia* (1598), a work directed against the two men who alone up to that time had made a bold and open protest against the witch mania and its abominations. Reginald Scot in his *Discovery of Witchcraft* (1584) took the view of a modern common-sense Englishman, that the whole thing is absurd, a mixture of roguery and false accusations. Weyer, on the other hand, his predecessor by twenty years, is a firm believer in the activity of the devil, whose object, however, is not to get possession of the souls of crazy old women, but by deluding them, to convert pious and learned lawyers and theologians into torturers and murderers.

Born about 1516 at Grave in Brabant, the son of a dealer in hops and faggots, Weyer was acquainted with the supernatural

[1] Lea, iii. 545, and references there given.

from his earliest years, for they had a domestic 'house cobold' or *Poltergeist*, who was heard tumbling the hop-sacks about whenever a customer was expected. At seventeen years of age the boy was sent to study medicine as apprentice to Cornelius Agrippa, an extraordinary man, long held to be a sorcerer, who had recently incurred yet stronger suspicion by his heroic and successful defence of a woman accused of witchcraft at Metz, and by his fondness for a black dog called 'Monsieur' which scarcely ever left him. The young Weyer used to take this animal out on a string, and soon became convinced, to use his own words, that it was 'a perfectly natural male dog'.[1] He next went to Paris and thence to Orleans, a university then famous for its medical school, where he took the degree of M.D. in 1537. He commenced practice in Brabant, became public medical officer at Arnheim in 1545, and in 1550 physician to Duke William of Cleves. In 1563 he published his great work *De praestigiis daemonum et incantationibus ac veneficiis*,[2] the object of which is to show that so-called witchcraft is usually due to delusions of demons, who take advantage of the weaknesses and diseases of women to bring about impious and absurd superstitions, hatreds, cruelties, and a vast outpouring of innocent blood, things in which they naturally delight.

He proposes to treat the subject under four heads corresponding to the four faculties, theology, philosophy, medicine, and law. In the first section he attempts to show that the Hebrew word *Kasaph* does not mean 'witch' but 'poisoner', or at any rate that Greek, Latin, and Rabbinical interpreters so vary, that no reliance can be placed upon them. Moreover the law of Moses was given to the Jews 'for the hardness of their hearts', and is by no means always to be used by Christians.[3] Magicians and sorcerers do indeed still exist, as in ancient Egypt, but these are always men, and usually rogues and swindlers, such as was Faust, of whom Weyer gives us one of the earliest and most authentic notices. Faust, he says, was once arrested by Baron Hermann of Batoburg, and given in charge of his chaplain, J. Dursten, who hoping to see some sign or wonder, treated him with much kindness, giving him the best of wine. But all he got out of him was a magic ointment to enable him to shave without a razor, containing arsenic, and so strong that it brought

1 *De praestigiis, &c.*, ii. 5.

2 The privilege for publication is dated November 4, 1562; three editions appeared before the end of 1564, and a sixth in 1583.

3 Op. cit., ii. 1.

not only the hair but the skin from the reverend gentleman's cheeks. 'The which he has told me more than once with much indignation.' [1]

Weyer, however, firmly believes that the devil may assist sorcerers, such as Faust, in some of their feats, though he does this chiefly by deluding the eyes of the spectators. He may also delude women into the belief that they have been at witch dances and caused thunder-storms, &c., but his greatest deception is to make men believe in the reality of witchcraft and so torture and murder the innocent.[2] Women are more liable to his deceptions owing to their greater instability both of mind and body, and the delusion may be favoured by the use of drugs and ointments, especially those containing belladonna, lolium, henbane, opium, and even more by herbs recently introduced from east and west, such as Indian hemp, datura, 'and the plant called by the Indians "tabacco", by the Portuguese "peto", and by the French "nicotiana"'.[3]

As for the supposed compact with the devil, it is an absurdity only surpassed by the belief in sexual intercourse with demons. This delusion, Weyer points out, may be explained medically by the phenomena of nightmare and the effects of certain drugs, and is not sanctioned by Scripture. For, though holy men such as Lactantius, Justin Martyr, and Tertullian have maintained that the 'sons of God' mentioned in Genesis vi. 2 were spirits, this interpretation is opposed by still more eminent theologians, such as Saints Jerome, Gregory Nazianzen, and Chrysostom, though he is obliged to admit that St. Augustine believed in *incubi* and *succubae*,[4] and that distinguished living theologians hold that Luther's father was literally the devil. This, however, says Weyer, is an unfair and prejudiced way of attacking the Lutheran heresy.[5]

People who fancy themselves bewitched are really possessed or assaulted by the devil, as were Job and the demoniacs of the New Testament. If these demoniacs had lived in our days, he remarks, they would probably have each cost the lives of numerous old women.[6] The strange objects vomited by such persons are either deceptions or put into the person's mouth by the devil, as is shown by there being no admixture of food, and the absence of pain or injury in spite of the size of the objects.[7]

[1] Op. cit., ii. 4. [2] iii. 6. [3] iii. 18. [4] iii. 21. [5] iii. 23. [6] iv. 1. [7] iv. 2.

A girl near Cleves fell into convulsions with clenched hands and teeth which, according to her father, could only be opened by making the sign of the cross. She also complained of pains for which it was necessary to buy a bottle of holy water from a priest at Amersfort, on drinking which she proceeded to vomit pins, needles, scraps of iron, and pieces of cloth. She spoke in an altered boyish voice, intended for that of a demon, and declared the whole was caused by an 'in my opinion honest matron', who was imprisoned with her mother and two other women.

Weyer undertook the case, 'whereupon she said in her boy's voice she would have nothing to do with me, and that I was a cunning fellow. "Look what sharp eyes he has."' Weyer opened her hands and mouth, without making the sign of the cross, 'not that I would in any way speak irreverently thereof'. He also showed that the objects produced, even soon after eating, were free from admixture of food, and had therefore never been farther than the mouth; and he thus obtained the release of the four women after a month's imprisonment.[1]

As for the stories of men changed into animals, they are partly poetic and moral allegories, as the sailors of Ulysses, and partly a form of insanity long recognized by physicians, and termed lycanthropy.[2]

Many think they are possessed when they are only melancholic, and others pretend to be so to excite interest and obtain money. Those who fancy themselves attacked by devils should, instead of accusing their neighbours, take to themselves the armour of God as described by St. Paul. Unfortunately, spiritual pastors, in their ignorance and greed, teach that not only diabolical possession, but even ordinary diseases are to be cured by charms, incantations, palm branches, consecrated candles, and an execrable abuse of scriptural words. Cures are, indeed, sometimes so produced, but are really due to the imagination.

Persons supposed to be possessed should first be taken to an intelligent physician, who should investigate and treat any bodily disorder. Should spiritual disorders be also present he may then send the patient to a pious minister of the Church, but this will often be unnecessary. The devil is especially fond of attacking nuns, who should be separated from the rest, and, if possible, sent home to their relations.[3]

[1] iv. 3. [2] iv. 23. [3] iv. 10.

Here Weyer inserts several instances in his own experience.

Philip Wesselich, a monk of Knechtenstein near Cologne, an honest, simple-minded man, was miserably afflicted by a spirit about the year 1550. Sometimes he was carried up to the roof, at others thrust in among the beams of the belfry, often carried unexpectedly through the wall (*plerumque per murum transferebatur inopinato*) and knocked about generally. At length the spirit declared he was Matthew Duren, a former abbot, condemned to penance for having paid an artist insufficiently for a painting of the Blessed Virgin, so that the poor man went bankrupt and committed suicide, 'which was true'. He could only be released if the monk went to Trèves and Aix and recited three masses in the respective cathedrals. The theological faculty of Cologne advised that he should do so, but the abbot Gerard, a man of firmness and intelligence, told the possessed man that he was a victim of diabolical deceptions, and that unless he put his trust in God, and pulled himself together, he should be publicly whipped. Whereupon the monk did so, and the devil left him and went elsewhere.[1]

A similar case was that of a young woman known to Weyer, who had convulsions in church whenever the 'Gloria in excelsis' was sung in German, and said she was possessed. It was observed, however, that she looked about for a soft place to fall on. She was therefore sent for by Weyer's friend the Countess Anna of Virmont, who said she was about to sing the chant, and that if the demon attacked her she would soon drive him out. The young woman fell in the usual fit, on which the countess, *prudens et cordata matrona*, with the aid of her daughter pulled up her dress and gave her a good whipping. 'She confessed to me afterwards that it completely cured her.' Extreme diseases, adds Weyer, require, according to Hippocrates, extreme remedies, but care should be taken to distinguish suitable cases.[2]

The last and most important section of the book treats of the punishment of witches, who are to be carefully distinguished from poisoners and magicians, such as Faust, who are often wealthy men and spend much money in travel, books, &c., to learn diabolic arts; or deceivers, such as the mason who buried wolves' dung in a cattle stall, and when the animals showed great excitement, said they were bewitched, and offered to cure them for a consideration. Such men, when proved to have done serious harm, are to be severely

[1] Op. cit., v. 34. [2] v. 35.

punished. The less guilty should be admonished, and among them are those who spread superstitious practices and persuade sick people that they are bewitched by some old woman.

This is all that the laws of Church or State require, and is a very different thing from seizing poor women possessed by diabolic delusions, or on the malicious accusations or foolish suspicions of the ignorant vulgar, and casting them into horrible dungeons, whence they are dragged to be torn and crushed by every imaginable instrument of torture, till, however guiltless they are, they confess to sorcery, since it is better to give their souls to God in innocence, even through flame, than longer endure the hideous torments of bloodthirsty tyrants. And should they die under torture or in prison, the accusers and judges cry out triumphantly that they have committed suicide, or that the devil has broken their necks.

Here follows a burst of indignant eloquence which would have cost Weyer dear had he fallen into the clutches of the witch-hunters, and which may be given in the terse vigour of the original:

'Sed ubi tandem is apparuerit quem nihil latet, Scrutator cordium et renum, ipsius abstrusissimae etiam veritatis Cognitor et Iudex, vestri actus palam fient, O vos praefracti tyranni, O iudices sanguinarii, hominem exuti et caecitate ab omni misericordia procul remoti. Ad ipsius extremi iudicii tribunal iustissimum vos provoco, qui inter vos et me decernet ubi sepulta et culcata Veritas resurget vobisque in faciem resistet latrociniorum ultionem exactura.'[1]

Their credulity almost equals their cruelty, as shown by the belief that a certain old woman caused the excessive cold of the preceding winter, and by the absurd swimming test. What effect can denial of faith, evil intentions, or a corrupt fantasy have upon a person's specific gravity, on which floating depends? Moreover, women usually float, since their specific gravity is less than that of men, as Hippocrates pointed out.[2] But nothing is too absurd for a witch inquisitor. Some fishermen at Rotterdam drew up their nets full of stones but fishless. This was clearly witchcraft, so they seized an unfortunate woman who confessed in her terror that she had flown out of the window through a hole the size of a finger-end, dived under the sea in a mussel-shell,[3] and there terrified the fishes

[1] vi. 4. [2] vi. 9.

[3] 'Mossel-scolp nostratibus dicitur.'

and put stones in the nets. The woman, says Weyer, was evidently mad or deluded by the devil, but they burnt her all the same. Treachery and cruelty go together. A priest, having failed to make a witch confess, promised that if she would admit some small act of sorcery, he would see that she was released after some slight penance. Thereupon she confessed and was burnt alive.[1]

In contrast to this, Weyer describes the method of dealing with witchcraft in the duchy of Cleves. In 1563 a farmer, finding his cows gave less milk than usual, consulted a witch-finder, who told him that one of his own daughters had bewitched them. The girl, deluded by the devil, admitted this and accused sixteen other women of being her accomplices. The magistrate wrote to the duke proposing to imprison them all, but the latter, probably at Weyer's instigation, replied that the witch-finder was to be imprisoned, the girl to be instructed by a priest and warned against the delusions of demons, and the sixteen women in no way to be molested.[2]

An old woman of eighty was arrested at Mons on charge of witchcraft, the chief evidence being that her mother had long ago been tortured to death on a similar charge. To make her confess they poured boiling oil over her legs, which produced blisters and ulcers, and her son hearing of it sent her a roll of lint to put round them. This was supposed to make magic bandages by the aid of which the woman might escape, and the son was promptly arrested. The mother was to be burnt in a few days, and her son would probably have followed, when Weyer, by permission of the Duke of Cleves, visited Count William of Mons and explained his views on witchcraft. He also examined the old woman, who was so broken down that she fainted several times, and finally obtained the release of both.[3]

Theologians (says Weyer in conclusion) may object that he is only a physician and bid him keep to his last. He can only reply that St. Luke was a physician, and that he is one of those who hope by the mercy of God and grace of Christ to attain that royal priesthood of which St. Paul and St. John speak. Finally he is ready to submit all he has said to the judgement of the Church, and to recant any errors of which he may be convicted.

The Church answered by putting his name on the *Index* as an *auctor primae classis*, that is, one whose opinions are so

[1] Op. cit., vi. 15. [2] vi. 16. [3] vi. 16.

dangerous that none of his works may be read by the faithful without special permission, while his book was solemnly burnt by the Protestant University of Marburg.[1] The Duke of Alva, then engaged in his notorious work in the Netherlands, used his influence to get Weyer removed from his position at the court of Cleves. In this he was aided by the duke's increasing melancholia and ill health, which were considered by many a judgement upon him for his protection of Weyer and neglect of witch-burning. In 1578 Weyer resigned his post to his son Galen, and in 1581 witch-hunting commenced in the duchy of Cleves. Weyer, however, as befitted the chivalrous defender of outraged womanhood, enjoyed the friendship and protection of Countess Anna of Techlenburg, at whose residence he died, 1588, aged seventy-two.

The work on *The Deceptions of Demons* has been aptly compared to a torch thrown out into the darkness, which for a moment brightly illumes a small space and then disappears. It made a temporary sensation, and was welcomed by a few of the more enlightened spirits of the time; it saved the lives of some unfortunate women (being successfully quoted the very year after publication in defence of a young woman at Frankfort, who confessed she had flown through the air and had intercourse with the devil), and it marks the beginning of an open and persistent opposition to the witch mania. Spee also has a curious story showing the influence of Weyer's book :

'A great prince invited two priests to his table, both men of learning and piety. He asked one of them whether he thought it right to arrest and torture persons on the evidence of 10 or 12 witches. Might not the devil have deceived them in order to make rulers shed innocent blood, as certain learned men had lately argued, "thereby causing us pangs of conscience"? The priest stoutly maintained that these pangs were needless, for God would never allow the devil to bring innocent men to a shameful and horrible death in this way; and so he (the prince) might continue the witch trials as usual. He persisted in this, till the prince said, "I am sorry, my father, you have condemned yourself and cannot complain were I to order your immediate arrest, for no less than 15 persons have sworn you were with them at the witch dances", and he produced the records of their trials in proof. Then the good man stood like butter in the sun in the dog-days, and had nothing more to say for himself.' [2]

[1] Diefenbach, p. 241.

[2] *Cautio Criminalis*, Dubium xlviii.

But it had little effect on the superstition itself, which reached its height during the following half-century; and the author is compelled by his religious beliefs to admit so much that his position is hardly tenable. Indeed, his premisses had already been granted by the witch-hunters themselves. The jurist Molitor, for instance, admits that much witchcraft is imaginary and due to the deceptions of demons, but while the physician argues that these deceptions are rendered possible by disease, and are themselves largely of the nature of disease, so that the victims deserve pity and medical treatment rather than burning, the lawyer asserts that a person can only be so deceived by his free will, and therefore a woman who believes she has made a compact or had intercourse with the devil is as deserving of punishment as if she had actually done so.[1]

Just over a century after the appearance of Weyer's book (1664)

'Sir Thomas Brown of Norwich, the famous physician of his time, was desired by my Lord Chief Baron [Hale] to give his judgement [in a case of witchcraft]. And he declared that he was clearly of opinion That the Fits were natural, but heightened by the devil co-operating with the malice of the witches at whose instance he did the villanies. And he added, That in Denmark there had been lately a great Discovery of Witches, who used the very same way of afflicting persons by conveying pins into them.'

The jury 'having Sir Thomas Brown's Declaration about Denmark for their encouragement, in half an hour brought them in guilty. . . . They were hanged maintaining their innocence.'[2]

Had Brown been better acquainted with *The Deceptions of Demons* he might have hesitated to make that 'Declaration about Denmark', but Weyer's early opponent, Bishop Binsfeld, has no difficulties. Quoting Origen (in Matt. xvii. 15) he exclaims, 'Physicians may say what they like, we who believe the Gospel hold that devils cause lunacy' and many other diseases.[3] But for a demon to cause disease or do other harm, two things are requisite, the permission of God and the free will of some malicious person, witch, or sorcerer. The physician, Weyer, has denied the possibility of a compact with the devil, but is easily refuted by

[1] U. Molitor, *Tractatus de lamiis*, 1561, p. 27.

[2] Hutchinson, *Historical Essay concerning Witchcraft*, London, 1718, pp. 40, 118, 120.

[3] Op. cit., Preludium, i.

Scripture and Church authority. Did not the devil try to make a compact with Christ Himself ? [1] Similarly he has no difficulty in showing that the Hebrew word for witch means much more than 'poisoner', and, given the almost universal beliefs of the age, it must be admitted that Brown and the bishop have the best of the argument.

In the opening chapter of his well-known work on rationalism, Lecky says that the decline of the belief in witchcraft 'presents a spectacle not of argument and conflict, but of silent evanescence and decay'; it was 'unargumentative and insensible'. Scot's work 'exercised no appreciable influence', and, so far as the result was concerned, he, Weyer, and their like might as well have kept quiet and waited for the change to be effected by 'what is called the spirit of the age', that is, 'a gradual insensible yet profound modification of the habits of thought' due to 'the progress of civilization'. This theory has been ably criticized elsewhere.[2] The truth it contains seems to be that argument would not have sufficed to change public opinion about witchcraft, without the aid of changes in other matters, and especially the development and success of scientific investigation. Such discoveries as the motion of the earth and circulation of the blood, when generally accepted (which was not till late in the seventeenth century), showed that the learned as well as the vulgar might be utterly mistaken in important beliefs supported by apparently good evidence, and that scientific methods of attaining truth differed widely from those of the witch-hunters.

The progress of civilization by practically abolishing the use of torture would alone have immensely diminished the number of victims, and of those 'confessions' on which the belief was fed. To use military language, the witch mania was an ugly and formidable redoubt connected with other forts and entrenchments. It suffered somewhat from the bombardment by Weyer and Scot, but could only be finally demolished by a general advance of the forces of science and civilization. But if every one had trusted to 'the spirit of the age' rather than disturb his neighbours' beliefs, we might still be burning our grandmothers.

Though born in what is now Holland and educated in France, German writers claim Weyer as their countryman and compare

[1] Preludium, vi.

[2] J. M. Robertson, *Letters on Reasoning*, London, 1905, cap. vi.

him with Martin Luther. The monk of Wittenberg is indeed a fine figure with his 'Here stand I; I cannot otherwise, God help me!' But he had half Germany behind him; both princes and populace were ready to protect him. Weyer stood practically alone, and if he escaped being burnt by jurists and theologians, had a fair chance of being lynched by an enraged mob as a sorcerer and protector of witches. There was little to save him from torture and death but the strength of mind of Duke William of Cleves, who came of an insane family and already showed signs of melancholia.

Weyer was happily spared such a trial of his fortitude, but none the less does he deserve our admiration as the chivalrous champion of womanhood, who first, with vizor up and lance in rest, greeted, alas! not, like the knights of legend, by prayers and blessings but by threats and imprecations, went forth to do open battle with the hideous monster which had so long tortured and slain the innocent and helpless.

وهذا كتاب الاسباب والعلامات

للحكيم موسى بن ميمون القرطبي الاسرايلي

العلامات ان كانت النزلة باردة فجة وكانت زكامية
كان في الانف او في احد منخريه سدد وفي الصوت غنة ويجد
العليل في الانف والجبهة ثقلا ويهر حاور [illegible] من المادة ماله غلظ
ولزوجة وان كانت حادة وكانت زكامية فحرارة في الوجه والانف
ورقة السيلان ولذعه وحدته وحرارة ملمسه وحمرة في الوجه وكذلك
علامة ما ينزل في الصدر منها ان كان حادا الذاعا في الحلق ودغدغة وحكة
وسعال العليل الدموي الحاد وان كانت غليظة باردة فثقل الصدر وظهور
النفث الغليظ وربما كان معها ضيق نفس ن العلاج الفصد عند
ابتدائها او الاسهال ان دعت الحاجة وتلطيف الغذاء والاقتصار على
حسو ماء النخالة ودهن اللوز والسكر الطبرزد وسويق الشعير واكل المزورة
من الماش او العدس مع حلبة اللوز وكذلك يلزم الحمام الى ان يظهر النضج
وينزل من الانف ما يخرج من المادة فيدخل حينئذ الحمام ويشطل [illegible] الماء الفاتر على
مقدم الراس ويستنشق بخار الماء المطبوخ فيه البنفسج والنيلوفر والبابونج
واكليل الملك ولا ينبغي ان يدخل الحمام الا بعد النضج الا ان يكون مع ذلك حمى
ولا ينبغي ان يقرب الحمام فان كانت النزلة حادة رقيقة يلطخ الخل على حجر محمى
ويستنشق بخاره او يطفى على الجمر نخالة وقد نقعت في الخل ويستنشق بخارها

Plate XLI. THE BODLEIAN MANUSCRIPT
MS. MARSH 379 fo. 73

THE 'TRACTATUS DE CAUSIS ET INDICIIS MORBORUM'

كتاب الاسباب والعلامات

ATTRIBUTED TO MAIMONIDES

By Reuben Levy

Among modern authorities on Arabian medicine, the opinion has been widely held that the position of Maimonides as a medical writer must depend mainly upon an unpublished work from his hand, known as the *Tractatus de Causis et Indiciis Morborum.*[1] It is here sought to demonstrate that the Bodleian MS. (Marsh 379), hitherto regarded as containing this work, is in reality by another author, while the Paris MS. (Bibliothèque Nationale, Ancien Fonds 411),[2] the only other alleged copy of the *Tractatus de Causis et Indiciis Morborum*, contains in fact no such work. Moreover, evidence will be adduced showing that it is not probable that Maimonides composed a treatise of this scope.

For their information concerning the *Tractatus*, the modern bibliographers evidently rely entirely on entries in the catalogues of the respective libraries. The 1739 Catalogue of Arabic and Hebrew MSS. in the Bibliothèque Nationale contains the following entry:[3] 'Codex bombycinus, Aleppo in bibliothecam Colbertinam anno 1673 illatus, quo continetur R. Mosis Maemonidae de morborum causis et illorum curatione tractatus, Arabice, charactere Hebraico.' Careful examination of the manuscript disclosed the fact that it contained no fewer than four works of

[1] See (*a*) H. Haeser, *Geschichte der Medizin*, Jena, 1875–82, vol. i, p. 596; (*b*) A. Hirsch, *Biographisches Lexico nder hervorragenden Aerzte*, Leipzig, 1884, art. 'Maimonides', vol. i, p. 178 f.; (*c*) K. Brockelmann, *Geschichte der arabischen Litteratur*, Weimar, 1897–1902, vol. i, p. 490.

[2] = No. 1211 in Zotenberg's Catalogue, Paris, 1866.

[3] Vol. i, p. 40, Cod. 411.

Maimonides, viz. on Poisons,[1] on Asthma,[2] the *Tractatus de Regimine Sanitatis*,[3] and the *Tractatus de Morbo Regis Aegypti*,[4] all bound together in confusion.[5] All these are known to be by Maimonides, and there is nothing besides them in the volume.

There has always been a good deal of confusion about the works *de Regimine Sanitatis* and *de Morbo Regis Aegypti*. The former is variously known as *de Regimine Sanitatis, de Cibo et Alimento, de Dietetica*, 'the letter to the Sultan', or as 'the Consultation concerning (the Sultan) Al Afḍal'.[6] The latter also has a number of titles, such as *de Causis Accidentium*,[7] *de Morborum Causis et Curatione*, and *Responsum ad Regem Raqqa*, in addition to its title of *de Morbo Regis Aegypti*. In 1514, in Venice the two treatises were printed together in Latin as one work.[8]

Leclerc [9] has made confusion worse confounded by saying that 'ce que l'on a désigné sous les titres, *De Morbo Regis Aegypti, De Causis Accidentium, De Causis et Indiciis Morborum, De Cibo et Alimento*, ne sont autre chose que tout ou partie du même ouvrage'.[10] No doubt he was led into making this statement

1 في السموم. Translated into Latin by Armengaud de Blaise of Montpellier; into French by J. M. Rabbinowicz, *Traité des Poisons de Maimonide*, Paris, 1865, and into German by M. Steinschneider, *Gifte und ihre Heilung, eine Abhandlung des Moses Maimonides.* Virchow's *Archiv*, LVII, vol. i, pp. 92–109.

2 في الربو. Unprinted. We hope shortly to issue this work.

3 في تدبير الصحة otherwise رسالة الافضليّة. 'Letter to [the Sultan] al Afḍal.' Printed in Latin at Florence, n. d.; Venice, 1514, 1521, &c.; Leyden, 1535; in the Hebrew translation of Moses ibn Tibbon edited by Jacob Saphir ben Levi, Jerusalem, 1885; and in German by Winternitz, *Diätetisches Sendschreiben des Maimonides*, &c., Vienna, 1843.

4 Printed in the Latin edition [Venice, 1514] of the *de Regimine Sanitatis* as Tractatus V of that work.

5 See L. Leclerc, *Histoire de la médecine arabe*, Paris, 1876, vol. ii, p. 60, and M. Steinschneider, *Die hebräischen Uebersetzungen des Mittelalters und die Juden als Dolmetscher*, Berlin, 1893, pp. 767, 772, 773.

6 رسالة الافضليّة.

7 في اسباب الاعراض and also في بيان الاعراض = on the diagnosis of accidents.

8 See note 4.

9 Op. cit., vol. ii, p. 61.

10 See Steinschneider, *Hebräische Uebersetzungen*, p. 770, and his *Catalogus Librorum Hebraeorum in Bibl. Bodl.*, Berlin (1852–60), p. 1921. In the *Zeitschrift der Morgenländischen Gesellsch.*, vol. xxx, p. 145, he makes the bare statement that the *Tractatus de Causis et Indiciis Morborum*—the *Hauptwerk* of Maimonides, as it is called by Haeser—rests upon an error. In his catalogue of Bodleian books (p. 1926) he puts the book down as a bookseller's fraud after what is obviously only a cursory glance. He says 'fraude bibliopolae ex variis opp. imperfectis confictus est, in quibus an Nostri sit aliquid non facile eruendum est'.

partly by the fact that Wüstenfeld [1] gives the title of *de Causis et Indiciis Morborum* both to the Bibliothèque Nationale MS. (which Leclerc knew as *de Causis Accidentium*) and to the Bodley MS.

The entry concerning the latter in Uri's Bodleian Catalogue of 1787 [2] reads as follows:

'Codex bombycinus, anno Hegirae 765, Christi 1363 exaratus, folia 116 implens. Comprehendit succinctum de omnium corporis humani morborum causis, signis et remediis tractatum ab Ibn Hobaish Hierosolymitano ex Hebraica lingua in Arabicam conversum, cui sectiones sex supra centum sunt. Initium fit a morbis capitis; finis in elephantiasi. Composuit Musa Ben Maimun Alcortubi, Israelita. [Marsh 379.]'

The MS. bears upon one of its pages the title

هذا كتاب الاسباب والعلامات الحكيم
موسى بن ميمون القرطبي الاسرايلي

'This is the book of the causes and symptoms, by the Doctor Mûsa ibn Maimûn the Cordovan, the Israelite.' (Plate XLI.)

Aa a matter of fact it is no such thing. This title, together with an extra title-page and colophon in the same hand, is a much later addition to the MS., which also has a fragment of some other medical work—at present unidentified—bound up with it. The folios of the MS. which deal with the *Tractatus* have been bound together in extreme disorder, but examination of them has shown that they really form a fragment of the second book of المختار في الطبّ, the *Delectus de Medicina*, by مهذّب الدين ابو الحسن على بن احمد البغدادي, Muhaddib ed Din Abu'l Hasan Ali Ibn Aḥmad of Bagdad.[3]

Ibn Abi 'Uṣaibia (1203–1269)[4] gives a life of this writer and a list of his works, which includes the *Delectus de Medicina*. Accord-

[1] H. F. Wüstenfeld, *Geschichte d. arabischen Aerzte*, Göttingen, 1840, § 198, No. 7.

[2] *Bibliothecae Bodleianae codicum manuscriptorum Orientalium . . . catalogus a Joanne Uri confectus*, Oxford, 1787, vol. i, p. 140, No. 594.

[3] Also known as الاخلاطي (of Akhlat) or التبريزي (of Tibriz) and as ابن هبل (Ibn Hubal).

[4] Ibn Abi 'Uṣaibia wrote an invaluable dictionary of the lives of the most noted physicians, entitled كتاب عيون الانباء في طبقات الاطباء (= The book of the sources of information concerning the various classes of physicians). It is especially full on the lives of Arab physicians. See the edition of A. Müller, Königsberg, 1884, vol. i, pp. 304–6.

ing to him, Muhaddib ed Din was born at Bagdad in A. H. 515 (= A. D. 1121), and after studying medicine and philosophy settled at Mosul. Later he became the physician of the Shah Arman, chieftain of Khalāt on Lake Van in Armenia, in whose service he amassed great wealth. He completed the *Delectus* at Mosul in the year A. H. 560 (= A. D. 1164), and died there in A. H. 610 (=A. D. 1213), with the reputation of being first physician of his time.

Another fragment of the same work of Muhaddib ed Din, which includes most of the contents of the Bodleian MS., besides a good deal of material which has been lost from the latter, exists in the British Museum.[1] The Leyden Library contains a unique copy of the work in three books. This is claimed to be complete by the Catalogue of the library,[2] although Bar Hebraeus [1226–1286] —Catholicus of the Jacobite (Monophysite) Church [3]—says that the work ran into four parts.[4] The three books of the Leyden MS. treat (i) of generalities (i. e. Anatomy, Physiology, and the general causes of disease), (ii) of medicaments, and (iii) of particular diseases and their treatment.

The Bodleian and British Museum MSS. contain part of the third book, which was probably in general use by itself as a dictionary of medicine. The British Museum copy has only lost the earlier chapters of this third part, but the Bodleian MS., although possessing a few more chapters at the beginning, is far less complete in the other portions.[5]

Wüstenfeld and the bibliographers that followed him have

[1] C. Rieu, *Supplement to the Arabic MSS. in the Brit. Mus.*, London, 1894, No. 796, II.

[2] Vol. iii, p. 242 of the Catalogue of Arabic MSS. compiled by P. de Jong and M. J. de Goeje, Leyden, 1865–6.

[3] Abu'l Faraj Gregory, Bar Hebraeus (Wüstenfeld, op. cit., No. 240).

[4] In his work entitled تاريخ مختصر في الدول, 'Compendious History of the Dynasties' (edited and translated by E. Pocock, Oxford, 1663), p. 457 f. of the Arabic and p. 300 of the Latin. Beyrout edition, 1890, p. 420.

[5] Two MSS. of the work are mentioned in the Catalogue of the Khedive's library, فهرست كتابخانه خديوية, vol. vi, p. 38. For further references concerning Muhaddib ed Din and his works, see (*a*) Wüstenfeld, op. cit., § 202; (*b*) Brockelmann, op. cit., vol. i, p. 490; (*c*) P. de Koning, *Traité sur le calcul*, Leyden, pp. 186–228. The more important Arab authors other than Ibn Abi 'Uṣaibia are: (*d*) Bar Hebraeus, Pocock's edition, p. 457 of theArabic part and p. 300 of the Latin part, Beyrout edition, p. 420; (*e*) Haji Khalfa, G. Fluegel's edition, Leipzig and London, 1835–58, vol. v, p. 436, No. 11584.

evidently derived their information concerning these MSS. from the catalogues of the Bodleian Library and of the Bibliothèque Nationale. No mediaeval bibliographer has up to the present been found who mentions this book of Maimonides.[1] Wüstenfeld's usual authority for his statements is the great thirteenth-century medical biographer, Ibn Abi 'Uṣaibia. But, though the latter gives a life of the Hebrew physician and a list of his writings,[2] he makes no mention of the *Tractatus de Causis et Indiciis Morborum.* Moreover, this *Tractatus* has no place in Haji Khalfa's admirable bibliography of Arabic works, which contains notices of four books bearing the title *De Causis et Indiciis Morborum*, not one of which is by Maimonides. Lastly, neither the historian Al Qifty in his *Classes philosophorum et astronomorum et medicorum*,[3] nor Bar Hebraeus, who is said to have plagiarized him,[4] notice the work in their sketches of the physician's life.

The Bodleian MS. alleged to contain the *Tractatus* is one of a collection of over seven hundred volumes bequeathed to the library on his death, November 2, 1713, by Narcissus Marsh, Archbishop successively of Cashel, Dublin, and Armagh. Most of his Oriental MSS. had been procured for him either in the East by Robert Huntington, Bishop of Raphoe and chaplain to the English merchants at Aleppo, or at the sale of Golius's library at Leyden in October 1696.[5] Golius was a Dutch orientalist, born at Leyden in 1596. He studied medicine and Oriental languages at the University of Leyden, and after leaving it he accompanied a French embassy to Morocco in 1622. He remained in Morocco for two years, and while there collected various MSS. On his return in 1624 he was appointed to the Chair of Arabic at Leyden, but was allowed a period of leave for travel in the East before taking up his appointment. He took with him a grant of money for the purchase of MSS., and these to the number of over two hundred are now deposited in the University

[1] See J. Pagel, 'Maimuni als medizinischer Schriftsteller', in the volume of studies on 'Moses ben Maimon' edited by W. Bacher and others, Leipzig, 1908, vol. i, p. 232.

[2] Op. cit., vol. ii, p. 117.

[3] طبقات الحكماء واصحاب النجوم والاطباء in MS. at British Museum (see Catalogue of Oriental MSS. at the British Museum, London, 1846, part II, No. 1503, p. 684), Leyden, Berlin, Escurial, and elsewhere. See Brockelmann, op. cit., vol. i, p. 325.

[4] See Leclerc, op. cit., vol. i, p. 5.

[5] See W. D. Macray, *Annals of the Bodleian*, Oxford, 1890, p. 270.

Library at Leyden. On several occasions during his travels in Arabia attempts were made by Arab chiefs to detain him for his medical knowledge, but he returned safely and later wrote a number of works mainly concerned with Arabic. He died in 1667.

Among the MSS. which Golius himself procured for the Leyden Library was that of the *Delectus*. It is at least unlikely therefore that such a profound Arabist, who was also a medical man, would have bought the Bodley fragment for a genuine work of Maimonides; the primary responsibility for the error thus probably rests with Huntington. However that may be, it was Uri, in his catalogue of the Bodleian MSS., who first published the error, and from him it was passed on to the modern bibliographers.

John Uri was a Hungarian who had studied Oriental literature under Schultens at Leyden, and was recommended to Archbishop Secker for the purpose of cataloguing the Bodleian Oriental MSS., by Sir Joseph Yorke, then ambassador in the Netherlands.[1] Many years were occupied in the preparation of the work, which appears to have commenced in 1766 and was not completed till 1787. In spite of the length of time which Uri occupied in his task, his successor, Pusey, found sufficient errors in it to fill sixty closely printed pages. In his preface to the second volume of the Catalogue, issued in 1835,[2] Pusey complains 'Urius vero MSS. haud raro negligenter exscripsit', and says that on re-examination of Uri's work he discovered, 'besides the errors which Uri himself would have admitted, that nearly all the purchasers of these books, Pocock alone excepted, had had spurious works foisted on them by wily Orientals. He therefore looked through all the books which Uri had enumerated, excepting the more common ones, to see if they corresponded to their titles or not. By doing this he discovered various irregularities. In some cases the titles had been covered over with paper or obliterated with ink, or practically erased with a knife. In others, by slight changes in the authors' names, more famous people were indicated as responsible for the works. Lastly, by changing the pagination in some of the volumes fragments were represented as complete

[1] See Macray's *Annals of the Bodleian*, p. 271, and the *Dict. of National Biography*.

[2] *Bibliothecae Bodleianae codicum manuscriptorum . . . catalogus*, vol. ii, ed. A. Nicoll and E. B. Pusey, Oxford, 1835, p. iv.

works, and a few pages of one work were even occasionally sewn on at the beginning of another.'[1]

Uri's errors will be the more readily condoned when it is remembered that he did not specialize on the Arabic MSS. alone, and that his work seeks to catalogue, *for the first time*, a two hundred years' accumulation of Oriental MSS., including Hebrew, Aramaic, Syriac, Aethiopic, Arabic, Persian, Turkish, and Coptic writings. Nevertheless, Uri's entry with reference to the present MS. deserves some of Pusey's criticism. The MS. has three parts, each written in a different hand, the first and most important part being the supposed *Tractatus de Causis et Indiciis Morborum*, which covers folios 2–87. The second part is a fragment of some as yet unidentified medical work (folios 88–115); and the third, consisting of the first and last folios, gives us an introduction and an end piece to the first part.

The alleged author and translator are named on the first page:

هذا كتاب موسى ابن ميمون الفه
للعموم قاطبا وقد نقله التميمي
الشيخ سليمان الحبشى المكنا بابن حبيش
في مملكة القدس الشريفة
تم

'This is the book of Mûsa ibn Maimûn which he put together as a compilation for general use. Al Tamimi, the sheikh Sulaiman the Abyssinian, known as Ibn Ḥubaish,[2] translated it in the noble city of Jerusalem. Finis.'

[1] 'Praeter errores enim quos ipse admiserit Urius, deprehendi omnibus fere horum librorum emptoribus, uno Pocockio excepto, libros supposititios pro veris subinde venditasse vafros Orientales. Codices ergo fere universos Arabicos, quos recensuit Urius (vulgatioribus quibusdam exceptis) oculis perlustravi, quo certius scirem titulisne responderent an non. Quo facto varias errorum formas deprehendi, titulis nunc charta coopertis, nunc atramento oblitis, nunc cultro paene abrasis; auctorum porro nominibus paullulum immutatis quo notiora quaedam referrent, numeris etiam quibus singula volumina signata sunt permutatis, quo quis opus imperfectum pro integro habeat, paginis denique pauculis operi alieno a fronte assutis.'

[2] Steinschneider (*Cat. Libr. Hebr. in Bibl. Bodl.*, p. 1926) says this title is invented and no doubt suggested by the name of Al Tamimi al Muqaddasi (the Jerusalemite), a doctor of the tenth century (Wüstenfeld, § 112) often praised by Maimonides in the Aphorisms, e.g. at the end of chap. 20. Pusey's only note on Uri's entry in the MS. is concerned with this title (vol. ii, p. 588): 'Translator in Cod. appellatur Alsheikh Soleiman Alhabashi, notus in terra Hierosolymitana nomine Ibn Hubaish. Opus autem A. D. 1363 ex Hebraico transtulit.'

On the next page there is an introduction to the book which commences :

بسم الله الرحمان الرحيم
قال موسى ابن ميمون القرطبى الاسرايلي الخ

‘In the name of God, the Merciful, the Compassionate.
‘So says Mûsa ibn Maimûn, the Cordovan, the Israelite,’ &c.

The whole of the passage is an extract from chapter vi of the Aphorisms of Maimonides, adapted as a kind of introduction, and runs as follows :

قد علمت في قولي هذا في قوة النفسيه والقوة الحيوانيه والقوة
الطبيعيه ولنسم الان في هذا الاصطلاح جميع افعال البدنيه
للانسان قول ان اشرف الافعال التنفس وبعده النبط والاحساس
واشرف الاحواس البصر ثم السمع وبعده الاحساس شهوة الطعام
والشراب وبعد ذلك الكلام وبعد ذلك التمييز اعني الذي بها التفل
والفكر وبعد ذلك الخلافه لساير الاعضاء علي المعتادة وهذه الرتبة (sic)
في شرف انما هي بحسب ضرورية الحيوة او صالحية فتعلم ان
الطبيعة اسم مشترك يقال علي معنى كثيرة كالقوة المدبرة (sic) الحيوان
ايضا طبيعية وما هو اشرف وتمسكت للاشرف في الاشراف وهذه
الاسباب الذي قد راينا ورتبناها وهو الابتداء في النزلات الزكاميه
من الراس ٭

Trans. ‘I teach in this discourse of mine concerning the animal power, the vital power, and the natural power, but we will here call all man’s bodily functions by one name. There is a saying that the noblest of the functions is breathing, next the pulse, and lastly the senses. Of the senses, the noblest is sight, which is followed by hearing. Following on the senses is the appetite for food and drink, after it being speech and then the mind ; I mean that which contains the reason and the intellect. Next comes the [?] allocation of [the various powers to] the other parts of the body according to the customary manner. This arrangement in order of nobility is only according to the requirements of life or [?] health.

‘You will recognize that “nature” is an equivocal term which can be used in many meanings. [One of these meanings,] for example, is “the motive power of animals”. So, too, is “natural”.

[?? . . .] ‘and that which is nobler. And you will retain the noblest of the noble [functions]. And these causes which we have noticed we have set down in their order; and the beginning is concerning catarrhal discharges from the head.'

Compare with this the real text of Maimonides:[1]

קד עלמת קול אלאטבא קוי׳ נפסאניה וקוי׳ חיואניה וקוי׳ טביעיה·
ולנפס אנא אלאן פי הדה אלאצטלאח גמיע אפעאל בדן אלאנסאן אלאפעאל
אלבדניה [ואשרף אלאפעאל אלתנפש ובעדה אלנבט[2]] ובעדה אל אחסאס·
ואשרף אלחואס אלבצר· תם אלסמע ובעד אלאחסאס שהוה̈ אלטעאם
ואלשראב· ובעד דאלך אלכלאם· ובעד דאלך אלתמייז· אעני בה אלתכייל
ואלפכר ובעד דאלך חרכה̈ סאיר אלאעצֹא עלי מעתאדהא והדה אלרטבה[3] (sic)
פי אלשרף אנמא הי בחסב צׄרוריה אלחיאה או צלאחיה̈ אסתמראהא״
ובעד הדה אלמקדמה̈ פלתעלם אן אלטביעה אסם משתרך יקאל עלי
מעאני כתירת ומן גמלה̈ תלך אלמעאני אלקוה אלמדברה̈ לבדן אלחיואן
פאנהא אלאטבא יסמונהא איצׄ טביעה והדה אלקוה הי איצׄ . . . . . .
פאן גלבת ען דלך בדלת מא הוא אשרף ותמסכת באלשרף פאלאשרף
ובחסב הדה אלתרתיב יעלם אלמרץ אלח׳·

'Thou knowest the opinion of the physicians [concerning] animal power, vital power, and natural power. But it is my intention here to call all the functions of man's body by the one name of "bodily functions". [The noblest of the functions is breathing, next the pulse,[2]] and lastly the senses. Of the senses, the noblest is sight, which is followed by hearing. Following on the senses is the appetite for food and drink, after it being speech and then the mind, by which I mean the thoughts and the intellect. Next comes the motion of the other parts of the body according to their customary manner. This arrangement in order of nobility is only according to the requirements of life or the health of its faculties.

'From this preface you will recognize that "nature" is an equivocal term which can be used in many meanings. One of these meanings [for example] is "the motive power in the bodies of animals" which the physicians call "nature" too. . . . And if you discover this, you will exchange that which is nobler and retain that which is noblest. By means of this process of arrangement, a disease can be recognized,' &c.

[1] From the text of the Aphorisms as given in the Bodleian MS. Pocock 319.

[2] Omitted from the MS. obviously by accident.

[3] No doubt for רתבה̈.

This introduction was added when the folios stood in a state of disorder different from their present one. The catchword at the bottom of the page [وهذا = and this] points forward to the title already mentioned,[1] which appears on folio thirty-nine of the present arrangement. The text below this title is part of the chapter on discharges and catarrh, so that the folio once followed immediately on the introduction, being then, too, out of its proper place.

The last page, written in the same hand as the introduction, bears a piece of some unidentified work and a colophon which reads:

وقد تم هذا الكتاب الشريف تاليف موسى ابن ميمون القرطبى
الاسرايلى رحمه الله مما الف وجرب هذا الكتاب المبارك، وعدد فصوله
مائة وست فصول للجميع (sic) امراض البدن مما رتبه على اوضاعه
تم الكتاب فى سنة ٦٥٢٧ (sic) سبع مائة وخمسة وستين *

‘This noble book is finished; the composiṭion of Mûsa ibn Maimûn the Cordovan, the Israelite, to whom God be gracious. This blessed book is part of that which he composed and tested. The number of its chapters is 106, dealing with all the diseases of the body, which he arranged in their proper order.

‘The book was completed in the year 765.’ [2]

The number 106, which according to the colophon is the number of chapters in the book, is really the number of titles in the MS. written in large hand. Fragments of many chapters whose titles are lost still remain in it however, while many of the chapters that have preserved their titles are no longer complete.

Again it may be pointed out that all the known medical works of Maimonides were written in Arabic and therefore did not need to be translated into that language as the Bodleian MS. claims to have been. The spurious title-page thus further betrays itself by saying that this work was translated from *Hebrew*.

Finally, the identification of the real contents of the Paris MS. disposes of the last foundation of the idea that Maimonides wrote any compendium of medicine known as كتاب الاسباب والعلامات (*Tractatus de Causis et Indiciis Morborum*), and clears up the confusion caused by the faulty entries in the Paris and Bodleian catalogues.

[1] See p. 227.

[2] = A. D. 1363. The numerals which accompany the written figures are equivalent to 6,527 and are meaningless.

SCIENTIFIC DISCOVERY AND LOGICAL PROOF

By F. C. S. Schiller

§ 1. Among the obstacles to scientific progress a high place must certainly be assigned to the analysis of scientific procedure which Logic has provided. This analysis has not only been inadequate in itself, but has set itself a mistaken aim. It has not tried to describe the methods by which the sciences have actually advanced, and to extract from their experience the logical rules which might be used to regulate scientific progress, but has treated scientific discoveries almost entirely as illustrations of a preconceived ideal of proof, and so has freely rearranged the actual procedure in accordance with its prejudices. For the order of discovery there has been substituted an order of 'proof', and this substitution has been justified by the assumption that if discovery had taken the ideally best course, it would have coincided with the process of proof. It followed, of course, that the same logic would do for both, and that this logic was already in existence.

The damage thus inflicted upon Science was twofold. Not only were the logicians given a plausible excuse for persisting in their profound misapprehension of scientific inquiry and rendered incapable of giving any help or guidance in the solution of actual problems, but, what was much worse, the scientists themselves were misled about the nature of their operations.

The precise value of the service which a correct logical analysis of its procedure might have rendered to Science is perhaps open to dispute, though it must surely be beneficial to operate consciously, and with a full understanding of their nature, the methods which have been hit upon empirically; but even if logicians have commonly been too unfamiliar with the details of scientific problems to offer much practical advice, it would be difficult to overrate the mischiefs which must have resulted from referring scientists to an incorrect analysis of their actual procedures. For the attempt to justify by such a false ideal what they had actually done was bound to divert their attention from the methods that were actually effective and fruitful to others which were impracticable and sterile, to waste energy upon false aims and impossible ideals, and so to hamper scientists fatally in the exercise of their scientific rights and powers.

Hence it is not too much to say that the more deference men of science have paid to Logic, the worse it has been for the scientific value of their reasoning, while the less they have troubled to know about the theory of Science, the better it has been for their practice.

Fortunately for the world, however, the great men of science have usually been kept in salutary ignorance of the logical tradition and left to their own devices, by the accident that the historical organization of academic studies nearly everywhere confined 'logic' to the literary curriculum. Nevertheless, the moral of this situation is not that it is right for science to neglect logic and for logic to despise science, but that science should appeal from logic as it is to logic as it ought to be, and should insist on being provided with a *reformed* logic. For surely if a scientific education is to be more than a narrow and technical specialty, and is to exert a 'liberalizing' and broadening effect on the mind, it *ought* to include a study of scientific method in its generality and a certain understanding of the intellectual instruments by which all others are operated and constructed.

The whole evidence for these contentions it will not, of course, be possible to marshal within the limits of this essay, but the systematic criticism to which the whole traditional logic has been subjected in my *Formal Logic* [1] may perhaps absolve me from the duty of substantiating them exhaustively. It may suffice to indicate the extent of the scientific grievance against 'logic' by drawing up a list of problems in the logic of science which the traditional logic has misconceived, and then to select for fuller treatment a palmary example of the radical discrepancy between the two.

The traditional logic may be convicted of having gravely misrepresented, (1) the value of classification and the formation of classes, scientific processes of which the real logic was only revealed by the Darwinian theory, (2) the function of definition, (3) the importance of analogy, (4) of hypothesis and (5) of fictions, (6) the incomplete dependence of scientific results on the 'principles' by which they are (apparently) obtained, (7) the formation of scientific 'law' and its relation to its 'cases', (8) the nature of causal

[1] Published by Macmillans, 1912.

analysis. Other important features of scientific procedure cannot be said to have been recognized at all, e. g. (9) the problem of determining what is *relevant* to an inquiry and what practically must be, and safely may be, excluded, (10) the methods and justification of *selection*, (11) the essentially *experimental* nature of all thought and consequent inevitableness of *risk*, (12) the necessity of so conceiving 'truth' and 'error' that it is possible to *discriminate* between them, and (13) the need for an inquiry into *meaning* and into the conditions of its communication.

I

§ 2. The most instructive, however, of the discrepancies between 'logic' and scientific procedure will appear if we compare the logical notion of *proof* with the scientific process of *discovery*, and examine how far it can afford any means of regulating, stimulating, or even apprehending the latter. We shall find that the logical theory of 'proof' has no bearing on the scientific process of discovery, is not related to what the sciences call proof, and can only have a paralysing influence on any scientific activities which try to model themselves upon it. On the other hand, the study of the process of discovery will point to an important correction in the notion of logic.

§ 3. The scientific uselessness of the traditional logic should not, however, excite surprise. For what reason was there to expect that the theory of proof should turn out to be adequate, or even relevant, to scientific procedure ? It had sprung from a totally different interest, proceeded on different assumptions, and aimed at different ends. It did not spring from interest in the exploration of nature, and did not aim at its prediction and control. Nor did it presuppose an incomplete system of knowledge which it was desired to extend and improve. It originated in a very special context, from the social need of regulating the practice of dialectical debate in the Greek schools, assemblies, and law-courts. It was necessary to draw up rules for determining which side had won, and which of the points that had been scored were good.

These were the aims Greek logic set itself, and successfully achieved. But the impress of this origin remains stamped all over it, and the accounts given of logical proof ever since have retained essential features of Greek dialectics.

Thus it was assumed that science could start from principles, as indisputable as are the current meanings of words in a dialectical debate, and the end of the whole theory of proof was always conceived as being to secure the conviction (ἔλεγχος) of one party to a dispute, who was to be definitely crushed by the triumphant cogency of a syllogistic demonstration, while the more real and fruitful analogy between scientific inquiry and debate, viz. that *there is always another side*, to which also it is well to listen, was unfortunately obscured by Aristotle's discovery of the syllogistic form and its show of conclusiveness. But for the purpose of apprehending scientific procedure the syllogism is a snare: by putting scientific reasoning into syllogisms, the difference between the true and the false views is made to appear qualitative and absolute, instead of being a quantitative question of more or less of scientific value. Thus dogmatism is fostered at the expense of progressiveness, and the mistake is committed of approaching the discovery of truth in a party spirit. Hence its dialectical origin has become *fons et origo malorum* for logic.

§ 4. It is true that this mistake is very old, and has grown deeply into the fabric of logic. For Aristotle had no sooner worked out the classic formulation of the rules of dialectical proof than he proceeded to extend their scope by applying them to the theory of science, in the *Posterior Analytics*. His instinct in so doing was sound enough; for there is no better verification of a theory than its capacity to bear extension to analogous cases. And of course if this extension had been successful, it would have supported the belief that the theory of discovery could profitably be amalgamated with that of proof.

Unfortunately, however, the verification only seemed to be successful. Aristotle chose to exemplify his theory of scientific proof from the mathematical sciences. His choice was natural enough, because they were the only sciences which had reached any considerable development in his day, and they had, moreover, an apparent necessity and universality and a fascinating appearance of exactness. But he had unwittingly chosen the most difficult and deceptive exemplification of scientific procedure. Because the mathematical sciences were in a relatively advanced condition they seemed to lend themselves to his design. He could there find terms whose meaning, and principles whose truth, was no longer in dispute. They could in consequence be argued from with as much assurance as debaters could assume the recognized meanings of words. And the fact that results seemed to follow from mathematical definitions and premisses which were not merely verbal, shed a delusive glory on the forms of dialectical proof by which they had been reached. Hence it easily escaped notice that the logical superiority of mathematics was an achievement, not a datum. Just because the mathematical sciences were very ancient, their origins had been forgotten, and with them the tentative gropings which had first selected, and subsequently confirmed, their principles. They had become immediately certain and 'self-evident', and no one was disposed to dispute them. On this psychological fact the whole theory of logical proof was erected.

Again, it was natural to suppose that the true nature of scientific knowing must be revealed in its most perfect specimens: no one stopped to reflect that even so the real difficulties of making a science are more keenly felt and more easily seen in the nascent stage than in one which has victoriously overcome them, and has rewritten its history in the assurance of its prosperous issue.

Lastly, the subtle ambiguity which pervades all mathematical reasoning, according as its terms are taken as *pure* or as applied, was overlooked entirely—with the disastrous result that the universality, certainty, and exactness pertaining (hypothetically) to the ideal creations of 'pure' mathematics were erroneously transferred to their 'applied' counterparts. To this day logicians are found to argue that real space is homogeneous because it is convenient in Euclidean geometry to abstract from the multitudinous deformations to which bodies moving through it are subjected, and to leave them to be treated by physics;[1] nor are they aware of any lack of 'exactness' and discrimination when they identify the ideal triangle with the figures they draw on the blackboard.

§ 5. After its apparent success in analysing mathematical procedure there was no more disputing the supremacy of the theory of 'proof'. The facts that its field of application was soon found to be much narrower than that of science, and that it failed egregiously to apply to the procedures of the (openly) empirical sciences, and *a fortiori* could not justify them, if they were noticed at all, were held merely to show that these sciences stood on a low level of

[1] Cf. Mr. H. W. B. Joseph's *Logic* ², p. 548.

thought, which from the loftier standpoint of logic could be contemplated only with contempt; if they required help and got none, so much the worse for them. Accordingly the whole theory of science was so interpreted, and the whole of logic was so constructed, as to lead up to the ideal of demonstrative science, which in its turn rested on a false analogy which assimilated it to the dialectics of 'proof'. Does not this mistake go far to account for the neglect of experience and the unprogressiveness of science for nearly 2,000 years after Aristotle?

§ 6. Yet the deplorable consequences of this error should not render us unjust. The influence of Aristotelian logic on the theory of science was natural, and in a sense deserved. For Aristotelian logic is perhaps the mightiest discovery any man has achieved single-handed. Its might is sufficiently attested by the length of its reign. Euclidean geometry alone is comparable with it, and Euclid owed far more to his predecessors than Aristotle. Moreover, the Aristotelian logic may be said to have achieved its purpose. It was able to regulate dialectical discussion. The syllogism did determine whether a disputant had proved his case, and for any one who had accepted its assumptions its decision was final, while even its severest critics had to admit that it was an indisputable fact, the interpretation of which was a real problem.

Unfortunately, there is not yet any agreement among logicians about the solution of this problem. Aristotle's own analysis did not go back far enough: he stopped short at the *Dictum de Omni* and the reduction of syllogisms in the second and third figures to the first. He did not penetrate to the ultimate assumptions which were implied in the dialectical purpose and social function of the syllogism. But the truth is that syllogistic reasoning presupposes quite a number of conventions which Aristotle did not state, and which can hardly be said to have been adequately recognized since.

§ 7. (1) The first of these may be called the *Fixity of Terms.* Syllogistic reasoning manifestly depends on the assumption that the terms occurring in it have meanings sufficiently stable to stand transplantation from one context to another; for only so can they establish connexions between one context and another. Thus a syllogism in *Barbara* argues that because *all M is P* and *all S is M, all S must be P*. But it can do this 'validly' only if *M*, its middle term, remains immutably itself, and is the same in both premisses. Doubt, dispute, or confute this assumption, and the cogency of the syllogism as a form of 'proof' is overthrown at once. If the sense in which *M is P* is not the same as that in which *S is M*, the syllogism breaks in two, and its conclusion becomes precarious. Raise the question of how far reality conforms to this assumption, and you get at once a subtle problem of the applicability of the syllogistic form to the case in hand, which is precisely analogous to the question whether a theorem of pure mathematics is applicable to the behaviour of a real thing. In either case the cogency of the 'proof' which establishes the conclusion is impaired and ceases to be unconditional. The conclusion of a 'valid' syllogism will only follow *if* the middle term can be known to be unambiguous, and if the objects designated by the terms do not change rapidly enough to defeat the inference. And that this is the case can usually be ascertained only by actual experience. The conclusion, therefore, cannot be simply deduced; it has actually to *come true*, before we can be sure that the reasoning *was* sound. Absolutely *a priori* proof thus becomes impossible, if the assumption of the fixity of terms is contested: all proof becomes, in a sense, empirical.

Nevertheless, experience shows that the fixity of terms, though not a 'fact', is *a valid 'fiction'*: in ordinary discussion the terms may usually be taken as fixed enough to render valid syllogisms common. An ordinary debate proceeds upon the assumption that the meaning of the terms involved is fixed, and cannot be varied arbitrarily. To science, however, this assumption does not apply without restriction. In a progressive science the meaning of terms often develops so rapidly that such verbal reasoning does *not* suffice. Hence the mere occurrence of verbal contradictions in a scientific reasoning is no proof that the argument is unsound. It may show merely that its terms are *growing*.

It should be observed further that this same assumption is implied in the fundamental 'laws of thought' on which the traditional logic rests. Indeed, the notorious 'Law of Identity' seems to be merely another statement of it. It is usually formulated as '*A is A*', but in its actual logical use it is really the assumption that 'everything *is* what it is called'. It is, of course, anything but self-evident that '*A*' *is A*, but *unless* the *S*, *M*, and *P* of the syllogism are *rightly* so called, the syllogism will not hold. Similarly, the Law of Contradiction collapses at once if the terms to which it is applied are allowed to change. The inability of '*A*' both to be *B* and not to be *B* vanishes if '*A*' is not fixed and may change its habits. And of course the real things known to science all change, and are fixed only by a fiction. Hence every application of the logical convention to real things may be challenged: it involves a fiction and takes a risk, and both of these may be bad. But the traditional logic ignores both the risk and the fiction and the lack of cogency in its attitude.

§ 8. (2) It is a further presupposition of the syllogism that the meaning of its terms is *known*. When a discussion is begun the parties to it are supposed to understand each other, and not to have first to find out and form the meaning of the terms they use. This assumption also is roughly true in ordinary debate, and its convenience is manifest. If things are rightly named, and if this feat has been accomplished once for all—presumably by Adam and Eve before they were turned out of Paradise for trying to know too much—we shall escape many of the most trying difficulties of scientific inquiry. We need no longer trouble whether the best names have been given, and whether a name good for one purpose is equally good for another, nor need we inquire whether our names may not unite what is alien on account of a superficial likeness, or separate what is akin on account of a superficial difference.

In science, on the other hand, the assumption that we know what meanings our terms can convey is not made as a matter of course. We may begin with roughly labelling objects of interest, and then inquiries may be conducted into, e.g., 'electricity', 'elements', 'life', 'species', &c., in the hope of settling what these terms *shall* mean, and of finding out *more* about their meaning, and without making the assumption that whatever new facts are discovered about them must conform to our preconceptions and confirm our nomenclature. Thus to a man of science it will not be cogent to argue that because an 'element' is (by definition) an ultimate form of matter which cannot be broken up, and 'radium' breaks up, 'radium' is not an 'element', or that because 'species' are eternal forms, and the Darwinian theory claims that they are not immutable, it can be dismissed as involving the 'contradiction' that a 'species' is not a species. Thus the best syllogisms lose their cogency so soon as a question is raised whether the verbal identity of their terms is an adequate guarantee of the real identity of the things they are applied to.

§ 9. (3) It is a further presupposition of the logician's conception of 'proof' that absolute truths exist, and that in the ideal demonstration they form the premisses from which the conclusion follows. This presupposition is not stated, and is not implied in the form of the syllogism. For a syllogism is no less 'valid' if its premisses are true only hypothetically, and not absolutely. Indeed, it is not thought to impair the 'validity' of a syllogism that its premisses should be utterly false. At any rate we can *reason* quite as well with hypotheses and probabilities as with absolute truths, and this is in fact what we usually do, whether or not we are aware that our premisses are conditional and hypothetical. This ordinary practice, however, is resented by the traditional logic. For if our premisses are only hypothetically true, how can they lead to conclusions which can be declared absolutely true? And if our conclusions are not absolutely true, how can they be certain? Are they not bound to remain infected with the doubts which beset their premisses? [1] As we value the certainty of our conclusions, therefore, absolutely true and certain premisses must be procured. If they cannot be procured, even the best formal proofs will remain hypothetical, and all truth will become dependent on experience. For if nothing is true absolutely, and every truth has originated humbly in a guess that has grown into a successful hypothesis, it can always be suggested that after all it may benefit by a little more verification. It may be true enough psychologically and for practical purposes, but it does not realize the ideal of 'logical certainty'.

§ 10. This ideal Logic has formulated from the first. Aristotle already was not content with merely analysing the form of reasoning; he aspired to formulate the norm of scientific demonstration. The 'demonstrative syllogism', which he held to be the form of truly scientific reasoning, differs from the formal syllogism in two essential respects. Its premisses are absolutely true, and its middle term states the real 'cause', which connects its terms and is not merely a *ratio cognoscendi*. The reasoning proceeds, therefore, from premisses which are unambiguous, true, and certain, i. e. *necessarily* true and *absolutely* certain. Nor does the conclusion lose any of this excellence. Logic puts on a fine air of modesty,

[1] Cf. §§ 10, 28.

and merely claims that the syllogistic form is a guarantee that no truth can be *lost* on the way from the premisses to the conclusion in a 'valid' argument. If, therefore, our thought is properly arranged, our conclusion will be as true and certain as were its premisses, and no man will be able to gainsay it. It is the great beauty and merit of the syllogistic form that it is an arrangement which gives us this guarantee.

It was natural, therefore, that throughout the history of logic enormous importance should be attached to the acquisition of unquestionable starting-points. For the possession of 'valid forms' was not enough. It only insured against loss of truth, it did not provide for its acquisition. It seemed, however, to imply that truth could only be generated out of truth, and handed down from the premisses to the conclusion. Hence the insistent demand for assured starting-points, self-evident 'principles', which the infallible method of syllogistic deduction might conduct to equally certain conclusions.

In reality, however, this demand for certainty was extralogical: it is not required for the purpose of analysing reasoning. For it is just as easy to reason from doubtful and probable premisses as from certainties, nor need the doubt in the reasoner's mind affect the form of the reasoning. If, however, there is an imperative desire for certainty, it must be somehow gratified by logic. And there seemed to be no way of doing so except by ascribing absolute truth and certainty to the initial principles of science.

Of course it was covertly assumed that certainty could only be reached by *starting* from certainty, and that no possibility of a growth of assurance in the progress of the reasoning could be entertained. In a sense this assumption was correct (cf. §§ 27, 28), because it is true that the gradual verification of scientific truths does not render them absolute; but it led to neglect of all methods which appeared to start with premisses initially doubtful and hardening into certainties by gradual confirmation. No doubt it was not strictly impossible to reason from premisses not known to be true, but such reasoning was despised as 'dialectical', and no inquiry was made into the frequency of its occurrence in actual science. Why, then, waste time upon so unworthy a procedure, instead of fixing one's whole attention upon the truly logical ideal, the absolute proof of absolute truth? Let us maintain, rather, the old Aristotelian[1] conviction that the truly scientific syllogism proceeds from premisses that are true and underivative (because 'self-evident') and inerrant, and demonstrates its conclusion with ineluctable necessity! Thus the attainment of *absolute truth* was unobtrusively smuggled in as the aim of reasoning, and became an integral feature of the ideal of 'demonstration'.

§ 11. From the standpoint of the scientific inquirer, however, this whole theory of proof is open to the gravest objections. He finds first that it is impracticable, being composed throughout of counsels of perfection with which he cannot comply, and then that, even if he could, they would be perfectly useless, and destructive of his aims.

(1) It strikes him at once that the Fixity of Terms is an obvious *fiction*. He will of course be aware, from his scientific experience, that fictions have their uses and are often indispensable; but he will know also that not all fictions are useful, and that the adoption of a fiction has in each case to be justified by its usefulness. Moreover, it is not so much its immediate and prospective use which justifies it, though this yields the usual motive for its adoption, as the ulterior uses ascertained *ex post facto* by experience.

He will ask, therefore, for evidence that an *absolute* fixity of terms is the vital necessity for logic it is declared to be. He will admit, of course, the familiar arguments for a certain *stability* of meanings which have come down from the days of Plato, but he will suggest that a *relative* fixity of terms is quite sufficient to content them. He will point out that in a progressive science any absolute fixity in its terms is precluded by the very progress of the science. For the terms in use must somehow manage to convey the *growing* knowledge they are employed to 'fix'. The term 'gas', for example, must not be tied down to the meaning Van Helmont desired to convey when he invented it; it must incorporate all that physics has discovered about 'gases' ever since. Similarly, when Darwinism transforms the notion of 'species', and the discovery of radio-activity that of 'atom', these developments of meaning must be recognized as perfectly proper. To object to these conceptions as modern science uses them, on the ground that, because to Plato and Aristotle species were eternal and immutable, a 'species' that changes cannot be truly a species, or that because an 'atom' is etymologically

[1] *Post. Anal.* i. 2. 71 b 20.

'indivisible', it becomes an impossible self-contradiction when it is made up out of 'electrons', will seem to him to reveal only the fatuous pedantry of an utterly unscientific mind.

§ 12. (2) If he is acquainted with psychology, he will perceive also that the fiction of the fixity of terms is subject to a further restriction. It is not only in science as such—for all sciences must be conceived as progressive—that the fixity of terms cannot be made absolute: a real fixity is strictly inconceivable for and in every human mind. For every term that is actually used to convey a meaning must be held to form part of a *new* truth,[1] i.e. of a truth that was not previously in being. It is not a question of principle whether the truth is supposed to be new only to the person to whom it is addressed, or claims to be new to all, i.e. to science. For no judgement would be made unless it had something new to say.[2] Hence *every real judgement*, as opposed to the verbal formulas which are called judgements in the logic-books, *more or less modifies the meaning of its terms*. If it succeeds in being a real judgement and a new truth, it establishes a new and previously unknown relation between its subject and its predicate. 'S' is henceforth an S-which-can-have-P-predicated-of-it, and 'P' a P-which-can-be-predicated-of-S. Thus both the psychological associations and the logical associates of S and P are changed. That logicians should not have noticed so obvious a fact can be attributed only to their inveterate habit of not using in their illustrations real judgements intended to cope with actual problems, but operating with their verbal skeletons, which are not being used by any one to convey his meaning, and so do not have any *actual* meaning.

Clearly, then, no science can interpret the fixity of terms quite literally. Or rather, it can only interpret it literally—as a matter of the literal integrity of the *words* that *may* convey a meaning. But in a scientific inquiry the convention of formal logic must be reversed; the fixity of terms must be *understood* not to be absolute, but to be merely *ad hoc* and sufficient to convey a definite meaning, which it is desired to develop. Accordingly it must always be assumed that the results of an inquiry are to modify its terms, and that it is permissible, and indeed inevitable, to develop their meaning, so long as they remain capable of expressing and conveying the new truth. We must come to every

[1] i. e. truth-claim. [2] Cf. *Formal Logic*, p. 173.

inquiry with a willingness to learn and to expand our terms. The Fixity of Terms, as it is tacitly presupposed in the traditional logic, is a scientific blunder of the gravest kind.

§ 13. (3) To renounce it, however, entails further consequences. It appears to undermine the whole notion of *formal validity*. For if we admit in principle that the meaning of terms depends vitally on that of the judgement in which they occur, how can we continue to rely absolutely on the mere verbal identity of its terms to hold together a syllogism ? In any syllogism the middle term, *M*, may have one shade of meaning in relation to *P*, another in relation to *S*. It may be quite right to call *M P* in one connexion, and to call *S M* in another ; and yet, when the two assertions are put together, they may lead to a conclusion which is an error or an absurdity. The man who (in his laboratory) would rightly declare that ' all salt is soluble in water ' and (at his dinner table) as properly hold that ' all Cerebos is salt ', could not combine these assertions to draw the conclusion that ' all Cerebos is soluble in water ', without finding that the facts confuted his anticipation.

No doubt, when this had happened, he might explain it, *ex post facto* (if he knew logic), by alleging a hidden ' ambiguity of the middle term '. We need not here discuss whether it is fair to treat as an inherent ambiguity what is really a juxtaposition of shades of meaning which were relative to different purposes and right in their original contexts, thus manufacturing a fallacy by selecting the premisses : the important thing is that the logician should be driven to admit that *any* middle term may become ambiguous in this way when a syllogism is constructed, and that this completely stultifies his assumption that the *verbal* identity of the middle guarantees the *real* identity of the objects to which it refers.[1] If we call two things, which are and must be different if they are to be two, both ' *M* ', we necessarily take the risk that the differences are irrelevant for the purpose of our argument. We may legitimately assume this, but if we do, our hypothesis

[1] Mr. Alfred Sidgwick has been pointing out for the past twenty years how fatal this difficulty is to the traditional notion of formal validity ; nor has any logician confuted his argument, or even shown that he apprehended its meaning and scope. It would seem, therefore, that the condition of formal logic is so precarious that its only chance of survival lies in hushing up all the vital objections to its stereotyped doctrines. But is not the policy of ignoring unanswerable objections the sure mark of a pseudo-science ?

has to be confirmed in fact ; it is naïve to think that the verbal identity of the terms is quite enough. If, then, actual identity cannot be absolutely guaranteed, if there is *always* a possibility that the same term when put into a syllogism and used in reasoning may develop an ambiguity and become effectively two, it is evident that no amount of formal validity will safeguard the truth of a conclusion, even when the premisses are in themselves severally true. The syllogistic form is convicted of *losing* truth which it started from, and this is the very thing it boasted it could never do. Moreover, its coercive 'cogency' is exploded : whoever wishes to deny a 'valid' conclusion after admitting its premisses, has merely to suggest that by putting the premisses together a fatal ambiguity has been generated in the middle term.

§ 14. (4) The assumption that everything has been named rightly, and is what it is called, will scarcely commend itself to the scientific researcher. He will know from much painful experience that language only embodies the knowledge which has been acquired up to date, and too often is only a compendium of popular errors. Hence in any research which really breaks new ground the existing terminology will always prove inadequate, and new technical terms have usually to be devised in order to embody the new knowledge. The reason is obvious. *Ex hypothesi* we are inquiring *farther* into the subject, because our knowledge is felt to be insufficient. Accordingly the probable defects of the terminology we are initially forced to use must be borne in mind : we may expect it to omit what is unknown, to misdescribe and to classify wrongly what is partially known, putting together what does not belong together and separating what does, emphasizing the unimportant and slurring over the important, and generally failing to provide the mind with words that give it a real apprehension of the objects under inquiry. Hence the tacit assumption of Aristotelian logic that the terms reasoned with are fully known, that adequate notions are already extant, that truth has merely to be *disentangled* by a verbal criticism of existing opinions, and has not to be discovered outright, is false ; nor can any argument from a verbal identity be taken as final.

§ 15. (5) But of all the assumptions lurking in the theory of proof, the belief that reasoning can and should start from certainty will seem the falsest and most pernicious to the man of science. For it means that we are committed to a search for absolutely certain premisses as a preliminary to every inquiry, and proscribes consciously hypothetical, i. e. truly experimental, reasoning altogether, or at least condemns it as incapable of leading to certainty. This search, however, will either be perfunctory and uncritical, if it accepts false claims to certainty; or else vain, if it is conscientious. For every attempt to prove a conclusion absolutely demands *two* absolutely true premisses; hence the more we try to prove, the more we have to prove, and our search grows the more endless and futile, the longer it is continued. An immutable basis of absolutely certain truths, therefore, for reasoning to start from, is nowhere to be found. In no science is it possible to *start* with truths that are absolutely certain. In every science the initial 'facts' are doubtful; they are alleged, but not yet approved. They embody only unsystematic observation and prescientific experience of the subject, and so are probably the products of inaccurate observation, bad interpretation, false preconceptions, and popular superstitions. To acquire any considerable scientific value, such material has to be thoroughly revised and refined.

The validity of methods and the certainty of 'principles' are no more assured than the 'facts', initially. Every science has to work out its own appropriate methods experimentally; even if it borrows methods from another, it has to find out how and how far they apply to a new subject. Neither does a science acquire its principles by divine revelation; even if they fell from heaven ready-made, it would insist on testing the authenticity of the revelation. But philosophers have been extremely reluctant to admit that the certainty of principles is a gradual growth: for over 2,000 years they have been endeavouring to discover some way of securing an infallibility to principles which would render them independent of the working of the sciences which use them. But if their labours have proved anything, it is that no such way can be found.

(*a*) They have recognized many principles as '*self-evident*', and equipped the mind with a variety of 'faculties', expressly invented to enable it to apprehend the 'self-evident' inerrantly. But they have not been able to agree upon a list of self-evident principles,[1] nor even to find any truth whose claims to self-evidence have not

[1] The latest I have noticed occurs in Abercrombie's *Inquiries concerning the Intellectual Powers* (1830); it reads very strangely now.

been denied by competent critics. Nor have they been able to define their notion of 'self-evidence' itself; they cannot discriminate between the sound 'logical' self-evidence, which they conceived to guarantee truth, and its merely 'psychological' 'mimic', which is certainly much commoner, and becomes more intense and extensive the more unsound is the mind that 'apprehends' it.[1] Hence an unprejudiced observer has no reason to put the 'intuitions' of philosophers and the 'faculties' which apprehend them on a higher cognitive level than those of women or even lunatics. They all impose themselves psychologically; but this proves nothing as to their logical value, and science has to test them just the same.

(*b*) The principles which are said to be *necessary* or *logical* '*presuppositions*' all turn out to be hypothetical when they are examined. They are needed, no doubt, to solve the problem in hand, *if* the particular way it is formulated is taken for granted. But if either the order or the formulation of problems is altered, they cease to be either 'necessary' or 'presuppositions'. For example, the 'axiom of parallels', *alias* 'Euclid's postulate', is a necessary presupposition of geometry, if the existence of parallels is assumed. But if we prefer it, we can just as well (with Aristotle) make it our axiomatic 'presupposition' that the interior angles of a triangle are equal to two right angles, and can then deduce the existence of parallels. I. e. Euclid might have deduced what he assumed, and assumed what he deduced. If, moreover, we do not desire to construct a Euclidean geometry at all, we can deny *both* presuppositions, and proceed from *alternative* postulates, which lead to the various metageometries. The only things, in short, which all scientific principles presuppose are the desire to construct a science, and the desire to construct it in a particular way, which is simplest, or easiest, or most systematic, or most in accordance with the reigning prejudices. But these desires are the very things which the logician's account of principles always omits to mention.

Again, the whole of Kant's scheme of *a priori* presuppositions in the theory of knowledge rests upon an arbitrary assumption, viz. that mental data are to be conceived as originally discrete and are therefore in need of 'synthesis'. But it is just as possible

[1] Controversially the criticism of 'self-evidence' has been met in the same way as that of the 'validity' of the syllogism, i. e. by total silence.

to conceive an analysis of knowledge which starts from the 'presupposition' of a continuum or flux, and proceeds to trace out the principles by means of which this continuum is broken up into a world of apparently distinct things and processes. Nor is it possible to say in advance of experience which of such 'presuppositions' is going to be more convenient and more conducive to scientific progress.

(*c*) It demands a high and rare degree of philosophic insight to perceive that very many principles are neither certain, necessary, nor probable, but simply *methodological.* Whether we think them true or not, we adopt them because of their eminent convenience. If they turn out to be false, candour compels us to call them *methodological fictions*; but they continue in use. Our belief in the trustworthiness of memory is a good example. For though we often find that our memory has played us tricks, we continue to accept as true what we 'distinctly remember'. If no limitations to the truth-claim of such assumptions are discovered, enthusiasts will probably insist on promoting them to the rank of indisputable 'axioms', and hail them as absolute truths. But their scientific value is not thereby enhanced, and the cautious will eschew such exaggerations. For there is no real reason why the scientific rank of principles should not rest openly and entirely on their actual services, and why a 'methodological assumption' should not rank higher than a 'self-evident truth'. For the latter is at most a fact of our mental organization which nothing has so far turned up in nature to set at naught, and as such a fact it is itself a thing to marvel at rather than an explanation of other things. The scientific spirit will always hesitate to acquiesce in the limits which are set to inquiry by sheer brute facts, and if the absolute truth of certain principles were merely an ultimate fact which could neither be impugned nor explained, this would go far to make these principles appear unintelligible and would be a constant challenge to dispense with them, or somehow to evade them. A principle, then, should always be prepared to state the reasons a science had for adopting it: only the reasons will appear from the actual working of the science. They will involve a reference *forward* to the facts it copes with, not *back* to higher principles or to any claim that proves itself by its self-assertion.

(*d*) Indisputable principles, then, are not consonant with the spirit of inquiry: it will gladly let them go, if it can attain truth and advance knowledge in other ways. It will not shrink even from repudiating the ideal of absolutely true and demonstrated truth, if it can be realized only by sacrificing the progressiveness of science; nor will it be dismayed to find that this ideal is unrealizable. For when the inquirer reflects upon his own procedure, he finds that it points to a radically different ideal, and that the existence of absolute truths would only be a hindrance and a restriction upon his endeavour (cf. § 28 (4)).

II

§ 16. Before, however, we attempt to delineate the logical ideal of the discoverer, it will be necessary to encounter a serious objection which protests on principle against such an undertaking, and urges that discovery by its very nature must elude logical treatment. It is contended, in the supposed interests of logic, that discovery is a process so inherently and incurably psychological that no logical account can ever be given of it. Discoveries are windfalls, and come as 'happy thoughts' to the gifted geniuses that make them, in a manner neither they nor any one else can account for or describe: they are therefore logically fortuitous, and to set forth the ideal of proof by which the truth of discoveries is tested is all that need, or can, be the concern of logic.

Certainly the great majority of deductive logicians have taken up some such attitude towards the process of discovery. Aristotle contents himself with a bare mention of 'sagacity' (ἀγχίνοια), which is defined as the instantaneous apprehension of the suitable middle term for constructing a demonstrative syllogism.[1] When one recollects the weary centuries of painful effort and continual failure which elapsed while the *élite* of the human race were seeking for clues to, e.g., the mysteries of disease and of physical happenings, before they hit upon the notions of microbes and the mechanical theory, this naïve underestimate of the most difficult and essential of scientific procedures sounds like a mockery. Yet the whole Aristotelian school pass over the problem as lightly. They all seem to believe that while it is merely low cunning to make a discovery, it is a real proof of mental capacity to arrange

[1] *Anal. Post.* i. 34.

it 'in logical order' after it has been made, and to show how far short it falls of the logical ideal. Even the inductive logicians may be said to have participated in this attitude. For they were not more anxious to propound methods of discovery than to contend that their conclusions were just as rigidly proved and just as formally valid as those of syllogisms. They did not see that they were thereby accepting the demonstrative ideal of proof and giving away their own; what they should have shown was that this ideal was utterly nugatory, and that their own methods could never conduct to 'proof', but only to something vastly superior.

§ 17. In spite, however, of this wonderful consensus of logicians the above argument depends essentially on a confusion. It has confused two things which are perfectly distinct, the actual procedure of the individual discoverer, and the generalized description of the attitude of mind and procedures of discoverers, as they appear to subsequent logical reflection. Both present problems to the logician, but the problems are not the same. To anticipate the process of actual discovery may well be left to the prophets; it will transcend the powers of logic and indeed of any science, unless it be individual psychology, if it exists, or history, if it be a science.[1] It may readily be admitted that anecdotes about the bath which fomented in the mind of Archimedes the idea of specific gravity, and the streets of Syracuse through which he ran and cried '*Heureka!*', or about the apple-tree which shed its fruit upon Newton's receptive head, and stimulated his brain to frame the law of universal gravitation, are beneath the dignity of science. Their narration belongs to history, which can go as deeply into their details as the scale of the history and the purpose of the historian demand; but the particular circumstances of a particular discovery may well be treated as 'accidental', and be smoothed out of the scientific record. But why does it follow that no common features can be traced in these histories of discovery, and that there cannot be compiled out of a sufficient number of them a generalized account of what appears to be the 'essential', i.e.

[1] It may be suggested that there is a similar confusion on this question: when history is called a science, it is often forgotten that its data are essentially such that they can only occur once, while the material of the other sciences is such that cases of 'the same' may always be found in it. But neither need it be denied on this account that history can, and should, be written in a scientific spirit.

really relevant, procedure of discoverers, which may serve as a guide and model to subsequent discoverers? Why should this be more difficult than to describe the method of lion-hunting from the records of lion hunts, or the treatment of a disease from the history of a number of cases? Indeed, it would seem that the thing has been done. Any discoverer may reflect upon his own discoveries, and, like Poincaré,[1] formulate the method he has found successful. And if discoverers are not all perfectly unique in their methods, important uniformities will probably be found by comparing the methods of a number of discoverers.

Why again should it be assumed that the general account thus extracted from a retrospective study of discoveries must at once coincide with the logical 'ideal of proof'? Why should it even point to this, or be related to it otherwise than by contrast? Surely the possibility should be discussed that there are *two* procedures for logic to consider, of which the one describes how human knowers, starting from what they believe themselves to know, set about it to fortify and extend their knowledge, while the other moves on a superhuman plane and describes, with Platonic fervour, how ideal demonstration, descending from absolutely certain principles, moulds into a closed and inexpugnable system all the truths which are deducible from these and alone intelligible. The two accounts must be distinct, for they have different starting-points and work upon different material. Nor need they ever have any point of contact. For it may well be that human knowing never attains to an absolute certainty and a completed system, while deductive proof never condescends to notice mundane fact.

This was certainly so in the first rapturous vision of *a priori* 'proof' which solaced Plato amid the elusiveness and opacity of the flow of happenings. The deduction of the intelligible order of the ideal 'Forms' from their supreme ground and (*sole!*) premiss in the 'Idea of the Good' stopped short of facts and events at the *laws* of minimum generality,[2] and recognized in all the happenings of the sensible world an ineradicable taint of 'not-being' which rendered their stability impossible and their prediction vain. Aristotle similarly distinguished between the procedure which started from the *notiora nobis*, the apparent facts

[1] *Science et Méthode*, ch. iii, L'Invention mathématique.

[2] *Republic*, 511 c.

of perception, and that which began with the *notiora naturae*, the self-evident principles which could form the ultimate premisses of demonstrations. But that these two methods must somehow coincide was assumed rather than proved, in a way that should have discredited the doctrine. For Aristotle also was not able to explain how 'science', being of 'universals', could apply to particulars, which nevertheless he would not with Plato stigmatize as 'unreal', while the ascent from the sensible fact to the 'universal', which was called the 'induction' of the 'principle', is hardly validated by the naïve allegation of a mental faculty of 'intuitive reason' (νοῦς) endowed with the special function of apprehending principles in their particular exemplifications. It is high time, therefore, that this whole assumption that a necessary congruity exists between the logic of discovery and of proof should be subjected to a thorough examination.

III

§ 18. Such an examination will speedily establish that the mental attitude of the discoverer is, and must be, quite different from that of the prover.

In the first place, the discoverer is not in possession of the knowledge he covets. It is for him a desire, an aspiration, an aim to be attained. Proof, on the other hand, presupposes knowledge. Not only must the demonstrator *know* the assured truths he uses as premisses, not only must he have a supply of absolutely certain truths if his proof is not to remain hypothetical (§ 9), but he must already *know* the conclusion he exhibits. He cannot be ignorant, like the discoverer, of the result he is to arrive at. He is not engaged in *discovering* new truth, he is only showing how it *follows from* old truths. His retrospective contemplation has merely to retrace the history of its attainment, or rather to rearrange it in the more pleasing order which he calls 'logical'. This order is not that in which it *was* discovered, nor even that in which it *could be* discovered. For there are such things as necessary errors, indispensable artifices, and indefensible fictions, and the way to a truth often lies through them. Thus from time immemorial mathematicians have represented the continuous by the discrete, quantities by numbers, knowing full well what fictions their practice involved. Again, mathematical calculation of shapes, areas, and motions necessarily presupposes the fictions that bodies have the ideal and regular forms to which they 'approximate', and that their 'mass' is concentrated at their (ideal) 'centre of gravity'. It is more than doubtful whether the notion of an 'evolution' of species could ever have been reached, except by starting from the false notion of the fixity of species, or whether the true nature of the mobility and development of meanings could have been understood except by correcting the Platonic theory of immutable and eternal 'universals'. To 'proof' all these incidents and accidents of the history of discovery are irrelevant; all that has to be done is to show that the new truth can be deduced from the old, and that a 'logical connexion' exists between them.

§ 19. Not only is this much easier to do than to make the discovery, but it is very much easier to follow. Any one can see the connexion once the data have been arranged in logical order. Hence the assumption that this order somehow represents the actual process in a perfected form is natural enough. But it leads to contempt for the procedure of discovery. The discovery is made to look so easy that it becomes impossible to appreciate its difficulty and its merit, and it seems astonishing that no one made it long before. For did not the 'facts' all but force it upon the dullest mind? Who could have failed to see that fossils must be (at least) as old as the rocks in which they are embedded, that obviously worked flints, similarly, attest the antiquity of man, that northern Europe is scratched all over with the marks of a gigantic glaciation? It is forgotten that these 'facts' were *not* there until there came a mind prepared to notice them. Hence none of these discoveries were in fact easy to make, and they were preceded by a long struggle of the human mind with false preconceptions and the illusory 'facts' which they had engendered.

Nor are discoveries easy to get recognized when they have been made. The persecutions to which discoverers of new truth are subjected always and everywhere (more or less) form as discreditable a chapter of human history as the persecution of moral reformers. Those may count themselves fortunate who are simply ignored. Hence everything has to be 'discovered' over and over again. Nothing new ever enters the world, just as nothing old ever passes away, without infinite pains and after a protracted struggle. One curious result of this inertia which deserves to rank among the great fundamental 'laws' of nature, is that when a discovery has finally won tardy recognition, it is usually found to have been anticipated, often with cogent reasons and in great detail. Darwinism, e.g., may be traced back through the ages to Heraclitus and Anaximander. Thus it is true that there is 'nothing new under the sun'; but only because when a new truth first appears it does not prevail: when after a hundred repetitions it is at length recognized, it is no longer strictly *new*. Accordingly, the 'discovery' of a truth is only the beginning of its career, the first step by which it makes its way in the world, and still very distant from the crowning 'proof' with which logic complacently adorns it *ex post facto*, when it has 'arrived'. The slowness and difficulty, then, with which the human race makes discoveries, and its blindness to the most obvious facts, if it happens to be unprepared or unwilling to see them, should suffice to show that there is something gravely wrong about the logician's account of discovery.

§ 20. Quite apart from the difficulties which the psychological constitution and social organization of man put in the way of innovators, the making of a new truth which formulates a new 'fact' is also intrinsically anxious work. It is not merely that its maker can have no assurance that his enterprise will succeed, that he cannot start with a feeling of certainty from established truths, and be wafted by an irresistible wave of logical necessity to the safe haven of a predestined conclusion. He *must* start with a consciousness of ignorance and an all-pervading feeling of doubt about every step of his inquiry. This doubt he should not, moreover, endeavour to disregard or to suppress; for it is the best guarantee that no way to the truth will be passed by in his explorations. Doubt, therefore, should be recognized on principle, and equipped with a technique of testing and experimentation: the inquirer should be proud that he has to feel his way in fear and trembling to the very end.

Yet his condition will not contravene Aristotle's dictum that all inquiry and research proceed from knowledge previously acquired.[1] In a sense he will still start from what he knows, or thinks he knows. For it is psychologically impossible to do anything else. The knowledge he believes himself to have cannot but affect all his ideas, and he cannot get away from it. His

[1] *Anal. Post.* i. 1.

boldest speculations, his most hazardous hypotheses, will have *some* relation, however subtle and recondite, to the knowledge at his disposal. It will influence all his thoughts and guide his guesses. As he cannot divest himself of his knowledge and the ideas it has rendered familiar to him, he has to accept its limitations. His only problem is to use it as effectively as possible.

But it is clear that he cannot regard his knowledge with the same sort and amount of confidence as the believer in demonstrative proof. He must conceive himself as an explorer, and his attitude must be tentative throughout. Knowing that his premisses are questionable and only doubtfully true, he will recognize that his inferences are only probable, and stand in need of confirmation. As a rule he can, no doubt, find accepted truths to argue from; but these being relative to the existing state of knowledge are known to be subject to correction. Even where he has started with premisses of the most superior kind, which are generally deemed absolutely self-evident and certain in themselves, he will still be conscious of a doubt whether they will prove to be the right premisses *for his purpose*. If they are not, their truth is irrelevant and will lead him astray. In no case, therefore, can he escape the responsibility of *choosing the right ones* from his limited stock of known truths and familiar ideas, as he contemplates the infinite expanse of possible discovery. In whatever direction he moves, the unknown lies before him; he may come upon surprises or be stopped by unsuspected obstacles. In short, there is nothing of the irresistible about his progress; it has not the faintest resemblance to the majestic march from inevitable premisses to a predestined conclusion which so fascinates us in the theory of proof.

§ 21. But, it may be said, all this is not enough. The differences in the attitudes assumed by the reasoner in discovery and in proof may be only psychological. They do not prove any real logical difference between them; the logician's account may still be what the discoverer would acknowledge to have been his best course, if he could have seen it. It has, therefore, to be shown that the differences in question arise out of, and develop into, differences which are indisputably logical.

Thus, the ignorance which the inquirer feels is doubtless a psychological fact, but the lack of knowledge which engenders it is surely a logical fact of some importance. In general, the feelings of doubt, expectancy, and perplexity which beset the mind of the inquirer, and contrast so distinctly with the feelings of confidence, knowledge, certainty, and necessity which accompany a 'proof', originate in a logical fact. Every inquiry starts from a *problem*, of which the solution is not yet known. An *inquiry* is, as the name implies, a *question*, put, not to nature at large and at random, but to some *part* of it, which is taken to be relevant and to contain a possible answer to the inquirer's question. Now this dependence of inquiry upon problems springs no doubt from the psychological fact that until there is something put before it the mind cannot get to work upon it; but it is surely a fact of the utmost logical significance, and it is astounding that the logical tradition should have slurred it over so completely.

Especially as in the very beginnings of logic some of the Greeks distinctly caught a glimpse of it. For, having started their reflection upon reasoning from a desire to regulate debate and to argue a case at law, they naturally noticed that there are two sides (at least) to every question. Accordingly, Protagoras appears to have taught systematically that there were always two reasonings (λόγοι) to be considered,[1] Socrates treated scientific inquiry as an extension of the art of cross-examination, and Plato conceived the search for ideal truth as a 'dialectical' process, as a sort of dialogue of the soul with itself. Now this whole doctrine is equally good as logic and as psychology. It is profoundly true of the inquirer's mind; he must be keenly alive, not only to the evidence *for*, but also to that *against* his working theory. But it is also true of the logical nature of inquiry that it is a process of determining *which* of the alleged 'facts' and of the theories to interpret them are real and true. Inquiry logically 'presupposes' a conflict between the data, and a dispute about them.

Unfortunately, however, the conception of scientific research as an inquiry lapses from the logical consciousness in consequence of Aristotle's work. His discovery of the forms and formulas of demonstration overshadowed it, and restored the reign of dogma which is so congenial to the authorities everywhere.[2] The true conception of inquiry does not revive again until our days, when Mr. Alfred Sidgwick and Professor John Dewey have endeavoured, not with the success they deserved, to reopen the eyes of logicians to the facts of the scientific situation.

[1] Diogenes Laertius, ix. 51.

[2] Cf. § 3.

§ 22. To conceive an inquiry as a question then is, we see, implicitly to conceive it as having a plurality of answers, all of which have to be examined. All these answers are initially hypotheses, and a choice has to be made between them. This renders the recognition of alternatives a paramount necessity for a logic of discovery, which can no longer dismiss them with a jejune chapter on 'disjunctive propositions'. Their existence is no longer to be treated as an annoying complication which delays the progress of science, but must be taken to inhere in the logical nature of problems, and to be essential to their proper elucidation.

Logic, therefore, should regard it as its duty to inquire (1) how the inquirer is furnished with an adequate supply of theories for analysing and testing the apparent facts of his subject, (2) what methods are used to sift hypotheses and to select the more valuable, and (3) if it can, to add some hints as to how theories and methods *ought* to be handled.

(1) To the first question there is no exhaustive answer. No logic can guarantee that *all* the possible theories which concern the facts under inquiry will be available. They may not yet have occurred to any human mind, and may never do so. This alone ought to be considered a fatal objection to all methods which presuppose exhaustiveness, and are pressed by the logician upon the man of science. It ought to dispose of methods which demand that *all* the facts should be assembled before theorizing is begun, or that *all* the alternatives should be stated and the true one extracted by the successive elimination of the false ones, or that define a 'cause' as reciprocating with its 'effect', and assume that the true cause has been discovered when no other has been thought of, or that if a theory works we may take it that it alone will do so and is (absolutely) true. All these notions demand an impossible exhaustion of the alternatives, and try to convert a (psychological) failure to think of any more into a logical proof that there are no more. And they all regard the plurality of alternatives as a hindrance to be got rid of, and not as a safeguard and a help to proper inquiry.

Hence the real difficulty was not perceived, viz. that there is no formal guarantee that the supply of hypotheses for use upon the facts in any inquiry will be adequate. It may well be that for lack of a good working theory to go upon, all the theorizing on a subject proves vain and sterile. In the beginnings of all the sciences this sort of condition always exists and often lasts for centuries, and it is a main reason why some sciences make little progress even now.

Nevertheless, the difficulty is not in practice as fatal as it looks on paper. It is probable that the inquirer will in fact usually have a supply of alternatives to start from. For (*a*) he will naturally select a subject in which there are disputed points. And (*b*), what is even more important, human minds are naturally various: they put, therefore, different interpretations on the same facts and value them differently. Some are attracted by novelty, others by orthodoxy; some incline to one type of theory and method of inquiry, others to another. Hence in any inquiry upon which a number of minds are actively engaged, there will always be differences of opinion, and these will be most marked in the rapidly growing regions of every progressive science, which, like the growing cells in the trunk of a tree, are always on the outskirts. There will always be a conservative and a liberal party, even in science, and the clash between their views will always provide alternative solutions of problems, the comparative merits of which the inquirer can examine. But the sciences owe their progress largely to the man who raises new questions, and should provide for him in their organization.

§ 23. It should be noted further that if this feature in discovery were properly recognized and emphasized, it would have important educational and ethical effects. At present the study of logic can hardly be said to liberalize and broaden the mind or to improve the temper. So long as its chief interest is in a theory of absolute proof and complete certainty, it will tend to breed pedants and bigots. The effect would be very different if an adequate logic of discovery had imbued the mind with an ever-present thought that every subject may and must be considered from several points of view, and that an inquirer should beware of letting his predilections and preconceptions blind him to possible alternatives. The logical attitude of inquiry, when fully understood, demands a tolerant and open mind, and excludes the narrow-mindedness and dogmatism which the theory of proof has fostered by its pretence of showing that there was but one truth and one inevitable way of reaching it. Moreover, the necessity of continually choosing between a number of alternatives should cultivate a judicial temper, conducing to fair-mindedness and consideration towards the views of others. For a mind which is in the habit of choosing between alternatives must be impressed by the facts that there is something to be said for the views it does not accept, that the view accepted is often not so very much superior to those rejected, and that new facts and new knowledge may always revive views which were supposed to be defunct.

Of course our natural dogmatism will take alarm at the flabby toleration of ideas which this attitude seems to imply. It will be objected that no one who can see the good and truth in beliefs he does not accept, can really be strenuous in upholding those he does. The full answer to this bigots' argument can only be appreciated when the attitude of progressive science is fully understood (cf. § 33), but in general it may be pointed out that a power of first weighing alternatives, choosing the best and acting upon it strenuously, is precisely what life demands of us at every step. It should not, therefore, be impossible to compass it in science.

§ 24. (2) To the second question of § 22, viz. what are the methods used by the inquirer in sifting the alternative hypotheses in the field, and picking out the most valuable, the answer is comparatively easy. It is substantially the answer given by the pragmatist analysis of knowledge. That theory is preferred, and tends to be accepted as true, which for the time being *works* best. The formula looks simple, but needs more thinking out than its critics usually bestow upon it.

(*a*) It implies, of course, that *all* the alternatives (before the mind) 'work' more or less. They must be (or appear) scientifically plausible, and proffer a more or less satisfactory explanation of some or all of the admitted 'facts'. This is why agencies like the Devil, who could once be extensively alleged to explain anything unusual, have dropped out of the purview of science.

(*b*) 'Working' must be conceived somewhat widely. Its *primary* appeal is to the accepted principles and recognized interests of the science; practically to 'work' means to conduce to the development of the science on the recognized lines, and the proper judges of what 'working' counts are the experts who cultivate each science.

(*c*) But there will often be complications due to certain disputable workings, of which the relevance is not yet established, and about these there will legitimately be differences of opinion. These should not be suppressed, but candidly argued out.

(*d*) Moreover, every *new* departure will be *pro tanto* disputable, because it will conflict more or less with the vested interests of the established doctrines. One great factor in the 'working' of a new truth is the extent to which it upsets, or is thought to upset, the old, and demands a reconstruction of beliefs, a correction of authorities, a revision of text-books, a renewal of plant, &c. Hence what works best in the abstract may not do so under the actual conditions. It may 'pay' a professor better to be 'orthodox' than to be an innovator, and he is usually quite alive to this, though it does not render him a good investment scientifically for the institution that appoints him. If then we looked at this side of the matter alone, the verdict would always go against the novelty. For very few new truths are fortunate enough to find the field free and unoccupied. Usually they have to spring up in a soil densely overgrown with a rank growth of prejudices, dogmas, and superstitions, to which the world is accustomed and even devoted. So they have to fight for an opening in which they can take root and grow up.

(*e*) The 'working', however, need not amount to a claim to represent 'the' truth. A discoverer may know that by reason of his deliberate use of fictions, his results have forfeited their claims to be strictly true; yet they may 'work' better than anything else in sight. The typical example here is, of course, mathematics. When physical objects are treated mathematically, they are identified by a fiction with the objects of pure mathematics, and it is only on this assumption that their behaviour can be calculated. They are, of course, vastly *more* than mathematical objects, but their surplus meaning becomes irrelevant wherever objects admit of mathematical treatment. And apart from the restriction of the claim to truth necessitated by the use of fictions, it should, of course, be recognized also that there are sound logical reasons for denying that truths which rest on their 'workings' can ever be 'absolute' (§ 26 *s.f.*). Their truth is pragmatic, and is *optimi iuris* only if pragmatism establishes that no other and no better truth exists.

(*f*) More specifically a very important form of working is the prediction of events. Knowledge of the future is an almost universal object of human desire, which men have sought to compass by fair means and foul, and the calculation of the future is the avowed aim of many scientific inquiries. Hence there is nothing more potent to dispose the mind to accept a theory than the success of the predictions it has led to. Yet here again this form of 'working' differs generically from 'proof'. It is clear that prediction is not strictly proof. For predictions may be made with considerable accuracy by the aid of hypotheses which turn out to be false or impossible. Thus eclipses and other celestial events were predicted for centuries by means of the Ptolemaic astronomy, and they cannot be predicted even now with absolute accuracy. Indeed, physically speaking, absolute accuracy is unthinkable. No instrument and no organ of observation can be conceived to measure to more than a finite degree of accuracy, and the *best value* for any physical 'fact' will always be the mean of a number of good observations after all the accessible sources of error have been allowed for.

At no point, then, does the test of 'working' conduct to the notion that absolute truth is discoverable. But the right inference may be, not that the test is worthless, but that absolute truth is a chimera.

§ 25. (3) It cannot then be seriously disputed either that alternative hypotheses are always (more or less consciously) present to the mind of the inquirer, or that the working of a theory is in fact used, in all the sciences, to test its claim to be true. But does it follow that logic should bow to scientific fact and recognize these practices? Should it set itself to devise a *technique* for regulating the formation of hypotheses and the establishment of their truth by their working? It is here that the traditional logic demurs, and disputes begin. Nevertheless, strong reasons may be advanced for answering both questions in the affirmative.

(*a*) An abundance of hypotheses is a guarantee of great logical value that all the important facts will be properly observed. For it is evident that every theory will produce a certain *bias* in the observer. It will direct his attention upon those facts and those features which are *relevant* to his theory, and, more particularly, which *support* it. This is usually an advantage, because it helps him to select what is relevant to his inquiry from the chaos of events; but it will *pari passu* blind him to whatever does not seem to be related to, and to fit into, his theory. He will, therefore, fail to observe and to appreciate what will seem to him to have little or no scientific interest. And in so thinking he may be quite wrong.

The old theory of ' induction ' thought to get over this difficulty by saying, ' Well, of course, *all* the facts must be observed '. It did not observe the fact that in practice this is impossible, and is never done. Nothing is observed but what the knowledge and preconceptions of the time make visible to the scientific eye. Of what is visible at any time only a small part seems worthy of the scientific microscope. Complete observation, therefore, of literally all the facts is scientifically impracticable.

As a logical ideal also this notion of all-inclusiveness is absurd. If no inquiry could ever begin until *all* the facts had been assembled, how could anything be discovered until omniscience had been achieved, i. e. when there was nothing left to discover ? For how are we to know that our assembly of ' facts ' really is complete ? And if literally all the facts have to be used as data in any inquiry, shall we not speedily find that every fact ramifies into infinity, and drags in the totality of reality, and a knowledge of all things present, past, and future ? This ' logical ideal ', therefore, renders inquiry impossible.

In point of fact the data of any inquiry are always a *selection*. They are such of the recognized facts as are thought to be *relevant*, i. e. to be truly ' facts ' for the purpose in hand. But being a selection they involve us in the risk that we may have selected wrongly, and omitted what is important while admitting what is not. *From this risk there is no escape.* For we cannot effect a compromise by including merely so much of the facts as we can lay hold of. Not only does this yield no guarantee that everything that is needed has been included, but it may be a positive hindrance to try to include too much. For if our data grow into an unwieldy mass, they will not seem susceptible of any order or principle, and even the most penetrating inquirer will lose his way.

It is better, therefore, to give up altogether the idea of securing formal validity by postulating an all-inclusive exhaustiveness. The obvious alternative is to operate simultaneously with a plurality of theories, each of which means a certain ordering of the ' facts ' relatively to what seems a relevant and promising point of view. Each will involve a selection and induce a bias ; but with any luck they will neutralize each other's bias, and so will increase the probability that no really relevant fact has escaped notice. This will not satisfy the logical ' ideal ', but in practice it means a good deal, and is enough for scientific progress. Of course it must be understood that the hypotheses employed are in a general way relevant to the problems and the condition of the sciences, and not random guesses. This proviso will cut down their exuberance even more than the limitations of the human imagination, which seems to be psychologically incapable of really departing very far from the suggestions of experience.

§ 26. When logic has recognized the use and value of 'working' as the test of truth, it must, however, make it clear to itself and to others both what precisely this test is, and what it can, and cannot, accomplish.

In the first place, it must be made clear that it is *not* a logical implication of the test that 'whatever works is true', and the reasons for disputing this dictum must be set forth. The fact is that we all have a strong psychological tendency to believe in the truth of what is found to work, without much criticism of the sort and extent of the 'working'. But the logician should carefully investigate the various sorts of working that occur, and take special note of those which either do not themselves lay claim to full truth, or do not (ordinarily) have their claim conceded.

For example, '*fictions*' are not supposed to be strictly true; but they may 'work' and be 'as good as true', or 'pragmatically true', or 'sufficiently true for the purpose in hand'. They work, in fact, within limits; but these limits are *known*, and so they are not confused with full-fledged truths, to the applicability of which there are no known limits.

The case of '*methodological assumptions*' is more difficult and instructive, and is usually misconceived. In their case the existence of limits to their 'working' is either not known or not relevant, because they owe their adoption to their use and convenience in analysing and organizing a subject of inquiry. Thus the principle of Causation, the assumption that every event has a cause which determines it fully, is properly to be regarded as methodological. It declares merely that if we desire to calculate the course of events, it is scientifically convenient to treat events as if they had 'causes', from which their occurrence could be predicted, whether or not they have them in fact. This assumption may be purely methodological; it need not, and should not, be turned into a dogmatic, metaphysical denial that there may be indeterminate happenings. There may even be good reasons to suspect their occurrence, and indeterminism may be ultimately true, and yet scientific method may rightly ignore this possibility, because it would render the calculation of events impossible.[1] Even an indeterminist then is fully entitled to reason *as if* events were determined, and to search for 'causes', for the purely methodological reason that this enables him to calculate events, and that after all they may be calculable. So long as they work for scientific purposes it is not, in the case of methodological principles, necessary to raise the question of their metaphysical truth.

The 'lie' again is a curious case of 'working'. A lie works, as a rule, only so long as it passes for truth, and is believed to have the meaning and value its author claims for it; when it is 'found out', it ceases to work. Hence it can both work and fail to work at the same time, according as it is, or is not, known to be a 'lie'. Clearly nothing can be made of the lie logically, until this double aspect inherent in its nature is recognized; if the logician refuses to distinguish between the *persons* concerned in its making, acceptance, and rejection, it remains (like 'error' to Plato) an insoluble 'contradiction'. It is, however, a mere prejudice to refuse to make these distinctions.

The 'working' of hypotheses is by no means simple and unambiguous. It admits of infinite gradations in amount and kind, and the 'truth' which is implicated in 'working' is nothing essentially but an index of its logical value, and may vary in quantity between values which cannot be *psychologically* discriminated from zero and from 100 % or 1 (= 'absolute' certainty). It is crude, therefore, to confront a scientific hypothesis with the rigid alternative 'either (absolutely) true, or (utterly) false'; its 'truth' really rests on its greater value, as compared with its competitors. Its value, then, is a question of more or less. The more extensively, conveniently, and economically a hypothesis works, the more value has it, i.e. the more likely is it to be called 'true', and to be supposed true absolutely: the more continuously and successfully the test of working has been applied to a doctrine, the greater the confidence and affection with which it is regarded, and the greater the presumption that it will continue to approve itself as true.

But, as we anticipated in § 24 (*s.f.*), it is vain to expect to establish any absolute truth by this method. It provides truth

[1] Or more difficult, if the indetermination is conceived as limited.

with ever-growing probability, but never with absolute certainty. For, however well a theory works, the thought that one may hereafter be found to work better can never logically be excluded. Even if every one alive were perfectly satisfied, and no one could imagine any improvement in an accepted truth—and these conditions are by no means often realized—such psychological considerations would not disprove the logical possibility that the best known was not the best absolutely, and logic would continue to distinguish between a truth that was absolute, and one liable to one billionth chance of error. The latter chance could be disregarded for all practical and scientific purposes, and would not have the slightest psychological effect on the confidence with which the truth was regarded; but logically it would still be there. Science, therefore, has to resign itself to the conclusion that its method cannot conceivably attain to absolute truth, and to make the best of it.

§ 27. Curiously enough this conclusion is fully confirmed by Formal Logic. It prides itself on pointing out that there is a formal fallacy involved in establishing truth by 'working'. The essence of this method is to argue that if a theory is found to work (after the proper precautions have been taken), it is true. If e. g. the events anticipated by a theory occur, and nothing occurs that could not be anticipated, it grows more and more probable until it convinces every one. But ought it logically to have done this? The logician declares emphatically, it ought not. For the argument suffers from an incurable flaw, which has been recorded as a 'fallacy' for over 2,000 years. It is a flagrant 'affirmation of the consequent'; symbolically, it argues that *if A is, B is, but B is, ∴ A is.* Now this is not 'cogent' or 'valid'. That *A is* can be proved only from the premiss '*only* if A is, B is', i. e. if A is the *only* theory which will account for the observed consequences. But this the fallacious method did not assert, and indeed could not assert. For that the best known is the best absolutely never can be proved (cf. § 26); and even if they happened to be identical, and we had somehow stumbled upon an absolute truth, we should never know that this was so.

§ 28. To the logician this fact only seems to prove the superiority of his conception of 'proof'. He infers, consistently enough, that no inductive reasoning from 'facts', no verification of hypotheses by events, can possibly amount to proof. What he seeks to impress upon his pupils is that *verification is not proof and can never lead to it.*

He considers himself entitled to look down upon science accordingly, its evidence, its methods, and its reasonings, and to contrast them with the absoluteness of his own ideal of demonstration. He upholds its validity in spite of all the failures of the sciences to realize it. As a rule he seems willing to grant that some mathematical proofs amount to logical demonstration;[1] but if pressed he would confess that scientific truth was only probable, whereas certain metaphysical truths, such as the law of contradiction, alone were absolutely certain.

The scientist, of course, is not in a position to deny that the nature of his truth is such as has been stated: but he should not attempt to do so. He should content himself with scientific truth, and contend that at its best it is good enough for any one. And he can carry the war into Africa by a vigorous counter-attack.

(1) He can deny—for the reasons stated in § 13—that the logician's formal 'proof' is as cogent and formally valid as the latter supposes, and show that after a conclusion has been '*proved*' true, it has still to *come true* before it can be trusted to be 'true'.

(2) He can point out that there is a serious *lacuna* in the logician's plea for his notion of 'proof'. The logician has assumed that the only alternative to his belief in absolutely certain premisses is complete scepticism, arguing that it must be possible to start from certainty, because otherwise no knowledge would be possible at all. He then urged 'but there clearly *is* knowledge—the sciences attest it', and consistently inferred that absolutely certain premisses must be obtainable. The more or less obvious failure of his attempts to explain their genesis by 'self-evidence', 'intuition', 'necessities of thought', &c. (§ 15), could not deter him from clinging to his belief, because the principles themselves seemed to him to be inevitable and to admit of no alternative.

In fact, however, there *is* a *via media* between scepticism and absolutism, and science safely pursues it, though logic has over-

[1] This we saw (§ 4) is really a mistake: mathematical proofs are really hypothetical, and deduced from the initial postulates and definitions. They hold of the ideal objects of mathematics, but that they can be advantageously applied to reality is merely an empirical fact, and it is not inconceivable that the world should grow *more* recalcitrant to mathematical treatment, though actually it has grown *less* so.

looked it. It is *not* necessary to start with absolutely certain premisses, because it is possible to adopt premisses hypothetically, to take them as true for the argument's sake and for the purposes of the inquiry, to experiment with them, and to revise them in the light of the results of such experiments. Thus their value may be judged and established, *after* their adoption, by the experimental results, and they may come to depend logically upon these, and not upon the processes (analogies, suggestions, guesses, fancies, &c.) which led to their adoption. If they show themselves capable of advancing the science and solving its problems, confidence in their 'truth' increases progressively, and their initial assumption is justified. They *cease* to be 'hypotheses' and become 'facts', and even 'principles' beyond dispute. If they fail to 'work', they may be discarded in favour of others which are tried in their turn and similarly tested. Hence it is not true that what is uncertain to begin with must always remain so, nor is it hard to understand that hypothesis, willingness to believe, and belief may be the psychological forerunners of logical proof, which, nevertheless, rests not upon them, but upon the solid value of the results subsequently reached by their means. The certainty of scientific premisses then admits of indefinite growth, which at some point or other will overpower even the most obstinately sceptical temper. This point naturally lies at a greater distance from the starting-point for some minds than for others, but when it is reached, and when the last doubts and scruples have been overcome, the triumphant truth will *feel* absolutely certain, and to all intents and purposes will function as such. But the 'practical certainty' thus achieved will still be distinguishable in thought from the absolute certainty which logical theory mistakenly demanded. And logicians, from Plato downwards,[1] will be convicted of having failed to allow for the possibility that the certainty of premisses and principles may be a fruit of continuous experience and experiment, and to perceive that this is the method the sciences have actually employed. In short, necessary (needed)

[1] In *Republic* vi his whole argument for the existence of metaphysical truth, culminating in a supreme 'Idea of the Good', depends on the assumption that the 'hypotheses' of the sciences, being insecure originally, remain so until they are deduced from a (self-proving) 'unhypothetical principle'. This assumes, of course, that they cannot be confirmed empirically by the results of their working, and exhibits the *lacuna* of logic in a typical way.

'truths' need *not* be regarded as '*a priori*', if it is seen how hypotheses are consolidated by experience.

(3) The scientist can deny that the ideal case, contemplated with so much satisfaction by the logician, can ever occur in actual knowing. He can point out that if the logical apparatus of demonstration is to work, it must be supplied with premisses that are absolutely true. But whence is the logician to obtain them? The 'self-evident' principles and 'necessary' axioms, for which so much has been claimed, have been shown (§ 15) to be highly disputable, and are themselves in need of support and verification. The truths which the sciences supply abundantly are all products of the method to which he takes exception. There are no scientific truths which have not to be, and have not been, verified, and if verification is logically vicious, and cannot amount to proof, they are not absolutely true. But if the premisses of a demonstration are not absolutely true, neither can its conclusion be. What then becomes first of the value, and ultimately of the 'validity', of an ideal of proof which can never be exemplified by actual reasoning, and serves only to condemn it?

(4) The ideal of absolute certainty may be repudiated altogether, even as an ideal, for sound scientific reasons. It may be shown that if it were possible it would be scientifically undesirable. For it would mean the creation of absolute bars to scientific progress. If truths existed which were absolutely certain, this would mean that nothing more could be learnt about them, and nothing could be done to strengthen their position. No experience, no inquiry, no experiment, could any longer affect them, and add to or detract from their value. They could not, therefore, form avenues to further knowledge. They would simply be stops which would arrest scientific inquiry. But how could such things form an ideal of scientific knowledge? How could it be in the spirit, and to the interest, of science to recognize them? They would merely be for science brute facts which it was forbidden to investigate. And must not science on principle hold out for the right to inquire into everything, to test every belief, however true it may seem? How, then, can it be the ideal of science to adopt an ideal which would stop inquiry?

Nor will it suffice in reply to point to the fact that the sciences continually assume the truth of the premisses they argue from. For though this is often a convenient assumption for the purpose in hand, it is one thing to assume the truth of premisses for the purposes of an inquiry, and quite another to assume it absolutely. For in the former case our assumption may be, and should be, accompanied by a consciousness that upon another and fitting occasion the premisses now assumed to be true may themselves be inquired into: to regard them, therefore, as absolute is to misinterpret their logical condition.

There are no good reasons, then, why the sciences should surrender to the arbitrary demands of the traditional logic, and sacrifice their practices which have been sanctified by the successes of 2,000 years to theories which sprang from a misunderstanding of scientific procedure, and have since lost all contact with it. The original mistake was pardonable, but it ought not to be regarded as an insult to logic to require it to understand the procedure by which the sciences actually progress.

§ 29. The scientist then should not be terrified by the charge that his 'truths' are 'only probable'. For it is better to be satisfied with probabilities than to demand impossibilities and starve. Moreover, a high degree of probability means 'practical certainty', i. e. confidence enough to move to action. Such certainty so convinces and satisfies the mind that it cannot feel more certain about anything; the logical gap between it and absolute certainty is psychologically negligible. We are sacrificing, therefore, nothing but a superstition, nothing that has any value for us, by renouncing the demand for absolute truth and demonstrative 'proof', and we gain in return a charter of liberty. For to admit the essential progressiveness of scientific truth and its indefinite capacity for improvement means unlimited freedom to research into truths which are infinitely perfectible, because they are never 'absolute'. The ideal of the infinite perfectibility of truth, and the infinite progressiveness of science, is more than an adequate substitute for the 'logical ideal' which is abandoned. For not only is it an ideal which works, but it really embodies a nobler aspiration than that which represented science as 'resting' in absolute perfection on fixed 'foundations' of 'eternal' truth. The sentiment which inspires this group of metaphors is given away by the word 'rest'. A science that desires to *rest* is one that is unwilling to *move* and unable to *advance*. Fixed 'foundations' are needed only for standing firm and standing still, and it turns out that what is strictly meant by 'eternal' is not that truths last for ever, but that they are not related to ' time ' at all, and so have really no application to ' events '.[1]

On the other hand, a science which sincerely desires to progress needs fixed foundations as little as fixed ideas, and firm ground as little as assurances to ' rest ' on. It needs only a starting-point, or jumping-off place, whence it can plunge into the unharvested seas of the unknown. Now the essence of a starting-point is to be a place you want to get away from, and its excellence lies in being such as to prompt you to leave it as easily and eagerly as possible. If, therefore, scientific ' principles ' (ἀρχαί) are really to be starting-points, they need not, and must not, be so comfortable and so deceptively similar to 'absolute' truths as to tempt the scientific spirit to repose. They should be tentative assumptions which are gladly abandoned in the hope of reaching something better, stepping-stones to farther and higher things, which are valued for their consequences, and logically dependent on the conclusions to which they formed the premisses. The logic of science, therefore, has no reason to postulate stability or solidity for its initial principles: the most indispensable of them are only principles of method, and even of the tried and tested principles it arrives at the 'validity' (= strength) demanded is merely that they should be able to float the accumulated wealth of knowledge down the stream of time.

IV

§ 30. It is clear, then, that the time has come when Science should break decisively with the logical tradition, and proclaim a logic of its own which has always been implicit in its procedure. It must definitely declare that what it needs is not a logic which describes only the static relations of an unchanging system of knowledge, but one which is open to perceive motion, and willing to appreciate the dynamic process of a knowledge that never ceases to grow, and is never really stereotyped into a system. To show that such a logic is not inconceivable will be the endeavour of the concluding sections of this essay.

We have already had occasion to note many of the most important features of this logic. We have seen that logical, i. e. critical, reflection upon discovery must start from, and be guided

[1] *Formal Logic*, ch. xxi, § 7.

by, the conception of a scientific *problem* with which the process of knowing *experiments* (§ 21). This problem has, of course, to be attacked with the existing resources of a science, i. e. with the knowledge it possesses up to date. These resources form the scientific *capital* which is necessarily *risked* in research if it is to yield interest. It comprises (*a*) approved principles, (*b*) known facts, and (*c*) established meanings of words. About each of them a little more may advantageously be said.

(*a*) We have seen (§ 15) that the principles of any science could not rightly be conceived as inscrutable, ultimate, absolute certainties of divine descent, and acknowledging no human ancestry. We saw that they could be understood only as hypotheses which reflection upon a problem had somehow suggested to an ingenious mind, which had been provisionally adopted in order to explore and organize a subject of inquiry, and had finally been verified and confirmed by their success (§ 15 (*c*), § 24).

The principles thus accepted by a science are often regarded as descriptive of fact when they are merely methodological and convenient,[1] but this is a point of secondary importance. And even the most amply verified principles never quite lose their hypothetical character. So long as they are used, their meaning, scope, and truth are not absolutely fixed. They can be extended, restricted, and modified by the working of the principles.

§ 31. (*b*) It is really obvious to any critical reflection that when a science appeals to 'facts', it is really appealing to the facts *as known*, or supposed to be known. It cannot from the first presume its knowledge to be absolute, and, *pace* some of our 'neo-realists', ignore the question whether the alleged facts are facts at all, and so pretend to start from 'the facts as they really are'. Such uncritical temerity would only conduct to insoluble pseudo-problems like that with which King Charles plagued the nascent Royal Society, as to *why* the weight of a bucket full of water was not increased when a fish was added to it. If, however, it is acknowledged that the 'facts' involved in a scientific inquiry are always relative to a definite state and date in the history of a science, several important corollaries follow.

(1) Being dependent on the condition of the science, the facts

[1] e. g. the 'accidental' distribution of variations in biology, for which see *Humanism*, pp. 146–50, and the postulates of causality and determinism in science generally (*Formal Logic*, ch. xx, § 6, and *Studies in Humanism*, ch. xviii, § 4).

of a science will not all be 'facts'. That is, not all that is relevant to the interest of the science will actually be within its cognizance, not all that turns out to be fact, and is antedated when it has been discovered, is as yet recognized as fact. It will be this fact, moreover, which constitutes the science a field for inquiry and renders it progressive.

(2) Though the 'facts' of the moment fail to include all the facts, they often manage to include too much. The 'facts' are not all fact. They include unknown, and often large, amounts of prejudice, illusion, error, superstition, and other remnants of the lurid past and stormy youth of every science. It is useless to repine at this inevitable consequence of past history, and childish to try to purge it away by defining as science only what *ex hypothesi* is free from such contaminations. To restrict the logical interest to science *qua* science, which is by definition infallible, is to forbid any logical treatment of the sciences we actually possess. But the logician should surely be encouraged to study the processes by which the sciences correct their initial errors and consolidate their acquisitions.

(3) It follows on both these grounds that the 'facts' of which a science takes cognizance will be subject to change. As the science grows, 'new' facts will come into it, and old facts will be discarded as erroneous. In particular, facts which at first were only inferred on theoretic grounds will be actually observed, even as 'Neptune' was the fruit of a theory about the perturbations of Uranus. Hence the antithesis of 'theory' and 'fact' must not be taken as absolute: they must be expected to play into each other's hands. It is the business of theories to forecast 'facts', and of facts to form points of departure for theories, which again, when verified by the new facts to which they have successfully led, will extend the borders of knowledge. Incidentally, however, this interaction between fact and theory often renders it difficult to decide whether a scientific doctrine is better regarded as a 'theory' or as a 'fact', and leads to differences of opinion. But it can hardly be wrong to advise the scientific mind to practise hospitality towards new facts, while it is no less fitting to show generosity towards old servants that have done their work and can now advantageously be retired. It is ungrateful to abuse them as 'errors', and to despise them with the lofty contempt of the higher knowledge to which they have conducted. And in both

cases the truly scientific attitude may be attained if an element of fanaticism is not imported into the conception of truth by attributing to it an absoluteness which no human truth in fact possesses.

(4) The same need for tolerance is emphasized by a further corollary of the conception of fact which has been advocated. It seems at first a paradox, but on reflection appears to be evident, that the 'facts' will not only *look* different but may really *be* different from different points of view and for different purposes. Once we permit ourselves to consider this possibility we shall easily perceive that there often are conflicts between 'facts', such that they cannot coexist for an abstract logic, while, nevertheless, each of the conflicting facts may be intelligible relatively to its own presuppositions and true under its own conditions, so that the 'contradiction' between them is generated merely because the logical statement has abstracted from the special circumstances of the case.

This situation is, of course, recognized very familiarly and universally in the case of *value-judgements.* We are all willing to admit that one man's meat may be another man's poison, that it is vain to dispute about tastes, and that the same mode of living does not suit all constitutions and all circumstances. We recognize, too, that profound differences of opinion and attitude exist, and always have existed, among men. The temperamental differences which make e.g. one man indolent another enterprising, one man daring another prudent, one a conservative another a radical, one an optimist another a pessimist, are so deeply rooted in human nature as to be, humanly speaking, ineradicable. And if so, must it not be conceded that situations occur which will inevitably, consistently, *and rightly,* be judged differently by these different persons?

Again, it should be noted that these differences in valuation are not merely subjective: they spring from objective differences in human nature, and are as objective as any other facts about it. For example, that certain persons dislike pork (because they cannot digest it), and hate cats (because their presence makes them feel ill), rests as much on a physiological fact of their constitution as that others suffer from 'hay fever'. Similarly, it is quite plausible to contend that 'every little boy and girl that is born alive, is born a little liberal or a conservative', and certainly the normal growth of conservatism as the individual mind ages is proof enough that changes of belief depend on psychological law, and are correlated with the hardening of tissue which is a general symptom of senescence. Again, is it possible to imagine a situation so bad or so good that it cannot be interpreted either optimistically or pessimistically? In most cases either interpretation is quite easy, and the choice between them is effected by sheer temperamental bias. If, then, we succeed in doing what the natural man will always find difficult, and regard such differences of opinion in a scientific and non-partisan way, must we not admit that *both* the conflicting standpoints are inevitable and justifiable? Neither can be pronounced wrong in general and *per se*, though in regard to a particular problem or occasion either may be. Let us conclude, then, that it may really be a 'fact' that the 'facts' justify one interpretation and attitude to one mind and another to another.

This argument is reinforced by the further consideration that even the most objective statements of fact involve *value-judgements* in their ultimate analysis. For they express, often explicitly and always implicitly, the choices and valuations by which a variety of pretenders to reality have been examined and sifted, and the most valuable have been declared 'truly real'. We have seen that in a scientific inquiry the 'facts' must always be taken as *alleged* facts, discovered up to date; hence a science must always be ready to defend the 'facts' it recognizes, when they are challenged, and to show wherein they excel conflicting allegations. The accepted 'facts' of a science, therefore, are always allegations which are thought to possess greater *value* than any known alternative; hence no sharp or absolute distinction between judgements of fact and judgements of value can be maintained. It becomes, moreover, quite possible that incompatible allegations of fact may in the actual state of a science be so nearly balanced that there is no convincing reason to prefer one to another, or at any rate none that could prevail against any ordinary temperamental bias. Consequently, in such cases the bias will condition the visibility of the 'fact'; it will be bathed in a 'subjective' atmosphere, and the 'eye of faith' will be necessary to perceive it. No doubt such situations are inconvenient, and repellent to the scientific spirit; but they do not occur only in the misty regions of religion and philosophy, and scientific alternatives like 'chance' or 'design', 'miracle' or 'law', 'mechanism' or ' vitalism', determinism or indeterminism are essentially of this order. There is no reason, therefore, why logic should not recognize them and acknowledge that the scientific 'facts' may be ambiguous, in the sense that further experience and experiments are needed to determine their character. As a rule, to judge by the past, further inquiry will resolve the ambiguity; but it may well be an illusion to assume that it must do so, and in some of the most important cases the decision will certainly be long in coming.

Thus the student of animal behaviour will probably long be left with a choice between minimizing the displays of animal intelligence and assimilating them to the human, while it will probably always be possible to put a pessimistic or an optimistic interpretation upon the facts of life as a whole.

A scientific logic therefore should radically disabuse the mind of any excessive trust in 'facts'. It is a superstition that 'facts' are plain, straightforward, and easy to discover; they are often subtle and recondite and relative to circumstances, changing their aspect to suit their scientific environment like any chameleon.

§ 32. (*c*) In considering the use of words in research, one cannot of course overlook the obvious fact that the employment of words is primarily determined by their established meanings, and that these greatly limit our freedom to use them as we please. Words naturally and inevitably suggest their established uses by their mere sounds, and should always be used with a proper respect for their past history and present meaning. To be sensitive to this appeal is the mark of the educated scholar; but it does not require the investigator to exhaust his energies in vain attempts to stereotype absolutely the current meanings, and so to deprive words of their essential function. For their essential function is after all to be instruments for the conveying of actual meaning, and actual meanings are always more or less new (cf. § 12). It occurs to a particular person in a particular situation to express and convey a meaning which has never in its full concreteness occurred before. If the novelty about this situation is appreciable and important, it may well be that the old words will not fully succeed in conveying the new meaning; and yet we shall always endeavour to use them, and select from the accumulated wealth of language the words which will suffice for our purpose. For the alternative is worse; we cannot always be coining new words for every new meaning we may desire to convey; they would not be understood or remembered, and even if they were, a science that employed nothing but technical terms, and was moreover compelled continually to change them, because it would not use them to convey new meanings, would speedily degenerate into an abstruse game, and could make no progress. How impracticable such a policy would be may be gauged by the grave inconvenience which even now systematists cause by so frequently changing the scientific names of plants and animals. It is indispensable, therefore, that words should retain a certain measure of plasticity, in virtue of which they can be transferred from old situations to new and be used to convey new meanings. Nor is there usually any difficulty about thus imposing new duties on the old terms; under the particular circumstances of the situation even wide departures from the established meanings may remain intelligible, and so the progress of science is not impeded.

The traditional logic, however, cannot treat the matter so lightly. For the plasticity of words may always engender a conflict between the old meaning and the new, between the scientific use of terms and the traditional conventions about their use. And this can always be represented as a defiance of the 'laws of thought'. For if the meaning of '*A*' may be altered by the growth of knowledge, it will no longer be true that everything once called '*A*' is truly *A*, nor that what was once incompatible with *A* will continue to be so for all time. Hence it is no longer necessarily true that '*A* is *A*', and that *A* cannot both be and not be *B*. It may be both in different senses, and in what sense '*A*' and '*B*' should be taken may be precisely the point at issue. Thus verbal contradiction ceases to be a clear proof of error; it may be only a much-needed warning that our terms have been developing new meanings. Hence, the 'laws' of Identity and Contradiction lose their last claims to be regarded as statements of fact, and have to be conceived as ideal postulates of just so much stability of meaning as is requisite for effective understanding.[1] They can be applied to reality only hypothetically, i. e. experimentally, to discover whether in a given situation the natural growth in the meaning of the terms may *rightly* be treated as irrelevant, and does not vitiate the conclusion which the reasoning forecasts. Now this problem can never be

[1] Cf. § 8 and *Formal Logic*, ch. x.

settled *a priori* by reasoning, but only by subsequent experience. Reasoning may forecast a result which experience fails to confirm; when we discover that comets' tails are not attracted by the sun but repelled, we do not declare the facts 'contradictory', but modify our notion of 'gravitation', and conceive it as inferior to 'light pressure' in its effects upon particles of a certain minuteness.

It follows that no merely logical scrutiny of the terms of an argument can ever settle a scientific question. If a 'contradiction' is real, it means either a difference of opinion between those who make the incompatible assertions, or, in the case of a real 'self-contradiction', the uttering of 'nonsense' and a failure to propound a meaning at all. But even the most glaring 'contradictions' may only be apparent, i. e. verbal: when we inquire into their actual meaning we may find that they refer to a context in which its terms are perfectly compatible. Thus the existence of a 'round square' may be predicated of London, and a 'triangle's' angles may equal or may exceed two right angles, according as it belongs to Euclid's geometry or to Riemann's.

§ 33. The problem of discovery, therefore, is never one of which the solution can be guaranteed in advance. The resources of a science are never sufficient to assure us of a prosperous issue of the research, though, rightly understood, they yield important safeguards. A recognition of the instrumental value of words as ancillary to meaning, and of the limitations under which they labour, will guard the inquirer against the terrible verbalism to which logic has been enslaved. A critical attitude towards allegations about 'facts' will enable him to minimize the dangers of error, deception, and bigotry. A conception of 'principles' as working hypotheses will discourage a servile and superstitious reverence for them, and justify the fullest freedom to experiment with whatever ideas hold out hopes of verification and of scientific progress. Together these three considerations will pretty thoroughly emancipate inquiry from the shackles of any mechanical scheme of 'proof'. Indeed, proof in the old formal sense will have become a chimera. It will no longer be possible to cherish the belief in a self-sufficing, self-satisfied form of absolute proof, of which the pure logician imagined himself the possessor and retailer.

Scientific proof, on the other hand, will be neither absolute nor formal. It will not be absolute, because it will always be relative to the actual condition of a science; it will not be formal, because it will never be absolute. It will only be the best known interpretation, and will always imply alternatives, to some of which it may wrongly have been preferred, while to others it may be destined to succumb (§§ 26, 27). It will be 'valid' so long as it is the strongest; but to it, as to the priest of Diana Nemorensis, as to Uranus and Cronus, will come the day when it is invalidated and superseded by a stronger and better, descended, it may well be, from itself. Scientific proof then will always be an *evaluation* of evidence, a making the most of the available resources of a science, a question of the *comparative values* of rival interpretations.

It stands to reason that such an evaluation cannot operate merely with the criteria of formal logic. Indeed, of the processes known to the traditional logic, only those which *cannot* be represented as 'formally valid' will be exemplified in scientific knowing. It will not be possible to find any genuine cases of absolute certainty or unconditional proof; but analogies, probabilities, hypotheses, alternatives, even fallacies and fictions, will abound, and will somehow have to be discounted. Clearly the evaluation of such things will be a delicate affair; it cannot be accomplished by reciting *Barbara Celarent* and crudely applying a few simple mechanical formulas. It will demand the energetic co-operation of the whole intelligence, and indeed of the whole personality, and cannot scorn the aid of psychological factors. For it is plain that the evaluation of a complicated scientific situation will require both expert knowledge of scientific detail and philosophic grasp of general principles and connexions; it will need also 'tact', 'judgement', an 'eye from experience', and a host of similar qualities that elude precise verbal formulation. It will no longer be practicable to flatter mediocrity and dullness, and to impede discovery, by proclaiming methods that dispense with imagination, ingenuity, originality, boldness, enterprise, and vainly endeavour to put genius for discovery on a par with mindless pedantry in applying stereotyped and sterile rules.

§ 34. But just because a logic that recognizes the actual process of discovery does not presume to dictate formal methods to the discoverer, and leaves him a very free hand, it does not relieve him of any of the responsibility for conducting his researches to a prosperous issue. As there is no longer any pretence that any logical machinery can be devised to guarantee success,

success and failure become his personal achievements. If he fails, he can no longer plead that it is not his fault, seeing that he has kept every letter of the law and broken no logical rule. This may be precisely *why* he failed. Perhaps he should have taken risks. He may have gathered such enormous masses of fact that he could no longer see through them, nor select the few that were relevant to his problem. He may have been so sensible of the need for caution that he dared not speculate or move. He may have devoted himself to unimportant problems or missed the important sides of important problems, or have wandered away into barren wastes of dialectics, or have got bogged in a mire of verbalism, or have pursued elusive phantoms of unverifiable speculations. For there are clearly many ways of failing. Only in whatever way he fails, his personal failure is *pro tanto* a failure of science to progress. Every science has somehow to get hold of a clue to guide it through the labyrinth of fact, and this clue has to lead it right, though it need not 'follow necessarily' from previous knowledge.

Nevertheless, if, and in so far as, a researcher succeeds in making a discovery, some of his personal credit is reflected upon his methods *ex post facto*. Their success does not, of course, establish their formal 'validity'; but it stops the mouth of those who argued that what is 'invalid' must be worthless. Methods that succeed must have *value*, a greater thing than 'validity', however far and however boldly they departed from the canons of formal proof. The success has shown that *in this case* the inquirer was right to select the facts he fixed upon as significant, and to neglect the rest as irrelevant, to connect them as he did by the 'laws' he applied to them, to theorize about them as he did, to perceive the analogies, to weigh the chances, as he did, to speculate and to run the risks he did. But only in this case. In the very next case, which he takes to be 'essentially the same' as the last, and as nearly analogous as is humanly possible, he may find that the differences (which always exist between cases) are relevant, and that his methods and assumptions have to be modified to cope with it successfully. But he should not be discouraged. For the ultimate ground of the whole cognitive procedure by which we analyse the flow of events is empirical. It is only an empirical fact that knowledge is possible, i. e. that the course of events is such that human minds can analyse it at all, that is, can pick out and construct cases of 'the same', of which the course can be predicted by means of the (verbally) stable formulas we call 'the laws of nature'. For logic at any rate these laws are neither supernatural behests nor metaphysical entities: they are forms for classifying happenings, in which the blanks have to be filled in with the variable values of the particular happenings. What the *right* values are, and even what is the *right* formula to apply, will always depend on the particular case which forms the actual problem. It is only the empirical fact that the differences between problems may so often be treated as irrelevant which generates the illusion that problems may be solved in advance by general formulas: in reality every problem in its full concreteness is unique, and we are never absolutely sure that it will submit to the rule we apply to it. Hence it is solved only when we come to it and find it amenable to our methods; in principle it eludes logical prediction, because it can be known as a 'case' of the successful 'law' only *after* the experiment has confirmed the forecast. To the inquirer, therefore, no result can seem certain until it has occurred; it is only *ex post facto* that the logician can describe it as an indubitable case of some law from which it follows of necessity. But in so doing he has changed it, and repudiated the duty of describing actual knowing. All he is doing is to rearrange a piece of knowledge, acquired without his aid by means he condemns as illicit, in the order he is pleased to call 'logical'. This order has a certain aesthetic value, but it is emphatically *not* the order of discovery, and throws no light on the process of acquiring knowledge.

§ 35. What function then can be assigned to the logician's reflection on the workings of science? In view of his failure to substantiate his claim to have provided a model for inquiry in his scheme of 'proof', it might seem that he was either useless or pernicious. Useless, if he merely devotes himself to constructing 'ideals of proof' which he admits to have no relation to the actual problems of science; pernicious, if he is prompted by these ideals to make demands with which no science can comply, and to deliver judgements which would paralyse the science that attempted to carry them into execution. Fortunately, he cannot enforce them, and the sciences actually go on their way, ignoring such 'logic'. The proper inference from his impotence is that he would do well to take up a position which is more useful and more influential, if less pretentious.

Let the logician then give up the pretence of dictating to the sciences and of judging the worth of scientific truth by rigid forms of absolute proof; let him abandon the vain pursuit of 'validity'. Nay, more, let him renounce the claim to determine the scientific value of an argument by a mere inspection of its logical character. Let him confess that what alone he can criticize is the incongruities in its verbal expression, and that its real value lies beyond his ken. If he will concede all this, his reward will be that he has vindicated for logic an important right of more real value than the claims he has abandoned. For he will have obtained the right of summoning the sciences to state their results in intelligible and consistent terms, and to confront them with a problem when they do not. Just because he does not presume to condemn them, and no longer ventures to declare that incompatible and verbally 'contradictory' results are necessarily wrong and worthless, but only urges that they are not intelligible as they stand, and need to be reworded or inquired into farther, he gains the right of *raising problems*, and stimulates the sciences to proceed to solve them.

It should be noted, moreover, that the problems thus raised are general, not special, i.e. are properly logical. The problem about 'contradictory' results is one about meaning, for contradictory assertions cancel each other's (apparent) meaning. This enables the logician to keep the sciences engaged upon the logical problem of solving the discrepancies between their results, so long as the sciences do not form one complete and congruous system, i.e. indefinitely.

Similarly the denial that truth is absolute is a general truth that affects all the sciences. It should stimulate them all, for it means that no statement is so perfect that it cannot be bettered and that no limits can be set to the progress of science.

Other topics which are 'logical', because they concern the general significance of scientific procedure and not the solution of particular problems, are the nature and importance of selecting 'facts' and the 'laws' they are taken to exemplify, the experimental attitude and the framing of hypotheses, the evaluation of probabilities and alternatives, the estimation of relevance and of verifications and of the amounts of the latter which are requisite and the sorts of it which are relevant. On all these points logic has hitherto had little or nothing to say, mainly because they did not lend themselves to formal treatment. Lastly, there are two extremely important subjects, which are so vital to the logic of discovery that a brief discussion of them may fitly conclude this essay. We may call them the problem of Novelty and the problem of Risk.

§ 36. In Logic we are not concerned with the metaphysics of Novelty, i. e. with the problem of whether there ever enter the world things that are really and truly unforeseen and unpredictable, that pop into it from nowhere, and if so, whether and how we can understand such things. This problem is deep and difficult, and so, until recently, philosophers have fought shy of it, and used to settle it off-hand by a flat denial that such things could be in a 'rational' universe. But now that M. Bergson has given us a radically new metaphysic, and that we are beginning to perceive that the principles used to dispose of the matter, viz. causality and the conservation of energy, are essentially methodological, the question has become an open one.

Logic, however, has no need to probe it; it can treat it more simply. For its purposes it can, and must, treat novelty as a real logical fact. It is a psychological fact, and logic must note it, that every moment of our life has for us a certain flavour of newness; it is also a fact that every real judgement that is ever made has a certain relation to novelty.[1] Its maker believes, either that it embodies a new truth, or that though known to him it is new to his hearers. If he did not believe this he would have no motive to make it. It would be stale repetition, devoid of interest or value alike to him and to others, whom he would merely bore by telling them what they, too, knew already.

So far, then, the logical nature of novelty seems simple. It gives rise to problems, however, when we consider the relation of the new truth to the old. It is clear, in the first place, that the new truth must affect the old. Even where we are willing to minimize its novelty, and to call it merely an 'extension' of what we already knew, it must modify it and change its value. For in the light of the new developments the old truth *means more*: it has relations in an enlarged field of knowledge. Moreover, the new truth is often not merely an extension but also a *correction*, and the effect of the correction may sometimes be revolutionary. It may even seem to upset the old beliefs altogether, though

[1] The 'novelty' which is claimed for the conclusion of a syllogism is only one case of this: in the traditional interpretation it is hopelessly at variance with the demand that it shall also follow from its premisses of necessity. Cf. *Formal Logic*, ch. xvi, §§ 8–10.

human ingenuity is far too fertile in building bridges (often only verbal) from the old to the new to allow this impression to be permanent. Still in all these cases there is more or less discrepancy between the new and the old.

The logician, however, should insist that this fact should not be blinked. He should recognize the discrepancy, and emphasize its significance, just because for other purposes it is usually convenient to ignore it. For it is not only the source of real ambiguity in the facts of science, and of the important differences of opinion among men and of their obstinate persistence, but the justification of the policy of open-mindedness and toleration which he regards as necessary to scientific progress. Inasmuch as of every discrepancy between the old truth and the new it will be possible to take two views, and either to cling to the old or to put one's trust in the new, there will always be a party of conservation and a party of innovation, or otherwise a conservative and a liberal bias, in science as in politics. It is, moreover, futile to discuss, in the abstract, which of them is right: for it would clearly be fatal to go all lengths with either. Science could make no progress, either if every novelty were at once condemned and suppressed because of its failure to conform with the accepted doctrine, or if everything new were hailed as true regardless of its concordance with the old truth, so that the course of science became a series of radical revolutions that had no consistent direction. In concrete cases of course both sides are sometimes right, though historically the stronger bias men have shown has been the conservative. What usually happens is that the new truth is first denounced as an immoral invention which is subversive of all intelligible order and cosmic rationality; it is then quietly assimilated and not infrequently converted in the end into the strongest support of the beliefs it was alleged to subvert. But it would be a real gain if logic, by viewing this natural feature of knowing in its generality, could induce men of science to take it more calmly. If it were generally recognized that every claim to new truth, however great the advantages it promises, necessarily entails certain inconveniences, because the old beliefs and notions have to be modified and readjusted, and this may involve too great an effort to be worth while, or an effort too great for certain minds, it would be seen that there are two sides to every question, and that both may be in a way legitimate. If, in addition, we recognize that the parties concerned usually have a bias which may render them dangerously blind to the case of the other side, and that both should be admonished to discount their bias duly, we shall have done not a little to secure fair-minded[1] consideration, reasonable discussion, and intelligent choice between the alternatives. And all this surely conduces to scientific progress.

It is clear, then, that the problem of relating the new to the old always exists, and has a vital influence on the fortunes of every science. But it is not capable of any formal or abstract solution *a priori*. Which is to be preferred is a matter which must be left to the expert who is cognizant of the circumstances of the case: logic can help only by broadening his mind, and putting him on his guard against his own personal bias, which might otherwise unconsciously determine his decision.

§ 37. To admit that scientific inquiries concern problems, and that to every problem (at least) two solutions may be propounded, between which a choice has to be made, is to admit that knowledge *must take risks* in order to progress. For there is always the risk of choosing the wrong solution of a problem, i.e. the one which works *less* well, just as there are always risks of choosing a bad problem and of selecting the wrong facts and the wrong theories to explain them withal. Nevertheless, we ought not to resent this fact. For the taking of risks is inevitable: we cannot escape it either by refusing to inquire or by refusing to decide. For in either case we run the risk of missing a valuable truth.

It is better, therefore, to recognize that every act of knowing must involve risks, just as every act of living does; and this for the simple reason that knowing is an activity comprised in living, and every judgement is an *act*, which might have been left undone, or for which another might have been substituted. The readiness of the new conception of logic to emphasize the existence of risks in all reasoning, and to sanction the willingness to take them,

[1] Usually, but wrongly, called 'dispassionate' or 'disinterested'. What is wanted is, not that the inquiring mind should take no interest in the conclusions it considers, but that, though it cares keenly and even passionately for one of them, it should yet be capable of sufficient self-control to consider fairly the case *against* the conclusion it favours. This mental attitude is probably best secured by caring more for truth than for a party victory, and is denominated a 'disinterested love of truth for its own sake'. But even so we love what we deem the truth, because it is the *best* thing to believe, and better (on the whole and in the end) than anything else that is propounded.

contrasts markedly with the vain efforts of the old logic to play for safety, and to make no move that was not absolutely necessary (cf. § 10). This was why it postulated absolutely certain premisses, and would contemplate nothing but 'valid' forms of reasoning. In its desire to elevate its proofs above the perplexities and vicissitudes of mundane problems, the old logic was expressing and comforting a deep-seated human craving: for life is so replete with the most hideous risks that it is a natural instinct to clutch at any promise of security. Hence the passionate and almost religious reverence with which formal logic has been regarded for over 2,000 years. Many philosophers still worship the syllogism, because it seems to them an incomparable exemplar of absolute security firmly fixed in the sphere of immutable necessity far above the flux of phenomena, which it illumines with its steady radiance. But to exalt in this way its ideal of proof, the old logic had to pay a heavy price. The price was cutting the ideal wholly adrift from the actual, contemplating exclusively a situation which could never occur in real life, and leaving all actual inquiry to its devices, unstudied, uncriticized, and unaided. Thus, the splendid aloofness of the logical ideal was purchased by a total repudiation of actual science. To many philosophic minds this price does not seem excessive. The more useless truth is made to appear, the purer and more admirable it seems to them. An ideal, they think, should be like Aristotle's 'god'; it should attract, without uplifting, and without running the risk of contamination by the dirty work of life.

These philosophers have always claimed for their attitude that it is philosophic *par excellence.* But their claim, besides being based on a somewhat rare personal idiosyncrasy, is not really sound. It is neither self-consistent nor a sound policy for life. An ideal which repudiates the actual, and yet professes somehow to be its exemplar, is left in the impossible condition of the Platonic 'Idea'. If it were as superhuman as it claims to be, no human mind could even speculate about it. And we have seen (§ 13) that it is not in the end possible to devise a form of proof which is bomb-proof against the attacks of experience and superior to verification.

Is it not wiser, then, to admit that life has its claims upon science, and science upon logic? We simply *must* have a science that can handle human life and meet human needs, and does not degenerate into a game with arbitrary and fantastic rules which depart from the actual conditions of life in any direction and to any distance unrestrained imagination carries them; and our logic must deign to study such a science. If to do so it has to 'scrap' its antique 'ideals', to abandon its pose of an inhuman, impassible, infallible aloofness, and to interest itself in the doubting, questioning, guessing, trying, risking, blundering, correcting, achieving that make up the sum of human knowledge, it will receive an ample reward in the gratitude of man for a logic that has entered his service, and in the salutary influence which it will exercise upon his actions.

Conclusions

(1) We have shown, negatively, that the notion of a form of proof, by which conclusions can be absolutely demonstrated by dint of pure logic alone, is a delusion. No such form can be constructed (§§ 13, 15), and if it could, it could neither find scientific material worthy of it (§ 28), nor contain the material which is fabricated by the sciences.

(2) We have thereby shown that formal logic cannot represent the logical nature of discovery or of any of the processes of actual knowing, and must condemn them all as 'invalid' (§§ 18, 20, 26, 28).

(3) We have seen that a logic which attempts to understand actual knowing cannot prescribe to the sciences how they are to solve their problems (§ 33).

(4) But it *can* grasp the general character of scientific procedure, appreciate its difficulties and dangers, understand the expedients for meeting them, and trace it to its roots in the constitution of the human mind and in the needs of life (§ 35).

(5) In virtue of its general grasp of the aim and method of the sciences a logic of science can at times offer advice to scientists: it may draw their attention to the general problems which their work involves, but which are apt to be overlooked by specialists, such as the claims of consistency and novelty and the regulation of risks (§ 36). Or, better still, if they will study it themselves, it may broaden their minds and enable them to handle these general problems for themselves far more effectively than a pure logician could do it for them.

(6) By abandoning its pretensions to rigour and conclusiveness logic does not really lose: it gains immensely by coming into contact with science and life, and becoming of use in the world.

INDEX

STUDIES

IN THE

HISTORY AND METHOD OF SCIENCE

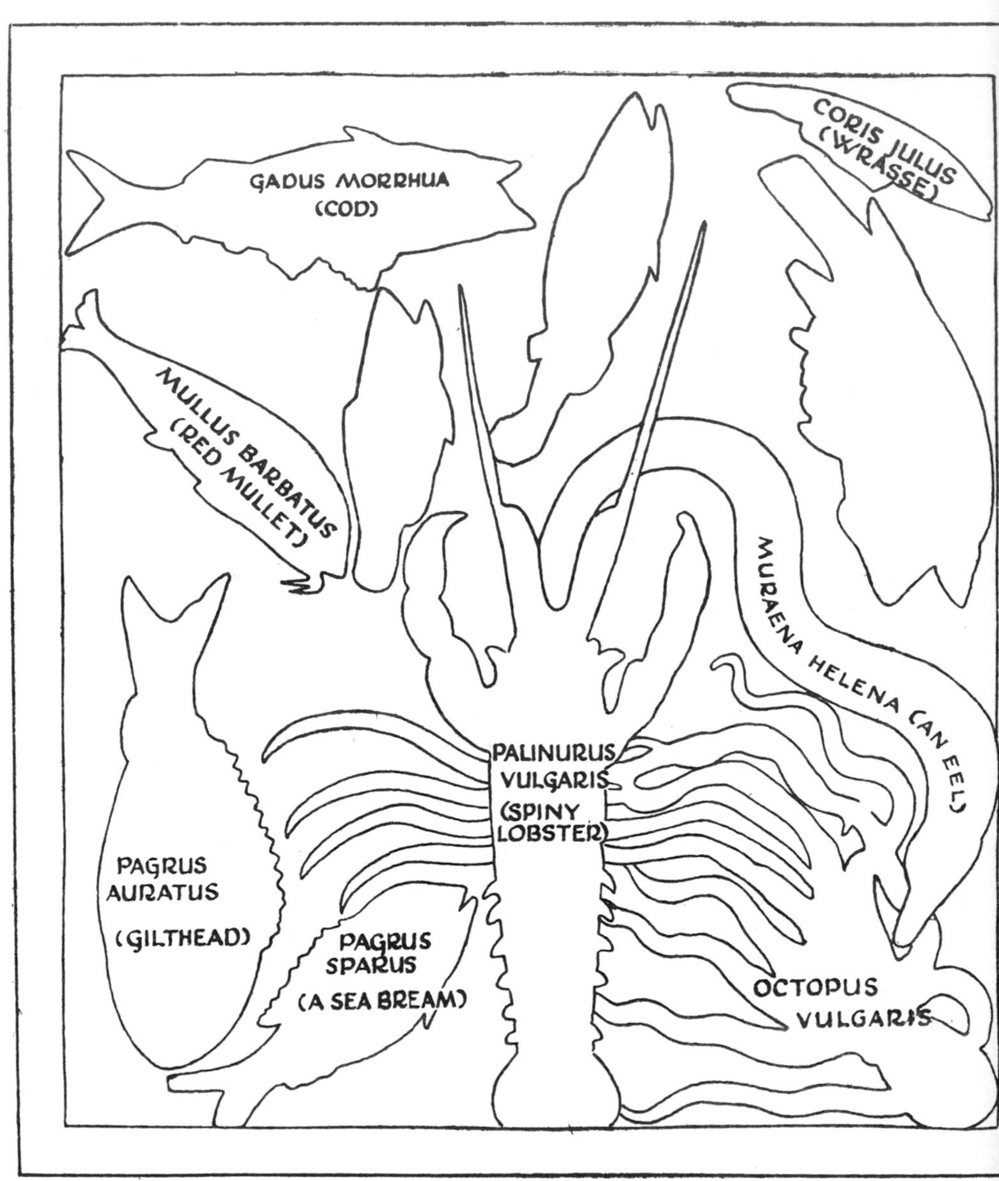
GADUS MORRHUA
(COD)
CORIS JULUS
(WRASSE)
MULLUS BARBATUS
(RED MULLET)
MURAENA HELENA (AN EEL)
PALINURUS
VULGARIS
(SPINY
LOBSTER)
PAGRUS
AURATUS
(GILTHEAD)
PAGRUS
SPARUS
(A SEA BREAM)
OCTOPUS
VULGARIS

Mosaic from Populonium (Etruria). Probably Ist Cent. A.D.

Now in the Victoria and Albert Museum.

OXFORD UNIVERSITY PRESS
LONDON EDINBURGH GLASGOW NEW YORK
TORONTO MELBOURNE CAPE TOWN BOMBAY
HUMPHREY MILFORD
PUBLISHER TO THE UNIVERSITY

STUDIES

IN THE

HISTORY AND METHOD OF SCIENCE

EDITED BY

CHARLES SINGER

VOL. II

OXFORD
AT THE CLARENDON PRESS
1921

TO THE MEMORY

OF

WILLIAM OSLER

PREFACE

The first volume of this series appeared in the autumn of 1917. The editor was unable to see it through the press owing to his absence from England on military duty. A Preface was accordingly provided by Sir William Osler, who, unhappily, has not lived to see the growing success that has attended the ideas he expressed there with so much force, and towards which he contributed life-long thought and effort.

The volume was received with an approval that far surpassed the hopes of its editor and the issue was rapidly exhausted. In the present collection an endeavour has been made to avoid some defects inevitable in the earlier volume. The undue prominence given to mediaeval studies will be found in part corrected and care has been taken to give more space to the evolution of the mathematical and exact sciences, though the balance is yet far from being fully redressed.

Among the many recent signs of public interest in this country in the History and Philosophy of Science special note must be made of the systematic course in the subject now organized at University College, London. The work is under the general direction of Dr. A. Wolf. Professors Sir William Bragg, Flinders Petrie, W. M. Bayliss, L. N. G. Filon, J. P. Hill, F. G. Donnan, E. J. Garwood, T. L. Wren, D. Orson Wood, and the Editor of this series are taking part in it. The course at Edinburgh in the History of Medicine, initiated by Dr. J. D. Comrie, has attracted a large and annually increasing number of students. Mr. A. E. Heath, Senior Lecturer on Education at Liverpool University, is giving systematic instruction there on the History of Science, and the Editor of this series has become Lecturer in the History of Medicine at University College, London, and University Lecturer in the History of the Biological Sciences at Oxford. Great service has been rendered by Mr. F. S. Marvin, who has not only done much to popularize the

History of Science, but has sought to integrate this important aspect of human development in the historical instruction of schools.

Considerable literary activity is also being exhibited. The English translation of Aristotle under the general editorship of Mr. W. D. Ross has continued to appear, as has also the much-needed publication of Roger Bacon's works under the editorship of Mr. A. G. Little, Mr. Robert Steele, and Dr. Withington. Sir Arthur Hort has rendered the *History of Plants* of Theophrastus into English, and a similar service has been done for Galen's short treatise *On the Natural Faculties* by Dr. A. J. Brock. A valuable work in a different department is Sir Norman Moore's monumental and learned *History of St. Bartholomew's Hospital.*

All historians of science look forward with lively interest to the appearance of Sir T. L. Heath's *History of Greek Mathematics*, Professor Dobson and Mr. Brodetsky's version of Copernicus, Mr. W. H. S. Jones's edition of Hippocrates, Professor William Wright and Mr. Foate's translation of Vesalius, and to further communications from the pen of Dr. J. K. Fotheringham, whose work peculiarly illustrates the aid rendered to modern scientific investigation by a wide and deep knowledge of ancient science. Less conspicuous but very real and useful will be the results of the tireless devotion which Dr. Withington has for many years bestowed on the vast mass of Greek scientific literature, with the object of rendering the forthcoming edition of *Liddell and Scott's Greek Lexicon* more complete in what has been hitherto a somewhat neglected department. It is a legitimate hope that his unique experience may now be utilized in rendering accessible to English readers the more important works of Galen.

Mrs. Singer's catalogue of early scientific MSS. in this country is now in a serviceable state and freely open to students. It is a card index with about forty thousand entries arranged by subject. The first section, Alchemy, is being printed as the opening fascicule of an international catalogue of Alchemical MSS., issued by the Union Académique Internationale under the general editorship of Professor Bidez of Ghent. The next of Mrs. Singer's sections to appear will probably be 'Anatomy' and 'Aristotle' by the Editor of this series.

Of foreign workers in our field mention must be made in the first place of the veteran scholar Professor K. Sudhoff of Leipzig, who has for many years been producing volume after volume of highly original researches, covering wide stretches in the history of science. His extraordinary fertility and equally great literary generosity lay all other workers in his department under a debt of gratitude impossible to repay. It is gratifying to observe that his venerable colleagues, Professor Eilhard Wiedemann of Erlangen and Professor J. Hirschberg of Berlin, are still prosecuting their researches in Oriental Science and the History of Optics, and that fascicules of the *Corpus Medicorum Graecorum* and *Corpus Medicorum Latinorum* continue to appear, though all too slowly. The recent output by Professor Max Wellmann on the science of classical antiquity has been very remarkable. Another useful work in the department of the History of Science is Professor E. O. von Lippmann's *Entstehung und Ausbreitung der Alchemie*.

In Austria the activities of the most philosophical of living medical historians, Professor Neuburger, have continued in spite of the terrible conditions through which the city of Vienna has passed and is passing. The first volume of his illuminating *History of Medicine* was rendered into English by Dr. Ernest Playfair, with the encouragement of Sir William Osler. The appearance of the second volume of this translation was a cherished wish, as it would be a suitable memorial, of that lover of learning.

The recent output in the History of Science of some of the smaller European States has of late years been truly remarkable, and the world owes to them a number of works of primary value for the subject, elaborately published at vast expense. The decipherment, transcription, translation, and publication of the beautiful Windsor *Quaderni d'Anatomia* of Leonardo have now been completed by Professors Fonahn, Hopstock and Vangensten, and the work has been published at Christiania. The Société Hollandaise des Sciences is proceeding with the monumental edition, collected from manuscript sources, of the *Œuvres complètes* of Christiaan Huygens. Denmark has given us the *Opera Philosophica* of Stensen, edited by Dr. Vilhelm Maar, the *Opera Omnia* of Tycho Brahe, edited by Dr. J. L. E. Dreyer, and the *Scientific Papers* of H. C. Örsted; all three are produced by the Carlsbergfond. Professor Heiberg of Copenhagen continues his series of fine works on Greek science, and his writings are an additional adornment of the Danish School. The Union Astronomique Internationale, founded at Brussels in 1919, has appointed a *Commission de réédition d'ouvrages anciens.*

In Italy the history of science has now been placed on a firm academic basis. The work of Professor Antonio Favaro on the physics and mathematics of the seventeenth century and of Professor Piero Giacosa on mediaeval science continues. The new *Archivio di Storia della Scienza* has completed its first annual volume under the editorship of Professor Aldo Mieli.

In France, since the appearance of Adam and Tannery's edition of the Works of Descartes and the termination by death of the labours of Henri Poincaré, the most important publication has probably been the fifth and posthumous volume of Pierre Duhem's very valuable *Le système du monde, histoire des doctrines cosmologiques de Platon à Copernic.* Mention must also be made of the very scholarly work on the *Commentaires de la Faculté de Médecine de l'Université de Paris* (1395–1516) by Dr. Ernest Wickersheimer, who has since become librarian of the University of Strassburg.

Nowhere, perhaps, has more general interest been taken in the History of Science than in the United States. Many courses of lectures have been devoted to the subject; much valuable original work in the department of mathematics has appeared from the pens of Professors L. C. Karpinski and Eugene Smith, in medicine from Colonel Garrison and Dr. E. C. Streeter, and in bibliography from Dr. A. C. Klebs. Among recent American publications are Mr. T. O. Wedel's work, *The mediaeval attitude toward astrology*, and a useful translation of Choulant's *History of Anatomic Illustration*, annotated and brought up to date by

Mortimer Frank, whose death at the age of forty-four is a real loss to learning.

Dr. George Sarton has been carrying on at Harvard the labours commenced before the war in Belgium. His journal, *Isis*, which first appeared in 1913 as a 'Revue consacrée à l'histoire et à l'organisation de la science et de la civilisation', has revived since the peace, and some form of joint publication is projected between this series and *Isis*. Moreover, in the future, the *Studies in the History and Method of Science* will appear regularly as an annual volume.

A central institute and library, devoted to the promotion of systematic investigation into the historical documents of science, is greatly needed in this country. Such a foundation would do much to place the subject on its proper academic basis and would rapidly react on the whole system of scientific education. It would help the teacher to present the sciences in their evolutionary relation to each other and to the course of history as a whole. It would especially help the teacher of science to develop his subject as the product of a progressive revelation of the human spirit rather than as a mere description and attempted explanation of the phenomena. We may well look to this new orientation of scientific teaching to counteract the effects of the regrettable but real decline in the study of the older 'humanities'.

CHARLES SINGER.

UNIVERSITY COLLEGE, LONDON,
February, 1921.

CONTENTS

LIST OF PLATES

GREEK BIOLOGY AND ITS RELATION TO THE RISE OF MODERN BIOLOGY

LEONARDO AS ANATOMIST

THE ASCLEPIADAE AND THE PRIESTS OF ASCLEPIUS

THE SCIENTIFIC WORKS OF GALILEO

THE HISTORY OF ANATOMICAL INJECTIONS

A SKETCH OF THE HISTORY OF PALAEOBOTANY

ILLUSTRATIONS IN TEXT

GREEK BIOLOGY AND ITS RELATION TO THE RISE OF MODERN BIOLOGY

A SKETCH OF THE HISTORY OF PALAEOBOTANY

GREEK BIOLOGY AND ITS RELATION TO THE RISE OF MODERN BIOLOGY

BY CHARLES SINGER

There is an extreme affecting of two extremities : the one antiquity, the other novelty ; wherein it seemeth the children of time do take after the nature and malice of the father. For as he devoureth his children, so one of them seeketh to devour and suppress the other ; while antiquity envieth there should be new additions, and novelty cannot be content to add but it must deface : surely the advice of the prophet is the true direction in this matter, State super vias antiquas, et videte quaenam sit via recta et bona et ambulate in ea. *Antiquity deserveth that reverence, that men should make a stand thereupon and discover what is the best way ; but when the discovery is well taken, then to make progression. And to speak truly,* Antiquitas saeculi iuventus mundi. *These times are the ancient times, when the world is ancient, and not those which we account ancient* ordine retrogrado, *by a computation backward from ourselves.*—Bacon's *Advancement of Learning*, v. 1.

I. THE COURSE OF ANCIENT AND OF MODERN SCIENCE COMPARED

IN the pages which follow we discuss certain elements in the exact, classified and consciously accumulated knowledge of living things possessed by the Greeks. This biological knowledge and the mode in which it was attained are well suited to the illustration of Greek scientific method, for the actual achievements of the Greek mind were no less remarkable and perhaps more characteristic in Biology than in other departments of physical science. As a preliminary to the discussion we may briefly consider the means available for forming an estimate of Greek science as a whole, and in doing this we shall inevitably compare and contrast the science of antiquity with that of our own time.

Ever since man has been man, he has had some control over nature through his power to adapt his instruments to make her serve his will, and it is possible to define science in terms of this power and of the knowledge that lies at the back of it. But the *conscious* formulation of theories to explain natural phenomena, and the *conscious* collection and record of data as a basis of these theories, come as a far later phenomenon in human development. It is this conscious and more sophisticated process to which, for our present purpose, we shall give the title *science*, and science so interpreted cannot be traced with certainty earlier than the speculations of the Ionian philosophers of the sixth century B.C. Greek science thus established continued its course of positive achievement until the second or third century of the Christian era. Then, from causes which we need not here discuss, it ceased to be original, having run a course of some eight hundred years.

Our effective record begins with the *Hippocratic collection.* Some elements in this are at least as early as the sixth century B.C., and it is therefore impossible that these earliest portions should be the work of Hippocrates himself who died in the first half of the fourth century. Nevertheless Hippocrates is almost certainly the first scientific writer of whom we have substantial remains. The latest original Greek scientific works were perhaps those of Galen and Ptolemy of the end of the second century of the Christian era, or, if we should include mathematics in our scheme, we may carry the period forward to Diophantus of the third or even to Theon of Alexandria of the fourth century.[1]

We may compare this course with the science of our own time. For a thousand years and more after the downfall of Greek science, the powers of observation and the scientific imagination of mankind seemed to sleep, a sleep broken only by disorderly dreams which either fitfully recapitulated the past, or conjured up what never was and never will be. At length, however, Man awoke to

[1] Or perhaps to Theon's daughter Hypatia, who survived to the second decade of the fifth century and is said to have made original mathematical investigations.

look around and to examine the world into which he had been born. During the long twilight of a new dawn he had been stirring in his slumber, but the year 1543 gave full proof that the night was over and he was at last awake. In that year appeared the two works which mark the real sunrise of modern science, the *De revolutionibus orbium celestium* of the Pole, Nicolaus Copernicus, and the *De fabrica corporis humani* of the Belgian, Andreas Vesalius. These two were the first great modern natural historians of the Universe and of Man, of the Macrocosm and the Microcosm, and if any single year be selected, 1543 has perhaps a better claim than any other to be regarded as the birth year of modern science, though we shall see good reason for assigning its conception to a much earlier period.

Beginning from 1543 we are thus near the end of the fourth century of modern science, and are now at about the middle of the total period of time that Greek science had to run. During these four hundred years a vast and ever-growing mass of original investigation has been recorded, and as time has gone on the stream has grown ever broader and fuller. Some idea of its enormous and unreadable bulk may be gained by a glance at the *International Catalogue of Scientific Literature*[1] which, while giving the titles alone of original articles, consists each year of seventeen closely printed volumes. The vast intellectual effort which this enormous output implies has gradually transformed our mode of life, our attitude to the world around us, and even our hearts and minds.

Now if we seek to compare this extraordinary movement and its results with its prototype of antiquity, we encounter difficulties at the very outset. These difficulties of comparison lie not so much in the relative scantiness of the Greek record—that in itself might be an advantage for our purpose—but rather in the character of that record. The differences in the mode of recording ancient and modern science become explicable when we consider certain differences in the history of the two systems, and to their history we therefore turn.

The earliest science, in the sense that we are using the word, arose in Asia Minor on the confines of the great Eastern

[1] Published for the International Council by the Royal Society of London. First annual issue 1902, last annual issue (with sequence disturbed by the intervention of the war) 1914–16. It is significant that the number of biological papers recorded in this enormous index is double that of the physical and mathematical combined.

civilizations. In the social systems of the valleys of the Euphrates, Tigris, and Nile there had accumulated a great mass of observations, and upon them rough generalizations had been erected. These generalizations seem in the main to have been an evolutionary product of the 'social consciousness', rather than the definite fruit of individual minds, and it is thus characteristic of the science of the ancient East that its products are anonymous. From all the centuries of intellectual activity of the civilizations of Babylon and of Egypt, hardly even the name of a scientific discoverer has come down to us. It was into a great *impersonal* heritage that the philosophers of the Ionian cities were fortunate enough to enter; with it as a basis they began to engage upon that active process of cosmical speculation that developed as Greek philosophy.

As time went on, knowledge accumulated, separate sciences or departments of knowledge were gradually differentiated and, in the course of centuries, these became more and more distinct from the parent stock of philosophy. Yet it is peculiar to Greek scientific thought that it never loses direct touch with the philosophic stem from which it sprang. Whether we look to such early traces of the scientific spirit as that of the sixth century B.C., when Pythagoras was contriving his first formulated conceptions of the relations of number to form, or whether we consider the last vitally original works of Greek science in the second century of the Christian era, when Galen and Ptolemy were giving forth those ideas on the structure of man and of the world that were to dominate Western thought for a millennium and a half, from end to end Greek science constantly betrays its descent from Greek philosophy.

Far different is the ancestry of modern science. The origin of modern science will be sought in vain in the lucubrations of the philosophers, who played but a subordinate part in the revival of letters. Copernicus and Vesalius were dead before the great philosophers of modern science, Francis Bacon and René Descartes, had been born. Nor is it a more fruitful task to attempt, as many have done, to draw a picture of our scientific system as but a rebirth of the wisdom of ancient Greece, for we must then seek its origin in the writings of the men who were the agents of that rebirth. Yet from them we get but little light. Science, as we understand the term to-day, was far from the minds of the men who made the New Learning. The scholars of the fourteenth and fifteenth centuries showed scant sympathy for the investigation of Nature

PLATE II

THEOPHRASTUS — ARISTOTLE

THEOPHRASTUS: From VILLA ALBANI. Copy (IInd cent. A.D.?) of earlier work

ARISTOTLE: From HERCULANEUM. Probably work of IVth cent. B.C.

PLATE III

LATE MINOAN GOLD CUPS. ATHENS MUSEUM

From VAPHIO about XVIth cent. B.C.

and the humanistic period dominated by them was, on the whole, backward or at best but retrospective in its scientific conceptions. Their thoughts were rather with the great past of literature and of art, which they sought to bring back to life.

It is certainly true that there were a few philosophical writers of the later part of this period in whom can be traced some consciousness of the value of the experimental method: Nicholas of Cusa (1401–64), Pomponazzi (1462–1525), Fracastor (1478 ?–1553) stand here to witness. But it is far from clear that their ideas on the mode of extension of natural knowledge were related to the re-discovery of the Greek texts and the diffusion of knowledge of the Greek language. These men, at best, were few and exceptional, and they come mainly in the late and academic period of the learned revival; their place is rather among the founders of modern science and they do not naturally fall into the series of the scholars of the classical Renaissance. It may, indeed, be claimed that the astronomical work of Regiomontanus (1436–76) and Purbach (1423–61) was dependent on their salvage of the text of Ptolemy.[1] But their recovery of the Greek original was the *result* of their scientific zeal and an incident in their scientific researches; it did not provide the primary stimulus for those researches.

If we turn to the revival of biological studies, we are encountered with the same phenomenon of dissociation from the learned revival of the Greek texts. The first modern records of the close scientific observation of plants or animals impinge on the intellectual orbit of the age either too early or too late to be explained as attracted thither by the new learning. Effective advance in zoological knowledge hardly begins until the second half of the sixteenth century, but it was preceded by the work of the anatomists whose activities we may trace back to the eleventh century. The records of botanical observation tell much the same tale. The familiar attempts of the 'fathers of Botany' are not encountered until well into the sixteenth century, but behind their work we can discern a long and slow evolution of first-hand plant-study reaching back to the twelfth century. However true it may be

[1] Purbach died before his project of obtaining the Greek text was attained. His *Theoricae novae planetarum*, Nuremberg, 1472, was published by Regiomontanus and relies on the text derived from Arabic. The *Epytoma Ioannis de monte regis* [i. e. Regiomontanus] *in Almagestum Ptolomei*, Venice, 1496, goes back however to the Greek.

that Greek thought is the final motive of these developments, that the desire to know is but the stirring again of the Greek spirit crushed and overlaid by barbarism and misunderstanding, it is yet clear that the actual recovery of the texts had no very close relation to the recommencement of direct biological observation. Above all, we need to distinguish mere *passive increase* of knowledge brought by the revival of the Greek language from the *active extension* of knowledge by direct observation that is the essence of the experimental method. This process of active extension began centuries before the learned Greek revival and received its great impetus long after it.

Thus modern science arose largely independent both of philosophy and of scholarship. She was, in truth, conceived in obscure and humble circumstances before the days of the new learning and in very different surroundings from those of her older sister. She did not take her origin, as did Greek science of old, among a group of philosophers thinking at large, and with little to check their investigations and their speculations save the limits of their own intellectual powers and a slowly accumulating mass of observations. The lines of modern physical science fell in far less pleasant places. Modern science entered on her first period of development in a highly sophisticated society, ruled intellectually by a most rigid tradition, limited by the claims of a priesthood, and constantly checked by an interpretation of Scripture which was only one degree more fettering than the scholastic presentation of such Greek philosophy as had reached that age. It was in the twelfth or perhaps the early part of the following century, with the flowing tide of the scholastic movement, that men began consciously but very slowly to modify by observation the details of a vast tradition concerning the structure of the universe and of man.[1]

[1] Some reservation, so far as Anatomy is concerned, must be made for the School of Salerno, where we have evidence that animals were being dissected at the end of the eleventh century. There is, however, no evidence that this example was followed elsewhere for generations, and perhaps the work of Salerno, so far as it is really scientific, is more logically regarded rather as the last phase of the ancient than as the first phase of modern science. The two earliest Salernitan anatomies are of the pig and they date respectively from about 1085 and 1100; they may be most conveniently consulted in Salvatore De Renzi, *Collectio Salernitana*, 5 vols., Naples, 1852–9, vol. ii, pp. 388 and 391. Even the earlier of these two tracts shows some trace of Arabian influence. Light is thrown on these early anatomical tractates by a recent excellent graduation thesis of a pupil of K. Sudhoff, F. Redeker, *Die 'Anatomia magistri Nicolai phisici' und ihr Verhältnis zur Anatomia Chophonis und Richardi*, Leipzig, 1917.

ATE IV Cotton Vitellius C. III fo. 19r. Anglo-Saxon Work c. 1000

LATO & CHIRON HAND MEDICAL ART TO AESCULAPIUS

ntrast typical Early English drawing of animals and plants with Classical work of frontispiece and Renaissance work of Plate XI

The process of development was perforce slow, for in the social and intellectual environment of the age speculation was impossible, and the only practicable advance was that of repeatedly verified experience. It was easy to confine a philosopher to his monastic cell for alleging aught new of the constitution of matter or of the structure of the heavens, but less easy, even in those days, to deny that his gunpowder exploded, when it did explode, or that his glasses magnified, when they did magnify. Princes and prelates had no need of the aid of Greek thought to discern the advantages that they might derive from the applications of gunpowder, nor did they await the re-discovery of Greek letters to become sensible of the uses of spectacles. If necessity is the mother of invention, experience is her father, and these two rather than Greek letters and Greek philosophy were the real begetters of the new experimental method.[1]

Perhaps it is in the order of nature that the root of the tree of knowledge is bitter though its fruit be sweet; it is at least true that the lesson of scientific verification needed to be well learnt before scientific speculation and theorization could become profitable. That lesson had at last been learnt, and we may be assured that the profound difference in the manner of setting forth the science of the modern and that of the ancient world is an expression of the historic difference in the origin of the two systems.

II. The Record of Ancient and the Record of Modern Biology

Consider examples of the two methods concretely and, turning to one of the thousands of files of journals that make up the bulk of any modern scientific library, compare its contents with one of the scientific works of Aristotle or of Theophrastus. Since we are discussing Biology, we will select a biological journal and briefly examine an average article in it, for such articles on special and narrow problems are a peculiar product of modern science and in these 'memoirs' modern science is most characteristically represented.

[1] In the history of scientific development mathematics and the mathematical sciences stand somewhat apart from the other departments of knowledge. It is a point that cannot be treated here, but it has been briefly discussed by the present writer in a pamphlet, *Greek Science and Modern Science, a Comparison and a Contrast*, London, 1920.

The author begins by pointing out a gap in knowledge. The structure or habits of some rare organism, he tells us, are inadequately known, or the development of some plant may be expected to throw light on the relationship between two groups of plants, or perhaps the function of an organ requires further elucidation. Having stated his problem, he reviews the efforts made by others to illumine this dark place in knowledge. He points out some of their errors or decides to accept their work and base his own upon it. Perhaps he distrusts their experiments or would like to re-interpret their results. Having summarized their labours he details his own experiments and observations.

But he is not able to tell us all of these. If he did, scientific literature would be far more bulky than it already is and science would quickly perish, suffocated under the dead weight of its own verbosity. Our author must, in fact, omit a great many of his mental processes. Space will not permit him to tell us how he embarked on many different lines of work and abandoned them as unprofitable or too difficult, nor anything of the months or years spent in merely repeating the experience of others. He says no word of how he acquired and improved his experimental skill and technical experience. He tells merely of those developments of his work that have yielded him results. But he does not tell us all even of these. When by gradual steps he had at last reached, or perhaps with the instinct of genius had more quickly discerned, a profitable direction for his investigations, he reached after a time those conclusions which his final line of work has verified and rendered more exact. It is this final process of verification that he mainly describes in his article, and it is the details of this that occupy the bulk, perhaps nineteen-twentieths or more, of all that he has to say. Then having described these verificatory experiments, he summarizes his conclusions in a short paragraph of a few lines.

That is a fair description of the average piece of scientific work as it is turned out to-day, and from vast accumulations of such work the generalizations of men of scientific genius have been mainly though sometimes unconsciously built up. The mass of scientific writings, bulky as they are, contain descriptions of only the verificatory experiments, and it is on account of this necessary limitation that it is impossible to understand scientific method from books without making independent observations.

How does the science of antiquity compare with material such as this ? The Greek work is of course less in quantity and often

OROBUS SP.
From MS. Bodley 130 fo. 16r 
Written 1120

BETONY
From MS. Ashmole 1462 fo. 12r 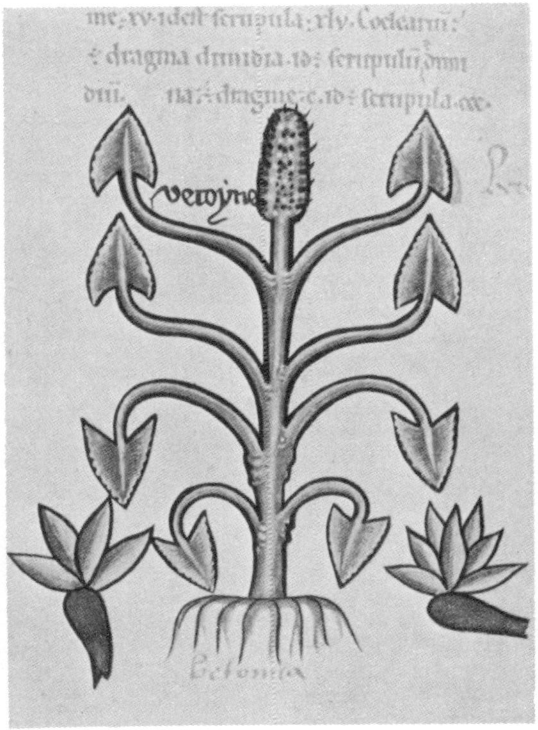
Early XIIIth Cent.

TEUCRIUM CHAMÆDRYS
From MS. Bodley 130 fo. 16r 
Written at St. Albans 1120

PLATE V

fragmentary in character, yet it is not that which makes comparison difficult. The difficulty arises from the habit of the Greek writers of setting down only their conclusions. Their methods of work, even their verificatory observations, they have almost completely hidden from us. It is as though we had a collection of the last few lines of a series of scientific articles. To grasp the nature of modern scientific method from a scientific article is difficult, since not all the mental processes involved are represented. In the case of Greek science the difficulty is far greater, for here we have only the conclusions with hardly any of the processes. There survives to this day a form of scientific literature, or rather of literature on scientific topics, which bears a distant analogy to some parts of the Greek record as it has reached us, though its motive is utterly different. It is the type of work known as the 'cram-book', which merely summarizes the known or recognized facts without reference to sources or methods. Such works present us with the final results of vast amounts of intellectual effort but they give little information as to how those results have been obtained.

Now it would appear that it is not the accidents of time that have reduced the Greek record to its present state. The responsibility rests far more with the Greeks themselves and with their method of research. The Greek scientific work lacks nothing in brilliance, the Greek scientist yields to none in keenness, the Greek record is at least the equal of our own in clearness. It is the constant solicitude for the exact *mode* of investigation, a solicitude characteristic of our own science, that we so often seek in vain among the Greeks. The method of modern scientific research is before all things and above all things a process of verification. 'The complexity of phenomena', says a critic of Aristotle, 'is as that of a labyrinth, and one wrong turn may cause the wanderer infinite perplexity. Verification is the Ariadne-thread by which alone the real issue must be sought.'[1] Yet the process of verification is slow and tedious and often difficult and dull, and Man is by nature lazy and impatient, hating labour yet eager for results. Hence Credulity. Credulity is the Pandora of Science, promising everything, yielding nothing. 'If you trust before you try, you may repent before you die' is as good a maxim in scientific as in human relationships. Μέμνησο ἀπιστεῖν would have been a salutary phylactery for the Greek scientific Pharisee, serene in

[1] G. H. Lewes, *Aristotle: a chapter from the history of science*, London, 1864.

the conviction of the power of his reason, to have bound upon his hand and to have worn as a frontlet between his eyes. It is true that the great Greek minds were singularly free from that baser credulity that we call superstition—from that they were preserved by their conviction that order reigned in Nature—but few indeed were the Greeks who showed an adequate scientific scepticism. The Greek often accepted data without scrutiny, induction without proof. His very brilliance was a source of weakness and he was often led to believe that the order of phenomena must perforce correspond to his own admirably clear conceptions.

But although Greek science failed to lay the enormous stress on verification that we in this last age have found to be the chief condition of scientific progress, it is yet certainly true that all Greek scientists were not equally deficient in this respect. Aristarchus of Samos, Archimedes of Syracuse, Hero of Alexandria, these are illustrious exceptions, and the lasting value of their results remains the justification of their method. Further, the process of verification is easier in some sciences than in others. Such are those purely observational studies classed under the general title of Natural History, for in them the phenomena are visible to all and the difficulty usually lies rather in the collection and arrangement of the vastly numerous data than in their verification. Moreover, biological phenomena are so exceedingly complex that it is very difficult to form general theories to explain any considerable proportion of them. The Greek naturalist was thus less tempted than the Greek physicist or astronomer to fit his facts to his preconceived theories.

Yet even in the field of natural history the Greek character had its own special sources of weakness. The Greeks were gifted with such a variety of talents that there may be a tendency to credit them with qualities that were not theirs, for although extraordinarily anxious to explain natural phenomena they were not remarkably observant of Nature except when their attention was specially drawn to it. It was the world *in relation to himself* and not as a mere objective complex of phenomena that interested and appealed to the Greek. Thus in his attention to purely external phenomena he ranks no higher than many peoples infinitely his inferiors in other respects. It is said that any Malay can distinguish some three hundred species of insects; set off against this the whole fauna of Aristotle with its total of but five hundred and forty forms. Again, Theophrastus, a professed botanist, mentions only three hundred plants, practically all cultivated; compare this to the record of barbarian Anglo-Saxon speaking tribes of whose literature the merest fragment survives, but a fragment containing about eight hundred plant names.[1]

But when the attention of the Greek was once fixed upon the structure or habits of living things, his success in elucidating or portraying them was unrivalled, for then the living things became part of his own world and not merely of the world around him, personal and not impersonal. With us it is quite otherwise, for it is just the impersonal or objective study that gives the hall-mark to our science and perhaps also to our art, and it is exactly in this objective or impersonal aspect that the Greek often failed. And he well knew his own weakness, which Plato has exposed for us with a sure hand: 'If we consider', he says, 'the works of the painter and the different degrees of gratification with which the eye of the spectator receives them, we shall see that we are satisfied with the artist who is able *in any degree* to imitate the earth and its mountains, and the rivers, and the woods, and the universe, and the things that are and move therein; and further, that knowing nothing precise about such matters, we do not examine or analyse the painting; all that is required is a sort of indistinct and deceptive mode of shadowing them forth. But when he paints the human form we are quick at finding out defects, and our familiar knowledge makes us severe judges of one who does not render every point of similarity.'[2] This criticism of Plato is well borne out by a study of Greek art.

If we examine the attempts to represent the forms of the animal and vegetable creations by early peoples, a very striking feature presents itself. A large number of ancient delineations of animals and plants have come down to us, and these figures show that the habits and forms of moving creatures fixed the attention of almost all races long before the same care was expended on plants. Animals are more like ourselves than plants, they move and feel and are subject to pleasure and passion, and so are capable of more personal treatment. The natural interest in the animal rather than in the plant might be illustrated by a hundred instances from the palaeolithic cave paintings downwards, but

[1] It is true that of these Early English plant names many are derived from Greek (through Latin) and many have only a literary use. But even allowing for these tendencies their number remains remarkable.

[2] *Critias.*

we will take our example from a pre-Hellenic people in the land of the Greeks. The best known of all the Minoan relics are perhaps the Vaphio cups (Plate III), and these betray the most careful study of the structure and movements of the bull. His anatomy is accurately shown and we can clearly discern the surface markings raised by the muscles which move the shoulders and the hind-quarters, as well as by those which support the head and control the ribs. Yet the representations of plants on these cups are very poor, so that the trees cannot be identified.

This neglect of plants, the beings least like ourselves, is characteristic also of the Hellenic art that succeeded and replaced the Minoan, and it has its analogue in Greek science. In the Greek scientific writings the interest in plants is usually practical; even Theophrastus, dealing almost exclusively with the domesticated varieties, frankly tells us that it was their medical application that had led to their more accurate study; the same is true of Dioscorides and of a number of minor Greek authors who have written on plants and their properties. We may search classical art in vain for any striking figures of plants, and the best representations come to us from a time when the creative force of Greece was dead. The most accurate Greek representations of plants are in a series prepared by artists of Constantinople as late as the sixth century of the Christian era (Plates VI and VII). The draftsmen of the Julia Anicia MS. of about 512 represent their originals faithfully and accurately, point by point, almost hair by hair, but with no trace of imaginative treatment.[1] These degenerate scions of the mighty race of Pheidias and Apelles were producing pictures of plants which are indeed no works of art but are yet accurate and clear and represent their subjects much as the illustrator of a modern scientific treatise might seek to do.

If we turn from the graphic representation of living things to scientific discourse about them we find ourselves in face of an extensive literature. Conrad Gesner, the most learned of biologists, estimated the number of Greek works with considerable bearing on Botany as over a hundred, and although many of these, it must be admitted, are very trivial, yet about half of them

[1] It is true that some of the figures in this MS. and possibly all of them are copied from earlier MSS. and not directly from nature. But even the originals, as shown below (p. 63, fig, 29), must have been long posterior to the best period of Hellenic, though contemporary with the so-called Augustan art with its remarkable treatment of plant forms.

PLATE VI

Julia Anicia MS. fo. 315 r Vth-VIth cent.

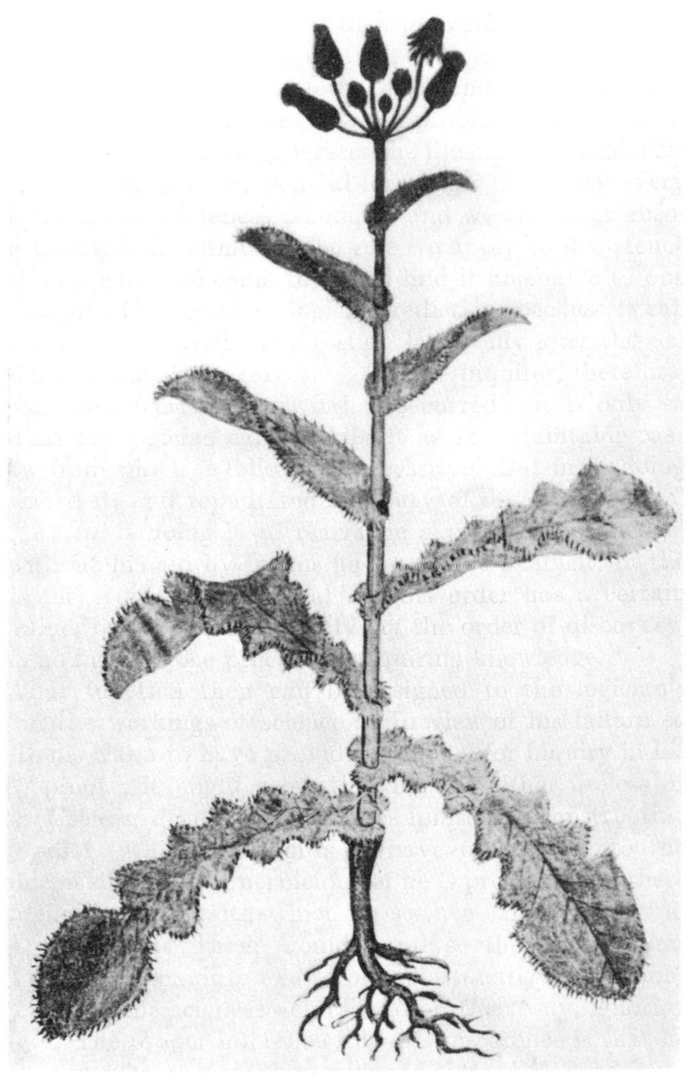

COΓKOC TPAXYC SOW-THISTLE

PLATE VII

Julia Anicia MS. fo. 370 v Vth–VIth cent.

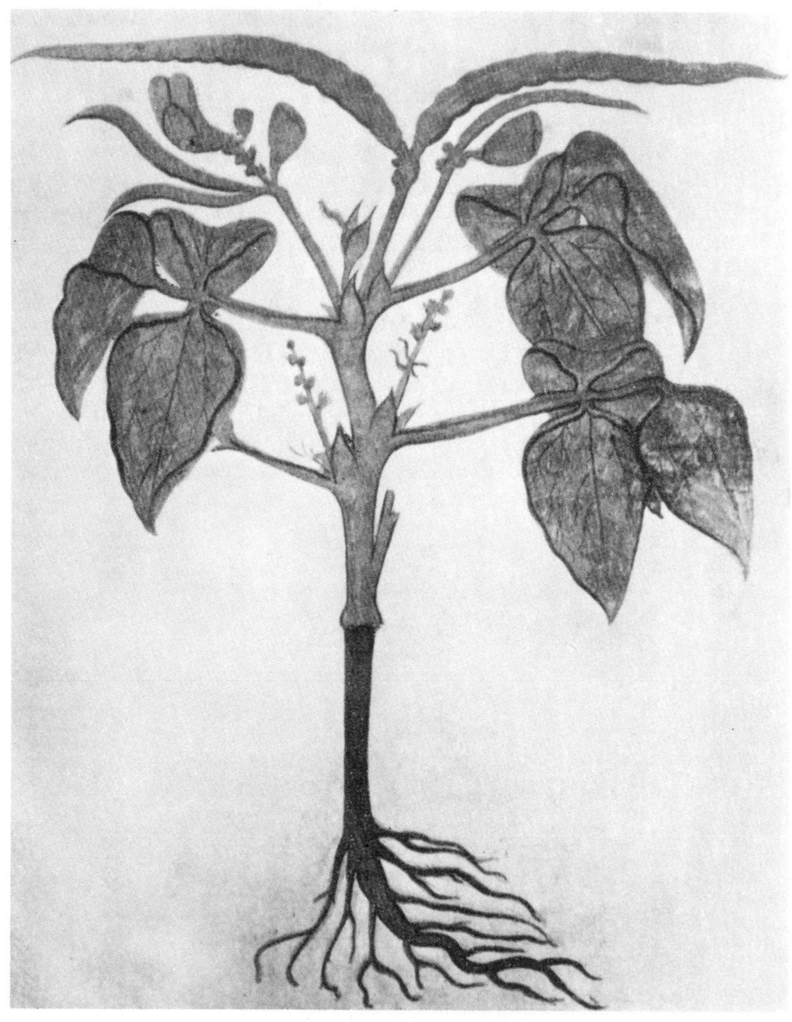

ΦΑCΙΟΛΟC SEEDLING BEAN

still find a place in the most exhaustive modern history of botany.[1] The number of works on the structure and habits of animals is also considerable. Now if the Greek was interested in men rather than things, as Plato tells us, how account for all this output ? What is it that thus fixed the attention of the Greek on animals and plants ? The answer is that these works have in the main a practical end ; the plants and animals are described for the use to which they can be put by man. But there remains a small residuum of works, chiefly those of the Lyceum, which have no such end in view. Why were these written, and where among the self-centred Greek people was the public interested in natural knowledge with no direct application to the circumstances of life ? The answer is that the best Greek biological opinion had come to regard Man himself as a natural product and was growing accustomed to look upon him as a member of a whole series of beings. These beings extended to the supra-mundane spheres, but the lower series also, plants and animals, partook of his essence in varying degrees, their resemblance to him increasing with their higher rank in the scale. Thus animals and plants, but especially animals, helped Man to understand *himself.*

III. The Bases of the Aristotelian Biological System

(*a*) *Classification*

Of the biological researches of the Lyceum we have the three great Aristotelian works, the *Historia animalium,* the *De partibus animalium,* and the *De generatione animalium,* and on plants the *Historia plantarum* and the *De causis plantarum* of Theophrastus, the pupil and successor of Aristotle, as well as the later and imperfect peripatetic work *De plantis,* probably composed by Nicholas of Damascus in the first century B.C.[2] There are also

[1] E. H. F. Meyer, *Geschichte der Botanik,* 4 vols., Königsberg, 1854–7.

[2] The history of this work is curious. The original work on plants by Aristotle was commented on by Nicholas in Greek. This commentary was translated by Hunein ben Ishâk into Syriac, and this translation was turned, by his son, into Arabic. In its Arabic dress it was then modified by Thâbit ben Curra. From the Arabic it was twice translated into Latin in the thirteenth century, on one occasion by the shadowy and elusive Alfredus Anglicus. An authoritative edition of the Latin text of Alfredus was published by E. H. F. Meyer, *Nicolai Damasceni de plantis libri duo Aristoteli vulgo adscripti,* Leipzig, 1841. See especially F. Wüstenfeld, *Die Uebersetzungen arabischer Werke in das Lateinische seit dem XI. Jahrhundert,* Göttingen, 1877.

two Aristotelian works on the movements of animals with which we are not here concerned, nor shall we take into consideration such points in Aristotle's works as deal with the structure of man, since these are best reserved for separate treatment. The Aristotelian and zoological writings may be considered first.

Whatever be concluded as to other works of Aristotle, it is probably true that any biologist who examines his zoological writings will accept what is known as the 'note-book theory'. The ill arrangement of much of the material and the gravity of some of the errors make it difficult to conceive that these works are in the state designed for publication by the master, with his genius for classification and undeniable powers of observation. The only explanation that will satisfy is that the more serious blunders are the mistakes of the student note-taker who had in his hands the rough-note books of the teacher. It is therefore probably true that if Aristotle's best biological observations are taken as samples of his scientific work we shall obtain the truest picture of what he himself was accustomed to teach.

There can be no doubt that through much of the Aristotelian writing there breathes a belief in a *kinetic* as distinct from a *static* view of existence. It is this aspect of his teaching which brings all living things into relation with man. In Aristotle's study of animal forms there are two departments where this kinetic view gains specially clear expression. These departments are respectively the Arrangement or Classification of Animals, and their Development, or, as we now call it, their Embryology.

It is now customary to summarize our knowledge of living beings in tabular form. As interpreted by a modern biologist these classificatory tables represent certain degrees of relationship. Modern systems of classification are not, however, as is often thought, closely comparable to genealogical trees, because two species may be very nearly allied and therefore close together in the table of classification, although they parted company far back in their history, or again historically allied species may become widely differentiated in comparatively few generations. Classificatory tables are rather intended to summarize structural similarities and structural differences, though, as a rule, the naturalists who draw them up have no exact quantitative conception of the amounts of differences signified by the degree of separation in their tables. Indeed this absence of a quantitative factor is among the weakest points of modern biology, and it is only in very recent times that effective attempts have been made to remedy it.

Now despite the fact that no classificatory table of Aristotle has come down to us, there can yet be no doubt that he formed general ideas of a classification based on the consideration of the structure and habits of animals. He uses the terms of classification and speaks of larger and smaller groups of animals which bear greater or less similarity to each other. It is thus in accordance with his meaning and is perhaps reproducing his method of teaching if we draw up a classification of animals from his works.

The primary basis of his classification would surely have been the method of reproduction, a subject to which he paid a vast amount of attention (cf. Table on p. 16). It is therefore necessary to examine some of his ideas on this subject.

He knew nothing of the mammalian ovum, and he regarded the mammalian embryo as a thing living from the first, and living in a higher sense than an egg can be said to live. The mammalia were thus for him viviparous internally and not merely externally in the sense in which the word viviparous is now used. The remaining *enaima*, sanguineous animals or 'vertebrates' as we now call them, were primarily oviparous, though some among them were viviparous in the external sense; that is to say, while the young in these cases were held to develop always from an egg, that egg might sometimes be hatched within the mother's body. The 'invertebrates', or *anaima* in Aristotle's notation, had on the other hand their own special methods of reproduction, among which the so-called *spontaneous generation* played an important part.

In considering the table of classification that we have drawn up from Aristotle's works and in comparing it with any modern system, the difficulties under which he was working must be recalled. He makes no attempt to produce a complete system, and he deals almost entirely with local forms. He exhibits knowledge of about 540 species of animals. When we consider that of insect species alone some half million are now recognized, a thousand times as many as his total fauna, we shall be more disposed to wonder that he has fastened upon so many points still regarded as of classificatory value, than to criticize his errors or the gaps in his knowledge.

ENAIMA. (Sanguineous and either viviparous or oviparous)=Vertebrates.

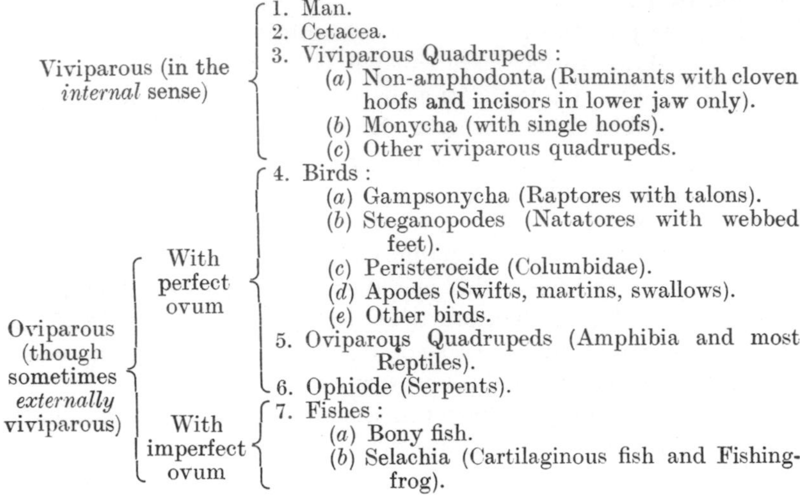

Viviparous (in the *internal* sense):
1. Man.
2. Cetacea.
3. Viviparous Quadrupeds :
   (*a*) Non-amphodonta (Ruminants with cloven hoofs and incisors in lower jaw only).
   (*b*) Monycha (with single hoofs).
   (*c*) Other viviparous quadrupeds.

Oviparous (though sometimes *externally* viviparous):

With perfect ovum:
4. Birds :
   (*a*) Gampsonycha (Raptores with talons).
   (*b*) Steganopodes (Natatores with webbed feet).
   (*c*) Peristeroeide (Columbidae).
   (*d*) Apodes (Swifts, martins, swallows).
   (*e*) Other birds.
5. Oviparous Quadrupeds (Amphibia and most Reptiles).
6. Ophiode (Serpents).

With imperfect ovum:
7. Fishes :
   (*a*) Bony fish.
   (*b*) Selachia (Cartilaginous fish and Fishing-frog).

ANAIMA (Non-sanguineous and either oviparous, vermiparous or budding) =Invertebrates.

With imperfect ovum . .	1. Malacia (Cephalopods). 2. Malacostraca (Crustacea).
With scolex	3. Entoma (Insects, spiders, scorpions, &c.).
With generative slime, buds, or spontaneous generation	4. Ostracoderma (Molluscs except Cephalopods, Echinoderms, &c.).
With spontaneous generation only	5. Zoophyta (Sponges, Coelenterates, &c.).

Aristotle's primary division into Enaima and Anaima, or as we call them Vertebrates and Invertebrates, is one still universally accepted. The two groups are now, it is true, regarded as incommensurate in evolutionary value, but this has only been recognized during the last generation or two, and the division survives as an effective dichotomy of our knowledge. When we examine his Enaima we see a division into groups which, with the forms known to him, could hardly be improved. The fish-like Cetacea are separated too widely from the other mammals, but Aristotle nevertheless knows of their mammalian character and recalls the fact that they suckle their young. He is in no danger of confusing them with fish. 'All animals', he says, 'that are internally and externally viviparous have breasts, as for instance all animals that have hair, as man and the horse, and the cetaceans, as the dolphin, the porpoise, and the whale, for these animals have breasts and are supplied with milk. Animals that are oviparous

PLATE VIII. Leyden MS. Voss Lat. Q. 9 Apuleius VIth cent.

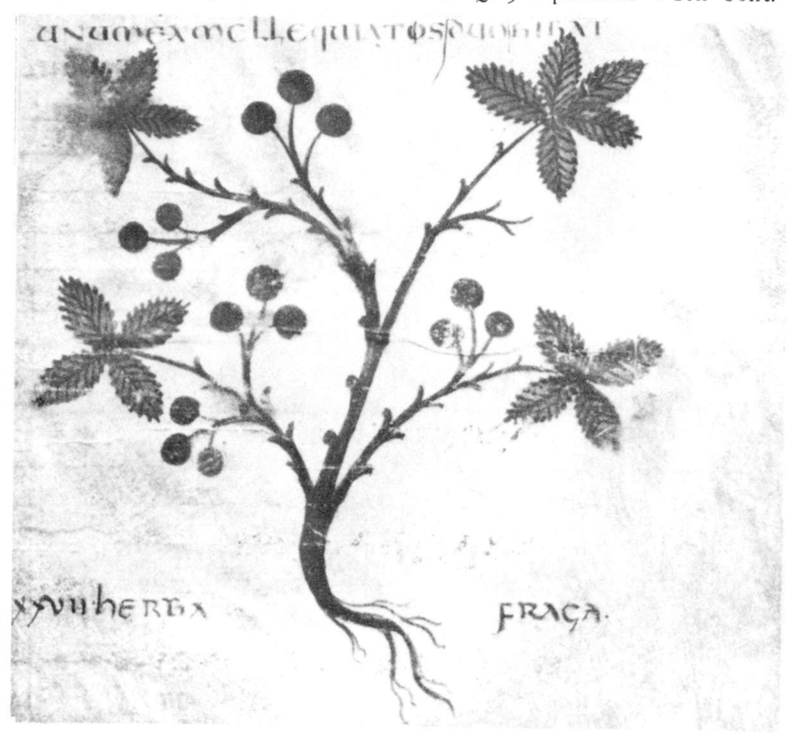

fo. 83 r *Fraga* = Blackberry

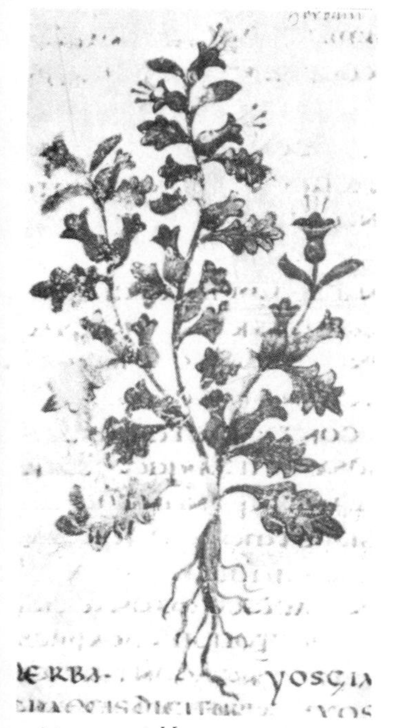

fo. 49 v *Hyoscyamus*

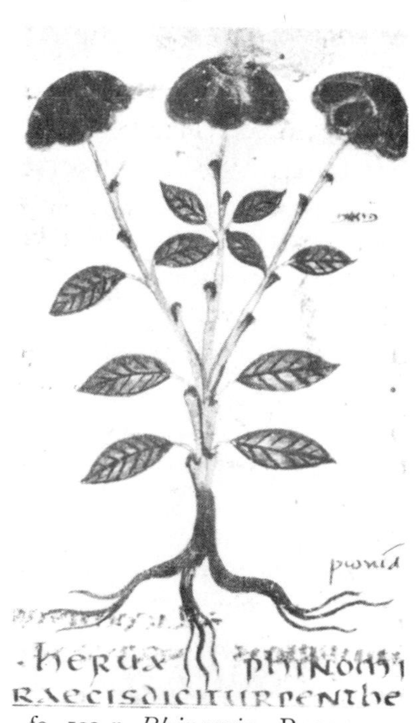

fo. 103 r *Phinomia* = Paeony

or only externally viviparous have neither breasts nor milk, as the fishes and the bird.'[1]

The passages in which Aristotle describes the Cetaceans are worth putting together as giving a general idea of his powers of zoological classification, and they provide an excellent illustration of the stage that he had reached in the study of comparative anatomy and physiology.

'Some animals are viviparous, others oviparous, others vermiparous. Among the viviparous are man, the horse, the seal, and all other animals that are hair-coated, and, of marine animals, the cetaceans, as the dolphin and the so-called Selachia.[2] . . . The dolphin, the whale, and all the rest of the Cetacea, all, that is to say, that are provided with a blow-hole instead of gills, are [internally] viviparous. That is to say, no one of all these is ever seen to be supplied with eggs, but directly with an embryo from whose differentiation the animal develops, just as in the case of mankind and the viviparous quadrupeds.[3] . . . All creatures that have a blow-hole respire and inspire, for they are provided with lungs. The dolphin has been seen asleep with his nose above water, and when asleep he snores.[4] . . . One can hardly allow that such an animal is terrestrial and terrestrial only, or aquatic and aquatic only, if by terrestrial we mean an animal that inhales air, and if by aquatic we mean an animal that takes in water. For the fact is the dolphin performs both these processes: he takes in water and discharges it by his blow-hole, and he also inhales air into his lungs; for the creature is furnished with these organs and respires thereby, and accordingly, when caught in the nets, he is quickly suffocated for lack of air. He can also live for a considerable while out of the water, but all this while he keeps up a dull moaning sound corresponding to the noise made by air-breathing animals in general; furthermore, when sleeping, the animal keeps his nose above water, and he does so that he may breathe the air. . . . For the fact is, some aquatic animals [as fish] take in water and discharge it again, for the same reason that leads air-breathing animals to inhale air; in other words, with the object of cooling the blood. Others [as cetaceans] take in water as incidental to their mode of feeding; for as they get their food in the water they cannot but take in water along with their food.'[5]

'The dolphin bears one at a time generally, but occasionally two. The whale bears one or at the most two, generally two. The porpoise in this respect resembles the dolphin. . . . The dolphin and the porpoise are provided with milk, and suckle

[1] *Historia animalium*, iii. 20; 521[b] 21.
[2] *Historia animalium*, i. 5; 489[a] 34.
[3] *Historia animalium*, vi. 12; 566[b] 2.
[4] *Historia animalium*, vi. 12; 566[b] 12.
[5] *Historia animalium*, viii. 2; 589[a] 31.

their young. They also take their young, when small, inside them. The young of the dolphin grows rapidly, being full-grown at ten years of age. Its period of gestation is ten months. It brings forth its young in summer, and never at any other season. Its young accompany it for a considerable period; and, in fact, the creature is remarkable for the strength of its parental affection.'[1]

The *Historia animalium* in which these passages occur became accessible in versions by Michael Scot (1175 ?–1294 ?),[2] by Albertus

La peincture de l'Oudre, que les Latins nomment Orca ou
Orcynum.

FIG. 1. GRAMPUS AND NEWLY-BORN YOUNG

The foetus is still surrounded by its membranes and the after-birth is in process of extrusion. From Pierre Belon, *Histoire naturelle des estranges poissons marins, avec la vraie peincture et description du daulphin et de plusieurs autres de son espèce*, Paris, 1551.

Magnus (1206–80),[3] and perhaps by William of Moerbeke (died *c.* 1281). The work was again rendered into Latin by Theodore Gaza, about 1450.[4] Yet the mammalian affinities of the Cetacea appear to have been generally overlooked in the Middle Ages and at the Revival of Learning until Pierre Belon (1507–64) repeated Aristotle's observations in the middle of the sixteenth century and published a description of the cetacean placenta

[1] *Historia animalium*, vi. 12; 566^{b} 7.

[2] Translated from the Arabic. Cf. F. Wüstenfeld, loc. cit., p. 101. Roger Bacon tells us that in 1230 'Michael Scot appeared [at Oxford] bringing with him the works of Aristotle on natural history and mathematics, with wise expositors, so that the philosophy of Aristotle was magnified among the Latins'. It appears that Scot produced two versions of the *De Animalibus*, one entitled *De Animalibus ad Caesarem* and the other *Tractatus Avicennae de Animalibus*. He also incorporated ideas from the *De Generatione Animalium* in his *Liber de secretis naturae.*

[3] An edition of Albert's commentary has been produced by H. Stadler, *Albertus Magnus de animalibus libri XXVI nach der Cölner Urschrift*, Münster i. W., 1916 and 1921, in Baeumke's 'Beiträge zur Geschichte der Philosophie des Mittelalters.'

[4] First printed Venice, 1476.

(Figs. 1 and 2). Belon is further worthy of commemoration as he was perhaps the first among the moderns to make an attempt at a *comparative* anatomy, for in one of his works he sets forth the homologues of the vertebrate skeleton along somewhat similar lines to those of Aristotle[1] (Fig. 3), a conception soon developed by Coiter well beyond the Aristotelian level.[2]

The classification of birds is to this day in an unstable state. We may say that Aristotle's grouping is substantially that which prevailed in scientific works till recent times and still remains as the popular division. His separation of the cartilaginous from the bony fishes, on the other hand, still stands in scientific works, and is a stroke of genius which must have been reached by means of careful dissection. It is marred only by the inclusion of one peculiar bony form, the fishing-frog, or *Lophius*, among the cartilaginous fishes, and investigation shows that the skeleton of this creature is, in fact, peculiarly cartilaginous. Aristotle himself regarded the Lophius as aberrant among cartilaginous fishes.

La peincture de l'Embryon d'vn Marsouin.

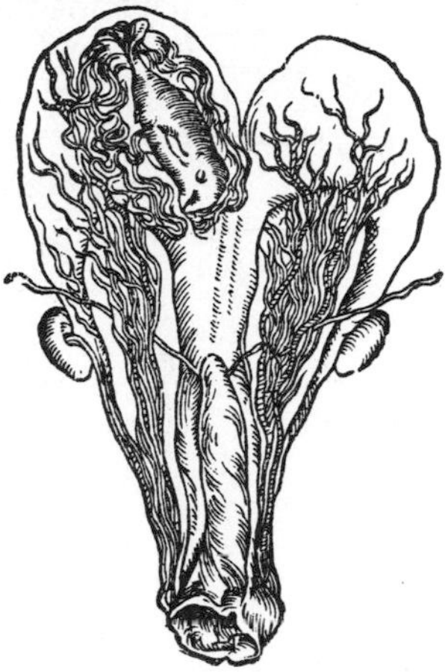

FIG. 2. THE UTERUS OF THE PORPOISE

Opened to show foetus attached by umbilical cord to placenta. From Pierre Belon, *Histoire naturelle des estranges poissons marins, avec la vraie peincture et description du daulphin et de plusieurs autres de son espèce*, Paris, 1551.

For the Anaima or Invertebrates even modern systems of classification are but tentative. There is an enormous number of species, and after centuries of research naturalists still find vast gaps even in the field of mere naked-eye observation. Nevertheless, with the instinct of genius, and with only some 240 of these forms on which to work, Aristotle has fastened on some of the most salient points. Especially brilliant is his treatment of the Molluscs. There can be no doubt that he dissected the bodies and carefully watched the habits of octopuses and squids, *Malacia* as he calls them. He separates them too far from the other Molluscs,

[1] The suggestion had already been made, though in a less complete form, by Vesalius in the *De fabrica corporis humani*, 1543. There are also traces of the conception of a comparative anatomy in the MSS. of Leonardo da Vinci.

[2] Volcher Coiter, *Lectiones Gabrielis Fallopii de partibus similaribus humani corporis ex diversis exemplaribus*, Nuremberg, 1575.

grouped by him as *Ostracoderma*, but his actual descriptions of the structure of the Cephalopods are exceedingly remarkable. (Cf. p. 39 ff.) His distinction between the *Malacostraca* or Crustacea,

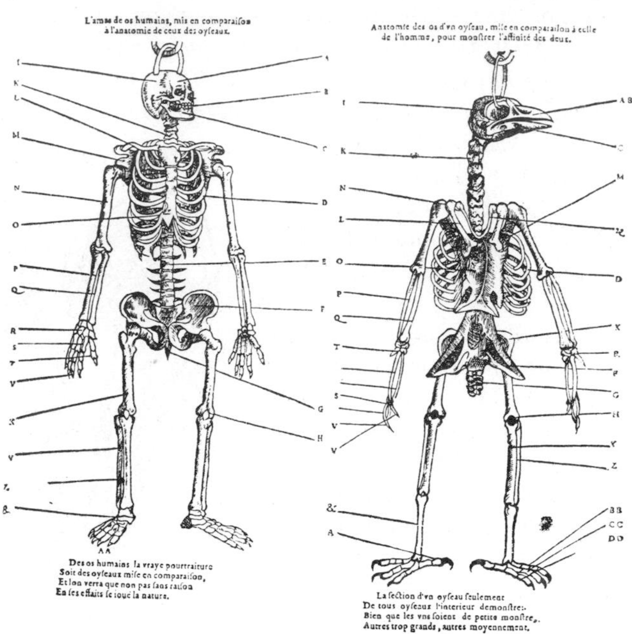

FIG. 3. THE SKELETON OF A MAN AND OF A BIRD COMPARED
From Pierre Belon, *L'histoire de la nature des oyseaux*, Paris, 1555.

Entoma, Sponges, and Jellyfish are also still of value, and these divisions remain along much the same lines as he left them.

(*b*) *Phylogeny*

Aristotle nowhere formally exhibits either a 'Scala Naturae' or a 'genealogical tree', devices in which naturalists have delighted for the last two centuries, but he constantly comes so near to such conceptions that there is no great difficulty in reconstructing his scale from his descriptions (Fig. 4).

'Nature', he says, 'proceeds little by little from things lifeless to animal life in such a way that it is impossible to determine the exact line of demarcation, nor on which side thereof an intermediate form should lie. Thus, next after lifeless things in the upward scale comes the plant, and of plants one will differ from another as to its amount of apparent vitality; and, in a word, the whole genus of plants, whilst it is devoid of life as compared with an animal, is endowed with life as compared with other corporeal entities. Indeed, there is observed in plants a continuous scale of ascent towards the animal. So, in the sea, there are certain objects concerning which one would be at a loss to determine whether they be animal or vegetable. For instance, certain of these objects are fairly rooted, and in several cases perish if detached.'[1]

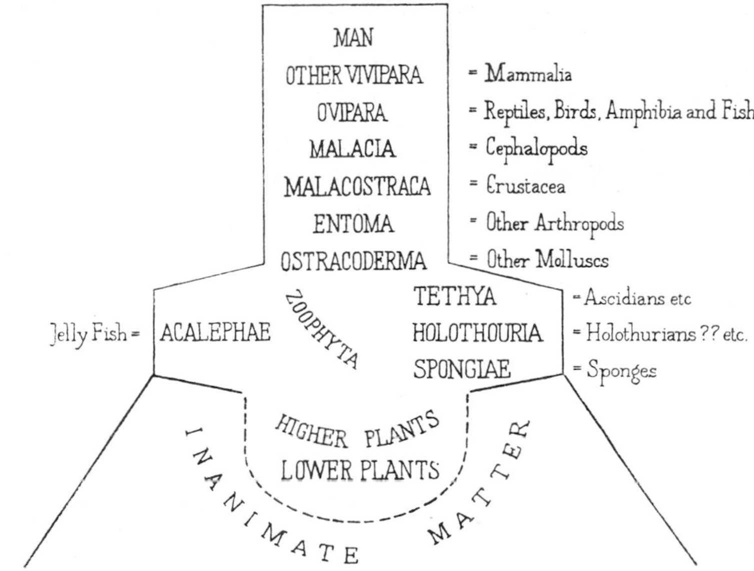

FIG. 4. The order of living things, put together from the descriptions of Aristotle.

'A sponge, in these respects completely resembles a plant, in that throughout its life it is attached to a rock, and that when separated from this it dies. Slightly different from the sponges are the so-called Holothourias and the sea-lungs, as also sundry other sea-animals that resemble them. For these are free and unattached, yet they have no feeling, and their life is simply that of a plant separated from the ground. For even among land-plants there are some that are independent of the soil . . . or even entirely free. Such, for example, is the plant which is found on Parnassus, and which some call the Epipetrum. This you may hang up on a peg and it will yet live for a considerable time. Sometimes it is a matter of doubt whether a given organism should be classed with plants or with animals. The Tethya, for instance, and the like so far resemble plants as they never live free and unattached, but, on the other hand, inasmuch as they have a certain flesh-like substance, they must be supposed to possess some degree of sensibility.[2]

[1] *Historia animalium*, viii. 1 ; 588^{b} 4.

[2] *De partibus animalium*, iv. 5 ; 681^{a} 15.

'The Acalephae or Sea-nettles, as they are variously called, . . . lie outside the recognized groups. Their constitution, like that of the Tethya, approximates them on the one side to plants, on the other side to animals. For seeing that some of them can detach themselves and can fasten on their food, and that they are sensible of objects which come in contact with them, they must be considered to have an animal nature. The like conclusion follows from their using the asperity of their bodies as a protection against their enemies. But, on the other hand, they are closely allied to plants, firstly by the imperfection of their structures, secondly by their being able to attach themselves to the rocks, which they do with great rapidity, and lastly by their having no visible residuum notwithstanding that they possess a mouth.'[1]

Thus 'Nature passes from lifeless objects to animals in such unbroken sequence, interposing between them beings which live and yet are not animals, that scarcely any difference seems to exist between two neighbouring groups owing to their close proximity.'[2]

'In regard to sensibility, some animals give no indication whatsoever of it, whilst others indicate it but indistinctly. Further, the substance of some of these intermediate creatures is flesh-like, as is the case with the so-called Tethya (ascidians) and the Acalephae (or sea anemones ?) ; but the sponge is in every respect like a vegetable. And so throughout the entire animal scale there is a graduated differentiation in amount of vitality and in capacity for motion. A similar statement holds good with regard to habits of life. Thus, of plants that spring from seed, the one function seems to be the reproduction of their own particular species, and the sphere of action with certain animals is similarly limited. The faculty of reproduction, then, is common to all alike. If sensibility be superadded, then their lives will differ from one another in respect to sexual intercourse and also in regard to modes of parturition and ways of rearing their young. Some animals, like plants, simply procreate their own species at definite seasons ; other animals busy themselves also in procuring food for their young, and after they are reared quit them and have no further dealings with them ; other animals are more intelligent and endowed with memory, and they live with their offspring for a longer period and on a more social footing.'[3]

(c) *Ontogeny*

So much for Aristotle's treatment of the kinds of living things. Evolutionary doctrine is also foreshadowed by him in his theories of the development of the individual. This fact is obscured, however, by his peculiar view of the nature of procreation. On

[1] *De partibus animalium*, iv. 5 ; 681ª 36.

[2] *De partibus animalium*, iv. 5 ; 681ª 10.

[3] *Historia animalium*, viii. 1 ; 588ᵇ 16.

this topic his general conclusion is that the material substance of the embryo is contributed by the female, but that this is mere passive formable material, almost as though it were the soil in which the embryo grows. The male contributes the essential generative agency, but it is not theoretically necessary for anything material to pass from male to female. The material that does in fact pass with the seed of the male is an accident, not an essential, for his essential contribution is not matter but *form* and *principle*. Aristotle, it appears, was prepared to accept instances of fertilization without material contact.

'The female does not contribute semen to generation but does contribute something . . . for there must needs be that which generates and that from which it generates. . . . If, then, the male stands for the effective and active, and the female considered as female, for the passive, it follows that what the female would contribute to the semen of the male would not be semen but material for the semen to work upon. . . . Now how is it that the male contributes to generation, and how is it that the semen from the male is the cause of the offspring ? Does it exist in the body of the embryo as a part of it from the first, mingling with the material which comes from the female ? Or does the semen contribute nothing to the material body of the embryo but only to the power and movement in it ? For this power is that which acts and makes, while that which is made and receives the form is the residue of the secretion in the female. Now the latter alternative appears to be the right one both *a priori* and in view of the facts.'[1]

There was, however, another view of generation, perhaps of Epicurean origin, that was prevalent in antiquity. According to this theory the foetus was a joint product of male semen and of some analogous factor secreted by the female.[2] Among later writers, from Galen onward, the Aristotelian and Epicurean views were often blended and confused. After the thirteenth century, however, the Aristotelian doctrine was that mainly held and it lasted on until quite modern times. Thus it profoundly influenced William Harvey in the seventeenth century. With the discovery of the spermatozoa, however, by Leeuwenhoek and Hamm, in 1677, it became substantially untenable. The view of Aristotle fell altogether into discredit in the nineteenth century, during the

[1] *De generatione animalium*, i. 21 ; 729ª 21.

[2] The theory is succinctly stated by Lucretius, iv, ll. 1229–31 :

> Semper enim partus duplici de semine constat,
> atque utri similest magis id quodcumque creatur,
> eius habet plus parte aequa.

long period of what may be called histological domination. We have now, however, entered on a new experimental period in biology, and recent work on mechanical stimulus of the ovum has demonstrated that it is indeed possible for development to proceed without passage of material from the male.

Aristotle's most important embryological researches were made upon the chick. He says that the first signs of development are noticeable after three days, the heart being visible as a palpitating blood spot whence, as it develops, two meandering blood-vessels extend to the surrounding tunics.

'Generation from the egg', he says, 'proceeds in an identical manner with all birds, but the full periods from conception to birth differ. With the common hen after three days and nights there is the first indication of the embryo. . . . The heart appears like a speck of blood, in the white of the egg. This point beats and moves as though endowed with life, and from it two vein-ducts with blood in them trend in a convoluted course . . . and a membrane carrying bloody fibres now envelops the yolk, leading off from the vein-ducts.' [1]

A little later he observed that the body had become distinguishable, and was at first very small and white.

'The head is clearly distinguished and in it the eyes, swollen out to a great extent. This condition of the eyes lasts on for a good while, as it is only by degrees that they diminish in size and collapse. At the outset the under portion of the body appears insignificant in comparison with the upper portion. Of the two ducts that lead from the heart, the one proceeds towards the circumjacent integument, and the other, like a navel-string, towards the yolk. . . .

'When an egg is ten days old the chick and all its parts are distinctly visible. The head is still larger than the rest of the body and the eyes larger than the head. At this time also the larger internal organs are visible, as also the stomach and the arrangement of the viscera; and the veins that seem to proceed from the heart are now close to the navel. From the navel there stretch a pair of veins, one towards the membrane that envelops the yolk and the other towards that membrane which envelops collectively the membrane wherein the chick lies, the membrane of the yolk, and the intervening liquid. . . . About the twentieth day, if you open the egg and touch the chick, it moves inside and chirps; and it is already coming to be covered with down, when, after the twentieth day is past, the chick begins to break the shell.' [2]

[1] *Historia animalium*, vi. 3; 561ª 4. [2] *Historia animalium*, vi. 3; 561ª 18.

To realize what must have been Aristotle's impressions on seeing the developing chick we should not only eliminate our own embryological prepossessions but should also divest ourselves of such modern conveniences of a laboratory as incubator, water-bath, and microscope, and even lens. For our purpose of comparison, better than the description of a modern text-book of embryology is the account of such a pioneer embryologist as Fabricius ab Aquapendente (1537–1619) whose work was done before the microscope had come into use (Fig. 5). Fabricius

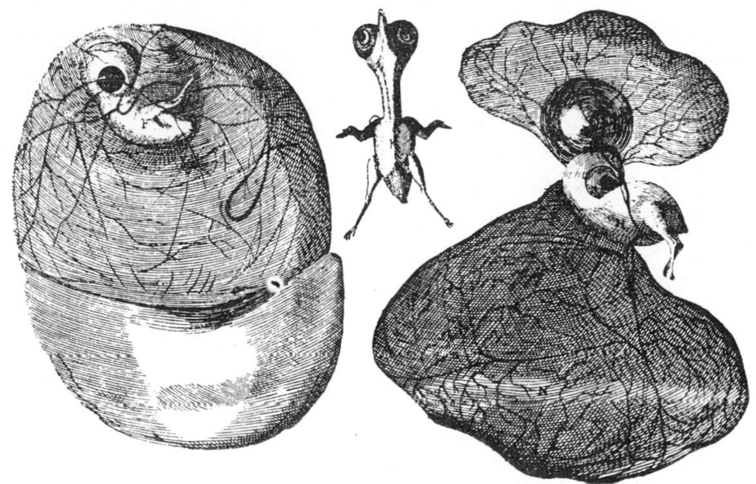

Fig. 5. THE YOUNG CHICK
From Fabricius ab Aquapendente's *De formatione ovi et pulli*, Padua, 1604.

was not, it is true, the first of modern writers to make embryological observations; for that position Leonardo da Vinci is perhaps the most likely candidate. Rieff (1500–58) also, in a work for midwives (1554), had written matter concerning the development of the human embryo that was not without point.[1] Moreover, Coiter (1534–1590) had already (1573) discussed the incubated fowl's egg[2] in a manner that displayed an understanding of the nature of the vitelline duct to which neither Fabricius nor even

[1] Jacob Rieff, *Trostbüchle*, Zurich, 1554. There is a Latin edition of this work (translated by Wolfgang Haller ?), *De conceptu et generatione hominis*, Zurich, 1554, and an excellent anonymous English translation, *The Expert Midwife*, London, 1637.

[2] Volcher Coiter, *Externarum et internarum principalium humani corporis partium tabulae, atque anatomicae exercitationes observationesque variae diversis ac artificiosissimis figuris illustratae*, Nuremberg, 1573.

Harvey altogether attained,[1] although it had been grasped by Aristotle.[2] But Fabricius was the first to imitate Aristotle in making extensive embryological observations under controlled conditions.[3]

Like Aristotle, Fabricius carried on his researches without means of magnification; like him he did his work almost without help from previous observers; following Aristotle, he held that there was a fundamental distinction between the male and female contribution to the formation of the embryo in the case of the viviparous placental animals, and as with Aristotle, the ovum of mammals was unknown to him, though its existence was suspected by his pupil Harvey. For these pioneers of embryology the *semen* or *sperm* was indeed literally the *seed*, the fertilizing principle—the word used for the seed of animals was the same as that used for plants both in Latin and in Greek—a seed which was sown in the soil of the mother's womb where it was nourished and where it grew. Harvey in his work on Generation suggested, without full evidence, that 'almost all animals, even those which bring forth their young alive and man himself are produced from eggs',[4] but it was not until the last quarter of the seventeenth century and the appearance of the work of de Graaf[5] (Figs. 6 and 7), Swammerdam,[6] and Stensen[7] that the notion came clearly into view that the so-

[1] In modern times the nature of the vitelline duct was first described by Coiter, but his work was overlooked until his observations were repeated by Stensen in his *De musculis et glandulis observationum specimen*, Copenhagen, 1664.

[2] Aristotle, *Historia animalium*, vi. 3.

[3] Hieronymus Fabricius ab Aquapendente, *De formatione ovi et pulli*, Padua, 1604.

[4] The phrase *omne vivum ex ovo*, sometimes attributed to Harvey, cannot, however, actually be found in his writings.

[5] Regnier de Graaf, *De mulierum organis generationi inservientibus tractatus novus, demonstrans tam homines et animalia, caetera omnia, quae vivipara dicuntur, haud minus quam ovipara, ab ovo originem ducere*, Leyden, 1672.

[6] Jan Swammerdam, *Miraculum naturae sive uter muliebris fabrica*, Leyden, 1672.

[7] In 1667 Stensen published his *Elementorum myologiae specimen . . . cui accedunt canis Carchariae dissectum caput, et dissectus piscis ex canum genere* at Florence, reprinted Amsterdam, 1669. In the third treatise of this work he maintained that the *testes* of women were analogous to the *ovaria* of oviparous animals and ought to be called by that name. In 1675 he published at Copenhagen in the *Acta Hafniensia* his *Observationes anatomicae spectantes ova viviparorum* in which he briefly described ova in a variety of viviparous animals. In an adjacent publication of the same year of the *Acta Hafniensia* entitled *Ova viviparorum spectantes observationes* he gives a diagram illustrating the analogy

called 'testes' of the female really contained eggs or ova comparable to the ova of oviparous creatures. In the absence of this knowledge Aristotle, like Fabricius, was unable to set forth an

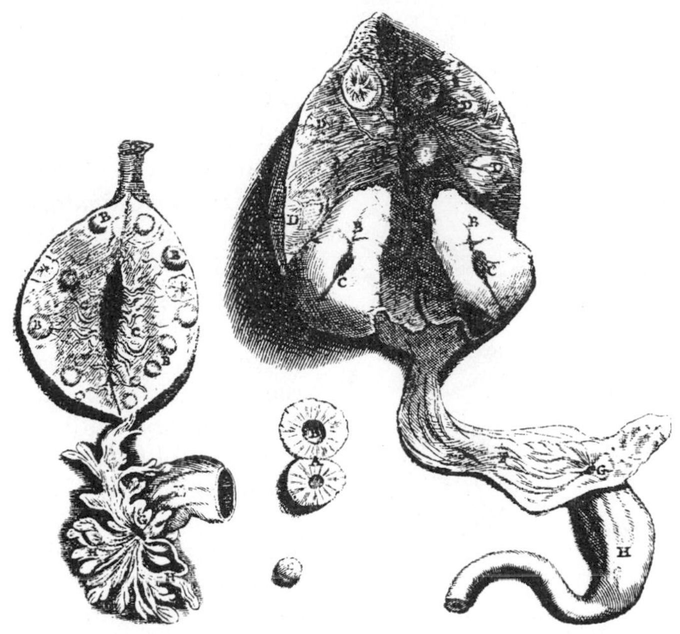

FIG. 6. From Regnier de Graaf's *De mulierum organis generationi inservientibus*, Leyden, 1672. To the left is the 'testicle or ovary', as he calls it, of a woman. It is cut open along the line A; BB are 'ova' of various sizes contained in the substance of the 'testis'; CC are blood vessels; D is the ligament of the ovary; E a part of the Fallopian tube and G its opening; H and I are the *ornamenta foliacea tubarum*. To the right is the ovary of a cow similarly cut open along the line AA; BB is the *Glandulosa substantia, quae post ovi expulsionem in testibus reperitur*, or as we call it the *corpus luteum*, CC being the 'almost obliterated cavity' in which the ovum was once contained; DD are ova; EE blood vessels; F, G, H the Fallopian tube. Between the two larger figures is the Graafian follicle AB of a sheep, C being the ovum removed from it.

explanation of the mechanism of generation that adequately covered the phenomena.

of the uteri of *vipara* and *ovipara*. Thus Stensen has the priority in the *suggestion* that the testes of the female mammal produce ova but de Graaf has the priority of *demonstration*.

The claim disseminated by A. Portal in his *Histoire de l'Anatomie et de la Chirurgie*, 6 vols., Paris, 1770–1773, that Jean Mathieu Ferrari da Grado (fl. 1450) was the discoverer of the ovarian nature of the female *testes*, is effectually disposed of by da Grado's collateral descendant H. M. Ferrari, in his *Une chaire de Médecine au XVᵉ siècle*, Paris, 1899, p. 115 ff.

Aristotle, true to the general gradational view that he had formed of Nature, held that the most primitive and fundamentally important organs make their appearance before the others. Among the organs all give place to the heart, which he considered 'the first to live and the last to die'.[1] Here again he was followed by the great investigators of the newly revived experimental method. Harvey has enshrined this idea in his work on the circulation. Indeed the conception of a hierarchy of the organs hardly departed from Biology until the observations of Caspar Friedrich Wolff (1733–94) and Karl Ernst von Baer (1792–1876) had been rationalized by Darwin, so that physiologists could at last turn from the consideration of the organ to a contemplation of the organism, and naturalists became enabled to think of the individual not as what it is but as what it has been and what it is becoming. Aristotle's kinetic view was at length vindicated.[2]

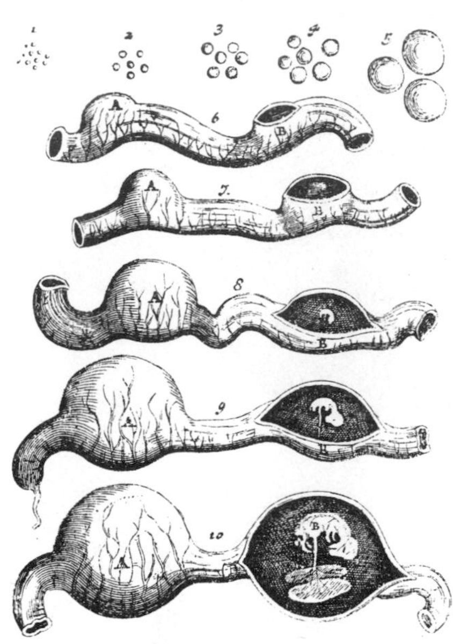

FIG. 7. From Regnier de Graaf's *De Mulierum organis generationi inservientibus*, Leyden, 1672. Illustrating the development of the rabbit's ovum: 1, ova on the third day after conception, 2 on the fourth, 3 on the fifth, 4 on the sixth, and 5 on the seventh day. The remaining figures show a section of the tube containing two embryos, 6 being on the eighth and 10 on the fourteenth day. The last figure shows the placenta.

[1] This is the sense of Aristotle, e.g. *De generatione animalium*, ii, 1 and 4; 735^{a} 25 and 738^{b} 16. The phrase. however, *primum vivens ultimum moriens* is, I think, first used in Latin translations of Averroes (1126–98), the commentator on Aristotle. There is a discussion of the origin of the phrase in the *Mitteilungen z. Gesch. der Med. und Naturwissenschaften*, xix, pp. 102, 219, and 305, Leipzig, 1920.

[2] There is a discussion of ancient embryological literature by Bruno Bloch, 'Die geschichtlichen Grundlagen der Embryologie bis auf Harvey', in the *Abhandlungen der kais. Leopold.-Carol. Akad.*, lxxxii, pp. 213–334, Halle, 1904. There is a shorter version of this same article in the *Zoologische Annalen*, i, p. 51, Würzburg, 1905.

IV. Some Aristotelian Zoological Observations and their Modern Counterparts

(a) *The Placental Shark*

We may now turn to observations in the Aristotelian writings on the habits of animals and on comparative anatomy. These are far too numerous for extended consideration here, and we therefore select a few that are of special historical interest for comparison with their modern counterparts.

Aristotle recognized a distinction in the mode of development of mammals from that of other viviparous creatures. Having distinguished the apparently viviparous animals as either truly and internally or merely externally viviparous, he pointed out that in the mammalia, a group regarded by him as internally viviparous, the foetus is connected until birth with the wall of the mother's womb by the navel string. These animals, in his view, produced their young without the intervention of an ovum. Such non-mammals, on the other hand, as are viviparous are so in the external sense only, that is, the young which arise from ova may indeed develop within the mother's body, but they do so out of organic connexion with her, so that her womb acts, as it were, but as a nursery or incubator for her eggs. It was thus a sort of accident whether in a particular species the ova went through their development inside or outside the mother's body. 'Some of the ovipara', he says, 'produce the egg in a perfect, others in an imperfect state, but it is perfected outside the body as has been stated of fish.' [1]

But it is exceedingly interesting to observe that although Aristotle regarded fishes as a whole as oviparous, he knew also of kinds that were externally viviparous and he knew, further, of one instance in which the manner of development bore an analogy to that of his true internal vivipara. 'Some animals', he says, 'are viviparous, others oviparous, others vermiparous. Some are viviparous, such as man, the horse, the seal and all other animals that are hair-coated, and, of marine animals, the cetaceans, as the dolphin, *and the so-called Selachia*. Of these animals, some have a tubular air-passage and no gills as the dolphin and the whale, others have uncovered gills, as the Selachia, the sharks and rays. . . . Of viviparous animals some hatch eggs in their own interior as creatures of the shark kind; others engender in their interior a live foetus, as man and the horse.' [2]

[1] *De Generatione animalium*, iii, 9 ; 758^a 37. [2] *Historia animalium*, i. 5 ; 489^b 35.

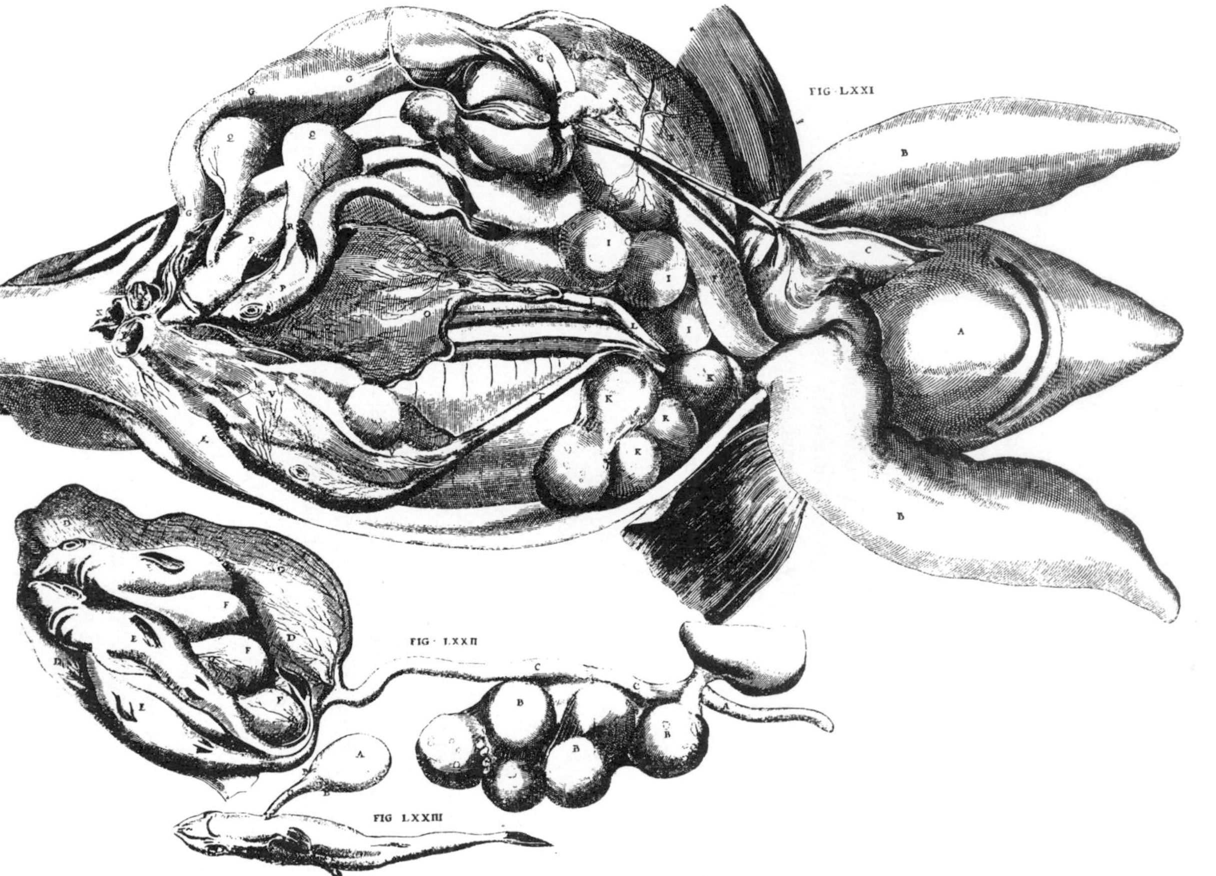

Fig. 8. From Fabricius ab Aquapendente, *De Formato Foetu*, Padua, 1604.

LXXI. Female *Galeus laevis* opened to show the gravid uterus. A, head of fish; BB, lobes of liver; C, gall bladder; D, bile duct; E, stomach; GG, intestine; II and KK, ova; LMNO, Fallopian tube; PP, foetuses in the uterus with yolk sacs, QQ, attached by the umbilical cord; RRSS; X and Y are the uterine orifices, and Z the rectal opening.

He even attempts to give an explanation of this peculiarity of the Selachians. His explanation may seem to modern ears to have little meaning, just as many of our scientific explanations will seem meaningless to our successors in a generation or two. But such explanations are worth recording not only as a stage in the historical development of biological theory, but also as illustrating the fact that, in those days as in these, while the function of science is the *description* of nature its motive is almost always the *explanation* of nature. Yet it is usually the descriptive, not the explanatory element that bears the test of time. 'Birds and scaly reptiles', says Aristotle, 'because of their heat produce a perfect egg, but because of their dryness, it is *only* an egg, the cartilaginous fishes have less heat than these but more moisture, so that they are intermediate, for they are both oviparous and viviparous within themselves, the former because they are cold, the latter because of their moisture; for moisture is vivifying, whereas dryness is furthest removed from what has life. Since they have neither feathers nor scales such as either reptiles or other fishes have, all which are signs rather of a dry and earthy nature, the egg they produce is soft; for the earthy matter does not come to the surface in their eggs any more than in themselves. That is why they lay eggs in themselves, for if the egg were laid externally it would be destroyed, having no protection.'[1]

This explanation is, of course, based on his fundamental doctrine of the opposite *qualities*, heat, cold, moistness and dryness that are found combined in pairs in the four *elements*, earth, air, fire, and water.

The intermediate character of the Selachians between the viviparous and the oviparous, as set forth by Aristotle, was well brought out by Fabricius ab Aquapendente[2] who described and figured young dogfish attached each to its own yolk sac and developing within the uterus of the mother (Fig. 8). But Aristotle had carried his investigation farther than Fabricius, for he knew of that small group of Selachian species in which the method of nourishment of the young presents remarkable analogies to that of the placental mammals. In this group of fishes the wall of the yolk sac becomes thickened at one point and attached to a corresponding thickening in the wall of the uterus. In this way a 'placenta' is formed very similar to, though not homologous with, the mammalian placenta, and the little developing fish derives nutriment from the mother's body through the placenta and navel-string much as in a mammal.

[1] *De generatione animalium*, ii. 1; 733^a 6.

[2] Hieronymus Fabricius ab Aquapendente, *De formato foetu*, Padua, 1604.

'The so-called smooth shark', says Aristotle—*Galeos* he calls it, and the name is still used by Greek fishermen—'has its eggs in betwixt the wombs like the dog-fish; these eggs shift into each of the two horns of the womb and descend, and the young develop with the navel-string attached to the womb, so that as the egg-substance gets used up, the embryo is sustained to all appearance just as in quadrupeds. The navel-string is long and adheres to the under part of the womb—each navel-string being attached as it were by a sucker, and also to the centre of the embryo in the place where the liver is situated. If the embryo is cut open, even though it has the egg-substance no longer, the food inside is egg-like in appearance. Each embryo, as in the case of quadrupeds, is provided with a chorion and separate membranes.'[1]

The attachment of the young Selachian to the womb of its mother was first observed in modern times by Pierre Belon

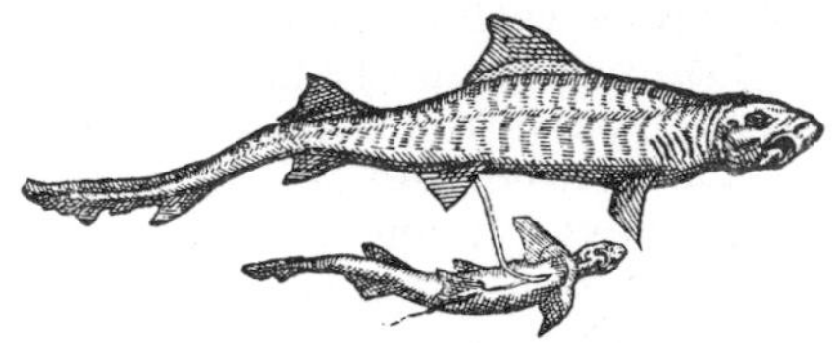

FIG. 9. *Galeus laevis*, from Rondelet's *De piscibus marinis*, Lyons, 1554.

'We have had an illustration made', says Rondelet, 'of the young attached by the navel cord to the mother so that it may be distinguished from sea-dogs, sea-wolves and the other sharks, as there is no other shark whose young is covered with secundines and membranes and attached to the mother by a navel string. I am aware', he continues, 'that there is another shark with a smoother skin than this: yet, as its manner of reproduction differs from that just described, I assert that it is not the *Galeus laevis* of the ancients but rather the *Galeus glaucus* of Aelian. [*De nat. animal.* i. 16.]'

(1517–1654) in 1553[2] and roughly figured by Guillaume Rondelet (1507–66) in 1554[3] (Fig. 9). A somewhat similar account was given by the missionary du Terte, in 1667.[4] The description of Belon was copied by Aldrovando (1522–1607) in a work published in 1613[5] and this came to the knowledge of Stensen, who definitely determined the relationship of the Selachian embryo to the wall of the maternal womb in 1675.

'In the *Galeus laevis*', says Stensen, 'each foetus has its own membrane which may be regarded as the amnion, since like the

[1] *Historia animalium*, vi. 10; 566^{b} 2.

[2] Pierre Belon, *De aquatilibus cum iconibus ad vivam ipsorum effigiem quod fieri potuit*, Paris, 1553, p. 69.

[3] Guillaume Rondelet, *De piscibus marinis*, Lyons, 1554.

[4] Jacques du Terte, *Histoire générale des Antilles habitées par les Français*, 4 vols., Paris, 1667.

[5] Ulisse Aldrovando, *De piscibus*, Bologna, 1613, p. 375.

PLATE IX. Bibl. nat. sup. grec 247 Nicander IXth cent.

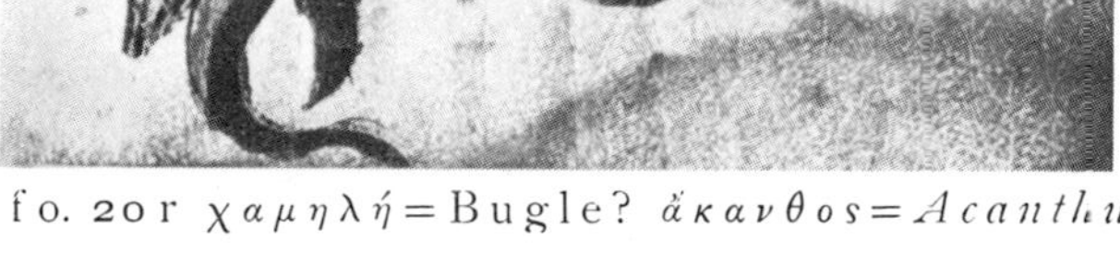

fo. 20r χαμηλή = Bugle? ἄκανθος = *Acanthus*

fo. 16v ἀλκίβιον = *Anchusa officinalis?*

Plate X. Paris Bib. nat. MS. grec 2179 Dioscorides IXth cent.

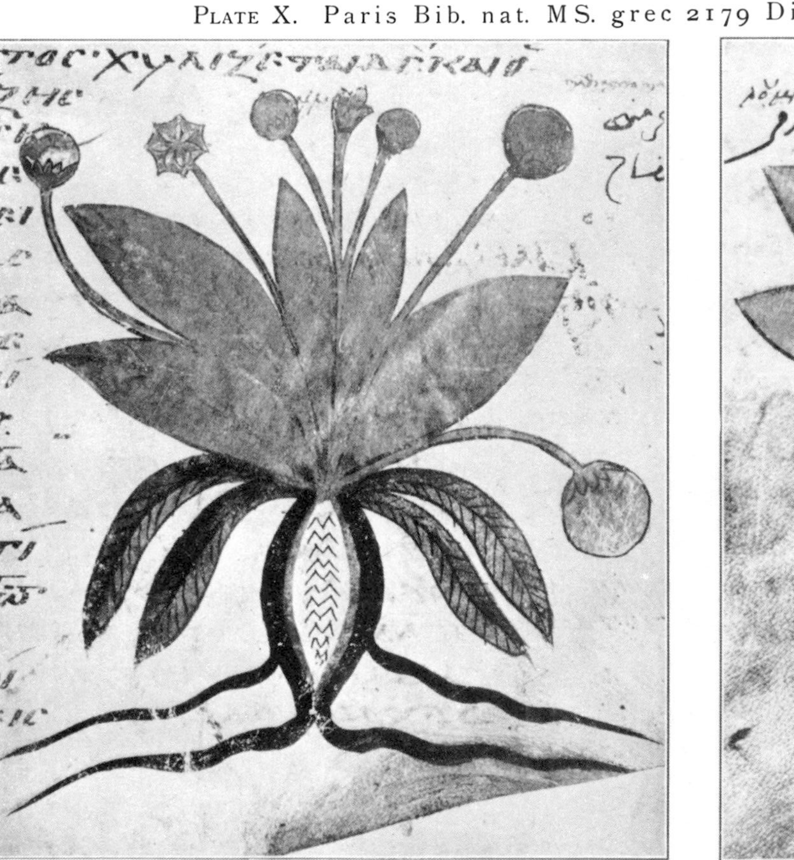

fo. 104 r 'MALE' AND 'FEMALE' MANDRAKES fo. 105 r

amnion it is the covering of the foetus and floats in the clear liquid. Yet it differs from the amnion in that it is united to the placenta in the way characteristic of the chorion (in mammals). . . . There is only a single very small placenta to each foetus, red in colour and situated close to the lower orifice of the yolk and a membrane drawn over forms a cavity.

'The umbilical vessels pass into the abdomen of the foetus by a channel beneath the diaphragm between the two anterior lobes. By following this duct I observed air bubbles floating in the intermediate liquid. On being propelled these disappeared into the intestine beyond. Next the intestine of a second foetus was inflated, and while I was moving it in different directions I opened a way for the air towards the placenta. Thus it was evident that a non-vascular tube was included among the *vasa umbilicalia*; of this vessel one extremity was joined to the spiral intestine within the abdomen, the other to the placenta where its upper surface forms a cavity with a thin membrane covering it. From the structure of the tube it is evident that nourishment is brought to the intestine from the cavity of the placenta in this fish as in birds from the yolk, as long as food is supplied to the foetus by the humours of the mother.'[1] (Fig. 10.)

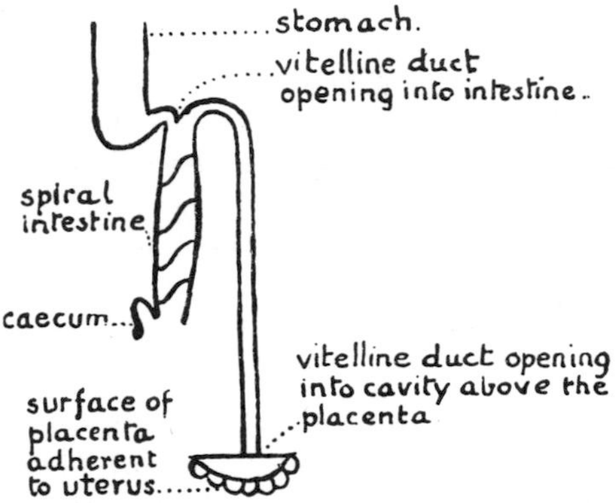

FIG. 10. After Stensen's diagram in *Ova viviparorum spectantes observationes*, Copenhagen, 1675, showing relation of yolk sac to umbilical cord and intestine.

The observations of Stensen were long disregarded. In 1828, Cuvier in his great work on fishes[2] remarked briefly that in Carcharias the yolk sac is attached to the uterus as firmly as a placenta. Neither Stensen nor Cuvier referred to Aristotle on this subject, and the importance of the ancient observations was unappreciated until the greatest of modern morphologists, Johannes Müller, took up the subject in 1839. Müller made it clear that there are at least two genera of Selachians in which this peculiar placental development takes place, namely *Carcharias* and *Mustelus* (Figs. 11–

[1] Nicolaus Stensen, *Ova viviparorum spectantes observationes*, in T. Bartholin's *Acta Hafniensia*, 1675. The works of Stensen have been made accessible by Vilhelm Maar in his *Nicolai Stenonis Opera philosophica*, 2 vols., Copenhagen, 1910. Cf. ii, p. 169.

[2] Georges Cuvier, *Histoire naturelle des Poissons*, Paris, 1828, vol. i, p. 341.

15). There can be little doubt from Aristotle's descriptions that his *Galeos leios* was not a large shark like *Carcharias* but a smaller dogfish answering to Müller's *Mustelus*. Müller further demonstrated the very peculiar fact that within the genus *Mustelus* one species (*M. laevis*) has the foetus firmly united to the uterus by means of a placenta, while in another closely allied species (*M. vulgaris*) the yolk sac is quite free and unattached. It is interesting to observe that the distinction between the two allied species was quite accurately made in 1554 by Rondelet (cf. legend to Fig. 9).

Johannes Müller describes the placenta of *Carcharias* (Figs. 12 and 13) in greater detail than that of *Mustelus* (Figs. 14 and 15).

'The *placenta foetalis* of these fish', he says, 'is formed by the folded yolk sac. The folds are much more complex in *Carcharias* than in *Mustelus laevis*. . . . In *Carcharias* the yolk sac, as usual, possesses two coats, an internal vascular coat, continuous with the intestine through the yolk duct, and an external non-vascular coat extending as a sheath over the yolk duct and the omphalo-meseraic vessels, and continuous with the skin of the foetus. . . . In the formation of the placenta both membranes are thrown into a mass of folds and the yolk sac thus converted into an irregular cavity. These folds on the side approximated to the uterus become closely involved with the wall of that organ and cannot be separated except by some force. On the other hand the part of the yolk sac on the side away from the uterus presents mere floating diverticula. Over the placental area the two walls of the yolk sac are in closest contact with each other, but elsewhere the membranes are separated from each other by a distinct interval. . . .

'The *placenta uterina* is formed by very prominent wrinkled folds of the inner membrane of the uterus which accurately correspond to those of the *placenta foetalis*. The folds of both are interposed between each other, and are as closely and firmly attached as the placenta uterina and placenta foetalis of any mammiferous animal. . . . The placenta uterina receives its blood-vessels from the uterine vessels, which are of large size, and run to the seat of the placenta at the lower part of the organ. The vessels of the placenta foetalis are the extraordinarily large *vasa omphalo-mesaraica* which are proportionally of as great size as the *vasa umbilicalia* of mammals. The organic relation of placenta foetalis and placenta uterina to one another is the same as in mammals.' [1]

[1] Johannes Müller, *Handbuch der Physiologie des Menschen*, 2 vols., Coblenz, 1840, vol. ii, p. 722. See also translation by William Baly, *Embryology, with the Physiology of Generation*, London, 1848, p. 1597, and Johannes Müller in *Monatsbericht der Akad. der Wissenschaften zu Berlin*, August 6, 1840, *Ueber den glatten Hai des Aristoteles*, Berlin, 1842, and in *Monatsbericht d. Berlin. Akad.*, 11th April, 1839.

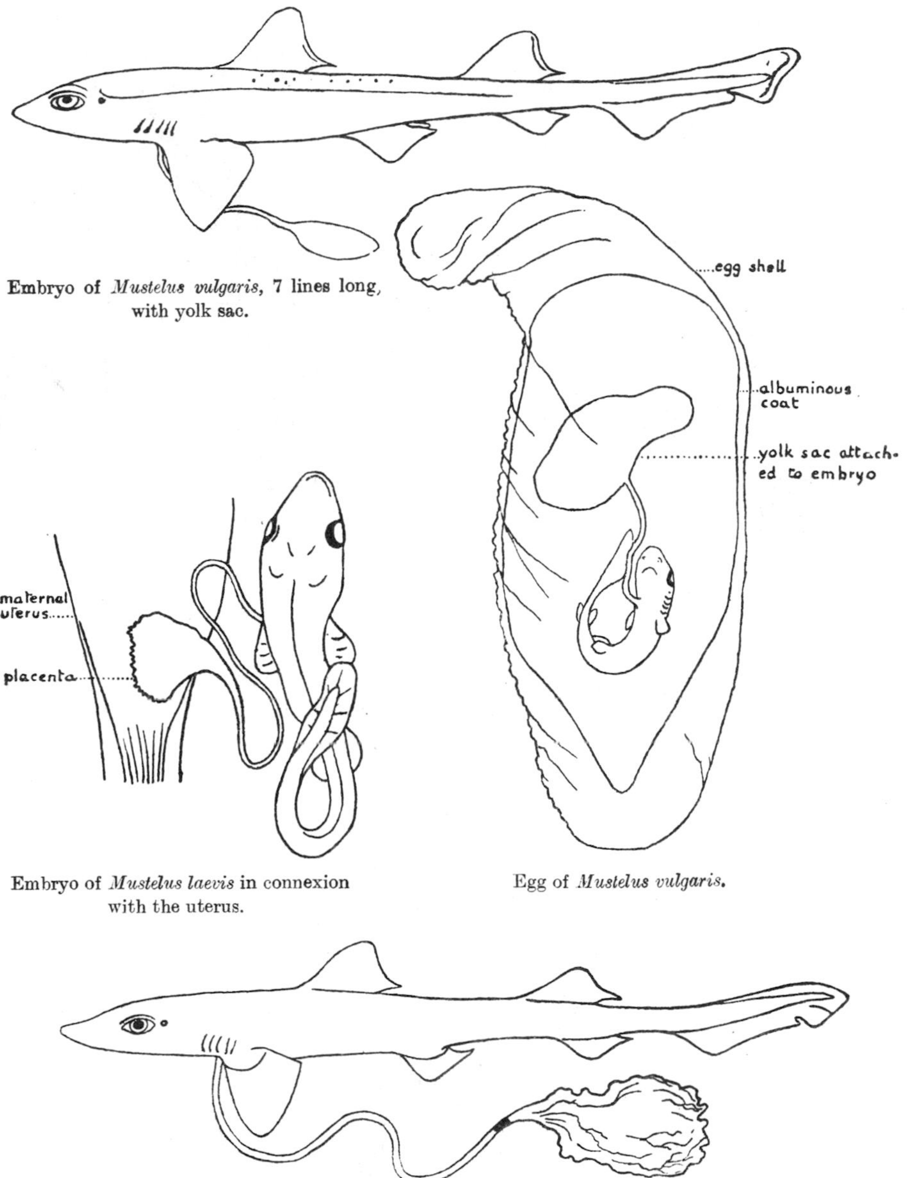

Embryo of *Mustelus vulgaris*, 7 lines long, with yolk sac.

Egg of *Mustelus vulgaris*.

Embryo of *Mustelus laevis* in connexion with the uterus.

Embryo of *Mustelus laevis*, 7 lines long, with the placental yolk sac separated from the uterus.

Fig. 11. Embryos of two species of *Mustelus*. From Johannes Müller, *Ueber den glatten Hai des Aristoteles*, Berlin, 1842.

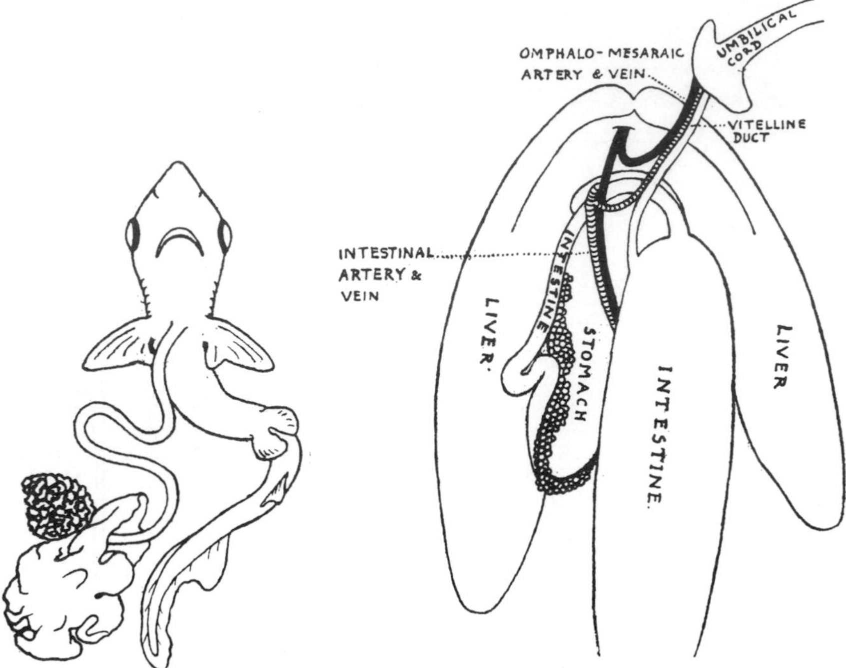

Fig. 12. Embryo of *Carcharias* with umbilical cord and placenta. From Johannes Müller.

Fig. 13. Dissection of umbilical structures of a foetal *Carcharias*, schematically represented. Modified from Johannes Müller.

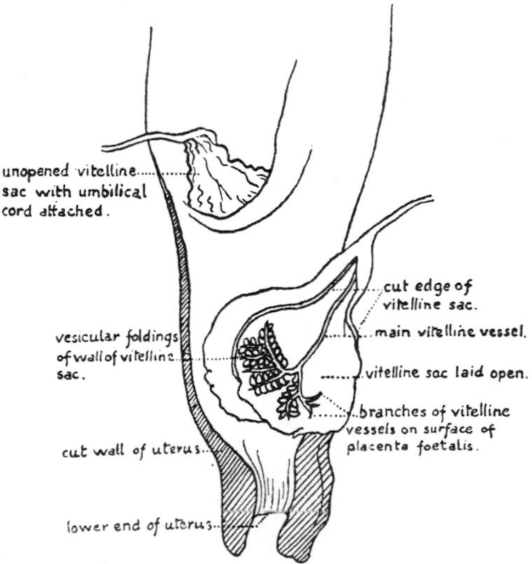

Fig. 14. A part of the uterus of *Mustelus laevis*, showing two placental attachments. From Johannes Müller.

Since Müller wrote, other observers have brought the phenomenon of the placenta of *Mustelus* and *Carcharias* more into line with what we know of allied viviparous forms. The embryos of many of these, when the yolk is nearly consumed, are nourished from other sources. In some the nutritive material is secreted

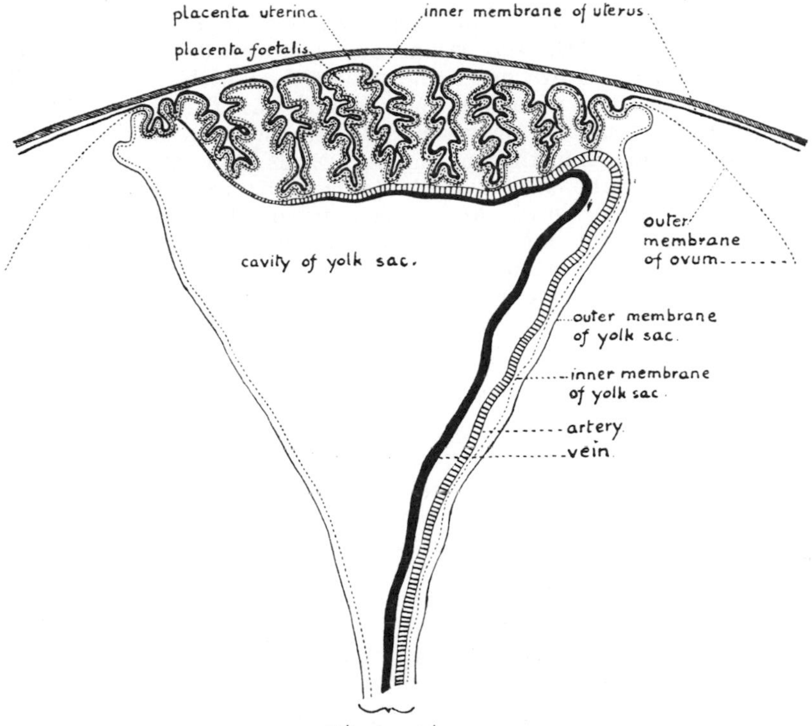

FIG. 15. DIAGRAMMATIC SECTION OF PLACENTA OF *MUSTELUS LAEVIS*. Modified from Johannes Müller.

into the cavity of the uterus and taken into the mouth of the embryo or absorbed by the blood-vessels of the yolk sac or of the gill slits. In others the wall of the uterus develops secreting villi which pass through the spiracles of the embryo into its pharynx. In yet others in which a placenta is actually developed the process of absorption is aided by villi which stud the umbilical cord throughout its course.[1]

[1] These accessory methods are described in the following papers which contain a bibliography of the subject: Franz Leydig, *Beiträge zur mikroskopischen Anatomie und Entwickelungsgeschichte der Rochen und Haie*, Leipzig, 1852;

(b) *The Ruminant Stomach*

Among the more remarkable of Aristotle's descriptions in the realm of comparative anatomy proper is that of the stomach of ruminants. He must have dissected these animals, for he gives a clear and correct account of the four chambers.

'Animals', he says, 'present diversities in the structure of their stomachs. Of the viviparous quadrupeds, such of the horned animals as are not equally furnished with teeth in both jaws are furnished with four such chambers. These animals are those that are said to chew the cud. In these animals the oesophagus extends from the mouth downwards along the lung, from the midriff to the *megale koilia* [*rumen*, or paunch], and this stomach is rough inside and semi-partitioned. And connected with it near to the entry of the oesophagus is what is called the *kekryphalos* [*reticulum*, or honeycomb bag]; for outside it is like the stomach, but inside it resembles a netted cap; and the kekryphalos is a good deal smaller than the megale koilia.'

The term *kekryphalos*, the reader may be reminded, was applied to the net that women wore over their hair to keep it in order.

'Connected with this kekryphalos', he continues, 'is the *echinos* [*psalterium*, or *manyplies*], rough inside and laminated, and of about the same size as the kekryphalos. Next after this comes what is called the *enystron* [*abomasum*], larger and longer than the echinos, furnished inside with numerous folds or ridges, large and smooth. After all this comes the gut.'[1] . . . 'All animals that have horns, the sheep for instance, the ox, the goat, the deer and the like have these several stomachs. . . . The several cavities receive the food one from the other in succession; the first taking the unreduced substances, the second the same when somewhat reduced, the third when reduction is complete, and the fourth when the whole has become a smooth pulp.'[2] . . . Such is the stomach of those quadrupeds that are horned and have an unsymmetrical dentition; and these animals differ one from

T. J. Parker and A. Liversidge, 'Note on the foetal membranes of *Mustelus antarcticus*', *Transactions of the New Zealand Institute*, xxii, p. 331, Wellington, 1890; J. Wood-Mason and A. Alcock, 'On the uterine villiform papillae of *Pteroplataea micrura*', *Proc. Roy. Soc.*, xlix, p. 359, London, 1891; J. Wood-Mason and A. Alcock, 'Further observations on the gestation of Indian Rays', *Proc. Roy. Soc.*, l, p. 202, London, 1892; A. Alcock, 'Some observations on the embryonic history of *Pteroplataea micrura*,' *Annals and Magazine of Natural History*, sixth series, x, p. 1, London, 1892; T. Southwell and B. Prashad, 'Embryological and Developmental Studies of Indian Fishes', *Records of the Indian Museum*, xvi, p. 215, Calcutta, 1919.

[1] *Historia animalium*, ii. 17; 507ᵃ 33.

[2] *De partibus animalium*, iii. 14; 674ᵇ 6.

Plate XI. From Brit. Mus. MS. Reg. 15 E III fo. 11 r

Le Livre des Propriétéz de Choses translated by Jehan Corbechon from Latin of Bartholomew de Glanvil, written at Bruges by Jehan du Ries in 1482. Frontispiece of Book XII 'On Birds'.

Plate XII. PAINTINGS BY EDWARD TYSON (b. 1650, d. 1708) made in 1687

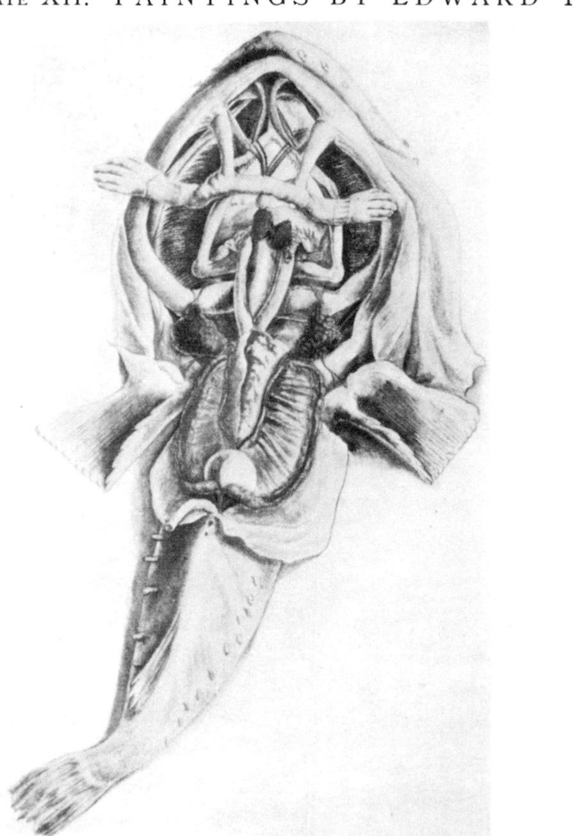

P. 41. Dissection of Lophius

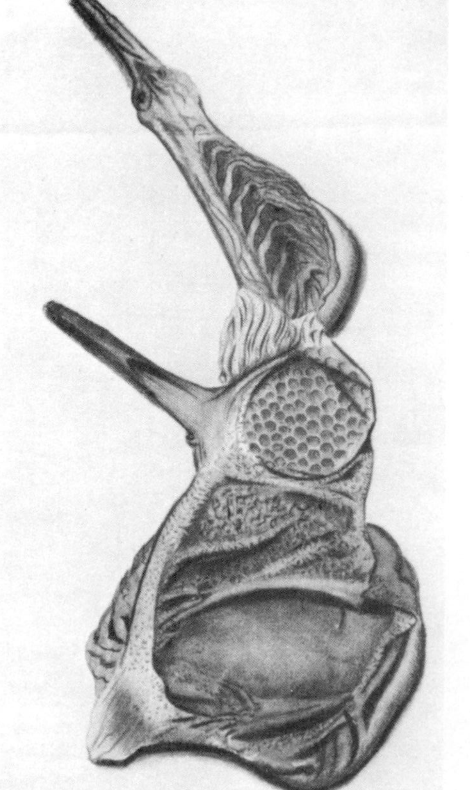

P. 92. Stomach of Gazelle

From a MS. at the Royal College of Physicians, London

another in the shape and size of the parts, and in the fact of the oesophagus reaching the stomach centralwise in some cases and sideways in others. Animals that are furnished equally with teeth in both jaws have one stomach; as man, the pig, the dog, the bear, the lion, the wolf.' [1]

The general appearance of the stomach of ruminants must always have been roughly known to butchers and its rediscovery cannot therefore be dated as can many of the biological observations of Aristotle that we have to recount. A fair scientific description of the organ was made by Aldrovando in 1613 [2] and by Fabricius in 1618.[3] The ruminant stomach was figured imperfectly by Severino in 1645 [4] (Fig. 16), and by Blasius in 1667.[5] There is a better figure by Grew of 1681 [6] (Fig. 17), and an excellent painting was prepared by Tyson, the earliest English comparative anatomist, in 1687 (Plate XII).

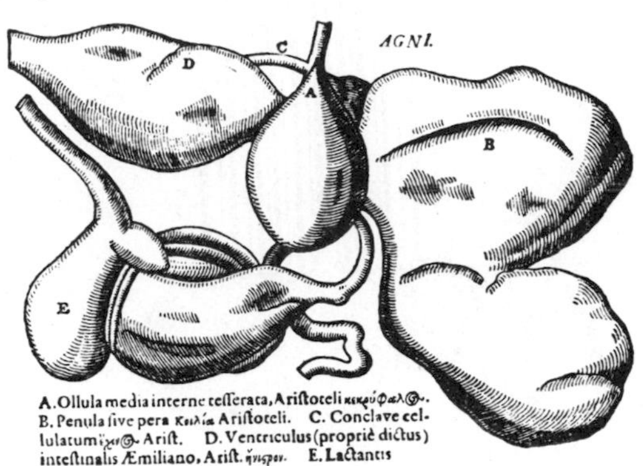

AGNI.
A
B
C
D
E
A. Ollula media interne teſſerata, Ariſtoteli κεκρύφαλος.
B. Penula ſive pera Κοιλία Ariſtoteli. C. Conclave cellulatum ἐχῖνος Ariſt. D. Ventriculus (propriè dictus)
inteſtinalis Æmiliano, Ariſt. ἤνυστρον. E. Lactantis

FIG. 16. THE FOUR CHAMBERS OF THE STOMACH OF A LAMB

From Marco Aurelio Severino, *Zootomia Democritea*, Nuremberg, 1645.

(c) *The Generative processes of Cephalopods*

Nowhere is the contrast between the ancient and modern method of setting out biological conclusions better brought out than in the investigation of the extraordinarily interesting generative processes of the Cephalopods. An examination of the modern accounts of the subject enables us to observe the slow emergence of a true conception of the actual nature of the phenomena. This

1 *Historia animalium*, ii. 17; 507b 12.

2 Ulisse Aldrovando, *Quadrupedium omnium bisulcorum historia*, Bologna, 1613.

3 Hieronymo Fabricius, *De gula*, Padua, 1618.

4 Marco Aurelio Severino, *Zootomia Democritea id est Anatome generalis totius animantium opificii, libris quinque distincta*, Nuremberg, 1645.

5 Gerhard Blaes, *Observata anatomia*, Amsterdam, 1676, p. 49.

6 Nehemiah Grew, *Catalogue of the rarities belonging to the Royal Society*, London, 1681.

process continues, in spite of many errors, because each observer records his actual observations and places them in such a form that the place, time and means of observation can be referred to at need, and the reader can himself separate what is observed from what is inferred and can grasp not only the nature of the observation itself but the means by which it was made. Thus the work of each writer can be criticized or modified by the next. In doing this both writer and reader are immeasurably aided by the use of figures. It is true that diagrams were used also by Aristotle,[1] but these appear to have been merely occasional devices rather than an intrinsic part of his method. Drawings or diagrams as routine aids to biological descriptions were probably uncommon until the first century B.C. In his account of the generative processes of the Cephalopods the ancient naturalist records only the final conclusions, and we hardly know which of the observations are his own and which are taken from

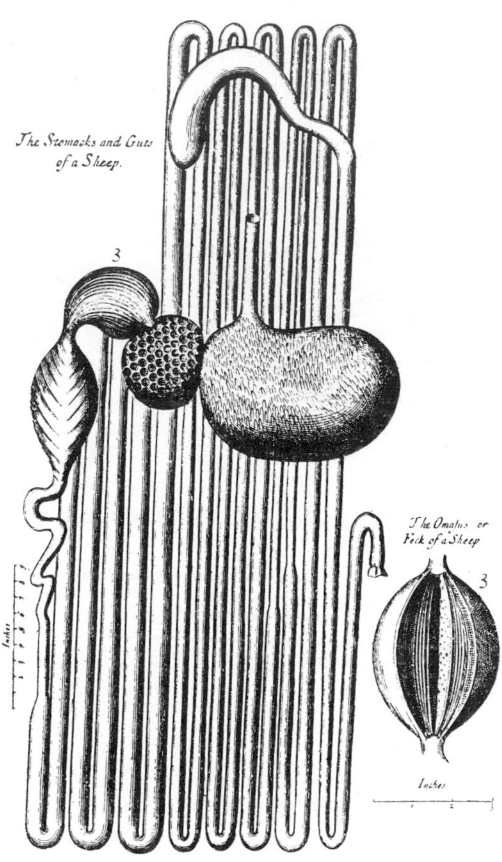

FIG. 17. THE FOUR-CHAMBERED STOMACH OF A SHEEP
After Nehemiah Grew, *The Comparative Anatomy of the Stomach and Guts Begun*, London, 1681.

[1] An interesting reference to the diagrams in Aristotle's lost work on Anatomy will be found in the *Historia animalium*, i. 17 ; 497ᵃ 33. Other references to anatomical diagrams are in the *De generatione animalium*, ii. 7 ; 746ᵃ 14, and the *Historia animalium*, iii. 1 ; 510ᵃ 29. The words used are σχήματα, διαγραφή, and παραδείγματα.

others, while he tells us nothing whatever of the conditions under which they were made. In spite of these faults in the record, his descriptions impel conviction that they are those of an acute and accurate observer, and that his work does not suffer from any lack in his powers as a naturalist. The main references to the reproduction of Cephalopods occur however in the ninth book of the *Historia animalium*, which is of more doubtful authenticity than the earlier parts.

With regard to the Cephalopod *Argonauta* Aristotle says that it is an

'octopus, but one peculiar both in its nature and habits.[1] . . . This polypus lives very often near to the shore, and is apt to be thrown up high and dry on the beach; under these circumstances it is found with its shell detached, and dies by and by on dry land.[2] . . . It rises up from deep water and swims on the surface. In between its feelers it has a certain amount of web-growth resembling the substance between the toes of web-footed birds; only that with these latter the substance is thick, while with the nautilus it is thin like a spider's web. (Cf. Plate XIII.) It uses this structure, when a breeze is blowing, for a sail, and lets down some of its feelers alongside as rudder-oars. (Cf. Fig. 18, p. 43.) If it be frightened it fills its shell with water and sinks. With regard to the mode of generation and the growth of the shell, knowledge is not yet satisfactory; the shell, however, does not appear to be there from the beginning, but to grow in their case as in that of the other shell-fish; neither is it ascertained for certain whether the animal can live when stripped of the shell.'[3]

The use of the membranes of the Argonaut as a sail and the arms as oars (Fig. 18) is now known to be pure myth, though many excellent naturalists, Vérany among them, have given colour to one or other of these ideas. It is but right to emphasize again that the ninth book of the *Historia animalium*, in which the statement occurs, is probably not the work of Aristotle himself.[4]

The questions that the Aristotelian treatise asks about the Argonaut can now at last, after many centuries, be answered in some fullness, although observation of the animal has been beset with numerous difficulties. It is a fact that the shell is *not* necessary to its life, and Lacaze-Duthier observed the animal recover the shells which had been taken from it. The shell when fully formed is in no organic connexion with the body of the

[1] *Historia animalium*, ix. 38; 622[b] 5. [2] *Historia animalium*, iv. 2; 525[a] 22.

[3] *Historia animalium*, ix. 38; 622[b] 6.

[4] On the question of the authenticity of the ninth book of the *Historia animalium*, see H. Aubert and F. Wimmer, *Aristotelis Thierkunde*, Leipzig, 1868, p. 11.

animal, and its function is but to support and aerate the developing eggs; it has therefore been aptly compared to a perambulator. The animal does not willingly sink below the surface of the water, and if forced to do so will rise again, and it is indeed doubtful if it is able to sink at all by its own effort. The act of congress has not been seen in this species, though since the appearance of the work of Heinrich Müller, the male, which is much smaller than the female, has been recognized. (Plate XIII.)

Turning now to the description given of the sexual processes of cephalopods by the naturalist of antiquity we read that

'The Malacia such as the octopus, the sepia and the calamary, have sexual intercourse all in the same way; that is to say, they unite at the mouth by an interlacing of their tentacles. When, then, the octopus rests its so-called head against the ground and spreads abroad its tentacles, the other sex fits into the outspreading of these tentacles, and the two sexes then bring their suckers into mutual connexion. Some assert that the male has a kind of penis in one of his tentacles, the one in which are the largest suckers; and they further assert that the organ is tendinous in character, growing attached right up to the middle of the tentacle, and that the latter enables it to enter the nostril or funnel of the female.' [1]

It is unfortunate that since he has given this accurate description Aristotle elsewhere contradicts it.

'In the cephalopods', he says, 'the same passage serves to void excrement and leads to the part like a uterus, for the male discharges the seminal fluid through this passage. And it is on the lower surface of the body where the mantle is open. . . . Hence the union of the male with the female takes place at this point. . . . But the insertion, in the case of the poulps, of the arm of the male into the funnel of the female, by which arm the fishermen say the male copulates with her, is only for the sake of attachment, and it is not an organ useful for generation, for it is outside the passage in the male and indeed outside the body of the male altogether.' [2]

We may now turn to the rediscovery in modern times of that peculiar sexual process of the octopods known as *hectocotylization* which explains Aristotle's description and even his self-contradiction concerning the cephalopods. The story may be given in the words of one of the most eminent investigators of these creatures, Richard Owen. The discussion concerning the nature of the process of fertilization arose chiefly in connexion with the octopod known as *Argonauta Argo*.

[1] *Historia animalium*, v. 6; 541^{b} 1.

[2] *De generatione animalium*, i. 15; 720^{b} 25.

PLATE XIII

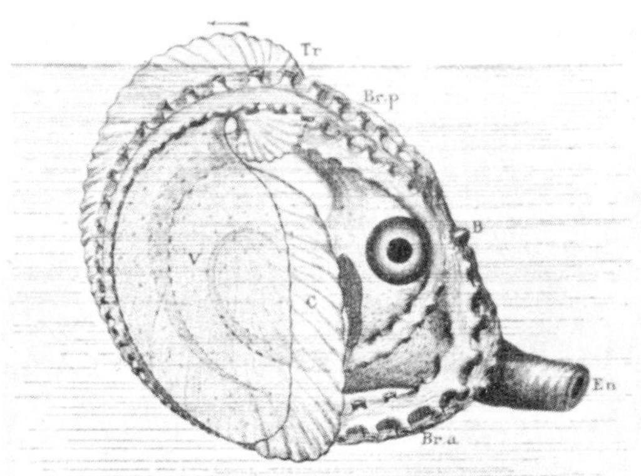

Female argonaut from H. de Lacaze-Duthiers, *Archives de Zoologie expérimentale*, 1892.

The animal is seen in profile and the arrow shows the direction of movement. *B*, mouth with parrot-like beak. *Tr*, the nidamental shell projecting beyond the two specially developed shell-carrying arms *Br.p.* If the outline of the shell is followed it will be seen to project beyond the shell membrane *V*, also at *C*. *En*, the funnel projecting between the two anterior arms, *Br.a.*

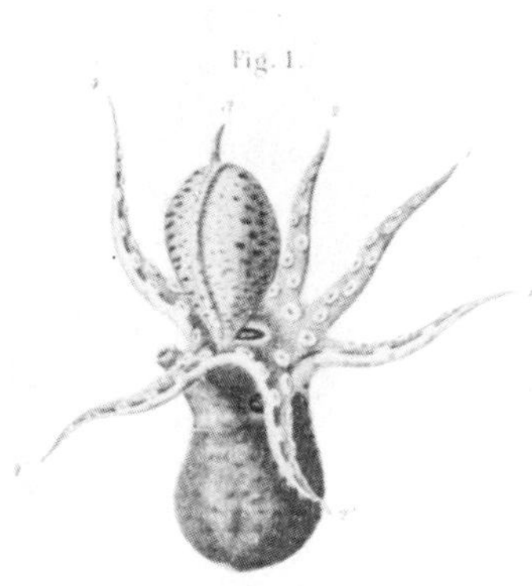

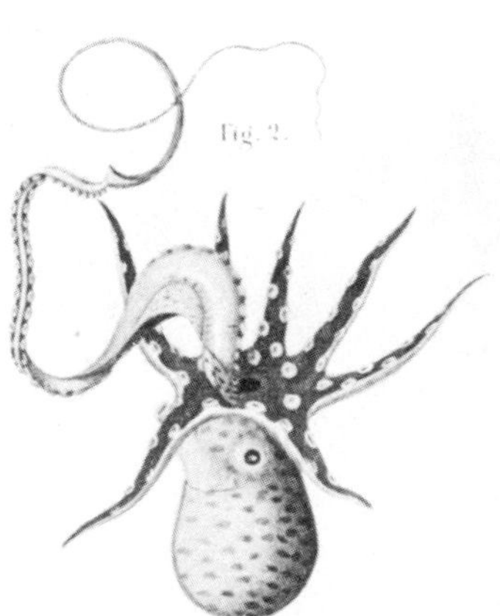

Male argonaut from Heinrich Müller, *Zeitschrift für wissenschaftliche Zoologie*, 1853.

1. The second and fourth arms of the left side are depressed to show the hectocotylus sac.
2. Hectocotylus unfolded showing how the sac is formed as a membrane developed at base of arm.

PLATE XIV

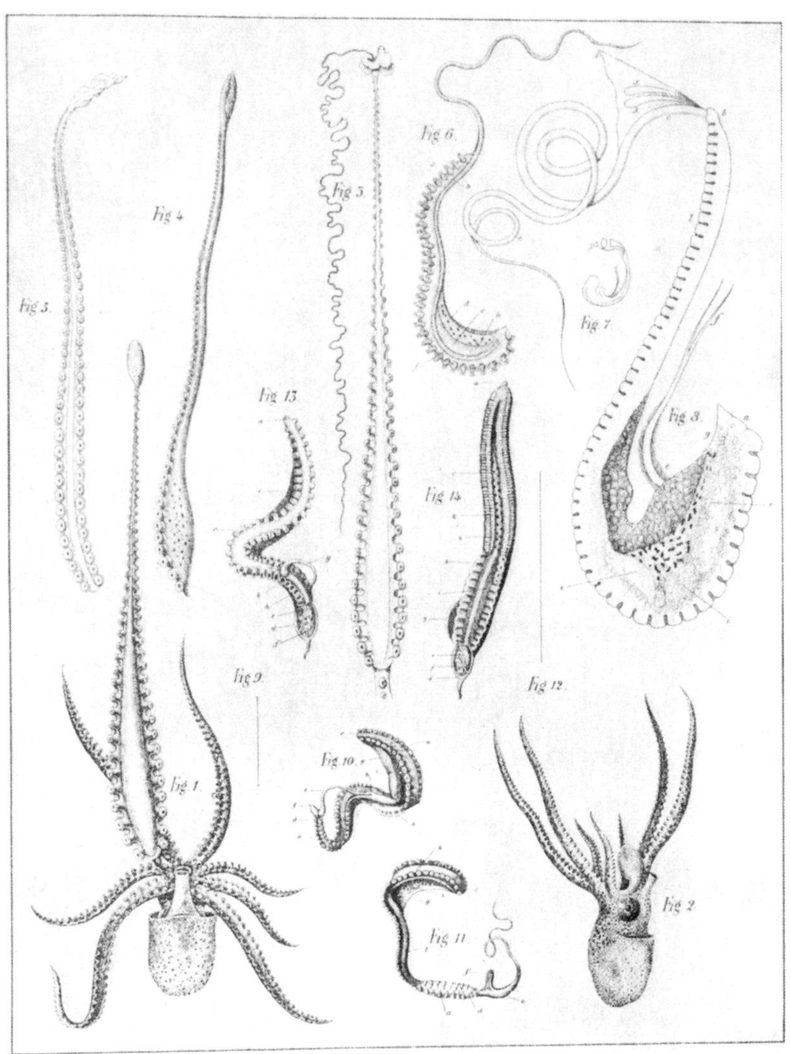

HECTOCOTYLIZATION

From J. B. Vérany, 1851; 1–3, *Octopus carenae* of Vérany; 1, Hectocotylus extended; 2, in sac; 3, removed from sac; 4–5, *Hectocotylus octopodis* of Cuvier, 1829; 6, *Tricho cephalus acetabularis* of Delle Chiaje, 1828; *l.b* alimentary canal, *c* ovary, *d* pigmented membrane, *ef* suckers; 7-8, *Alleged male of Argonaut* from Costa, 1841; *ab* trunk, *cc* terminal appendix, *ef* tentacular cirrhus, *l* suckers, *ii* mucous sac, *d* membrane with special strands *xx*; 10–11, *Hectocotylus Argonautae* of Kölliker, 1849 (see text figs. 19–21); 13–14 Hectocotylus of *Tremoctopus violaceus* of Vérany after Kölliker, figured by the latter as a separate animal, *e* spermatic duct, *f* testis, *g* penis.

'The cumulative experience of numerous observers since 1839', says Owen, writing in 1855, 'had led to the conviction that the Argonauta with the expanded arms and shell was the female form of the species. The discovery of the male has been attended with difficulties. . . .

'Delle Chiaje first (1828) figured and described an organism which he found attached to the female Argonaut,[1] and which he believed to be a parasite, describing it under the name of *Trichocephalus acetabularis* (Plate XIV, item 6) on account of the number

Portraict du Nautillus, lequel Pline nôme Pôpilus ou Nauplius.

FIG. 18. THE PAPER NAUTILUS

Argonauta Argo. From Belon's *Histoire naturelle des estranges poissons*, Paris, 1551. The animal is drawn as though using its arms as oars and its membrane as a sail.

of suckers with which it was beset. In the following year, Cuvier, having received a similar organism which Laurillard had detected in a cephalopod called *Octopus granulosus*, also believed it to be a parasitic worm for which he proposed the name of *Hectocotylus Octopodis*, assigning the name *Hectocotylus Argonautae* to the previously observed species.[2] (Plate XIV, items 4 and 5.) In 1842 Kölliker, having detected the same organism apparently parasitic on the female Argonaut, carefully scrutinized its structure and found that of the skin, with its complex pigment cells and that of the acetabula, identical with the same parts in the Argonaut. He detected, moreover, in a dilated hollow part of the organism a quantity of spermatozoa . . . and came to the bold conclusion

[1] Stefano delle Chiaje, *Memorie sulla storia e notomia degli animali senza vertebre del regno di Napoli*, 4 vols. and 2 atlases, Naples, 1828, vol. ii, Plate 16.

[2] Georges Cuvier, 'Mémoire sur un ver parasite d'un nouveau genre (*Hectocotylus octopodis*)', in the *Annales des sciences naturelles*, xviii, Paris, 1829, p. 147, plate 11.

that it was the long-sought-for male of the Argonaut, arrested in its development and subsisting practically parasitically on the female, like the diminutive males of the Rotifera, Epizoa, and Cirrhipedia. It may serve as a wholesome warning against entering upon

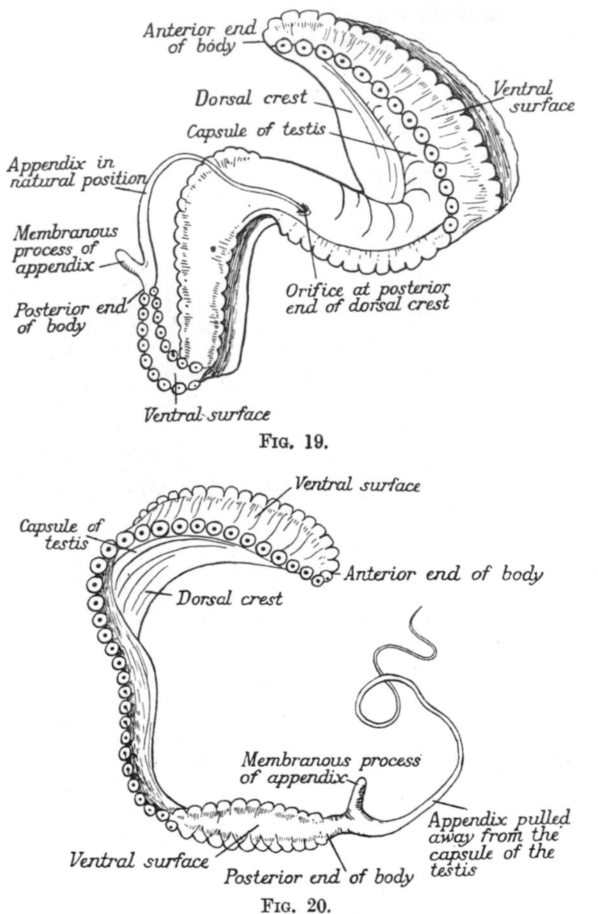

FIG. 19.

FIG. 20.

FIGS. 19, 20. DRAWING FROM KÖLLIKER OF THE 'MALE ARGONAUT'

a scrutiny of parts while prepossessed by a foregone conclusion to remember that the acute and usually accurate observer describes the digestive, circulatory, and respiratory organs of the same supposed independent individual male animal.'[1] (Figs. 19–21 and Plate XIV, items 10, 11, 13, and 14.)

[1] Albrecht von Kölliker, 'Hectocotylus Argonautae D. Ch. und Hectocotylus Tremoctopodis Köll., die Männchen von Argonauta Argo und Tremoctopus

'Vérany first had the good fortune to discover the Hectocotylus or presumed parasitic male Argonaut, forming one of the arms singularly modified and developed, of a little octopod which he figured under the name of *Octopus carenae* (Plate XIV, items 1–3).[1] Müller and others were not slow in demonstrating that this, or a similarly modified octopod, was really the male of the Argonauta [2] (Plate XIII).

'Certain species of the Octopod family thus have the male apparatus extended into one of the cephalic arms. . . . In *Octopus granulosus* and *Argonauta Argo* the spermatic duct is continued from the testis . . . into the base of the sexual arm and opens into a dilated reservoir at the termination of that singularly modified member. It is somewhat longer than the longest of the unmodified arms, and is much thicker. The acetabula are larger and more numerous, but retain the arrangement in a double row. . . .

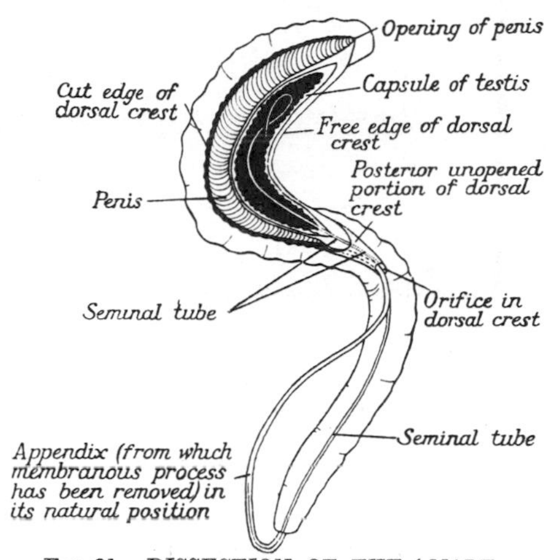

FIG. 21. DISSECTION OF THE 'MALE ARGONAUT'. After Kölliker.

'One presumes that the arms of the two sexes being interlaced, as described by Aristotle, the expanded receptacular end of the modified arm, with the spermatozoa, is introduced into the funnel of the female . . . and that . . . the modified arm is snapped off and left adhering to the female by the suckers where it long retains the power of motion. Such is the conclusion of the long mooted questions of the Argonaut, its shell, the use of the brachial membranes as 'sails', and the true sexual distinctions of the male and female.' [3]

violaceus D. Ch.', in the *Berichte von der zootomischen Anstalt in Würzburg für das Schuljahr* 1847–8, Leipzig, 1849, p. 67. This article is exceedingly rare, but there is a copy in the collection of T. H. Huxley at the Royal College of Science, London.

[1] J. B. Vérany, *Mollusques méditerranéens*, Genoa, 1851, and 'Mémoire sur les hectocotyles de quelques Céphalopodes' in the *Annales des sciences naturelles*, xvii, Paris, 1852.

[2] Heinrich Müller, 'Ueber das Männchen von Argonauta Argo und die Hectocotylen' in the *Zeitschrift für wissenschaftliche Zoologie*, vol. iv, Leipzig, 1853, p. 1.

[3] Richard Owen, *Lectures on the Anatomy and Physiology of the Invertebrates*, 2nd edition, London, 1855, pp. 628 ff.

It remains only to add that the actual process of congress of an octopus, though not of *Argonauta* itself, has been watched by Racovitza (Fig. 22) and the hectocotylized arm of the male has been observed to be inserted through the mantle and into the

FIG. 22. CONGRESS OF OCTOPUS VULGARIS
as observed in a tank by É. Racovitza, *Archives de Zoologie expérimentale*, Série ii, Paris. 1894. The male is seen to the right. He has fixed himself by the bases of his arms to the glass of the aquarium and has introduced the extremity of the hectocotylized arm into the pallial cavity of the female to the left.

funnel of the female.[1] There is still a gap in our knowledge however, for the passage of the fertilizing elements from the testis into the arm of the male has not yet been observed.

(*d*) *The Habits of Animals*

i. *The Frog-fish and Torpedo*

Aristotle is perhaps at his best and happiest when describing the habits of living animals that he has himself observed. Among his most pleasing accounts are those of the fishing-frog and torpedo. In these creatures he did not fail to notice the displacement of the fins associated with the depressed form of the body.

'In the Torpedo and the Fishing-frog,' he says, 'the breadth of the anterior part of the body is not so great as to render locomotion by fins impossible, but in consequence of it the upper pair [*pectorals*] are placed further back and the under pair [*ventrals*] are placed close to the head, while to compensate for this advancement they are reduced in size so as to be smaller than the upper ones. In the Torpedo the two upper fins [pectorals] are placed in the tail, and the fish uses the broad expansion of its

[1] Émile Racovitza, 'Notes de biologie' in the *Archives de zoologie expérimentale*, series 3, vol. ii, Paris, 1894, p. 25.

body to supply their place, each lateral half of its circumference serving the office of a fin.' [1]

'In marine creatures one may observe many ingenious devices adapted to the circumstances of their lives. For the account commonly given of the frog-fish or angler is quite true; as is also that of the torpedo. The frog-fish has a set of filaments that project in front of its eyes; they are long and thin, like hairs, and are round at the tips; they lie on either side, and are used as baits.[2] . . . The little creatures on which this fish feeds swim up to the filaments, taking them for bits of seaweed such as they feed upon.[3] Accordingly, when the animal stirs himself up a place where there is plenty of sand and mud and conceals himself therein, it raises the filaments, and when the little fish strike against them the frog-fish draws them in underneath into its mouth. The torpedo narcotizes the creatures that it wants to catch, overpowering them by the force of shock that is resident in its body, and feeds upon them; it also hides in the sand and mud, and catches all the creatures that swim in its way and come under its narcotizing influence. This phenomenon has been actually observed in operation. . . . That the creatures get their living by this means is obvious from the fact that, whereas they are peculiarly inactive, they are often caught with mullets, the swiftest of fishes, in their interior. Furthermore, the frog-fish is unusually thin when he is caught after losing the tips of his filaments, while the torpedo fish is known to cause a numbness even in human beings.' [4]

The Fishing-frog is well known in northern waters and several of the early naturalists give recognizable figures of it (Fig. 23). It is remarkable how in most of them a curious error has crept in as regards the structure of the fins. The pectoral limbs of the Lophiidae are, in fact, specially developed for crawling about on the sea floor. In the older pictures, however, the fins are usually represented as divided into digits, a mistake perpetuated in the careful paintings of dissections of this creature that have been left by Edward Tyson (1650–1708) [5] (Plate XII).

The peculiar action of the Torpedo described by Aristotle has been the subject of a number of classical investigations and has been brought into line with widely distributed phenomena. It is characteristic of muscle substance that at the moment of contraction it produces an electric disturbance. In ordinary

[1] *De partibus animalium*, iv. 13; 696ª 26.

[2] *Historia animalium*, ix. 37; 620ᵇ 10.

[3] This passage only doubtfully refers to the fishing-frog. We have transposed it from ix. 37; 620ᵇ 30.

[4] *Historia animalium*, ix. 37; 620ᵇ 15.

[5] Drawings of these dissections are found in MSS. of Tyson at the Royal College of Physicians of London, and at the British Museum.

muscular tissue this can only be detected by means of delicate instruments, but in the Torpedo-fish there are two tracts of musculature which show a reduction of the function of contraction accompanied by a great increase in the power of producing electric disturbance. These electric organs are two large kidney-shaped structures occupying the greater portion of the forepart of the animal and are controlled by a special set of nerves which have their origin in a characteristically developed lobe of the brain (Fig. 24).

PETRI BELLONII CENOMANI

Βάτραχος θαλάττιος, Græcis: Rana marina, Latinis: *Diauolo marino*, Italis: *Bauldroy*, Massiliensibus: *Pescheteau*, Burdegalis.

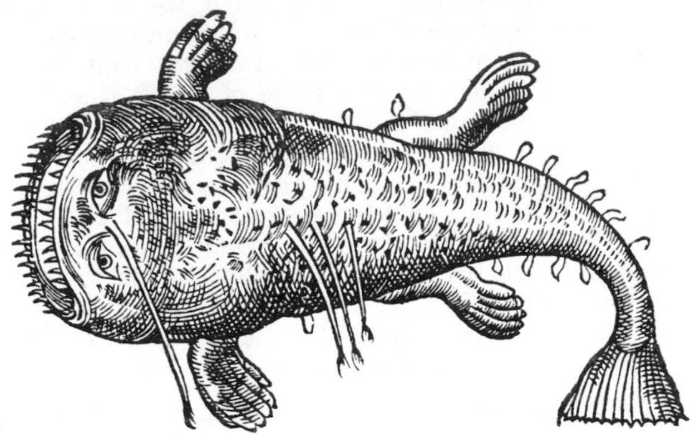

FIG. 23. THE FROG-FISH
From Pierre Belon's *De aquatilibus*, Paris, 1553.

Aristotle was by no means alone among the ancients in his knowledge of the electric action of the torpedo. The animal was commonly eaten and is referred to as suitable for invalids in several of the Hippocratic treatises (fifth–fourth cent. B.C.).[1] Torpedo and fishing-frog were both known to Pliny (*c.* A. D. 23–79), and it is worth repeating his account of these fishes, copied evidently from Aristotle, in Philemon Holland's inimitable translation:

'I marvel', he writes, 'at them who are of opinion that fishes and beasts in the water have no sence. Why, the very Crampe-fish Tarped (i. e. *Torpedo*) knoweth her oune force and power, and being herselfe not benummed, is able to astonish others. She lieth hidden ouer head and ears within the mud unseene, readie to catch those fishes, which as they swim ouer her, bee taken with

[1] In the *De victus ratione*, a work in part older than Hippocrates, in the spurious but ancient *De internis affectionibus*, and in the equally ancient and spurious *De mulierum morbis*.

a nummednesse, as if they were dead. Also the fish called the sea Frog, Diable de Mer, (and of others, the sea Fisher) is as craftie everie whit as the other: It puddereth in the mud, and troubleth the water, that it might not bee seene: and when the

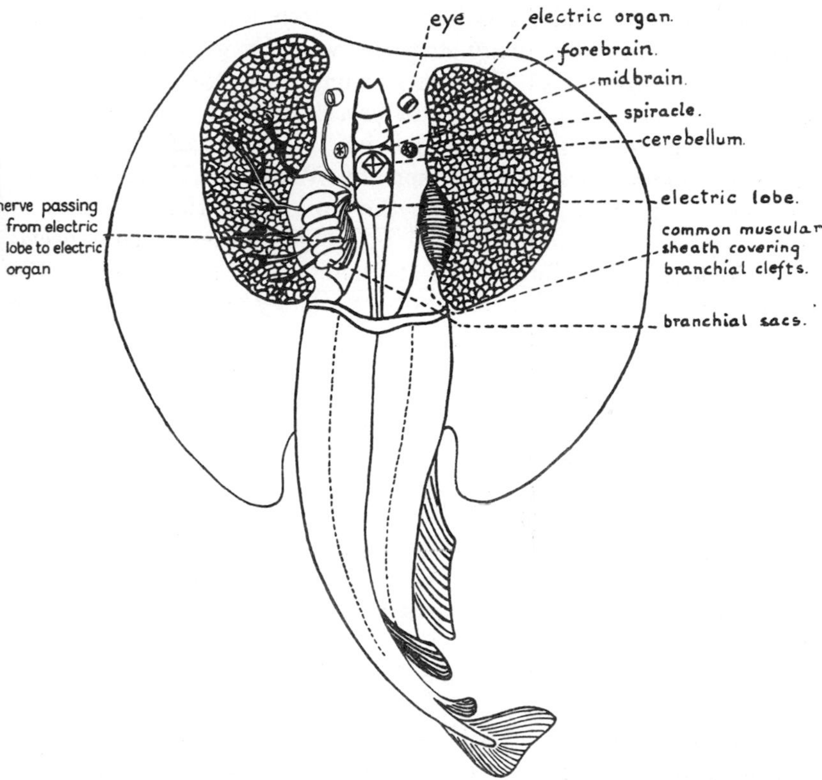

Fig. 24. THE ELECTRIC ORGAN OF THE TORPEDO

little seely fishes come skipping about her, then she puts out her little hornes or Barbils which shee hath bearing forth under her eies, and by little and little tilleth and tolleth them so neere that she can easily seaze upon them.'[1]

The torpedo must have been somewhat extensively experimented upon in antiquity. Galen (A. D. 131–201) knew of its power, which he compared to the magnetic effect of a lodestone;[2] he noted that it produced numbness or anaesthesia of parts that

[1] *The Historie of the World, commonly called, the Naturall Historie of* C. Plinius Secundus, *Translated into English by Philemon Holland Doctor in Physicke,* 2 vols., London, 1601. The quotation is from book ix, chapter 42, of Holland's notation.

[2] Galen, *De locis affectis,* Lib. vi. Kühn viii. 421.

it touched [1] and he therefore recommended the application of the living fish to the head for headache.[2] Similar remedies are referred to by Scribonius Largus (A. D. 47), Marcellus Empiricus of Bordeaux (fifth century), and Aetius of Amida (sixth century).

The Torpedo and its powers have been known continuously from Greek times, but many centuries went by without any elucidation of its structure. The first to suggest the true nature of the shock of electric fish was Muschenbroeck in 1762, who compared their shocks to those of a Leyden jar.[3] A series of experiments on the Torpedo was undertaken by Walsh in 1772,[4] and by Ingenhousz in 1775,[5] and about the same time the structure of the electric organ was described by John Hunter.[6] The special electric lobe of the brain and its homologies and those of the nerves that arose from it were not completely elucidated until the work of Letheby in 1843.[7]

ii. *Bees*

Aristotle's account of bees is remarkable for its extent. He has given more attention to these than to any other group of insects. Yet it must be confessed that although his account contains many accurate and striking observations, it also contains numerous errors and is obviously largely drawn from hearsay and from secondhand accounts. We have omitted many erroneous elements, and have put together some of his best passages on these insects.

Of the structure of the bee Aristotle tells us very little, nor does that little contain any evidence of close observance of the actual parts. We read that 'some insects have the part which serves as tongue inside the mouth as with ants . . . while in others

[1] Galen, *De locis affectis*, Lib. ii, Kühn viii. 72, and *De symptomatum causis*, Lib. i, cap. 5, Kühn, vii. 109.

[2] Galen, *De simplicium medicamentorum temperamentis ac facultatibus*, Lib. xi, cap. 11, Kühn, xii. 365.

[3] Pieter van Muschenbroeck, *Introductio ad philosophiam naturalem*, Leyden, 1762.

[4] J. Walsh, 'On the Electric Property of the Torpedo', *Philosophical Transactions*, London, 1772.

[5] J. Ingenhousz, 'Experiments on the Torpedo', *Philosophical Transactions*, London, 1775.

[6] John Hunter, 'Anatomical Observations on the Torpedo', *Philosophical Transactions*, July 1, 1773.

[7] H. Letheby, 'An account of a second *Gymnotus Electricus*, together with a description of the electrical phenomena and the anatomy of the Torpedo', *Proceedings of the London Electrical Society for 1843*, p. 512, London, 1843.

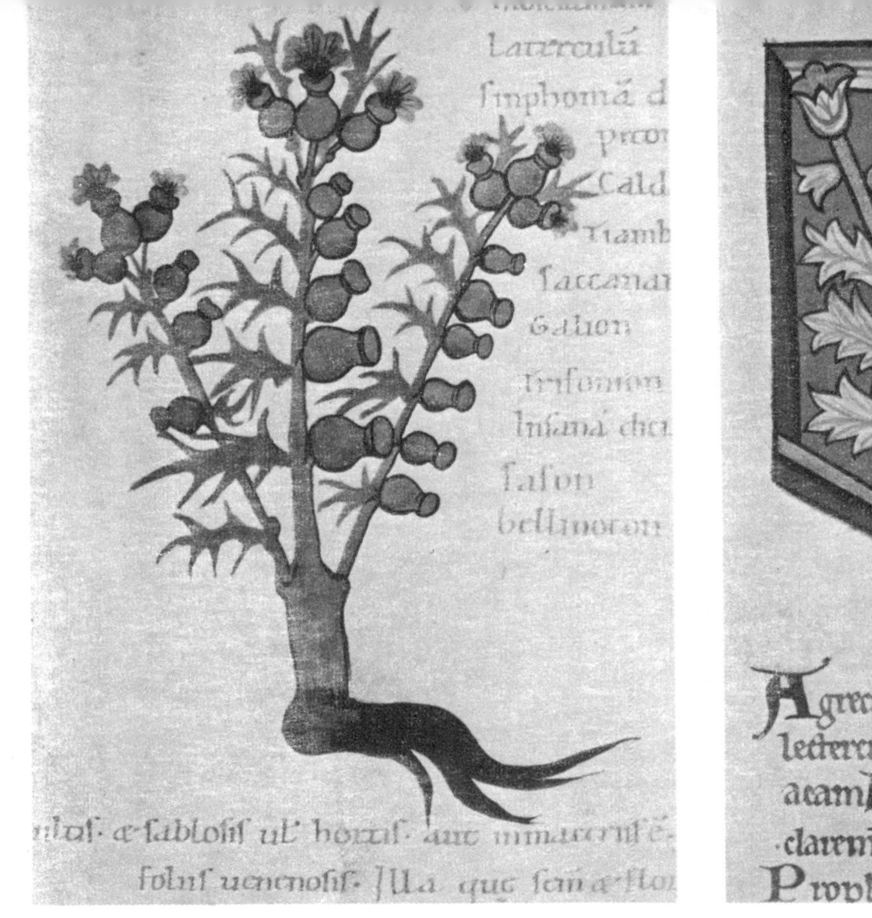

MS Bodley 130 fo. 37r Written at St. Albans 1120

MS Sloane 1975 fo. 15r Written in England Early XIIIth Cent.

it is placed externally. In this latter case it resembles a sting, and is hollow and spongy, so as to serve at one and the same time for tasting and for the taking up of nutriment. This is plainly to be seen in flies and bees.'[1] Thus in these animals, as he tells us, the proboscis protrudes from the mouth,[2] and he generalized correctly that in insects the mouth parts are either biting or sucking. Beyond this stage Western biology hardly advanced until the time of Mouffet (1553–1604), who prepared beautiful and accurate paintings of a large number of insects. Unfortunately his best figures of bees have greatly deteriorated. We reproduce, however, his drawing of the hornet (Plate LV).[3] Definite improvement of Mouffet's work was made by the members of the first Academy of the Lynx, a small body that came into existence in 1603 and fell to pieces with the death of their president, Federigo Cesi, in 1628. For the researches of the Academy the microscope was used, and as the instrument was hardly available until 1611 (vide Article No. 10 of this volume), the period during which their microscopic work was done is further narrowed.[4] The figure that we reproduce is the earliest drawing known for which the microscope was certainly used (Fig. 25). It was probably made about 1625.

Aristotle's description of the habits of bees is much more entertaining and complete than his account of the structure of these creatures. It is impossible, however, to give more than a fraction of what he has to say on the subject.

'There are nine varieties [of bee-like creatures] of which six are gregarious—the bee, the king-bee, the drone-bee, the annual wasp, . . . the anthrene [or hornet], and the tenthredo [or ground-wasp]; three are solitary—the smaller siren, of a dun colour, the larger siren, black and speckled, and the third, the largest of all, that is called the humble-bee.[5] . . . The little bees are more industrious than the big ones: their wings are battered; their colour is black, and they have a burnt-up aspect. Gaudy and showy bees, like gaudy and showy women, are good for nothing.'[6]

[1] *De partibus animalium*, ii. 17; 661^a 15.

[2] *De partibus animalium*, iv. 5; 678^b 15.

[3] The figures in the printed work of Mouffet are much inferior to the drawings in his MS. See his *Insectorum sive minimorum animalium Theatrum*, London, 1634, English translation in E. Topsell's *History of four-footed beasts and serpents*, London, 1658, and compare with Brit. Mus. MS. Sloane 4014 (Plate LV).

[4] Early microscopes and microscopic observations are described in two papers by Charles Singer, 'Notes on the Early History of Microscopy,' *Proc. Roy. Soc. of Med.*, London, 1914, vol. vii (sect. of Hist. of Med.), p. 247, and 'The Dawn of Microscopical Discovery,' *Journal Roy. Microscopical Soc.*, London, 1915, p. 317.

[5] *Historia animalium*, ix. 39; 623^b 7.

[6] *Historia animalium*, ix. 40; 627^a 12.

'Ants never go a-hunting, but gather up what is ready to hand; the spider makes nothing, and lays up no store, but simply goes a-hunting for its food; while the bee . . . does not go a-hunting, but constructs its food out of gathered material and stores it away, for honey is the bee's food. . . . They have also another food which is called bee-bread; this is scarcer than honey and has a sweet, fig-like taste; this they carry, as they do the wax, on their legs.

'Very remarkable diversity is observed in their methods of working and their general habits. When the hive has been delivered to them clean and empty, they build their waxen cells, bringing in the juice of all kinds of flowers and the "tears" or exuding sap of trees. . . . With this material they besmear the ground-work, to provide against attacks of other creatures; . . . they also with the same material narrow by side-building the entrances to the hive if they are too wide. . . . They begin building the combs downwards from the top of the hive, and go down and down building many combs connected together until they reach the bottom. The cells, both those for the honey and those also for the grubs, are double-doored; for two cells are ranged about a single base, one pointing one way and one the other, after the manner of a double goblet. The cells that lie at the commencement of the combs and are attached to the hives, to the extent of two or three concentric circular rows, are small and devoid of honey; the cells that are well filled with honey are most thoroughly luted with wax.[1] The ordinary bee is generated in the cells of the comb, but the ruler-bee in cells down below attached to the comb, suspended from it, apart from the rest, six or seven in number, and growing in a way quite different from the mode of growth of the ordinary brood.[2]

'The drones, as a rule, keep inside the hive; when they go out of doors, they soar up in the air in a stream, whirling round and round in a kind of gymnastic exercise: when this is over, they come inside the hive and feed to repletion ravenously.[3] . . . The kings are never themselves seen outside the hive except with a swarm in flight: during which time all the other bees cluster around them.[4] . . . These rulers have the abdomen or part below the waist half as large again, and they are called by some the "mothers" from an idea that they bear or generate the bees; and, as a proof of this theory of their motherhood, they declare that the brood of the drones appears even when there is no ruler-bee in the hive, but that the bees do not appear in his absence. Others, again, assert that these insects copulate, and the drones are male and the bees female.[5]

'Bees scramble up the stalks of flowers and rapidly gather the beeswax with their front legs; the front legs wipe it off on to the middle legs, and these pass it on to the hollow curves of the hind legs; when thus laden, they fly away home, and one may see plainly that their load is a heavy one. On each expedition

[1] *Historia animalium*, ix. 39; 623ª 13.
[2] *Historia animalium*, v. 22; 553ᵇ 2.
[3] *Historia animalium*, ix. 40; 624ª 22.
[4] *Historia animalium*, ix. 40; 625ᵇ 7.
[5] *Historia animalium*, v. 21; 553ª 27.

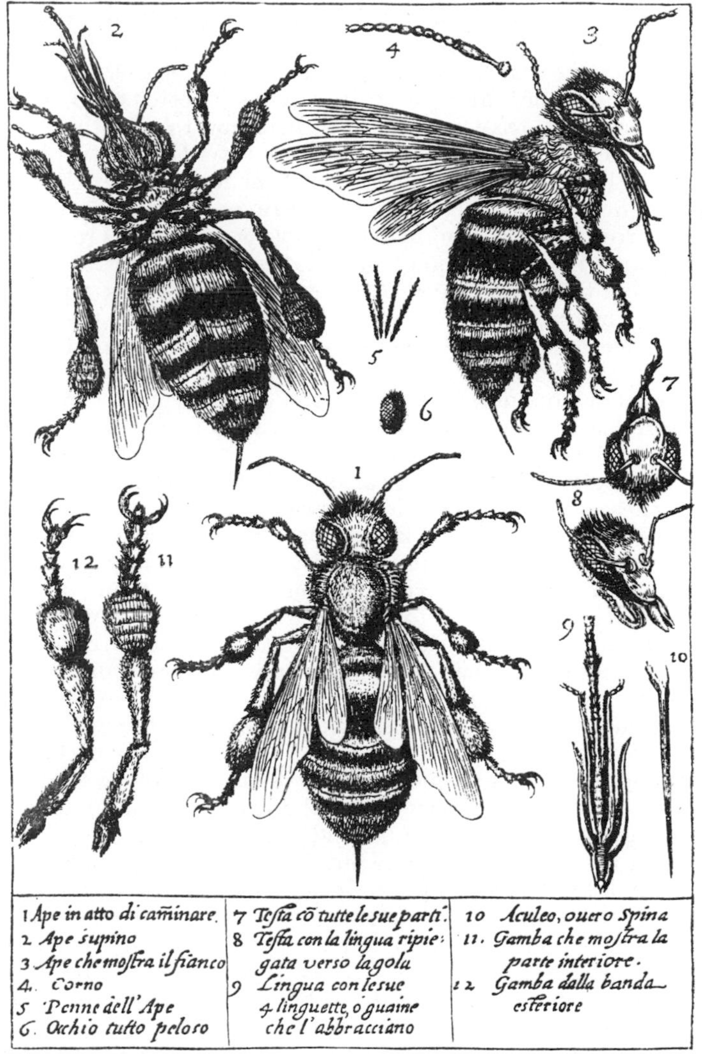

FIG. 25. ENLARGED FIGURES OF THE BEE

From Francesco Stelluti's *Persio tradotto*, Rome, 1630. The plate is the product of the Accademia dei Lincei. It was based on the work of Cesi (d. 1628), drawn by Fontana (probably about 1625), and contains observations by Johannes Faber and Francesco Stelluti. It is probably the earliest printed figure drawn with the aid of the microscope.

the bee does not fly from a flower of one kind to a flower of another, but flies from one violet, say, to another violet, and never meddles with another flower until it has got back to the hive; on reaching the hive they throw off their load, and each bee on his return is accompanied by three or four companions. One cannot well tell what is the substance they gather nor the exact process of their work.[1] . . . Bees feed on thyme: and the white thyme is better than the red.[2] . . . Bees seem to take a pleasure in listening to a rattling noise: and consequently men say that they can muster them into a hive by rattling with crockery or stones.[3] . . . Whenever the working-bees kill an enemy they try to do so out of doors.[4] Bees that die are removed from the hive, and in every way the creature is remarkable for its cleanly habits; in point of fact, they often fly away to a distance to void their excrement because it is malodorous; and, as has been said, they are annoyed by all bad smells and by the scent of perfumes, so much so that they sting people that use perfumes.[5]

'When the flight of a swarm is imminent, a monotonous and quite peculiar sound made by all the bees is heard for several days, and for two or three days in advance a few bees are seen flying round the hive; it has never as yet been ascertained, owing to the difficulty of the observation, whether or no the king is among these. When they have swarmed, they fly away and separate off to each of the kings; if a small swarm happens to settle near to a large one, it will shift to join this large one, and if the king whom they have abandoned follows them, they put him to death. So much for the quitting of the hive and the swarm-flight. Separate detachments of bees are told off for diverse operations; that is, some carry flower-produce, others carry water, others smooth and arrange the combs. A bee carries water when it is rearing grubs. No bee ever settles on the flesh of any creature, or ever eats animal food. . . . When the grubs are grown, the bees put food beside them and cover them with a coating of wax; and, as soon as the grub is strong enough, he of his own accord breaks the lid and comes out. . . .

'When the bee-masters take out the combs, they leave enough food behind for winter use; if it be sufficient in quantity, the occupants of the hive will survive; if it be insufficient, then, if the weather be rough, they die on the spot, but if it be fair, they fly away and desert the hive. Away from the hive they attack neither their own species nor any other creature, but in the close proximity of the hive they kill whatever they get hold of. Bees that sting die from their inability to extract the sting without at the same time extracting their intestines. True, they often recover, if the person stung takes the trouble to press the sting out; but once it loses its sting the bee must die.[6]

The diseases that chiefly attack prosperous hives are (*a*) first of

[1] *Historia animalium*, ix. 40; 624ᵃ 35. [2] *Historia animalium*, ix. 40; 626ᵃ 20.
[3] *Historia animalium*, ix. 40; 626ᵃ 15. [4] *Historia animalium*, ix. 40; 625ᵃ 31.
[5] *Historia animalium*, ix. 40; 626ᵃ 23. [6] *Historia animalium*, ix. 40; 625ᵇ 6.

fo. 21v Wagbræde=Waybroad=Meadow Plantain

fo. 23v Hennebelle=Henbane=*Hyoscyamus reticulatus*, a Mediterranean species

all the *clerus*—this consists in a growth of little worms on the floor, from which, as they develop, a kind of cobweb grows over the entire hive, and the combs decay.[1] . . . Bees brood over the combs and so mature them; if they fail to do so, the combs are said to go bad and to get covered with a sort of spider's web. . . . When the combs keep settling down, the bees restore the level surface, and put props underneath the combs to give themselves free passage-room; for if such free passage be lacking they cannot brood, and the cobwebs come on.[2] . . . (*b*) There is another insect resembling the moth, called by some the *pyraustes,* that flies about a lighted candle: this creature engenders a brood full of fine down. It is never stung by a bee, and can only be got out of a hive by fumigation. (*c*) A caterpillar also is engendered in hives, of a species nicknamed the *teredo,* or " borer ", with which creature the bee never interferes.[3] . . . (*d*) Another diseased condition is indicated in a lassitude on the part of the bees and in malodorousness of the hive.'[4]

The account of the diseases of hives described by Aristotle is paraphrased by Pliny.[5] These diseases can be identified with comparative confidence.

(*a*) The *Clerus* is held to be the *Trichodes apiarius,* a red and blue beetle which is known to destroy the larvae of the honey bee. When honeycomb is destroyed from this or any other cause a mould is liable to grow upon it. It is possible that Aristotle was confusing this growth with the cocoons described under the next heading.[6]

(*b*) The *moths that inhabit hives* are *Galleriidae,* a small group, many of which, *Galleria mellonella* especially, live in and consume bee hives. The spider's web described above is probably, in part, the cocoon of this species.

(*c*) The *caterpillars called Teredo* are the larvae of species of the *Galleriidae,* some of which have the habit of spinning their cocoons into a mass which is perhaps described by Aristotle as the 'cobweb' and also as the 'brood full of fine down'.

(*d*) This is some form of *foul brood,* a condition first described in modern times by Schirach in 1769,[7] and more minutely and carefully in the island of Syra by Della Rocca[8] towards the end of the eighteenth century. The actual nature of the disease—in the causation of which a number of spore-bearing organisms are

[1] *Historia animalium,* ix. 40; 626b 16. [2] *Historia animalium,* ix. 40; 625a 6.

[3] *Historia animalium,* viii. 27; 605b 12. [4] *Historia animalium,* ix. 40; 626b 18.

[5] Pliny, xi. 20.

[6] C. J. Sundevall, *Die Thierarten des Aristoteles,* Stockholm, 1863.

[7] A. G. Schirach, *Der Sächsische Bienenmeister, oder kurze Anweisung für den Landmann zur Bienenzucht, nebst beygefügtem Bienenkalender,* Löbau, 1769.

[8] L'abbé Della Rocca, *Traité complet sur les Abeilles avec une méthode nouvelle de les gouverner, telle qu'elle se pratique à Syra,* Paris, 2 vols., 1790.

involved—has been more fully cleared up only in quite modern times, beginning with the work of Ferdinand Cohn (1828–98).

Despite the figures of Mouffet prepared late in the sixteenth century,[1] and of the Academy of the Lynx early in the seventeenth,[2] the knowledge of the various forms of bee advanced but slowly. The male, female, and neuter were not adequately distinguished for what they were until the work of Goedart and de Mey in 1662.[3] The knowledge of the habits of these creatures was even more behindhand, and although some advance was made by Charles Butler (died 1647) in 1609,[4] yet Rusden in 1689[5] was still describing the queen bee as a *king* in the language of Aristotle. The whole subject had, however, by that time, been put on a firmer basis by the microscopical researches of Swammerdam (between 1662 and 1675)[6] and Malpighi[7] (about 1680).[8]

V. The General Course of Botanical Knowledge

(a) *Botany among the Greeks*

The history of Botany is more fortunate than that of its companion branch of Biology in that the material exists for telling it as an almost continuous story. Thus for the observation of plants our own age is linked with, and not separated from, that of the Greeks. Much of the material, especially for the period from the sixth to the twelfth century, is difficult of access, and this we have sought to present here in greater detail than its quality would otherwise demand.

Among uncivilized peoples the knowledge of plants is by no means confined to their culinary uses. Even the most primitive races have also a herb-lore which instructs them what plants to use and how to use them for the treatment of disease. Much effort has been expended in attempting to demonstrate the

[1] British Museum MS. Sloane 4014, written before 1589.

[2] Francesco Stelluti, *Persio tradotto*, Rome, 1630, illustrates this work. The more important *Apiarium* of Federigo Cesi, Rome, 1625, is excessively rare, and I believe there is no copy in this country.

[3] Jean Goedart, *Metamorphosis et historia naturalis insectorum cum commentario Iohannis de Mey*, Middelburg (1662–7).

[4] Charles Butler, *The Feminine Monarchie or a Treatise concerning Bees and the due ordering of Bees*, Oxford, 1609.

[5] Moses Rusden, *A Further Discovery of Bees*, London, 1689.

[6] Jan Swammerdam, *Bybel van de Natuur*, Leyden, 1737.

[7] Marcello Malpighi, *Opera Omnia*, London, 1686.

[8] A useful article on the knowledge of bees displayed by the Aristotelian writing is by J. Klek and L. Armbruster, 'Die Bienenkunde des Aristoteles und seiner Zeit', *Archiv für Bienenkunde*, i, Abt. 6, Freiburg im Breisgau, 1919.

therapeutic value of such drugs, but it would appear that folk herb-lore is no more rational than other departments of folk medicine and the majority of the ingredients of all pharmacopoeias, save perhaps the most modern, are in fact without appreciable physiological action and certainly are no cures for the conditions for which they are given.

The recipients of the herb-lore traditions among the Greeks were the *rhizotomists*. These, as a class, were ignorant men, corresponding in a measure to our herbalists, and they occupied themselves with gathering herbs, sometimes for the use of the physicians, sometimes in order that they might themselves usurp the functions of the physicians. They were superstitious and practised a complex ritual in obtaining their drugs. Fragments of this ritual have survived,[1] and we can detect in it ceremonies still closely followed by European peasantry engaged in similar practices. Rhizotomists were often of evil reputation, and Sophocles, who was contemporary with Hippocrates, wrote a play, now lost, entitled 'The Rhizotomists', the word having been used by him as almost equivalent to *poisonmongers*. The profession and to some extent the tradition of the rhizotomists extended from Greek into mediaeval and even modern times, and they and their work are not infrequently illustrated in the manuscripts (Fig. 26).

We learn little of the botanical knowledge of the sixth and fifth centuries from the works that make up the Hippocratic collection. About three hundred plants are mentioned as of value for medicinal purposes, and this implies considerable botanical knowledge, but we can glean almost nothing of the plants themselves from the Hippocratic writings. With the fourth century botany as a separate study comes into full view, and there is evidence that the subject may have been systematically studied at the Academy even before it was taken up by the Lyceum. From the latter we possess documents of first-class importance in the works of Theophrastus, the so-called 'father of botany', whose achievements are considered in greater detail below (p. 79).

But Theophrastus is the father of botany only in the sense that he is the first botanist whose writings have come down to us. There is nothing primitive about his work, nothing to suggest

[1] e. g. in Theophrastus, *Historia plantarum*, ix. 8. Other instances can be found in Pliny and have been collected by J. J. Mooney in his *Hosidius Geta's Tragedy 'Medea'*, Birmingham, 1919. Yet further material from both Greek and Latin sources may be culled from A. Abt, *Die Apologie des Apuleius von Madaura und die antike Zauberei*, Giessen, 1908.

that he is treading paths that none have trod before. On the contrary he is a thoroughly sophisticated writer, evidently the product of generations of thought and even of research. We know, moreover, that his work was preceded by a treatise on plants by Aristotle which is now lost, though fragments of it have perhaps survived.[1] But the writings of Theophrastus are specially valuable as presenting, after Hippocrates, the first substantially complete works of Greek observational science extant. Viewed in relation to the completeness and ordered sequence of the *History of Plants*, the biological works of Aristotle, in the form in which they have reached us, seem mere disarranged note-books. Like the works of Aristotle, those of Theophrastus are scientific compositions; that is to say, they are written, in large part at least, for the purpose of describing nature and not for the direct applicability of such knowledge to the needs and amenities of life. In this respect they stand almost alone among Greek botanical works, for the rest of Greek botanical history has to be pieced together mainly from poetical and pharmaceutical writings.

From the second century B. C. we have in Greek the *Alexipharmaca* and the *Theriaca* of Nicander, two works on poisons and their antidotes which deal with a number of plants from the point of view of their special topics and not from the scientific aspect. The *De re rustica* of Cato the Censor is contemporary with these works and is even more devoid of the Greek scientific spirit. For the first century B. C. we are better provided, for we have the literary works of Virgil and the agricultural treatises of Varro and Columella, supplemented by what can be gleaned from the valuable medical compendium of Celsus.

The first century of the Christian era is richer in important botanical works than any period from the fourth century B. C. until the sixteenth century. In Pliny's *Natural History* we have a collection of current views on the nature, origin, uses, and treatment of plants such as we might expect from a very intelligent, industrious, and honest member of the landed class who was devoid of critical or special scientific skill. More valuable and ranking second only to Theophrastus in importance is the treatise

[1] Put together by F. Wimmer, *Phytologiae Aristotelicae Fragmenta*, Breslau, 1838 (unfinished). The spurious *De Plantis* of Aristotle is attributed to Nicholas of Damascus by E. H. F. Meyer, *Nicolai Damasceni de plantis . . . Aristoteli vulgo adscripti ex Isaaci ben Honaici versione Arabica Latine vertit Alfredus*, Leipzig, 1841. It probably contains elements in the Aristotelian tradition. See note, p. 13.

FIG. 26. This drawing is traced from a facsimile on Plate XXII of the Atlas to Piero Giacosa's *Magistri Salernitani,* Turin, 1901. The MS. from which Giacosa's facsimile was taken has since perished. It bore the arms of Savoy and was work of the fifteenth century. The figure represents herbalists at work on a mountain, the slopes of which are covered with a variety of plants. One of the herbalists is climbing an oak tree to secure the mistletoe growing on it. The other is digging up plants and placing them in a receptacle.

on *Materia Medica* of Dioscorides. It consists of a series of careful descriptions of plants, and of their uses in medicine, arranged, it is true, almost without reference to the nature of the plants themselves, but quite invaluable for its terse and striking descriptions which often include habits and habitats. Its history has shown it to be the most influential botanical treatise ever penned.

After Dioscorides Greek botany declines. The monumental intellect of Galen in the second century of the Christian era hardly applied itself to plants, and his pharmacopoeia, copious though it is, gives but a scant glimpse of contemporary botany. Such Greek botanical writers as followed were little but copiers of Galen and Dioscorides, and the history of Greek botany is from now on a story of the continued disintegration of knowledge.

But, retracing our steps, there yet remains for consideration a writer of the first century B. C. the full importance of whom has only recently been realized. There is evidence that Crateuas, the most intelligent and instructed of the rhizotomists who lived during this century, occupied himself not only in collecting but also in drawing plants. He is thus the father of the very important botanical department of plant illustration.[1] Crateuas was the attendant of Mithridates VI Eupator, but was also an author, and though his works have perished, fragments have survived in a Vienna codex.[2] This manuscript, to which the name of Julia Anicia is attached, was prepared in Constantinople a little before the year 512 as a wedding gift to that lady, the daughter of Flavius Anicius Olybrius, Emperor of the West, and it is not improbable that many of the figures in it are copies of Crateuas's own drawings or of pictures prepared under his supervision in the first century B. C. He is perhaps represented at work in one of the more damaged miniatures of this great manuscript (Fig. 27), and we have what are perhaps portraits of Dioscorides himself in this and other miniatures of the same volume (Fig. 28).[3]

[1] Pliny, *Hist. Nat.* xxv. 4.

[2] Max Wellmann, 'Krateuas', *Abh. der kgl. Gesellschaft der Wissenschaften zu Göttingen, philologisch-historische Abt.*, Neue Folge, Bd. ii, No. 1, Berlin, 1897; and M. Wellmann, 'Das älteste Kräuterbuch der Griechen', in *Festgabe für Franz Susemihl*, Leipzig, 1898.

[3] These miniatures are represented in their damaged state in Rendel Harris, *The Ascent of Olympus*, Manchester, 1917. The entire MS. has been reproduced in two luxurious but excessively inconvenient elephant folios by J. de Karabacek, Leyden, 1906. The original work has been removed from Vienna to St. Mark's Library at Venice.

Thus even before the Julia Anicia MS., botanical illustration had had a long history, and surviving fragments of a herbal on papyrus of the second century [1] show that a strong tendency to diagrammatize the forms of plants had already set in among the

Fig. 27. Restored from the Julia Anicia MS., fo. 5 v. about A.D. 512. Dioscorides writes while Intelligence ('Επίνοια) holds the mandrake for the artist, Crateuas (?), to copy.

Greeks. But the tradition represented by the Julia Anicia MS. was that which for many centuries largely controlled the whole art of plant illustration in Greek-speaking lands. The history of this manuscript is thus itself the history of an important line of botanical knowledge during the Middle Ages. The Julia Anicia MS. is not exactly a text of Dioscorides but is a composite docu-

[1] J. de M. Johnson, 'A botanical Papyrus with Illustrations' in the *Archiv f. Gesch. d. Naturwissenschaften und der Technik*, iv, p. 403, Leipzig, 1913.

ment of which the nature and origin is represented in the diagram opposite (Fig. 29).

Even in the seventh century the sources of the Julia Anicia figures were still being imitated with considerable accuracy and

Fig. 28. Restored from the Julia Anicia MS., fo. 4 v. about A.D. 512. Discovery (Εὕρεσις) presents a mandrake to the physician Dioscorides. The mandrake is still tethered to the hound whose life is sacrificed to obtain it.

skill,[1] but as the centuries went by and the figures were copied and recopied without reference to the original they moved further and further from the facts until at last they remained as mere diagrams. By this time the gift of naturalistic representation had left the Greeks, but traces of some power to imitate classical models of

[1] As in the so-called Codex Neapolitanus of the seventh century at St. Mark's, Venice. A facsimile of a page of this work has been reproduced by the New Palaeographical Society, ii, Plate 45. Copies of the Julia Anicia itself, of which several exist, are much later, e. g. that described by O. Penzig in his *Contribuzione alla storia della Botanica*, Genoa, 1904. There is a similar copy in the University Library at Cambridge. Press Mark Ee. 5. 7 (Browne 1385).

plant paintings may still be traced in a fragment of a herbal of the eighth century discovered in the cover of an Armenian book [1] (Fig. 30), in the famous eighth-century Vatican MS. of Cosmas Indicopleustes,[2] in a very peculiar little manuscript of Nicander of the ninth century [3] (Plate IX), in the Paris and Cheltenham

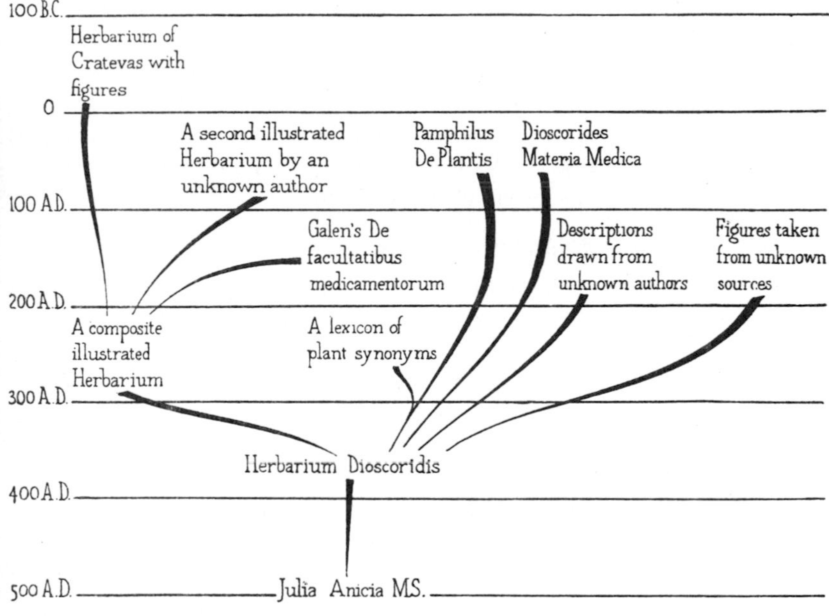

FIG. 29. THE GENEALOGY OF THE EARLIEST MANUSCRIPT OF DIOSCORIDES

Dioscorides MSS. of the ninth and tenth centuries respectively (Plates X and XVIII), in the Smyrna Physiologus of the eleventh century,[4] and in many other works of Byzantine origin. For the rest, however, the story of botany is mainly in the West and not in the East.[5]

[1] Discovered by F. C. Conybeare. Facsimile in MS. Bodley E. 19 (31528).

[2] Vat. Gr. 699. Reproduced in facsimile by C. Stornajolo, Rome, 1909.

[3] Bibl. nat., Supplément grec, 247.

[4] Josef Strzygowski, 'Der Bilderkreis des griechischen Physiologus', in the *Byzantinisches Archiv*, Heft 2; Leipzig, 1899.

[5] We do not here touch on the very interesting group of Dioscorides herbals of Persian origin, which are known from the twelfth century onwards. Chinese herbals are also known.

A few words may be devoted to the Greek text and manuscripts of Dioscorides. Modification and interpolation set in early and on an extensive scale. This process was encouraged by the alphabetical rearrangement of the paragraphs according to the name of the drug the properties of which are described. Such a rearrangement was made in the fourth, or perhaps in the third century. The oldest codices are of this type, and are thus of comparatively little *textual* value, though the illustrations that adorn them have the highest interest, alike for the history of Botany and of Art. On the other hand the non-alphabetical manuscripts are less ancient and, on the whole, less beautifully illustrated, but some of them are of high textual importance.

Alphabetical Greek Codices of Dioscorides

The alphabetical Greek codices fall naturally into three classes:

I. This class contains the two most ancient manuscripts. Both are elaborately illustrated, text and figures being derived from one common archetype of the fourth century. The text is alphabetically arranged from beginning to end. It draws on the *Materia Medica* alone among the works of Dioscorides, but also contains certain works of other authors (as shown in Fig. 29). The figures illustrating the text of these manuscripts preserve the tradition of Crateuas. This class includes only:

(*a*) The *Julia Anicia*, written in capitals before 512, and known as the *Constantinopolitanus*. This manuscript was formerly at the Vienna Hofbibliothek, where it was numbered Med. Gr. I. It is now in St. Mark's Library at Venice (Plates VI and VII). The MS. is accessible in a beautiful photographic facsimile.

(*b*) The so-called *Neopolitanus*, written in half uncials in the seventh century. This manuscript was also formerly at the Vienna Hofbibliothek, where it was numbered Suppl. Gr. 28. It is also now in St. Mark's Library at Venice.[1]

II. This class contains a text borrowed not only from the *Materia Medica*, but also from the other works of Dioscorides. The text is divided, according to the subject, into five sections or books, within each of which the order is alphabetical. These sections treat respectively of (1) plants, (2) animals, (3) oils, (4) trees, (5) wines and stones. In (*c*) and (*d*) the text is com-

[1] There is an excellent but antiquated article on the *Constantinopolitanus* by L. Choulant 'Ueber die HSS. des Dioscorides mit Abbildungen', *Archiv für die zeichnenden Künste*, I, p. 56, Leipzig, 1855. In R. Dodoens' posthumous *Stirpium Historiae Pemptades sex*, Antwerp, 1616, figures on pp. 123, 126, 149, 288, 372, 377, 439, and 572 marked *e cod. Caesareo* are taken from the *Constantinopolitanus*.

bined with certain spurious works of Nicander. The figures, like those of class I, preserve the tradition of Crateuas with certain well marked differences. There is evidence that they spring from an archetype of not later than the fourth century.

(*c*) Cheltenham, Phillipps, 21975, tenth cent., now in the Pierpont Morgan Library, New York (Plate XVIII).

(*d*) Mount Athos, Laura Monastery, twelfth cent.

(*e*) Venice, St. Mark's Library, No. 92. Cotton paper, thirteenth cent.

(*f*) Escurial Σ.T. 17. Paper, fifteenth cent.

III. This class contains only the *Materia Medica* of Dioscorides

FIG. 30. 'Chamaepitys', from a fragment of an eighth-century Greek Herbarium [Bodl. E. 19]. The plant is probably *Ajuga reptans* Linn., the Creeping Bugle, or some allied species.

and has an alphabetical text not divided into books. No manuscript of this class is earlier than the fourteenth century.

(*g*) Milan, Ambros. A95 Sup., fourteenth cent.

(*h*) Rome, Vatican Urbinas 66, fifteenth cent.

(*j*) Venice, St. Mark's 272, fifteenth cent.

(*k*) Venice, St. Mark's 597, fifteenth cent.

and certain sixteenth-century manuscripts at Berlin, Paris, and the Escurial.

Non-Alphabetical Greek Codices of Dioscorides

The subject-matter of the remaining manuscripts of Dioscorides is not in alphabetical order. These manuscripts are very numerous and we need only mention the more interesting. They fall into two classes, of which the first is the more important for fixing the text as it contains only the genuine work of Dioscorides. In the second class this text is mingled with other material.

I. Manuscripts containing only works of Dioscorides. Among the most important are the following:

(*l*) Paris Bibl. nat. MS. gr. 2179, ninth century (Plate X).

This is the most valuable of all the manuscripts for fixing the text. It is beautifully illustrated and it is interesting to observe that the figures of the mandrake, that are anthropomorphic in all the other manuscripts, are not here given any human attributes. Thus these figures may contain elements in an even earlier tradition than the *Constantinopolitanus* and *Neopolitanus* and may take us back beyond the fourth century and perhaps beyond Crateuas. The manuscript is badly preserved and partially illegible, but is supplemented by

(*m*) Venice, St. Mark's 273, twelfth century, and

(*n*) Florence, Laurentian Plut. 74, 17, twelfth century.

(*m*) and (*n*) are parts of one manuscript copied from (*l*) or from a closely similar manuscript.

(*o*) Escurial III, R 3, imperfect, eleventh cent.

(*p*) Florence, Laurentian Plut. 74, 23. The only perfect manuscript of this class, fourteenth cent.

(*q*) Rome, Vatican Pal. 77. The older leaves of this manuscript represent the same tradition as (*o*), fourteenth cent.

II. The Greek manuscripts of Dioscorides that contain other works besides his are very numerous, but of little importance for the text. A few, however, have interesting figures. Among these are:

(*r*) Paris, Bibl. nat., MS. gr. 2180, fifteenth cent.

(*s*) Paris, Bibl. nat., MS. gr. 2182, written 1481.

The relation of the tradition of the figures to that of the text has not been investigated. It is a subject for a special research, and would be of great importance for the history of Art. From a first examination we may say that the line of descent of the figures may be traced as far as the thirteenth century along lines parallel to those of the text.[1]

To sum up the history of Greek botany, we may say that it probably arose with other sciences in the sixth century B. C., but that it does not come into clear view until the fourth century B. C. It was then firmly established as a scientific discipline, botanic gardens had been established, much research had been made, lectures were being given, and the work of one important writer has survived. By the first century B. C. scientific botanical

[1] The classification of the manuscripts given above is largely taken from M. Wellmann's masterly article on 'Dioskurides' in Pauly-Wissowa's *Real-Encyclopädie der klassischen Altertumswissenschaft*, vol. v, Stuttgart, 1905. Much information may also be obtained from H. Diels' 'Die Handschriften der antiken Ärzte, II. Teil. Die übrigen griechischen Ärzte ausser Hippokrates und Galenus', in the *Abhandlungen der königl. preuss. Akademie der Wissenschaften*, Berlin, 1906.

principles had been absorbed into agriculture and medicine, and their application is happily illustrated by works which survive from the two centuries which follow. Then begins a process of decay, but as late as the seventh century A. D. very careful and accurately illustrated herbaria were still being prepared. From then on only decadent imitations are known. Some of these, however, are remarkably accurate, and a study of them aids us in forming an idea of the botanical attainments of the earlier and the limits in the botanical knowledge of the later centuries.

(b) *Botany in the West from the sixth to the twelfth century (The Dark Ages)*

During the Dark Ages, and even until the twelfth or thirteenth century, there were no effective additions to botanical knowledge. Up to that time we may say that such codified botanical knowledge as existed in the West was contained in a very small group of Latin works of which countless manuscripts existed. The versions of these works, their illustrations and their general appearance, present remarkable similarities in spite of their wide distribution. Three of them especially were often compounded in various proportions to form a single text. These three works are:

(i) The *Materia Medica* of Dioscorides.
(ii) The *Herbarium* of 'Apuleius'.
(iii) The pseudo-Dioscoridean *De Herbis Femininis.*

(i) The work of Dioscorides had already been translated into Latin in the time of Cassiodorus (490–585), who recommended the study of illustrated copies to such of his monks as were unable to read Greek.[1] The earliest surviving manuscripts of the Latin translation of Dioscorides are, however, of the ninth century.[2] Several of the manuscripts of the Latin Dioscorides contain remarkable illustrations.[3] The text has been printed in modern times.[4]

(ii) The *Herbarium* bearing falsely the name of Apuleius

[1] Cassiodorus, *Institutio divinarum litterarum*, c. 31 'Si vobis non fuerit graecarum litterarum nota facundia imprimis habetis herbarium Dioscoridis qui herbas agrorum mirabili proprietate disseruit atque depinxit.'

[2] The two oldest manuscripts of the Latin Dioscorides are at the Bibl. Nat., where they are numbered 9332 and 12995.

[3] Notably the Munich MS. Monac 377 of the tenth century in the characteristic Beneventan script. See H. Stadler, *Der lateinische Dioscorides der Münchener Hof- und Staatsbibliothek*, Janus iv, p. 548, Leyden, 1899.

[4] In Vollmöller's *Romanische Forschungen*, i, x, xi, and xii, Erlangen. Cf. also V. Rose in *Hermes*, viii, p. 38.

Platonicus was compiled in Latin perhaps during the fifth century. It describes very briefly the character and habitat, and at greater length the virtues of herbs, and it is possible that the name of Apuleius was attached to it because he was the author of a *Liber floridarum*, which, however, is of a totally different character. The *Herbarium* of Apuleius is in part taken from the real Dioscorides and in part from the work known as the *Medicina Plinii*. At an early date the text became associated with a lexicon of synonyms similar to that found in the alphabetical Greek manuscripts of Dioscorides. Some have seen in the character of these synonyms evidence of an African origin of this pseudo-Apuleian document. The earliest manuscript is of the sixth century,[1] and already exhibits these peculiar synonyms. There are a number of other manuscripts of the pseudo-Apuleius which contain a remarkably persistent group of illustrations becoming progressively more conventionalized.[2] The pseudo-Apuleian *Herbarium* has been printed a number of times, usually under the title *Apuleii Platonici de virtutibus herbarum*. The earliest edition was printed at Rome before 1484 by Philip de Lignamine. The printed editions of this work, like the manuscripts, vary greatly in their contents.

(iii) The *De herbis femininis* of the pseudo-Dioscorides was also written or rather compiled in Latin. It is only about half the length of the pseudo-Apuleius, and consists of a description of seventy-one herbs and their properties without the synonyms. It is drawn from Dioscorides, pseudo-Apuleius and Pliny, relying perhaps on a separate Latin version of the first of these. The earliest manuscript is of the ninth century,[3] but the work was probably put together no later than the sixth century, and perhaps in Italy during the domination of the Goths (493–555).[4] Manuscripts of this work in its pure form are not very common.[5] It has also been printed in modern times.[6]

[1] Leyden, Voss. lat. Q 9.

[2] Among the most ancient of these are the *Codex Hertensis* of the ninth century, described by Sudhoff in the *Arch. für Gesch. der Med.*, x, p. 265, Leipzig, 1916, and the *Codex Fuldensis* of Sudhoff of the tenth century at Cassel, where it bears the pressmark *Phys. et hist nat.*, *fol.* 10.

[3] Rome, Barberini, ix. 29.

[4] M. Wellmann, 'Krateuas', in *Abh. d. kgl. Gesellschaft d. Wissenschaften zu Göttingen, Philol.-hist. Klasse*, Berlin, 1897.

[5] Ten manuscripts of this work (one of which has been destroyed) are recorded by H. Diels, *Die Handschriften der antiken Ärzte*, Berlin, 1906.

[6] H. F. Kästner, 'Pseudo-Dioscoridis de Herbis feminis', in *Hermes*, xxxi, p. 578, Berlin, 1896.

More interesting than these botanical works themselves are the paintings of plants with which a number of them are provided. These were copied from manuscript to manuscript, century after century, without direct reference to the objects intended to be represented, and thus arose definite traditions which can be traced in the representation of plants from age to age. The traditions of botanical illustration fall along somewhat different lines to those of the general history of manuscript illumination, and the figures of plants as they are encountered in manuscripts therefore merit a special study which has hardly yet been accorded them. In what follows we shall deal mainly with English developments which we have had most opportunity to study. There is evidence that the evolution of the herbal in other countries took a somewhat different course.

Our earliest illuminated Latin herbal is a sixth-century manuscript (Plates VIII, XX, and XXI) probably written in southern France. Most of its figures are already stylized and far removed from nature drawing. This model remained little altered until the tenth century, and herbals of that period from the Rhineland and from Italy still preserve the same traditional pictures.[1] By then, however, another treatment of plants becomes traceable in northern France. The figures in this new style have become symmetrical and heavy and some of them approach the manner which we describe below as Romanesque (Plate XXV).

An examination of a series of manuscript herbals of between the eleventh and thirteenth centuries reveals the fact that the figures fall in the main into two divisions, one of which we may call *Naturalistic* and the other *Romanesque.* Both these divisions, like other manifestations of Western culture, were doubtless derived from Italy, but it would appear that they reached the West by different routes, or at least with different cultural waves.

The naturalistic tradition of these Western herbals of the eleventh, twelfth, and thirteenth centuries probably took its rise in the south of Italy, perhaps in that region where developed the peculiar script to which the name of Beneventan has now become

[1] To this group belong Hertensis 192 written in Westphalia (?) in the ninth century (cf. Sudhoff, *Archiv für Gesch. d. Med.* x, p. 265, Leipzig, 1916), the Laurentian 73, 41 written in southern Italy in the ninth century, and Cassel Phys. Fol. 10 written at Fulda in tenth century. Munich Lat. 337, Italian work of the tenth century, belongs however to a quite different tradition.

attached. It drew on Greek models which were already beginning to undergo the change known as ' Byzantinism '. In line with this view of the southern origin of certain herbal illustrations is the fact that a medical work in the Anglo-Saxon language circulating in England in the twelfth century[1] was actually a translation of a known Salernitan document, and English leech-craft of the eleventh and twelfth centuries was profoundly influenced by Salerno.[2] It is of interest for our purpose to analyse the medical manuscripts in the Beneventan script that have actually survived. Some 600 Beneventan manuscripts (including fragments) are known, and nearly half of these are still lodged at the Benedictine monastery of Montecassino. The manuscripts vary in date from the early ninth to the mid-thirteenth century and thus represent the period during which Salerno was an important medical school. Of the Beneventan MSS., 14 (13) or about 2 per cent. are medical.[3] Of the 14 surviving early southern medical manuscripts we have adequate details of 10, and all of these are of works which were in fact influential in establishing the main currents of Dark Age medicine and nearly all of them contain Herbaria. The most important perhaps was the recently destroyed Turin codex of the eleventh century. The figures in this manuscript, a few of which have been fortunately preserved in photographs, were of a definitely Byzantine cast and must have been derived from an earlier Greek original, and the influence of this manuscript, or one very closely resembling it, can be traced in some English herbals.[4] (Fig. 31.)

The first herbals brought to this country had however probably been prepared in northern France (cf. Betony in central figure of Plate XXV with the same plant of Plate XVII). But the models from which the artist worked must have betrayed the fullest evidence of an origin from a region further south, for traces of a Mediterranean flora may still be clearly discerned in the English copies (cf. Henbane in Plate XVI). During the tenth, eleventh, and twelfth centuries there developed a very characteristic English school of draftsmanship and this new manner was not without its influence

[1] Max Löweneck, 'Περὶ διδάξεων, eine Sammlung von Rezepten in englischer Sprache', in the *Erlanger Beiträge zur englischen Philologie*, ix, Erlangen, 1896.

[2] Charles Singer, 'A Review of the Medical Literature of the Dark Ages with a new text of about 1110', *Proceedings of the Royal Society of Medicine* (*section of the History of Medicine*), x. 102, London, 1917.

[3] The known Beneventan MSS. are recorded by E. A. Lowe in *The Beneventan Script, a study of the South Italian minuscule*, Oxford, 1914.

[4] Especially in the MS. Harley 5294.

PLATE XVII. Paris, Bib. nat. MS. lat. 6862 Apuleius Xth cent. French work

fo. 20 v *Vetonica* = Betony

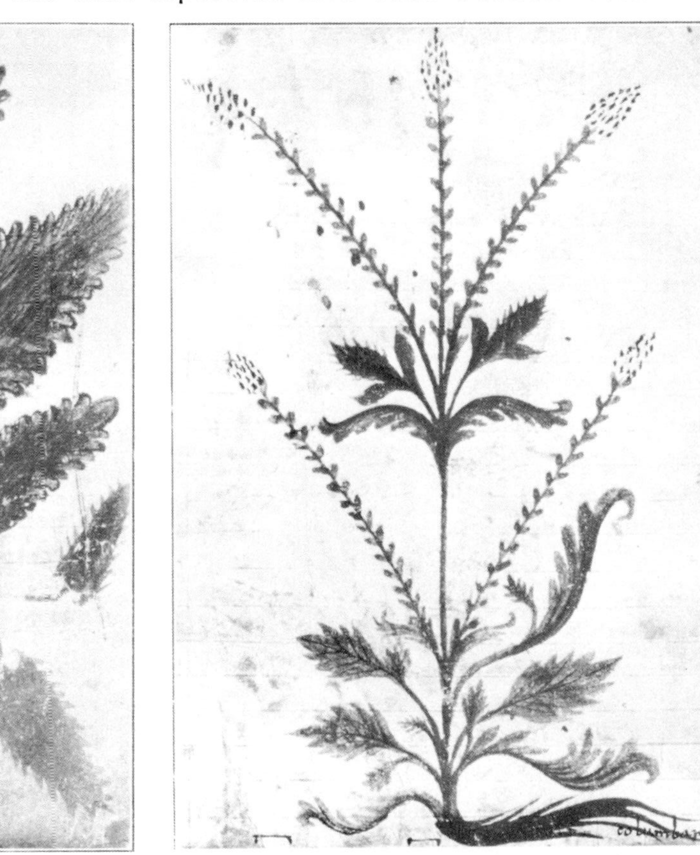

fo. 26 v *Verminacia* (*columbaris*) = *Verbena officinalis*

on the paintings that adorned the herbals (cf. figures of Orobus and Teucrium on Plate v and of Ivy on Plate xxiii).

But this native style was, in turn, largely replaced by the Romanesque. The change in the figures was accompanied by a well-marked alteration in the character of the script. These changes took place in England a generation or two after the

Fig. 31. OUTLINES OF THE MANDRAKE AND ITS GATHERERS

Traced from a twelfth-century English Herbarium in the British Museum and a contemporary Italian Herbarium to illustrate the close similarity.

Conquest, which introduced a revolution into art and letters at least as great as that to which it subjected the social system of this country.

On the Continent the Romanesque method of plant illustration had established itself at an earlier date. This style originated or at least emanated from northern and north-eastern France. Its ultimate origin appears to be a debased style of Roman art proceeding from northern Italy and perhaps modified in Carolingian times by English influence. Its characteristics, so far as plants are concerned, are a replacement of free drawing by symmetrical design (Plate v, central figure), while an increased formalism is ultimately obtained by the enclosure of the picture in a rectilinear frame (Plate xxv, lateral figures). As the style develops, a background at first simple and afterwards more elaborate is provided. The human body is often drawn half naked or even completely nude, and the drapery arranged along characteristic lines (Plate xxiv).

Comparing the styles we may characterize the Italian as more free and the Romanesque as more formal. Both styles are trace-

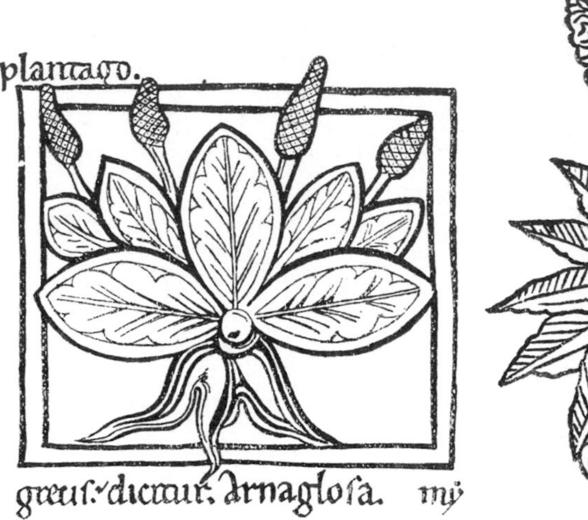

Fig. 32. THE PLANTAIN
From a thirteenth-century Herbarium in Romanesque style prepared in England. Sloane, 1975, fo. 12 v.

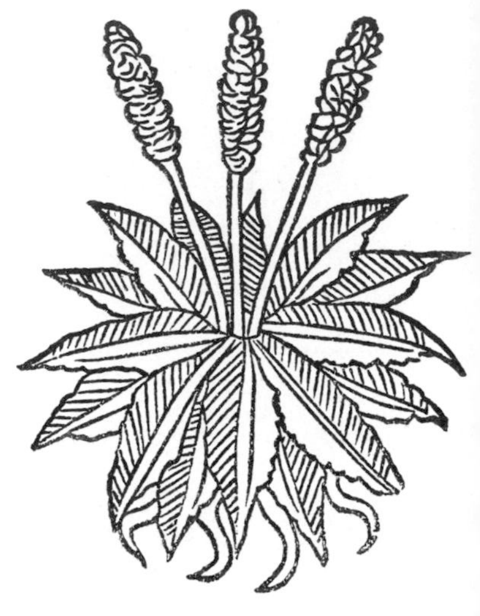

Fig. 33. PLANTAIN
From the *Herbarius latinus* printed at Mainz in 1484.

able even in the later manuscripts and are carried over into the early printed herbals. Some manuscripts even contain specimens of both styles, and in all herbals the centaur holding a plant is a favourite device. It is noteworthy that in none of the English herbals is there any trace of the influence of the characteristic Celtic school of illumination, while the peculiar and well-known Anglo-Saxon style of draftmanship is far less obtrusive in the books of plant pictures than in ecclesiastical documents. The influences that produced the herbals were, in fact, mainly of a lay character, but nevertheless a certain characteristic English manner of plant representation can be detected in them. This English manner had completely disappeared by the beginning of the thirteenth century (cf. Sloane 1975, Plates xv and xxv).

(c) *Botany in the West from the twelfth to the fifteenth century (The Middle Ages)*

After the long depression of the Dark Ages the study of plants began to take an upward turn at a much earlier period than the sister study. The revival begins with literary rather than scientific effort, for neither the poem called that of Macer Floridus composed by Ódo of Meune (died 1161), nor the account of plants in the later

FIG. 34. WALL-FLOWER WITH DODDER Cf. Orobus on Plate v. FIG. 35. YELLOW FLAG
From the German version of the *Hortus Sanitatis*, printed at Mainz in 1485.

Subtilitatum diversarumque creaturarum liber (early thirteenth century), wrongly attributed to Hildegard of Bingen, contains evidence of new or even direct observation. More scientific in spirit is the somewhat later work *De vegetabilibus* of Albertus Magnus (1206?–80). This is primarily a compilation based on the peripatetic work on plants by Nicholas of Damascus (first century B.C.), of which only fragments remain. Albert's treatise is essentially a learned product and its author is hampered by his desire to fit the nature of plants into an ill-thought-out system, while his work is marred by the scholastic doctrine that '*philosophia* is concerned with generalities, not particulars'. It is a phrase which itself explains the failure of scholasticism to erect an enduring scientific structure. But in spite of such a handicap the work of this extraordinary man contains evidence of a certain amount of careful first hand observation. His botanical treatise gives a not unpleasing impression of the common sense and open-eyedness of the great mediaeval scholar who has a place among the grandfathers, if not among the fathers, of modern botany. We translate one of his chapters as a good specimen of mediaeval descriptive botany.

'Of the Oak and its qualities.

'The oak is a very large and tall tree with broad branches; it has many roots which go deep down and when old has a very rough bark, but the young tree is smooth. It has great breadth and size in its branches; when it is thriving its leaves are thickly set, broad, and hard. The leaves are wholly surrounded by triangles, the bases of which are upon the leaf and the angle at the exterior. Many leaves are attached to it, but they fall off. When they are dried up, however, they still sometimes cling in large numbers. Its wood increasing by straight layers is composed of straight pores, can be split to a line and can be hewn and retains well the shapes of large incisions, but in this box-wood surpasses it. The outer zone is of a pale colour but towards the centre it shades into a reddish tint. If it is put in water, at first it swims, but at last sinks owing to its earthy nature and then grows black. Its fruit is called acorn (*glans*); it is not joined by a stalk of its own to the branch on which it grows, but small cups spring out from the branches and in these the acorn is formed. The acorn also has outside a hard shell in which it is enclosed; this resembles well-polished wood, shaped like a column except that its apex is not a plane superficies but a hemisphere and has a point in the middle to represent the pole. Below is the base of the acorn through which it draws nourishment from the little cup; that also is not simply a plane superficies but is somewhat flattened at the pole; this depression is formed by the weight of the acorn, for if it were an exact hemisphere there would be no place for the reception of nourishment except the point, and through that it could not receive enough. The acorn within the sheath is surrounded by rind, not hard but soft, which is formed from the excretion of the acorn; the acorn is twisted round itself and divided down the middle as a column might be cut lengthways by a plane surface. At the apex, however, is the life germ and what is beneath is of a floury substance and is to be regarded as matter and food for the germ. The little cup in which the acorn itself sits is concave and evenly formed as if it were smoothed on a lathe inside. The bottom is somewhat levelled since from it the acorn draws its nourishment; on the outside it is rough

PLATE XVIII. From the Phillipps Dioscorides MS. Xth c.

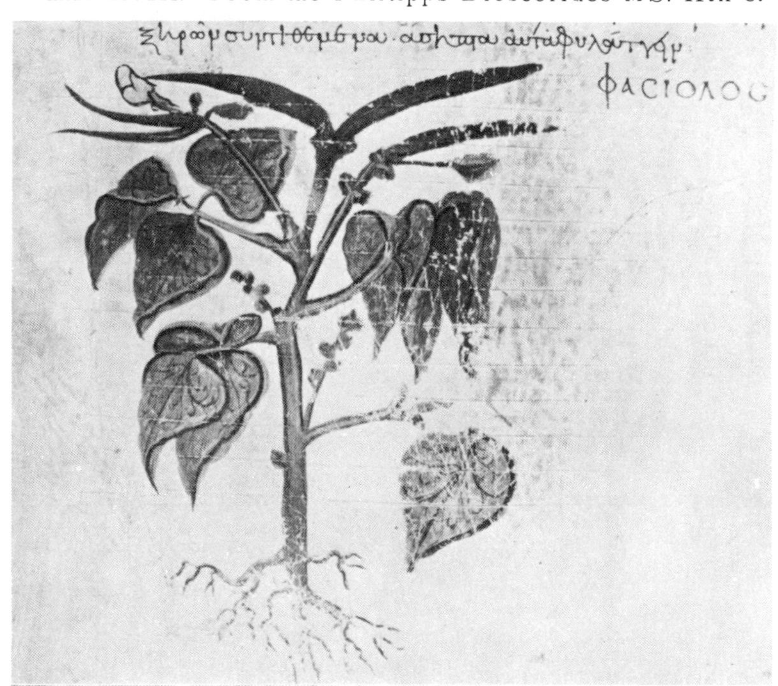

fo. 186 v ΦΑCΙΟΛΟC SEEDLING BEAN. Cf. PLATE VII

fo. 34 v ΓΟΓΓΥΛΗ TURNIP

Now in the Pierpont Morgan Library--New York

HYOSCYAMUS

From MS. Harley 1585 fo. 19 v

English, about 1200

PLANTAIN

From British Museum MS. Add. 17063 fo. 4 r

Italian, of about 1450
Compare Text Fig. 36

DRACUNCULUS

From MS. Harley 1585 fo. 22 v

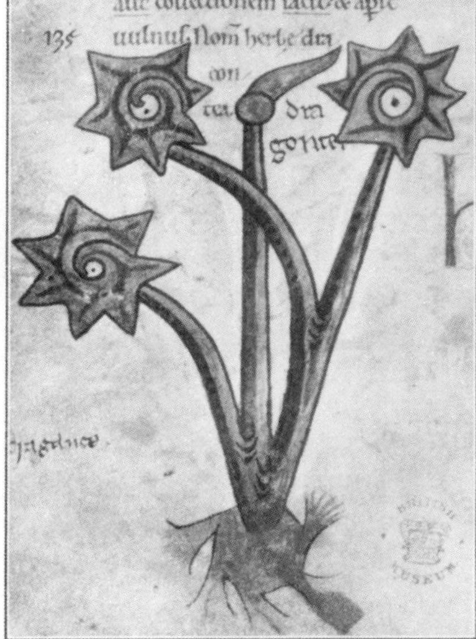

English, about 1200

PLATE XIX

because of its own earthy nature which is expelled from the material of the acorn. It is not joined by any sort of stem to the branch but sits immediately upon it. This is to prevent the acorn from being too far distant from the branch, because if it had to draw its food a long way, it would become hard and cold and would do no good, especially as the juice of this tree is very earthy.

On the leaves of the oak often grow certain round ball-like objects called galls, which after remaining some time on the tree produce within themselves a small worm bred by the corruption of the leaf. If the worm exactly reaches the midst of the gall apple weather prophets foretell that the coming winter will be harder: but if it is near the edge of the gall they foretell that the winter will be mild. . . . Galls have a juice pure in itself as long as the apple is green and moist; but when it is rubbed against a flat clean piece of iron it immediately is transformed into a kind of very black encaustic.

Fig. 36. THE PLANTAIN
From Lignamine's Apuleius, Rome, 1483, printed probably from metal blocks.

'The leaves of the acorn are extraordinarily astringent but less dry. The acorn resembles the chestnut in that both are astersive and cause flatulence in the lower bowel; both strengthen the limbs and both are good food especially for pigs. Galen says that the acorn as well as the chestnut is good for nourishment and deserves more praise than all the fruits of growing trees: but the chestnut is more nutritious than the acorn on account of its greater sweetness. But the food they afford lacks the praise of men because it is too astringent, but if chestnuts are mixed with sugar they make good food; taken otherwise they will be of slow digestion but the oak is even slower. The astringency in the inner bark of the acorn is greater than in the acorn itself. The leaves of the oak ground to powder and laid upon wounds make the flesh unite. The acorn is of as much value as the chestnut as a remedy against poison.

'The juice of galls darkens the hair: powder of them gets rid of superfluous flesh and warts. Galls are helpful also if placed on decayed spots of the teeth and in many other medical operations which must be determined in (books on) simple medicines.' [1]

[1] Translated from Albert's *De vegetabilibus*, Lib. VI, Tractatus 1, *De arboribus*, heading *Quercus*. There is a good appreciation of Albert as a botanist in E. H. F. Meyer's *Geschichte der Botanik*, vol. iv, p. 28. Königsberg, 1857.

In this chapter Albert alludes to the relationship of galls to insects. This was not mentioned by Aristotle and hardly suspected by the ancients, being but faintly referred to by Pliny. The subject had to await the microscopic researches of Malpighi to gain further elucidation. In the same passage Albert distinguishes the two cotyledons of the acorn, gropes after a botanical nomenclature adequate to describe plant forms, and succeeds in giving a description which would convey some definite picture to one who had never seen the plant. We can at least say that in the hands of Albert botany has begun to move on the upward grade toward the level at which Theophrastus left it.

Yet the scholastic movement, to the furthering of which Albert's chief efforts were directed, was of its nature inimical to the first-hand study of plants and animals. The thunders of the contest between Nominalism and Realism might well drown the still small voice with which Nature calls for direct observation. The great scholastic centuries from the twelfth to the fifteenth are hardly more fertile than the previous period in botanical writing exhibiting any first-hand knowledge. In one department of intellectual activity, however, there was some clear revival of the spirit of the naturalist. The artistic spirit early showed its kinship with the scientific by the closeness with which some illuminators of manuscripts sought to imitate nature. The herbarium itself remained a fixed text unaltered from that inherited from the preceding age, but its illustration underwent a definite development in the direction of increased naturalism. A close study of some of these beautiful works shows that the early printed herbals had predecessors, and that already in the thirteenth century the older merely stylistic method of plants was giving place to a real attempt to represent nature.

To explain the development of illustration in the herbarium in the later mediaeval centuries some reference must be made to the mode in which these volumes were prepared. The text was usually written before the figures were inserted, and writing and illumination were the work of different hands. Thus the figures are as a rule a little later and may occasionally be much later than the text. Sometimes the provenance of the model can be determined from an examination of the figures. Thus in one of the figures from the Anglo-Saxon herbarium of about the year 1000, the plant representing the henbane, or to call it by its earlier English name, *hennebelle*, is not our familiar *Hyoscyamus niger* but *Hyoscyamus reticulatus*, a Mediterranean form not found in this country (Plate XVI). The illustrator must have had an herbarium of a far southern tradition before him from which he copied his figures. Another proof of the way these plant illustrations were copied has been afforded by Fig. 31, which shows the mythical mandrake with its roots shaped in human form. The form on the left is traced from a manuscript prepared in southern Italy, that on the right from a manuscript made in England in the twelfth. The English figure is clearly based on an Italian original.

Now it must be remembered that those who used these herbaria had no idea of plant distribution. The conception that flora had local peculiarities had been familiar enough to Aristotle and Theophrastus, and traces of it can be found in Pliny and Dioscorides, but the idea had been almost lost in the Middle Ages and remained obscured until revived by Euricius Cordus. When the scribe copied his text he was accustomed to leave a space of a particular size and shape into which the illustrator could then fit his figure, and many herbals have come down to us in which the illustrator has either not completed or not begun his work, so that these spaces remain blank. The gaps might be filled in later, sometimes centuries later, according as the owner of the book had the talent or the financial resources at his disposal. Sometimes the original model was not available when the later figures were inserted, so that these do not fit the spaces left for them.

Such figures of plants were usually copied from earlier figures and therefore became further removed from nature at each stage. But the degradation of the copied herbarium had its limits, and those limits were reached when the figures had so utterly deteriorated that no semblance to an indigenous plant could be discerned by the native scribe or owner of the book. At this point it was necessary to return to nature and to give some impression of a real and local plant, though not necessarily that originally intended by the author of the text. The point of lowest illustrational degradation appears to have been generally passed with the full development of the Romanesque manner at the end of the twelfth or the beginning of the thirteenth century, and from this period we can trace the rise towards modern botany. But individual instances can be adduced of the onset of the change at an even earlier date and there are figures of plants of the first half of the twelfth century which show a definite upward tendency (Bodley

130, Plates v, xv, and xxiii). The movement was continuous and in manuscripts of the fourteenth and fifteenth centuries we can distinguish beautiful attempts to imitate nature comparable in their degree to the work of the artists who employed their talents on grander schemes (Plate xi, especially margins). Such painters of plants as Leonardo da Vinci and Albrecht Dürer had, therefore, their humbler craftsmen predecessors.

These general statements require some modification, for not only did the naturalistic school of plant illustration produce good models in the twelfth century, as we have already shown, but also in the thirteenth, fourteenth, and fifteenth centuries, side by side with real artistic efforts, are still to be found the crudest and most wooden imitations of the outworn models. So it is also with the earlier printed herbals. Some of these are mere repetitions of ancient diagrams, some are real attempts to represent plant life as it is, some are a mixture of the two types of illustration, and some contain illustrations which are themselves a mixture of the two types. The early printed herbals therefore present a stage of development that can be paralleled in the manuscripts, and it is thus perhaps unfortunate that historians of botany have usually elected to begin their accounts with these printed works. Botany is perhaps alone among the sciences in that it is possible to tell its history as an almost continuous tale, and the invention of printing introduces no specially new element into that tale nor does it mark an important period in it.

Among the purely traditional pictures in the incunabula we may class that of the plantain from the Latin Herbarium printed at Rome in 1484, which we can exactly parallel from an Italian manuscript of somewhat earlier date (Fig. 32, p. 72, and Plate xix). As a naturalistic representation we may place that of the wallflower surrounded by dodder from the German version of the *Hortus sanitatis* of Mainz, 1485 (Fig. 34, p. 73), beside a figure of an Orobus from a herbarium written and illustrated in England early in the twelfth century (Plate v). As an intermediate form we may regard the plantain from the Latin *Herbarius latinus* of Mainz of 1484 (Fig. 33, p. 72) and compare it with the Anglo-Saxon effort of about 1000 on the one hand (Plate xvi) and the stereotyped thirteenth-century forms on the other (Fig. 32, p. 72). An instructive series is also provided by the plants henbane (Plates viii, xv, xvi and xix), Betonica (Plates v, xvii, and xxv), or Dracunculus (Plates xix, xxi, and xxii).

PLATE XX. From Leyden MS. Voss Lat. Q.9 Apuleius VIth cent.

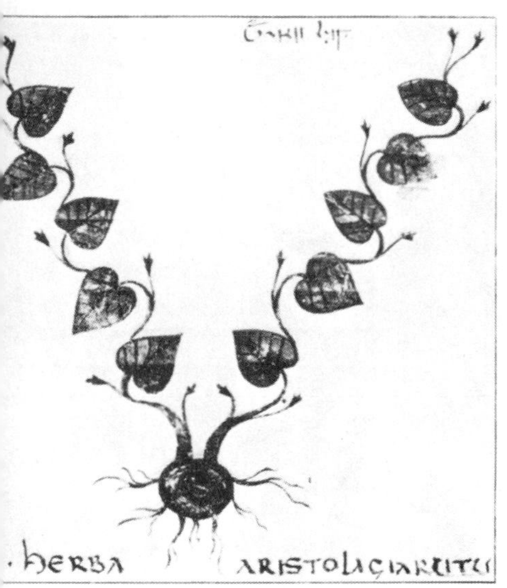

fo. 66 v *Aristolochia*

. 92 v *Heliotropia* = Forget-me-not

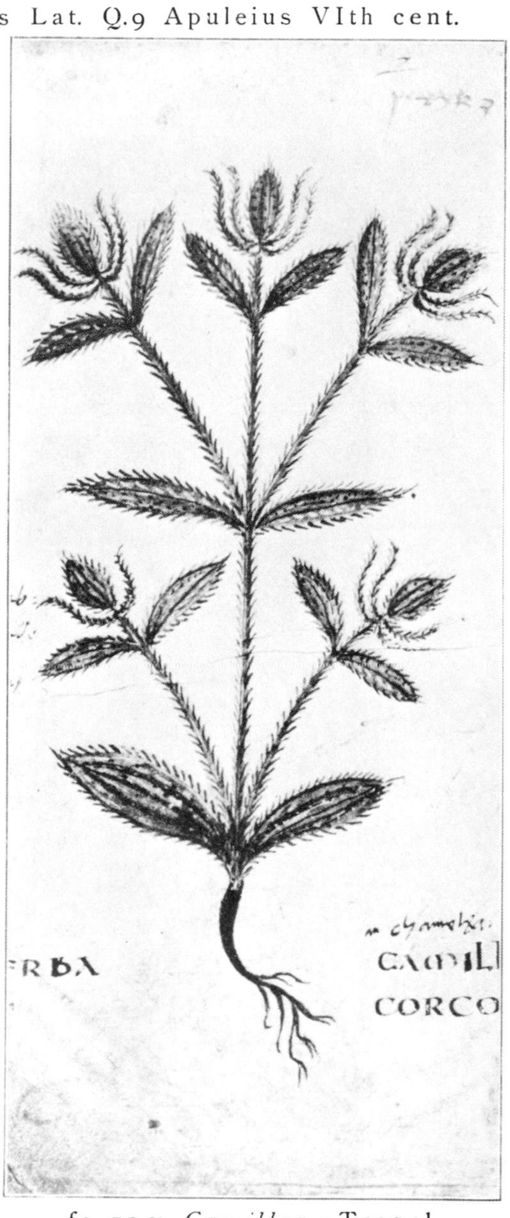

fo. 72 v *Camillea* = Teasel

Plate XXI. Leyden MS. Voss Lat. Q. 9 Apuleius VIth cen

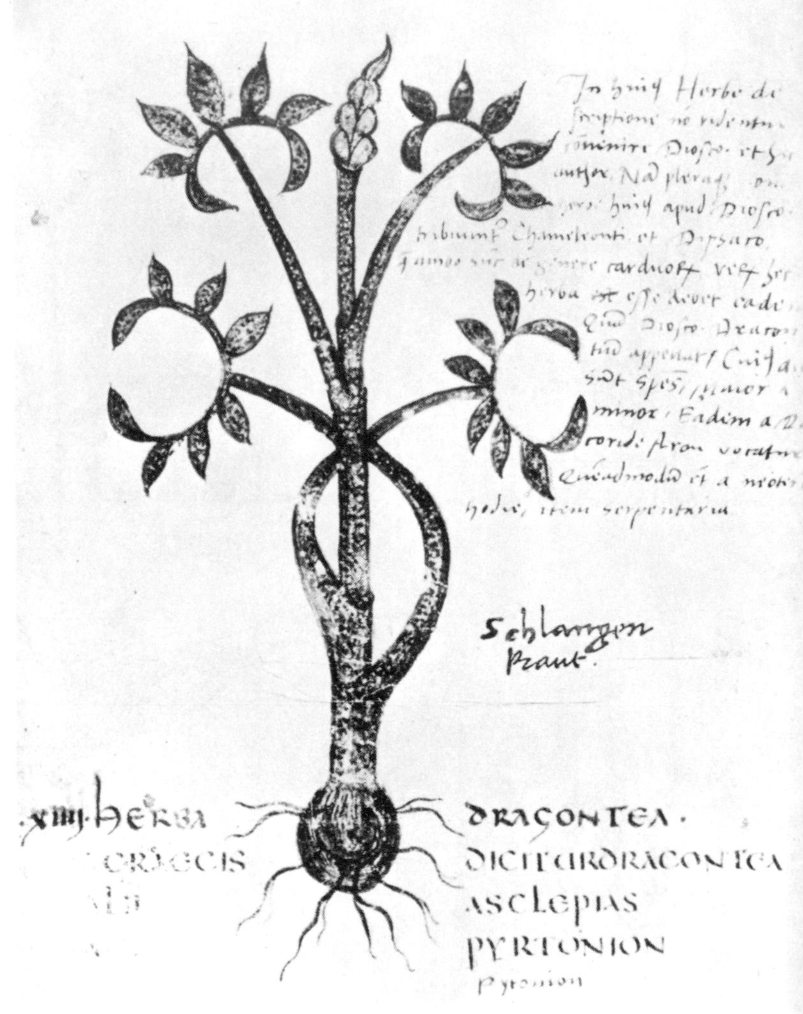

fo. 60 v *Dracontea = Dracunculus vulgaris*
A Mediterranean Species

From the beginning of the sixteenth century the development of the Herbarium may be followed in several readily accessible works, and we need therefore pursue it no further.[1]

VI. The Botanical Results of Theophrastus compared with those of Early Modern Botanists

(a) *Nomenclature and Classification of Plants*

We may now return to glance at the botanical work of the Lyceum. If we would realize the course of ancient botany we must mentally sever two ideas which we have inherited in combination, viz. the description of plants and the system of classification of plants. Unlike investigators of animal forms, botanists of the third and fourth centuries B.C. had not developed a system of classification. It is true that some would discern the idea of plant families in the descriptions of Theophrastus, but it is hardly possible to draw from his works a botanical table such as the zoological table that we have extracted from Aristotle. Still less shall we find in Theophrastus a definite technical nomenclature. Even Dioscorides who lived four hundred years later and whose names for plants still form the larger part of the popular English botanical vocabulary is scarcely more advanced in this respect.

In the absence of any adequate nomenclature or classification the work of Theophrastus gives at first a confused impression. It is a descriptive treatise seldom illumined by the philosophic flashes characteristic of the Aristotelian biological writings, and its author, unequipped with the exact terms with which we can now describe plants and parts of plants, seems to be working under insuperable disadvantages. Yet if we take trouble to comprehend his method we shall see that he has produced a very scientific and thorough piece of investigation. Not only are the descriptions almost always accurate and the illustrations apposite, but the writer is unusually careful to distinguish his own observations from those which he has merely heard, and to separate hypotheses from the records of observations. In this respect Theophrastus is unequalled among Greek biologists. But his work, like all Greek biology, is marred by the almost complete absence of any account of the processes of investigation. Compared to a modern

[1] Agnes Arber, *Herbals* (Cambridge, 1912), and J. F. Payne, 'The Herbarius and Hortus Sanitatis' (*Transactions of the Bibliographical Society*, vi. 63, London, 1903). A. C. Klebs, 'Herbals', in *Papers of the Bibliographical Society of America*, xi, p. 75, Chicago, 1917.

scientific work it is therefore but a fragment containing the conclusions only.

We have said that Theophrastus has no system of classification. It would perhaps be better to say that he suggests many systems but that he has discovered no natural system, and is fully aware of this. He develops, however, several tentative methods of dividing the kinds of plants. The one which he found in practice most effective was a division into trees, shrubs, under-shrubs, and herbs. Other distinctions are the common popular divisions into wild and cultivated, flowering and flowerless, fruit-bearing and fruitless, or again aquatic, terrestrial, marsh-living, and marine. In all this there is no effective and permanent scheme of arranging plant forms, and this defect he shares with all the older botanical writers.

Yet the methods of arranging plants adopted by Theophrastus, imperfect as they are, were hardly improved upon for nearly two millennia. In the illustrated manuscript herbals there is seldom any grouping of plants according to their structure. They are usually either in alphabetical order or placed according to their uses in healing or in some other artificial manner. In Dioscorides, however, plants are roughly grouped in some places according to their form, and occasionally he presents us with a series belonging to the same family, e.g. the *Compositae*, the *Labiatae*, or the *Leguminosae*. The same tendency is also encountered in the fine Anglo-Saxon Herbarium of about A. D. 1000 extracted from the Herbaria of Dioscorides and Apuleius. In this work there is a real grouping of Umbelliferous plants, and in other respects there are traces of a rudimentary attempt at a system of classification. It is a point that does not perhaps appear so clearly from the printed text as in the manuscript itself, but there is every reason to suppose that such arrangement as exists cannot be placed to the credit of the Anglo-Saxon leech, but to the compiler of the work from which he was translating.

In general it may be said that we encounter no real classification until towards the end of the sixteenth century. Among the sixteenth-century writers the tendency to group plants according to their physical characteristics advanced only with extreme slowness. The herbal of Brunfels (1484–1534) that appeared in 1530 [1] is no more systematically arranged than Dioscorides, while that of

[1] Otto Brunfels, *Herbarum vivae icones*, 3 parts, Strasburg, 1530–40; also the *Contrafayt Kreüterbůch*, Strasburg, 1532.

Fuchs (1501–66) dated 1542 [1] is merely alphabetical. Indeed the rudimentary classificatory method exhibited by Theophrastus is hardly attained even by the work of Bock (= Tragus, 1498–1554) of 1546.[2] The herbal of Bock is divided into three parts, the first and second containing the smaller herbs, the third the shrubs and trees. In Bock's work the feeling for relationship is confined to smaller groups as with Theophrastus, from whom indeed he inherits them.

No satisfactory basis of classification was in fact forthcoming until the appearance of a description of a few plants in the *Historiae Stirpium* by Valerius Cordus (1515–44), published posthumously in 1561.[3] Cordus was the first to suggest the structure of the flower as a basis of classification, but his work was largely disregarded until reprinted in the eighteenth century, embedded in the botanical writings of Conrad Gesner (1516–65),[4] who had adopted and developed the views of Cordus. As regards precedence of publication, the first modern botanist to attain even to the low classificatory level of Theophrastus was probably Charles de l'Ecluse (Clusius, 1526–1609), the books of whose *Rariorum plantarum historia* of 1576 [5] are divided to some extent according to the plants of which they treat, the first of trees, shrubs, and undershrubs, the next of bulbous plants, the third of scented flowers, the fourth of scentless flowers, the fifth of poisonous, narcotic, and acid plants, and the sixth of a group containing plants with milky juice, *Umbelliferae*, ferns, grasses, *Leguminosae*, and some Cryptogams.

In de l'Obel (1538–1616) [6] and Cesalpino (1519–1603) [7] we encounter at last two writers who have a preponderating interest in the arrangement of plants according to their natural affinities. The primary divisions of de l'Obel are the traditional ones, trees, herbs, &c., and Monocotyledons and Dicotyledons are dis-

[1] Leonhard Fuchs, *De historia stirpium commentarii insignes*, Basel, 1542.

[2] Hieronymus Bock, *New Kreutter Buch*, Strasburg, 1539; 2nd edition, with figures, Strasburg, 1546.

[3] Valerius Cordus, *In hoc volumine continentur Valerii Cordi Annotationes in Pedacii Dioscorides . . . eiusdem Val. Cordi historiae stirpium. . . . Omnia Conr. Gesneri collecta*, Strasburg, 1561.

[4] Conrad Gesner, *Opera botanica*, 1751.

[5] Charles de l'Ecluse, *Rariorum aliquot stirpium per Hispanias observatarum Historia*, Antwerp, 1576.

[6] Mathias de l'Obel, *Plantarum seu stirpium historia*, Antwerp, 1576.

[7] Andrea Cesalpino, *De plantis*, Florence, 1583.

tinguished about as clearly as by Theophrastus. His system of classification, like much of Theophrastus, is largely based on leaf form, but he shows real advance in the synoptical tables that he constructed for the diagnosis of plant forms. Many of these tables betray a knowledge of true natural relationships. Cesalpino brings us into another realm; he pays much attention to the fruit, seed, and flower for purposes of classification, and distinguishes clearly though with insufficient emphasis between monocotyledons and dicotyledons. He has at last passed definitely beyond Theophrastus.

In the absence of any adequate system of classification almost all botany until the seventeenth century consisted primarily and mainly of descriptions of species. To describe accurately a leaf or a root in the language in ordinary use would often take pages and overwhelm the reader by its bulk. Modern botanists have invented an elaborate terminology which, however hideous to eye and ear, has the crowning merit of helping to abbreviate scientific literature. Botanical writers previous to the seventeenth century were substantially without this special mode of expression. It is partly to this lack that we owe the persistent attempts throughout the centuries to represent plants pictorially in herbals, manuscript and printed, and thus the possibility of an adequate history of plant illustration.

Theophrastus seems to have felt acutely the need of botanical terms, and there are cases in which he seeks to give a special technical meaning to words in more or less current use. Among such words are *carpos*=fruit, *pericarpion*=seed vessel=pericarp, and *metra*, the word used by him for the central core of any stem whether formed of wood, pith, or any other substance. Thus he speaks of 'the seed belonging to the *carpos* (fruit): by *carpos* is meant the seeds bound together with the *pericarpion* (seed-vessel)'.[1] It is from the usage of Theophrastus that the exact scientific definition of fruit and pericarp has come down to us.[2] We may easily discern also the purpose for which he introduces the term *metra*, a word meaning primarily the *womb*, into botany and the vacancy in the Greek language which it

[1] *Historia plantarum*, i. 2, i.

[2] Though it is possible that Theophrastus derived it from Aristotle. Cp. *De Anima*, ii. 1, 412[b] 2. In the passage τὸ φύλλον περικαρπίου σκέπασμα, τὸ δὲ περικάρπιον καρποῦ in the *De Anima* the word does not, however, seem to have the full technical force that Theophrastus gives to it.

PLATE XXII. THREE FIGURES OF *DRACONTEA = DRACUNCULUS VULGARIS*

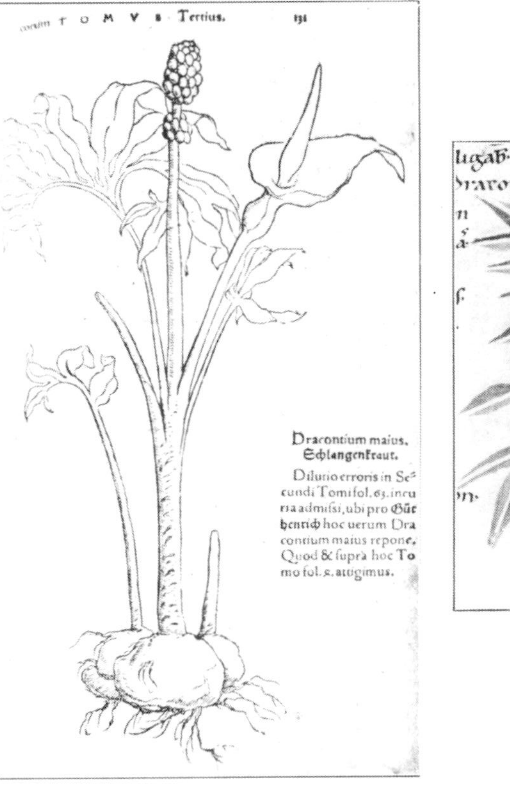

O. Brunfels, *Herbarum Vivae Icones*, Vol. III, p. 131
Strasburg, 1536

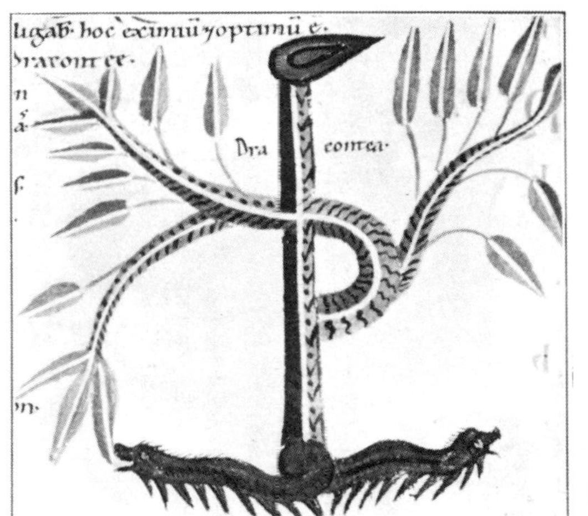

MS. Harley 4986 fo. 7 v
German work of end of XIIth cent.

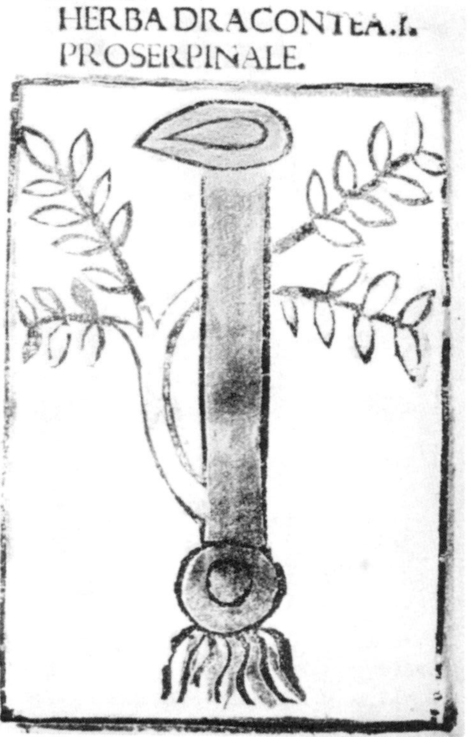

Apuleius, printed Rome, 1483, from metal (?) blocks by
De Lignamine

PLATE XXIII

MS. Bodley 130 written in St. Albans 1120

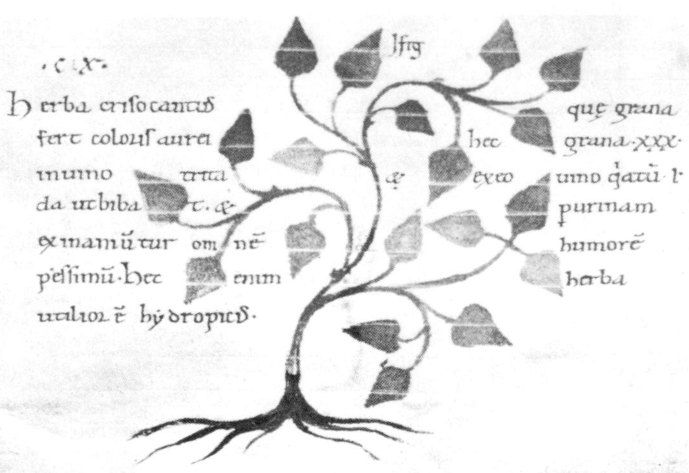

fo. 55 r *CRISOCANTUS*=IFIG=IVY (Flowering form)

fo. 55 r *CYSSON*=*EDERA*=YVYE=IVY (Climbing form)

was made to fill. '*Metra*', he says, 'is that which forms the middle of the wood, being third in order from the bark and corresponding to the marrow in bones. Some call this part the *cardian* (heart), others call it the *enterionen* (inside), others again call only the inner part of the metra itself the *cardian*, while others distinguish this as marrow.'[1] He is clearly inventing a word to cover all the different kinds of core and importing it from another study. This is the method of modern scientific nomenclature which hardly existed for the sixteenth-century botanists. The real foundations of our modern nomenclature were laid in the later sixteenth and in the seventeenth century by Cesalpino and Joachim Jung.

(*b*) *Generation and Development of Plants*

Theophrastus understood the value of developmental study, a conception that he must have derived from his master Aristotle. 'A plant', he says, 'has power of germination in all its parts for it has life in all its parts, wherefore we should regard them not for what they are but for what they are becoming.'[2] The various modes of plant reproduction are correctly distinguished as they would be by any farmer. 'The ways in which trees and plants in general originate are these: spontaneous growth, growth from a seed, from a root, from a piece torn off, from a branch or twig, from the trunk itself; or again from small pieces into which the wood is cut up.'[3] The spontaneous origin of living things was taken for granted by Aristotle and was hardly questioned until the sixteenth century, when, in a flash of genius, Fracastor (1478 ?–1553) suggested that the supposed spontaneous generation was really a process arising from undiscovered seeds;[4] the suggestion did not gain demonstration until the experiments of Redi (1626–94) in the seventeenth century.[5] It is therefore the more interesting to find an inkling of this idea in the mind of Theophrastus. 'Of these methods', by which plants originate, he says, 'spontaneous growth comes first, one may say, but growth from seed or root would seem most natural; indeed

[1] *Historia plantarum*, i. 2, vi.

[2] *Historia plantarum*, i. 1, iv.

[3] *Historia plantarum*, ii. 1, i.

[4] Girolamo Fracastoro, *De contagionibus et contagiosis morbis*, Venice, 1546. See also Charles and Dorothea Singer, 'The Scientific Position of Girolamo Fracastoro,' *Annals of Medical History*, i, New York, 1917.

[5] Francesco Redi, *Experimenta circa generationem insectorum*, Amsterdam, 1671.

these methods too may be called spontaneous; wherefore they are found even in wild kinds, while the remaining methods depend on human skill or at least on human choice.' [1]

There are other passages in which Theophrastus expresses some doubts as to the existence of spontaneous generation.[2] He quotes the view of Anaxagoras (*c.* 450 B. C.) who thought that the air contained seeds (σπέρματα) of all the things, plants among them, that make up the visible universe, and contrasts this theory with that of other philosophers who held that plants and animals were generated *de novo* from special combinations of the elements, and then he observes that 'this kind of generation is somehow beyond the ken of our senses. There are other admitted and observable kinds, as when a river in flood gets over its banks . . . and in so doing causes a growth of forest in that region that by the third year casts a thick shade', an event which he assures us took place at Abdera. The same results may be brought about by heavy rains.

'Now, as the flooding of a river, it would appear, conveys seeds of fruits of trees . . . so heavy rain acts in the same way; for it brings down many of the seeds with it, and at the same time causes a sort of decomposition of the earth and the water. In fact the mere mixture of earth with water in Egypt seems to produce a kind of vegetation. And in some places, if the ground is merely lightly worked and stirred, the plants native to the district immediately spring up.' [3]

The process of germination of seeds is one that must have awakened admiration from a very early date. Even the Egyptians, a people by no means observant of the minute phenomena of plant life, were struck by it, and in a bas-relief, put up to record one of the Syrian expeditions of Tethmosis III (about 1500 B.C.), we may discern a series of figures illustrating the development of seedlings (Fig. 37). It is thus by no means remarkable that the process should have impressed Theophrastus, who has left on record his views on the formation of the plant from the seed.

'In germinating some of these plants produce their root and their leaves from the same point, some separately from either end of the seed. Wheat, barley, spelt, and in general all the cereals produce them from either end, in a manner corresponding to the

[1] *Historia plantarum*, ii. 1, i.

[2] Theophrastus seems, nevertheless, to accept fully the doctrine of spontaneous generation in the *De causis plantarum*, i. 2.

[3] *Historia plantarum*, iii. 1, v.

Fo. 23r Centaur holding Centaury

Fo. 26r Mercury brings 'Electropion' to Homer

position of the seed in the ear, the root growing from the stout lower part, the shoot from the upper part; but the part corresponding to the root and that corresponding to the stem form a single continuous whole. Beans and other leguminous plants do not grow in the same manner, but they produce the root and the stem from the same point, namely, the point at which the seed is attached to the pod, which, it is plain, is a sort of starting-point of fresh growth. In some cases there is a process, as in beans, chick peas, and especially lupines, from which the root grows downwards, the leaf and stem upwards.

'There are then these different ways of germinating; but a point in which all these plants agree is that they all send out their roots at the place where the seed is attached to the pod or ear, whereas the contrary is the case with the seeds of certain trees, as almond, hazel, acorn, and the like. . . . In certain trees the bud first begins to grow within the seed itself, and, as it increases in size, the seeds split—for all such seeds are in a manner in two halves; and those of leguminous plants again all plainly have two valves and are double—and then the root is immediately thrust out; but in cereals, since the seeds are in one piece, this does not occur but the root grows a little before the bud.

YOUNG DRACUNCULUS

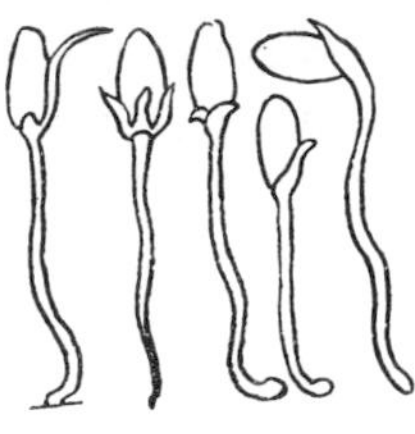

SEEDLING BEANS

FIG. 37. Seedlings from the 'Syrian Garden' of Tethmosis III (about 1500 B.C.) at Karnak.[1]

'Barley and wheat come up with a single leaf, but peas, beans, and chick peas with several. All the leguminous plants have a single woody root, and also slender side roots springing from this . . . but wheat, barley, and the other cereals have a number of fine roots whereof they are matted together. . . . And there is a sort of contrast between these two classes; the leguminous plants which have a single root, have many side-growths above from the stem . . . while the cereals which have many roots, send up many shoots, but these have no side-shoots.'[2]

There can be no doubt that this is a piece of first-hand and minute observation of the behaviour of germinating seeds. The distinction between dicotyledons and monocotyledons is accurately

[1] These tracings were made from photographs kindly taken for the purpose by Captain Engelheart of the Department of Antiquities of the Egyptian Government. Cf. also Mariette-Bey's *Karnak*, Leipzig, 1875, Pl. xxxi.

[2] *Historia plantarum*, viii. 1, i.

set forth, though the stress is laid not so much on the cotyledonous character of the seed as on the relation of root and shoot. In the dicotyledons the root and shoot are represented as springing from the same point and in the monocotyledons from opposite poles in the seed.

No effective work was done on the germinating seed until the invention of the microscope, and the appearance of the work of Highmore (1613–85)[1] (Fig. 38), and the much more searching investigations of Malpighi (1628–94)[2] (Figs. 42 and 43) and Grew (1641–1712)[3] after the middle of the seventeenth century. The observations of Theophrastus are, however, so accurate, so lucid, and so complete that they might well be used as legends for the plates of these writers two thousand years after him.

Much has been written as to the knowledge of the sex of plants among the ancients. It may be stated that of the sexual elements of the flower no ancient writer had any clear idea. Nevertheless, sex is often attributed to plants, and the simile of the *Loves of Plants* enters into works of several of the poets, agricultural authors, and writers of fiction. Among these Achilles Tatius, the fifth-century author of the romance *The Adventures of Leucippe and Clitophon*, offers a good example.

> ‘Plants’, we there read, ‘fall in love with one another and the palm is particularly susceptible. . . . If the female be planted at any considerable distance, the loving male begins to wither away. The gardener realizes what is the cause of the tree's grief, goes to some slight eminence in the ground, and observes in which direction it is drooping (for it always inclines towards the object of its passion) . . .’ Then ‘he takes a shoot of the female palm and grafts it into the very heart of the male. This refreshes the tree's spirit, and the trunk, which seemed on the point of death, revives and gains new vigour in joy at the embrace of the beloved; it is a kind of vegetable marriage.’[4]

Plants are frequently described as male and female in ancient biological writings, and Pliny goes so far as to say that some considered all herbs and trees were sexual.[5] Yet when such passages can be tested it will be found that these so-called males and females are usually different species. In a few cases a sterile

[1] Nathaniel Highmore, *A History of Generation*, London, 1651.

[2] Marcello Malpighi, *Anatome plantarum*, London, 1675.

[3] Nehemiah Grew, *Anatomy of Vegetables begun*, London, 1672.

[4] Achilles Tatius, i. 17.

[5] Pliny, *Naturalis historia*, xiii. 4.

variety is described as the male and a fertile as the female.[1] In a small residuum of cases dioecious plants are regarded as male and female, but with no real comprehension of the sexual nature of the flowers. There remains a minute group, which cannot be extended beyond the palms, in which the knowledge of plant sex had advanced a trifle further.

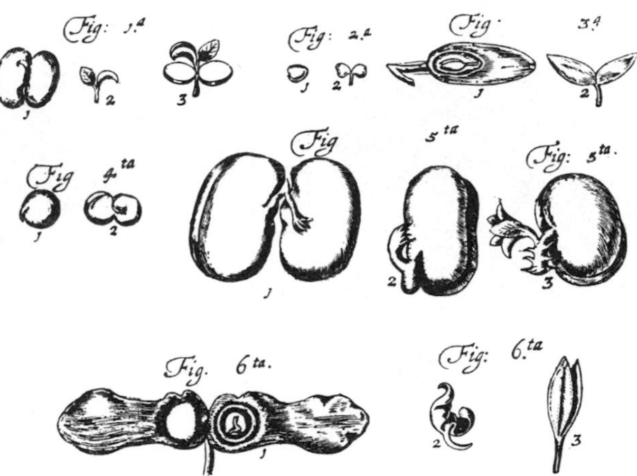

FIG. 38. GERMINATION OF SEEDS. From Nathaniel Highmore's *History of Generation*, London, 1651.

'The first figure, of the first Table, shews the Kidny Bean opened; in which is a little crooked leaf folded up, which being displayed, shews itself, as in the second; and when, being set, it arises above ground, it is such a Plant as the third shews, with the very same leaves and no other.

'The second figure shews a Colewort seed: the first shews both leaves, with the stalk folded up, as they lie in the husk of the seed: the second shews it come up out of the ground.

The Third Figure hath the small germen of an Ash; lying with his two leaves in the kernel of an Ash, both in the husk inclosing them. The second shews him sprung up above the Earth, at his first coming abroad.

'The fourth delineates the young germen of the Pease in the midst of the grain, and its breaking forth.

'The fifth shews the young Plant in the midst of the Bean: with the manner of his putting forth, with the same leaves displayed in the third, which are wrapt up in the first and second.

'The sixth Figure displayes the young Maple wrapt up in his husk; and how he lies, as in the first: The second shews him a little unfolded, when it is taken out of the husk. The third shews him gotten from his shell, and the surface of the Earth.'

'Common to all trees', Theophrastus tells us, 'is that by which men distinguish the "male" and the "female", the latter being fruit-bearing, the former barren in some kinds. In those

[1] A good collection of references to plant loves and plant sexes in classical writings can be found in R. J. Thornton's sumptuous *New illustrations of the sexual of Carolus von Linnaeus*, London, 1807.

kinds in which both forms are fruit-bearing, the "female" has fairer and more abundant fruit, however, some call these the "male" trees—for there are those who actually thus invert the names. This difference is of the same character as that which distinguishes the cultivated from the wild tree.'[1] The description

FIGS. 39 and 40. Supernatural figure from Nineveh holding male inflorescence o date palm (from Layard).

by Theophrastus of the fertilization of the date palm is, however, quite clear. 'With dates it is helpful to bring the male to the female; for it is the male which causes the fruit to persist and ripen, and this process some call by analogy (!) *the use of the wild fruit.* The process is thus performed; when the male palm is in flower they at once cut off the spathe on which the flower is, just as it is, and shake the bloom with the flower and the dust over the fruit of the female, and, if this is done to it, it retains the fruit and does not shed it.'[2] The fertilizing character of the spathe of the male date palm was familiar in Babylon from a very early date, and is represented by a frequent symbol on the monuments in which a divine figure is represented as holding

[1] *Historia plantarum*, iii. 8, i.

[2] *Historia plantarum*, ii. 8, iv.

the male inflorescence of the palm and fertilizing the female tree by shaking it (Figs. 39, 40, and 41).

The description of caprification by Theophrastus may here be appropriately related. He tells us that there are certain trees, the fig among them, ' which are apt to shed their fruit prematurely and remedies are sought for this. In the case of the fig the device adopted is caprification. Gall insects come out of the wild

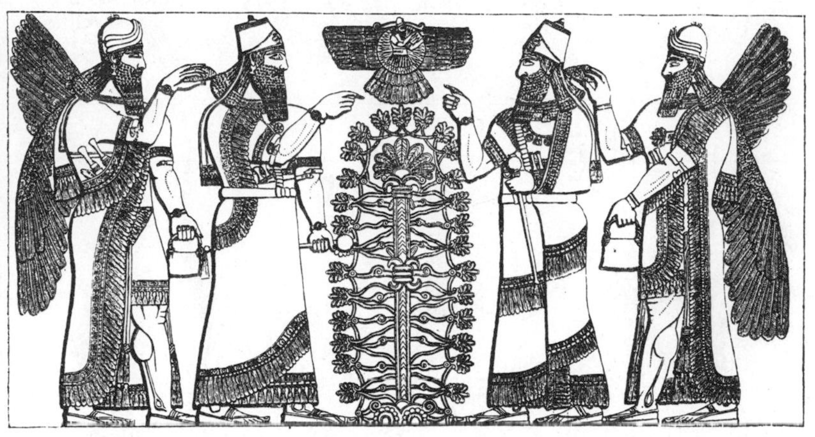

FIG. 41. Assur-nasir-pal, King of Assyria, about 885–860 B.C., with winged attendants holding male inflorescence of date palm, performing ceremony of fertilization before conventionalized tree. Above is the symbol of the god Assur. From a bas-relief on the walls of the palace of Assur-nasir-pal, discovered at Calah (Nimrûd), now in the British Museum.

figs which are hanging there, eat the tops of the cultivated figs, and so make them swell '.[1] These gall-insects 'are engendered from the seeds. The proof given of this is that, when they come out, there are no seeds left in the fruit; and most of them in coming out leave a leg or wing behind. . . . A fig which has been subject to caprification is known by being red and parti-coloured and stout, while one which has not been so treated is pale and sickly '.[2] That Theophrastus had no idea whatever of the true nature of this caprification he shows a little farther on by observing that ' in the case both of the fig and of the date it appears that the " male " renders aid to the " female "—for the fruit-bearing tree is called female—but whilst in the case of the fig there is a union of the two sexes, in the case of the palm the result is brought about somewhat differently '.[3]

[1] *Historia plantarum*, ii. 8, i.
[2] *Historia plantarum*, ii. 8, ii.
[3] *Historia plantarum*, ii. 8, iv.

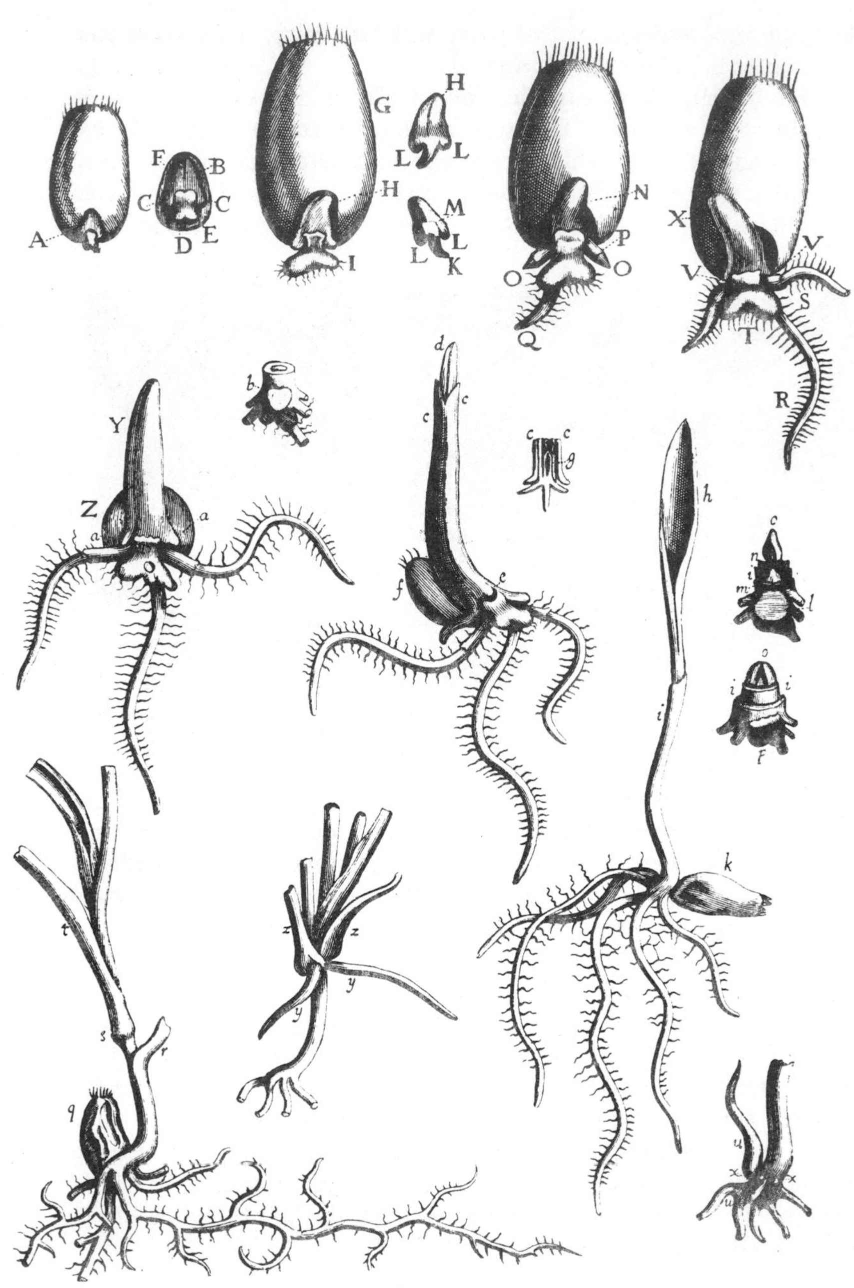

Fig. 42. THE GERMINATION OF WHEAT
From the *Anatome plantarum* of Malpighi, London, 1676.

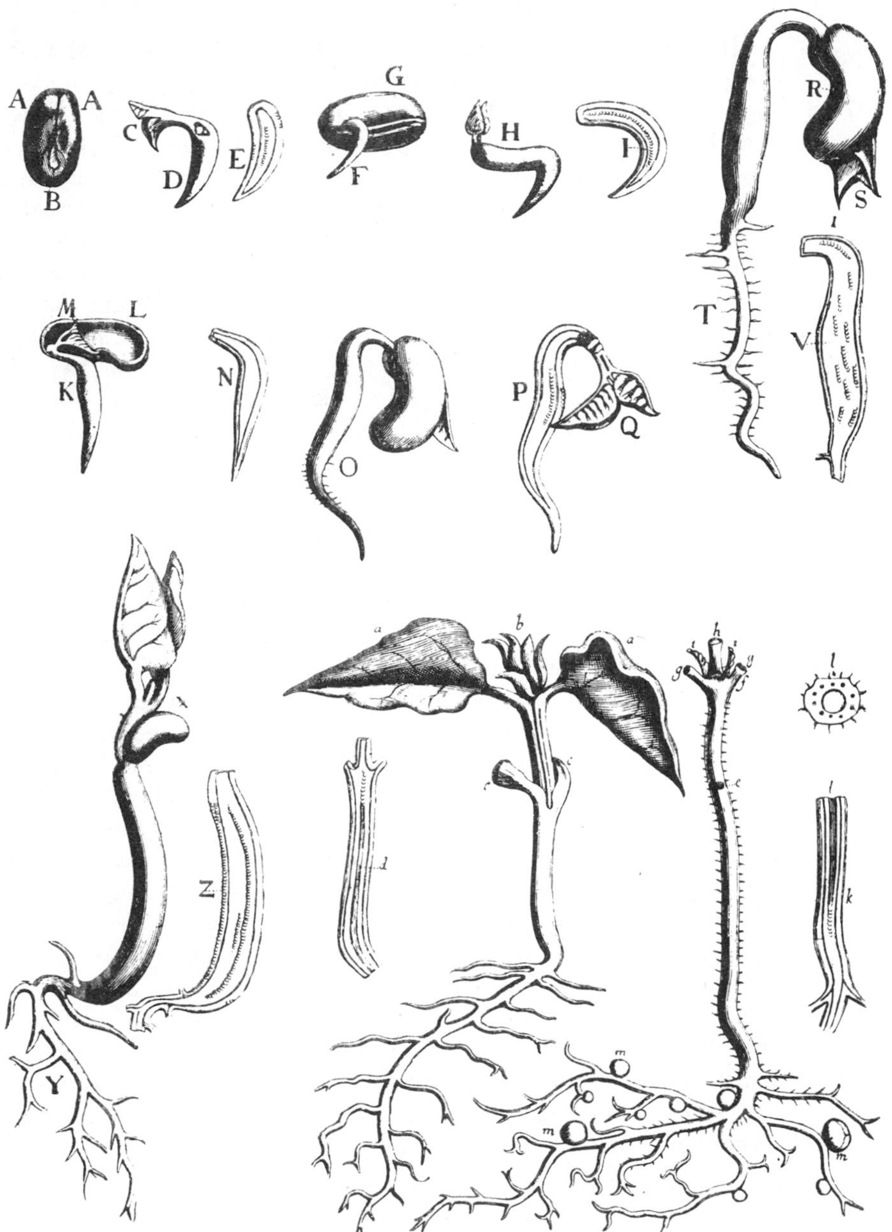

Fig. 43. THE GERMINATION OF THE BEAN
From the *Anatome plantarum* of Malpighi, London, 1676. In one of the lower figures the root nodules may be seen.

It is interesting to observe that Herodotus (about 500 B. C.), describing the fertilization of the date palm in Babylon, compares the process with that of the fig. 'The people of Babylon', says Herodotus, 'have date-palms growing over all the plain, most of them fruit-bearing, and to these they attend in the same manner as to fig-trees, and in particular they take the fruit of those palms which the Hellenes call male-palms, and tie them upon the date-bearing palms, so that their gall-fly may enter into the date and ripen it and that the fruit of the palm may not fall off; for the male palm produces gall-flies in its fruit just as the wild fig does.'[1]

Neither Theophrastus nor any ancient author ever saw the flowers of the fig. These were first distinguished by the youthful botanical genius Valerius Cordus in the first half of the sixteenth century.[2]

(c) *Form and Structure of Plants*

Theophrastus was not successful in distinguishing the nature of the primary elements of plants, though he was able to separate root, stem, leaf, stipule, and flower on morphological as well as to a limited extent on physiological grounds. For the root he adopts the familiar definition, the only one possible before the rise of chemistry, that it 'is that by which the plant draws up nourishment',[3] but he shows by many examples that he is capable of following out its morphological homologies. Thus he knows that the ivy regularly puts forth roots from the shoots between the leaves, by means of which it gets hold of trees and walls,[4] that the mistletoe will not sprout except on the bark of living trees into which it strikes its roots, and that the very peculiar formation of the banyan tree is to be explained by the fact that 'this plant sends out roots from the shoots till it has hold on the ground and roots again: and so there comes to be a continuous circle of roots round the tree, not connected with the main stem but at a distance from it'.[5] He did not succeed, however, in distinguishing the real nature of such structures as bulbs, rhizomes, and tubers, but regards them all as roots. Nor was he more successful in his discussion of the nature of stems.

[1] Herodotus, i. 493.

[2] The passage in Valerius Cordus in which the flowers of the fig are described presents certain difficulties. It is discussed by E. L. Greene, *Landmarks of Botanical History*, Washington, 1909, p. 292.

[3] *Historia plantarum*, i. 1, ix.

[4] *Historia plantarum*, iii. 18, x.

[5] *De causis plantarum*, ii. 23.

As to leaves he is more definite and satisfactory though wholly in the dark as to their function. In speaking of them he is especially handicapped by the absence of a botanical nomenclature, so that his descriptions are seldom more than comparisons to well-known objects. 'Leaves', he says, 'differ in their shape, some are round, as those of pear, some rather oblong, as those of apple; some come to a sharp point and have spinous projections at the side, as those of smilax . . . some are divided and like a saw, as those of silver fir and of ferns. To a certain extent those of the vine are also divided, while those of the fig one might compare to a crow's foot. Some leaves again have notches, as those of elm, filbert, and oak, others have spinous projections both at the tip and at the edges, as those of kermes-oak, oak, smilax, bramble, *Paliouros* [Christ's thorn], and others. . . . Again there is the difference that some leaves have no leaf-stalk, as those of squill and purse-tassels, while others have a leaf-stalk. . . .'[1] He was well on the way, however, towards arriving at a correct idea of the nature of certain pinnate leaves. Thus of the mountain ash he says that 'the leaves grow attached to a long fibrous stalk, and project on each side in a row, like the feathers of a bird's wing, the whole forming a single leaf but being divided into lobes with divisions which extend to the rib; but each pair are some distance apart, and when the leaves fall, these divisions do not drop separately but the whole wing-like structure drops at once.'[2] Again of the elder, 'the leaf is composed of leaflets growing about a single thick fibrous stalk to which they are attached at either side in pairs at each joint; and they are separate from one another, while one is attached to the tip of the stalk',[3] and of the Terebinth, 'the leaf is made up of a number of leaflets, like bay leaves, attached in pairs to a single leaf-stalk. So far it resembles the leaf of the [mountain ash], there is also the extra leaflet at the tip, but the leaf is more angular than that of the [mountain ash] and the edge resembles more the leaf of the bay'.[4] These passages on the nature of pinnate leaves are the more remarkable when we recall that they remained neglected until similar distinctions and observations were published by Johann Vaget in 1678 in his edition of the work of his master Joachim Jung (1587–1657).[5]

[1] *Historia plantarum*, i. 10, v, vi and vii.

[2] *Historia plantarum*, iii. 12, vi.

[3] *Historia plantarum*, iii. 13, v.

[4] *Historia plantarum*, iii. 15, iii.

[5] Joachim Jung, *Isagoge phytoscopica*, Hamburg, 1678.

In spite of his frequent use of the terms 'male and female' as applied to plants, Theophrastus, as we have already said, had no correct idea of the nature of sex in flowers. His description of flowers is thus almost entirely morphological. 'Some flowers', he says, 'are hair-like, as that of the vine . . . some are "leafy" as in almond, apple, pear, plum. Again some of these flowers are conspicuous, while that of the olive, though it is "leafy", is inconspicuous'. The flowers of annuals are usually, he says, 'two coloured and twofold. I mean by *twofold* that the plant has another flower inside the flower in the middle, as with rose, lily, violet. Some flowers again consist of a single leaf, [i.e. are gamopetalous or sepalous], having merely an indication of more, as that of bindweed. For in the flower of this the separate *leaves* are not distinct; nor is it so in the lower part of the narcissus, but there are angular projections from the edges. . . .' [1]

Notwithstanding his lack of insight as to the nature of sex in flowers, he attained to an approximately correct idea of the relation of flower and fruit. 'Some plants', he says, 'have the flower close above the fruit as vine and olive; in the latter, when the flowers drop off, they are seen to have a hole through them, and this men take for a sign whether the tree has blossomed well; for if the flower is burnt up or sodden, it sheds the fruit along with itself, and so there is no hole through it. The majority of flowers have the fruit case in the middle of them, or it may be the flower is on the top of the fruit case as in pomegranate, apple, pear, plum, and myrtle . . . for these have their seeds below, beneath the flower, and this is most obvious in the rose because of the size of the seed vessel. In some cases again the flower is on top of the actual seeds as in pine, thistle, safflower, and all thistle-like plants. . . . In some other plants the attachment is peculiar as in ivy and mulberry, and in these the flower is closely attached to the whole fruit-case.' [2] Thus Theophrastus, while never finally defining a flower, really comes gradually to abandon his first suggestion that a flower is but a whorl of specially coloured leaves, and almost comes to regard as the essential floral element its relation to the fruit. He has, moreover, succeeded in distinguishing between the hypogynous, perigynous, and epigynous types of flowers.

In spite of the discoveries of Valerius Cordus and the use of flowers for classification by de l'Obel and by Cesalpino in the

[1] *Historia plantarum*, i. 13, ii.

[2] *Historia plantarum*, i. 13, iii.

sixteenth century, the sexual character of flowers remained very obscure for a hundred years after their time. Grew (1641–1712) in 1682 distinguished the stamens, or *attire* as he called them, from the outer floral whorls. He watched the anthers or *semets* bursting and scattering their pollen. Grew almost ignores the pistil, but 'in discourse with our learned *Savilian* Professor Sir Thomas Millington, he told me, he conceived, That the *Attire* doth serve, as the *Male*, for the *Generation* of the *Seed*. I immediately reply'd, That I was of the same opinion and gave him some reasons for it. . . . But withall, in regard every *Plant* is *Male* and *Female*, that I was also of opinion, That it serveth for the *Separation* of some *Parts* as well as the *Affusion* of others '.[1] Ray (1627–1705) spoke in somewhat similar indefinite terms,[2] and the sexual character of flowers was only cleared up by the work of Jacob Camerarius (1665–1721) in the last decade of the seventeenth century.[3]

(*d*) *Habits and Distribution of Plants*

Theophrastus had a perfectly clear idea of plant distribution as dependent on soil and climate. ' Differences in situation and climate, he says, affect the result. In some places, as at Philippi, the soil seems to produce plants which resemble their parent ; on the other hand a few kinds in some few places seem to undergo a change, so that wild seed gives a cultivated form, or a poor form one actually better.' [4] These changes he acutely contrasts to the metamorphoses of animals.

' In pot-herbs ', he says, ' change is produced by cultivation ; for instance, they say that if celery seed is trodden and rolled in after sowing, it comes up curly ; it also varies from change of soil, like other things. . . . It would seem more surprising if such changes occurred in animals naturally and frequently ; some animals do indeed seem to change according to the seasons, for instance, the hawk, the hoopoe, and other similar birds. . . . Most obvious are certain changes in regard to the way in which animals are produced, and such changes run through a series of creatures ; thus a caterpillar changes into a chrysalis, and this in turn into the perfect insect. . . . But there is hardly anything abnormal in this, nor is the change in plants, which is the subject of our inquiry, analogous to it. That kind of change occurs in trees and

1 Nehemiah Grew, *The Anatomy of Plants*, London, 1682, p. 171.

2 John Ray, *The Wisdom of God manifested in the Works of the Creation*, London, 1691.

3 R. J. Camerarius, *Ephem. Leopold. Carol. Acad.*, 1691, and *De sexu plantarum epistola*, Tübingen, 1694.

4 *Historia plantarum*, ii. 2, vii.

in all woodland plants generally, as was said before, and its effect is that when a change of the required character occurs in the climatic conditions, a spontaneous change in the way of growth ensues.' [1]

He is very clear as to the difference between the vegetation of mountain and plain, and gives formal lists to illustrate it.[2]

'Again the character of the position makes a great difference as to fruit-bearing. The *persea* of Egypt bears fruit . . . but in Rhodes it only gets as far as flowering. The date-palm in the neighbourhood of Babylon is marvellously fruitful; in Hellas it does not even ripen its fruit, and in some places it does not even produce any.' 'That each tree seeks an appropriate position and climate is plain from the fact that some districts bear some trees but not others; the latter do not grow there of their own accord, nor can they easily be made to grow, and that even if they obtain a hold, they do not bear fruit. . . . Thus in Egypt there are a number of trees which are peculiar to that country, the sycamore, the tree called *persea* . . . and some others. Now the sycamore to a certain extent resembles the tree which bears that name in our country. Its leaf is similar, its size, and its general appearance; but it bears its fruit in a quite peculiar manner . . . not on the shoots or branches, but on the stem; in size it is as large as a fig.' [3]

At times Theophrastus seems to be on the point of passing from a statement of climatic distribution into one of real geographical regions. Thus:

'Among the plants that grow in Arabia, Syria, and India the aromatic plants are somewhat exceptional and distinct from the plants of other lands; for instance, frankincense, myrrh, cassia, balsam of Mecca, cinnamon, and all other such plants. . . . So in the parts towards the East and South there are these special plants and many others besides.

'In the northern regions it is not so, for nothing worthy of record is mentioned except the ordinary trees which love the cold and are found in our country.' [4]

The general question of plant distribution long remained at, if it did not recede from, the position where Theophrastus left it. The usefulness of the manuscript and early printed herbals was marred by the retention of plant descriptions prepared for the Greek East or Latin South, and these works were saved from complete ineffectiveness only by an occasional appeal to nature.

[1] *Historia plantarum*, ii. 4, vi. and vii.
[2] *Historia plantarum*, iii. 1, i.
[3] *Historia plantarum*, iii. 3, v.
[4] *Historia plantarum*, iv. 1, v.

Sloane 1975 fo. 10v

Early XIII Cent.

Cotton Vitellius C.III fo. 20r

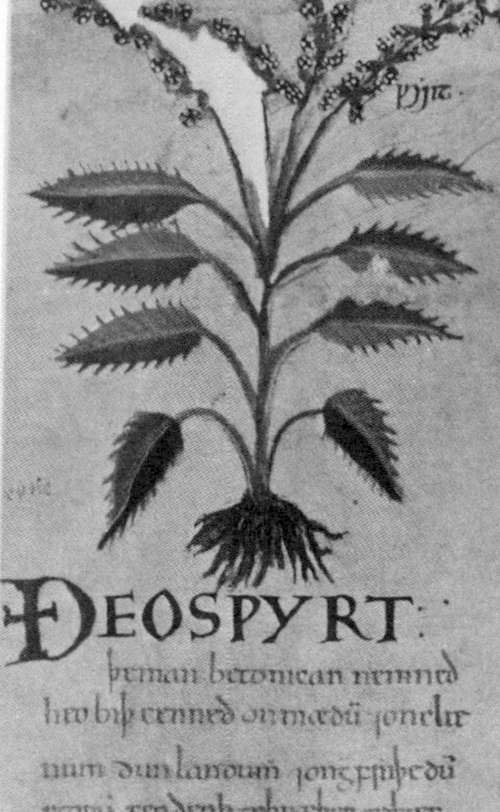

About 1000

Harley 1585 fo. 14r

About 1200

Euricus Cordus (1486–1535) was perhaps the first to realize that there was only a general and not a specific correspondence between the plants of middle Europe and those of the Mediterranean region known to Dioscorides. With the extension of travel in the preparation of works on exotic botany such as those of Oviedo, Clusius and Acosta, this point gradually came into clearer view, but the conception of botanical regions was a very slow growth.

Theophrastus does not abandon the doctrine of plant metamorphosis and he applies this view especially to the transformation of corn into tares. 'They say', he says, 'that wheat and barley change into darnel, and especially wheat; and that this occurs with heavy rains and especially in well watered and rainy districts.' It is a fact that in damp fields the *Lolium* is liable to spring up, and this event, it is said, is still held by farmers to be of the nature of a metamorphosis. Theophrastus, though he does not deny that this is a metamorphic process, expresses his doubts. 'That darnel', he says, 'is not a plant of the spring like others (for some endeavour to make this out) is clear from the following consideration: it springs up and becomes noticeable directly winter comes; and it is distinguished in many ways; the foliage is narrow, abundant, and glossy.'[1]

In the case of the *Lolium* Theophrastus rightly expresses his disbelief in a dimorphism or transformation that we now know does not exist. There is another case in which a plant exhibits two stages of development that were almost invariably regarded by the ancients as separate species. The habit of the ivy is to develop a climbing shoot well provided with rootlets which hold it to its support. The climbing shoot is beset with dark angular foliage. After reaching the top of the object on which it is climbing there sets in an extensive production of free terminal branches bearing somewhat paler and larger ovate leaves. On these terminal twigs the flowers are born, and often the production of the terminal growth becomes so extensive as to conceal the climbing portion. In ancient times, as in the Middle Ages and indeed into quite modern times, these two stages in the development of the ivy were regarded as two species (Plate XXIII). The climbing plant was named by the Greek *helix* and the bushy terminal growth *cittus* or *chrysacanthus*. Theophrastus, however, ventures to traverse this view.

'The *helix*', he says, 'presents . . . differences in the leaves, which are small, angular, and of more graceful proportions, while

[1] *Historia plantarum*, viii. 7, i.

those of the *cittus* are rounder and simple; there is also difference in the length of the twigs, and further in the fact that this tree is barren. For as to the view that the helix by natural development turns into the *cittus* some assert that this is not so, the only true ivy according to these being that which was ivy from the first; whereas if, as some say, the helix invariably turns into [cittus], the difference would be merely one of age and condition and not of kind.' [1]

We may terminate our discussion with a summary of the botanical position of Theophrastus:

1. He distinguished the external organs of plants, naming them in regular sequence from root to fruit, and attained in many cases to a really philosophical distinction.
2. He definitely set forth the leaf homology of the perianth members of flowers but attained to no real knowledge of their sexual nature.
3. He established the first rudiments of a botanical nomenclature.
4. He watched the development of seeds and was able to some extent to distinguish between dicotyledons and monocotyledons.
5. He established a relationship between structure and habits and approaches the conception of geographical distribution.
6. He saw the need for a general classification of plants and made some attempt at a system though he failed to produce one which was in fact workable.
7. He perceived a general relation between structure and function in plants, and thus laid the basis of scientific botany.

NOTE.—I have to thank many friends for much kind help during the preparation of this essay.

Mrs. Agnes Arber and Professor D'Arcy Thompson read the work in manuscript and made a number of corrections and emendations.

Professor Sudhoff has most generously handed over to me a large collection of photographs of early herbals which must have taken years to put together. It is a source of great regret to me that these photographs reached me too late for full use in illustration of my text. I have, however, been able to include figures of the Leyden Apuleius and it is a pleasure to me to

[1] *Historia plantarum*, iii. 18, vii.

acknowledge that I owe my acquaintance with that important document and many others, including the frontispiece of this book, entirely to his good offices. In a later publication I hope to display more fully the rich vein of treasure which the generosity of this eminent scholar has placed at my disposal.

I am grateful to Mr. Henry Balfour, Dr. A. H. Church, Professor F. J. Cole, Professor Clifford Dobell, Dr. Claridge Druce, Professors Ernest and Percy Gardner, Mr. E. A. Lowe, Mr. Eric Maclagan, Mr. F. S. Marvin, Professor F. W. Oliver, Mr. C. Tate Regan, Mr. R. R. Steele, and Dr. W. J. Turrell for a number of suggestions that they have made.

For the loan of the blocks of Plates VI and VII illustrating the Julia Anicia MS. I have to thank Mrs. Arber and the Cambridge University Press. Plate XVIII of the Phillipps Dioscorides is from photographs taken for me by Mr. E. A. Lowe by kind permission of the then owner, Mr. Fitzroy Fenwick. The Paris Bibliothèque nationale MSS. lat. 6862 and gr. 2179 were examined for me by Miss A. Anderson who has helped me also with many details of the work. Miss C. Hugon has redrawn many of the text figures and has been of great assistance in the preparation of the plates.

In quoting passages from the works of Aristotle I have used the translations of Professor D'Arcy Thompson, Professor A. Platt, and the late Dr. Ogle in the *Oxford Aristotle.* The only deviation I have made has been the restoration, in a few cases, of the Greek term for that used by the translator. In every case I have given the reference by page and line to Bekker's Greek text. For the *History of Plants* of Theophrastus, the translation by Sir Arthur Hort has been followed. In one instance, however (*Historia plantarum*, i. 1, iv, p. 83 of my essay) I have ventured to diverge from it. To all these writers and to their publishers, the Clarendon Press, and the Loeb Classical Library, my thanks are due for permission to avail myself of these translations. There is no English version of the *De causis plantarum* and for that I have used Wimmer's text.

The Appendix contains a list of Aristotelian MSS. abstracted from the *Catalogue of Scientific and Medical MSS. in the British Isles* that is being prepared by my wife. She has helped me in innumerable ways and the revision of the proofs has been her work.

APPENDIX

MSS. IN ENGLISH LIBRARIES OF ARISTOTELIAN BIOLOGICAL WORKS

A list of the MSS. as they exist in the public libraries of this country is here given. For convenience of reference the Greek MSS. have been included in the list but not in the summary at the end. All MSS. are Latin unless otherwise described. Many give abridged versions of the works.

The time distribution of the MSS. is significant. It does not suggest any revived interest in Aristotelian biology with the revival of Greek learning in the fifteenth century. On the contrary, the number of Aristotelian biological MSS. of the fifteenth century is but a third of that of the fourteenth. The fall in numbers in the fifteenth century cannot be explained by the advent of printing, since the number of medical and scientific MSS. of almost every class of the fifteenth century is, in fact, greater than of the fourteenth century.

Historia animalium

1. British Museum : Royal 9 A XIV. Imperfect . . . . . . 13th century.
2. British Museum : Royal 12 C XV. Translation by Michael Scot . . 13th „
3. British Museum : Royal 12 F XV. Translation by Michael Scot. . 13th „
4. Oxford : Balliol 250 . . . . 13th „
5. Oxford : Balliol 252 . . . . 13th „
6. Cambridge : Gon. & Caius 109. Translation by Michael Scot . . 13th „
7. Cambridge : Peterhouse 121 . . . . 13th „
8. Cambridge Univ. Lib. Ii. III. 16. Translation by Michael Scot . . 13th „
9. Salisbury Cathedral 111 . . . . 13th „
10. British Museum : Harley 4970 . . . . 14th „
11. British Museum : Royal 7. C. I. Translation by Michael Scot . . 14th „
12. Oxford : All Souls 72. Avicenna's paraphrase . . . . . 14th „
13. Oxford : Merton 278. Translation by Michael Scot . . . . 14th „
14. Cambridge Univ. Lib. Dd. IV. 30. Translation by Michael Scot . . 14th „
15. Oxford : Bod. Can. misc. 418 . . . . . . ? early 15th, 14th „
16. Oxford : Bod. Barocci 95. Excerpts. GREEK . . . . . 15th „
17. Lincoln Cathedral B. 6. 4. Begins imperfectly . . . . . 15th „

De partibus animalium

1. Oxford : C.C.C. 108. GREEK. Ends imperfectly . . . . late 12th century.
2. British Museum : Royal 9 A XIV. . . . . 13th „
3. British Museum : Royal 12. F. XV. Ends Imperfectly . . . . 13th „
4. Cambridge : Peterhouse 121 . . . . 13th „
5. Cambridge Univ. Lib. Ii. III. 16 . . . . 13th „
6. British Museum : Harley 4970. . . . . 14th „
7. British Museum : Royal 7. C. 1. Translation by Michael Scot . . 14th „
8. Oxford : Merton 270 . . . . 14th „
9. Oxford : Merton 271 . . . . 14th „
10. Oxford : Bod. Can. misc. 412 . . . . early 15th „

De generatione animalium

1. Oxford : C.C.C. 108. GREEK . . . . late 12th century.
2. British Museum : Royal 9. A. XIV . . . . 13th „
3. Cambridge : Peterhouse 121 . . . . 13th „
4. Cambridge Univ. Lib. Ii. III. 16 . . . . 13th „
5. British Museum : Harley 4970 . . . . 14th „
6. British Museum : Royal 7. C. I. Translation by Michael Scot . . 14th „
7. Oxford : Merton 270 . . . . 14th „
8. Oxford : Merton 271 . . . . 14th „
9. Oxford : Bod. Can. misc. 412 . . . . early 15th „
10. Oxford : New Coll. 226 . . . . 15th „

De incessu animalium

1. Oxford: C.C.C. 108. Greek . . . . late 12th century.
2. Oxford: Balliol 250 . . . . 13th ,,
3. Cambridge: Peterhouse 121 . . . . 13th ,,
4. Cambridge: Peterhouse 190 . . . . 13th ,,
5. Cambridge: Fitzwilliam. 155 . . . . 13th ,,
6. Oxford: Balliol 232 A . . . . 14th ,,
7. Oxford: Trinity 67. . . . . 14th ,,
8. Oxford: Merton 271 . . . . 14th ,,
9. Cambridge Univ. Lib. Ii. II. 10. . . . . 14th ,,
10. Oxford: Bod. Can. misc. 418 . . . . early 15th ,,
11. Oxford: New Coll. 226. Greek . . . . 15th ,,
12. Cambridge Univ. Lib. Mm. III. 11. Extracts . . . . . . 15th ,,

De motu animalium

1. Oxford: Balliol 250 . . . . 13th century.
2. Oxford: Balliol 250. Fragment . . . . 13th ,,
3. Cambridge: Peterhouse 12 . . . . 13th ,,
4. Cambridge: Fitzwilliam 154 . . . . 13th ,,
5. Cambridge: Fitzwilliam 155 . . . . 13th ,,
6. British Museum Add. 19582 . . . . 14th ,,
7. Oxford: Bod. Can. Lat. auct. 290 . . . . 14th ,,
8. Oxford: Merton 270 . . . . 14th ,,
9. Oxford: Balliol 232 A . . . . 14th ,,
10. Oxford: Trinity 67. . . . . 14th ,,
11. Cambridge Univ. Lib. Ii. II. 10 . . . . 14th ,,
12. Oxford: Bod. Can. misc. 412 . . . . 15th ,,
13. Oxford: Bod. Digby 44. Extracts . . . . 15th ,,
14. Oxford: New Coll. 226. Greek . . . . 15th ,,
15. Cambridge Univ. Lib. Mm. III. 11. Extracts . . . . . . 15th ,,

De plantis

1. Oxford: Bod. Auct. F. 5. 31 . . . . 12th century.
2. Oxford; C.C.C. 114. . . . . 13th ,,
3. Cambridge: Gon. & Caius 409. Fragment . . . . . . 13th ,,
4. Cambridge: Gon. & Caius 452 . . . . 13th ,,
5. Cambridge: Gon. & Caius 506 . . . . 13th ,,
6. Cambridge: Fitzwilliam 154 . . . . 13th ,,
7. British Museum: Harley 3487 . . . . 14th ,,
8. British Museum: Add. 19582 . . . . 14th ,,
9. Oxford: Bod. Can. Lat. auct. 291 . . . . 14th ,,
10. Oxford: Balliol 232 A . . . . 14th ,,
11. Oxford: C.C.C. 111. . . . . 14th ,,
12. Cambridge Univ. Lib. Ii. II. 10 . . . . 14th ,,
13. Durham Cathedral C. III. 17 . . . . 14th ,,
14. Durham Cathedral C. IV. 18 . . . . 14th ,,
15. British Museum: Sloane 2459 . . . . 15th ,,
16. Oxford: Bod. Bodley 675 . . . . 15th ,,
17. Oxford: C.C.C. 113. Greek . . . . 15th ,,
18. Cambridge: Peterhouse 184 . . . . 15th ,,

Summary of Latin Texts

	No. of MSS.
12th century	1
13th ,, ...	30
14th ,, ...	32
15th ,, ...	12

MEDIAEVAL ASTRONOMY

By J. L. E. Dreyer

The study of Mediaeval Cosmology and Astronomy has hitherto not attracted many students. This is perhaps partly due to the fact that the period in question is one of stagnation, during which astronomy made absolutely no progress in Christian countries, while the high state reached by science at Alexandria had gradually to be won back. But the chief reason of the neglect is that many interesting writings from the Middle Ages have never been printed and have therefore to be looked for among the manuscript treasures in great libraries, particularly in the Bibliothèque Nationale at Paris. The great work of M. Pierre Duhem, *Le Système du Monde, Histoire des Doctrines cosmologiques de Platon à Copernic,*[1] is therefore particularly welcome, and it is quite up to the high standard of excellence of his previously published historical works, *Études sur Léonard de Vinci* and *Les Origines de la Statique.* So far, five volumes of more than five hundred pages each have been published, and it is remarkable that so great a work should have appeared in France during the terrible struggle in which that country was then involved. The five volumes reach to the beginning of the fifteenth century; and how far the work will be continued may be doubtful, as the death of the author was announced in 1916. It was stated in the Paris Academy in December 1916[2] that the work was to have been completed in ten volumes, and that the fifth and sixth had been entrusted by M. Duhem's daughter to the Academy. The fifth volume appeared in 1917. But even if not completed according to the original plan, the work will be of exceptional interest on account of the great number of manuscripts which M. Duhem has examined, and from which he has given lengthy extracts.

The work is (so far) divided into three parts. 'Greek Cosmology' occupies the first volume and four-fifths of the second

[1] Paris, A. Hermann et Fils, 5 vols., 1913–17.

[2] *Comptes rendus*, December 18, 1916.

one. 'Latin Astronomy in the Middle Ages' reaches to the middle of the fourth volume, and is followed by 'The rise of Aristotelism', which at the end of the fifth volume is carried as far as Thomas Aquinas. As Greek Cosmology has been dealt with in two works published in England during the last fifteen years, that part of M. Duhem's work does not call for special notice in this place. Neither do the chapters of Part I dealing with Arabian astronomy (which the author considers as a mere continuation of Greek science) contain anything new. As a rule, the author shows a thorough acquaintance with the literature of his subject, though we have in a few cases failed to find references to important works, such as the *Liber Jesod Olam* of Isaac Israeli or *Le livre de l'Ascension de l'esprit* of Abu'l Faraj. It is particularly the chapters dealing with Latin Astronomy in the Middle Ages which will be of permanent value, as they give accounts of many manuscript treatises never before described.

The last great astronomer of the Alexandrian school, Claudius Ptolemy (about A. D. 140), wrote a complete compendium of ancient astronomy as finally developed by Hipparchus and himself. During the 270 years which had elapsed since the days of Hipparchus astronomy had certainly not stood still, but we know next to nothing about the progress made, as Ptolemy gives very little historical information. The details given by Pliny about the situations of the apsides of the excentric orbits of the planets show, however, that Ptolemy had more than the work of Hipparchus to build on. Pliny's source was no doubt the book *De novem disciplinis* by M. Terentius Varro, which is unfortunately lost. It seems to have been a sort of condensed encyclopaedia, and was superseded by writings of a similar kind from the fourth and fifth centuries which have come down to us. These are, the commentary to Plato's *Timaeus* by Chalcidius, the commentary to Cicero's *Somnium Scipionis* by Macrobius, and the encyclopaedic book *De nuptiis Philologiae et Mercurii* by Martianus Capella. Being written in Latin they were more readily accessible to Western readers than the lengthy Greek works of Proklus and Simplicius; and during the first half of the Middle Ages they were, together with Pliny's *Natural History*, the only books from which some knowledge of Greek science might be derived by students in the West.

The first feeble light after the dark night of the patristic writers came from Isidore, Bishop of Seville, who died in 636. When dealing with dangerous topics such as the figure of the world and the earth he does not lay down the law himself, but quotes 'the philosophers' as teaching this or that, though without finding fault with them. In this manner he repeatedly mentions that heaven is a sphere rotating round an axis and having the spherical earth in its centre. The water above the firmament mentioned in the first chapter of Genesis had of course to be brought in, and Isidore states that the Creator tempered the nature of heaven with water, lest the conflagration of the upper fire should kindle the lower elements. Isidore gives as his authorities Hyginus (author of a versified description of the constellations), Clement of Alexandria, and the patristic writers, but does not mention Pliny, so that it is no wonder that his knowledge is very fragmentary.

The Venerable Bede, who lived a century later (he died about 735), was better informed. The contents of his treatise *De Natura Rerum* are taken from Pliny, often almost verbatim; and the spherical form of the earth, the order of the seven planets circling round it, the sun being much larger than the earth, and similar facts are plainly stated. But the water around the heaven and the usual explanation of its existence could not be kept out of the book, even though Pliny did not mention it and though Bede had stated that the earth was a sphere. Another and much larger book on chronology (*De Temporum Ratione*) shows a fair knowledge of the annual motion of the sun and the other principal celestial phenomena. It is deserving of notice that Bede from his study of Pliny and from personal observation knew a good deal about the tides, and was the first to show that the 'establishment' of a port (or the mean interval between the time of high water and the time of the moon's previous meridian passage) is different for different ports. But the sphericity of the earth was still rather unpopular among ecclesiastics, and even in the first half of the ninth century Hrabanus Maurus, Archbishop of Mainz, thought it best to say nothing about it. He merely says that the earth is in the middle of the world, and tries hard to reconcile the roundness of the horizon with the four corners of the earth alluded to in Scripture. His statement that the heaven has two doors, east and west, through which the sun passes, also looks as if his point of view was much the same as that of the patristic writers.

But he was the last prominent author of whom this may be said, and from about the ninth century the spherical figure of the earth and the geocentric system of planetary motions were reinstated in the places they had held as facts ascertained with certainty among Greek philosophers twelve hundred years earlier.

Among the writers of the ninth century who paid any attention to the construction of the Universe, the most remarkable was John Scotus Erigena. In his great work *De Divisione Naturae* he shows that he is acquainted with Chalcidius and Martianus Capella, and for the first time we perceive a very curious influence which these rather inferior writers exercised throughout the Middle Ages. In the fourth century B. C. Herakleides of Pontus, struck with the fact that Mercury and Venus are never seen at a great distance from the sun, had come to the conclusion that these two planets move, not round the earth, as the sun and the other planets were supposed to do, but round the sun, so that they are sometimes nearer to us and sometimes farther off than the sun. But this idea was coldly received; it was quite ignored by Ptolemy and is only mentioned by Theon of Smyrna and Macrobius (without alluding to Herakleides), and by Martianus Capella and Chalcidius, who give the credit to Herakleides. Theon was not known in the Middle Ages, but the three other writers were held in high repute; and this led to the planetary system described by them being known to many mediaeval writers, though to most of them rather confusedly, as if they did not quite understand it. Thus Erigena says: 'As to the planets which move round the sun, they show different colours according to the quality of the regions which they traverse; I speak of Jupiter, Mars, Venus, and Mercury, which incessantly circle round the sun, as Plato teaches in the *Timaeus*. When these planets are above the sun, they show us clear aspects, they look red when they are below it.' Plato says nothing at all about this; but perhaps Erigena had only read Chalcidius and assumed that what he said was also to be found in the *Timaeus*. Chalcidius only mentions Venus as moving round the sun, but as he had already described the apparent motions of the two inferior planets, he probably made no distinction between them. M. Duhem seems to consider it highly creditable to Erigena that he extended the system of Herakleides to Mars and Jupiter, which nobody else did for fully seven hundred years, till Copernicus and Tycho Brahe let all the five planets move round the sun. But with regard to Jupiter, it is simply absurd to imagine that it is sometimes above the sun (i. e. more distant than the sun), sometimes below it (or nearer). As to Mars, that planet is certainly at opposition nearer to the earth than the sun is, but there is no reason to think that the astronomical knowledge of Erigena included that fact, as it is in other directions scanty enough and is confined to carelessly copied scraps from his few authorities. We need only mention his statement that half the circumference of the earth is equal to its diameter!

Probably also from the ninth century is another book about the Universe, *De mundi caelestis terrestrisque constitutione liber*, formerly ascribed to Bede, but quite certainly of much later date, since there are several allusions to the chronicles of Charlemagne. The author has a fair knowledge of the general celestial phenomena such as could be gathered from the above-mentioned sources, but no more. It is interesting to see that he favours the old idea sometimes met with among the Greeks, that the planets do not really travel from west to east, but from east to west, only more slowly than the sphere of the fixed stars do; so that Saturn, which comes to the meridian about eight seconds later every night, is the quickest planet, and the moon, which takes fully three-quarters of an hour longer than the fixed stars do, is the slowest—contrary to the usual idea, that Saturn, which takes 29½ years to go round the heavens in its orbit, is the slowest planet, and the moon, going round the heavens in 27 days, is the fastest.[1] We shall see presently that this primitive idea obtained many adherents towards the end of the Middle Ages. As to Mercury and Venus, the writer's opinion is, that they are sometimes above the sun and sometimes below it, as it is recorded in the *Historia Caroli* that Mercury was visible for nine days as a spot on the sun, though clouds prevented both the ingress and the egress being seen. But he does not say that they move in orbits round the sun. The writer shows himself somewhat independent of his authorities by adding a good deal of astrology and suggesting various rationalistic theories about the unavoidable 'supercelestial waters'.

Passing over the extremely elementary *Imago Mundi* of

[1] See for instance Plato, *Timaeus*, pp. 38–9, Leg. 821 sq.

doubtful age and authorship, we must next mention another work formerly counted among the writings of Bede, entitled Περὶ διδάξεων *sive elementorum philosophiae libri* IV. It was written by William of Conches, a Norman of the first half of the twelfth century.[1] It is strange that it should ever have been attributed to Bede, as it shows a freedom of thought which would have been impossible early in the eighth century. But his astronomical knowledge is often confused and erroneous. For instance, he knows that the orbit of the sun is a circle excentric to the earth, but he imagines that the great heat in summer is caused by the sun being at that time nearer to the earth than in winter. He is aware of the difference of opinion among the ancients as to the position of the solar orbit, whether it was just outside the lunar orbit (according to the Pythagoreans, Plato, Eudoxus, and Aristotle); or between the orbits of Venus and Mars, as taught by Archimedes and all subsequent writers, including Ptolemy. William of Conches thinks that this difference of opinion is caused by the fact that the periods of Mercury, Venus, and the sun are nearly equal, so that their circles must also be nearly equal and therefore are not contained one within the other but intersect each other. He therefore did not grasp the real meaning of the system of Herakleides, but merely conceived the three orbits to be nearly equal in size with their centres at short distances from each other and in a line with the earth; and his description agrees with a diagram given in an anonymous manuscript of the fourteenth century, copied by M. Duhem.

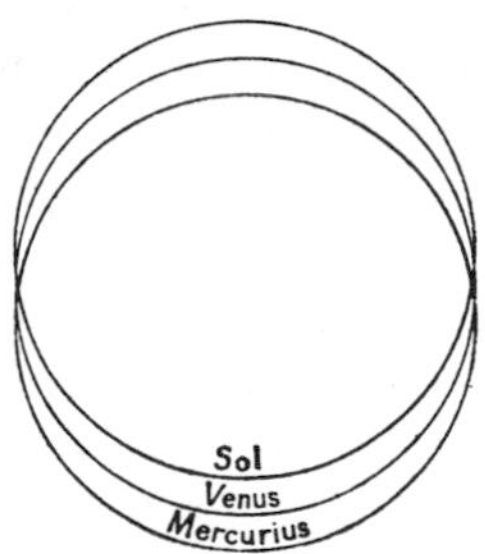

FIG. 1.

The question of the orbits of Mercury and Venus continued to crop up now and then, as long as Chalcidius and other late authors continued to be considered as great authorities by some writers who were much behind their own time. The Rabbi Abraham ben Ezra of Toledo (1119–75) in several of his astrological writings, which were printed in 1507 at Venice in a Latin translation, alludes to the orbits of Mercury and Venus being between those of the moon and the sun; but in one place he says

[1] Two manuscripts in the Bibliothèque Nationale give the author's name as William of Conches, and there are other proofs from other undoubted writings of his.

that the two planets are sometimes above and sometimes below the sun. The same expression is used in the following century by Bartholomaeus Anglicus in his encyclopaedic work *De proprietatibus rerum* (*c.* 1275). The only time that it is clearly and distinctly stated that Mercury and Venus travel round the Sun is in an astrological manuscript in the Bibliothèque Nationale of the year 1270 by an anonymous astrologer to the last Latin Emperor at Constantinople, Baldwin of Courtenay. After saying that the orbits of moon, sun, and three outer planets surround the earth, he continues: 'Li cercles de Vénus et de Mercure ne l'environent mie. Ainz corent environ le Soloil et ont lor centre de lor cercles el cors del Soloil; mes Mercurius a le centre de son cercle el milieu del cors del Soloil, Vénus l'a en la souraineté del cors del Soloil; et por ce sunt il dit épicercle, qu'il n'environent mie la terre, si cum j'ai dit desus des autres.' The author might have lived hundreds of years earlier; he knows nothing of Ptolemy or of the Arab writers on Ptolemaic astronomy, who lông before the time he wrote had become known in the west of Europe.

For while Europe had been content to pick up a few crumbs here and there, the East had been feasting on the intellectual repast left by the Greeks. Works on Philosophy and Science had been translated into Arabic, and Mohammedan authors had written text-books founded on them and had continued the work of the Greeks in Mathematics and Astronomy. Arabic authors began to be known in the West from about the year 1000; Gerbert (Pope Sylvester II) probably wrote a book on the astrolabe founded on Arabic writings, and several tracts on the same subject were written in the eleventh and twelfth centuries in France, especially at Chartres, at that time the principal seat of learning there. Translations were also made in Italy by Plato of Tivoli; but it was in Spain, where science was still under the protection of powerful Arabian kings, that the work of translation was chiefly carried on. The first translations were the work of a college of interpreters established at Toledo; an Arabian scholar translating a book into the vulgar tongue, and a Spaniard afterwards turning this into Latin. Ptolemaic astronomy became known about the middle of the twelfth century through the medium of the books of Al Battani and Al Fargani. The original work of Ptolemy, the *Syntaxis* or the *Almagest,* as it was generally called in Latin countries from a corruption of part of the Arabic title

(Al-$\mu\epsilon\gamma\acute{\iota}\sigma\tau\eta$), was first translated about the same time by Gherardo of Cremona, who died about 1184 at the age of 73. He seems to have spent most of his life at Toledo, where he went to find the Almagest. Seeing what a great number of valuable works in Arabic were to be found there, he learned Arabic and is said to have translated no less than seventy-four different works, both by Greek and Arabian authors. But it took a very long time before people could be found capable of mastering the great work of Ptolemy.

That there were some people in the middle of the twelfth century anxious to spread knowledge of astronomy may be seen from a manuscript in the Bibliothèque Nationale, examined by M. Duhem. The name of the author of the 'Tables of Marseilles' is not known; from internal evidence it appears that they were prepared about the year 1140. The author says that students of astronomy were compelled to have recourse to worthless writings going under the name of Ptolemy and therefore blindly followed; that the heavens were never examined, and that any phenomena not agreeing with such books were simply denied. He therefore decided to transform the astronomical tables of Al Zarkali, which were computed for the meridian of Toledo and adapted to Arab years, so as to arrange them for the meridian of his native city and according to years dated from the birth of our Lord. This attempt to make the Toledo tables known in Latin countries did not bear fruit immediately; but early in the thirteenth century imitations of the tables of Marseilles began to appear, adapted to the meridians of Paris, London, Pisa, and Palermo, even for that of Constantinople, at that time ruled by Latin Emperors. The London tables date from the year 1232; the author mentions Ptolemy, but evidently only knows his work by name.[1] This is certainly also the case with the celebrated little book on the Sphere by John of Holywood (Joh. de Sacrobosco), written in the first half of the twelfth century, which continued to be a favourite text-book for three hundred years and was repeatedly printed. He only had his wisdom from Al Fargani and Al Battani, for he copies a mistake made by them and omits what they omit.

[1] Similar tables, founded on those of Al Zarkali, were made at Montpellier towards the end of the thirteenth century by the Jew Jacob ben Makir, generally called Profatius.

But astronomical books were far from being the only ones transmitted through the Arabs. The philosophical books of Aristotle and of his commentators, as well as neoplatonic and Arab speculations, also crossed the Pyrenees. At first they were not welcomed by the Church, and at a provincial council held at Paris in 1209 it was decreed that neither Aristotle's books on natural philosophy nor commentaries on them should be read either publicly or privately in Paris. In 1215 this prohibition was renewed in the statutes of the University of Paris. But this resistance wore off by degrees; better translations both of Aristotle and of Arab astronomers were produced by Michael Scot; while Guillaume d'Auvergne, Bishop of Paris from 1228 (died 1248), lent his powerful aid to the spreading of knowledge. He was a prolific writer, and was the first to make serious use of Greek and Arab philosophy, rejecting what was contrary to the Christian faith and combining the rest with what the Church taught, to compose a philosophical system acceptable to the Christian world. Among his writings was a treatise *De Universo*, which stands half way between the old works of Isidore, Bede, Pseudo-Bede, Honorius, and the later encyclopaedias of Albertus Magnus and Vincent of Beauvais, containing more philosophy and less theology than the former. But his opinions on celestial motions are very confused. For instance, he thinks that one can 'by means of astronomical instruments and certain geometrical instruments' determine the distance of the earth from each of the fixed stars and from each of the planets. The waters above the moving spheres are neither fluid nor in a state of vapour; they form an ethereal mass, perfectly transparent and immobile, separating those spheres from the Empyrean.

The introduction of Aristotelian natural philosophy in the Universities of Paris and Oxford brought about a prolonged strife between Aristotelian ideas of the construction of the Universe and the Ptolemaic system of the world; or rather a revival in France and England of an old dispute which had existed first in the Hellenistic and then in the Mohammedan world. Aristotle had adopted the 'homocentric spheres' of Eudoxus to account for the motions of the planets; but though this would to some extent explain the chief irregularities in these motions, continued observations soon showed that the system was insufficient to 'save the phenomena', particularly as it could not account for the variable distance of a planet from the earth. A totally different system had therefore been developed at Alexandria in the course of nearly four hundred years, until it was completed by Ptolemy. According to this, a planet moved on the circumference of a circle (the epicycle), the centre of which travelled on a larger circle (the excentric or deferent) the centre of which was at some distance from the earth; but in such a manner that its motion was uniform, not with regard to the centre of the deferent, but as seen from another point, the *punctum aequans*. The centre of the deferent was midway between that point and the earth. Further complications had to be introduced to account for the motion in latitude of the planets. From a mathematical point of view this system was perfect, as it really could 'save the phenomena', that is, represent the actually observed motions with an accuracy nearly corresponding to that attainable by the crude instruments then in use. But it was totally at variance with Aristotelian Physics, the adherents of which viewed the movements around points outside the centre of the world with extreme disfavour. Long before Ptolemy's time attempts had therefore been made to reconcile the two systems. This was simple enough, as long as the deferent was assumed concentric with the earth. The epicycle might then be conceived to be the equator of a solid sphere, rolling between two solid concentric spheres. This idea is described by Theon of Smyrna (soon after A. D. 100), but it must be much older. It became untenable, as soon as the deferents became excentric circles. In the *Syntaxis* Ptolemy merely alludes to planetary spheres when describing the order of the various orbits (ix. 1); and his attitude with regard to the equivalence of the epicyclic and excentric theories shows that he had broken with the idea described by Theon and did not attribute any reality to the multiple motions of the *Syntaxis*, but merely considered them as geometrical means of representing the real motions. But in a later work, *Hypotheses of the Planets*, or rather in the second book of it, Ptolemy's ideas are quite different.[1] Here he proposes to do for the complicated theories of the *Syntaxis* what the system of Theon did for the simple epicyclic motion, producing not a mere model but a real representation of the constitution of the universe,

[1] The Greek original is lost, but a translation into Arabic has been preserved, from which a German translation was printed in 1907 (*Claudii Ptolemaei Opera*, ed. Heiberg, T. ii).

as real as that described in Aristotle's *Metaphysics*. The epicycle-sphere now fits between two excentric spherical surfaces which touch two other surfaces (an inner and an outer one), in the common centre of which the earth is situated. This system of the world does not seem to have been a success ; in the neoplatonic schools the theories of the *Syntaxis* appear to have been more valued, although the old Platonic and Aristotelian dogma, that every celestial motion must be circular and uniform round the centre of the earth, still found partisans.

Towards the end of the eighth century Mohammedan nations began to become acquainted with Alexandrian astronomy, in the first instance through the medium of northern India, where a knowledge of Greek science had spread in the first couple of centuries after the conquests of Alexander the Great.[1] The system of spheres seems to have appealed strongly to Eastern minds ; and throughout the time when astronomy continued to be successfully cultivated in the Mohammedan world we find that various combinations of spheres were proposed by people who could not be satisfied with the Ptolemaic system of circles, while the latter was accepted and used by professional astronomers. The first to describe the spheres was Tâbit ben Korrah, in the second half of the ninth century. He seems to have been the first to fix the number of spheres at nine, and he was followed by the 'Brethren of Purity' in the tenth century and by Ibn al Haitham (*c.* A. D. 1000).[2] The necessity of introducing a ninth sphere above the eighth sphere (the sphere of the fixed stars) was due to the imaginary phenomenon of trepidation or oscillatory movement of the equinoxes. This dates back to the time before Ptolemy, but he quite ignored it and taught that the Precession of the Equinoxes is uniformly progressive, while Tâbit (though speaking with a certain reservation) accepted the phenomenon of trepidation as real. The Arabian combinations of spheres were mainly borrowed from Ptolemy, though with modifications. It

[1] From what M. Duhem says (vol. ii, p. 213) it looks as if he thought that Arabian astronomy was founded on indigenous Indian knowledge. But it is quite certain that the Indians derived all their knowledge of planetary motion from the Greeks. See J. Burgess, in *Journal of the R. Asiatic Society*, October 1893, pp. 746 sq., and Dreyer, *Hist. of the Planetary Systems* (Cambridge, 1906), pp. 240 sqq.

[2] Known in the West as Alhazen, author of a celebrated book on Optics.

was particularly in Spain that the opposition to the Ptolemaic system of excentrics and epicycles came to the front, being intimately connected with the rapid rise of Aristotelian philosophy in that country in the twelfth century, which culminated in the work of Averroes, the greatest philosopher of Islam. An ingenious attempt at reviving the principle of homocentric spheres in a perfectly novel manner was made by the astronomer Al Bitrugi (Alpetragius), though the leading idea was probably due to the philosopher Ibn Tofeil.

This system of homocentric spheres differs from that of Eudoxus and Aristotle by assuming that the prime mover (the ninth sphere) everywhere produces only a motion from east to west, the independent motion of the planets from west to east being rejected. We have already mentioned that this was a very old idea which had been revived in Europe by Pseudo-Bede. But Al Betrugi saw that this was not sufficient, as not only is the pole of the ecliptic different from that of the equator, but the planets do not even keep at the same distance from the pole of the ecliptic but have each their motion in latitude as well as a variable velocity in longitude, all of which had to be accounted for. This is done by letting the pole of each planet's orbit describe a small circle round a mean position (the pole of the ecliptic) in the synodic period of the planet.[1] But the system was not worked out in detail; and only philosophers who wanted nothing more than a representation of the principal phenomena could be satisfied with it. In the eyes of astronomers it had many faults, the greatest being (as in the case of the system of Eudoxus) that it assumed a planet to be always at the same distance from the earth.

When Michael Scot about the year 1230 had produced his translations, the attacks of Averroes and Al Betrugi on the epicyclic system spread rapidly among the Scholastics. Though people who only desired to account for the apparent motions of the planets as seen projected on the celestial vault continued to follow the rules of Ptolemy, philosophers were greatly concerned

[1] The account given by M. Duhem of this historically important system is most unsatisfactory. For further details see *Hist. of the Planetary Systems*, p. 265, a book which seems to be unknown to M. Duhem. It is curious to see how astronomical historians have fought shy of explaining the system; see e. g. Delambre, *Hist. de l'Astr. du Moyen Age*, p. 174.

about the contradiction between Aristotle and Ptolemy. During the whole of the second half of the thirteenth century this agitated the two rival orders of Dominicans and Franciscans, who dominated the University of Paris. Among the former Albertus Magnus was at first the most prominent. He was much attracted by the system of Al Bitrugi, which he thought was very simple, because he ignored the small circles and thereby made it quite useless. It was in this simplified form only that the system continued to be known to most of the Scholastics, which sufficiently characterizes their superficial knowledge of celestial motions. Yet Albert was quite aware of the fatal objection to every form of homocentric system, and he finally declared for Ptolemy. The same was the case with his disciple Thomas Aquinas. For the simplified system of Al Bitrugi he substituted the idea that celestial bodies are animated by two movements; the first is a uniform rotation from east to west, a principle of eternal duration; the second is a rotation from west to east round the poles of the ecliptic, a principle of generation and transformation, a movement in which the different orbs take part to a lesser extent the more noble they are. Like Albert he ends by abandoning Averroes for Ptolemy, chiefly on account of the change of distance.

If we turn to the Franciscans we are met with more hesitation as to the choice between the rival theories. Bonaventura, the Doctor seraphicus, does not seem to have devoted much attention to astronomy. According to him there is a ninth heaven, the aqueous one, which is the primum mobile; some philosophers have perceived that the firmament has a proper motion of 1° in a hundred years,[1] but whether this is true or not it is certain that Doctors of Theology admit that there is a moving heaven without stars.

But in Roger Bacon we meet at last with a man who was thoroughly acquainted with the astronomical writings both of Greeks and Arabs. At Oxford he was under the influence of Robert Grosse-Teste, afterwards Bishop of Lincoln, who had devoted some attention to astronomy and was a follower of Ptolemy, except when he wanted to be a metaphysician and had to follow Al Bitrugi. But it was not till Bacon went to Paris (about 1235) that he was able to study scientific problems seriously. M. Duhem gives a lengthy account of a manuscript

[1] The amount of precession according to Ptolemy.

in the municipal library of Amiens, containing several series of questions on the *Physics* and *Metaphysics* of Aristotle. They are probably written by pupils of Bacon at Paris, at latest about the year 1250. The questions on the *Metaphysics* must be the earliest; the teacher does not appear to know Ptolemy's works, and only to have heard of Al Bitrugi. The questions on Aristotle's *Physics* show more knowledge, especially of the system of Eudoxus and of Precession. The subsequent writings of Bacon show how he persevered in his study of cosmology and astronomy; but he continues all his life to hesitate between the two systems of the world. He studies the *Almagest*, he borrows from Al Fargani what he says about the dimensions of the planets and of their orbits; on the other hand he makes himself thoroughly acquainted with the theory of Al Bitrugi, the details of which hitherto had scared readers in France. In his *Opus tertium* he gives a carefully written summary of that theory, and then gives an account of the system of solid orbs described by Ptolemy in his *Hypotheses* and taken up by Ibn al Haitham. Bacon acknowledges that it does away with many of the objections formulated by Averroes against the epicyclic system, but he thinks that there are too many questionable hypotheses in it. His main objection is, that the two bodies between which each deferent is comprised have in their various parts different thickness,[1] but this cannot be the case in celestial bodies on account of their simple and homogeneous nature. He also thinks that a celestial body cannot be supposed to be devoid of all motion. Other objections made by Bacon seem to show that he cannot have been acquainted with the second book of Ptolemy's *Hypotheses*, the Greek original of which was probably lost before that time; and that he only had before him an incomplete résumé of the systems of spheres

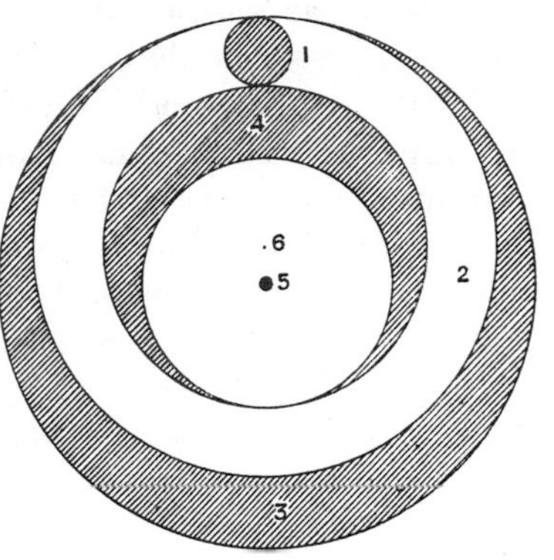

FIG. 2

1. Epicycle Sphere. 2. Excentric Sphere. 3. The Surrounding Sphere. 4. Complement of Surrounding Sphere. 5. The Earth. 6. Centre of Excentric Sphere.

[1] Compare the diagram in the *History of the Planetary Systems*, p. 259 (reproduced above), the surrounding sphere and its complement.

of Ptolemy and Ibn al Haitham.[1] Bacon therefore never reached any conclusion as to which system of the world was the true one.

But a younger contemporary of Bacon among the Franciscans, Bernard of Verdun, made up his mind to break with Aristotle and Al Bitrugi. In a *Tractatus optimus super totam Astrologiam*, of which two manuscripts are known to exist, and which shows familiarity with Bacon's works, he begins by counting up the facts which must be explained: the change of velocity of the planets; the variability of the moon's diameter, the moon being more or less completely eclipsed even in the same point of the ecliptic; the upper planets (particularly Mars) being brightest at opposition to the sun, while Mercury and Venus are of greater brightness after leaving the sun and moving eastward than when they leave it to move westward. All these facts distinctly contradict all homocentric theories, and Bernard therefore finally rejects the system of Al Bitrugi, with which he shows that he is well acquainted. In the system of spheres of Ptolemy or its imitation by Ibn al Haitham he sees the means of shielding the Ptolemaic system from Aristotelian attacks; and as the theory of excentrics and epicycles is the only one which is able to produce tables of planetary motions and 'save the phenomena', Bernard does not hesitate to proclaim it as the true system. The above-mentioned objection raised by Bacon he brushes aside by saying that the surrounding spheres, &c., are not proper celestial bodies, but like those parts of a cithern which do not give any sound.

This treatise by Bernard of Verdun seems, at least for the University of Paris, to mark the epoch when the Ptolemaic system began to reign absolutely among students of astronomy; the adherents of Al Betrugi had to give up the struggle. There are various other tracts in existence from the time around the year 1300 which confirm this result, while they show at the same time that the question about the precession of the equinoxes, whether it was a steady progressive motion or only an oscillation, was still unsettled, which involved uncertainty as to the number of spheres above those of the planets. The Alfonsine Tables, prepared under the direction of King Alfonso X of Castille, were finished about 1270, but they were probably not at once issued to

[1] The fact that he calls the system *Ymaginatio modernorum* also points to his only having known an Arabic account of it.

the public. At any rate it is certain that they were not known at Paris until towards the end of the thirteenth century, the tables of Al Zarkali (the tables of Toledo) being still in use there. The time had come however when some people among the Christian nations had begun in a small way to occupy themselves with practical astronomy, instead of merely speculating whether there were nine or ten heavens, or considering whether an astronomer or a philosopher was the best guide among the stars. There is a codex in the Bibliothèque Nationale which contains several tracts from the end of the thirteenth and the beginning of the fourteenth century bearing witness to this change of study. Among these are two by Guillaume de St. Cloud, one being an Almanac for twenty years from 1292, the other a Calendar, the date of which is 1296, which gives the time of entry of the sun into each of the twelve signs for two hundred years before and after the year 1296. At the beginning of the Almanac it is stated that the tables of Ptolemy of Alexandria, Tolosa (?), and Toledo do not agree with observations; there is no mention of the Alfonsine Tables. From observed solstitial altitudes of the sun in 1290 the obliquity of the ecliptic is found to be 23° 34′ and the latitude of Paris 48° 50′. Guillaume also determines the time of the spring equinox of 1290 (March 12, 16^h) and corrects the errors of the Tables of Tolosa by the simple expedient of correcting the mean motions thus: of Saturn by $-1° 15'$, of Jupiter by $+1°$, of Mars by $-3°$.[1] M. Duhem does not explain what the Tables of Tolosa are, but a fragment of a tract by Guillaume de St. Cloud quoted by Nicolaus de Cusa gives the clue to this riddle.[2] The fragment gives the errors of the Alfonsine Tables for the three planets, and the amounts are exactly the same as those given in the codex as the errors of the Tables of Tolosa. This shows that the Alfonsine Tables had become known at Paris by the year 1292.

The same interesting codex also contains a tract by Johannes de Muris, a native of Normandy. He quotes the determination of the equinox of 1290 and describes how he repeated it at Evreux in 1318 (March 12, $16^h 40^m$); he says this agrees with King Alfonso's and Guillaume de St. Cloud's observations. Joh. de Muris and Firmin de Belleval wrote a report on the reform of the calendar, by order of the Pope, which if adopted would have settled this question more than two hundred years before the time when the

[1] Duhem, T. iv, p. 18. [2] Ibid., p. 23.

Gregorian reform was actually carried out. Firmin wrote a little book on Meteorology which was printed at Venice in 1485.[1]

Another astronomer of the first half of the fourteenth century who must not be passed over here, though not connected with Paris, was a Jew, Levi ben Gerson of Avignon, who died in 1344. He wrote against the system of Al Betrugi and (what is more important) he invented (or at least introduced) the instrument known as Baculus Jacobi or Cross Staff for measuring the angular distance between two stars ; and better still, he applied a diagonal scale to it. The latter invention (not mentioned by M. Duhem) appears to have remained unnoticed for about two hundred years.[2] Practical astronomy was also cultivated by Johannes de Lineriis (Jean de Linières), from whose hand there are several manuscript treatises in the Bibliothèque Nationale, among them a guide to the use of the Alfonsine Tables and *Theorica Planetarum*, anno Christi 1335. The latter is an account of the Ptolemaic System from the *Almagest* (without spheres) ; there is a chapter on the motion of the eighth and ninth spheres, in which it is shown that Tâbit's theory of oscillation must be rejected, as the equinox has now receded more than that theory allows; the author is inclined to adopt the theory proposed by Alfonso X combining progressive and oscillatory motion. Jean de Linières also produced a catalogue of positions of forty-seven stars, the first attempt in Europe to correct some of the star-places given in Ptolemy's Catalogue.[3]

[1] *Opusculum repertorii pronosticon in mutationes aeris.* M. Duhem (T. iv, p. 42) says that according to a manuscript copy in the Bibl. Nat. there was an interval of 68 years between the Alfonsine Tables and the epoch of certain tables ; this gives the date of the book 1252 + 68 = 1320. The sentence quoted occurs in the printed book on f. 12v at the foot, but the interval is 86, which gives 1338. The context shows that 86 is a misprint ; yet compare f. 3v, where 1338 is given as the epoch of some star-places brought up from Ptolemy's catalogue.

[2] M. Duhem (T. iv, p. 40) says that the use of the baculus was introduced among Portuguese navigators by the German scientist Martin Behaim towards the end of the fifteenth century. But it has been conclusively shown by Joaquim Bensaude (*L'Astronomie nautique au Portugal à l'époque des grandes découvertes*, Berne, 1912) that the baculus was known in Portugal long before the time of Behaim, to whom and his compatriots the Portuguese owed nothing. M. Duhem quotes this book in a footnote without noticing that it demolishes what he has just stated in the text.

[3] See a paper by G. Bigourdan in the *Comptes rendus*, December 1915 and January 1916. The catalogue was printed in Riccioli's *Astronomia Reformata*, i, p. 216.

Other manuscripts from the fourteenth century show that various teachers at Paris continued to expound the Ptolemaic system, with or without spheres. At the beginning of the century Aegidius, a native of Rome, wrote a Hexaëmeron (printed several times in the sixteenth century) in which he proposes to let the epicycle-sphere lie, not between two spheres, but in a cavity in the celestial matter of the form of an anchor ring. Some of these writers had not a very clear idea of what they wrote about; e.g. Albert of Saxony in a Commentary to Aristotle's *De Caelo* (also printed several times) says that the moon does not move in an epicycle; for if it did, the figure of a man in the moon carrying a bundle of sticks would sometimes be seen upside down! He means of course that the moon would not always turn the same side to the earth.

Though M. Duhem several times alludes to the Alfonsine Tables, he has evidently not examined any codices of them. He has thus missed the interesting fact, that the tables first published at Toledo about the year 1270 were totally different from those first printed in 1483. The latter were a modification of the original tables made at Paris, probably by Jean de Linières. At Oxford, where there seems to have been a good deal of interest taken in astronomy from the beginning of the fourteenth century, the tables in their original form remained in use longer than anywhere else.[1]

While science had thus begun to be earnestly cultivated in France and England, Italy had remained behind. Though Gherardo of Cremona had translated the *Almagest* about the year 1175, more than a century passed before a single Italian studied it, while the text-book of Al Fargani was the usual guide. But it is remarkable how badly even that book was understood, the most extraordinary blunders being made by the best of the Italian writers. Dante is an honourable exception; there are no blunders in his cosmological ideas, neither in the *Divine Comedy* nor in the *Convivio*. The latter shows the craze for astrology which prevailed in Italy, while this branch of learning was at Paris always of secondary importance. The Italians took no part in the dispute whether a system of Physics deduced from peripatetic principles

[1] Cf. Dreyer, 'On the original form of the Alfonsine Tables', in *Monthly Notices of the Royal Astronomical Society*, vol. lxxx, pp. 243–62.

or a science constructed to agree with observed facts should conquer. Only Petrus de Abano (about 1300) alludes to it, but the dispute had been settled before he went to Paris. In the Universities of Spain and Portugal the fight did not begin till the middle of the fifteenth century and lasted about a hundred years, during which time objections were raised which had been demolished elsewhere long before.[1]

The second half of M. Duhem's fourth volume and the whole of the fifth deal with Arabian and Scholastic Philosophy. His account of mediaeval cosmology and astronomy reaches to the end of the fourteenth century, except that it does not deal with any English writers after Roger Bacon. The revival of astronomy in Europe has hitherto been supposed to date from the middle of the fifteenth century, and to have commenced with the labours of Cusa, Peurbach, and Regiomontanus in Germany. We know now from the researches of M. Duhem that the revival began in France fully a hundred years earlier, while the studies of M. Bensaude have shown that the scientific light spread by the Arabs in Spain and Portugal had never been put out, so that the navigators who found the way to the Indies and to the New World had nothing to learn from German astronomers. We shall look forward with great interest to the promised sixth volume of M. Duhem, while deeply regretting that his early death should have prevented him from completing this vast monument of learning and perseverance.

[1] In the south of France a last attempt to substitute spheres for epicycles and excentrics was in the fourteenth century made by Levi ben Gerson in his work *Milchamot Adonaï*, of which an account was published by Carlebach in 1910 (Duhem, v, pp. 201 sqq.).

ROGER BACON AND THE STATE OF SCIENCE IN THE THIRTEENTH CENTURY

By Robert Steele

The real value of a man's work can only be estimated with any approach to accuracy when it is seen against the background of the intellectual life of his time: when his contribution to the world's thought is confronted with the ideas of his contemporaries, whether they run together in harmony or diverge into independent ways of development. The position of Roger Bacon in the history of science depends not only on the actual contributions to the stock of knowledge he made, but also, and more especially, on his mode of approach to the problems of his age, and on the foresight with which he selected the objects of his study.

The story of his life has often been told, and is tolerably familiar to most readers. Born about 1214, the year before Magna Carta, he first appears in history, if it be indeed the same, as a young clerk at the court of Henry III in 1233. Soon afterwards he left Oxford and went to Paris, at least for a time, before 1236. He seems to have travelled in Italy, dedicated a book to Pope Innocent IV, returned to Paris, lectured there as a Master in the University, and returned to Oxford about 1251. He joined the Franciscans about 1254, and towards 1257 was sent to their Paris convent, where he remained ten years. In 1266 he received the commands of Pope Clement IV to send him a fair copy of his works, but Clement's death in 1268 frustrated any hopes from him, and in 1278 he was condemned for teaching 'suspected novelties' and, tradition says, imprisoned till 1292, in which year he composed his last work, and shortly afterwards died.

The public life of Roger Bacon, then, extends from 1236 to 1272, the middle years of the great thirteenth century, almost the same as that of Louis IX, St. Louis of France, 1236–70. He was a contemporary of two great Popes, Gregory IX (1227–41) and Innocent IV (1242–54), and of the Emperor Frederic II, the new 'Stupor Mundi', who died in 1250. Two great Councils were held in his time at Lyons, and two Crusades were undertaken, while the Latin Empire of Constantinople came to its end and the Greek rule there was restored. The Mongols had spread out and conquered from China to Silesia. Europe from end to end was filled with unrest and criticism of established authority, while authority was asserting itself with varying degrees of success. The strife between Pope and Emperor ended in the defeat of the Hohenstauffen and the extinction of his line, leaving the Imperial crown vacant for a score of years. Heresy stoutly fought a losing battle in the south of France, while in the north free cities established their rights under the royal protection. In England the controversies begun under John culminated in the summoning of a Parliamentary assembly by Simon de Montfort in 1265 before his defeat and death at Evesham. Among the masses this feeling of unrest manifested itself in opposition to the clergy, usually by a more or less good-humoured incredulity towards their personal character and teaching, but sometimes showing itself, as in the case of the revolt of the Pastoreaux, in the determination to destroy the clergy, suppress the monks, and turn on the knights and lords, as the Custos of the Franciscans of Paris puts it.

Civilized Europe was in those days comparatively small. Roughly speaking, from the Black Sea to the North Sea the northern boundary of civilization lay along the valleys of the Danube and the Rhine, and its common language was some sort of French, from England to Constantinople, and along the Mediterranean. French nobles ruled in Athens and Thebes, a French king gained the crown of the Sicilies, and everywhere there was a movement to and fro which made for internationalism. With the end of the thirteenth century this particular unifying influence had decreased almost to the vanishing-point, but others remained, only less potent. Every magnate of Europe, lay or clerical, had at some time or another of his career business to transact with the Papal Court: and this caused a continual stream of suitors journeying to Rome or returning from it to their own lands. Every prelate had to visit, at least once, the threshold of the Apostle, every ecclesiastical lawsuit found there its final court of appeal. The existence of the great Universities was another influence making for the international spirit. The readiest way to high promotion in the Church, and often in the State, for a poor man, was eminence in the law, Canon or Civil, and this eminence could only be attained in the schools of Bologna, while Theology, the Queen of Sciences, had her chosen seat at Paris, and welcomed students from every country to her bosom. Lastly, modern commerce was beginning to appear in the great cities situated on the routes between the East and the West; and Florence, Genoa, and Venice were extending their relations in every direction within the ring-fence of Christendom, and sending outposts to the ends of the known earth. Interpenetrating all this was the Jewish community, whose solidarity and love of learning make the exact amount of its influence on Christendom hard to estimate.

The temperament of this great commonwealth is reflected in its popular literature, especially that of France. The two great vernacular works of Bacon's period are *Reynard the Fox* and the first part of the *Romance of the Rose*, the former a satiric criticism of life—of Church and State, not political but human; the second, gathering up all the poetry of its predecessors and veiling it in a symbolism instinct with study. No one can be said to have any real conception of mediaeval thought to whom these books do not appeal. The passionate love of beauty and the tolerance for human weaknesses of the age reveal themselves too in the architecture and the sculpture of such a church as Rheims was as much as in the literature of the day: they are a background which must be borne in mind while our attention is directed to other aspects of the thought of the day.

As the distinguishing feature of the late twelfth century was a revival of literature, so the thirteenth century was distinguished by a revival of science. The immediate cause of this was the introduction of the scientific works of Aristotle to western Europe, in nearly every case through the medium of Arabic translations. Up to the middle of the twelfth century the only works of Aristotle known were the *Categories* and *De Interpretatione* in the translation of Boethius, which, with the *Isagoge* of Porphyry and the *Timaeus* of Plato, represented to Western Europe the sum total of the scientific work of the Greeks. Not that Greek was unknown in these lands; on the contrary, a knowledge of it seems to have been fairly common, but the interest of the learned of the Byzantine Empire and of Magna Grecia lay in mysticism and devotional theology, while the ordinary speaker of Greek was about as fitted to translate Aristotle as an English merchant in an Asiatic port would be to translate modern philosophy or biology into the language of his surroundings.

It was in the Arabic-speaking countries that the new scientific movement began. Its great names are Avicenna (d. 1057), Algazel (d. 1111), and Averroes (d. 1198). Their commentaries on the text of Aristotle represented the highest point of the efforts of the battle between reason and faith, between the philosophers and the theologians of Islam. The battle spread from Islam to the Synagogue, but though the Mohammedan theologians finally crushed the school of Arabic Aristotelianism, its Jewish disciples carried on the tradition, and played a not unimportant part in its transfer to the Christian community.

The texts used by the Arabs were translations, in some cases made from Syriac versions, in others directly from the Greek; from manuscripts not always perfect, as may be seen from the phrase 'album in Greco'—a blank space in the Greek—so often met with in early Latin manuscripts of Aristotle. When these translations were turned into Latin, it was usually by means of two persons, a Jew or native who knew Arabic and gave the sense of the passage in Latin or the vernacular, and a writer who set down the passage in a more or less coherent and intelligent form. A good deal of our information as to these versions comes from the pen of Bacon himself, who knew personally some of the translators, and was markedly sensible of the faults in their work.

The spread of knowledge in the thirteenth century lay through three distinct channels, the Universities, the schools of religious houses both of the older orders and of the new friars, and through little groups and individual scholars working on their own account without any organization to back them up. The Universities were the chief means of spreading the study of philosophy and theology; the religious houses more or less preserved the chain of literary study and of historiography, in addition to their normal devotional writing; while the study of astronomy, cosmogony, alchemy, and such experimental science as existed was in the hands of small groups such as those of Pisa, Marseilles, and London, who issued astronomical tables for the calculation of the position of the sun, moon, and planets; of the alchemists of Italy, France, and, perhaps, England; and of such experimentors as Peter de Maricourt, the 'magister experimentorum', and others whose name and work are lost to us by their isolation and obscurity. Bacon himself for many years of his life seems to have belonged to this class.

It is, however, to the Universities that we must turn for the main influences on European thought, and more especially to Paris and Oxford. It was to them that the new Aristotelianism came, at first possibly as summaries of the text with the commentaries of Avicenna and Algazel, and, as curiosity was aroused, by more or less complete versions of the text of 'The Philosopher' himself (Aristotle) and of the works of 'The Commentator' (Averroes). The first of these versions that we can date with certainty were made in Spain about 1150 at Toledo, which long continued to be a centre of translation. It was from there that the new text-books came, according to Giraldus Cambrensis, 'libri quidam tanquam Aristotelis intitulati Toletani Hispanie finibus nuper inventi et translati, Logices quodam modo doctrinam profitentes', which led to the adoption of so many 'novitates et hereses' that their use was prohibited in Paris in 1210. These must have caused inquiry at Constantinople after its capture by the Latins in 1204, for in 1209 we learn 'legebantur Parisiis libelli quidam de Aristotele ut dicebantur compositi, qui docebant Metaphysicam, delati de novo a Constantinopoli et a Greco in Latinum translati, qui . . . iussi sunt omnes comburi', at the council next year.

Thus reaction triumphed for a while, and in the earliest statutes of the University of Paris (1215) dealing with the course of study we find that the only works of Aristotle studied were the old and new *Dialectics* and the *Nichomachean Ethics* (four books) with the *Topics* of Boethius. The prohibition was renewed by Gregory IX in 1231, but, long before, the *Physics* in one form or another were studied. Giraldus Cambrensis, in his old age (*c.* 1220), complains of young men arriving at the professorial chair (the M.A. degree) in three or four years instead of the twenty years of his own time, and devoting themselves to logic and physics instead of the poets and historians who used to be studied. William of Auvergne, who became Bishop of Paris in 1228 and was still writing in 1248, quotes the *Metaphysics*, the *De Anima*, the *Physics*, the *De Caelo et Mundo*, the *Meteorics*, the *De Animalibus*, the *De Somno et Vigiliis*, and the *Ethics*. Everything goes to prove that the end of the first quarter of the thirteenth century saw a nearly complete Aristotle in the hands of the Western world. But in practice these translations, passing through the hands of unnumbered and anxious copyists, soon grew to be wofully corrupt. Even Bacon himself, lecturing at Paris, is led into accepting as a dictum of Aristotle the exact opposite of what now appears in the mediaeval texts, and we may perhaps attribute to this unfortunate experience a part of the criticisms which he lavishes on the ignorance of the translators.

It was not, however, the study of Aristotle that first aroused the mind of western Europe to the pursuit of science. Only one branch of study at that period could present in the fullest way the essential requisites for a complete science, the study of Astronomy. Its two branches, theoretical and experimental, allowed for close and detailed observation, the construction of theories founded on that observation, the prediction of future phenomena from theory, and the verification of these predictions, in a way that no other study could begin to approach: the only other scientific studies of that time, Medicine and Alchemy, being at the stage of empirical arts. Astronomy was another of the heritages of Europe from the Arabs. Since the time of the Greeks its greatest advances had been made in the country of its origin, at Bagdad. The *Almagest* of Ptolemy was translated into Arabic about A. D. 800 and reached Europe at the end of the twelfth century. The observatory at Bagdad was a centre of study, and here Thebit was led to his theory of the trepidation of the equinoxes to explain discrepancies between the calculated and observed positions of the heavenly bodies. From here a constant influence was exercised west and east from Granada and Cairo to Samarkand. With better observations, more correct tables were from time to time calculated and issued, and these tables were seized on by the Western world when they came into their hands, and adapted to their use. With each advance in precision came improvements in geometrical methods, more accuracy in the determination of periods like the length of the solar year or the relation of its length to the lunar month, to name no others. But in all this Arab science there was no fundamental difference from its Greek source, nothing was changed by the additions made, so that when it came into Europe it harmonized well with the traditional astronomy handed down from the Latins by Isidore of Seville and his copyists.

The science of Mathematics had grown with the needs of Astronomy: the art of the algorists had been improved by the publication of the *Liber Abaci* of Fibonacci in 1202, and was now perfectly adequate not only for the ordinary needs of commerce but for the intricate calculations demanded by the observations of astronomers: a general method of extracting roots had been obtained, and was taught in the ordinary work of the schools. The works of Euclid had been translated by Adelard of Bath in the twelfth century and were revised by Campanus of Novara in 1246, and a sufficient amount of spherical geometry had been obtained for the needs of the day, while a sort of algebra and the elements of trigonometry only awaited a scientific nomenclature to be recognized as new branches of the science. What was needed to make it a vigorous study were better teachers and a wider outlook.

The science of medicine received a new impulse in the early part of the thirteenth century by the translation of the medical work of Avicenna, made by the order of Frederick II—a translation which dominated the teaching of medicine till the seventeenth century, and long survived its usefulness. It furnished a complete body of medical teaching, theoretical and practical, and, though Avicenna did not differ greatly from his Eastern predecessors, it introduced some new theoretical principles as a foundation or as an explanation of medical practice. But this translation was only one of the services rendered to medical science by the court of Sicily: a far greater one was the insistence on the value of dissection in the course of medical study. In 1231 the Emperor issued an ordinance forbidding any one to practise surgery, unless he had studied at least a year in human anatomy, and he is said (on, we are afraid, insufficient grounds) to have ordered the medical schools of Salerno to dissect publicly a human corpse at least once in five years. But it cannot be said that this practical side of the study of medicine has left much trace on the writings of the thirteenth century, perhaps because surgeons were of an inferior class to physicians, regarded as handicraftsmen simply obeying their orders, perhaps because the needful stimulus to original thought was drowned by the vast mass of literature thrust upon it.

Later in the century the University of Bologna was a great centre of medical teaching. An English bishop of Durham in 1241, Nicholas of Farnham, had been a professor of medicine in Bologna, but the formation of a scientific school there dates from about 1260 and is associated with the name of Thaddeus of Florence, who seems to have founded his teaching on Avicenna.

It is to Bologna later on at the end of the century that we owe the close connexion between medicine and astrology. Surgery was well taught there, judging by the many famous names of its pupils—Theodoric of Lucca, William of Saliceto, and Roland of Parma among others. Anatomy was taught by public annual dissections to which medical students of over two years' standing were admitted.

Of the Experimental Science of the early thirteenth century we know almost nothing except that it existed, and that some teachers of it were famous enough to be called ' Experimentator '. Thomas of Cantimpré quotes one such. Villars d'Honcourt, the thirteenth-century architect, has left us a sketch of a perpetual motion machine he had designed, the properties of the magnet had been studied, and it is difficult to believe that the fine architecture of the period was arrived at without an exact knowledge of mechanical law.

Chemistry and Alchemy were indistinguishable, but metallurgy was sufficiently advanced to ensure that the properties of the commoner metals and ores were known experimentally and that a theory to account for their more obvious properties should be evolved. One has only to consider the number of treatises quoted by Albertus Magnus to realize the mental activity expended on this subject. But no scientific study was possible until a much wider and deeper knowledge was attained, sufficient to indicate the points to which inquiry should be devoted. In the thirteenth century only empirical knowledge could be hoped for.

In this world of awakening science Roger Bacon began to take part somewhere about 1235, after having been at Oxford for some years, and having possibly attended the teaching of Grosseteste, who was lecturing in the Franciscan schools from about 1230 to 1235. There is no doubt that Grosseteste (*maximus naturalis et perspectivus quem vidi*) was the man from whom more than any other Bacon derived, not only much of his knowledge and many of his theories, but above all his love of mathematics and his insistence on the value of a mathematical demonstration. But certain features of his early teaching of astronomy in the notes of his lectures preserved by the Amiens manuscript lead us to the conclusion that Bacon's early studies were in the hands of more conservative teachers, whose astronomy was derived from that current before the invasion of Mohammedan science.

However that may be, Bacon seems to have come to Paris with a reputation good enough for him to be taken up by the Chancellor of the University, who advised him in the direction of further studies. Bacon was recommended to study medicine, and to apply himself to the composition of a work on some branch of it. The subject he selected was the relief of old age, and his *Epistola de accidentibus senectutis,* dedicated and sent to Innocent IV (1243–54), was the result. In the preparation of this work Bacon visited Italy 'in partibus Romanis' and made himself acquainted with those works of Mohammedan medicine which had been introduced into Europe. His book is largely a compilation, perhaps one of the most complete on its subject that has been made, from Avicenna, Rhazes, Haly, Isaac, and Damascenus, and contains little that would mark out its writer as an independent or innovating spirit. It was after this visit that Bacon began the study of Greek, then a spoken and written language in southern Italy, as his treatise makes it pretty evident that he knew no Greek, Hebrew, or Arabic at the time, his knowledge of Greek medicine—Galen and Dioscorides—being obtained from Italy.

We owe to the piety or the industry, it would seem, of one of the pupils of Roger Bacon a report of several of the courses given by him as a regent master at Paris. They consist of two courses on the *Physics* of Aristotle and three on the *Metaphysics,* with another on the *De plantis.* An examination of the contents assures us that we have in the questions on the *Physics* the report of two distinct courses, not two reports of the same lectures, and the three sets of questions on the *Metaphysics* must belong to at least two courses and most probably three. As the *Physics* and *Metaphysics* were lectured on at the same time in 1245, and most likely earlier, we thus have records of three years' teaching—ending most probably in or before 1247. We put this date because Bacon, writing in 1267 or 1268, speaks of the twenty years in which he had laboured specially in the study of wisdom, after abandoning the usual methods.

These lectures bear the mark of youth—a youth of genius. They have been examined at some length by the late M. Duhem, who has made it plain that the principal subjects dealt with in them are those which preoccupied Bacon throughout his career, and that his treatment shows a progressive growth in knowledge and boldness of views. Even at this early stage of his career he is an innovator, and it is to his exposition of the question as to the existence of a vacuum, that mediaeval science owed the theory which held sway for centuries till new facts were observed which showed its unsoundness, a theory crystallized into the expression 'Nature abhors a vacuum'. It is probable that the earlier course represents the views current at Oxford, the later was certainly given at Paris.

The style of these early works has little in common with that of twenty years later, and this offers a warning to us, forbidding the rejection of works attributed to Bacon merely on account of their style, provided that their subjects and treatment permit of their being placed among the 'multa in alio statu . . . propter iuvenum rudimenta' written by him, or the 'aliqua capitula nunc de una scientia nunc de alia . . . aliquando more transitorio'. There is, it is true, none of the fierce and outspoken criticism which marks his later work, but it must be remembered that few of the abuses he attacks had come into existence. His period of lecturing must have coincided with the last teaching of Alexander of Hales, who died in 1245, and the first appearance of Albertus Magnus in the Jacobin convent at Paris in the same year, and he had retired from his chair before St. Thomas came to lecture at Paris in 1252. His teaching compares favourably with that of Albert, as preserved in the Commentary on Aristotle's *Physics*, written most probably at or soon after this period.

It is in astronomy that these lectures show best the fluid state of Bacon's knowledge. In the earlier of them, that on the eleventh Metaphysics, his theory is a development of the Aristotelian reasoning in the *de Caelo et Mundo*. He seems entirely ignorant of the Ptolemaic theories, and to found his teaching on the Commentator of the *Timaeus* rather than on Alpetragius. His theories are those of the twelfth-century school of Chartres. In his second lectures on the *Physics* his knowledge of contemporary astronomical theory is greatly enlarged. He now finds the necessity of supposing nine orbs instead of eight, to account for the precession of the equinoxes, and though he is not yet prepared to acknowledge the opposition between the homocentric system of Aristotle and that of the school of Ptolemy, he is evidently on the road to this position.

We cannot with any certainty give a reason for Bacon's withdrawal from his work as a master at Paris. It may have been due to the public affront he received, when, bungling over a strange word in the text of the *De plantis* he was explaining, and calling it Arabic, one of his students replied that it was Spanish. Certainly it is a curious fact that the reported lectures on this book break off just before the point at which this passage occurs. At any rate we may date from this period his resolution to become a master of the languages in which the fundamental documents of religion and science were written.

It was about this time that a revival of the teaching of mathematics in the University of Paris took place, following the republication of the Euclidean corpus by Campanus of Novara in 1246. Bacon tells us that he attended the lectures of many new masters on mathematics, but found that they did not understand the terminology of their subjects, so that he supplied them with geometrical proofs of positions from Aristotle and Averroes which none of the official critics in the disputations could attack. About this time (1248), Bacon seems to have learnt from Grosseteste his theory as to the origin of the tides and now probably began that intensive study of natural science of which his work bears such witness. He writes in 1267 that during the last twenty years he had spent more than 2,000 livres (a sum equal in purchasing power to something under £10,000 at present) in purchasing secret books, on various experiments, languages, and instruments, and astronomical tables, in forming friendships with the wise, and in instructing helpers in languages, figures, numbers, tables, instruments, and the like.

The experiments to which he devoted his attention may be gathered from his account of the mysterious Peter de Maricourt, whom he seems to have met about this time. The account of him in the *Opus Tertium* seems to be Bacon's ideal of a student. 'He makes no account of speeches and wordy conflicts but follows up the works of wisdom and remains there. He knows natural science by experiment, and medicaments and alchemy and all things in the heavens or beneath them, and he would be ashamed if any layman, or old woman or rustic, or soldier should know anything about the soil that he was ignorant of. Whence he is conversant with the casting of metals and the working of gold, silver, and other metals and all minerals; he knows all about soldiering and arms and hunting; he has examined agriculture and land surveying and farming; he has further considered old wives' magic and fortune-telling and the charms of them and of all magicians, and the tricks and illusions of jugglers. But as honour and rewards would hinder him from the greatness of his experimental work he scorns them.' We learn that in 1267 he had just completed a concave mirror capable of acting as a burning-glass, after three years' work, at an expense of 100 livres.

A reflection of this acquaintance may be found in the rather puzzling work known under the name of *De mirabili potestate artis et naturae*, consisting of ten or eleven chapters, of which the first six must have been written about this time, prompted, there is little doubt, by a recent denunciation by William of Auvergne, Bishop of Paris, of dealings in magic. The remaining chapters are of a different character and may possibly have formed part originally of another work.

The principal subject of these chapters is a delimitation of the wonders that can be wrought by an application of scientific principles from those which are mere trickery or those which may be due to the powers of evil. He seems to have gone thoroughly into the question of magic, and to have attended the thirteenth-century spiritualistic seances. 'When inanimate objects are quickly moved about in the darkness of morning or evening twilight, there is no truth therein but downright cheating and cozenage.' He allows a certain utility in charms, for example in the hands of a physician, that the patient may be induced to hope and confidence of a cure, quoting Avicenna as to the effect of the mind on the body. He then proceeds to give a general explanation of the way in which one person or thing can act on another, confusing what we now recognize as infection, sympathy, and hypnotism into one class of action at a distance, and while repudiating such books as pay worship to evil, warns his reader that many books reputed magical contain much useful knowledge.

He then proceeds to describe a few of these wonders, which give us some idea of what were really the mechanical problems of the time. The first is a ship, without men rowing yet moving faster on rivers or the sea than a galley. This seems to point either to an improved form of sails, or to something like a paddle moved with cranks. The car moving without animals to draw it may have depended on the same thing, the real obstacle to its use being the want of good roads. The flying machine with wings he acknowledges has only existed in plans. The small instrument for lifting weights is evidently a sort of jack, unless it is a system of pulleys; similarly, the instrument by which one man can pull a thousand toward him. Walking under the sea, building suspension bridges, and so on, are also mentioned by him as possible at the time. He himself lays much stress on the properties of mirrors. Some of these appear obvious enough to us, such as the arrangement by which images of an object are repeated indefinitely, which it is difficult to believe had not been observed before, though he uses it to explain the existence of parhelia and mirages, and proposes its utilization in war. He also speaks of an arrangement of lenses by which distant things may be brought near and small things made visible: 'possunt enim sic figurari perspicue ut longissime appareant propinquissima, et e converso.' This is undoubtedly a description of the telescope without the enclosing tube. Such an arrangement is known to have been made by Leonard Digges, who died in 1571, from his study of Bacon's works. The magic-lantern is not obscurely hinted at, in a description of some optical illusions. But Bacon reserves his greatest praise for the use of lenses and mirrors as burning-glasses, and for the construction of a self-moving celestial globe, probably dependent on the use of a magnet, as was that of Peter de Maricourt. He further describes some inflammable mixtures and an explosive mixture which later on proves to be gunpowder.

We do not know how much of this period of investigation was spent at Paris, and everything points to the fact that Bacon returned to Oxford about 1251 and entered the Franciscan Order about 1253 to 1256. But during these years of absence from the University schools another and a rival influence was establishing itself there.

Albert the Great, the first doctor of the Dominican Order, had come to Paris in 1245; he was some years older than Bacon and brought with him a great renown as a teacher. He was accompanied by his pupil and socius Thomas Aquinas, now in his eighteenth year, whose fame was destined to put his own in the shade. Their task was to explain Aristotle and to reconcile him with dogma. In some respects this was difficult. Aristotle tells of no Creator, no Adam, no Providence, no eternity of the individual soul after death; and his commentators had to pass over these defects in silence, sometimes to deny the plain meaning of what he wrote, and occasionally to throw his teaching overboard altogether.

Albert's encyclopaedic work on the *Physics* and *Metaphysics* of Aristotle, begun perhaps before 1245 in view of his coming to Paris, was concluded by about 1256. His *Logic* and his *Summa* are parts of another scheme. It was written as a result of a demand from the Dominicans for a compendious account of physics which would enable them to understand the works of Aristotle on the subject. Albert felt himself unequal to the task, but after repeated entreaties set to work to write a Commentary for the use of the friars and of all those who desired to attain to natural science—a commentary many times longer than the original text. He goes on to describe the plan on which he worked, and to explain that there are three divisions of real philosophy; that is, philosophy not caused in ourselves by our own work like Moral Philosophy. These three parts are Natural Philosophy or Physics, Metaphysics, and Mathematics; and he promises to deal with each in turn. As a matter of fact he never did deal with Mathematics, and it is probable that he was not in a position to do so, but he dealt fully with everything included in the thirteenth century in Physics and Metaphysics.

Albert was, beyond doubt, one of the most learned men of his age, as Bacon, his severest critic, admits. Writing to the Pope in 1267 he mentions Albert and William de Shyrwood as the two most famous scholars of the day, and challenges comparison with them. In the *Opus Minus* he says of Albert, 'truly I praise him more than all the crowd of students, for he is very given to study and has seen an infinite number of things, and has spent much, and so has been able to collect many useful facts from an ocean of authors'. But Bacon insists that being a self-taught student he is not thoroughly grounded, seeing he was never trained by hearing, lecturing, and disputing in the ordinary arts course. Moreover, he does not know languages, and so cannot check the errors of his texts, and as he is ignorant of the laws of perspective he cannot fully explain any part of natural philosophy.

The dates of Albert's treatises are not well established, and the order in which they were written is only partly known, so that we cannot say how far Bacon's statement as to his ignorance of Greek and Arabic was true at the time. There seems no doubt that Albert consulted Greek texts and probably Arabic ones, though he may have had expert help. He certainly was a very keen and diligent observer; his tracts on mineralogy contain many proofs of it. Perhaps this was because he could not find the complete book of Aristotle *de Mineralibus*: 'que diligenter quesivi per diversas mundi regiones . . . exul enim aliquando factus fui longe vadens ad loca metallica ut experiri possem naturas metallorum.' He goes on to describe his investigations among the alchemists, theoretical and practical, and compares his impressions with his own knowledge of how gold, silver, and other metals are found native or as ores. Of course these studies may have followed Bacon's criticisms and resulted from them. We know that Albert was sensitive to them.

But Bacon's criticism of Albertus, however strong, was, from the point of view that we share with him, well founded. The ideal science of Albertus Magnus was a science based on words and books, not on the phenomena of nature. He looked to logic for advance in knowledge. 'Et hoc est quod per investigationem rationis ex cognito devenitur ad cognitionem incogniti, hoc enim fit in omnia scientia quocunque modo dicta, sive sit demonstrativa', he says in asserting the claims of logic to be a science. 'Investigatio enim sive ratio investigans ignotum per notum, speciale quoddam est.' It is true that he says elsewhere, 'Dicendum quod scientie demonstrative non omnes facte sunt, sed plures restant adhuc inveniende', but these demonstrative sciences are further branches of logic, not sciences founded on observation of nature. Similarly the statement, 'Oportet experimentum non in uno modo sed secundum omnes circumstancias probare', does not apply to experiments in natural science but to words. Summing up our impressions, his influence on his time is on the whole bad; his value for us is that he presents us with a *résumé*, from the scholastic point of view, of the knowledge already acquired by western Europe at the end of the first half of the thirteenth century.

Albert's great pupil, St. Thomas, enters late into the scheme of Bacon's surroundings. He may have been the 'Doctor Parisius' on whom an attack is made, but his treatment of the Aristotelian system of nature seems to have been a little later in date than the *Opus Tertium*, and certainly to have avoided some of Bacon's criticisms by an attempt to obtain a new version made directly from the Greek. That this version was defective, too, was not his fault. Thomas was lecturing in Paris in 1252, after having spent three years there as a pupil of Albertus in 1245–8, and he took his divinity degree at the same time as Bonaventura in 1257. His commentaries on the *Naturalia* of Aristotle are said to have been begun in 1263, and he lectured in Paris again from 1268 to 1272, when he left to take up a post in the reorganized University of Naples. The 'duo moderni gloriosi' of the *Communia Naturalium* must have been Albert and Thomas, but it is not till later that Thomas is mentioned (1271) in the *Compendium Studii*, with Albert as an example of the 'pueri duorum ordinum studentium', and it must be about this time or a little earlier (1268–9) that the strictures in the later part of the *Communia Naturalium* on his doctrines were written. 'Temporibus autem meis non fiebat mencio de istis erroribus.' But while these discussions have lost much of their meaning to us of this day, Thomas remains the greatest of the schoolmen, and his supremely intellectual attitude has carried over his theological work to our own days almost unaltered.

The work of Albert was not the only attempt to sum up the totality of human knowledge at the time. Several veritable encyclopaedias were composed about 1250, large and small, including all that was known or believed about the universe. Of these the most notable was that of Vincent of Beauvais, a Dominican like Albert and St. Thomas, who wrote between 1240 and 1264 his colossal *Imago Mundi* divided into three sections—the *Speculum Naturale*, *Speculum Doctrinale*, and *Speculum Historiale*, to which was later added by other hands the *Speculum Morale*. Perhaps Bacon had Vincent in mind when he wrote to Pope Clement that his own scheme could only be carried out by the aid of the Pope or of some great prince like the King of France, for it was to St. Louis that Vincent was indebted for the unrestricted use of the royal library, containing, it is said, some 1,200 manuscripts, and for the large staff of copyists and helpers necessary for the completion of the work. His method is to make extracts from the various authorities on each subject, and when necessary to add a few remarks of his own. It has been computed that he quotes from some 450 writers by name, some of whom we only know by his extracts.

The *Speculum Naturale* is divided into 32 books and 3,718 chapters, and is intended to be a full description of all created living beings, from angels down to fishes and plants. Book 15, for example, treats of astronomy and the calendar; books 23 to

27 deal with the psychology, physiology, and anatomy of man. In book 6, cap. 7, the question is discussed as to what would happen if a stone were dropped down a hole passing through the centre of the earth, and he decides that it would stay there. The *Speculum Doctrinale* treats of sciences and arts, beginning with the trivium, ethics, the economic arts, housekeeping, and agriculture, and so on. The eleventh book treats of the mechanical arts, which include architecture, navigation, alchemy, and metallurgy, weaving, smith-work, commerce, and hunting. Books 12 to 14 deal with medicine, theoretical and practical, and book 16 with mathematics. It is a most useful summary of the knowledge and belief of the time. The *Speculum Historiale* is, as its name implies, a gossipy abstract of the history of the world.

Another encyclopaedia, the most popular of all, was the *De Proprietatibus Rerum* of Bartholomew Anglicus, a Franciscan friar, in nineteen books. It is of the same general character as the *Speculum Doctrinale*, but shorter. It exists in hundreds of manuscripts and several early translations were made, while its popularity in the early days of printing was very great. It appears to have been written somewhere about 1250, for use, not in University circles, but among the general body of friars. The taste for encyclopaedic manuals spread to the laity, and we have a number of compositions in the vulgar tongue such as *L'image du monde* of Gautier de Metz, written about 1245, treating of cosmogony, astronomy, and geography, and the *Fontaine de toutes sciences*—a dialogue between King Boetus and the philosopher Sidrach on things in general—written a little earlier. The *Trésor* of Brunetto Latini, written in French about 1265, was a book of a superior sort, with as wide a range.

Bacon after these years of study, after repeated entreaties from his friends, and, it may be, criticisms from opponents, determined in his turn on the composition of an encyclopaedia which was to be a compendium of real knowledge, not a commentary on the works of Aristotle or on the compilations of other and lesser minds than his. The great work was to be founded on a reasoned scheme of mental progress, leading up to the supreme science of the mind, for him as for every one else in that century, the science of theology.

The scheme of his great encyclopaedia, as he conceived it, included four principal divisions: the first volume comprised

Grammar and Logic, the second Mathematics, the third Natural Science (Physics and Experimental Science), the fourth Metaphysics and Morals. He worked at this scheme until his task was arrested by his condemnation in 1277, only breaking off during the year devoted to the compilation of the *Opus Maius*, the *Opus Minus*, and the *Opus Tertium*, much of these works being obviously composed of material ready to hand. He seems to have worked at them in a desultory way, writing now on one subject now on another. 'Aliqua capitula de diversis materiis', 'nihil continuum est' are his words. This accounts for the great number and diversity of the smaller works that pass under his name, destined to take their place in a larger frame, and often unrelated to their fellows. Attempts have obviously been made since his death to effect this synthesis; the fine Royal MSS. of Bacon have been through the preliminary stages of such an attempt. The Oxford MS. of the *Opus Maius* is evidently an aggregation of several tractates to the original work, and the Mazarine MS. of the *Communia Naturalium* offers undoubted proof of having been made up at the end of the fourteenth century from a number of independent fragments written for the *Opus Tertium*, and added to the work as it left the pen of Bacon.

The first volume of this encyclopaedia has not been handed down to us in any complete form. Several fragments of it remain: a nearly complete Greek grammar, a fragment of a Hebrew grammar, a *Summa grammatice*, and the tract published by Brewer under the title *Compendium Studii Philosophiae*. Most of them, in the form we have them at any rate, appear to have been written after 1267, though a part of what we find concerning Grammar and Logic in the *Opus Tertium* was probably taken from pre-existing fragments. Bacon returns to the subject in his latest work the *Compendium Studii Theologie*, written in 1292.

The second volume was intended to deal with Mathematics, pure and applied, of which nine divisions were recognized at the time, five speculative and four operative. This division is due to Alfarabi; he divides geometry, arithmetic, astronomy (to use our modern nomenclature), and music into two sections each, theoretical and applied, and precedes them by a general study of principles 'de communibus mathematice'. We have a part of this treatise written before 1267, and a later and enlarged form of it written some time after, together with some shorter additional tracts on Geometry. Astronomy was relegated to the third volume.

The third volume dealt with Natural Philosophy, the modern Physics. It was intended to include a treatise on the first principles of physics, the nature of rest and motion, of the four elements, of the theory of compounds both inanimate and animate, of generation and corruption, alteration, growth, and diminution. After this he proposed to treat of seven special sciences: Perspective, Astrology, the Science of Weights, Alchemy, Agriculture, Medicine, and Experimental Science. His scope is fully explained in the surviving parts of this volume, two forms of which exist, one written before 1267, the other after. We have in addition to those parts formally included in this volume a number of detached works, which would naturally fall into their place in the scheme, on Alchemy, Medicine, and Experimental Science.

The fourth volume of the *Compendium* dealt with Metaphysics and Morals. Of this we have a number of detached parts, a separate treatise on the lines of the *Metaphysics* of Avicenna, and a treatise on the propagation of force treated mathematically, which has been printed with the *Opus Maius*. There is no doubt but that he had completed a part of his treatment of the subject long before 1266, and two important treatises written at this period, showing the development of his theories, are still unprinted. A number of references to this work are to be found in the *Opus Maius*. The Morals would probably correspond largely to the seventh part of that work.

Not only had this great encyclopaedic compendium been undertaken, but Bacon had given considerable attention to a matter which was becoming rather urgent, the reform of the Calendar. The slightest knowledge of the beginnings of our history will remind us of the importance attached by early Christianity to the observance of Easter on the proper date. By the middle of the thirteenth century the Julian Calendar had amassed so great an error that it was perfectly possible that Easter might be celebrated a month later than the date on which, according to the spirit of the rule, it should fall.

Two works of Bacon on the subject are known. The large *de Computo* is an historical and general treatise on the various divisions of time and on the methods of using tables for the calculation of festivals or reducing Arabic to Christian dates.

That included in the *Opus Maius* is more controversial. It exposes the faults of the Julian Calendar, by the observance of which Easter is celebrated on the wrong date in the third and fourteenth years of the lunar cycle. He points out the superior advantage of using the Arab system of 30 years of 12 lunations, making 10,631 days, or of adopting the Hebrew system which is nearly the same. In this attack on the established order of things Bacon stands nearly alone in his time; and though his arguments were repeated in the next two centuries several times, it took three hundred years to effect this simple reform.

In connexion with his study of this work Bacon seems to have written several treatises, one *de Temporibus* which once existed in the Austin Friars' Library at York but is now lost; the long work, the *Computus,* written in 1263–4; and a shorter treatise dealing with times and seasons in 1266. He sums up his arguments in the fourth part of the *Opus Maius,* returning to the point in the *Opus Tertium.*

The turning-point in Bacon's career was in 1266. Probably about 1264, Cardinal Guy de Foulkes heard of Bacon's writings from a clerk in his service, Raymond of Laon, whom he commissioned to obtain them. In 1265 the Cardinal became Pope under the title of Clement IV. In March 1266 Sir William Bonecor was sent by Henry III on a special mission to the new Pope, and he seems to have carried with him a communication from Roger Bacon, which Professor Little conjectures, with some probability, to have been the *Metaphysica de viciis contractis in studio theologi.* On receiving this the Pope replied in a letter, first printed by Martene, asking him to indicate the remedies he would propose for the evils he had pointed out, and ordering him to send the writings he had previously asked for, secretly and without delay, any Constitution of his Order to the contrary notwithstanding. This letter is dated June 22, 1266.

As a result of this Papal command we possess the most widely known of Bacon's works: the *Opus Maius*—a preliminary work he calls it—a 'tractatus preambulus' in contradistinction to his great encyclopaedia, and with it three treatises, summaries and supplements to it: the introductory letter and summary found in the Vatican Library by Cardinal Gasquet; the *Opus Minus*—a summary with additional remarks on the causes of error in the church, Biblical textual criticism, and a short theory of alchemy;

and the *Opus Tertium*, a work of which the exact size is still uncertain, which was probably never completed. The *Opus Maius* and *Opus Minus*, with some other treatises, were probably sent to the Pope early in 1268, but as he died in the same year, and his successor was not elected till 1271, no answer was received by Bacon and no result followed from them.

Of the work of the succeeding years there is little to tell. Bacon seems to have worked on his *Compendium*, to have completed his introduction to the *Secretum Secretorum*, and about 1272 to have written the work published by Brewer. He was evidently in residence at Oxford at this period, to which may be referred the legends as to his construction of magical mirrors and perspective glasses referred to in the foundation for Greene's play of *Friar Bacon and Friar Bungay*. He was imprisoned in 1278 by the Minister-General of his Order for some 'suspected novelties', and it is believed, on no definite grounds, that he remained in prison till 1292, the date of composition of his last fragment, the *Compendium Studii Theologie*, published by Rashdall—a work which was intended to be completed in seven parts, including much of his earlier writing.

It would be wrong to infer, because his name is rarely quoted in the few works of the half-century succeeding him, that his influence was negligible. Nearly all the manuscripts of his works we possess must have been written after his death. They range in date from the end of the thirteenth century to the middle of the fifteenth, and their form attests a constant attempt to edit and rearrange them according to the author's intention. In the face of his express condemnation, the wish that his 'dangerous teaching might be completely suppressed', the absence of his name is easily understood, but the influence of his thought is clearly discernible in Oxford teaching up to the time of the Renaissance.

In estimating Bacon's position among the men of his own time it is important to remember, first of all, the complete originality of his scheme. His great work, unfinished though it most probably was, and almost beyond the powers of any one unaided scholar to complete, the *Compendium Studii Philosophie* or *Theologie*, as the case may be, was as distinct in kind as in form from the works of his great contemporaries. As we have already said, it was an age of encyclopaedias, but none of them were independent in form. Albert's life-work consisted in a series of comments on the words of Aristotle, following the order of his text with occasional excursuses when the subject seemed to suggest them, but breaking no new ground, and making no rearrangement of his matter. The great work of St. Thomas is conditioned by the *Sentences*, following its order with more originality and power than Albert, but, after all, adopting another man's scheme. His physics are mere commentaries on the text of Aristotle. On the other hand, the encyclopaedia of Vincent of Beauvais owes nothing of its arrangement to Aristotle, and comparatively little to its early forerunners, and was, so far, original. But there its originality ceased : it was merely a collection of facts and dicta collected from approved authors by the industry of a small college of clerks, and arranged under convenient headings. It added nothing to human knowledge, no inspiration for the progress of thought.

Bacon's schematic arrangement was not only unparalleled among the writers of his time ; it was absolutely new. Nothing like it had been devised since the time of Aristotle, and it had the advantage of not being obliged to combat a large number of exploded hypotheses. The whole system of human thought was re-cast, and the plan simplified to an extraordinary degree to meet the necessities of the age. It may be that the framework of his scheme owed something to Al Farabi's *de Scienciis*, or to Avicenna, but in its conception and execution its originality is manifest. His plan has already been described, its execution was to be marked by the most rigorous economy. Everything superfluous or unnecessary to the development of the argument was to be cut away ; all the excrescences which dialectical skill had embroidered on the simplest notions were to be discarded. Bacon did not exclude the notion of special treatises ancillary to the main lines of his course, but he did not regard them as necessary to every pupil. He thought it more necessary, for example, that the results of geometry should be known as a whole, than that the pupil should be able to prove the fifth proposition and be ignorant of the sixth book of Euclid.

The foundation of his system was laid on an accurate study of language : 'Notitia linguarum est prima porta sapientie.' Latin, of course, was taken for granted like English or French (is it, by the way, noticed that English is spoken by a Norman family as early as 1270 ?). But Latin alone was no sufficient key to the treasury of knowledge : its vocabulary was too limited, the masterpieces of the world's teaching were either not translated into it, or were so badly translated that they were wholly misleading to such teachers and taught as attempted to profit by them. We need not labour the point as Bacon had to do; the fact remains that no mediaeval translation of Aristotle or his commentaries is now consulted, except as a curiosity. Bacon's study of languages, *Grammatica*, included not only the modern Philology but also a strictly utilitarian Logic and Dialectic, sufficient for the study of mediaeval science which was to follow: 'Grammatica et logica priores sunt in ordine doctrine.' We have a great part of Bacon's teaching on the subject, scattered over many writings. His main work was destructive; the schools were lumbered with inefficient text-books and antiquated errors. After that came reconstruction, 'secundum linguas diversas prout valent immo eciam necessarie sunt studio Latinorum'; Greek, Hebrew, and Arabic. He recognized three stages in the knowledge of languages, looking at the question from the point of view of actual life, not of a teacher. The first was that of a man who recognizes that a word is Greek, Hebrew, &c., can read it and even pronounce it, knows approximately what grammatical forms it may take, and so on: something like a man who can read enough German to know the names of the streets and the directions on the tramcars, but cannot speak the language. This is the sort of knowledge of Hebrew or Greek which Bacon pledged himself to impart in three days, when writing in his *Opus Tertium* and elsewhere. The second stage of knowledge was the power to read and understand in an ordinary way. 'Certes,' he says, 'this is difficult, but not so difficult as men believe.' His third stage is only reached when the student can talk and write and preach in the foreign language as in his own. The only grammars actually preserved to our own day which have been identified are the Greek and a fragment of the Hebrew. It is doubtful whether he ever knew Arabic or composed a grammar of that language. He certainly knew Chaldaic enough to explain its relation to and difference from Hebrew. His Greek Grammar is a very remarkable one for the time, even if it be true that it is founded on a Byzantine original. As we have it, the work is not complete, and the study of Greek made no progress in this country for nearly two centuries and a half.

This, then, is his first distinction in the study of languages, that he laid down the principle that every language has an individuality of its own, and that the grammar appropriate for one of them, say Latin, would not lend itself to the study of another, such as Greek. It is a principle which was only recognized in practice towards the end of the nineteenth century, and is still disregarded by many writers. But he was also distinguished as a critic.

All human knowledge was, in his day, assumed to lead up to the study of theology—the queen of sciences. Whatever progress was made on the great lines of modern thought was made, not with that aim, but with the intention of facilitating the formation of right conclusions as to the relation between God and man. If we are to undervalue our predecessors on this account, we need not study the thought of the Middle Ages. Bacon's criticism of his contemporaries was actuated by the thought that they were bad teachers because they were insufficiently taught, that they taught in error because they were unable to test the truth of the maxims they repeated. His zeal for truth, his application of the tests of truth, remain of value if his conception of the highest truths no longer satisfies us. His criticism was applied to the text of the Vulgate; the principles of textual criticism he laid down are of universal application.

A large part of Bacon's published work on the subject is taken up with the proofs of the corrupt state of the Vulgate text, made worse by the number of correctors, for the most part ignorant of both Greek and Hebrew. More than that, many of these corrections would not have been made if the correctors had even consulted a good Latin grammar. The scheme he proposes is an official attempt to restore the genuine text of the Vulgate as issued by St. Jerome from a comparison of the oldest manuscripts, which were to be collected, examined, and compared, while the readings were to be judged by the original Greek or Hebrew from which St. Jerome had made his translation. Here for the first time in the Middle Ages were the true principles of textual criticism laid down, principles valid for the work of every editor since his time.

After the science of language had been thoroughly mastered, so that the student was able to read the principal documents of scholarship in the original and follow their train of thought, Bacon next directed his attention to Mathematics. As we have already pointed out, this science was already reviving in western Europe. In Italy Leonardo Fibonacci had published his *Liber Abaci*, Campanus had re-edited Euclid and written on the sphere, de Lunis had begun the study which was to become algebra. In France Alexandre de Villedieu had written the *Carmen de Algorismo*, the popular treatise on Arithmetic ; in Germany Jordanus Nemorarius on the theory of numbers, the geometry of the sphere and on triangles; while the English John of Halifax wrote *c.* 1232 the *De sphera mundi*, the most popular text-book on the subject of the Middle Ages, and his *Tractatus de arte Numerandi* ; and Peckham's *Perspectiva communis* and Grosseteste's semi-mathematical tractates were also published.

Bacon's own reading, as evidenced by quotations in his *Communia Mathematica*, was considerable. Besides the general scholastic learning of his day he quotes from all the works of Euclid, the *Almagest* and *Aspects* of Ptolemy, Theodosius on the Sphere, Apollonius, Archimedes, Vitruvius, and Hipparchus. Boethius is his main stand-by. The Arabic writers were well known to him, and early mediaeval writers such as Adelard of Bath, Jordanus, Anaricius, Bernelius, Gebert, and others are often quoted ; indeed, we learn of a hitherto unknown work by Adelard from his writings. It would seem, however, that the general interest in mathematics of his time was strictly utilitarian. 'The philosophers of these days,' says he, 'when they are told that they ought to know perspective or geometry, or languages, and many other things, ask derisively, "What good are they ?" asserting that they are useless. Nor will they listen to any account of their utility, and so they neglect and despise the sciences of which they are ignorant.'

His own view of the value of mathematics in education was a very high one. Like Plato, he saw in it the master key to all correct reasoning and all progress in knowledge. 'Mathematica est omnino necessaria et utilis aliis scientiis.' 'Impossibile est res huius mundi sciri, nisi sciatur mathematica.' 'Oportet ut fundamenta cognitionis in mathematica ponamus.' It was not so much mathematics for its own sake, as mathematics a handmaid to the natural sciences and theology. We have already referred to his classification of the subject as speculative and practical, following a study of the elements of the science. His remaining work is largely taken up by discussions of the meaning of continuity, infinity, dimensions, axioms, postulates, definitions, and the like, and he spends much time on the various ratios, arithmetical, geometrical, and harmonical. But in pure mathematics he was not an originator, he made no discoveries in geometry or the theory of numbers: his title to remembrance as a mathematician is his sympathy with it, his wide knowledge, and his insistence on its value as the foundation of a liberal education.

Bacon's work on optics was really a part of a larger scheme in his own mind—a study of the propagation of force at a distance. This he treated geometrically and used one variety of force as an example which was susceptible of measurement—light. Thus is explained the emphasis laid on the science of optics, or perspective, as he called it. His main treatises on it, the *Perspective* and the *Multiplication of Species*, were written before the *Opus Maius*. The scheme of the *Perspective* was not entirely new, of course; it had to include much that had been treated by his predecessor Alhazen, and, as a summary, to omit many of his detailed geometrical extensions of theorems. But on the other hand, it carried on the science a considerable way, as, for example, by proving that a concave spherical mirror would bring the reflected rays from different parts of its surface to a focus on different parts of its axis, and that to obtain a single focus the mirror must have a surface produced by the rotation of a parabola or hyperbola.

His study of the theory of optics went hand in hand with practical work; he caused to be constructed concave mirrors for use as burning-glasses, time after time, remarking on the diminishing cost as the craftsman grew more skilful; he was familiar with the use and properties of a convex lens both for magnification of objects, i.e. the simple microscope, and as a burning-glass; and there is every reason to suppose that he was acquainted with the combination of lenses which makes up the telescope, though he only used it for terrestrial objects and did not make it portable.

It is this combination which lies at the base of the legend of Bacon and Bungay's magic mirror which had grown up by 1385. 'Friar Roger Bacon took such delight in his experiments that instead of attending to his lectures and writings he made two mirrors in the University of Oxford: by one of them you could light a candle at any hour, day or night; in the other you could see what people were doing in the uttermost parts of the earth.

The result was that the students either spent their time in lighting candles at the first mirror instead of studying books, or, on looking into the second and seeing their relations or friends dying and lying ill, left Oxford to the ruination of the University—and so both mirrors were broken by the common counsel of the University.' This legend refers evidently to the burning-glass and telescope. The latter is shown to have existed by a statement printed in 1579 that Leonard Digges, then dead, 'was able by Perspective Glasses duely situate upon convenient Angles, in such sort to discover every particularitie of the Countrie round about, wheresoever the Sunne beames might pearse . . . which partly grew by the aid he had by one old written book of the same Bakon's experiments, that . . . came to his hands'. This work of Bacon's is no longer known to exist.

We have already spoken of his devotion to Astronomy in the modern sense of the word. His account of the science in the *Opus Maius*, the *De Celestibus*, and the fragment of the *Opus Tertium* published by Duhem, not only shows that he was abreast of the best work of his time, but also forms the best epitome of the state of knowledge at the day. His continual labour in the construction of astronomical tables bore fruit in his attempt at the reformation of the calendar, of which we have given some account. His work on Chronology, sacred and secular, is closely connected with this subject.

Bacon also takes rank as one of the earliest mediaeval writers on Geography, and part of his treatise was reprinted for the first time by Purchas in 1625 from the *Opus Maius* manuscript. His study was founded in the first place on Ptolemy, checked by modern travel, and his first consideration is an approximate determination of the relative amounts of land and water on the globe. It was a passage of this part of his work which had a leading part in deciding Columbus to make an attempt to reach the Indies by the Atlantic route, as shown by his letter from Haiti to Ferdinand and Isabella. Bacon's description covers the known world with the exception of western Europe, and he insists on the habitability of the earth south of the equator, and on the extension of Africa to the south. We have only to compare this treatise with Albert's *De natura locorum* to understand the great advance Bacon has made.

His position in the history of Chemistry has yet to be fully investigated. The very large number of alchemical tracts which pass under his name show that his influence upon the students of the next century was very great. We are, naturally, at a loss to do more than form a reasonable conjecture as to the extent of his practical acquaintance with the operations of chemistry or alchemy, since his writings are devoted rather to theoretical than practical considerations. There was, of course, a large number of industries which depended on chemical reactions for their methods ; brewing, dyeing, enamel-making, glass-making, metallurgy, lime-works, are but a few examples ; but the eyes of inquirers were rarely turned towards these, and a superficial reason was given for the effects produced. What was really being sought by students was the general formula of the universe, adopting more modern terms, its integral equation, which, once found, would resolve any particular case by substituting suitable values for its constants. Whether he had made up his mind as to the existence of a universal primary matter in our modern sense, taking up the properties which made it a distinct material, is doubtful. The theories he held as to the action of the celestial bodies on this earth swayed him first one way, then another, while the dicta of Aristotle and Avicenna that no change can happen 'nisi fiat resolutio ad materiam primam', which he accepted without questioning, led him towards its acceptance. His doctrine of the multiplication of species was one of the most fruitful in his theory of alchemy. Just as celestial fire produced fire by means of a lens, so the celestial bodies might act on a suitable primary matter to produce their cognate metal, if their influence were as great. The science of weights was a branch of alchemy, because what distinguished the four elements in changeable matter from those in the super-celestial regions was their combination with the qualities of heaviness and lightness.

Alchemy, according to Bacon, was either speculative or practical. Speculative or theoretical alchemy treats of the generation of materials from their elements inanimate or animate. His list of inanimate things comprises metals, gems, stones, colours, salts, oils, bitumen, &c. ; his animate things include vegetables, animals, and men. Alchemy was for him linked with Physics and Medicine in a chain of development. Among the treatises which give us the clearest views of his thought are the *Opus Minus* fragments and the *Opus Tertium*: by these the others are to be tried and accepted or rejected. A striking example of the effect of his teaching is to be found in the treatise *De lapide philosophici*, attributed to St. Thomas Aquinas but really written by Fr. Thomas, chaplain to Robert, son of Charles of Anjou, in 1296.

Nothing has yet been said of his relation to the peculiar science of his scholastic contemporaries—speculative philosophy. This was almost a creation of his own time, due to the fuller study of Aristotle now possible. The meaning of matter, form, and substance, the struggle between realist and nominalist, the question of species and individual, all involved questions of the highest religious importance, pantheism or theism. Here Bacon, leading the Oxford schoolmen of later years, rejects much of the controversy as useless. There is no answer to the question what causes individuality or what universality: God makes things as their nature requires. The second part of the *Communia Naturalium* alone is quite enough to place Bacon in a high place among mediaeval schoolmen: its clear treatment shows solid thinking as well as sound criticism.

Looking back on the whole activity of this remarkable scholar we may try to sum up the interest he has for the modern world. Perhaps to himself the question would have been otiose: he was, like many men of science to-day, prepared to accept results and methods without lingering over the history of how they came into being. But on the other hand, to the large and increasing number to whom the history of scientific thought and method is often almost as important as its results, Roger Bacon stands out prominently as the first English leader of scientific thought. More still, he has the special English quality of fighting a lone battle for his views, unsupported as he was by his own order, attacking its opponent's chiefs, and remaining unshaken to the end. His works, though not entirely neglected, have usually been treated as curiosities, while those of his two great contemporaries have been held in reverence for centuries, and even to-day are receiving the full honours of scholarship in new editions from the manuscripts. The publication of his remains would be invaluable, if only as marking the development of a mediaeval thinker, ranging as they do over a period of forty years' activity from his early lectures at Paris. In them we can trace the process of emancipation from established ruts of thought and the entrance of new conceptions.

We can follow his attempts to make a theory to explain the whole body of natural phenomena, the gradual elaboration of a mathematical theory of action at a distance, which, unfruitful at the moment, reappears in a fuller form in modern science. We see him as a pioneer of textual criticism, a critic of established authorities in whom the spirit of *Reynard the Fox* and the *Fablaux* is incarnated, a critic of received doctrines who applies to them in an ever-increasing degree the test of common sense and experiment. The work of such a one should be available to all the world of scholars: more than half of it in bulk is still locked up in single manuscripts difficultly legible and almost inaccessible.

PLATE XXVI

Leonardo da Vinci (1452–1519)

From a crayon portrait by himself at the Royal Library

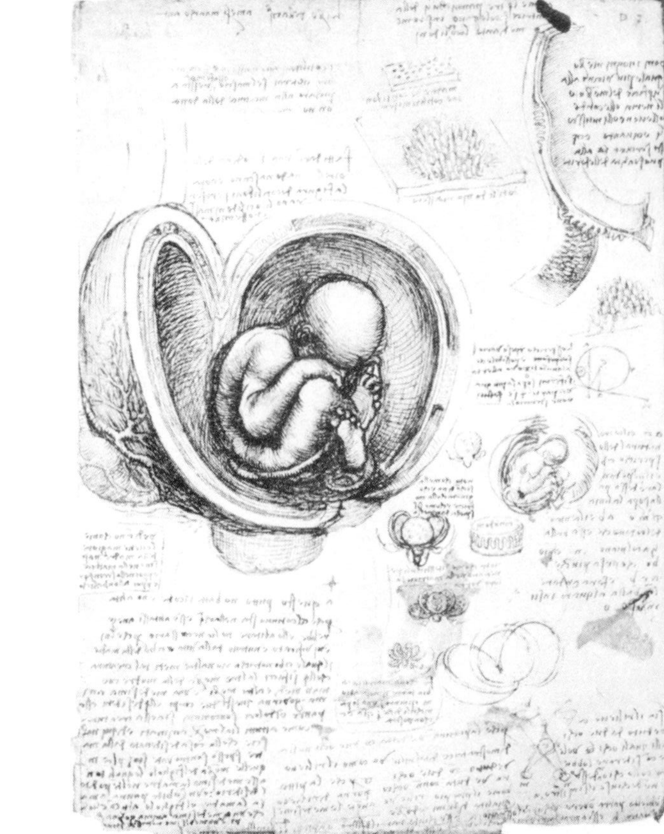

Quaderni V fo. 8 r

Foetus in utero and relations of membranes to uterine wall

LEONARDO AS ANATOMIST

By H. Hopstock

TRANSLATED FROM THE NORWEGIAN BY E. A. FLEMING

> Wisdom is the daughter of Experience,
> Truth is only the daughter of Time.
>
> Leonardo.

The greater part of Leonardo da Vinci's anatomico-physiological manuscripts are preserved at Windsor. Sixty leaves of these, with altogether about 400 drawings, have been published in facsimile with a diplomatic transcription and French translation as *Fogli A*, Paris, 1898, and *Fogli B*, Turin, 1901, the two together constituting the edition of the Russian Sabachnikoff and the Italian Piumati, with a preface by the French anatomist Duval. The remaining Windsor manuscripts of 129 leaves, with altogether about 1,050 drawings, have been published in facsimile with diplomatic transcription and English and German translations as the six volumes of the *Quaderni d'Anatomia* by Vangensten, Fonahn, and Hopstock, Christiania, 1911–16.

The facsimiles in the *Fogli* show Leonardo's drawings without colour. The *Quaderni* contain an exact reproduction of his manuscripts, showing the various tones of the paper, and the shades of the red chalk, pencil, ink, and other pigments which he used. The drawings and text of the Fogli, with very few exceptions, treat of anatomical and physiological questions only. Of the Quaderni, three-quarters are concerned with these subjects, whilst a quarter of the drawings, and much of the text, deals with other matters, especially mathematics, geometry, physics, and art.

The Quaderni show more clearly than the Fogli the manner in which Leonardo carried out his anatomical researches, and the period during which these developed. The pages of the Fogli, on the whole, reveal Leonardo as an independent and confident anatomist, especially with regard to his drawings. These manuscripts must therefore have been written in his later years. They discuss osteology, myology, the peripheral nervous system, the blood-vessels, the abdominal organs, and so on, all in a comparatively fluent and clear style. The Quaderni, on the other hand, cover a very long period in Leonardo's investigations, from his very first anatomical studies in 1489 up to his latest years. His language in the Quaderni is consequently not infrequently uncouth and involved, so that it is difficult to make out the meaning. The style, however, is extremely characteristic throughout.

Leonardo apparently at first made extensive use of old anatomical literature and diagrams. He attempted to elucidate and explain these, but as his authorities were inaccurate, so also were his delineations indefinite and clumsy. He then began to pursue his own studies, cautiously and tentatively at first, and then with more ease and definite purpose, until at last, emancipated from tradition, he stands forth as what he is, a great biologist.

He carried out his own precept to dissect the same part repeatedly. This is evident from the Quaderni, where sometimes on the same page, sometimes on different folios, numerous sketches appear of the same organ, sketches which show his progress from a hesitating student to a confident and independent investigator.

The Quaderni show also that Leonardo dissected animals. He has in part applied these discoveries to man, and it is evident from some embryological drawings and sketches of the processes of reproduction that there are instances in which he rested satisfied with these, though for the most part he finally portrays conditions as they occur in man. Yet strange to say, in the midst of his best topographical work, he occasionally takes an illustration of a single organ from an animal; it would seem that he has done this but as a trial or experiment, for in other cases the organ is correctly reproduced.

Certain departments of anatomy, only lightly touched on in the Fogli, are made the subject of careful study in the Quaderni. Such, for example, are the study of embryology, the structure of the generative organs, the form and functions of the diaphragm, the lungs, the brain cavities, and especially the heart and vascular system, as well as surface anatomy, and the study of proportion. It is indeed evident from the Quaderni that Leonardo undertook his dissections, not only in order to obtain anatomical data, but also, with the aid of this knowledge, to arrive at a clear understanding of physiological processes. He therefore makes his anatomical and physiological researches conjointly, expending not infrequently more work on the latter.

Great physicist and mathematician as Leonardo is, he corroborates his physiological investigations by experiments and by proofs and tests drawn from physics and mathematics; several of his dicta emphasize this fact: 'He who is not a mathematician according to my principles', he says, 'must not read me';[1] and again, 'Oh, students, study mathematics, and do not build without a foundation'.[2] For Leonardo the naturalist it was a matter of course to seek the principles of movement in animals in the laws of mechanics.

Under the heading *On Machines* he indicates four primary natural forces: (1) local movement which is produced by the three other forms of movement, (2) natural weight, (3) force and (4) *percussion*.

'We shall therefore', he says, 'first describe this local motion and how it produces and is produced by each of the three other Powers. Then we shall describe the natural weight, although no weight can be termed otherwise than accidental; but so it has pleased (us) to call it, to distinguish it from the force which is, in all its operations, of the nature of weight and is therefore called accidental weight; and this is set up as the third Power of Nature or the one produced by Nature. The fourth and last Power shall be called percussion, i. e. end or impediment of motion. And we shall first mention that every local involuntary motion is produced by the voluntary motor, like the counterpoise of a clock lifted up by its motor, Man.'[3]

Thus in his opinion every local or involuntary movement is, in the ultimate analysis, produced by a motive power that is itself voluntary, just as the involuntary movements of a clock depend ultimately on the voluntary movements of him who raises the clock weight.

'Furthermore, the Elements mutually repel or attract each other, as one sees that water expels the air and the fire entered as heat into the bottom of the cauldrons and escapes through the bubbles on the surface of the boiling water. And again the flame attracts the air, and the heat of the Sun draws up the water in the form of moist vapour, which afterwards falls down as heavy rain; but percussion is the immense Power of things which is produced in the Elements.'[4]

[1] Q. iv, f. 14 v. [2] Q. i, f. 7 r. [3] Q. i, f. 1 r. [4] Q. i, f. 7 r.

Leonardo's general idea of the proper method for the investigation of the human body and its parts is given in his *Plan for the Book*:

'This my exposition of the human form', he says, 'shall be demonstrated to you not otherwise than if you had the real man before you; and the reason is, that if you want to know thoroughly the parts of a dissected person, you must turn him, or your eye, examining him from different aspects, from below, from above, and from the sides, turning him, and investigating the origin of each member, and in this way the natural dissection has satisfied you as to your knowledge. But you must know that such knowledge does not satisfy you on account of the great confusion of pannicles (membranes) with veins, arteries, nerves, tendons, muscles, bones, and blood, which colours each part with the same colour, and the vessels which empty themselves of blood are not recognized on account of their diminution; and the integrity of the membranes is broken by the examination of the parts which are enclosed in them, and their transparency, tinged by blood, prevents you from recognizing the parts covered by them on account of the similarity of their blood-colour; and you cannot learn to know these parts without confounding and destroying the others. It is therefore necessary to do several dissections. . . . Thus each part and each whole will become known to you by my diagrams with the aid of demonstrations from three different aspects of each part. . . . Accordingly, the cosmography of the microcosm (*minor mondo*, i. e. man) will be demonstrated to you here through 15 full figures in the same order as has already been adopted before me by Ptolemy in his cosmography of the macrocosm (i. e. the world).'[1] And Leonardo adds: 'You must in your anatomy, depict each phase of the parts from man's conception until his death and till the death of his bones, stating which part of them decays first, and which part of them lasts longer.'[2]

Leonardo was a well-read man, conversant with the anatomical writings of Galen, Avicenna, Mondino, and Benedetti, but his opinion of authors is apparent from many passages in his works.

'I do not understand', he says, 'how to quote as they do from learned authorities, but it is a much greater and more estimable matter to rely on experience, their masters' master. These men go about puffed up, and boasting, adorned, not with their own qualifications, but with those of others, though they will not admit mine. They scorn me who am a discoverer; yet how much more do they deserve censure who have never found out

[1] Q. i, f. 2 r.

[2] Q. vi, f. 22 r.

anything but only recite and blazon forth other people's works. . . . Those who only study old authors and not the works of nature are stepsons, not sons of Nature, who is mother of all good authors.'[1]

The depth of feeling which animates Leonardo during his work of dissection can be gauged from the following passage:

'O searcher of this our machine, you must not regret that you impart knowledge through the death of a fellow creature; but rejoice that our Creator has bound the understanding to so perfect an instrument.'[2]

And what demands he makes on the dissector are seen in the following:

'And if you have love for such things you may be prevented by nausea; and if this does not hinder you, you may be prevented by fear of living during the night hours in the company of these quartered and flayed corpses, hideous to look at; and if this does not deter you, perhaps you lack the good art of draughtsmanship, which is essential for such demonstrations, and if you have the art of drawing, it may not be accompanied by the sense of perspective, and even if it is, you may lack the order of geometrical demonstrations, and the method for calculating the force and strength of the muscles; or perhaps you lack patience, so that you will not be painstaking. . . .

'As to whether all these things have been in me or no, the hundred and twenty books written by me will furnish sentence, yes or no, for in these I have not been hampered by avarice, or by negligence, but only by time. Vale.'[3]

Leonardo's nomenclature is very deficient. Bones, muscles, nerves, and vessels, have, as a rule, no definite names but are indicated by letters or some such means. In the case of the bones, the old names that were in use by mediaeval writers often occur, for example, *adiutorium* for humerus, *furcula* for clavicle, *focile maius* and *minus* for the ulna and radius; the muscles are also indicated by their origins and insertions, thus *pars domestica* and *pars silvestris* describe the palmar and dorsal sides of the extremities, *rascetta* and *pecten manus* indicate carpus and metacarpus; and then there are the mediaeval Arabic terms *meri* for oesophagus, *sifac* for peritoneum, and *mirac* for abdomen.

Leonardo's representation of embryological conditions is naturally by no means complete. He seems to have examined the

[1] Quoted from Oswald Sirén's *Leonardo da Vinci*, Stockholm, 1911.

[2] Q. ii, f. 5 v.

[3] Q. i, f. 13 v.

embryos of animals, hens, and calves, before he studied the human foetus. He says in fact: ' But you must first dissect the hatched egg before one shows the difference between the liver in the foetus, and the fully developed human.' [1] One of his figures [2] is evidently taken from a bird's egg, and he counsels one to observe ' how the bird nourishes itself in the egg '.[3] He remarks that chickens can be hatched by the warmth of an oven.[4]

' Ask the wife of Biagin Crivelli how the capon rears and hatches the eggs of a hen, when he is intoxicated. Her chickens are given into the care of a capon, which is plucked on the underside and then rubbed with nettles and set under a basket; then the chickens go in under it, and it feels it is being tickled by the warmth and likes it, for which reason it afterwards leads them and fights for them, jumping into the air against the goshawk in ferocious defence.' [5]

He seems also to have sought after the cause of difference of sex, and considers he has found it in that ' eggs that are round-shaped produce males, and those that are long-shaped produce females '.[6]

Leonardo has a series of very beautiful drawings of the human foetus lying in the uterus [7] (Plate XXVI). The position of the foetus is correct, and apparently he must have had opportunity to dissect a gravid uterus, but much of his work must have been done on the foetal calf. The foetus is surrounded by three membranes, the ' animus ', ' alantoydea ', and ' secondina ', probably corresponding to the amnion, allantois, and chorion of our notation. He does not describe a placenta, but on the other hand his drawings show how the chorion connected in several places with the inner surface of the uterus by cotyledons, the *male cotyledons* on the chorion embracing the *female cotyledons* on the uterus.

' The child in the uterus ', he says, ' has three panniculi which surround it, of which the first is called *Animus*, the second *Alantoydea*, the third *Secondina*; with this Secondina (chorion) the uterus is conjoined by means of the cotyledons, and all join in the umbilical cord, which is composed of vessels.[8] . . . How the three panniculi of the uterus bind themselves together by means of the cotyledons . . . female and male cotyledons. . . . Let some one give you the secondina of a calf when it is born, and observe the form of the cotyledons whether they retain the male or female

[1] Q. i, f. 10 r. [2] Q. iii, f. 8 v, fig. 3. [3] Q. iii, f. 9 v. [4] Q. iii, f. 7 r.
[5] Ibid. [6] Ibid. [7] Q. iii, ff. 7 r. and 8 r. [8] Q. iii, f. 8 v.

cotyledons; I observe how the fetal membranes are joined to the uterus, and how they loosen themselves from it.'[1]

The word *secondina* is thus used by him in various senses: as chorion, as secundine, and as foetal membrane generally.

The foetus does not breathe in the uterus, for it would drown if it did so, lying, as it does, in water.[2] Leonardo considers as the reason for this arrangement that heavy things weigh less in water than in air,[3] and that the foetus does not require to breathe because it is animated and fed by the mother's life and nourishment.[4] He denies the truth of the old story that the foetus cries or wails in the uterus, and considers that any sounds that seem to emanate from a gravid uterus must be caused by maternal flatus.[5] He frequently remarks that 'one soul governs (the) two bodies', and that what the mother eats, and the impressions she receives, leave their mark on the foetus.[6] He defines the length of the full-grown foetus as a *braccio* and the length of an adult as three times that of the full-grown foetus.[7] He has examined a foetus which was less than half a *braccio* in length at nearly four months.[8] He remarks 'how in four months the child is half of its length, i. e. eight times less in weight than when it is born.'[9] He draws attention to the fact that the foetus in the uterus grows three times as quickly as the new-born infant, and that a year after birth a child has not yet attained to twice the length of a nine-months-old foetus.[10] He has examined the viscera of the foetus, and draws attention to the relatively greater size of the left lobe of the liver at this stage of development, and points out that it diminishes after birth.[11] He seems to have observed the conversion of the umbilical vein of the foetus into the round ligament of the liver of the child.[12]

As a result of his study of conception and of the growth and birth of the foetus, Leonardo comes to the conclusion that man and his works are in no way things isolated in nature, but are only a part of one great whole, a single link in one vast chain, so that man is subject to the same 'necessità' as all other living things. He formulates these thoughts in the following words: 'Every seed has an umbilical cord which breaks when the seed is

[1] Q. iii, f. 8 r. [2] Q. iii, f. 7 r. [3] Q. iii, f. 1 v. [4] Q. iii, f. 8 v.
[5] Q. iii, f. 7 v. [6] Q. iii, ff. 3 v. and 8 v. [7] Q. iii, f. 7 r.
[8] Q. iii, f. 7 v. [9] Q. i, f. 10 r. [10] Q. iii, f. 7 v.
[11] Q. iii, f. 8 v and Q. i, f. 10 r. [12] Q. iii, f. 10 v.

mature. And similarly they have matrix and secondina as the herbs and all the seeds which grow in pods show.'[1] In this connexion we meet one of Leonardo's characteristically abrupt transitions of thought. While sketching a foetus in the intra-uterine position with elbow bent and hand prone, he suddenly asks himself which muscles flex the elbow-joint, and, immediately under the drawing of the foetus, makes two rough sketches of the muscles of the arm with pronated hand and the following legend attached. 'Demonstrate here only those muscles which serve to bend the arm to a right angle and those which cause it to turn the hand back and forward. Do not concern yourself with anything else, but demonstrate only the functions performed by those muscles which rise immediately from the bone of the said humerus.'[2]

Leonardo differs from many of the mediaeval writers who preceded him in representing the uterus with only one cavity. The tubes go outwards and upwards and the ovaries lie to the side of the uterus. On the right the artery of the ovary is seen to come from the aorta and the vein to go to the vena cava inferior (Plate XXVII). The ovaries are designated as 'seed-vessels (*vasi spermatici*) in the form of testicles, and her seed is first blood like that of the male'.[3] He is aware too that there is a difference between the male and the female pelvis. 'Measure how much less the woman's pubic bone is than the man's. It is for the sake of the space between the lowest part of the pubic bone and the point of the coccyx in view of parturition.'[4]

It is apparent from various drawings that Leonardo had a fairly good knowledge of the structure and relations of the testis, vas deferens, vesiculae seminales, vasa spermatica and nervi spermatici interni (Plate XXVII), 'nerves, originating in the vertebral column, which join the vein of the testicle',[5] although here also he partly transfers his findings from animals to man.

Leonardo discusses the results of conception from the union of white and black parents, and remarks that the colour of the offspring is not conditional upon the influence of the sun but of the parents' colour, and states 'that the mother's seed has an influence on the embryo equal to that of the father'.[6]

A number of general observations on Osteology are found in

[1] Q. iii, f. 9 v. [2] Q. iii, f. 7 v. [3] Q. iii, f. 1 v.
[4] Q. iii, f. 4 v. [5] Q. iii, f. 3 r. [6] Q. iii, f. 8 v.

PLATE XXVII

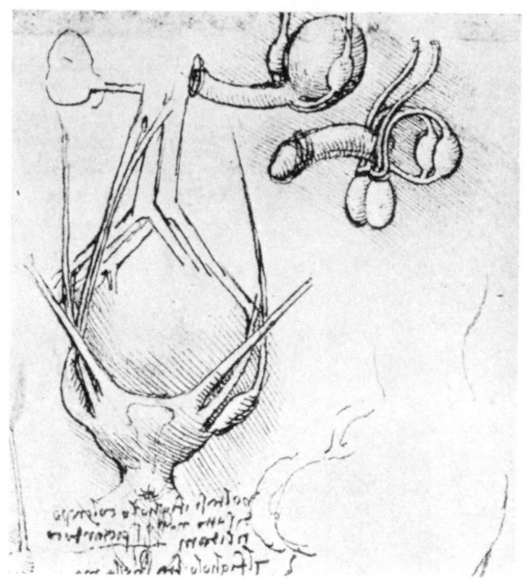

Quaderni III fo. 1 v

(Left) General structure of uterus and sources of its blood supply.
(Right) Male organs.

Quaderni V fo. 18 r

Topographical anatomy of neck and shoulder in a thin aged individual.

the Quaderni, which could only have been written by one well versed in the subject. In the *Plan for the Book* occurs this passage:

'It is necessary to make three dissections for the anatomy of the bones, which must be sawn through to demonstrate which is perforated and which is not, which is medullary and which is spongy, and which from without inwards is thick and which is thin, and which at one place has great thinness, and at one place is thick and at one is perforated or full of bone, or medullary or spongy, and thus all these things will sometimes be found in the same bone, and there may be a bone which has none of them.'[1]

In another place, Leonardo writes: 'Bone is of inflexible hardness adapted for resistance, and is without feeling. It terminates in cartilages at its extremities. And the medulla is composed of sponge, blood and soft fat covered with the finest veil. Spongy bone is a substance composed of bone, fat and blood.'[2]

Leonardo's treatment of the hands in his paintings is well known. Under the heading *The Hand from the Inside* he demands that the bones shall first be studied in order, so that their number, shape, and position should be learnt, and afterwards they should be further examined by being sawn through; they must then be put together according to the articulations; then the muscles that connect carpus with metacarpus and the tendons which move the first, second, and third joints must be demonstrated; next the nerves, arteries and veins, then the hand as a whole with the skin must be examined, and lastly the measurement of the hand and its parts must be given. A similar systematic procedure must be adopted in dealing with the back of the hand.[3]

With the exception of two delicate sketches in red chalk of the bones of the lower limbs set at the correct inclination to the pelvis (Plate XXXV), the Quaderni—in contrast with the Fogli—contain no osteological drawings of great interest, but only a few rough sketches, mostly of the bones of the extremities. A drawing of the cranium, the cervical, and part of the thoracic vertebrae has no close relation to the actual facts.[4] This drawing must date from the earliest period of Leonardo's anatomical studies, before he had begun to dissect, and when his fantasy had free reign, working rather on information gained from books, and possibly from old drawings that have now disappeared than on actual observation.

[1] Q. i, f. 2 r. [2] Q. ii, f. 18 v. [3] Q. i, f. 2 r. [4] Q. ii, f. 5 v.

As regards the morphology of the muscles, Leonardo writes:

'Muscles are of many kinds, some without tendons, like the trabeculae in the right ventricle of the heart, and others similar. Some are round like the above-mentioned and isolated (*musculi papillares*), being connected only by tendons (*chordae tendineae*) with that of the flexible part. . . . Some are broad and thin, some broad and thick, some long and narrow, others long and thick; some are thin and oval, some shaped like a fish, others like a lizard, some are twisted and some straight. Some have tendons along one side only, others at both ends, others are divided by several tendons, as for instance the longitudinal muscles. Some may move the part from either end, others from one end only, another moves behind its tendon, others draw their tendons towards themselves.' [1]

Leonardo states that muscles move longitudinally,[2] and he speaks generally of muscles with several heads.[3] In a passage on *Definition of the Instruments* he writes: 'The muscles, the functionaries of the nerves, draw to themselves the tendons which are connected to these members. . . . The tendons are mechanical instruments which in themselves have no feeling, but carry out whatever is imposed upon them.' [4]

In order to demonstrate the structure of the lower limb and the actions of its muscles he gives three illustrations of models formed from copper wire (Plate XXVIII). These are in the main correct and to them Leonardo attaches the following text:

'How many muscles originating in the hip are formed for the movement of the femur? Present the leg in full relief and make the cords of red-hot copper wire, and bend them on it to their natural position, and when you have done this you will be able to sketch them from four sides, and place them as they are in nature, and discuss their functions. When you have finished with the bones of the lower limb, give the number of all the bones, and having completed the tendons, give the number of these tendons, and you must do the same with the muscles and the nerves, the veins and the arteries, stating: the thigh has so many, the leg so many, the foot so many, and the toes so many, and then you must say: so many muscles spring from a bone and end in a bone, and there are many that spring from a bone and end in another muscle, and in this manner you can describe each detail of every part of the body, and especially the ramifications made by some muscles which produce various tendons.' [5]

[1] Q. ii, f. 15 r. [2] Q. iv, f. 6 r. [3] Q. iii, f. 9 v.
[4] Q. ii, f. 18 v. [5] Q. v, f. 4 r.

PLATE XXVIII

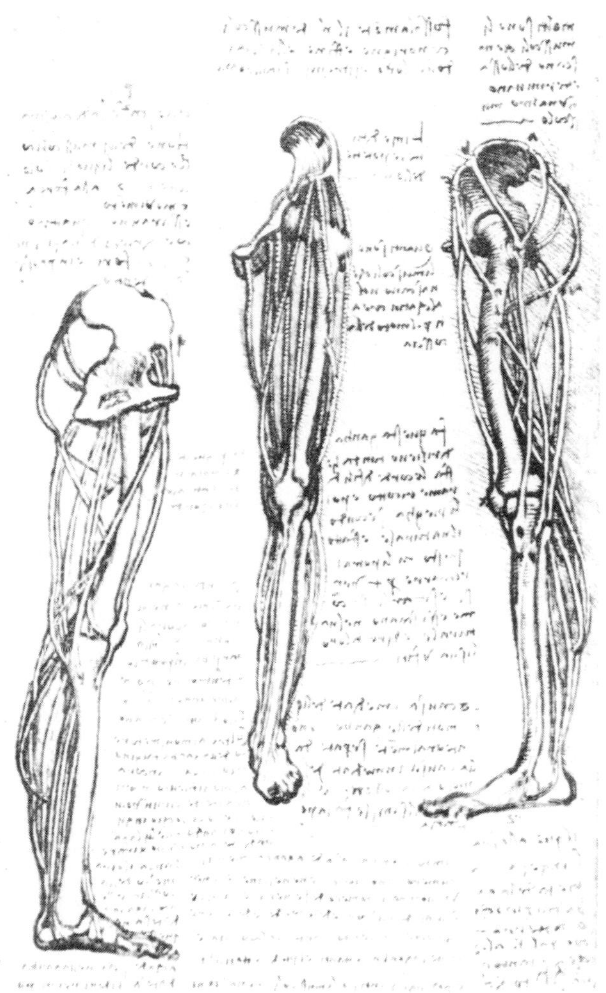

Quaderni V fo. 4 r

Bones of lower limb to which wires are fitted to illustrate lines of muscular traction

Leonardo emphasizes the fact that with the knee flexed the action of the sartorius muscle gives rise to internal rotation and the biceps to external rotation, but when on the other hand the knee is extended, rotation of the limb can only occur at the hip-joint. He adds:

'Nature has attached all the muscles which control the movements of the toes to the bones of the leg and not to the thigh, because these muscles, when the knee is bent, would, if they were attached to the thigh-bone, contract and become locked under the knee-joint, and would not be able without great difficulty and effort to work the toes.' [1]

The *biceps brachii* is described as a flexor and supinator, the *brachialis anticus* as a powerful flexor only, the *pronator radii teres* as pronator and antagonist of the biceps.[2] All these are compared with the cords of the 'trepan' which serve to pronate and supinate the hand. The ulna is characterized as 'non-rotating' in contradistinction to the radius, and it is specified that pronation and supination take place 'without alteration of the bone which is called adjutorium (humerus)'.[3] It is stated that when the arm is bent at the elbow the flexor muscles contract whilst the extensors stretch as the angle of the bend becomes more acute.[4] The three parts of the *deltoid* muscle and their functions, together with those of the *pectoralis major* and *teres major*, are correctly reproduced.[5]

The muscles of the back are stronger than those of the front, for, since one can bend farther forward than backward, more power is required to raise oneself after bending forward than backward.[6]

The topographical dissections carried out by Leonardo, of the throat and adjacent parts, are significant.[7] Of these he has made a series of drawings ranging from rough sketches to illustrations which reproduce carefully performed dissections in great detail. With these may be classed a pair of delicate silver-points where, through the thin skin of aged subjects, we discern the fossae of the throat with the underlying muscles (Plate XXVII). In another drawing Leonardo has topographically reproduced the lower section of the face, the column of the neck, the *hyoid* bone and its connexion to the *styloid* process, the larynx, *trachea, sterno-cleido-mastoid, trapezius, splenius, scaleni, levator anguli scapulae*, the

[1] Q. vi, f. 17 r. [2] Q. iii, f. 9 v. [3] Q. iv, f. 14 r. [4] Q. vi, f. 20 r. [5] Q. vi, f. 13 r. [6] Q. iv, f. 6 r. [7] Q. v, ff. 15–18, and 20.

jugular vein with its tributaries, the carotid artery, the *vagus* with the *superior laryngeal* and the *hypoglossal* nerves and what seems to be the sub-maxillary gland, though it is perhaps a lymph node (Plate XXX). It is remarkable that in these regional drawings of the human throat he should have taken his model for the hyoid bone and larynx not from a human subject but from an animal, perhaps a dog. His serial sections of the lower extremities are also noteworthy. Leonardo seems to have been the first to make topographical dissections and serial sections to illustrate the structure of the parts (Fig. 1).

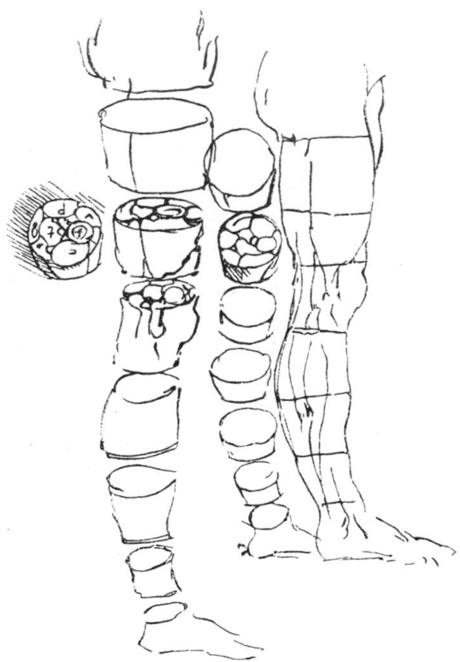

FIG. 1. Leonardo's use of serial sections

Leonardo's papers on comparative anatomy also display his great skill as a dissector, and his understanding of anatomical conditions. As commentary to a finished drawing of the abdomen and lower limbs of a muscular man, he writes:

'In order to make the comparison you must draw the leg of a frog which has great resemblance to that of a man in the bones as in the muscles. This must be followed by the hare's hind leg, which is very muscular, with conspicuous muscles unimpeded by fat.'[1]

On a folio of red-brown paper, between exquisite drawings of muscular men and of the skeleton of the pelvis and the lower limbs of a man and a horse, in which some muscles are marked out by cords (Plate XXXV), Leonardo thus writes:

'The union of the fleshy muscles with the bones without any tendon or cartilage—and you must do the same with several animals and birds. Represent the man on tiptoe so that you can

[1] Q. v, f. 23 r.

more easily compare him with other animals. Draw the man's knee bent like the horse's. In order to compare the skeleton of a horse with that of a man, you must present the man on tiptoe, when portraying the bones. On the affinity, which the conformity of bones and muscles of animals have with the bones and muscles of man.' [1]

Leonardo's great dissecting skill is seen again in the most perfect way in certain figures which display the muscles and tendons on the distal portion of the leg and on the foot, together with a part of the crucial ligament as well as the vaginal sheaths of the tendons. In these figures the toes are armed with claws of some cat-like animal, perhaps a lion [2] (Plate XXX). These drawings are by no means faultless in details, since the inner and outer edges of the foot have been interchanged, but they show, by the combination of the foot of a man with an animal's claws, one of Leonardo's most outstanding traits, his highly imaginative and artistic creative spirit. The treatment of these sketches with silver-point and colour, united to the author's power as a dissector and artist, give them a peculiar charm. They are without text, and Leonardo must have felt that they conveyed a sufficiently obvious and clear meaning.

The study of the diaphragm is a subject which Leonardo pursues with especial predilection.

'It is shaped', he says, 'like a deeply hollowed spoon.[3] . . . If it were not arched so that it could receive the stomach and other viscera into its concavity it could not afterwards contract . . . and exert pressure on the intestines, and drive the food from the stomach into the intestines, nor could it help the abdominal muscles to expel the faeces, nor could it by contracting enlarge the thoracic cavity and compel the lung to expand, so that they may inspire air to refresh the veins coming from the heart.' [4]

Leonardo points out that the diaphragm has four functions, primarily it is a respiratory muscle, secondly it presses on the stomach and drives its contents into the intestine, thirdly it aids the abdominal wall in the act of defaecation, and fourthly it divides the 'spiritual parts' (the lungs and heart) from the 'natural parts' (the abdominal organs).[5] All these functions are brought into action by the rise and fall of the diaphragm. The way in which the movements of the diaphragm and the abdominal wall

[1] Q. v, f. 22 r. [2] Q. v, ff. 11–14. [3] Q. i, f. 5 r.
[4] Q. i, f. 4 v. [5] Q. i, f. 5 r. and v.

alternate, like ebb and flow, is described and illustrated by an outline drawing.[1] Leonardo states that the muscles outside the ribs (*serrati*) must regulate them when the diaphragm contracts, as it would otherwise draw down the ribs, it being attached to them at its margin.[2]

Leonardo's treatment of the cerebral ventricles clearly shows the development of his investigations from vague and groping efforts, based on ancient and erroneous views (Plate XXIX, upper figure) to his own independent study of the phenomena (Plate XXIX, lower figure). By means of sagittal and horizontal sections through the head, he sketches the cerebral ventricles as three small vesicles lying behind each other and nearly equal in size, the foremost of which is, by means of *canals* (i. e. nerves), connected with the eye and ear.[3] In all this he has simply followed earlier authors. On the same sheet is drawn a cross-section of an onion. After he has calculated the layers he successively cuts through in bisecting the head, he says:

'If you cut an onion through the middle you will be able to observe and count all the circular layers and cases which cover the centre of the onion. In the same way, if you wish to bisect a human head, you will first cut through the hair, then the skin, then the muscular flesh and the peri-cranium, then the skull, and inside that the dura mater and pia mater, and the brain, thereupon (i. e. at the base) again the pia and dura mater, and *rete mirabile* and the base, the bone.'

But on another folio he approaches much nearer the actual conditions and illustrates almost perfectly by horizontal sections and profile drawings the inner aspect of the base of the skull, the base of the brain, the surface of the cerebral hemispheres, and the cerebral ventricles and their relations to each other.[4] He enumerates three ventricles and the drawings show that he counts both the lateral ventricles, which are connected, as one (Plate XXIX). He designates the lateral ventricles as *impressiva,* the third cerebral ventricle as *sensus communis,* and the fourth as *memoria*; this last seems to continue downwards into the spinal cord as a thin tube (? the central canal). The fourth cerebral ventricle is mentioned as the source or meeting-point of 'all nerves which give feeling'.

In order to acquire a correct idea of the ventricles, Leonardo performed the following experiment. Into a brain removed from the cranium, he injected melted wax through a hole in the fourth

[1] Q. i, f. 6 v. [2] Q. iv, f. 1 r. [3] Q. v, f. 6 v. [4] Q. v, f. 7 r.

PLATE XXIX

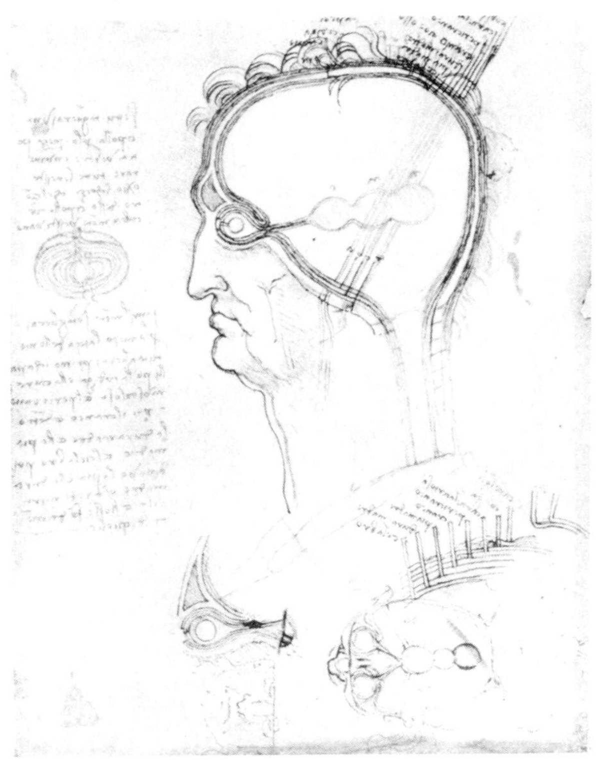

Quaderni V fo. 6 v

Ventricles and layers of head and eye in section

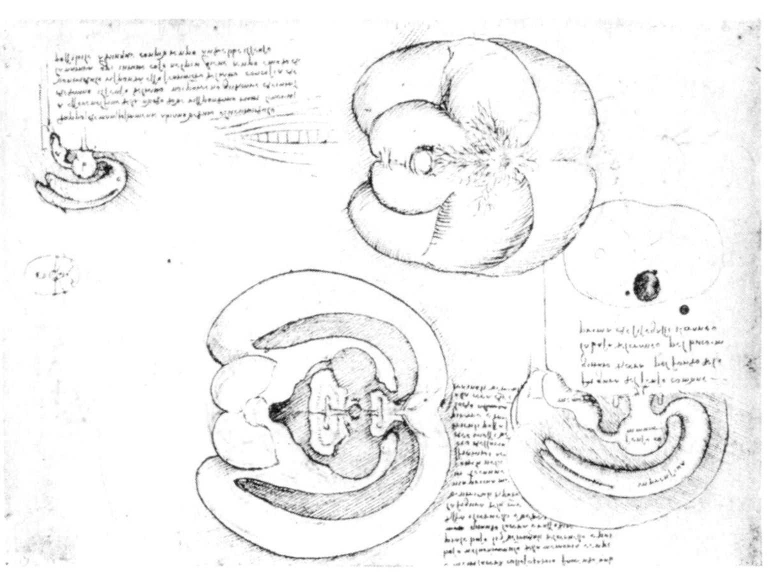

Quaderni V fo. 7 r

Casts of cerebral ventricles

ventricle, having already made an opening in both the lateral ventricles and inserted a tube so that 'the air can stream out'. He then removed the brain matter from the wax so as to display the shape of the casts formed in the ventricles. He made a similar experiment with a brain without removing it from the cranium, injecting the wax through a hole bored through the base of the skull, which probably led up through the infundibulum. His words are—

'Make two air-holes in the horn of the larger ventricle and inject the melted wax into it, at the same time making a hole in the *memoria* and fill through such a hole the three ventricles of the brain; and then when the wax has hardened, remove the brain and you will see the exact form of the three ventricles. But first insert the fine tubes into the air-holes, so that the air in the ventricles can stream out, giving place to the injected wax. The shape of the sensus communis filled with wax through the hole M at the bottom of the basis cranii, before the cranium was sawn through' (Plate XXIX).

These operations of filling the soft brain cavity with a solidifying substance are fraught with many difficulties, and it is not an easy matter to get casts true to nature. As is to be expected, therefore, the figures show certain deviations from the actual state of the parts. He was, however, the first to conceive the idea of injecting a solidifying substance into the cerebral cavities, and he was the first to give a fairly correct representation of those cavities. Yet as recently as the twentieth century the claim of priority in this method has been made[1] for a modern investigator, though it was in use nearly four hundred years before his time.[2]

On several pages in the Quaderni are drawings of a number of cerebral nerves and of the spinal cord.[3] In one such drawing of the base of the brain we see the olfactory nerves, and behind them the optic tract with chiasma and optic nerve and bulbs, behind these again are shown branches to the superior maxilla from the trigeminal, next the vagi, and farthest back the spinal cord. Elsewhere the vagi are sketched in their length, and shown passing from the thorax into the abdomen, where they obviously

[1] Cf. Regius on the Raube-Welche drawings, *Biologische Untersuchungen*, 1911.

[2] Cf. Holl, 'Leonardo da Vinci', in the *Archiv für Anatomie und Physiologie*, 1911.

[3] e. g. Q. v, f. 8 r.

ramify.[1] On other pages are seen the hypoglossal, and the vagus, with the superior laryngeal.[2] Leonardo often mentions the inferior laryngeal, *nervi reversivi*, as he calls them. 'The recurrent nerves are bent upwards only because they would be torn asunder in the great movement which the neck makes in extending itself forward and further because it partly carries with it the trachea and such nerves.'[3]

In one outline drawing Leonardo probably intended to represent the medulla oblongata.[4] To this sketch, indistinct and obscure, are attached some passages which show that he had grasped the importance of the spinal cord by experiments on frogs. 'The frog retains life for some hours after the head, heart, and intestines are removed. And if you perforate this cord, it immediately shrivels and dies. All the nerves of the animals derive from here. And if you perforate the said string it suddenly twitches and dies.' And on the back of the sheet the description of these experimental investigations continues thus: 'The frog instantly dies when its spinal cord is perforated. And formerly it lived without head, without a heart, or any entrails or intestine or skin. It thus seems that here lies the fundamentum of motion and of life.' Similar outline drawings are found in others of Leonardo's manuscripts.[5]

Both in the Fogli and the Quaderni, a cord-like structure is seen on either side of the spinal cord stretching down from the brain to the foramina in the transverse processes of the vertebrae. There are several connexions shown between these cord-like structures and the spinal cord, and they are also connected with the brachial plexus. These structures are products of his imagination: they cannot be the vertebral arteries. It may be that an explanation of them would result from a further investigation of his sources.[6]

In the Fogli much of the peripheral nervous system is correctly reproduced, but in the Quaderni it is on the whole cursorily treated. The brachial plexus can be traced sometimes to the elbow, sometimes to the wrist and hand.[7] The lumbar plexus is seen to consist of three lumbar nerves from which issue the femoral

[1] Q. i, f. 13 v. [2] Q. v, ff. 16 r. and 17 r. [3] Q. i, f. 13 v.

[4] Q. v, f. 21 r, fig. 5. [5] e. g. Fogli B, f. 4 r. and v. and f. 23.

[6] Cf. Holl, 'Leonardo da Vinci: Quaderni d'Anatomia', v and vi, *Archiv f. Anatomie und Physiologie*, 1917.

[7] Q. v, ff. 19 r., 21 r. and v.

PLATE XXX

Quaderni V fo. 16 r

Dissection of triangles of neck

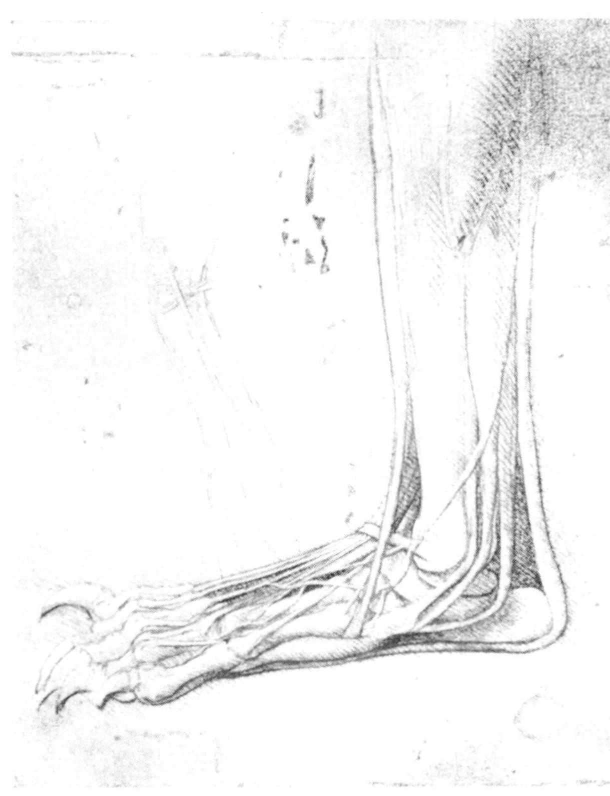

Quaderni V fo. 11 r

[Dissection of foot. Nails replaced by claws.

nerve with a branch to the inside of the leg besides a number of other branches,[1] and the sacral plexus is represented with branches of the ischiadic nerve distributed to the pelvis,[2] high up on the thigh [3] and above the knee.[4]

Leonardo devotes a large portion of his researches to the consideration of the lungs and respiration. His portrayal of the lungs seem to be based on examination of animals [5] as well as of man.[6] He knows the shape of the lungs and their lobes.[7] Their substance is dilatable, extensible and spongy, and they are enclosed in a delicate membrane (the pleura) which interposes itself into the spaces between the ribs when they expand.[8] He shows how the pleura covers the inner side of the ribs and the surface of the diaphragm [9] and he discusses if there be air in the pleura. 'Whether between the lungs and the chest, at any part, a quantity of air interposes itself or not.' [10] Later he decides against this view 'because there can be no vacuum in nature, the lung, which touches the ribs on the inside, must follow their dilatation '. [11]

The bronchi ramify in the lung substance, and gradually diminish until they become blind tubes which develop minute vesicles when inflated. These endings are drawn and described, both inflated and empty, '*trachea minima* uninflated and again inflated which redoubles its capacity in its increasing ',[12] and throughout the lungs the bronchi are accompanied by the blood-vessels and the finest ramifications of the bronchi are in close touch with the most minute branches of the blood-vessels.[13] Whilst in many of Leonardo's drawings the two bronchi enter the lung near its apex, others show approximately the actual conditions (Plate XXXVI), and here he says, 'first describe the entire ramification of the trachea in the lung, and then the ramifications of the veins and arteries each separately, and then all three together '.[14]

On the findings of his experiments with inflation of the lungs, Leonardo considers it impossible that air, as such, can reach the heart from the bronchi; on the contrary he holds that it is the pulmonary arteries that receive 'the freshness of the air' from the bronchi. 'To me it seems impossible that any air can penetrate

[1] Q. v, ff. 9 r., 20 v., and 21 r. [2] Q. iv, f. 9 r. [3] Q. v, f. 9 v.
[4] Q. v, f. 15 r. [5] e.g. Q. ii, f. 1 r. [6] e.g. Q. ii, f. 7 v.
[7] Q. iii, f. 4 v, fig. 7. [8] Q. ii, f. 1 r. [9] Q. iv, f. 3 r.
[10] Q. ii, f. 7 v. [11] F. A, f. 15 v. [12] Q. ii, f. 1 r. and f. 2 r.
[13] Q. ii, f. 1 r. [14] Q. iii, f. 10 v.

into the heart through the trachea (i. e. the bronchi) for he who inflates these does not expel any air from any part of these, and this occurs because of the dense panniculo, with which the whole ramification of the trachea is coated, which ramification goes on dividing into the minutest branches together with the minutest branches of the veins.'[1] And 'the lung is unable to transmit air to the heart . . . and further the air which is inhaled by the lung continually enters dry and cool, and leaves moist and hot. But the arteries which join themselves in continuous contact with the ramification of the trachea distributed through the lung are those which take up the freshness of the air which enters such lung.' Leonardo thus represents the bronchi as a tubular system, terminating blindly, from the expanded ends of which the inhaled air passes to the pulmonary blood-vessels. 'The dilatation of the lungs occurs in order that the lungs may inhale the air with which the veins which the heart sends into them can refresh themselves.'[2]

Leonardo frequently affirms that the diaphragm is the most essential respiratory muscle; but in deep inhalations such as a yawn or sigh, the contraction of the diaphragm is insufficient; then the serratus posticus superior comes into action. This he describes as made up of six muscles, three on either side, which stretch from the vertebral column to the uppermost ribs. The mode of action of these muscles is demonstrated by levers.[3] He further states that the scaleni and the serratus anterior are inspiratory muscles, and that they prevent the diaphragm from drawing the costal cartilages inwards. This is also illustrated by drawings.[4] Both in text and drawings he defines the internal intercostal muscles as expiratory, stating that they proceed obliquely upwards and forwards, that the external intercostal muscles are inspiratory and run in the opposite direction, and that the intercostal nerves, arteries and veins pass between the ribs.[5] The thorax expands on account of the oblique disposition of the ribs and of the bending of the costal cartilages,[6] the lower ribs move more than the upper; but in the case of irregular breathing Leonardo draws attention to the intervention of the abdominal muscles with action on the intestines, which again act on the diaphragm.[7]

Light sketches appear of the oblique plane of the superior

[1] Q. ii, f. 1 r.

[2] Q. i, f. 4 v. The old 'vena arterialis' = arteria pulmonalis.

[3] Q. i, f. 2 v. [4] Q. i, ff. 5 r. and 8 r. [5] Q. iv, f. 9 r. [6] Q. ii, f. 6 v.

[7] Q. ii, f. 16 v, p. 35.

PLATE XXXI

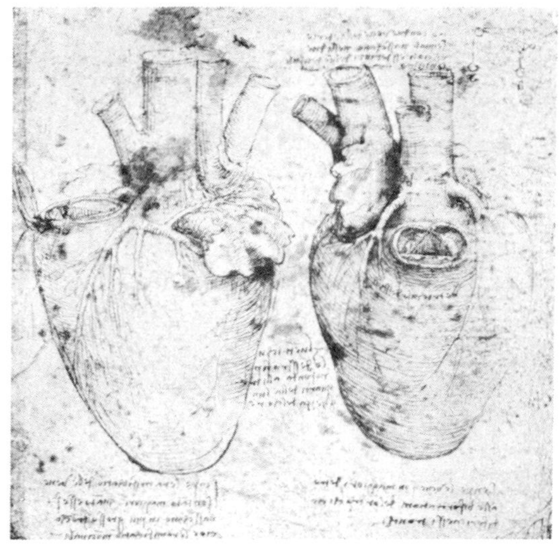

Quaderni II fo. 3 v
Dissection of coronary vessels

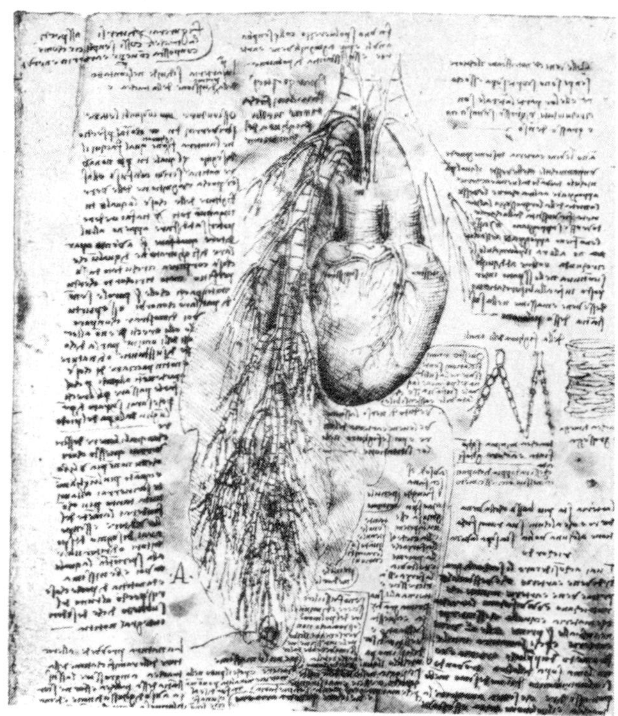

Quaderni II fo. 1 r
Dissection of bronchi and bronchial vessels

thoracic aperture, with the union of the costal cartilages with the sternum,[1] and also of the cartilages of the lower ribs which form the costal arch and are ranged one below the other, like a part of a cable, so that the skin may more easily glide over the cartilages when they move. The ribs are 'pivot-shaped' at their attachment to the vertebral column for the benefit of the respiration.[2]

Leonardo has observed that the shoulders move in breathing but 'the raising of the shoulders does not always force the lungs to inhale'. The thorax is the receptacle for the spiritual organs, the abdomen for the natural. The lungs expand and contract continuously in all directions, but mostly downwards.[3] Leonardo has observed a calcified focus in a lung and meditates—as this unique observer always does when he meets anything unknown to him—over the causes of the process.[4] Emphasizing the recuperative power of Nature, he concludes that Nature prevents a break in the bronchi, by the thickening of the substance which becomes cartilaginous, and forms an 'incrustation like the shell round a nut'. Inside the focus is found 'dust and watery humor'. From the passage containing this description, he draws a line to the focus on the drawing. In another manuscript which also treats of the lungs he remarks that 'dust is injurious',[5] so that he seems to have fixed on the idea of diseases of the lung as originating from this cause.

Leonardo points out that the cartilaginous ring of the trachea, which is elastic as a spring, is incomplete at the back, where the oesophagus enters.[6] 'But the trachea contracts in the epiglottis in order to condense the air which seems animated from the lungs in order to form various tones of voice.'[7] He considers that the relation of the trachea to the formation of the voice must be studied and he describes which and how many muscles act on the larnyx in phonation.[8] 'And thus you must not give up this study of the voice and of the trachea and its muscles until you have acquired full knowledge of all the parts contiguous to the larynx and of their functions made by nature for the modulation of this voice. And of all this you must make a special drawing, sketching and discussing the various[9] parts.' Then resuming his experiments he deals with various phonetical problems.[10] He now busies himself with the development of sound and shows, both in text and

[1] Q. ii, f. 6 v. [2] Q. iv, f. 1 r. [3] Q. iv, f. 3 r. [4] Q. ii, f. 1 r.
[5] Q. iii, f. 1 v. [6] Q. 1, f. 9 r. [7] Q. i, f. 5 v. [8] Ibid.
[9] Q. i, f. 9 r. [10] Q. iv, f. 10 r. and v.

drawings, that the palate divides the air current, so that it goes partly through the nose, partly through the mouth. He describes the position of the lips and the part played by the soft palate in the formation of vowels and arranges these in conjunction with various consonants in tables. He compares the trachea to an organ-pipe, and thinks that the voice can be modulated by its lengthening and shortening, its expansion or contraction.

Although in other manuscripts Leonardo has reproduced the larynx very clearly, it is uncertain how exactly he understood its function.[1] He does not use the word larynx, but calls it sometimes the 'trachea', sometimes the 'upper part of the trachea', or its '*ring*', sometimes the 'epiglottis' or the 'fistula or flute, i. e. the place where the voice is formed'.[2] He states that by the dissection of animals he will make an experiment of pressing the air in and out of a lung, 'contracting and dilating the fistula, the generator of their voices'.

In dealing with the trachea and larynx Leonardo also brings in the tongue, which consists of 28 or 24 muscles, but he 'notes how they transform themselves into six in their formation in the tongue . . . and further it must be demonstrated where such muscles have their origin, that is to say from the cervical vertebrae, where they join the oesophagus, and some from inside from the maxilla, and some from the outside from the side of the trachea'.[3] The tongue takes part in the enunciation and articulation of syllables, sets the masticated food in motion, cleanses the mouth and teeth, 'and its principal functions are seven, i. e. extension, retraction and attraction, thickening, shortening, dilation and straightening'. On account of its great mobility, Leonardo frequently compares it to the penis: 'but here you might perhaps argue with the definition of the membrum which receives in itself so much natural heat, that it, besides its thickening, lengthens very much'. The surface of the tongue in the cat and bovine tribes is very rough, and as an illustration of this, Leonardo relates the following: 'I once saw a lamb being licked by a lion in our town of Firenze where there are always twenty-five or thirty of them, and where they are bred; this lion removed with a few strokes the whole skin of the lamb, and, thus denuded, ate it up.'[4]

[1] Fogli A, f. 3 r.

[2] See C. L. Vangensten, 'Leonardo da Vinci og fonetiken', *Videnskabs-selskabets Forhandlinger*, No. 1, 1913.

[3] Q. iv, ff. 9 v. and 10 r.

[4] Q. iv, f. 9 v.

PLATE XXXII

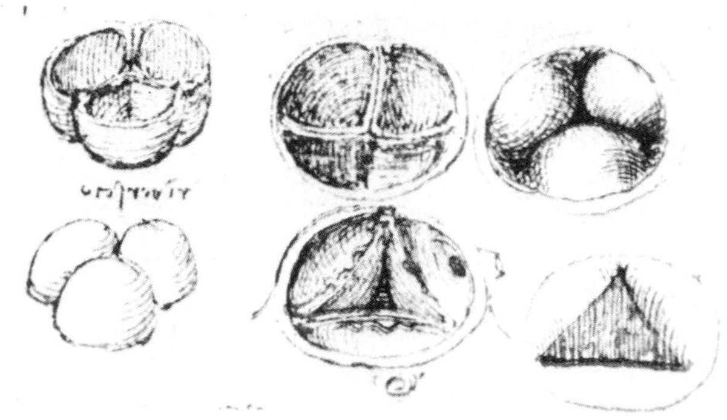

Quaderni II fo. 9 v

The Semilunar Valves

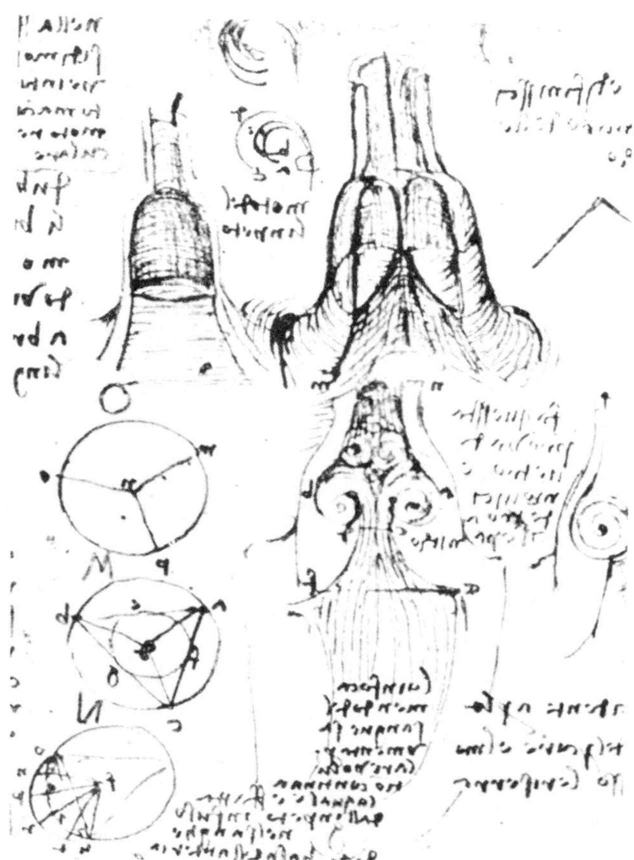

Quaderni IV fo. 11 v

Above. Glass casts with valves to illustrate action of semilunar valves.

Below. Diagrams of semilunar valves; to the right the eddies of the blood are shown.

Here again Leonardo makes a deviation in his train of thought. After dealing with the rough surface of the leonine tongue on one page, on the next his thoughts turn for a moment to Florence, for here appears a lightly sketched representation of that town [1] in the form of two inscribed circles joined by lines which go to the centre, and on which are written the names of its eleven gates.

In no less than three of the Quaderni, besides other manuscripts, Leonardo pursues, with surpassing skill, the study of the heart and the movement of the blood in it and in the larger blood-vessels. A number of drawings and paragraphs in one of the Quaderni [2] suggest that these manuscripts must have been written in his early days as an author, whilst others show that he had dissected and made much progress in comprehension of the vascular system.[3] It is obvious from these drawings that the representations of the heart are, for the most part, based on investigations of the hearts of animals, principally cattle. Occasionally however drawings of the human heart appear.[4] In several places Leonardo mentions the veins and arteries. He frequently indicates by the word 'veins', both veins and arteries—in other words, the blood-vessels. In some manuscripts he must mean arteries when he writes 'veins', and occasionally it is doubtful whether veins or arteries are implied. He draws the outer surfaces of the heart (Plate XXXI), and by longitudinal and transverse sections, he demonstrates its cavities, their form and projections (Plates XXXIII and XXXIV), with the trabeculae, the pectinati muscles with the depressions between the septum, the papillary muscles, the cordae tendineae, the valves, the columnae carneae, and again in a transverse section of the base [5] he shows the venous and arterial openings with their similarly disposed valves (Plate XXXIII, upper figure).

Leonardo draws the heart in the shape of a cone, with the base upwards and to the right. On the surface of the heart are seen the coronary vessels, arteries as well as veins. These 'lie together, the arteries deeper than the veins, but some of the arterial ramifications lie above those of the veins',[6] and 'with regard to the third vein, I have not yet seen whether it has an artery with it, for which reason I shall make a dissection (really peel off the flesh) to satisfy myself'.[7] The coronary vessels are

[1] Q. iv, f. 10 v. [2] Q. i. [3] Q. ii. and iv.
[4] e. g. Q. ii, f. 14 r., fig. 1. [5] Q. iv, f. 14 r.
[6] Q. ii, f. 1 r. [7] Q iv. f. 13 v.

surrounded with fat, and covered by the pericardium which covers half the width of the vessels, the other half is covered by the flesh.[1] The arteries feed the heart's substance,[2] and spring from the aorta;[3] they 'issue from both the outer openings of the left ventricle'. By this is perhaps meant Leonardo's *hemicycles*, later known as the sinuses of Valsalva. The coronary veins are called '*vene nere*',[4] probably because they are dark in colour on account of their venous blood. Leonardo describes how the outer surface of the heart is visibly divided by the coronary vessels[5] and these anastomose at the apex.[6]

He states that the heart has four *ventricles*, two larger on the right, and two smaller on the left. The two lower ones lie in the heart's substance, and the two upper ones outside it. He opposes those who say that there are only two cavities in the heart, and maintains that if one means that the two right and two left ventricles each only form one cavity, then the room and ante-room which are separated by a small door are only one room.[7] The upper or outer ventricles should in this case answer to the ante-rooms and this word is indeed often used to designate them.[8] In consequence of the fact that most of Leonardo's material on the subject of the heart is derived from animals, his outer or upper ventricles are sometimes to be taken not as atria but as auricles, and he occasionally says this himself. 'The heart has four ventricles, that is to say, two upper, called heart-ears ('orechi'), and below them the two lower ones called the right and left ventricles.'[9] He says that these *ears of the heart* are of the nature of expanding pockets, so as to receive the percussion on the movement made by the blood when it is forcibly driven out of the ventricles when these contract, and he compares them to the bundles of wool and cotton placed on the bulwarks of a ship to soften the impact of shots from enemy bombardment.[10]

It seems reasonable to assume that Leonardo, who has dissected animal as well as human hearts, often meant atria when he used the words 'upper ventricles', and 'ears'. This is also obvious from a drawing and observation he made of an open *foramen*

[1] Q. ii, f. 1 r. [2] Q. ii, f. 4 r. [3] Q. ii, f. 3 v. [4] Q. ii, f. 4 r.

[5] Q. iv, f. 13 v. [6] Q. iv, f. 14 v. [7] Q. i, f. 3 r.

[8] Holl points out that Leonardo was the first to introduce the designation ventricles for the cardiac cavities: before Leonardo, various other expressions were in use, as *sinus*, *vacuitas*, or *concavitas cordis*.

[9] Q. ii, f. 17 v. and f. 3 v.

[10] Q. ii, f. 3 r. Here is one example of the many images in which Leonardo's work is so rich.

ovale. 'I have found from (a) in the left to (b) in the right ventricle a perforation which I here note to ascertain whether it also occurs in other hearts.'[1]

The muscles of the heart consist of longitudinal and transverse fibres. In all the four ventricles the interior muscles, which are adapted for contraction, are of a similar nature. The surface muscles on the other hand serve only the lower ventricles, but the upper ones have a continuous membrane, dilatable and contractable; elsewhere[2] he states that the upper ventricles could not empty themselves if they could not fold together, and if they had not longitudinal, oblique, and transverse muscles, able to contract.

Leonardo considers and represents the wall of the left ventricle as much thicker than that of the right,[3] and the apex of the heart to be formed mostly by the left ventricle[4] (Plate XXXIV, lower figure, and Plate XXXVI, left). To obtain a correct idea of the shape and contour of the heart's cavities, they must be inflated before dissection.[5] One can then find 'cells' or 'cavernosities' separated by rounded walls (i. e. the trabeculae and pectinati muscles). 'If you inflate the auricle you will find out the shape of the cells.'[6] The cavity of the heart is divided into two parts by a septum in which are pores, *meati*, for the passage of blood from the right to the left ventricle. As a rule, Leonardo makes the septum quite solid, occasionally with indications[7] of the *meati*, but these he admits he was unable to find himself, for he refers to them as invisible.[8]

Musculi papillares, which Leonardo calls the *muscles of the heart*, merge 'near the valves' into the cordae (tendineae), and the valves are held firmly by these cords. These *muscles of the heart* divide into two parts, each having its own cordae, and Leonardo has noticed that the cordae are fastened to different parts of the cusps of the valves.[9] He has made a very beautiful drawing of the papillary muscles with cordae, fixed to the tricuspid and bicuspid valves[10] (Plate XXXIII, upper fig., and Plate XXXIV, upper fig.).

In some of his drawings of ventricles Leonardo sketches the intraventricular moderator band,[11] which takes its origin in the septum and is attached at the base of a papillary muscle or a trabecula[12] (Plates XXXIV and XXXVI) This band he called

[1] Q ii, f. 11 r. [2] Q. i, f. 4 r. [3] Q. ii, f. 11 r.
[4] Q. ii, f. 4 r. [5] Q. iv, f. 13 r. and v. [6] Q. iv, f. 13 v.
[7] e. g. Q. i, f. 3 r. [8] Q. iv, f. 11 v. [9] Q. ii, f. 3 r.
[10] Q. iv, f. 13 r. and v., f. 14 r. [11] Q. i, f. 14 r; Q. iv, f. 13 r.
[12] Q. iv, f. 13 r.

catena,[1] and in his opinion it serves to prevent the heart dilating more than necessary, for without this structure the heart would draw too much blood from those vessels into which it had previously thrown it [2] (Plates XXXIII and XXXIV).

In one of his drawings the vena cava superior and inferior are distinctly seen to enter the right auricle separately.[3] The other drawings of these vessels and their relation to the heart seem to be taken from animals. To the pulmonary artery and vein he gives the old names of *vena arterialis* and *arteria venalis* respectively.

The aorta (*vena aorto* or *arteria aorto*) comes from the left ventricle.[4] 'The right ventricle has two orifices, one in the vena aorto, and when the heart dilates in the left ventricle its base contracts to close the door of the arteria aorto.' [5] The other orifice is 'the arteria venalis (pulmonary vein),[6] and goes from the heart to the lung'. Leonardo is here indefinite. He does not mention the left auricle.

On one occasion Leonardo calls the pulmonary artery 'the door of the lung'. Under the heading *On the names of the vessels of the heart* he says, in fact, 'the door of the lung, and it is called vena arterialis, and it is named *vena*, because it conveys the blood to the lung.' Leonardo has here erased these last words *to the lung*, but that he meant that the pulmonary artery carries blood *to* the lung seems obvious from the continuation of the passage: 'And it has three valves which open from within outwards (valvulae semilunares) with perfect closure, and these are in the right ventricle.' [7] The dominating position attributed by him to the aorta is seen from the following passage: 'And in the middle of the base of the heart is the source or base of the aorta, founded in the centre of the heart's base, having power over the state of this heart's base as the latter has power over the animal's life.' [8]

Leonardo deals often with the valves of the heart. The atrio-ventricular valves are formed, according to Leonardo, by the endocardium above and the cordae tendineae below; these may

[1] Q. ii, f. 4 v.

[2] Holl points out that Leonardo was the first to observe these fibres passing through the ventricular cavities and suggests that, in honour of their discoverer, they should be called Leonardo da Vinci's columnae carneae. Tawara in 1908 first pointed out that these fibres form bridges through which the fibres, the atrio-ventricular bundle of His, reach from the septum to the papillary muscles.

[3] Q. ii, f. 14 r.

[4] Q. ii, f. 2 v. Holl points out that Leonardo must here have written right for left.

[5] Q. ii, f. 13 v.

[6] Q. ii, f. 2 v.

[7] Q. ii, f. 2 v.

[8] Q. iv, f. 14 v.

stretch from one of the papillary muscles to two of the cusps of the valve.[1] Valvulae semilunares are portrayed both open[2] and closed.[3] Leonardo describes how the inside of the artery (i. e. either of the aorta or pulmonary artery) is covered by a thin membrane (intima) which merges into and forms one side of the valves, the other side of the valve being formed of another layer of the pannicle.[4] A similar pannicle is found in the ventricles (endocardium) and in the pericardium. The shape of the closed semilunar valves seen from above and below are so beautifully reproduced that they must have been copied after the large vessels had been filled up with a solidifying substance. Leonardo indeed here remarks: ‘ but first pour wax into the ports of an ox heart, so that you can see the true form of these doors.’ [5]

In his opinion the valves and the roots of the great arteries are enclosed in the substance of the heart, so that the blood when it presses against the closed valves will not destroy these, but will break its impetus against the walls of the blood vessels by dilating them.[6] And in dealing with the closing of the semilunar valves when the blood passes over them, he finds, from his physico-mathematical reflections that three aorta-valves are more satisfactory than four,[7] for if their number were more than three, their angles or triangles would be weaker than those [8] formed by three valves.

It is difficult to judge how Leonardo pictured to himself the working of the heart, the movement of the valves and of the blood. He is not in the least clear in his pronouncements concerning them, and his remarks on these points are spread over many leaves of manuscript. He often says that when the ante-chambers, the upper ventricles (i.e. auricles), contract, the heart-chambers, the lower ventricles, expand, and vice versa, e.g. ‘ On the two lower ventricles situated in the root of the heart; their dilatation and contraction are made at one and the same time through the flux of the flood, and the reflux of the blood is made at one and the same time, succeeding the first, through the reflux in the upper ventricles, situated above the root of this heart.’ [9] The movement of the blood at the alternating dilatation and contraction of the upper and lower ventricles is thus compared with ebb and flow. By the contraction of the auricles, the blood is driven through the atrio-ventricular openings into the ventricles which open, causing

[1] Q. ii, f. 3 r.
[2] Q. ii, ff. 3 v. and 4 r.
[3] Q. ii, f. 9 v.
[4] Q. iv, f. 14 v.
[5] Q. ii, f. 12 r.
[6] Q. iv, f. 14 v.
[7] Q. iv, f. 12 r.
[8] Q. iv, f. 12 v.
[9] Q. ii, f. 4 v.

the semilunar valves to close. When the ventricles contract, some blood returns to the ante-chambers before the atrio-ventricular valves have closed entirely; [1] the latter when so stretched increase somewhat in size and approach one another, causing ultimately a complete closure both of the right and left atrio-ventricular orifices.[2] With the systole of the right ventricle, another portion of the blood goes through the pulmonary artery to the lungs, while a third portion goes through the septum into the left ventricle.[3] Thus less blood is driven back from the right ventricle to the right auricle than goes from the auricle to the ventricle, and the right auricle has its quantity of blood made good from the vena cava inferior, 'through the liver, the treasurer, the generator of blood'. 'The blood', which comes to the lungs, 'gives without hindrance the necessary nourishment to the pulmonary veins, where the blood, after it has been refreshed in the lung, returns for the most part to refresh the blood left in the ventricle where it divided.' [4] What Leonardo means by this is obscure, as he does not mention through which vessels the blood returns to the heart. 'In the lungs the arteries which are connected with the minute branches of the bronchi absorb the freshness of the air entering the lungs.' [5]

Elsewhere he asks: 'Whether the pulmonary veins send back the blood to the heart when the lung contracts at the expulsion of the air',[6] and he says that the contraction of the diaphragm compels the lungs to expand, 'which occurs in order that they may absorb the air with which the veins (vena arterialis = pulmonary artery) proceeding from the heart may refresh themselves.[7] The blood is heated in the heart's cavity, part of it evaporating (spiritus vitalis), and this vapour, mingled with dense moisture, is excreted through the farthest ends of the capillaries ('vene chapillari') from the skin in the form of perspiration.' [8] When the left ventricle contracts, the blood goes through the aorta [9] ('the upper vessel'), and the wave of blood thus formed goes through all the arteries,[10] and with the pulsation of the blood in the heart, which closes the valves, 'a tone is created which goes through every artery and which the ear often hears in the temples'.[11]

By drawings as well as descriptions, Leonardo discusses the flow of the blood from the left ventricle through the opening of

[1] Q. ii, f. 3 v. [2] Q. ii, ff. 3 r., 8 v., 11 r., and 12 r. [3] Q. ii, f. 17 v.
[4] Q. ii, f. 4 v. [5] Q. ii, f. 11 r. [6] Q. i, f. 5 r. [7] Q. i, f. 4 v.
[8] Q. ii, f. 11 r. [9] Q. ii, f. 17 v. [10] Q. ii, f. 13 v. [11] Q. ii, f. 3 r.

PLATE XXXIII

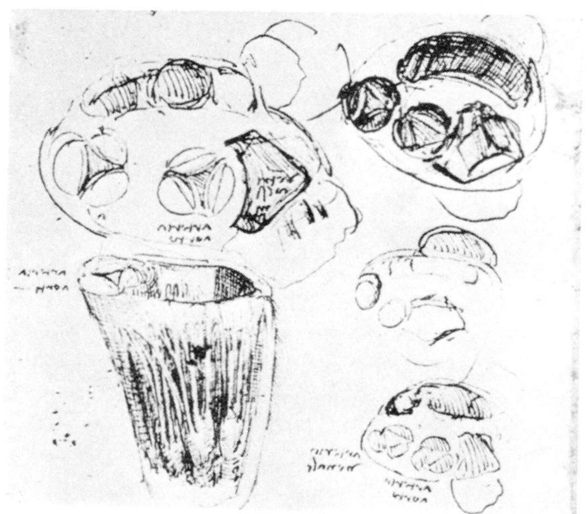

Quaderni V fo. 14 r
Details of cardiac anatomy

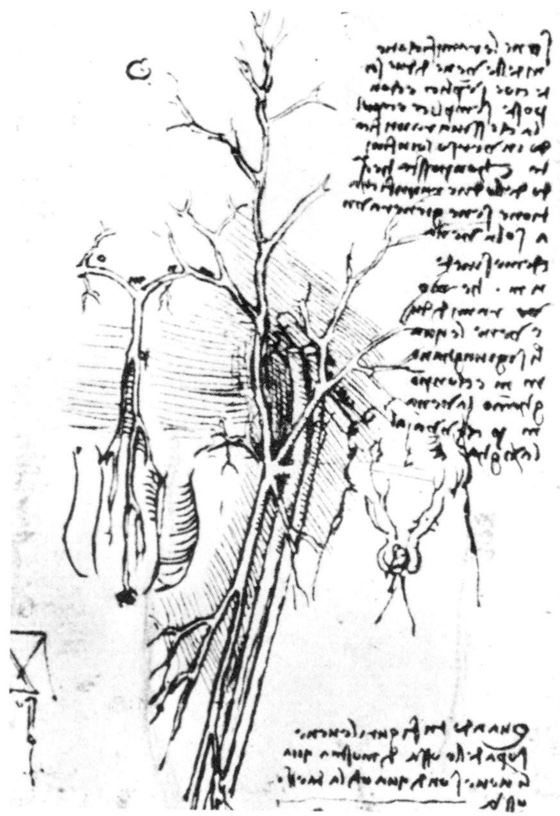

Quaderni IV fo. 8 r
Blood-vessels in inguinal region

the aorta and its branches.[1] The semilunar valves open at the ingress of the blood and close with its withdrawal. He theorizes as to how far the arterial ostia open only with the central part of the semilunar valves, and to what extent the openings of the heart could have closed by mere muscular action without valves: he comes to the conclusion that the closure of the heart's openings proceeds both better and more quickly with the aid of the valves than if it occurred by action of the heart's substance.

When the left ventricle contracts, the blood, as already stated, flows into the aorta; its speed is there varied proportionately to the calibre of the vessel. When the wave of blood enters the aorta, the centre part of the wave which goes directly upwards is higher than the sides, the impetus of which dissipates itself by the lateral motion. Leonardo proves that in this the blood acts like other fluids, demonstrating both by words and drawings the action of water when it runs out of a vertical and a horizontal tube.[2]

During his deliberations over the movement of the blood in the *ostium aortae*, and above its semilunar valves, Leonardo made several experiments with models. His wax castings of the heart have already been mentioned,[3] and in this connexion he says: 'A form of gypsum to be inflated, and a thin glass within, and then break it from head to foot'; and: 'The form of the glass, to see in the glass what the blood does in the heart when it shuts the openings of the heart'.[4] In order clearly to understand the movement of the blood in the heart, Leonardo thus first made a wax cast of the ventricles and their vessels, over this he made a gypsum cast, and from this a glass cast. Through this glass cast he has examined the vortices made by the blood when it is driven out by the systole into the aorta and pulmonary artery, as shown in some of his drawings[5] (Plate XXXII). The semilunar valves close during these vortices, and the walls of the blood-vessels protrude into the 'semiventricles' or 'hemicycles' (sinuses of Valsalva).

Leonardo, however, finds it difficult to gauge to what extent this actually occurs. He says: 'It is doubtful whether the percussion caused by the forcible movement in the front of the upper arch of the hemicycle divides into two parts, of which one goes upwards and the other backwards, and this doubt is subtile

[1] Q. iv, f. 11 r. and v, f. 12 r.
[2] Q. iv, f. 11 r.
[3] Q. ii, f. 12 r.
[4] Q. ii, f. 6 v.
[5] Q. ii, ff. 12 r. and 13 v.

and difficult to elucidate.'[1] He then makes another experiment: 'Make this experiment in a glass and move . . . the panniculae (i.e. the valves) about in it'[2]. And now his doubt has vanished; he has completed his experiments and has come to the following conclusions: When the blood enters the hemicycles, it strikes the wall of the aorta and divides at the topmost edge of the hemicycle into an ascending and a descending stream. The descending part makes a spiral curve, follows the concavity of the hemicycle, and percolates through its base; it then follows the surface of the semilunar valve, stretches the valve and closes it against the other valves, the stream then turns upwards in a retrograde movement and ends in a reflex vortex.[3] The ascending part of the blood-stream also makes a whirling motion, but in the other direction. This vortex again forms other vortices, which gradually decrease until 'the impetus consumes itself'. When the blood by the systole of the left ventricle is driven through the mouth of the aorta, it strikes against the blood over the semilunar valves, and 'this concussion shakes all the arteries and pulses distributed throughout the body'.[4] The systole collapses simultaneously with the concussion of the apex and the thorax, also with the heat of the pulse, and the entry of the blood into the auricle[5] (Plates XXXII and XXXIV).

One must investigate the relations of the recurrent nerve to the heart, he tells us, to see whether this nerve lends movement to the heart, or whether the heart's movement is spontaneous.[6] Its dilatation and contraction are spontaneous, he concludes, and these movements occur on the heart's longitudinal axis.

In the ventricular hollows, between the trabeculae and the pectinati muscles, to which we have already referred, the blood is driven round in a whirling movement,[7] and because it does not meet any edges or corners this movement acquires 'its vertiginous impetus', which causes heat, and this can become so great as to cause suffocation. Leonardo says that he has witnessed such an occurrence in the case of a man whose heart 'broke' as he fled from the enemy, a blood-stained sweat exuding from all the pores of his skin. Leonardo's thought then turns to the general and vital purpose of heat, and he states: 'And so heat gives life to all things, just as one sees that the warmth of the hen and turkey hen gives

[1] Q. ii, f. 13 v. [2] Q. iv, f. 11 v.

[3] It seems, to judge from fol. 11 r., par. 11, and fol. 11 v., par. 11, that Leonardo means the sinus of Valsalva. [4] Q. iv, f. 11 v.

[5] Q. iv, f. 11 r. [6] Q. iv, f. 7 r. [7] Q. iv, f. 13 r., par. iii.

PLATE XXXIV

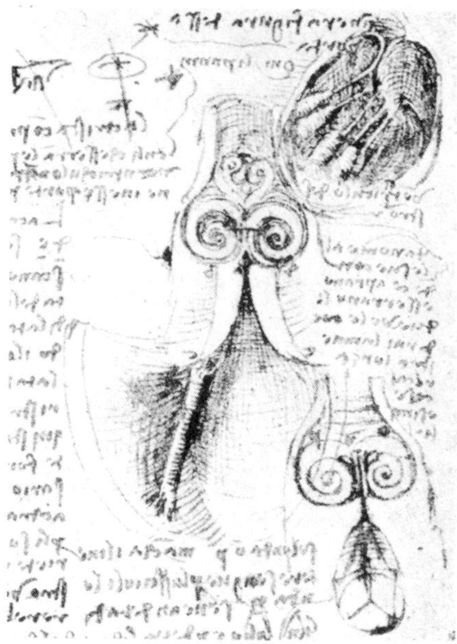

Quaderni II fo. 12 r

Right ventricle pulmonary artery and *musculi papillares.* The eddies of blood are shown around the semilunar valves.

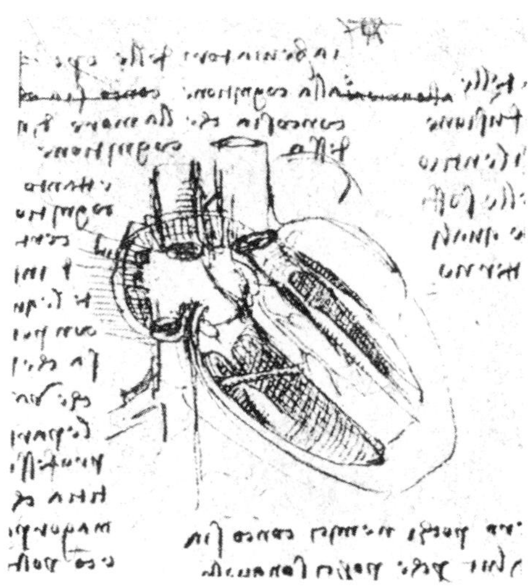

Quaderni II fo. 14 r

Ventricles, right auricle, and great vessels

life and birth to their chickens, and the sun when it returns gives life and blossoming to all fruits.'[1]

Leonardo endeavoured also to study the movement of the living heart. He relates that when pigs were killed in Tuscany, the animal was turned on its back, fastened securely, and an instrument called a 'spillo', used to draw wine from casks, is thrust into its heart.[2] He observes that if it enters the heart the instrument begins to move: at the diastole its point goes upwards, and its handle downwards, with the shortening of the heart,—the contrary with its lengthening. At last, when the heart has ceased to move, the handle becomes stationary in the exact middle of the two extremes. During the experiment Leonardo estimated the length of the movements. 'And I have seen this several times and taken such measurements, and left such an instrument in the heart until the animal was cut up.' Leonardo demonstrates the experiment by drawings, and points out that the movements of the instrument do not continue equal in extent, giving reasons for this (Fig. 2).

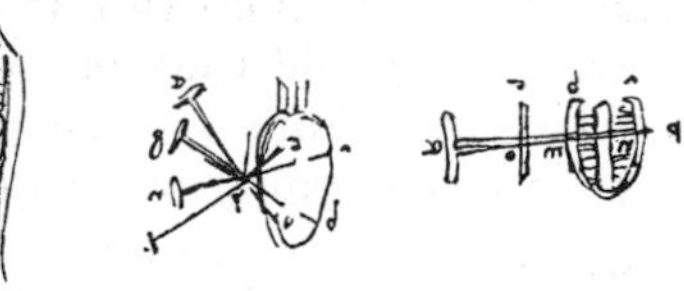

FIG. 2. An experiment of Leonardo on the heart.

Of all the bodily movements Leonardo evidently found that of the heart and vascular system the most perplexing. To this question he returns again and again, expending on it much of his time, his art of dissection, his keen observation, and his knowledge of the laws of physics. He seeks to elucidate this theme with question and counter-question directed from every point. He controverts actual authors and fancied antagonists, and his drawings, experiments, and deliberations tell their tale of his unceasing efforts to reach the ultimate truth concerning this abstruse problem. The fact that more than a quarter of all the anatomical and physiological drawings in the six Quaderni deal with the heart and its cavities shows how intense was his concentration on this subject, and it is evident from the following

[1] It is extremely interesting to follow Leonardo's train of thought in this manuscript. Par. iii reaches nearly to the edge of the manuscript; it is therefore evident that he did not intend to write anything further, but his thoughts turned to the general value of heat, and he made a note of them (par. iv) in the narrow margin at the side of par. iii. From this par. iv he drew a line to the foot of par. iii to indicate the order in which they should be read. See Plate XXXVI.

[2] Q. i, f. 6 r.

passage how necessary he thought it to use drawings as well as words for the purpose of demonstration. 'With what words will you describe this heart so that you do not fill a book; and the more minutely and elaborately you describe it, the more you will confuse the mind of the listener, and you will ever require a commentator or to revert to experience, which in your case has been very short, and explain few things concerning the subject in its entirety about which you wish full knowledge'; [1] and it is in connexion with this subject that he exclaims: 'Give an address on the shame, which is necessary for the students, impeders of anatomy and abbreviators thereof,[2] nay not abbreviators but destroyers should they be called who curtail such a task as this.' [3]

To judge from the Quaderni, the main results of Leonardo's research concerning the working of the heart may be stated as follows: At the contraction of the auricles the blood flows through the venous ostia into the ventricles. At the systole of the ventricles the blood goes out into the pulmonary artery and the aorta. By means of the pulmonary artery the blood goes from the right ventricle to the lungs, from where it returns 'refreshed' to the heart—by which vessels is not stated. From the left ventricle the blood is driven into the aorta and from it into all the arteries. Towards the skin it passes into the 'capillary veins'. Blood enters the right auricle by the vena cava.

Leonardo has not given any clear description of the circulation, his comments on the subject being disconnected and incomplete. He was not able entirely to emancipate himself from the old idea of the passage of the blood from the right ventricle through the septum into the left ventricle, although he himself seems doubtful about the truth of it, for he says that the pores in the septum are invisible. He likewise maintains that with the systole of the ventricles, some blood returns to the auricles until the atrio-ventricular valves have closed entirely. On the other hand, one must remember that Leonardo has not given any systematic exposé of the subject; his remarks are distributed over many years and many manuscripts. Nor is it surprising that Leonardo, who was no physician, and who occasionally fell short in the matter of proportion, of which as an artist he made daily use, should be incomplete and obscure in dealing with one of the most difficult physiological problems. It is also to be noted that Leonardo did not seek to publish the results of his research—it is only comparatively recently that they have been found in

[1] Q. ii, f. 1 r. [2] Q. i, f. 4 v. [3] Q. i, f. 4 r.

PLATE XXXV

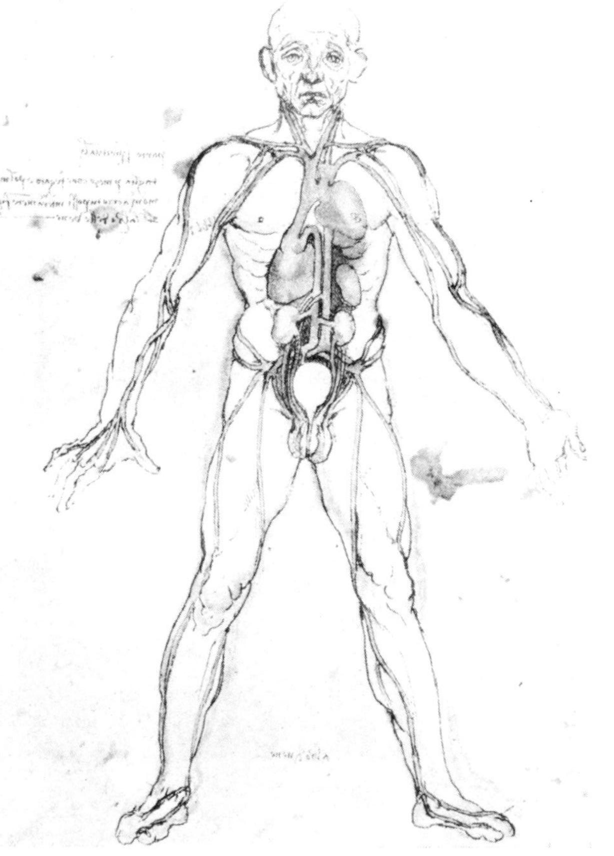

Quaderni V fo. 1 r

The 'Vessel-Tree'

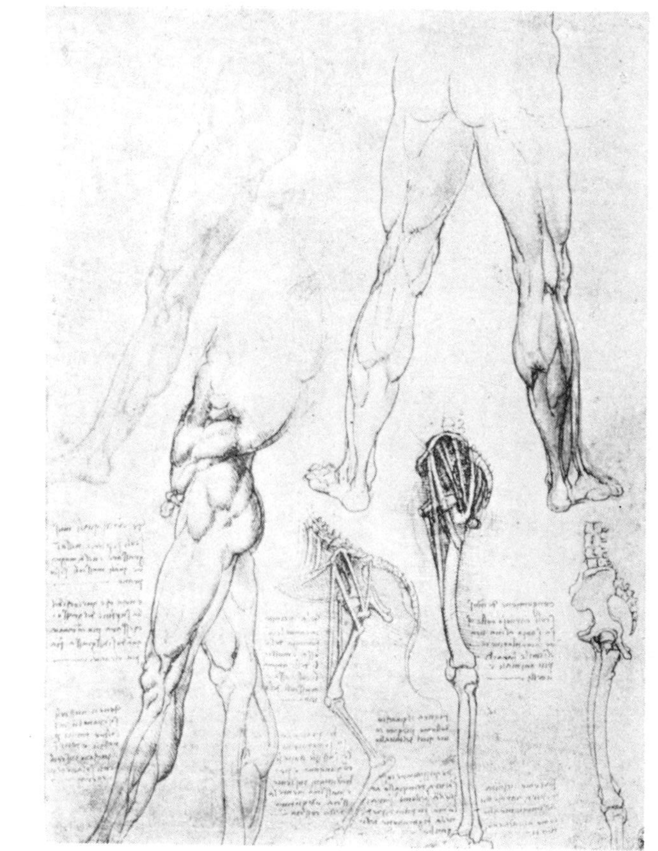

Quaderni V fo. 22 r

Surface Anatomy. Lower limbs of man and horse

his manuscripts. It is therefore possible, in spite of errors and omissions, that he had a fairly correct conception of the circulation. It is certain that in the Quaderni he beheld the Promised Land—that he began to enter it seems evident when one compares the Quaderni with the following statements in the Fogli:

'By the ramification of the veins in the mesentery, the food is drawn from the corruption of the aliments in the intestines, and eventually it returns by the ultimate ramifications of the artery to these intestines . . .'; and again, 'the origin of the sea is the contrary of that of the blood, for the sea receives in its bosom all the rivers, which are produced only by the vapours of water, risen into the air: the sea of the blood is the cause of all the veins. The aorta is only one which subdivides into as many principal branches as there are principal parts to be nourished, branches which continue to ramify *ad infinitum*.'[1]

From this it may be inferred that Leonardo came very near to the conception of the circulation of the blood.

The drawings of blood-vessels in the Fogli are so beautiful that Leonardo must have prepared the vessels by injections. Many of those in the Quaderni date from an earlier period. A large and very fine full-page drawing in one Quaderni,[2] named the 'Vessel-tree' and 'Spiritual Parts', is, however, not very accurate, although the three great blood-vessels springing from the arch of the aorta are correctly rendered (Plate XXXV). His advice on the study of the blood-vessels is, 'Bisect the heart, liver, lungs, and kidneys, so that you may be able to portray the complete ramification of the blood-vessels'. Below a very rough sketch of the vascular system he writes, under the heading *Anatomia venarum*: 'The vascular system must be treated as a whole as Ptolemy represented the world in his cosmography. Later the blood-vessels of each part must be described separately and from various aspects. Study the ramifications of the blood-vessels from the back, the front, and the sides, otherwise you cannot give the true information as to their ramifications, form, and position.'[3] Elsewhere he gives a beautiful and quite correct representation of the subcutaneous

[1] Fogli A, f. 4 r. Here is written 'la vena', but there is no reason why *vena* should not here mean vessel, that is *aorta*, as it does in Q. i, f. 1 r., for example, where Leonardo calls the abdominal aorta and the vena cava inferior *le vene massime*, and in Q. ii, f. 2 v., where he deals among other things with the aorta under the heading '*de nomi delle vene del chuore*', which is translated as 'On the names of the vessels of the heart'. In the same place Leonardo calls the aorta 'Vena aorto'.

[2] Q. v, f. 1 r. [3] Q. v, f. 2 r.

veins of the groin.[1] One sees here the union of the *vena saphena magna* and the *vena femoralis*, which latter is depicted as lying mesial to the arteria femoralis. To this figure the following passage is subjoined: 'From the soft parts (anguinaie) of the arms and thighs go veins, which, branching from the main veins, traverse the body between the skin and the flesh. And remember to note how the arteries part from the company of the veins and nerves.' The *vena dorsalis penis* also appears on the drawing, and the remark: 'There are two kinds of vein ramifications, simple and compound; it is the simple one which continues dividing itself indefinitely. A compound vein is one which is formed by two branches, as one sees n.m. and m.o. branches from two veins which unite at m. and form the vein m.p. which goes to the penis.' This sketch of Leonardo's, like many others of his anatomical drawings, might well be placed in a modern text-book (Plate XXXIII).

One would expect Leonardo's study of proportion[2] to be discussed in *Trattato della Pittura*, a work which is composed of extracts from various of his manuscripts, but it is not. In that work the fact that Nature never makes two individuals exactly alike is emphasized, the inference being that one cannot draw all one's figures from the measurement of a single subject. Only a few special measurements appear in this book. The manuscripts constituting the opening section of one of the Quaderni[3] contain the most important part of Leonardo's studies of the proportion of an adult.[4] These are sometimes rather obscure, some of the points of measurement being represented by letters

[1] Q. iv, f. 8 r.

[2] The study of proportion is the teaching concerning a harmonious relation between the body and its parts stated in figures formulated for practical as well as artistic requirements. From ancient times artists have sought a basic measurement, a module, by which they could establish a standard, i. e. the normal length and breadth of the body and its parts, and the schedule used has been called a 'canon'. The length of the head, the face, the hand and the foot, have all been used for the purpose, and the length of the body and the limbs deducted from these. The Egyptian canon seems to have been calculated from the length of the middle finger, the length of the body being 19 middle fingers. Judging from the Doryphorus of Polycleitus, the Greek canon may be taken from the Egyptian, or as some think, from the length of the head which should go eight times into the length of the body. With Albrecht Dürer the length of the body ranges from six and a half to eight times the length of the head. According to Michael Angelo's canons the length of the body seems to be between nine and ten times the length of the head.

[3] Q. vi, ff. 1–12.

[4] Leonardo states (see Richter, *The Literary Works of Leonardo, &c.*, ii, p. 109) that he will describe the proportions of the adult man and woman, but no orderly exposition of these is known to have been written by him.

which do not appear in the adjacent sketches. He also uses for purposes of measurement points which vary in different individuals, for instance, the margin of the hair, and some of his text and drawings are difficult to interpret. He is also liable to designate the same part by different terms, thus the base of the nose is named *fine di soto del naso*, *principio del naso*, and again, *nasscimento di sotto del naso.*[1]

Like many contemporaries, as well as earlier and later artists, Leonardo made for himself a canon of proportion, and as a basis for this he took sometimes the length of the head, sometimes that of the face or of the foot, but he also made use of others. Thus he tells us that the entire height is four times the breadth of the shoulders, and again, that four times the breadth of the shoulders equals only the distance from the sole of the foot to the base of the nose. In one manuscript the head is the length of the hand, in another it is the face that is of this length. One gathers from such contradictions that Leonardo has measured various subjects of different heights and proportions.

It is thus by no means easy to follow Leonardo's study of proportion. The manuscripts containing this material are evidently a collection which he intended to revise, for here—as elsewhere in the abundant material that he has left behind him—he deals several times with the same subject, probably at different periods and with different models, and jots down his findings without putting them in order, so that we do not know which of his measurements are individual and which general. A *résumé* of the partially worked-up sections of Leonardo's study of proportion may, however, be attempted.

Bases. The height of the head is the distance from the underside of the chin to the highest point of the head. The length of the face from the lower edge of the chin to the edge of the hair. The length of the foot is from the back of the heel to the point of the big toe or second toe. The width of the shoulders is to be measured between the contours of the deltoid muscles, though occasionally he indicates the 'shoulder-joints' as points of measurement.

The proportions of the head. Half the length of the head is from the crown to the inner canthus, from the inner canthus to the under edge of the chin, or from the under edge of the chin to the angle of the jaw, also from the top of the ear to the crown of the head ; the width of the throat from back to front is also equal to half the height of the head. The distance between the mouth

[1] The quotations are given in Leonardo's own orthography, which is not always consequent (e. g. *soto* in one place, *sotto* in the next).

and the edge of the hair, and between the chin and the nape of the neck is three-quarters the length of the head, as is also the greatest width of the face.

The *face* is divided into three equal parts, namely, from the chin to the base of the nose, from here to the root of the nose, 'where the eyebrows begin', and thence to the commencement of the hair. Half the length of the face is from the middle of the nose to the chin, quarter from the lower edge of the chin to the orifice of the mouth, from the back of the ear to the nape, from the most prominent part of the chin to the throat. The width of the mouth is also one-quarter the length of the face. From the labio-mental furrow to the edge of the hair is five-sixths the length of the face, one-sixth from the labio-mental furrow to the under side of the chin, one-seventh from the edge of the hair to the crown, and from the base of the nose to the orifice of the mouth, one-twelfth of the length of the face from the labio-mental furrow to the orifice of the mouth (Plate XXXVII).

The height and its proportions. The total height equals eight times the length of the head. From the edge of the hair to the ground is nine times the length of the face, three times the distance between the wrist and the top of the shoulder, four times the width across the shoulders, four times the distance from the centre-line of the body to the elbow of the stretched and abducted arm. Any of these measures equals four ells, one ell (chupido) being the distance from the elbow to the point of the middle finger with stretched arm, or the distance from the point of the elbow to the point of the thumb with bent arm. Again, the height may be expressed as six times the distance from the hair edge to the pit of the throat, 12 times the width of the face, 12 times the distance from the mouth to the edge of the hair, 15 times the diameter of the throat in profile, 15 times the distance from the chin to the eye, 16 times the distance from the point of the chin to the angle of the jaw, 16 times the distance from the chin to the inner corner of the eye, 16 times the distance from the top of the ear to the crown, 18 times from the upper bend of the throat to the pit of the throat, 42 times from the front to the back of the arm at the wrist, and 54 times the distance from the labio-mental furrow to the underside of the chin.

If one compares some of these measurements of the body with those for proportions of the head, one finds that the body varies between $7\frac{1}{2}$ and sometimes nine times the length of the head.

By kneeling, the height is lessened by a quarter. When in this attitude the hands are folded on the breast, the navel is the centre-point and the elbows are on a level with the navel. In the sitting posture the lower margin of the shoulder-blades and the breast will be on the same plane and both at equal distances from the seat, and from the crown of the head, while from the seat to the crown 'will be as much more than half of the man as the thickness and length of the testicles'. In an upright position, the aural orifice, the top of the shoulder, the great trochanter, and outer ankle bone will lie in a line (Plate XXXVII).

The proportions of the trunk. The width of the shoulders equals the distance from the pit of the throat to the navel, and is twice the height of the head; from the navel to the root of the penis is equal to the length of the head. From the nipple to the navel is one foot, as is also the distance from wrist to elbow, and from elbow to armpit. The width across the shoulders is equal to the distance from the great trochanter to the knee and from here to the ankle. The distance between the arm-pits is equal to the width of the hips, and to the distance from the shoulder-joint to the top of the hip, and from here to the lower extremity of the buttocks. The waist lies midway between the shoulder-joint and the lower extremity of the buttocks.

The proportions of arm and leg. From the point of the longest finger of the hand to the shoulder-joint there are four hands or, if you like, four heads, also three feet; one foot from the wrist to the elbow, also from here to the arm-pit, and with bent elbow, the distance from the top of the shoulder to the point of the elbow, and from here to the base of the fingers both equal two heads; from the finger-points to the arm-pits is equal to the distance from the top of the hip to the knee-cap, and from here to the sole of the foot, and each of the distances is two feet; two feet is also the distance from the sole of the foot to the front of the knee bent at a right angle, and from here to the back part of the buttocks; the distance from the great trochanter to the knee, and from here to the ankle, are equal.

It is probable that Leonardo's study of anatomy first sprang from his wish to represent the surface anatomy as perfectly as possible. In this he carried out his motto 'to know is to see'. He gradually became enthralled by his work of investigation and so decided to continue until he became complete master of the subject. His success was achieved through his intuition, his power to handle a variety of subjects, his tireless research, and his unparalleled skill as a draughtsman. He knew how to reproduce as he saw, and he saw perfectly. On account of his intimate knowledge of nearly every part of the human frame, and of his delicate and artistic treatment of them, his drawings acquired a rare and hitherto unsurpassed beauty. He excels in his many delineations of surface anatomy, and in them we see portrayed a life vigorous in all its many richly varied phases and a power of observation that pierces all the layers of the body. His plastic colour drawings of the body in different positions, of which a number on red-brown and blue-grey paper appear in the Quaderni, show clearly his greatness alike as anatomist and as artist.

Under the heading[1] *On the human shape* he asks, 'Which part of the man is it where the flesh never increases when the man gets fat, Which is the part which, when a man gets thin, never becomes emaciated with too noticeable an emaciation? Among the parts which become fat, which become the fattest? Among the parts which become thin, which become most thin? Among those men who are very strong, which muscles are of greater thickness and development?' And he continues under the heading *On Painting*:

> 'Which are the muscles which, in old age, or in a young person who is becoming thin, separate? Which are the parts of the human limbs where the flesh never increases on account of some quality of the fat, and where also the flesh never decreases, on account of some quality of the fat, and where also the flesh never decreases on account of some degree of thinness? What one seeks in all these questions will be found at all the surface joints, as the shoulder, the elbow, the joints of the hands and fingers, the hips, the knee, ankles, and toes, and such things which will be discussed at their places.'

The following singular remark shows Leonardo's speculative turn. He states: that Nature has disposed in front all the parts most liable to be bruised, for example, the brow, the nose, the skin. If these parts were not highly sensitive, they would certainly be destroyed by the many blows to which they are exposed.[2]

Fearless and imperious, zealous and grave, Leonardo reproaches those who scorn the mathematical sciences 'in which the true knowledge of things is contained', those who satisfy themselves with superficial acquirements, those who deliberately abbreviate authors, and those who think they can dissect the spirit of God.

[1] Q. vi, f. 22 r.

[2] Ibid.

PLATE XXXVI

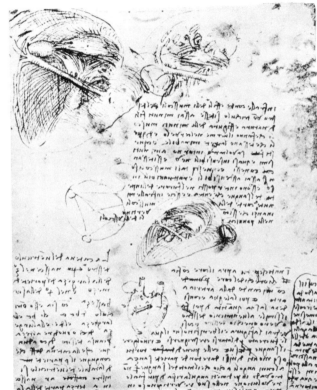

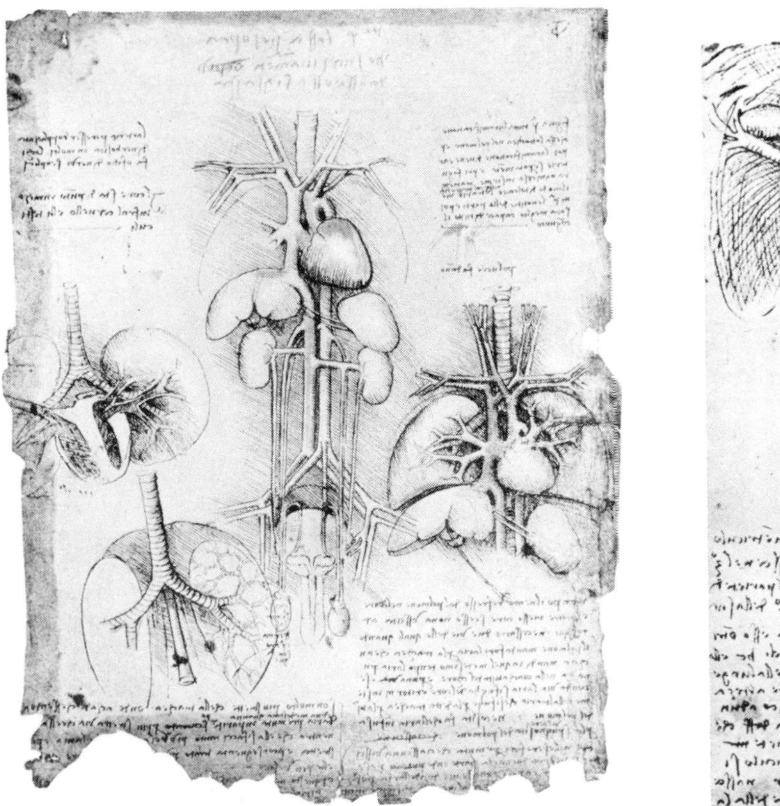

Quaderni III fo. 10 v

Heart, great vessels, bronchi, &c.

Quaderni IV fo. 13 r

The intraventricular muscle band

He scorns sophists and despises human folly. He condemns those who make a god of their stomach, and the betrayers of the weak and innocent. He counsels humility and the recognition of genius, and warns against persecuting them:

'And if any one is found to be *virtuoso* and good, do not drive him away, honour him so that he will not flee from you, and retire to deserts or holes or other solitary places to escape your snares. And if any one of these is found, honour him for these, our earthly gods, merit statues, portraits, and marks of honour. But I had better remind you that you must not eat their portraits, as is done in some parts of India, where, when their portraits perform some miracle, the priests cut them up when they are made of wood, and give a piece to each inhabitant—and not without compensation; and each shaves his piece and strews it on the first food he eats; and in this way they think they have eaten his virtue, and believe that he thereafter protects them from all dangers. What think you, man, of your species, here: Are you as wise as you think? Are these things which ought to be done by men?'[1]

How his steadfast will stands out. 'Obstacles', he says, 'do not deter me; every obstacle can be overcome by resolution.' Leonardo approaches with awe and reverence the anatomical problems he sets himself, but he meets them with energy and ardour, and it is in the elucidation of these scientific difficulties that his fascinating personality, his creative imagination, most spontaneously appear; it is then that one meets the full force of this powerful mind.

An illegitimate child, disinherited by his father, hated by his step-brothers, and often misjudged by his contemporaries, who regarded him as a mystic and one possessed of many secrets which he would not divulge—he naturally felt forsaken and misunderstood. He became meditative, and he commends solitude, for it alone gives time for study. 'When you are alone, you are yourself absolutely; but if you are attended by a single companion, you are only half yourself—and if you serve two masters, giving yourself sometimes to society, sometimes to thoughts of art—then I assure you that in this you will fail.'[2]

Although Leonardo was without doubt the greatest naturalist of the fifteenth century, and must have realized the importance of the diverse discoveries he made, yet he remained an unassuming man. 'As I see', he says in a *proemio*, 'that I cannot acquire

[1] Q. ii, f. 14 r. [2] Quoted from Sirén.

any more useful or more attractive matter for study since those who have gone before me have secured all that is most useful and necessary, I shall imitate the man who, through poverty, went last to the market, and being unable otherwise to get aught else, seized those things thrown aside for their worthlessness—trivial and despised wares, the leavings of many customers. These I shall lay on my miserable pack-ass and with them wander, not through the big cities, but through the poor villages, distributing them there and earning a recompense in keeping with my deeds.' [1] This attitude may also be taken as an outcome of his loneliness. A genius, whose creative impulses never left him any peace, Leonardo hardly felt himself held by any of the common bonds which force other mortals to remain where they have settled down. He seeks service with various princes—Ludovico Moro, Caesare Borgia, Louis XII, and François I, wherever he thought he might find suitable conditions for the development of his talent and his plans, and for the appeasement of that force which continually drove him onward, his insatiable desire to learn, to understand, and to create.

Leonardo seeks, line by line, to trace the lineaments of Nature in all her moods. He has the whole always in view, regarding Nature as one. He arrives at the conclusion that there is but one Natural Law which governs the whole world, and that Law is *Necessity*. Necessity is Nature's master and guardian; it is Necessity that makes the eternal laws.[2]

All Leonardo's study of Nature, a research which has been called a divine service—is visibly inspired by a great love for everything in Nature, from man to the meanest creature, and fortune always follows his researches, whether on the Alps at Mombosa [3] he formulates the first scientific hypothesis regarding glaciers, or whether in his study he is examining the minute parts of the smallest insect or plant.

'Great love is born of great knowledge of the objects one loves. If you do not understand them you can only admire them lamely or not at all—and if you only love them on account of the good you expect from them, and not because of the sum of their qualities, then you are as the dog that wags his tail to the person who gives him a bone. Love is the daughter of knowledge, and love is deep in the same degree as the knowledge is sure—Love conquers all things.' [4]

[1] *Codice Atlantico*, f. 119 r.
[2] Richter 1135.
[3] Probably Monte Rosa.
[4] Quoted from Sirén.

The man who says: 'When I think I have learned to live, (then) I will learn to die,' sees clearly the vanity of all our desires.[1]

'Man who with ceaseless longings awaits the new festive spring, the new summer, coming months and coming years—man imagines that all this lingers too long on the way, and does not perceive that it awaits his own dissolution. But', he adds, 'this desire is the quintessence, the true spirit of the elements which feed themselves through the soul, imprisoned in the human form and is always demanding to return—and I wish you to understand that this longing is that quintessence which is nature's ally, and that man is the model for the whole world.'[2]

Leonardo's significance as an anatomist judged from the Windsor manuscripts can thus be shortly summed up: No one before him, so far as is known, made so many dissections on human bodies nor did any understand so well as he how to interpret the findings. His account of the uterus was far more accurate and intelligible than any that preceded him. He was the first to give a correct description of the human skeleton—of the thorax, the cranium and its various pneumatic cavities, of the bones of the extremities, of the vertebral column, of the correct position of the pelvis and the corresponding curvatures of the column. He was the first to give a correct picture of practically all the muscles of the human body.

No one before him had drawn the nerves and the blood-vessels even approximately as correctly as he, and in all probability he was the first to utilize dissections of a solidifying mass in research on the blood-vessels. Nobody before him knew and depicted the heart as Leonardo. He was the first to describe the intraventricular moderator band. Whether and to what extent he understood the circulation of the blood is still an open question.

He was the first to make casts of the cerebral ventricles. He was the first who employed serial sections. No one before him, and hardly any one since, has given such a marvellous description of the plastic surface anatomy, nor had any one before him brought forward that wealth of anatomical details which he observed, nor given such correct information as regards topographical and comparative anatomy.

Most of Leonardo's anatomical and physiological drawings must be said to belong to the domain of science rather than of art. An *artist* has no use for any knowledge of the cerebral

[1] *Codice Atlantico*, f. 252 r.

[2] Quoted from Herzfeldt CXX.

ventricles, of the anatomy and physiology of the heart, of the situation of the nerves and deeper blood-vessels, of the ramification of the bronchi and their relation to the pulmonary vessels. The man who takes interest in such things and makes them the object of his study and research is surely a *scientist.* It was in drawing that he found the most satisfactory medium for dealing with such problems, and in these drawings he is scientist and artist alike. He has drawn even the most intimate anatomical details with the accurate objectivity of a scientist and yet with a virile sense of beauty which seems unparalleled.

Although Leonardo has not left a complete and systematic descriptive anatomy nor even dealt with all its chapters, although he has not always given an exhaustive description of such chapters as he has dealt with, and although in his descriptions we can trace his dependence on old and traditional ideas, yet in the history of science Leonardo will rank as *the first to have illustrated anatomy by drawings from the object, the first of the moderns to have treated anatomy in a methodical and scientific way by means of independent research and post-mortem dissections.* His singular form of observation and concentration, his careful treatment of the scientific problems before him, his intuitive powers and his ability draw right conclusions and prove them by experiment, all these resulted in the many discoveries which he made in human anatomy and physiology, discoveries that placed him centuries ahead of his contemporaries in knowledge and in thought.

He placed anatomy in close relation to physiology, he regarded the human organism as well as that of animals and plants as ruled by the general laws of Nature; he is, in short, *a modern biologist in the disguise of a mediaeval artist.*

When we bear in mind that Leonardo's writings and drawings have shared the fate of his pictures—that most of them have been lost and the remainder consists of fragments only—when we further contemplate what a wealth of observation, what a sum of natural science these fragments contain, when we realize that Leonardo throughout his life was busied with numerous different occupations, then we can only marvel at the gigantic energy and genius that found time for such intimate and painstaking anatomical research of a kind foreign to the ordinary artist. No one has ever, with the same right as Leonardo, been able to apply the quotation from Horace which in a somewhat free translation

PLATE XXXVII

Quaderni VI fo. 8 r
Proportions of trunk

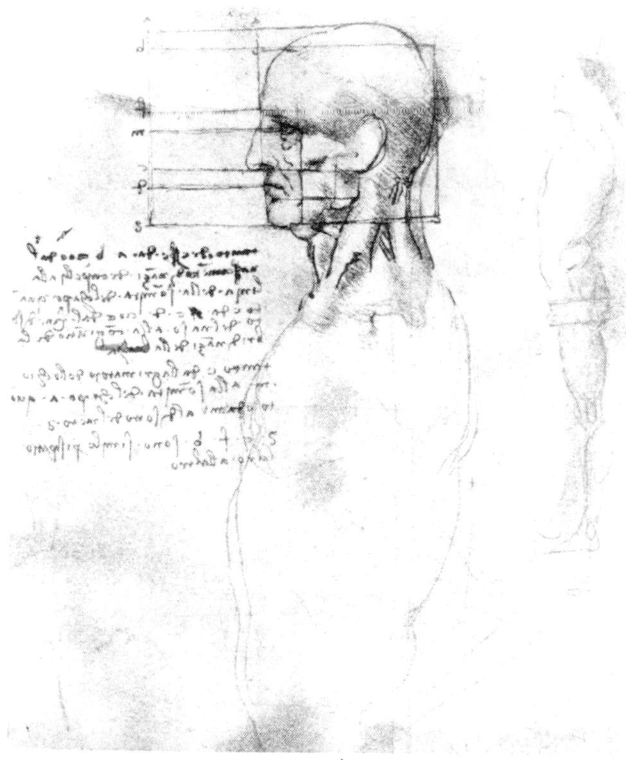

Quaderni VI fo. 1 r
Proportions of head

we find in one of his manuscripts, 'God sells us everything good at the price of fatigue.'[1]

One may ask perhaps what do we learn from Leonardo's anatomical writings and drawings? They can have had no influence on the development of anatomy, for the fragments were not collected, interpreted, and published till centuries after his death. Is it not a sad tale of labour lost, a tragedy of genius working in vain without influencing the development of science? And yet Leonardo's anatomical investigations have not been altogether vain; they will for all time stand out as a proof of what human genius is capable, and Leonardo will always remain a glorious example of an unconquerable will, of continuous and intense application to work, of reverent self-restraint, of love for created things—a glorious example whose influence on posterity has not been lost with the majority of his works. Truly he is to be regarded as a genius of the highest order that the world has ever known.

To have had the opportunity to try to track his way of reasoning, the development of his theories, his methods of research, has been a privilege which has given more pleasure and joy than one can easily express. Gratitude is the predominant and overwhelming sentiment of one who has had the opportunity of laying even a single stone to the building now being reared out of the remnants of the products of one of the world's greatest investigators and thinkers, a work of restoration which yet will for ever remain unfinished since many of its fundamental stones will be for ever wanting.

NOTE.—In the English translation of the manuscript of Leonardo's *Quaderni* an attempt was made to maintain some of his quaint personal style of writing. The quotations given above, however, have not been given the identical form of that publication but have been altered into a more readable English.

[1] Q. v, f. 24 r. Horace, *Satire* ix, Liber i, vers. 59:

Nil sine magno
vita labore dedit mortalibus.

The phrase is doubtless derived from Epicharmus in Xenophon's *Memorabilia*, ii. 1. 20:

τῶν πόνων πώλουσιν ἡμῖν πάντα τἀγάθ' οἱ θεοί.

THE ASCLEPIADAE AND THE PRIESTS OF ASCLEPIUS

By E. T. Withington

In the ancient history of medicine there are two epochs of special interest: its apparently sudden appearance as an art, or rather the art *par excellence*, with its rules and method laid down in the Hippocratic treatises in the fifth century B.C., and the transmission of this Greek medicine in modified form to the nations of the west and north in the early Middle Ages. The obscurity which covers both processes tempted early historians to seize upon any plausible explanation, and they declared that the first had its origin in the Asclepieia or temples of Asclepius, and the second in the monasteries of St. Benedict.

It was soon objected to the former theory that there is nothing priestly about the Hippocratic writings, and that no ancient author ever calls Hippocrates a priest, or represents any physician as practising in a temple. The controversy thus begun still continues, and we are told by two distinguished authorities in the same work that:

(1) 'Born of a family of priest physicians, and inheriting all its traditions and prejudices, Hippocrates was the first to cast superstition aside.... His training was not altogether bad, though superstition entered largely into it. . . . There is every reason to believe that the various "asclepia" were very carefully conducted hospitals, possessing a curious system of case books in the form of votive tablets left by the patients. . . . One of his great merits is that he was the first to dissociate medicine from priestcraft.'[1]

(2) 'The priests of Asclepius were not physicians. Although the latter were often called Asclepiads, this was in the first place to indicate their real or supposed descent from Asclepius, and in the second place as a complimentary title. No medical writing of antiquity speaks of the worship of Asclepius in such a way as to imply any connexion with the ordinary art of healing. The two systems appear to have existed side by side, but to have been distinct. . . . The theory of a development of Greek medicine from

[1] Art. 'Hippocrates', Sir J. B. Tuke, *Encyclopaedia Britannica*, ed. 1911.

HIPPOCRATES PLATE XXXVIII AESCULAPIUS

British Museum IInd or IIIrd cent. B.C.

British Museum IVth cent. B.C.

the rites of Asclepius, though defended by eminent names, must be rejected.' [1]

Sixty years earlier, Adams, the British translator of *Hippocrates*, and his French translator, Daremberg, were contradicting each other in almost the same terms. Among German authorities on Greek antiquity, von Wilamowitz opposes Thraemer, who, though less famous as a scholar, has written the articles on Asclepius in *Roscher's Lexicon* and the *Pauly-Wissowa Cyclopaedia* as well as that on Health Deities in the *Cyclopaedia of Religion and Ethics*, now in progress, and continues a stubborn upholder of the priest-physician theory. Finally, the great name of Littré, always mentioned with reverence by students of Hippocrates, must be quoted on the side of the priests, though we may hope that with our present knowledge he would have thought differently.

It is therefore not surprising that, though the priest party were supposed to have received a knock-out blow by the discovery in 1883 of the Epidaurus inscriptions,[2] with their amazing mixture of miracles, dogs, snakes, and bare-faced quackery, at the head centre of Asclepius worhsip, they should have rapidly recovered, and that most casual references to the subject assume rather than defend the priest theory, while a beautiful book by Dr. Aravantinos,[3] with the prestige of being written in Greek, supports the same view with copiousness and ingenuity.

It is not intended to discuss here the whole of this wide and intricate subject, but merely to contribute certain munitions in the way of arguments (some perhaps not used hitherto) to the side represented in the second extract.

The most important quotations are:

(1) *Strabo. Geography*, xiv. 2. 'They say Hippocrates was trained in the knowledge of dietetics by the cures dedicated there' (the temple at Cos).

(2) *Pliny. Natural History*, xxix. 1. 'Hippocrates is alleged to have copied out the accounts written by the patients cured in the temple of this god.' He proceeds to connect this with a story of the burning of the temple, apparently by Hippocrates himself

[1] Art. 'Medicine,' Dr. F. Payne, *Enc. Brit.*, ed. 1911.

[2] *Ephemeris Archaeologike*, 1883–5. For English translations see Hamilton, *Incubation*, London, 1906, 17 ff.; Withington, *Medical History*, London, 1894, appendix 2.

[3] Ἀσκληπιὸς καὶ Ἀσκληπίεια. Leipzig, 1907.

to hide the source of his wisdom. A similar tale had been told three centuries earlier by the physician, Andreas,[1] who, however, makes him burn not the temple of Cos, but the library of Cnidus.

(3) *Pausanias. Tour*, ii. 27, says he saw six pillars at Epidaurus 'inscribed with the names of men and women healed by Asclepius, as well as the disease from which each suffered, and how he was cured'. Two of these pillars were discovered in 1883.

(4) *Strabo*, viii. 6. 'Epidaurus is famous for the manifestation (epiphany) of Asclepius and he is relied upon to heal all kinds of disease, and has a temple there which is always full of patients and dedicated tablets on which the cures are inscribed just as at Cos and Tricca.'

The first of these is the main argument for the priest theory. Pliny is later and less reliable, and the story of the deliberate burning of the temple is universally discredited. If good for anything, it rather favours the other side, for it may indicate that no 'cures' of scientific value were to be found in the existing temple, while Andreas, who by age and profession was nearer to the facts, may have known that no temple ever contained records that could have been of much use to Hippocrates, and so substituted a library. But Strabo is a good authority, and his statement was accepted by Littré,[2] who even declared that the two treatises known as *Prorrhetics I* and The Coan *Prenotiones* are probably collections of temple records—an assertion frequently repeated.

When a great scholar makes what appears an amazing error it is well to get another great scholar to point it out. Daremberg in his translation of *Hippocrates* not only discusses the general subject, but addressing Littré on this particular point, asks, 'Why not say also that the Aphorisms are derived from these tablets; except that, from the little known of the temple inscriptions, the supposition is still more improbable? Strabo speaks of votive tablets at Epidaurus describing treatment. Now the Coan *Prenotiones* contain only prognostic statements, treatment is rarely mentioned. They have, then, nothing to do with Asclepius, nor with his priests, nor with the patients they treated'.[3]

Littré made his assertion before editing the treatises, and

[1] Soranus, *Vita Hip.*, in Kühn's *Hippocrates*, iii. 851.

[2] Hippocrates, i. 48.

[3] *Œuvres choisies d'Hippocrate*, Paris, 1855, Introduction, p. 85.

afterwards admitted himself partly converted.[1] But his original statement is still repeated by adherents of the old theory, though the treatises can now be read in several languages, and are obviously attempts to discover a natural history of diseases and their probable terminations. These, in a large proportion of cases, are disastrous—' they have a painful death ' . . . ' they are carried off quickly ' . . . ' this symptom is pernicious ' . . . ' that is fatal '. How could this sort of thing be derived from votive tablets intended to encourage future suppliants, and advertise the gratitude of the patients and power of the deity ? For, though there were doubtless sceptics in the background, so far as our records go no one (except possibly Aristophanes, the precise limit of whose ridicule is hard to define) doubted the fact of supernatural interference. It is the god who works every cure throughout the thousand years of his known activity. In the earlier period he is alleged to have cured most frequently by immediate miracles, later he often prescribed natural remedies, but the prescription remained oracular, and was admitted to be so by all writers, medical and lay. No one is ever described as going to an Asclepieion to benefit by change of air, &c., or for the cure of a disease obviously amenable to ' the art ', any more than a Catholic patient goes to Lourdes for the sake of the climate, or to get a tooth stopped. They all hoped, if highly favoured, to get an immediate miraculous cure, or, failing that, supernatural advice in a dream ; and unless we reject a vast amount of evidence, they nearly always got the dream, and sometimes the instantaneous cure.

It seems improbable that any one should even begin to study the natural course of disease, the great object of Hippocratic medicine, in conditions where supernatural intervention was continually expected.

We may add a subsidiary argument which seems to have escaped notice. The 34th Prorrhetic says, ' Muttering delirium with trembling of the hands and carphology are strong indications of phrenitis, as with Didymarchus in Cos '. Strangely enough, though many localities are mentioned in the Hippocratic Collection, this is the only patient said to have been treated ' in Cos '. But if these are the records of the local Asclepieion, they were obviously all treated there, and the remark is quite uncalled for. Further, a large proportion of the patients evidently suffered from

[1] *Op. cit.* viii. 628.

very acute disease, but our records give hardly a single case of a patient carried to a temple ; they could almost all walk.

Disturbed by the discovery of the pillars at Epidaurus, later upholders of the Strabo-Littré view attempt to distinguish between pillars and tablets in quality as well as size. They even suggest that the tablets were like our hospital charts put at the head of the patient's bed, with a diagnosis of the disease and daily remarks ;[1] but this is inconsistent both with the words of Strabo and all else we know of the tablets. The first case on the pillars is that of Cleo, who was safely delivered of a five years' boy, 'and he promptly washed himself and walked about with his mother. Now when this had happened to her, she wrote on a votive tablet,

"Marvel not at the size of this tablet but at the occurrence,
Five years Cleo was pregnant ; she slept and the god made her whole ".'

This lady evidently did things on a large scale, but even her tablet only contained a couplet. There is no reason to suppose that those of others contained more than a brief mention of how they were cured by the god, and when, in the late Roman period, we meet with something like a clinical history in the record of M. J. Apellas,[2] it requires much more than a tablet.

Still, there must be some explanation of Strabo's statement, and the following suggestion, though novel, may be worth notice. The four books 'On Diet', ascribed to Hippocrates, end with the statement that the author has investigated the rules of diet as well as he could 'with the help of the gods'. This common phrase occurs nowhere else in the Collection. It would therefore be noticeable, and might have been used by votaries of supernatural as opposed to secular treatment somewhat in this way: 'Hippocrates himself admits he got his knowledge of dietetics by the aid of the gods. He must evidently have meant Asclepius in particular, and how could he learn his treatment except from the temple cures ?'

Here is a possible origin for Strabo's 'They say'—which is after all a qualification, while his statement that there was no difference between the tablets at Epidaurus and those at Cos, which the priest party now tries to distinguish, is categorical.

[1] Aravantinos, *op. cit.*, p. 164.

[2] See the Epidaurus inscriptions.

To pass to wider questions: when we try to find dates for the definite establishment of the therapeutic dream oracle they strike us as unexpectedly recent. Getting advice in dreams is, of course, as old as mankind. Some anthropologist has said that when primitive man was in a difficulty he first asked advice of his friends; if they were no good he went to the old men, and if they failed him, to those still older, the dead. He slept on the tomb of his ancestors and got advice in a dream, as Herodotus tells us was the custom of certain Africans. But for incubation in Asclepieia the oldest definite date is 420 B.C., when the worship of the god was established at Athens, and his method of cure received with ridicule by the sturdy conservative Aristophanes. This, no doubt, shows that the custom had been long in use, and there is one example of possibly earlier date in a fragment of Hippys of Rhegium, who is supposed to have flourished 'during the Persian wars', about 480 B.C. This case also occurs on the pillars, which were set up two centuries later, but the following is the Hippys version:

'A woman had a worm, and the ablest physicians gave up curing her. So she went to Epidaurus and besought the god for deliverance from the domestic plague. The god was not there, but the attendants made the patient lie down where the god is wont to heal the suppliants, and she went to rest as directed. But they, in the absence of the god, began her treatment and cut off her head. Then another put his hand down and took out the worm, a great thing of a beast, but they were no longer able to put the head back and fit it accurately to the old joining. Then the god came, and was angry with them for undertaking a task beyond their wisdom, but he himself by power invincible and divine put the head back on the trunk, and raised up the guest.'[1]

On the pillars the story is told with variations and additions. The woman was Aristagora of Troizen, and she incubated there, but Asclepius was at Epidaurus and had to be sent for—a hint that if one wants prompt supernatural aid, Epidaurus is the place.

Admitting the date of Hippys, this story not only shows the existence of lay physicians before Hippocrates, of which we have ample proof, but also indicates a clear distinction acknowledged to exist between human medicine and the supernatural aid of the Asclepieia.

It is fair to add that Dr. Aravantinos sees in this case a pretended tracheotomy by medical student deacons. Then

[1] Aelian, *H.A.* ix. 33.

enters the priest physician and administers a strong emetic or vermifuge. In the pillar version Asclepius extracts the worm after gastrotomy. The patient meanwhile is under narcosis from mandragora, or the like, and dreams of decapitation followed by gastric disturbance. Her story, though not accurate, was considered edifying, and was published accordingly. The priest physicians kept their real treatment secret and performed sham superficial operations to mislead and increase the wonder of the public.[1] There is something to be said for this, but its bearing on our present problem lies in the application.

Apart from the above, Greek literature of the fifth century has nothing to say about the therapeutic dream oracle. It is perhaps not strange that Athenian writers should ignore it, and we may notice that they always speak of medicine as a secular art. Thucydides, describing the plague at Athens,[2] contrasts the human art of medicine with prayers, divinations, and the like, saying that they were equally useless. Sophocles, in the famous ode in the *Antigone*,[3] puts the healing of diseases among the things achieved by unaided man. Even Aeschylus classes medicine with agriculture and navigation as a natural art, and if he gives it a demigod as founder, he is not Asclepius but Prometheus [4]—a myth which the Hippocratics with their emphasis on prognosis would have readily explained.

But the silence of Herodotus is remarkable. He came from the immediate neighbourhood of Cos, from a Dorian colony founded by Troizen, as was Cos by Epidaurus, though, like the Hippocratic Asclepiadae, he writes in Ionic. He was interested in medicine, giving us vivid glimpses of certain physicians and mentioning medical schools, as to which it is noteworthy that while the oldest Asclepieia are all in Greece proper, the oldest medical schools are all outside.[5] He was still more interested in the gods and their strange oracles; yet he has no word of Asclepius or the great therapeutic oracle supposed to exist close to him and to be a main source of the medicine of the age; and that though he mentions other dream oracles both among Greeks and barbarians.

[1] *Op. cit.*, p. 118. [2] ii. 47. [3] Line 364. [4] *Prom.* 460, 484.

[5] There is, perhaps, as much to be said for the suggestion that the prominence of the dream oracle in Greece proper hindered the development of medical schools there as for the theory that it gave origin to the actual schools of Cos, Cnidus, Rhodes, Cyrene, and Croton.

Most amazing is the silence of the Hippocratic writers themselves. Omitting the late and spurious 'Letters', the collection contains literally one word on the subject; the name, Asclepius, coming after Apollo in the Oath. And even the Oath gives some evidence against the priest theory, for it is taken by visiting practitioners. 'Into whatever houses I enter I will go for the good of the patients.' These Asclepiadae are great travellers (περιοδευταί), and if they settle anywhere they practise not in a temple but a surgery (ἰατρεῖον), concerning which there is a special treatise. There are also two works of great antiquity and excellence, *On the Art* and *On Ancient Medicine* dealing with its nature and history. Here surely we might expect something about Asclepius, his priests and his temples, but there is nothing, nor do the writers show the least consciousness that the Art is being, or has been, freed from superstition.

This Art makes no claim to supernatural aids or inspired knowledge, but 'clearly has and ever will have its essence in causation and the power of foretelling'. In other words, relying on the uniformity of causal sequence in nature we can, from observation of like cases, foretell the probable sequence of events in diseases, and the modifications which may be produced by changes in diet and other agencies.[1]

The author of *Ancient Medicine*, perhaps Hippocrates himself, also makes diet the main source of the art, but knows nothing of temple cures. It was observed of old, he says, that if healthy men ate the coarse uncooked food of animals they were taken ill, and, similarly, if the sick followed the diet of the healthy they became worse, and combining these observations with reason men proceeded to establish by experiment the rules of dietetics. This he calls a great invention elaborated and perfected with no mean display of intelligence and observation.[2]

The healing art, he adds, 'possesses all things from of old, a principle and a beaten track along which in the course of ages many splendid discoveries have been made, and along which the rest will be discovered, if competent men, knowing the things already discovered, set out thence on further inquiries. He who, rejecting these, takes another road and claims to have discovered something is deceived and deceives himself, for it is impossible'.[3]

This idea that the method of medical progress by observation

[1] *De Arte*, § 6, Littré, vi. 10. [2] § 4, Littré, i. 580. [3] § 2, Littré, i. 572.

and reason has long been established is repeated elsewhere,[1] and though 'charms, amulets, and other such vulgarity' may be used by some, there is no suggestion that true votaries of the art have ever been deluded by these or other superstitions. What is objected to is not priestcraft, but theorizing philosophy.

The *Ancient Medicine* contains the one intimation that the art was thought worthy to be ascribed to a god.[2] We look out hopefully for Asclepius, but the writer tells us instead that this saying is justified because medical discovery depends on the application of reason to the nature of man.

Physicians are called δημιουργοί,[3] 'practisers of a public art' as in Homer,[4] and Plato, our best authority on Hippocrates, classes the physician with Pheidias and Polyclitus as an 'artist'. If you pay any of these men money they will teach you their art, medicine, or sculpture.[5] Similarly, Socrates is made to draw his comparisons with about equal frequency from the arts of the cobbler, the pilot, and the physician.

The Collection also contains a treatise *On Dreams*.[6] Certain diviners are said to be skilled in predicting from dreams things about to happen to cities and individuals, but when they try to interpret dreams foreboding bodily affections they get hopelessly muddled owing to their ignorance of physiology. The writer then makes a laudable attempt at a naturalistic theory of dreams, connecting them with bodily states, but there is not a word about therapeutic dreams. Various gods are mentioned who may be prayed to for preservation of health and prevention of disease, but Asclepius is not among them.

Diviners come off badly in these treatises. Ignorant physicians who quarrel with one another and give contradictory advice are said to bring scandal on the art, 'and almost make it resemble the art of divination, for this is how diviners act'.[7] Another treatise holds up to reprobation certain 'cheating diviners' who persuade young women recovered from hysteric affections to dedicate their best dresses 'and many other things' to Artemis, and hints that they would be better used to get married in.[8]

The forty-two clinical histories of the 'genuine' Hippocrates

[1] Littré, vi. 342; *Loc. Hom.*, § 46.
[2] § 14, Littré, i. 600.
[3] § 1, Littré, i. 570.
[4] *Od.* xvii. 384.
[5] *Protagoras*, 311 B.
[6] Littré, vi. 640.
[7] Littré, ii. 242, *Morb. Acut.* 3.
[8] Littré, viii. 468, *Morb. Virg.* 1.

correspond in number with the forty-two 'Cures of Apollo and Asclepius' on the pillars, but in everything else it is impossible to imagine a greater contrast. Most of the Hippocratic cases end fatally, and the treatment of those who recover is rarely considered worth mention. The temple cures are instantaneous or nearly so; 'when day appeared he went away healed' is the usual ending, the few instances of delay being due to want of faith or over hastiness of the attendants, as with Aristagora.

To sum up: if there was a therapeutic dream oracle at Cos in the fifth century, the Hippocratic writers will have nothing to do with it, they will not even mention it. As von Wilamowitz says, 'The Epidaurian colonists carried Asclepius with them to Cos, but not medicine. Where medicine is concerned 'the god disappears. Instead of the priest comes the physician; instead of the dream oracle, science; instead of Asclepius the Asclepiad'.[1]

A partial explanation of this remarkable silence may be found in the practical certainty that the Asclepiadae were not only not priests but not Dorians. Our earliest information about them is a chapter-heading from Theopompus. 'Concerning the physicians in Cos and Cnidus, how they are called Asclepiadae, and how they first came from Syrnus', a town in Caria. They came, then, from the East, not West, and belonged to an older stratum of Greek emigrants. They wrote in a dialect more Homeric than Doric. and held the Homeric view of Asclepius, that he was a mere man, the founder of their guild and ancestor of its more prominent members. They were, perhaps, astonished by the appearance of their patron as a Dorian earth-god followed by mystic snakes; and though they may soon have established a *modus vivendi* with the invaders, a difference of view as to Asclepius remained permanent, for both Galen[2] and Pausanias[3] tell us that it was disputed even in their time whether Asclepius was a deified man or 'a god from the beginning'.

There is evidence of this difference at Cos. The temple certainly acknowledged the primacy of Epidaurus, whence it got its snakes, but, according to another story, it was founded direct from Tricca, the home of the Homeric and purely human Asclepius. A deified hero might appear and give advice at his tomb, as did Amphiaraus, but an epiphany of Asclepius, like that of the high gods, might

[1] *Isyllos von Epidaurus*, Berlin, 1886, pp. 37 and 103.
[2] i. 22. Kühn's edit., *Protrept.* 9. [3] ii. 26.

occur anywhere, for he was also an embodiment of the great earth-spirit in his capacity of counsellor and healer. Possibly there was first a simple shrine at Cos, where the Asclepiadae celebrated the memory of their patron but did not expect him to appear; while the dream oracle and the snakes came later from Epidaurus, and were for a time deliberately ignored by the medical school.

So much for the silence of Hippocrates, or rather, of the Asclepiadae, the importance of which as regards our problem, especially when combined with what they do say, has perhaps been overlooked.

Neither Celsus nor the author of the brief outline of medical history in the *Introduction*, formerly ascribed to Galen,[1] mention priest-physicians, and the former tells us that Hippocrates separated medicine from philosophy.

For the bearing on the question of inscriptions concerning medical men and priests the reader should refer to the work of Pohl (*De Graecorum medicis publicis*, Berlin, 1905), where full references are given. The writer sums up his conclusion thus: 'In the latest age, then, the true art of medicine and that of the priests of Asclepius got mixed, after arising from widely separated sources and continuing a long time distinct.'

The slight modification we would make in this statement may be illustrated from another point of view, not yet sufficiently studied, the attitude of later medical writers towards the oracle.

We have seen that the earlier ignore it, and our first available reference is from Rufus of Ephesus, called by Oribasius 'the Great', whose fragments reveal a man of much ability and sober-mindedness.

A certain Teucer, afflicted with epilepsy, went to the Asclepieion at Pergamus and besought the god to heal him. Asclepius appeared, as usual, in a dream, and asked whether he would like another disease instead. Teucer replied, this was not his most earnest desire, in fact he would rather be healed entirely; but if that was impossible, and the other disease less troublesome, he would accept it. The god replied that it was less troublesome, and was also the best cure for his complaint. Thereupon he was attacked by a quartan fever, but was delivered from his epilepsy.[2]

[1] xiv. 674, *Introductio seu Medicus*.

[2] Oribasius, xlv. 30, Daremberg's edition, iv. 86, Paris, 1862.

This is told in illustration of the Hippocratic Aphorisms that fevers generally,[1] and quartans in particular,[2] relieve spasms, a principle which, according to Rufus, had been successfully utilized in medicine.

We pass to *Galen* with great expectations, for he was a native of Pergamus. He calls Asclepius his ancestral god (though he was not an Asclepiad), and tells us he saved his life at least once by curing a disease, and perhaps again by warning him not to go with Marcus Aurelius to the Danube. He therefore became a special θεραπευτής or votary of the god. Yet the vast bulk of his extant works affords us only four or five cases, including his own. This, he says, was ' a pernicious abscess ',[3] but he describes it elsewhere as a chronic pain between the liver and diaphragm.[4] He was then a young man and thought he was going to die, but Asclepius in a dream recommended bleeding from an artery and he rapidly recovered. The god ordered the same treatment in the case of another ' votary ' with a chronic pain in his side, and it was partly owing to these ' cures ' that Galen became convinced of the value and safety of arteriotomy, of which he previously had doubts.

A third case is that of the corpulent Nicomachus,[5] whose treatment unfortunately is not given, but may be connected with a remark made elsewhere that patients will obey the directions of Asclepius even to the extent of going entirely without drink for fifteen days, which they would never do on the orders of physicians.[6]

A fourth cure is that of ' elephas ', supposed to be leprosy.[7] Here the patient was summoned by the god from Thrace to Pergamus, and the treatment ordered was the internal use of *theriac,* a famous medicine containing viper's flesh, combined with viper's-flesh ointment. The disease was thereby converted into ' lepra ', which was cured ' by drugs ordered by the god '. According to Galen, the treatment of elephas by viper's flesh had been discovered accidentally some time before, and had been successfully used by himself. There is also evidence that it had been employed by Archigenes half a century earlier (Aetius, xiii. 121).

[1] iv. 57, Littré, 4, 522.
[2] v. 70, Littré, 562.
[3] xix. 19, Kühn's edition, *Libr. Propr.*, 2.
[4] xi. 314 ff., *Venesect.* 23.
[5] vi. 869, *Dif. Morb.* 9.
[6] xvii *b*. 137, *In Hipp. Epid.* vi. 4, 8.
[7] xii. 315, *Simpl. Med.* xi. 1.

Finally, we have a recommendation in a dream of a local application for swollen tongue, confirming Galen's advice against that of other physicians, but this is not clearly attributed to Asclepius.[1]

In these cases the treatment, so far as known, contains an element of recognized medicine, but the theurgic side remains prominent. It is the god who cures, and the patients attribute their recovery to his supernatural interference. So, too, Rufus and Galen, as will be clearer to those who look up the context, show not the slightest doubt of the supernatural character of the intervention. They may point out the medicinal nature of the means, and some circumstances which favour success, but the cure is essentially divine even when it gives divine sanction to human methods. There is no true mixture: the 'Cures' of Asclepius are separated to the last from those of the art by a strong element of miracle: they are two distinct things, and medical writers show no consciousness that they ever had been, or were tending to become, the same. The latest recorded temple cure is as purely miraculous as the earliest, though in a different way.

Archiadas of Athens had a daughter afflicted by a painful and incurable disease. So he besought Proclus the philosopher to intercede with Asclepius for her cure. Proclus, whose benevolence equalled his wisdom, at once went to the Asclepieion, 'fortunately not yet plundered', and prayed according to the ancient rites. The patient felt a sudden change in her condition and great relief, and when the philosopher, having ended his petitions, came to see the effect, he found her 'delivered from her pains and in perfect health'. This was in the latter part of the fifth century.[2]

The closing of the Asclepieia was followed almost immediately, or even preceded, by the establishment of dream oracles in Christian shrines, especially that of SS. Cyrus and John at Alexandria. Seventy of their cures are recorded by Sophronius, Patriarch of Jerusalem (d. 640), who had himself been healed by them. They closely resemble those of Asclepius, the chief difference being that the saints are represented as indignant at any suggestion that their cures are not miraculous throughout,

[1] x. 971, *Meth. Med.* xiv. 9.

[2] Marinus, *Vita Procli*. For translation see Taylor, *Proclus*, London, 1788, i. 26. The god also appeared to Proclus on his deathbed, 'and would probably have cured him had he desired it,' *op. cit.* i. 27.

and inflict severe punishment on a rash physician who declared that some of their prescriptions might be found in *Galen* and *Hippocrates*.[1] In fact secular and sacred medicine are in open opposition.

To return to the fifth century B.C. and the origin of Hippocratic medicine. While not professing to solve that obscure and interesting problem, it is hoped that some additional reasons for rejecting the superficial view that 'it originated in the health temples' are given in these notes. What can be said in favour of that view may be found in the works noticed at the beginning. It may seem to some that the arguments there adduced do not outweigh the single fact (the most certain in medical history) that in the fifth century B.C. medicine was known throughout Greece as an art, the chief of arts, too long to be learnt in this short life. For arts are not learnt in temples by observing real or supposed supernatural intervention, but, as the Hippocratic writers tell us, by experience and the application of reason to the natures of men and things.

[1] Mai, *Spicilegium*, vol. iii, case 30, Rome, 1840.

THE SCIENTIFIC WORKS OF GALILEO (1564—1642)

WITH SOME ACCOUNT OF HIS LIFE AND TRIAL

Being a Review of Favaro's *Edizione nazionale delle Opere di Galileo* (1890–1909)

BY J. J. FAHIE

I. THE GALILEIAN RESEARCHES OF ANTONIO FAVARO

IN the current year 1920, Professor Antonio Favaro, of Padua University, completes forty-four years of Galileian studies. The result is monumental; besides editing the National Edition of Galileo's Works in twenty large quarto volumes, Favaro has published over 450 separate studies on matters relating to the life, times, and activities of the great master. The vastness of the labour involved in this kind of literary work can be appreciated only by those who have themselves engaged in historical researches.

Antonio Favaro was born on the 21st of May, 1847, in the Villa Favaro at Fiesso d'Artico, on the Brenta, and about half-way between Padua and Fusina. He was educated at the Universities of Padua, Turin, and Zürich, and, after a few years of private teaching, was elected in 1872 to the newly created chair of Graphical Statics in the University of Padua, being then only twenty-five years of age. A few years later he was appointed Director of the *Scuola d'applicazione per gli Ingegneri*. In 1877 he published his first work in Padua, his *Lezioni di Statica grafica.* This same year was destined to be of the first importance in the ordering of his life, for at this time, stimulated in the first instance by a suggestion from the poet Giacomo Zanella, Favaro began to devote himself to the study of the life and work of Galileo.

PLATE XXXIX

Devot:mo et Oblig:mo Ser:re

Galileo Galilei

From a portrait in the Bodleian brought from Florence in 1661

After four years of close study and research, chiefly amongst the 303 volumes of Galileian MSS. and papers in the National Library at Florence, Favaro came to the conclusion that a new and complete edition of Galileo's works was necessary. After six years of strenuous advocacy he had the satisfaction of seeing his ideas crowned with the highest recognition.[1] On the 20th of February, 1887, King Umberto issued a Royal Order decreeing the publication of a new and complete edition of the works of Galileo, at the cost of the State, and under the care of the Minister of Public Instruction. Favaro was appointed Editor-in-Chief, and there were nominated as his coadjutors Professor Isidoro Del Lungo, of the Accademia della Crusca, who was to occupy himself with all that concerned the care of the text, and Professors Genocchi, Govi, and Schiaparelli to assist in surmounting the scientific difficulties that might present themselves.[2] The first volume appeared in 1890, and the twentieth and last in 1909. The series represents a magnificent tribute by the Italian people to the greatest scientific genius that their nation has produced.

In our exposition of the science of Galileo, we shall follow the chronological order adopted in the National Edition, prefixing to our presentment in each case a note indicating all the places in the National Edition where the particular matter is dealt with.

We desire to express our obligations to Mr. John Murray for kindly allowing us to utilize part of the material of our book *Galileo; His Life and Work*, London, 1903.

II. Early Manhood of Galileo, 1564–92

1. *Training and Education*

Galileo Galilei was born of Florentine parents at Pisa on the 15th of February, 1564. Here he passed the first ten years of his life, and he received his early education, partly at the school of one Jacopo Borghini, and partly at home, where he was helped in the study of Greek and Latin by his father, who was a good scholar and mathematician, and an authority on the theory and practice of music.

At the age of twelve Galileo was sent to the monastery of Vallombroso, near Florence, for a course of the 'Humanities',

[1] *Intorno ad una nuova edizione delle Opere di Galileo*, Venezia, 1881.

[2] *Per la Edizione nazionale delle Opere di Galileo—Esposizione e Disegno*, Firenze, 1888.

the literary education then considered indispensable for a well-born youth. Here he made himself acquainted with the best Latin authors, and acquired a fair command of Greek. For the course on logic, as then taught, he had little taste, contenting himself with what scraps of elementary science and philosophy he could pick out of the lessons.

From early boyhood Galileo was remarkable for mental aptitudes of various kinds, coupled with a very high degree of mechanical skill and inventiveness. His favourite pastime was the construction of toy-machines or models, such as wind and water mills, boats, and other common mechanical contrivances. When unable to supply some essential part, he would adapt the machine to an entirely new or quaint purpose, never resting satisfied until he saw it work.

As he grew he imbibed from his father something of the theory and practice of music, and became skilful with the lute. He was also, it is said, a creditable performer on the organ and one or two other instruments. Music, and especially the lute, gave him pleasure through life and afforded the greatest solace in later years, when blindness was added to his other afflictions. His talent in drawing and painting was equally striking. In later life he used to tell his friends that, had circumstances permitted him to choose his own career, he would have decided to become a painter.[1] Galileo was also very fond of poetry, and his essays on Dante, Tasso, and Ariosto, as well as some verses and the fragment of a play, bear witness to a cultivated taste.[2]

In 1581, when seventeen and a half years old, Galileo was sent to study medicine and philosophy at the University of Pisa, a course which his father, who was in straitened circumstances, regarded as likely to prove lucrative.

2. *On the Pulsilogia*[3]

About a year after his matriculation (1582–3), Galileo made his first discovery—that of the synchronism of the oscillations of

[1] Among the circumstances which assisted the rise and progress of the reformed Florentine School, Lanzi includes 'the readiness with which the celebrated Galileo imparted to artists his discoveries and the laws of perspective'. *History of Painting*, Bohn's edition, vol. i, p. 210.

[2] These are collected in vol. ix of the National Edition of his works.

[3] Cf. Nat. Ed., vol. x, p. 97; vol. xix, pp. 603, 648.

the pendulum, timing the excursions by his pulse. This was the first attempt at accurate measurement of any bodily function, as well as the basis of the modern clocks. He was not, however, then thinking of clocks, but only of the construction of an instrument which should mark with accuracy the rate of the pulse and its variation from day to day. He quickly gave form to his idea, and it was welcomed with delight by physicians, and was long in general use under the name of Pulsilogia. Sancto Santorio, Professor of Medicine at Padua, was the first to give diagrams of the Pulsilogia in his *Methodus Vitandorum Errorum in Arte Medica* (Venice, 1602), three of which we reproduce (Figs. 1–3).

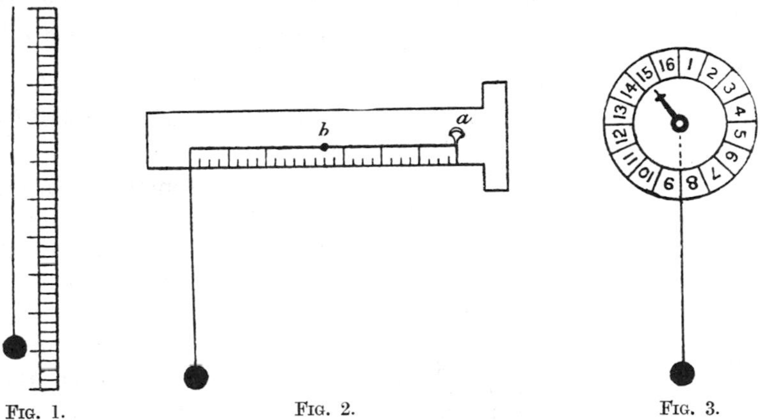

FIG. 1. FIG. 2. FIG. 3.

Fig. 1 shows a weight at the end of a string held at the top of a scale, which can be graduated so as to show the number of pulsations per minute. The string is gathered up in the hand till the oscillations of the weight coincide with the beats of the patient's pulse. Then, a greater length of string, i. e. a longer pendulum, would indicate a slower pulse, and a shorter length a more lively action. In Fig. 2 an improvement is made by connecting the scale and string; the length of the latter is regulated by turning the peg *a*, and a bead *b*, on the string indicates the rate of pulsation. Fig. 3 is still more compact, the string being adjusted by winding (or unwinding) upon an axle or drum at the back of the dial plate.

In many of his subsequent experiments and investigations, Galileo utilized this principle, as in his innumerable experiments on motion, his long-continued observations on the periods of Jupiter's satellites, and, just before he died, in the design of a pendulum clock.

Vincenzio Viviani, a pupil and the earliest biographer of Galileo, tells us that the young Galileo's attitude from the first in the philosophical classes was not at all to the satisfaction of his teachers, owing to his habit of examining every assertion to see what it was worth. In consequence of this he soon acquired a reputation among the professors and students for bold contradiction, and was dubbed 'The Wrangler'. His eager questioning, especially of the dictates of Aristotle, found no favour in their eyes.

The first volume of the National Edition opens with what the editor calls Juvenilia, written in or before the year 1584, and entirely in Galileo's own hand. They are commentaries in Latin on the *De Caelo* and kindred tractates of Aristotle, and are almost entirely transcripts of lectures he had attended. Beside their personal interest these Juvenilia are of value for the light they throw on the scholastic character of the teaching then prevalent in Italy.

3. *Galileo and the State of Mechanics in Italy at the end of the sixteenth century*

Up to Galileo's time the study of mathematics, although included in the *Rotoli* or lists of University lectures, was practically neglected in all the Universities of Italy, though Commandino (1509–75) and Maurolico (1494–1575) had recently revived a taste for the writings of Euclid and Archimedes; and Vieta (1540–1603), Tartaglia (1500–59), Cardano (1501–76), and others had made considerable progress in algebra. Guido Ubaldi del Monte (1540–1607),[1] soon to become a warm friend and patron of Galileo, Besson (d. *c.* 1580),[2] Ramelli (1531–90),[3] and one or two others had done something towards the application of statics, the only part of mechanics as yet cultivated. Thus Guido Ubaldi's *Mechanicorum Liber* of 1577 consists of little more than bald descriptions of the mechanical powers—the balance, the lever, the pulley, &c. He explains combinations of pulleys at great length, and reduces their theory to that of the lever, but his solution of the problem of equilibrium is very erroneous. The other contemporary works on mechanics consist solely of descriptions of actual and imaginary machines, with no reference to general principles. There was, in fact, no general recognition of the necessity of applying mathematics to the study of mechanisms.

[1] *Mechanicorum Liber*, Pisauri (now Pesaro), 1577.

[2] *Theatrum Instrumentorum*, Lyon, 1582.

[3] *Le diverse ed artificiose macchine*, Paris, 1588.

To the close of his nineteenth year, Galileo knew nothing of mathematics, and appears to have been first led to study geometry by his fondness for drawing and music, the underlying principles of which he wished to understand. His father tried to prevent these studies, but the natural bent of the young man's mind was not to be denied.

During the winter and spring of 1582–3 the Court of Tuscany was at Pisa, and among the suite was one Ostilio Ricci, mathematical tutor to the Grand-Ducal pages, and a friend of the Galilei family. The tutor and the young student became friends. Going on one occasion to pay Ricci a visit, Galileo found him lecturing to the pages on some problem in Euclid. He did not enter, but, standing by the door, followed the instruction with rapt attention. This was the awakening of a new craving of the intellect, under the influence of which he found himself drawn repeatedly to the class-room. Each time, concealing himself behind a door, he listened, Euclid in hand, to the teacher's demonstrations. At last, taking courage, he begged the astonished tutor to help him, which Ricci readily consented to do. Henceforth, mathematics were more studied than medicine, for which, truth to say, he never showed any relish. For some unknown reason, however, after nearly four years' residence, and without taking the doctor's degree, Galileo withdrew from the University, and went home to Florence about the middle of 1585, where, still under the guidance of Ricci, he devoted himself heart and soul to mathematics and physics. From Euclid he passed to Archimedes, whose work in mechanics he was destined to continue, and for whom he then conceived a life-long veneration.[1]

4. *The Hydrostatic Balance*[2]

In 1586, when fresh from the study of the great Syracusan's work on hydrostatics, Galileo constructed a hydrostatic balance (La Bilancetta) for finding with accuracy the relative weights of any two metals in an alloy. In describing it he refers to the popular account of the way in which Archimedes detected the fraud committed by the goldsmith in the making of Hiero's crown. He doubts the correctness of that oft-told story, the results of which, he says, are fallacious, or at least little exact, and he

[1] Nat. Ed., vol. i, pp. 231–42. [2] Cf. Nat. Ed., vol. i, pp. 211–28.

believes that his own 'most exact method' was that really employed by Archimedes himself.

Take a lever A B (Fig. 4) at least a yard long (the longer it is, the more accurate its indications) and delicately suspend it from its centre C, and at the end B let there be means of attaching the body to be tested, say an alloy of gold and silver, and its counter-weight at the other end A. First, take a piece of pure gold and weigh it in air; now immerse it in water; it will seem to be lighter, and the counterpoise D must be moved from A to, say, E to obtain a balance. Then, as many times as the space C A contains the space A E, so many times is gold heavier than water. Proceed in the same way with a piece of pure silver. When placed in water it will seem to lose more of its weight than

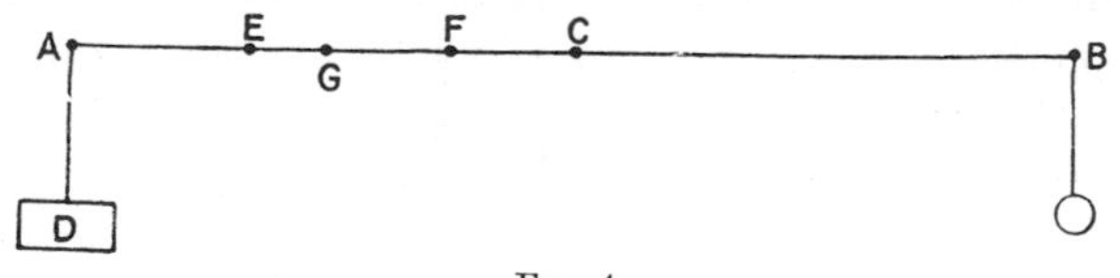

FIG. 4.

the gold, and its counterpoise will have to be moved to, say, F, showing that silver is specifically less heavy than gold in the ratio A E to A F. Now taking the alloy, it is clear beforehand that it will weigh less than an equal volume of pure gold, and more than an equal volume of pure silver. Weigh it in air, and then in water, when it will be found that its counterpoise must be moved to some point between E and F, say, G. From this we conclude that the weight of the gold in the alloy is to that of the silver as F G is to E G.

5. *The Centre of Gravity in Solid Bodies* [1]

After this work Galileo passed to the investigation of the centre of gravity in solid bodies. The results are embodied in an essay first circulated in manuscript, and only printed in 1638, as an appendix to the Leyden edition of his *Dialoghi delle Nuove Scienze*. His first intention was to prepare an exhaustive treatise in completion of the work of Commandino on the same subject. There chanced, however, to fall into his hands the book of Luca Valerio (1552–1618),[2] the Neapolitan mathematician, in which he

[1] Cf. Nat. Ed., vol. i, p. 182; vol. viii, p. 313; vol. xvi, p. 524.

[2] *De Centro gravitatis solidorum*, Rome, 1604.

found the matter treated so fully that he discontinued his investigations, although, as he tells us, the methods he employed were quite different from those of Valerio. This study, however, attracted the attention of the Marquis Guido Ubaldi del Monte of Pesaro, himself an able mathematician, who introduced the author to the notice of Ferdinando I, Grand Duke of Tuscany.

Late in 1587 Galileo made his first journey to Rome, probably with the object of finding there some opening, or in furtherance of his designs on the chair of mathematics at Bologna, which had been vacant since 1583. The visit led to his acquaintance with Father Cristoforo Clavio (1557–1612) of the Society of Jesus, a celebrated mathematician, to whom mainly the world is indebted for the reform of the calendar in 1582. In later years he was a stout opponent of the new astronomical doctrines of the younger man, though before his death on the 6th of February, 1612, he became one of Galileo's most distinguished converts.

Evidently while in Rome the two new friends had been discussing mathematics, for, on his return to Florence, Galileo sent a letter to Clavio, under date 8th January, 1588 (the earliest of his letters known), respecting his theorem on the centre of gravity of a rectangular conoid, or truncated pyramid. 'Those', he says, 'to whom he had already submitted it, were not satisfied, and, therefore, he could not be so himself.' In this dilemma he solicits the learned Father's opinion, adding that, 'if it, too, was unfavourable, he would not rest until he had found such a demonstration as would be convincing to all.'

Galileo's efforts to secure employment in Bologna were frustrated, and he was equally unfortunate in successive applications for similar posts at Padua, Pisa, and Florence. Disappointed by repeated rebuffs, he and a young Florentine friend decided, towards the end of May, 1589, to seek their fortunes in the East. They had reached Milan, and were setting out for Venice *en route* to Constantinople, when the mathematical chair at Pisa became again vacant. Once more he applied for the post; and this time was successful, through the joint influence of his Maecenas, the Marquis Guido Ubaldi del Monte, and Cardinal Francesco del Monte, brother of the Marquis. He was still barely twenty-five and a half years old. The salary itself was insignificant, only sixty scudi per annum, or about £14 of our money; moreover, the appointment was for three years only, though renewable. But in his needy circumstances even this meagre opportunity was not to be rejected, and the office would after all enable him to add something to his means by private tuition.

6. *Certain Mathematical Problems*

No sooner was he settled in his new office than he resumed his physico-mathematical investigations. In the first year he carried to greater length his studies on the centre of gravity of solids, and arrived at results which excited afresh the admiration of Guido Ubaldi del Monte. At about the same time he undertook the investigation of that particular curve which is described by a point in a circle (as a nail in the rim of a wheel) while it revolves along a surface. To this curve thus described he gave the name Cycloid. This curve (known as Aristotle's wheel) and its properties had long been a puzzle to geometers, and had passed into a proverb —' Rota Aristotelis quae magis torquet quo magis torquetur.' He attempted the problem of its quadrature, and guessed that the area contained between the cycloid and its base is three times that of the describing circle, but he was unable to show this geometrically —a task which his disciple Torricelli (1608–47) achieved soon after his death.[1] Galileo recommended the curve as a form of arch for bridges, and it is said to have been applied in the construction of one across the Arno at Pisa (? the present Ponte di Mezzo).

Side by side with these studies Galileo was at this time steadily revolving those novel ideas on motion which were to be the basis of his latest and greatest work. In pursuance of these ideas, he now began a systematic experimental investigation of the mechanical doctrines of Aristotle.

Galileo was not the first to call in question the authority of Aristotle in matters of science. Cardinal Nicholas de Cusa (1401–64) was among the earliest to enter the lists.[2] The philosopher, Peter Ramus (1515–72), was another early opponent, and suffered the penalty for his convictions in the massacre of St. Bartholomew. Leonardo da Vinci (1452–1519) also held some very correct views on mechanics, and even anticipated Galileo in a few instances.[3]

[1] *Opera Geometrica*, Florence, 1644. Cf. Nat. Ed., vol. xviii, p. 153. Huygens applied the tautochronic property of this curve to the better regulation of pendulum clocks.

[2] Cusa, *De Docta Ignorantia*, Paris, 1514.

[3] His writings, mostly short notes and memoranda, were not known in Galileo's time. They remained in MS., practically lost to the world, till 1797, when Venturi brought them to light in his *Essai sur les Ouvrages physico-mathématiques de Léonard de Vinci*, Paris, 1797. Cf. Favaro's 'Leonardo da Vinci e Galileo' (*Estratto dalla Raccolta Vinciana*, July 1906), and 'Léonard de Vinci a-t-il exercé une influence sur Galilée et son École ?' (*Scientia*, December 1916).

Rizzoli,[1] again, in a posthumous work, had condemned the peripatetic philosophy in forcible terms, declaring that, although there were many excellent truths in Aristotle's *Physics*, the number was scarcely less of false, useless, and ridiculous propositions. Giovanni Battista Benedetti,[2] another sixteenth-century writer, had written expressly to confute several of Aristotle's mechanical problems, and so clearly expounded some principles of statical equilibrium that he may be regarded as a precursor of Galileo. Tartaglia [3] had discussed the theory of projectiles, and Varrone wrote on the force of inertia, and recognized the power of gravity as the cause of accelerated motion in falling bodies.

But while Galileo was thus by no means the first to question the authority of Aristotle, he was undoubtedly the first whose questioning produced a profound and lasting effect in men's minds. The reason is not far to seek. Galileo came at the fitting moment, but, above all, he came armed with a new instrument—*experiment.*

7. *Sermones de Motu Gravium* [4]

The results of these earlier isolated investigations on the foundations of dynamics are given at great length in the treatise *Sermones de Motu Gravium,* written in 1590, and, as was then the custom of Galileo and for many years after, the work was first circulated in manuscript. It did not appear in print until two hundred years after his death,[5] and then in an incomplete form. These *Sermones* consist chiefly of objections to Aristotelian doctrines, but a few of the chapters are devoted to an entirely new field of speculation. Thus, the 11th, 13th, and 17th *Sermones* relate to the motion of bodies along planes inclined at various angles, and of projectile motion ; the 14th enunciates a new theory of accelerated motion ; while in the 16th his assertion, that a body falling naturally for however great a time would never acquire more than an assignable degree of velocity, shows that he had already formed accurate notions of the action of a resisting medium.[6]

[1] Rizzoli, *Antibarbarus philosophica,* Frankfurt, 1674.

[2] Benedetti, *Speculationum liber,* Venice, 1585.

[3] Tartaglia, *Quesiti et Inventioni diverse,* Venezia, 1546.

[4] Cf. Nat. Ed., vol. i, pp. 245–419.

[5] In Alberi's *Le Opere di Galileo,* Florence, 1842–56.

[6] Most of these theorems were afterwards developed and incorporated in his larger work, *Dialoghi delle Nuove Scienze,* Leyden, 1638, and they can best be studied in that admirable compendium.

Galileo did not content himself with writing and circulating his *Sermones*, but as soon as he succeeded in demonstrating the falsehood of any Aristotelian proposition he did not hesitate to denounce it from his professorial chair.

8. *Public Experiments on Falling Bodies*, 1590–1 [1]

From professorial denunciation he proceeded to those public experiments with which the leaning tower of Pisa has become for ever associated. Aristotle had said that, if two different weights of the same material were let fall from the same height, the two would reach the ground in a period of time inversely proportional to their weights. Galileo maintained that, save for an inconsiderable difference due to the disproportionate resistance of the air, they would fall in the same time. The Aristotelians ridiculed such 'blasphemy', but Galileo determined to make his adversaries see the fact with their own eyes. One morning, before the assembled professors and students, he ascended the leaning tower, taking with him a 10 lb. shot and a 1 lb. shot. Balancing them on the overhanging edge, he let them go together. Together they fell, and together they struck the ground.

Neglecting the resistance of the air, he now boldly announced the law that all bodies fall from the same height in equal times. The correctness of this law was easily, if roughly, established by the leaning tower experiments, but, as the vertical fall was too rapid to admit of exact measurement, he made use of the inclined plane—a long straight piece of wood, along which a groove was accurately made, and down which bronze balls were free to move with the least friction. With this he proved that, no matter what the inclination of the plane, and, consequently, no matter what the time of fall, the movement of the balls was always in accordance with his law. In these investigations he utilized his discovery of the isochronism of pendulum oscillations as a measurement of time.

It might have been thought that such experiments would have settled the question. Yet with the sound of the simultaneously fallen weights ringing in their ears, the Aristotelians still maintained that a weight of 10 lb. would reach the ground in a tenth of the time taken by a weight of 1 lb. A temper of mind like this could not fail to produce ill-will, and we learn that, with the exception of the newly appointed Professor of Philosophy,

[1] Cf. Nat. Ed., vol. i, p. 249; vol. xix, p. 606.

Jacopo Mazzoni, the whole body of the teaching staff now turned against their young colleague.

> Stimano infamia il confessar da vecchi
> Per falso quel che giovani apprendero.
>
> Viviani, after *Horace.*

Soon a wholly unforeseen circumstance came to their aid, and led to Galileo's retirement from Pisa. Giovanni de Medici, natural son of Cosimo I, was at the time Governor of Leghorn. He was not unskilled as an engineer and architect, and had himself just designed a monster machine which he wished to use in cleaning the harbour. A model was submitted to the Grand Duke, and Galileo commissioned to examine and report on it. He did so, and declared it useless—an opinion which subsequent trial fully confirmed. Smarting under this failure, the inventor was induced to combine with the Aristotelians, to whose machinations were now added intrigues at Court. The position became intolerable, and Galileo resigned his post before the three years' term had expired, and once more returned to Florence about the middle of 1592.

III. Life Work, 1593–1632

1. *Early Years at Padua*

Galileo had not long to wait for a new post, for on the 22nd of September, 1592, he was appointed to the mathematical chair in Padua. Here he displayed at once extraordinary and varied activity. Besides the routine lectures on Euclid, the Sphere, and Ptolemy's Almagest, he gave special courses on Military Architecture and Fortifications,[1] on Mechanics, and on Gnomonics. On these and other subjects he prepared treatises which long circulated in manuscript among his pupils. Some were printed many years afterwards; others, like the treatise on Gnomonics, are lost; while others again found their way into the hands of persons who did not scruple to claim and publish them as their own.

The treatise on the Sphere, first published in Rome in 1656,[2] is supposed by some to be apocryphal, as it teaches the Ptolemaic cosmogony, placing the earth immovable in the centre of the

[1] Cf. Nat. Ed., vol. ii, pp. 9–146. In this there is nothing very original, his object being to lay before the student a compendium of the most approved principles of military science as then known.

[2] *Trattato della Sfera di Galileo*, &c., by Urbano Daviso, under the pseudonym of Buonardo Savi. Cf. Nat. Ed., vol. ii, pp. 205–55.

universe, and adducing the usual orthodox arguments. But we have it from Galileo's own hand that for many years he taught the Ptolemaic system in his classes in deference to popular feeling, although at heart he was even then a follower of Copernicus.[1]

2. *On Mechanics* [2]

The treatise *Delle Meccaniche* was written in 1594, and deals with the powers of the balance, the lever, the windlass, the screw, the pulley, cogged wheels, and the endless screw; and concludes with a note on 'The Force of Percussion', added later, apparently in 1599. Galileo had not yet discovered the principle of the decomposition of forces, but in his present treatment of these subjects he ingeniously uses the theory of the screw, reducing the screw to the inclined plane, the inclined plane to the pulley, and the pulley to the simple lever. Here for the first time is mentioned that condition of equilibrium, regarded from an entirely new aspect, to which modern mechanics owes all its splendid achievements. This condition of equilibrium is now called the principle of virtual velocities, and the credit of its discovery is with Galileo, who undoubtedly perceived its importance.

3. *Machine for Raising Water* [3]

While carrying on his professorial duties, giving private lessons, and writing learned tracts, Galileo was occupied in giving practical form to his ideas on the proper application of mechanics to machinery. Thus, towards the end of 1593, he devised a machine for raising water, of small dimensions but of great power, so constructed that one horse could raise the water and distribute it through twenty channels. The Venetian Government gave him the monopoly of this invention for a period of twenty years, though it does not appear to have been used in a practical way.[4]

[1] See his letter to Kepler dated August 4, 1597. His lectures on the New Star (p. 220, *infra*) may be regarded as his first *public* note of antagonism to the ruling astronomy. Cf. Favaro, *Galileo e lo Studio di Padova*, vol. i, pp. 148–67.

[2] Cf. Nat. Ed., vol. ii, p. 149; vol. viii, pp. 216, 321.

[3] Cf. Nat. Ed., vol. xvi, p. 27; vol. xix, p. 126.

[4] The early biographers of Galileo, Viviani and Gherardini, state that he was often employed 'to his great honour and profit' in the construction or superintendence of other machines for use in the Venetian State, but repeated searches in the archives of Venice and Padua do not afford any ground for this statement. Cf. Favaro, *Intorno ai servigi straordinarii prestati da Galileo alla Republica Veneta*, Venezia, 1890.

4. *The Geometrical and Military Compass* [1]

In 1597, however, he invented a more profitable instrument, and one that came into very extensive use, the Geometrical and Military Compass. This instrument is now known as *Gunter's Scale*, or the *Sector*. It consists of two straight rulers connected by a joint so that they can be set to any required angle. On one side are four sets of lines:

Arithmetical lines, which serve for the division of lines, the solution of the Rule of Three, the equalization of money, the calculation of interest.

Geometrical lines, for reducing proportionally superficial figures, extracting the square root, regulating the front and flank formations of armies, and finding the mean proportional.

Stereometrical lines, for the proportional reduction of similar solids, the extraction of the cube root, the finding of two mean proportionals, and for the transformation of a parallelopiped into a cube.

Metallic lines, for finding the proportional weights of metals and other substances, for transforming a given body into one of another material and of a given weight.

On the other side of the instrument are:

Polygraphic lines, for describing regular polygons, and dividing the circumference of the circle into equal parts.

Tetragonical lines, for squaring the circle (approximately) or other regular figure, for reducing several regular figures to one figure, and for transforming an irregular rectilineal figure into a regular one.

Joined lines, used in the squaring of the various portions of the circle and of other figures contained by parts of the circumference, or by straight and curved lines together.

There is joined to the compass a quadrant, which, besides the usual divisions of the astronomical compass, has transversal lines for taking the inclination of a scarp of a wall.[2]

5. *The Thermometer* [3]

To a somewhat later period may be referred his invention of the air thermometer. The date is uncertain, for while Viviani asserts that the instrument was designed during the first term of

[1] Cf. Nat. Ed., vol. ii. pp. 337–601; vol. xix, pp. 167, 222.

[2] The treatise on this subject, *Le Operazioni del Compasso geometrico e militare* (1606), is Galileo's first printed work, and was entirely set up in his own house in Padua.

[3] Cf. Nat. Ed., vol. xvii, p. 377.

his professorship at Padua (1592–8), other evidence takes us back only to about 1602. Thus Benedetto Castelli (1577–1643), writing to Ferdinando Cesarini, on the 20th of September, 1638, says:

'I remember an experiment which our Signor Galileo showed me more than thirty-five years ago. He took a small glass bottle about the size of a hen's egg, the neck of which was about two palms long [about 22 inches], and as narrow as a straw. Having well heated the bulb in his hand, he inserted its mouth in a vessel containing a little water, and, withdrawing the heat of his hand from the bulb, instantly the water rose in the neck more than a palm above its level in the vessel. It is thus that he constructed an instrument for measuring the degrees of heat and cold.'

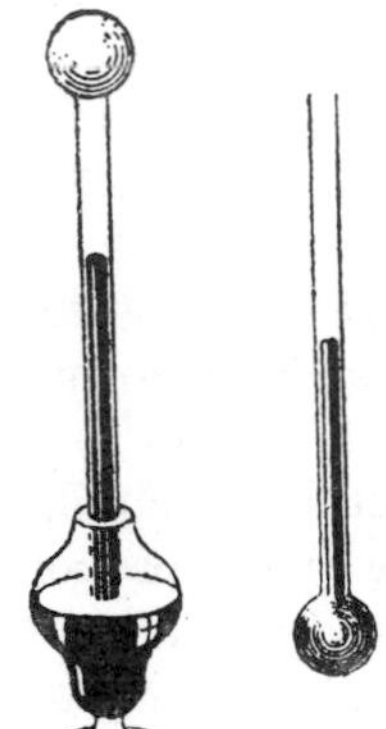

FIG. 5. Galileo's thermometer.

In this instrument different degrees of temperature were indicated by the expansion or contraction of the air which remained in the bulb; so that the scale was the reverse of that of the thermometer now in use, for the water would stand at the highest level when the weather was coldest. So long as the orifice of the tube remained open, this instrument could not be an efficient measurer of temperature, for it would be impossible to distinguish the effects of heat and cold from the effects of varying atmospheric pressure. It was, in truth, a barometer as well as thermometer, although Galileo did not recognize this (Fig. 5).

His friend Sagredo of Venice (1571–1620) was the first to divide the tube into 100 degrees in 1613. Sagredo also appears to have experimented with closed tubes from about 1615; but it was not until many years later (1653), after the death of Galileo, that the practice of closing the orifice, after exhausting the air, was introduced.[1]

6. *New Star of* 1604 [2]

In 1604 a new star appeared with great splendour in the constellation Serpentarius. Maestlin, one of the first to notice it, says:

'How wonderful is this new star! I am certain that I did not

[1] The credit of this capital improvement is due to Leopoldo de' Medici, brother of Ferdinando II, who adopted the plan of filling the tube with spirit, boiling it, and sealing the orifice whilst the contained spirit was in an expanded state, thus obtaining a partial vacuum, and depriving the instrument of its barometrical property.

[2] Cf. Nat. Ed., vol. ii, pp. 269–305, 526.

see it before the 29th of September, nor, indeed, on account of several cloudy nights had I a good view till the 6th of October. Now that it is on the other side of the sun, instead of surpassing Jupiter as it did, and almost rivalling Venus, it scarcely equals the Cor Leonis, and hardly surpasses Saturn. It continues, however, to shine with the same bright and strongly sparkling light, and changes colour almost every moment, now tawny, then yellow, presently purple and red, and, when it has risen above the vapours, most frequently white.'

Galileo appears to have noticed the new star about the 15th of October. The appearance of the new phenomenon had given rise to the most bewildering statements. Some said it was a light in the inferior regions of space, in 'the elementary sphere', that is, in that sphere of the four elements below that heavenly and incorruptible region where the Aristotelian school placed the heavenly bodies. Others thought that it was an old star hitherto unnoticed; others again believed that it was a new creation, while the astrologers deduced from it the wildest forebodings.

After observing the new star for some time, Galileo expounded his views upon it in three extraordinary lectures, which were delivered to the public in the great hall of the University in December, 1604. Unfortunately, only fragments of these lectures have come down to us, but from these and one or two other references,[1] we learn that his purpose was to indicate the position of the new phenomenon as 'far above the sphere of the Moon'.

Now, unlike his contemporaries, Tycho Brahé and Kepler, who thought that new stars were temporary conglomerations of a vapour-filling space, Galileo had suggested that they might be products of terrestrial exhalations of extreme tenuity, at immense distances from the earth, and reflecting the sun's rays—an hypothesis which, as we shall see later on, he also applied to comets. From the absence of parallax he showed that the new star could not be, as the current theory held, a meteor engendered in our atmosphere, and nearer to us than the moon, but that it must be situated beyond the planets and among the remote heavenly bodies. This was inconceivable to the Aristotelians, who

[1] Cf. *Difesa contro alle Calunnie di Baldassare Capra*, Nat. Ed., vol. ii, and *Postille al Libro d'Antonio Rocco*, Nat. Ed., vol. vii.

thought of the outer sphere of the Universe as an incorruptible and unchangeable heaven, subject neither to growth nor to decay, where nothing was created or destroyed.

7. *On Magnetism* [1]

Galileo's study of the magnetic properties of the loadstone dates from about 1600, apparently after he had seen the *De Magnete* of William Gilbert of Colchester (1544–1600), which was published in London in 1600. Gilbert's book had very great attraction for him, firstly, because its arguments traversed many of the principles of the Aristotelian school, and secondly, because it contained a number of original experiments, coupled with philosophical reflections of a far-reaching kind which appealed to his own daring spirit. Both Sagredo and Paolo Sarpi (1552–1623), his Venetian friends, tell us that at this time (1602) Galileo had not only repeated many of Gilbert's experiments, but made new ones of his own, especially on the best method of arming loadstones.

During the long vacations at Padua Galileo was wont to return to Florence, where he gave lessons to the young Prince Cosimo de' Medici. In the summer of 1607 he had evidently been rehearsing the wonderful properties of loadstones, and on returning to Padua he sent his pupil a stone which he had picked up in Venice, about ½ lb. in weight, not elegant in form but very powerful. In the letter which accompanied it he speaks of another stone belonging to Sagredo.

'This is elegantly shaped, and weighs about 5 lb. I have made it capable of sustaining 5½ lb. of iron, and I think I can make it do more before it leaves my hands. Much diligence is required in finding the true poles of these stones, that is, where their virtue is most robust, and where, consequently, their full sustaining power is manifested. This sustaining force depends as much on the quality of the armature as on the stone itself. Not every piece of iron of any size and shape is equally sustained, but well-made steel of a particular size and figure is most powerfully attracted. Then again, a slight shifting of the armatures from their true positions causes a great variation in the sustaining force. In the last four days I have so arranged that the stone now bears 1 lb. more than its owner was ever able to make it carry, and I hope, after preparing some pieces of the finest steel, to make it sustain still more.'

[1] Cf. Nat. Ed., vol. iii, p. 279 et seq.; vol. x, pp. 89, 185 et seq.; vol. xiii, p. 328.

He then describes a curious case of what is now called superposed magnetism :

'I have also observed in this stone another admirable effect which I have not met with in any other, namely, that the same pole repels or attracts the same piece of iron according to distance. Thus, placing an iron ball on a smooth and level table, and quickly presenting the stone, at a distance of about one finger, the ball moves away, and can be chased about at pleasure. But now, when the stone is sharply withdrawn to a distance of about four fingers, the ball moves *towards* it, and, with a little dexterity, can be made to follow it about.'

Ultimately Galileo was able to secure this stone also for Prince Cosimo, and in an accompanying letter to Chief Secretary Belisario Vinta, dated 3rd May, 1608, he writes :

'I send for his Highness the loadstone which, after many experiments, I have finally made to sustain a weight of 12 lb., or more than double its own, and I am certain that, had I more time and more suitable tools at my disposal, I could have done better still. I have fashioned the two armatures in the form of anchors, in allusion to the fable of a stone so large and powerful as to hold securely a ship's anchor. The form is also convenient for attaching weights to measure its holding capacity. I have purposely not made these anchors equal to the largest weights that I know the stone will carry, first, to leave a margin of safety, so that it will hold them firmly when presented, and secondly, because I am of opinion that its power may vary according to locality with respect to the poles of the great magnet, the earth ; for, whereas along the equinoctial line both poles will be of equal strength, one may be more powerful than the other in the northern hemisphere, and vice versa in the southern. Hence I am led to believe that the more powerful pole sustains here in Padua somewhat more than it can in the more southerly latitude of Florence or Pisa.[1] . . .

'I have marked the poles, so that one can readily see where the anchors or armatures should be applied ; that with the greater weight should be attached to the more robust pole, and the less heavy one to the weaker pole, taking care to apply them at the same instant, for I have found, not without great astonishment, that the stone more willingly supports two weights together than one alone. Thus, a piece of iron, so heavy as not to be supported when applied alone, will be held if at the same time another piece is applied to the other pole. . . . I have in mind some other

[1] Here he confuses two properties of the magnet, (1) its portative force, and (2) its directive force or 'inclination'.

artifices to render the stone still more marvellously powerful, and I am certain that I shall not fail. I believe I can make it sustain four times its own weight, or 20 lb., which, for such a large stone, is something very admirable. Indeed, I have no doubt that with proper cutting it can be made to support more than 30, and even 40 lb. I have noticed in this stone, not only that it never tires of holding a weight, but that with time it invigorates itself the more.'[1]

The 'other artifices' referred to in the letter consisted in breaking up large stones, shaping the best pieces so as to bring out their maximum of polarity, and providing them with suitable armatures, with the result that the portative force of the selected pieces far exceeded his own first and Gilbert's achievements in the same direction. Thus, in the *De Magnete*,[2] the English philosopher speaks of a stone (weight not given) which normally could sustain 4 oz. of iron, but which, when capped with steel, could support 12 oz. 'But', he goes on, 'the greatest force of a combining, or rather united, nature is seen when two stones, armed with iron caps, are so joined by their concurrent (commonly called contrary) ends that they mutually attract and raise one another. In this way a weight of 20 oz. of iron is raised, when either stone unarmed would only allure 4 oz.'

Galileo went far beyond this, since he was able to fashion small stones of extraordinary power. Of such a one he speaks in a letter to Cesare Marsili, dated 27th June, 1626. It weighed 6 oz., and unarmed could only support 2 oz., but when armed it was able to sustain 160 oz., or twenty-six times its own weight. He had this by him when writing his Dialogue of 1632, where he speaks of it as still in his possession. Later he appears to have presented it to the Grand Duke of Tuscany, Ferdinando II, as we gather from Castelli's *Discorso sopra la Calamita* (*circa* 1639–40).[3]

[1] These two stones were lost in after years, for in 1698 Leibnitz searched for them in vain.

[2] Chapter XVII, book 11.

[3] This stone is now preserved in the Tribuna di Galileo, Florence. The weight is in the form of a tomb (*sepolcro*), a form which was probably suggested by the legend of Mohammed's coffin suspended in the air by loadstones.

The Editor of the Florentine edition of Galileo's Works, 1718, mentions another small stone fashioned by Galileo, and of still more extraordinary power. It weighed only three-tenths of a grain, yet could support 121 grains (vol. i, Preface, p. xlvi). This, if true, beats Sir Isaac Newton's stone, which he wore set in a ring. It weighed 3 grains, and was able to support 746 grains.

8. *Development of Galileo's Ideas on Mechanics*, 1590–1609 [1]

In the years 1590 to 1609, Galileo completed, or, at least, laid the foundations of, his *Dialoghi delle Nuove Scienze* of 1638. In the 6th Day of these dialogues, he speaks of Paolo Aproino as assisting at a great number of experiments in Padua on diverse problems in mechanics, and especially, 'on the marvellous problems of percussion'. On the 29th of November, 1602, we find him writing to the Marquis Guido Ubaldi on the fall of bodies through two successive chords of a quadrant (discussed in 3rd Day), and on pendulum oscillations of different amplitudes. In 1604–6 he further studied the properties of the inclined plane, and took up the subject of naturally accelerated motion (discussed in 3rd Day). During 1609 he was occupied with the strength of beams and their resistance to fracture (discussed in 2nd Day). In the same year he investigated the motion of projectiles as applied to artillery, using the experiences derived from his many visits to the arsenal in Venice (discussed in 4th Day), and the coherence of the particles of solid bodies (discussed in 1st Day). In fact, about May, 1609, he was intending to publish an account of all these studies as a complete system of mechanical philosophy, when the telescope and consequent astronomical work intervened, and turned his energy in another direction.

Writing of this period, 1602–9, Favaro says:

'In truth the house of Galileo in Padua was not only a place for genial intercourse; not only a school to which flocked students, Italians and foreigners from every country in Europe, but more than this, it was a laboratory, where his marvellous mechanical talent knew how to devise ever new expedients. It was an academy in the true sense of the word, where the gravest problems in physics, in mechanics, in astronomy, and in mathematics were discussed with perfect freedom, and where it was possible to submit the deductions of reason to the salutary test of experiment, and the results of experiments, in their turn, to reasoning and calculation. Thus it may be said that the principal problems in the *Dialoghi delle Nuove Scienze* were raised and discussed within the walls of Padua.' [2]

[1] Cf. Nat. Ed., vol. ii, p. 259; vol. x, pp. 97, 228, 244, 248.

[2] *Galileo e lo Studio di Padova*, 2 vols., Firenze, 1883, vol. i, p. 314. Much the same may be said of his Dialogue of 1632, where many of the arguments are drawn from his experiments in Padua.

Galileo had long been desirous to return to his native Tuscany, and he now (1609) opened negotiations with the Chief Secretary of the Grand Duke, explaining that what he especially sought was leisure to pursue his researches without the distraction and strain of teaching and lecturing.

'The works which I have to finish', he wrote in May 1610, 'are chiefly: (1) two books on the system or structure of the Universe, an immense undertaking full of philosophy, astronomy, and geometry; (2) three books on local motion, a science entirely new, no one, ancient or modern, having discovered any of the many admirable consequences which I demonstrate in natural and violent motions, so that I may with reason call it a new science invented by me from its very first principles; (3) three books on mechanics, two of them on the demonstration of principles, and one of problems. Although others have treated this subject, no one either in quantity or quality has done a quarter of what I am writing on it. I have also treatises on natural subjects such as on sound and speech; on light and colours; on the tides; on the composition of continuous quantity; on the movements of animals; and others. I have also an idea of writing some books relating to the military art. . . .

'Then I need not say what an amount of labour will be required to fix the periods of the four new planets, a task the more laborious the more one thinks of it, as they are separated from one another by very brief intervals, and are all very similar in size and colour.'

Some of the treatises named in this letter are now lost, partly through the accidents of Galileo's stormy life, and in transport from place to place, and partly, through the extraordinary negligence and criminality of custodians. The loss of the essay on Continuous Quantity is particularly to be regretted. It is to his early disciple Buonaventura Cavalieri (1598–1647), who refused to publish his own book [1] so long as he hoped to see Galileo's printed, that we owe *The Method of Indivisibles*, which is recognized as containing some of the germinal ideas of Newton's Fluxional Calculus. The treatises on *Sound and Speech*, and on *Light and Colours*, were probably never completed, but we find fragments of them in later works, as, *Il Saggiatore* and the *Dialogues* of 1632 and 1638. Similarly, of the movements of animals we have the fragment *Intorno al camminare del cavallo.*[2]

1 *Geometria indivisibilibus continuorum nova quadam ratione promota*, Bologna, 1635.

2 Nat. Ed., vol. viii, last section.

PLATE XL. From the Galileo Museum at Florence

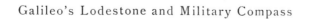

Galileo's Lodestone and Military Compass

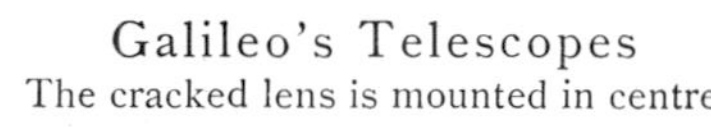

Galileo's Telescopes
The cracked lens is mounted in centre

9. *The Invention of the Telescope*[1]

Leaving aside the vexed question of the priority of invention of the telescope, and of the reports of it which circulated in Italy from about November 1608,[2] we pass to Galileo's own account of his share in the discovery. Of this there are three slightly discordant versions—one to his brother-in-law, Landucci, dated 29th August 1609, the second in *Sidereus Nuncius*, published in March 1610, and the third in *Il Saggiatore*, 1623. In the first version (probably the most reliable as being nearest in time) he says:

'I write now because I have a piece of news for you, though whether you will be glad or sorry to hear it I cannot say, for I have now no hope of returning to my own country, though the occurrence which has destroyed that hope has had results both useful and honourable. You must know then that about two months ago (i. e. about June 1609) a report was spread here that in Flanders a spy-glass had been presented to Prince Maurice, so ingeniously constructed that it made the most distant objects appear near, so that a man could be seen quite plainly at a distance of two miles. This result seemed to me so extraordinary that it set me thinking, and, as it appeared to me that it depended upon the laws of perspective, I reflected on the manner of constructing it, and was at length so entirely successful that I made a spy-glass which far surpasses the report of the Flanders one. As the news had reached Venice that I had made such an instrument, six days ago I was summoned before their Highnesses the Signoria, and exhibited it to them, to the astonishment of the whole Senate. Many of the nobles and senators, although of a great age, mounted more than once to the top of the highest church tower in Venice, in order to see sails and shipping that were so far off that it was two hours before they were seen without my spy-glass, steering full sail into the harbour; for the effect of my instrument is such that it makes an object 50 miles off appear as large as if it were only five miles away.

'Perceiving of what great utility such an instrument would prove in naval and military operations, and seeing that his Serenity the Doge desired to possess it, I resolved on the 24th inst. to go to the palace and present it as a free gift. On quitting the presence-

[1] Cf. Nat. Ed., vol. iii, pp. 18, 60, 869; vol. x, pp. 250–3; vol. xix, p. 587.

[2] Favaro discusses these questions very fully in *Galileo e lo Studio di Padova*, vol. i, ch. xi, and in *La Invenzione del Telescopio secondo gli ultimi Studi*, Venezia, 1906. Cf. Drinkwater's *Life of Galileo*, 1833, pp. 20–6; Grant's *History of Physical Astronomy*, 1852, pp. 514–37.

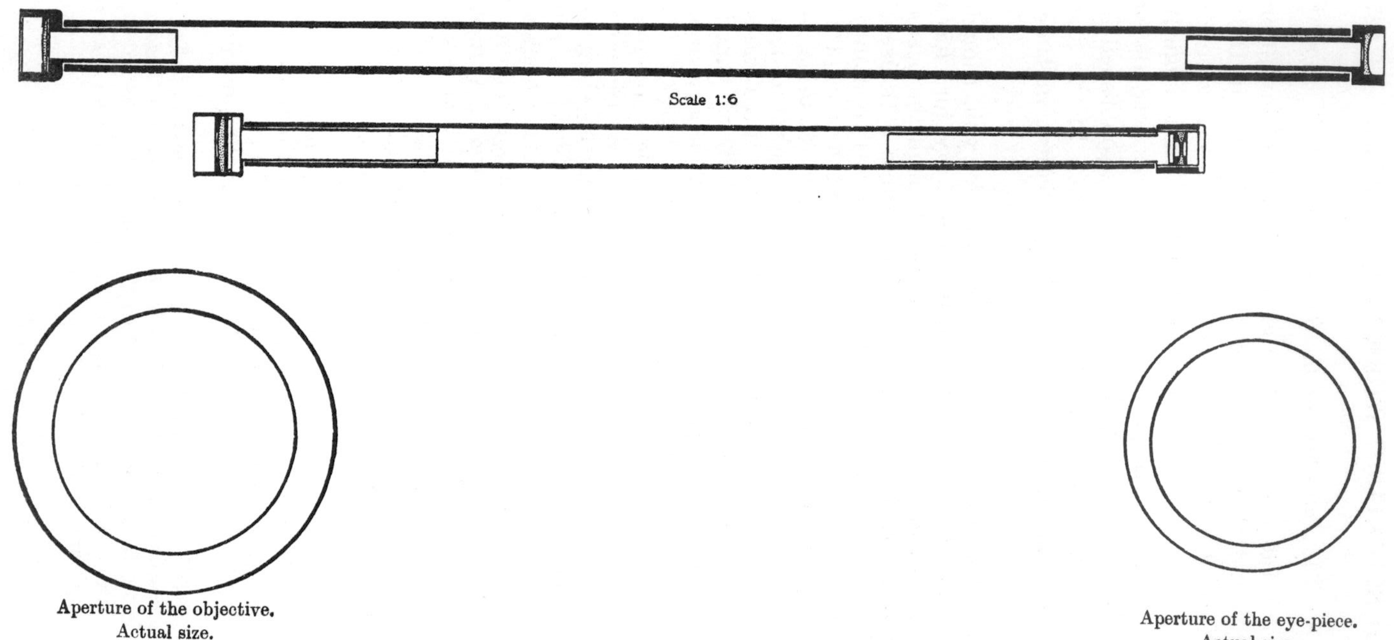

FIG. 6. DIAGRAM TO ILLUSTRATE THE PROPORTIONS OF GALILEO'S FIRST TELESCOPES

chamber I was commanded to bide awhile in the hall of the Senate, whereunto the Procurator, Antonio Prioli, one of the heads of the University of Padua, came, and, taking me by the hand, said that the Senate, knowing the way in which I had served it for seventeen years at Padua, and being sensible of my courtesy in making it a present of the spy-glass, had ordered my election (with my goodwill) to the Professorship for life.'

This telescope is unfortunately lost. It is mentioned with some particulars in his *Sidereus Nuncius*, and consisted of a leaden tube, with a plano-concave eye-glass and a plano-convex object-glass, and had a magnifying power of three diameters. From other sources we learn that the tube was 2·39 metres long and about 42 millimetres diameter (see Fig. 6, p. 228). This instrument was shown for the first time in public on the 21st of August, 1609, from the top of the campanile of San Marco, when the farthest object that could be seen clearly was the campanile of the Church of San. Giustina in Padua, distant about 35 kilometres.

10. *Invention of the Microscope* [1]

The invention of the telescope could hardly fail to lead to the disclosure of the principle of the microscope, which at first was but a telescope adapted. John Wodderborn, a Scotch student who attended Galileo's lectures at Padua, in a defence of his master (published in 1610),[2] states that he 'heard Galileo describe in what manner he perfectly distinguishes with his telescope the organs of motion and of the senses of the smaller animals, especially in a certain insect which has each eye covered by a rather thick membrane, which, perforated with seven holes, like the visor of a warrior, allows it sight '.[3]

In 1614, the Frenchman, Jean Tarde, called on Galileo, whom he found ill in bed, and he says that

'Galileo told me that the tube of a telescope for observing the stars is no more than 2 feet in length; but to see objects well, which are very near, and which on account of their small size are hardly visible to the naked eye, the tube must be two or three times longer. He tells me that with this long tube he has seen flies which look as big as a lamb, are covered all over with hair,

[1] Cf. Nat. Ed., vol. xiii, pp. 36, 40, 199, 201, 208.

[2] John Wodderborn, Scotobritannus, *Quatuor problematum quae Martinus Horky contra Nuntium Sidereum de quatuor planetis novis disputanda proposuit confutatio*, Padua, 1610.

[3] Nat. Ed., vol. iii, pp. 151–78.

and have very pointed nails, by means of which they keep themselves up and walk on glass, although hanging feet upwards.'[1]

In *Il Saggiatore* and in Viviani's biographical sketch there are similar references, the latter adding that in 1612 Galileo presented a microscope to the King of Poland. All this shows that he was well acquainted from the first with the use of his invention *qua* microscope. For many years, however, he gave the matter little attention—not, indeed, until his visit to Rome in 1624, when he found the microscope discussed as a novelty which nobody could understand.

An optician of Middelburg had invented a form of microscope about 1590, in which objects were seen inverted. One of these instruments he presented to the Archduke Charles Albert of Austria, who in turn gave it to Cornelius Drebbel, a Dutchman, then living in London. For many years after, the instrument was practically forgotten; but about 1621 Cornelius Drebbel appears to have resumed its manufacture.[2] In the following year Jacob Kuffler, a relative of Drebbel, brought a specimen to Rome, a present from Nicholas Fabri de Peiresc of Paris (1580–1637) to one of the Cardinals. Unfortunately, Kuffler died before he had time to explain the management of the instrument, and so it remained a mystery. Two years later two other specimens arrived, also sent by de Peiresc, with brief instructions as to their use. One of these was nothing more than a magnifying glass, but of the other, consisting of two glasses, nobody in Rome could make anything, 'although they had the help of mathematicians'.

At this moment Galileo arrived. The instrument was shown him, and he at once told his friends that he had himself made a somewhat similar instrument many years previously, 'which magnified things as much as 50,000 times, so that one sees a fly as large as a hen'. He made some specimens, showing objects *erect*, which he sent to his friends, and soon his microscopes were in as great request as his telescopes. Amongst others, he sent one to Prince Federigo Cesi, on the 23rd of September, 1624, with the following letter:

[1] Cf. Nat. Ed., vol. xix, p. 589. Tarde's voyages in Italy are in MS. at the Bibliothèque nationale, Paris.

[2] Charles Singer, 'Notes on the Early History of the Microscope', in the *Proceedings of the Royal Society of Medicine, Historical Section*, 1914, vol. vii, p. 247, and 'The Dawn of Microscopical Discovery,' *Journal of the Royal Microscopical Society*, 1915, p. 317, and the article by him in this volume.

'I send your Excellency a little spy-glass (*occhialino*) for observing at close quarters the smallest objects, which I hope will afford you the same interest and pleasure that it has to me. I delayed sending it because my first attempts were imperfect by reason of the difficulty in fashioning the lenses. The object is placed on a movable circle (at the base of the instrument), which can be turned in such a way as to show successive portions, a single pose being unable to show more than a small part of the whole. As the distance between the lens and the object must be precisely adjusted in order to see things that are in relief, it is necessary to bring the glass nearer to or farther from the object according to the parts to be examined. Therefore the little tube is made adjustable on its stand or guide. The instrument should be used in a strong light, or even in full sunlight, so as to illuminate the object as much as possible.

'I have examined with the greatest delight a large number of animals, amongst which the bug is most horrible, the gnat and the moth very beautiful. I have also been able to discover how the fly and other little animals are able to walk on window-panes and ceilings feet upwards. But your Excellency will now have the opportunity of observing thousands of other details of the most curious kind, of which I shall be glad to have an account.

'P.S.—The little tube is in two pieces, so that you may lengthen it or shorten it at pleasure.' [1]

11. *The Siderius Nuncius*

First Telescopic Discoveries in the Heavens [2]

After exhibiting his telescope in Venice, Galileo returned to Padua, and at once constructed a third instrument, of which he only says that 'it made objects appear more than sixty times larger', equivalent to a magnifying power of about eight diameters. But in a very few days he had a much better telescope which enlarged four hundred times. With this in the autumn of 1609 he made his first discoveries in the heavens—an immense number of fixed stars, more than tenfold the number at that time catalogued. He noticed the property of irradiation common to all

[1] The only relics (two tubes) of these instruments now known to exist are preserved in the Tribuna di Galileo, Florence. The lenses are missing, and the genuineness of the tubes themselves is doubtful. For much interesting information on this subject see Prof. Govi's 'The Compound Microscope invented by Galileo', in *Journal of the Royal Microscopical Society*, 1889, pp. 574 et seq., and the two articles by Charles Singer quoted above.

[2] Cf. Nat. Ed., vol. iii, Part 1, *passim*; vol. x, pp. 273, 410; vol. xix, pp. 229, 610.

luminous bodies, and the scintillation of the fixed stars, which differentiated them from the planets. Upon directing the telescope to the more conspicuous star-clusters, he was astonished to find that they contained a great number of other stars too faint to be recognized by the naked eye. Thus, the number of the Pleiades, from six or seven, now rose to thirty-six; while in Orion, instead of thirty-seven, he now counted as many as eighty stars. Next, examining portions of the Milky Way and other nebulous patches he resolved them into congeries of stars of various magnitudes. Turning to the moon, he brought it to a distance of less than three semi-diameters of the earth, making it appear about twenty times nearer and four hundred times larger than when seen by the unaided eye.

Early in January, 1610, Galileo had constructed a still more powerful telescope, which showed objects more than thirty times nearer and nearly one thousand times larger. With this instrument he not only verified and completed the observations begun the previous autumn with his fourth telescope, but he also discovered Jupiter's moons.

Writing to a friend at the Tuscan Court on 30th January, 1610, he thus alludes to this series of discoveries:

'I give thanks to God, who has been pleased to make me the first observer of marvellous things unrevealed to bygone ages. I had already ascertained that the moon was a body very similar to the earth, and had shown our Serene Master, the Grand Duke, as much, but imperfectly, not then having such an excellent spy-glass as I now possess, which, besides showing the moon most clearly, has revealed to me a multitude of fixed stars never before seen, being more than ten times the number of those that can be seen by the unaided eye. Moreover, I have ascertained what has always been a matter of controversy among philosophers, namely, the nature of the Milky Way. But the greatest marvel of all is the discovery of four new planets. I have observed their motions proper to themselves and in relation to each other, and wherein they differ from the motions of the other planets. These new bodies move round another very great star, in the same way as Mercury and Venus, and, peradventure, the other known planets, move round the sun. As soon as my tract is printed, . . . I shall send a copy to his Highness, the Grand Duke.'

This tract is the *Sidereus Nuncius*, the preface of which is dated 4th March, 1610. In this epoch-marking treatise he gives the results of his observations. He speaks first of the moon. The discovery of spots, added to those already visible to the naked eye, and observations on the changes of light on those spots, led him to the conclusion that the surface of the moon, far from being smooth and polished according to the then accepted opinion, was rough with deep depressions and high mountains. The brilliant parts he inferred were land, while those which remained obscure—the permanent spots—he regarded as water. The illuminated edges of the moon showed themselves smooth and without those indentations which one would expect from the inequalities of the surface. Galileo explained this appearance (1) by supposing that the mountainous parts masked each other as it were, so that at the distance of the earth the intervening depressions were not discernible, and (2) by the existence of a lunar atmosphere of a density such as to reflect the solar rays while not obstructing the vision. From the appearance of illuminated mountain-tops in the dark part of the moon, at some little distance from the broken line along which sunrise or sunset was general, he was able to judge of the height of some of the mountains, and his calculation agrees very well with the modern estimate. The higher mountains were found to rise four or five miles above the general level.

Galileo remarked the feeble light,[1] which, in the first and last quarters of the moon, makes visible to us that part of its disk which is no longer illuminated directly by the sun. After showing that the light did not originate in the moon itself, that it was not caused by sunlight passing through its body, that it was not reflected there from Venus, he concludes that it can only be due to the sunlight reflected from the earth to the moon, and thence reflected back to our eyes. He contended therefore that our earth shines, like the moon and planets, by light from the sun; that it far exceeds the moon in luminosity; and that since it is a moving planet it is thus fully comparable to the other heavenly bodies.

In using the telescope to examine the fixed stars and comparing them with the planets, Galileo observed a remarkable difference. While the planets showed themselves as disks, like little moons, the stars appeared but little larger than with the naked eye, just bright specks sending forth twinkling rays. He explained this apparent failure of the telescope to enlarge in proportion to its magnifying power as due to the effect of irradiation. In virtue

[1] Now known as *earthshine*. Previously noticed by Pythagoras (*c*. 580–504 B.C.), by Plato (428–347 B.C.), and by Leonardo da Vinci (1452–1519). In 1640 Fortunio Liceti held that the moon is a phosphorescent body like the Bologna Stone. This drew from Galileo a reply—his last great effort. See Nat. Ed., vol. viii, pp. 467–556.

of this principle, a light projected upon a dark ground is dilated in all directions, so as to appear larger than it really is, and the greater its luminosity the greater is the irradiation, and the larger the light appears to be. A star, then, as seen by the naked eye is a luminous point plus its dilated corona. Now, the telescope has the property of cancelling or masking, more or less according to the degree of luminosity, this false light, so that what we see is the resultant of two opposite effects, (1) the star bereft more or less of its corona, and (2) the star as enlarged by the power of the glass; and, since the resultant enlargement is but little, we must conclude, not that the telescope fails in such cases, but that the corona, though diminished, is still able to nearly counterbalance the telescopic enlargement, with the result that the apparent size is little larger than that exhibited to the naked eye.[1]

When Galileo turned his fourth telescope to the planets he saw them as little moons. Jupiter's disk was of considerable size, but in no other way did he differ from the other planets. Now, on the 7th of January, 1610, directing his fifth and more powerful glass towards Jupiter, his attention was drawn to three small but bright stars in his vicinity, two on the east side and one on the west. He at first imagined them to be fixed stars, and yet there was something in their appearance which he thought curious, and they were all disposed in a right line parallel to the plane of the ecliptic. Happening, by mere accident, as he says, to examine Jupiter again on the next night, he was surprised to find these stars now arranged quite differently. They were all on the west side, nearer to each other than on the previous evening, and at equal distances apart. He therefore waited for the following night with some anxiety, but he was disappointed, for the heavens were enveloped in clouds. On the 10th of January he could see only two stars, and they were both on the east side! He suspected that the third might be concealed behind the disk of the planet. Those visible appeared as before in the same right line, and lay in the direction of the ecliptic. Unable to account for such changes by the motion of the planet, and being fully assured that he always observed the same stars, he concluded that the motions must be referred to the stars themselves and not to the planet.

On the 11th of January he again saw only two stars, still on the east side, but the outer one was now nearly twice as large

[1] He has a great deal more on this subject in his letters to Griemberger on Lunar Mountains, in his works on Sun-spots, and in *Il Saggiatore*, which see *infra*. Irradiation was first treated as a general principle by Kepler in 1604.

as the other, although on the previous night they were almost equal. This fact, taken in connexion with the constant change of the relative positions of the stars, and the total disappearance of one of them, now revealed to him their real character. He concluded that there are in the heavens three stars revolving round Jupiter in the same way as Venus and Mercury revolve round the sun. On the 12th of January he again saw three stars, two on the east side, and one on the west. The third began to appear about three o'clock in the morning, emerging from the eastern limb of the planet ; it was then very small, and discernible only with great difficulty. On the 13th of January he saw four stars, three on the west side and one on the east. They were all in a line parallel to the ecliptic, with the exception of the central one of the western group, which was a little towards the north. They were all about the same size, and shone with a much greater lustre than fixed stars of the same magnitude. January the 14th was cloudy, but next night he saw all four stars to the west of the planet, all nearly in the same right line, and increasing in size and brilliancy, according to their distance from Jupiter. And so he continued nightly, up to the 2nd of March, 1610, to make these observations, sixty-six of which are figured and described in the *Sidereus Nuncius*.[1]

The persistence of the relative distances between these four bodies and Jupiter in all their changes left no room for doubt that they accomplished with him, and in about twelve years, a revolution around the sun as a centre. Their own orbits round the planet were unequal in time, those nearest moving more rapidly than those more remote ; while the most remote of all appeared to complete its revolution in about fifteen days.

During the Easter recess of 1610 at Padua, Galileo had an express invitation from the Tuscan Court, then at Pisa, to explain to the Grand Duke his discovery of the four satellites of Jupiter, which, in honour of the reigning family, he proposed to call Medicean Stars, after the four brothers Cosimo II, Francesco, Carlo, and Lorenzo. Cosimo II, who all his life showed a sincere attachment to his tutor, asked for and obtained the gift of the instrument with which this discovery was made ; but Galileo quickly repented of his generosity. He evidently could not part with his 'old

[1] There are also (1) Diagram to illustrate principle of the telescope, (2) five drawings of the moon's superficies, (3) two diagrams on lunar measurements (4) Orion, (5) Pleiades, (6) Nebulosa Orionis, and (7) Nebulosa Praesepe.

discoverer ', as he affectionately called it in after years ; so, while always reserving it for the Grand Duke, he kept it near himself till his death, when it was handed over to Prince Leopoldo, brother of Ferdinando II.

Of its subsequent history little is known with certainty. It would appear that in Galileo's last years the instrument was accidentally broken. Then, in 1675, there is a record in the inventory of the effects of Cardinal Leopoldo de' Medici of a ' broken object-glass with which Galileo discovered the four new planets ' ; and in 1677 another record of its having been set in an ivory frame. It is now preserved, together with two telescopes, said to have been made by Galileo, and certainly of his time, in the Tribuna di Galileo at Florence, with many other precious relics of the period. Accurate measurements of it have been made quite recently by Professor Roiti of the University of Florence, as follows : Focal distance 1·70 metres, diameter 0·056 metre. One face has the curvature of a sphere with radius of 0·935 metre, and the other face is practically plane, having just a trace of convexity.[1]

12. *On Saturn* [2]

After completing his study of Jupiter, Galileo turned his glass to the other planets to see if they also had attendant moons. On the 25th of July, 1610, he was rewarded by another brilliant discovery—the phenomenon that we now describe as the ring of Saturn. To him, however, it did not appear as a ring, but as a triple star of which the central part was about three times larger than the laterals, and all three almost touching, and in a plane parallel to the zodiac. He made further observations in the autumn, when Saturn was well above the horizon, and, fearing that some one might forestall him, he announced the discovery in a brief letter, dated Padua, 30th July, 1610, to Belisario Vinta, at Florence, but begged him to keep it secret for a while. As a further precaution lest his claims should be forestalled he sent to friends in Italy and Germany a cryptogram of thirty-seven letters as follows :

SMAISMRMILMEPOETALEUMIBVNENUGTTAVIRAS

Kepler and other friends puzzled long over this anagram, the former thinking it had some reference to his favourite planet

[1] Cf. Favaro's *Intorno ai Cannocchiali costruiti ed usati da Galileo*, Venezia, 1901.

[2] Cf. Nat. Ed., vol. x, pp. 410, 474 ; xi, p. 439 ; xviii, p. 238.

Mars. At length, Giuliano de' Medici, Tuscan ambassador at the German Court, was charged by the Emperor Rudolph II to ask for the solution, to whom Galileo, replying 13th November, 1610, gave the following startling solution:

'Altissimum Planetam Tergeminum Observavi.'

'I have observed', he says, 'with great admiration that Saturn is not a single star but three together, which, as it were, touch each other. They have no relative motion, and are constituted in this form [see Fig. 7, I], the middle being much larger than the lateral ones. They are not strictly in the line of the zodiac, but rather parallel to the equinoctial line. . . . I have already discovered a court for Jupiter, and now there are two attendants for this old man, who aid his steps and never leave his side.'

The learned world had not yet had time to digest the surprising facts announced in the *Sidereus Nuncius*, when this asserted triple nature of Saturn again contravened the prevailing Aristotelian ideas. Continuing his observations, Galileo found that the lateral bodies did not retain the same apparent magnitudes. In fact, they had been gradually diminishing, although they appeared to be immovable, both with respect to each other and to the central body. They continued to grow less and less during the next two years, and towards the close of 1612 they vanished altogether! Horrified at this extraordinary phenomenon, and full of alarm for the consequences to himself when his Aristotelian opponents should come to hear of it, he thus wrote to Welser on December 1st, 1612:

'Looking at Saturn within these last few days, I found it solitary without its accustomed stars, and, in short, perfectly round and defined like Jupiter, and such it still remains! Now what can be said of so strange a metamorphosis? Are, perhaps, the two smaller stars consumed like spots on the sun? Have they suddenly vanished and fled? Or has Saturn devoured his own children? Or was the appearance, indeed, fraud and illusion, with which the glasses have for so long mocked me and many others who have observed with me?'

He continued, however, to conjecture that the two attendant stars would reappear after revolving with the planet, and that, by the summer solstice of 1615, they would be not only again visible, but more luminous and larger. And by the middle of 1615 he was able to verify his prediction, for the lateral stars were now reappearing (Fig. 7).

No change calling for special comment was noticeable until the summer of 1616, when he made a new observation relating to Saturn. In August of that year, writing to Prince Cesi, Galileo says :

'I cannot rest without signifying to your Excellency a new and most strange phenomenon observed by me in the last few days in Saturn. Its two companions are no longer two small and perfectly round globes, as they have hitherto appeared to be, but are now bodies much larger, and of a form no longer round, but, as shown in the annexed figure (see Fig. 7, III), with the two middle parts obscured, that is to say, the very dark triangular-like spaces contiguous to the middle line of Saturn's globe, which latter is seen, as always, perfectly round.'[1]

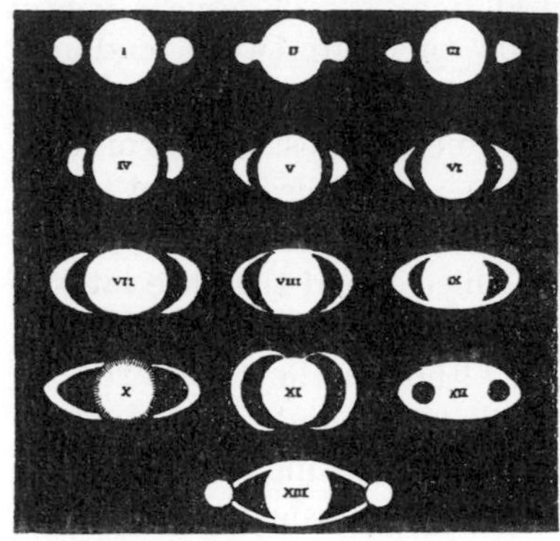

FIG. 7. EARLY DRAWINGS OF SATURN
From the *Systema Saturnum*.

Up to the last, Galileo made no announcement as to the precise nature of Saturn's appendages. He contented himself with describing what he saw, and, recognizing the incompleteness of his knowledge, and, perhaps, the inadequate power of his glass, he left it to the future to solve the problem. This was done by Christian Huygens in 1655. Working with a refracting telescope, magnifying 100 diameters, this astronomer not only saw and described the ring as a ring, but discovered one of Saturn's satellites.

13. *Venus, Mercury, and Mars* [2]

These discoveries stimulated yet further the interest of Galileo's grand-ducal pupil, and in June 1610 Cosimo II nominated him 'First Mathematician of the University of Pisa, and First Mathematician and Philosopher to the Grand Duke'. Galileo now returned to Florence. Here, on September 30, he made another astounding discovery in the heavens, namely, the occasional

[1] Cf. Favaro, 'Intorno all' Apparenza di Saturno osservata da Galileo nell'Agosto 1616' (*Atti del Reale Istituto Veneto*, February, 1901).

[2] Cf. Nat. Ed., vol. x, pp. 483, 499, 503 ; vol. xi, pp. 11–12.

crescent form of the planet Venus. After satisfying himself, by three months' observations, of its correctness, he announced the fact in a letter (11th December) to his friend Giuliano de' Medici at Prague, but concealed it again in an anagram as follows: 'Haec immatura a me iam frustra leguntur o y.' He did not, however, leave his friend long in perplexity, for on the 1st of January, 1611, he sent him the solution: 'Cynthae figuras aemulatur mater amorum.' 'That is, *Venus rivals the appearance of the moon*; for, being now arrived at that point of her orbit in which she is between the earth and the sun, and with only a part of her enlightened surface turned towards us, the telescope shows her in a crescent form, like the moon in a similar position.' Following her through the visible portion of her orbit, he had the satisfaction of seeing the illuminated part assume successively the crescent forms appropriate to his hypothesis.

It was with reason, therefore, that he laid stress on the importance of these observations, which established yet another fact obnoxious to the Aristotelians—a further resemblance between the earth and moon and one of the principal planets. As he had shown in *Sidereus Nuncius* that the earth, like the moon, is luminous only where exposed to the sun's rays, so this change of figure in Venus demonstrated that she and, probably, all the other planets were not luminous of themselves, but reflected the sun's light. Thence he concluded that they must all revolve round the sun—'a fact surmised by Pythagoras, Copernicus, Kepler, and their disciples, but that could not be proved by ocular demonstrations'. For it had always been a formidable objection to the Copernican theory that Venus and Mercury did not exhibit the same phases as the moon, which they should do if they revolved round the sun, and Copernicus himself had endeavoured to account for this by supposing that the sun's rays passed freely through the body of the planets.

Of similar changes in Mercury, the existence of which he inferred by analogy, he could observe nothing, because that planet's orbit does not take him far from the sun, and, in consequence, his small disk is always so resplendent that not even the best telescope could deprive him of his factitious rays.[1]

[1] The revolution of Mercury about the sun, which Galileo assumed, was confirmed twenty years later. Just before his death in 1630, Kepler predicted a transit of Mercury for the next year, and it was duly observed, on November 7, 1631,

The orbit of Mars being exterior to that of the earth, he is not subject to phases like the inferior planets Venus and Mercury, but in certain positions he assumes a gibbous appearance, like that of the moon a little before and after the full. Galileo recognized this feature, and, after four months' careful observation, he announced that ' when Mars is in quadrature, or the middle points of his path on each side of the sun, his figure varies slightly from a perfect circle. I dare not affirm that I can observe phases, but, if I mistake not, I already perceive that he is not always perfectly round.' He also observed that the apparent size of the planet varied according to its distance from the sun, being sixty times larger when in opposition than when in conjunction.

14. *On Sun-Spots* [1]

In consideration of the intense interest, friendly and otherwise, excited by these discoveries in Rome, Galileo thought it desirable to go there himself, and acquaint at first hand the *savants* and dignitaries of the Church with his work. It was not till March 23, 1611, that he was able to set out, provided with many letters of introduction, amongst them one from Michelangelo the younger (nephew of the great sculptor and painter) to Cardinal Barberini (afterwards Pope Urban VIII). He was received with distinction by princes and all the Church dignitaries, as well as by the learned laymen. Even those who discredited his discoveries, either through obstinacy or through fear of their results, were as eager as the true friends of science to see and hear this wonder of the age.

After exhibiting on several occasions all his recent discoveries, or ' celestial novelties ' as they were called, a commission of four scientific members of the Roman College was appointed to examine them. Their report of April 24 was favourable on all points, and was considered as equivalent to an official *Imprimatur*. Pope Paul V granted him a long audience, and assured him of his unalterable goodwill ; high dignitaries of the Church followed suit, and the Accademia dei Lincei elected him a member.

Immediately after the publication of the report of the com-

by Gassendi, who followed Kepler's instructions. Our own countryman, Horrocks, was the first to observe a transit of Venus, in 1639.

[1] Cf. Nat. Ed., vol. v, pp. 10–260 ; vol. vii, p. 372 ; vol. xiv, p. 299.

SIDEREVS NVNCIVS

MAGNA, LONGEQVE ADMIRABILIA
Spectacula pandens, suspiciendaque proponens
vnicuique, præsertim verò
PHILOSOPHIS, atq; ASTRONOMIS, quæ à

GALILEO GALILEO
PATRITIO FLORENTINO

Patauini Gymnasij Publico Mathematico

PERSPICILLI

Nuper à se reperti beneficio sunt obseruata in LVNÆ FACIE, FIXIS INNVMERIS, LACTEO CIRCVLO, STELLIS NEBVLOSIS,
Apprime verò in

QVATVOR PLANETIS

Circa IOVIS Stellam disparibus interuallis, atque periodis, celeritate mirabili circumuolutis; quos, nemini in hanc vsque diem cognitos, nouissimè Author depræhendit primus, atque

MEDICEA SIDERA

NVNCVPANDOS DECREVIT.

VENETIIS, Apud Thomam Baglionum. M DC X.

Superiorum Permissu, & Priuilegio.

Fig. 8. Title-page of Sidereus Nuncius.

mission, Galileo announced yet another new discovery in the heavens, namely, dark spots on the body of the sun, which, towards the end of April 1611, he showed to several prelates and men of science in Rome. Describing these phenomena,[1] he states that at first he was undecided whether to explain the ever-changing form and position of the spots by supposing that the sun revolved, or by imagining that other and hitherto unknown planets revolved about the sun, and were visible only as spots on his disk. Further observation, however, led him to abandon the latter supposition and to announce positively that the spots were in contact with the body of the sun, where they were continually appearing and disappearing much as clouds about our earth. These observations were, in their consequences to Galileo, particularly unfortunate, as he thereby became embroiled with the powerful Jesuit party whose influence was one of the chief causes of his subsequent misfortunes.

A Jesuit father, Christopher Scheiner, Professor of Mathematics at Ingolstadt, claimed priority in the discovery of the sun-spots, asserting that early in 1611 he first noticed them and showed them to his pupils. Scheiner stated his case in three open letters addressed to Mark Welser, Chief Magistrate of Augsburg, though it is clear from these letters that at first he attached no importance to these appearances, and even thought them due to defects in his glasses. The spots he supposed to be caused by multitudes of little planets, revolving round the sun in an orbit inside Mercury, and producing the appearance of spots in crossing his disk.

On the publication of Scheiner's letters, Welser sent a copy to Galileo, requesting to be favoured with his opinions of the phenomena therein described. He replied in three letters dated respectively 4th May, 14th August, and 1st December, 1612. In the first letter (the autograph copy of which is now in the British Museum) he begins by saying that the phenomena are not illusions produced by the glasses, but veritable facts, which he himself had observed eighteen months before in Florence, and which he had shown to many people in Rome in April of the past year.[2] He

[1] *Discourse on Floating Bodies*, 1612, which see *infra*, p. 249.

[2] At the end of *Il Saggiatore* (1623), and in the Dialogue of 1632 (3rd Day), he states that he first observed the spots while still in Padua, and that he showed them to some friends. This would take the date back to the summer of 1610. The claim is supported by Fulgenzio Micanzio and Viviani. Galileo explains his

then proceeds to combat Scheiner's various and often contradictory assertions as to the nature of the spots, and their places and movements in relation to the body of the sun.

'I would say', he writes, 'that they are formed on the sun's superficies, that they are carried round with him in his rotation, remaining visible for about one-half month; and that they may be something of the nature of our own clouds. Certainly, if our earth were self-luminous and surrounded by clouds, it would seem to a far-distant observer to have spots like those we see on the sun, now uniting, now separating, and now dissolving. They would follow her in her rotation, appearing very large at the centre of her disk, where their motion would be most rapid, and contracting towards the edges, where they would be smallest, and where also their velocity would be least.'

This he recognized as an effect of fore-shortening which would result if, and only if, the spots were on or very near the sun.

In the second letter he restates his views, adding some further particulars as to the constant, slow, and irregular changes in the form of the spots, and their varying density, being very dark at the centre, and less so towards the circumference—'a manifest proof of the sun's sphericity'. They are confined to a zone about the sun's equator, extending 28 or 29 degrees (in his third letter he says 29 or 30) on each side, beyond which they are never seen; and, finally, they all have a common motion of rotation. From all these facts, and from the additional one, that often the same spot disappearing at one side reappears at the other, he concludes that the sun is a sphere, that it rotates on its axis from west to east, and that it performs one such rotation in about a lunar month. In an appendix he gives forty sketches of spots as observed from day to day during June, July, and August, 1612.

In his third letter, dated 1st December, 1612, Galileo notices some observations of Scheiner on Venus and the moon, and shows the falsity of his 'facts' and deductions; he recurs to the sun-spots and adduces a further proof of the correctness of his own hypothesis in the behaviour of some very bright spots (*piazzette*, now called *faculae*). Some parts of the sun's disk are perceived

silence as to these earlier observations thus: 'Having regard to the extraordinary nature of the phenomena, so contrary to the received opinions, I judged it more prudent to wait until I had convincing proof. I prefer to be the last to produce a true conception than to anticipate others at the risk of having to unsay what I was in a hurry to affirm' (Nat. Ed., vol. v, p. 94).

to be brighter than the rest, and these parts appear to traverse the disk just as the other spots do. Now, were these very bright spots planets, as Scheiner would have it, they ought sometimes to appear beyond the sun's limb, but this they never do—an irrefragable proof that they are part and parcel of the sun himself.

After referring to various subjects, as the inhabitability of the planets, the supposed crystalline and transparent substance of the moon, the diversity of figure amongst the planets, and the periods of Jupiter's satellites (of which Scheiner had recently 'discovered' a fifth), he returns once more to the sun-spots and their general resemblance to clouds or smoke. We can, he says, imitate them in various ways, as, for instance, by dropping on a red-hot iron plate bits of bitumen. He supposes that the sun's light (and heat) may be sustained by a constant supply of new *pabulum*, which, like the bitumen, first gives off black smoke, which we see as spots. In a later letter, 23rd March, 1615, to Piero Dini, he refers to this idea. 'I suggested', he says, 'that these spots might be part of that pabulum (or rather the *débris* of it), of which, according to certain ancient philosophers, the sun has need for his sustentation.' [1]

These letters were ultimately published at Rome in 1613, at the expense of the Accademia dei Lincei, and under the title *Istoria e dimostrazioni intorno alle macchie solari.*[2]

15. *On Lunar Mountains* [3]

Soon after his return to Florence in June, 1611, Galileo wrote a series of letters on *The Inequalities of the Moon's Surface,* in defence of the views expressed in his *Sidereus Nuncius.* The moon was with him a stock subject for observation, the results of which he utilized in his astronomical works, or communicated in long letters to friends, notably to Griemberger and Gallanzoni in Rome, and to Welser and Bernegger in Germany. His last astronomical

[1] Newton and Buffon conjectured that comets might be the aliment of the sun, and, at present, a nearly similar explanation finds favour, viz. that streams of meteoric matter, varying in volume, are constantly pouring into the sun from the regions of space. Professor Turner of Oxford is the latest exponent of this hypothesis. See his paper in *Monthly Notices of R.A.S.*, December, 1913. Cf. Mayer, *Beiträge zur Dynamik des Himmels*, Heilbronn, 1848.

[2] Nat. Ed., vol. v, pp. 75–260.

[3] Cf. Nat. Ed., vol. iii, pp. 301, 313 ; x, pp. 461, 466 ; xi, pp. 141, 178.

discovery, towards the close of life and just before he became blind, was connected with the moon.

It had been asserted that, as the full moon always presented a well-defined outline, whether viewed with the naked eye or through a telescope, it was impossible that there could exist any inequalities around her circumference. Galileo maintained that the irradiation of the moon's light might be great enough to mask the asperities around her edge, and so effectually conceal the real nature of her surface. With respect to irradiation generally, he remarked that it increases with the brightness of the object. It is from this cause that the planets near the sun have a greater irradiation than those more remote. So intense is the irradiation of Mercury that it is impossible, even with the most powerful telescope, to deprive him of his brilliant corona. The same is true, though in a less degree, with respect to Mars. On the other hand, Jupiter, and especially Saturn, being more feebly illuminated by the solar light, lose their irradiation in the telescope, and disclose their true figures.

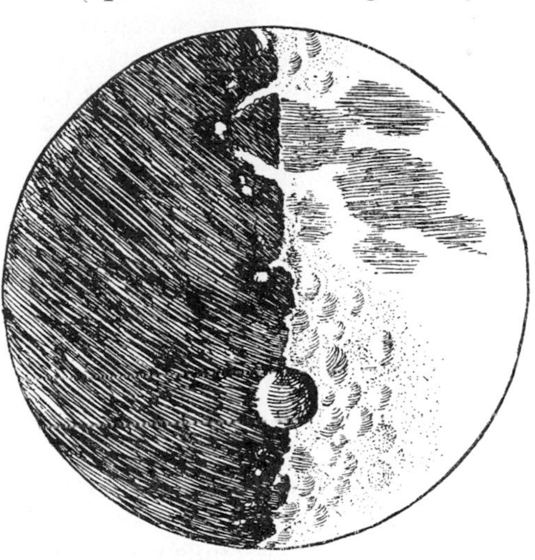

FIG. 9. The moon as seen by Galileo, 1609–10. From *Sidereus Nuncius.*

With respect to Venus, when she is near her inferior conjunction, she in reality resembles the new moon; but such is the effect of her irradiation that she appears to the naked eye round like any other star. In this position, as the extent of the illuminated surface is small and the light is at the same time enfeebled by the obliquity of the surface, it is possible by means of a telescope to discern the real crescent appearance of the planet. When, however, she is near her superior conjunction, she presents a complete hemisphere of vivid light towards the earth of such intensity that even the most perfect telescope does not reveal to us her true figure. Galileo therefore contends that it is probable that even the telescope will fail to efface the irradiation of the moon enough to disclose the eminences and cavities which may be situated near the edge of her disk.

His peripatetic opponents next tried to reconcile the old doctrine of a polished and perfectly plane surface with these new observations. Father Clavio doubted at first the reality of the inequalities, and thought that the appearances were due to inequalities in the reflecting power of the moon's substance. Other Aristotelians, as delle Colombe and Lagalla,[1] supposed that every part of the moon, which to us appears hollow, is, in fact, filled with clear crystal matter, thus preserving a round and smooth superficies, and it is this diversity of *substance*, with its more or less transparency, which gives the impression of inequality of *form*.

16. *Discussion of Habitability of Moon and Planets* [2]

Among the many burning questions to which the *Sidereus Nuncius* gave rise was that of whether the moon and planets were inhabited. This was openly discussed in Rome from 1611 onwards, and its ' manifest absurdity ' was used as an argument against the Copernican theory in general and Galileo's lunar observations in particular. If the moon, it was said, is so like the earth with land and water, mountains and valleys, and surrounded by an atmosphere, we may suppose that she too is the home of beings like ourselves. Again, if our earth be not the centre of the universe, but one of a number of planets, and a small one at that, then the other planets are inhabited like ours. The arguments thus resolved themselves into the syllogism—all planets are alike, the earth is a planet and is inhabited, *ergo* all planets are inhabited. The peripatetic philosopher, Lagalla, maintained this thesis in a public discourse in Rome in 1612. The same was gravely adduced by Scheiner in one of the numerous digressions in his letters on Sun-spots, and he insinuates that Galileo must hold the belief as a necessary consequence of his observations.

Yet far from admitting this view, Galileo took pains to show its impossibility. In his third letter to Welser on Sun-spots, he points out that for fifteen days continuously the moon is exposed to the scorching sun-rays, and for another fifteen consecutive days is

[1] Nat. Ed.. vol. iii, Part I.

[2] Cf. Nat. Ed., vol. xii, p. 240.

plunged in cold and darkness. A day of glare and heat equal to fifteen of ours, and a night of cold and darkness of equal length were, he pointed out, impossible conditions for life such as ours.[1]

17. *On Finding the Longitude at Sea*[2]

Very soon after he had discovered the satellites of Jupiter in 1610, Galileo began a work the difficulty and fatigue of which he has himself indicated by comparing it with the labours of Atlas. It was a series of observations on the periods of the satellites, 'with a view to drawing up tables so as to be able to predict all particulars of their situations, relations, and eclipses, and thus to have the means of determining at any hour of the night the longitude of the place of observation'. Kepler thought this enterprise so difficult as to be wellnigh impossible, and certainly Galileo did not find it easy. Notwithstanding many hundreds of observations in the next twelve months, repeated often twice and sometimes three times in a night, he had made little or no progress up to April, 1611. It was not until another year and more had elapsed that he was able to announce satisfactory results in his 'Discourse on Floating Bodies'. To show their close approximation, we give them here side by side with the modern figures.

PERIODS OF REVOLUTION

	Galileo.	*Modern.*
Innermost satellite . .	1 day 18 hrs. 30 mins.	1 day 18 hrs. 29 mins.
Second ,, . .	3 days 13 ,, 20 ,,	3 days 13 ,, 18 ,,
Third ,, . .	7 ,, 4 ,, 0 ,,	7 ,, 4 ,, 0 ,,
Fourth ,, . .	16 ,, 18 ,, 0 ,,	16 ,, 18 ,, 5· ,,

Our moon had already been suggested for the same purpose: it changes its position amongst the stars continuously, and, if at specified times throughout the night that position can be predicted, the mariner is able to determine his longitude. But, at the beginning of the seventeenth century, tables of the moon's positions were very inaccurate; and even its proximity to the earth was a disadvantage, for an observer at sea would get

[1] Cf. his letter to Giacomo Muti, February 28, 1616; Dialogue of 1632, first Day.

[2] Cf. Nat. Ed., vol. iii, Part II, *passim*; vol. v, pp. 415–25; vol. viii, p. 451; vol. xi, p. 321; vol. xii, pp. 256, 289, 311, 358, 392; vol. xiii, pp. 17, 370; vol. xiv, pp. 53, 91, 202, 349, 374.

a different view of it from one on land, and this difference in the moon's position amongst the stars would have to be allowed for. The errors due to these defects would, it was thought, be avoided if Jupiter's satellites were used instead of the moon. The much greater distance, the frequency of their eclipses (more than 1,000 yearly), and (it was expected) their suddenness seemed to promise success to Galileo's method. But in practice difficulties cropped up. First, there was the difficulty of observing such small objects as the satellites from a moving ship, and secondly, there was the want, common to both methods, of accurate time-keepers. To obviate the first he contrived what he called the *Celatone* or *Testiera*:

'I made', he says, 'for the use of our navy a kind of cap, fitted to the head of the observer, and supporting a telescope in such a way that it always points in the same direction as the free eye, so that an object viewed by the latter is also seen by the other eye through the telescope. A similar apparatus could be made and fixed on the shoulders and chest of the observer, to support a telescope of a power sufficient to show the satellites of Jupiter, and adjustable as in the case of the *Celatone*. When, then, the free eye is turned towards Jupiter, the other eye sees through the telescope not only the planet but its satellites.'[1]

With this contrivance and a chair for the observer, hung like a ship's compass on a binnacle, he hoped to overcome the first difficulty of unsteadiness; while to remedy the second he had hopes of utilizing his early observations on the pendulum and applying it as an exact measurer of time.

In September, 1612, Galileo offered his method to the Spanish Government for use in their navy, but the proposal was not well received, and for the next four years he took no further steps in the matter. Now, during his visit to Rome in 1616, he reopened the negotiations through Count di Lemos, the Spanish Viceroy of Naples. Di Lemos was fully alive to the importance of the proposal, and promised to submit it to his Government. This was done, after much unaccountable delay, in March, 1617, and, in the same leisurely way, the proposal was discussed by the King in Council. To the various objections advanced Galileo replied, and even offered to go himself to Spain, but he could not, with all his enthusiasm, bring the Spanish Court to a decision. His

[1] Letter to Lorenzo Realio, June 6, 1637. From this it is clear that Galileo did *not* propose a *binocular* telescope as has sometimes been supposed.

disappointment was mitigated by his own sovereign taking up the method for use in the Tuscan navy. Its practical application, however, proved to be beset with so many difficulties that it soon fell into disuse.[1]

18. *On Floating Bodies* [2]

During the summer of 1611 the subject of floating bodies had been debated at one of the scientific parties which the Grand Duke liked to assemble round him. The general opinion was that of Aristotle, that the sinking or floating of a body in water depended upon its shape. Galileo undertook to show this view to be untenable,[3] and he embodied his arguments in a famous treatise, published in Florence in 1612, ‘ Discorso intorno alle cose che stanno in su l'Acqua, o che in quella si muovono ’. In this work he restores the true principles of hydrostatics as laid down by Archimedes, alludes to the so-called hydrostatical paradox, first noticed by his contemporary Stevin of Bruges, and explains it on the principle of virtual velocities, as first clearly enunciated by himself in his treatise on the mechanical powers in 1594.

In the course of the discussion it was asserted that condensation is the effect of cold, and ice was mentioned as an example. Galileo retorted that ice is rather water rarefied than water condensed, the proof of which is that it always floats upon water. His opponents rejoined that this was due, not to the lightness of the ice, but to its incapacity, owing to its flat shape, to overcome the resistance which the water opposed to its sinking. Galileo denied this, and asserted that ice of any shape would float, and that if a flat piece were forced to the bottom it would, when left to itself, rise again to the surface.

The behaviour of ebony was then instanced, which in the shape

[1] In August, 1636, Galileo offered his method to the States General of Holland, but here again its practicability was questioned and for the same reasons as above. The negotiations dragged on wearily, and by the middle of 1640 came to an infructuous end, except for a collar of gold, as a mark of the Dutch Government's esteem, which Galileo refused, or, as is more likely, was not allowed by the Inquisition to accept.

[2] Cf. Nat. Ed., vol. iv, *passim* ; vol. xi, pp. 176, 304, 317.

[3] Cardinals Gonzaga and Maffeo Barberini (afterwards Pope Urban VIII) were among the guests, and the latter took Galileo's side in the discussion against the peripatetics led by Gonzaga.

of a ball sinks, but as a thin board floats when gently placed on the surface. To this he replied :

'The diversity of figure given to any solid cannot be the cause of its floating or sinking in water, though the breadth of the figure may indeed retard its velocity as well of ascent as of descent, and more and more in proportion to the breadth and thinness. If you examine carefully your thin boards of wood you will see that they have part of their thickness under water ; and, moreover, you will see that shavings of ebony, stone, and metal, when they float, have not only broken the continuity of the water, but are *with all their thickness* under the surface, and this more and more according to their specific gravity.'

To show more clearly the non-resistance of water to penetration, he directs a cone to be made of wood or wax, and asserts that when it floats, either with its base or its apex in the water, the solid content of the part immersed will be the same, although the apex is by reason of its shape better adapted to overcome the resistance of the water to division. Shape, then, cannot be the cause of the buoyancy. He goes on :

'Now, let us return to the thin plate of gold or silver, or the thin board of ebony, and lay it lightly upon the water, so that it may stay there without sinking, and observe the effect. It will be seen that the board or plate is lower than the surface of the water, which rises up and makes a kind of rampart round it. But if it have already penetrated the water, and is of its own nature heavier than the water, why does it not continue to sink ? My answer is, because in sinking till its surface is below that of the water, it carries with it the air, so that that which descends is not merely the board of ebony, but a compound of ebony and air, from which results a body no longer specifically heavier than the water, as was the ebony alone. But, gentlemen, we want the same matter ; you are to alter nothing but the shape, and therefore have the goodness to remove this air. This may be done by washing the upper surface of the board, for the water, having once got between the board and the air, will run together, and the ebony will sink to the bottom. To demonstrate how truly the air does support these solids, I have found that when one of these bodies (which float when placed lightly on the water) is thoroughly bathed and sunk to the bottom, by carrying down to it a little air, without otherwise touching it in the least, I am able to raise and carry it back to the top, where it floats as before. To effect this I take a ball of wax, and with a little lead make it just heavy enough to sink slowly to the bottom, taking care that its surface is quite smooth and even. This, if put gently into the water, submerges almost entirely, there remaining outside only a very little of the top, and, so long as it is thus joined to the air, the ball floats; but if we take away the air, by wetting this top, the ball sinks to the bottom. . . . There is, therefore, a certain affinity between air and other bodies, which holds them united, so that they separate not without a kind of violence, just as between water and other bodies, for, in drawing such bodies wholly out of the water, we see it follow them, and rise sensibly above its level before it quits them.'

There is a confusion here between the phenomena of hydrostatic pressure and of capillary attraction or surface tension; and Galileo would, perhaps, have carried conviction more readily had he realized himself that the floating plate of metal indicated a natural property of liquids which deserved special investigation.

This book, like all his other works, encountered violent opposition from the 'book philosophers'; and it was in reference to this controversy that we have one of his fine *obiter dicta*. 'Ignorance', he said, 'had been the best teacher he ever had, since, in order to be able to demonstrate to his opponents the truth of his conclusions, he had been forced to prove them by a variety of experiments, though to satisfy his own mind alone he had never felt it necessary to make many.'

19. *First Encounter with the Inquisition* [1]

The uncompromising boldness with which Galileo published and supported his opinions, as we have seen, raised crowds of enemies against him. The Aristotelian professors, the Jesuits, the political churchmen, and those timid and respectable persons who at all times dread innovation, were drawn together against the man who threatened them with the penalties of too much knowledge. No longer able to combat his observations and deductions by asserting that the former were due to faults in his glasses or to apparatus 'devilishly designed to produce them', and that the latter were vainglorious and philosophically absurd, his enemies now took their stand on theology. After some months of underground agitation, Father Caccini, of the Dominican convent of San Marco, was the first to declare war openly, in a sermon from the pulpit of Santa Maria Novella in Florence. Preaching on the fourth Sunday in Advent (December 21, 1614), and selecting as his text Joshua x. 12, 13, and Acts i. 11, he opened with the words:

[1] Cf. Nat. Ed. vol. v, pp. 264, 281, 291, 309, 351 : vol. xii, pp. 123, 183, 244, 277 ; vol. xix, pp. 272–421.

'Viri Galilaei, quid statis aspicientes in caelum?' Galileo explained and defended his position in long letters to Castelli, to Piero Dini, and to the Grand Duchess Cristina (December, 1614, to June, 1615), which together constitute a powerful *Apologia.*[1]

The pith of his argument is contained in the saying of Cardinal Baronius, which he quotes—'The Holy Spirit intended to teach us in the Bible how to go to heaven, not how the heavens go'. One or two passages from his letter to the Grand Duchess may be quoted:

'Methinks that in the discussion of natural problems we ought not to begin with the authority of passages from Scripture, but with sensible experiments and necessary demonstrations. . . . Nature being inexorable, acting only through immutable laws which she never transgresses, and caring nothing whether her reasons and methods of operating be or be not understandable by men, I hold that our conception of her works, which either sensible experience sets before our eyes, or necessary demonstrations prove, ought not to be called in question—much less condemned upon the testimony of Scriptural texts which may conceal under their words senses or meanings seemingly opposite. . . . To command professors of astronomy that they see to confuting their own observations and demonstrations is to ask the impossible. . . . As to opinions which are not directly articles of faith, certainly, no man doubts that his Holiness hath always an absolute power of admitting or condemning them; but it is not in the power of any creature to make them to be true or false otherwise than as they are.'

All through the year 1615 the agitation went on with unabated violence, denunciations were sent to the Holy Office, and the Inquisition began to make secret inquiries. At length the situation became so threatening, not only for our philosopher himself, but for science in general and the Copernican theory in particular, that Galileo, with the advice of friends, decided to take himself to Rome. Accordingly, on December 3, 1615, he set out, provided with cordial letters from the Grand Duke to his Ambassador Guicciardini, to Cardinals del Monte and Orsini, and others. After many weeks of alternating hopes and fears, the matter came officially before the Inquisition, with the result that on February 26, 1616, Cardinal Bellarmine, who had sat on the commission that investigated his discoveries on his former visit to Rome, was directed 'to summon before him the said Galileo, and admonish him to abandon the said opinions, and, in case of

[1] Cf. his letter to Francesco Ingoli, Nat. Ed., vol. vi, pp. 504–61.

refusal, the Commissary is to intimate to him, before a notary and witnesses, a command to abstain altogether from teaching or defending the said opinions, and even from discussing them. If he do not acquiesce therein, he is to be imprisoned '.

Here the process ended so far as Galileo was concerned, but on March 5 a decree was issued prohibiting and suspending certain writings. Amongst them the book of Copernicus (*Revolutions of the Celestial Orbs,* first published 1543) was ordered to ' be suspended until corrected '. Galileo appears to have taken the admonition with a very bad grace, if we are to believe Guicciardini, who was no friend of his. Writing to the Grand Duke on May 13, 1616, he says :

' Galileo seems disposed to emulate the monks in obstinacy. . . . It may be heard at any moment that he has stumbled into some new abyss or other. However, the heat will probably drive him from Rome before long, and that will be the best thing that can happen to him.' [1]

20. *The Tides* [2]

Yet amidst all these cares and worries Galileo's teeming mind was busy with many scientific problems. He had not been many days in Rome when a suggestion from Cardinal Orsini was enough to start him on a treatise on the Flux and Reflux of the Sea. This problem had from the earliest ages deserved its name—' The grave of human curiosity '. Some supposed the rise of the waters to be due to the influx of rivers ; others supposed the existence of subterraneous fires which periodically made the sea to boil up ; while others again attributed this boiling effect to changes of temperature in the sun and moon, or to variations in the amount of their light.

The ancient philosophers had vague ideas that the moon's attraction was the cause. ' The flow ', says Pliny, ' takes place every day at a different hour, according to the rising of the moon, which, with greedy draught, drags the seas along with it.' In modern times, Gilbert of Colchester had speculated on this connexion. ' There are ', he says, ' two primary causes of the motion of the seas—the moon and the diurnal revolution of the earth.

[1] Cf. letter from Mgr. Antonio Querengo to Cardinal d'Este, January 20, 1616 ; and, on the other side, Cardinal del Monte's letter to the Grand Duke, June 4, 1616.

[2] Cf. Nat. Ed., vol. v, pp. 373–95.

The moon does not act on the seas by its heat rays, or by its light. How, then? Certainly by the common or mutual effort of the bodies, or, to explain it by something similar, by their magnetic attraction.' The Jesuits of the celebrated college of Coimbra, as well as Marc Antonio de Dominis, once Catholic Archbishop of Spalatro, later Protestant Canon of Windsor, and finally himself a martyr of the Inquisition, and Kepler, the law-giver of the planets, all held much the same views. Kepler's words are worth quoting, as they embody his ideas of that universally mutual gravitation, which Borelli, Wallis, and Hooke after him saw more clearly, and which it was the glory of Newton to establish:

'Gravity is a mutual affection between cognate bodies towards union, similar in kind to the magnetic virtue, so that the earth attracts a stone more than the stone attracts the earth. Assuming the earth to be the centre of the world, then, wherever it may be, or wheresoever it may be carried by its animal faculty, heavy bodies will always incline towards its centre. If two stones be placed in any part of the universe near each other, and beyond the influence of a third cognate body, they, like two magnetic needles, will come together at an intermediate point, each approaching the other by a space proportionate to its mass. If the moon and the earth were not retained in their orbits by their animal force, or some other equivalent, the earth would mount towards the moon by one fifty-fourth part of the distance between them, and the moon would fall towards the earth by the other fifty-three parts, i.e. in the inverse ratio of their masses, and assuming their substances to be of the same density. If the earth should cease to attract its waters to itself, all the waters of the sea would be raised and flow towards the moon. This attractive virtue of the moon extends as far as the earth, and entices its waters, but, as she flies rapidly across the zenith, and the waters cannot follow so quickly, a flow of the ocean is occasioned in the torrid zone and towards the west.' [1]

Galileo's theory is that the tides are the visible effects of the terrestrial double movement, since they are the combined result of (1) the earth's daily rotation on its axis, and (2) the inequality of the absolute velocities of the various parts of the earth's surface in her revolution round the sun. In the whole range of the sciences over which Galileo left indelible marks of his genius, he

[1] *Astronomia Nova*, Prague, 1609. Ten years later Kepler abandoned these correct ideas, and depicted the earth in his *Harmonice Mundi* as a living monster whose whale-like mode of breathing occasioned the rise and fall of the oceans in recurring periods of sleeping and waking dependent on solar time.

made very few fundamental errors, and this is perhaps his most serious one.[1]

21. *On Comets and 'Il Saggiatore'* [2]

In August, 1618, three comets appeared, and the very brilliant one in the constellation of the Scorpion—one of the most splendid of modern times—especially attracted the attention of astronomers. Although this comet was visible until January, 1619, Galileo had little opportunity of observing it, as he was confined to bed nearly the whole time. However, on a small basis of observation he reflected much and imparted his views to his friends, amongst others to Mario Guiducci, a Florentine disciple. In May of the same year, 1619, Guiducci, in a Presidential address to the Accademia Florentina, gave the views of the Master—'not as demonstrative truth, but as plausible conjectures in a matter so abstruse'.

This address was published immediately under the title *Discorso delle Comete* (Florence, 1619), and from it we learn that Galileo did not regard comets as heavenly bodies analogous to the planets, but as atmospheric phenomena—columns of vapour which, rising from the earth to great heights, far beyond the moon (*spazi celesti*), became there temporarily visible by reflection of the sun's light. In fact, he classed them in the same category as rainbows and mock suns, thus for once agreeing with Aristotle, and opposing himself to the more correct views of his contemporaries, Tycho Brahé and Kepler.[3] Referring to some proposed observations for parallax he pointed out the difference in this respect between a fixed object, the distance of which may be calculated by two angular observations at a known distance apart, and atmospheric appearances like rainbows, which are simultaneously formed in different drops of water for each spectator, so that two observers in different places are, in fact, viewing different objects. He then

[1] Between 1630 and 1637 Galileo would seem to have changed his view and suggested that the moon's librations may be the cause of the tides—'which by the common consent of all philosophers are ruled by the moon'. See his letters to Fulgenzio Micanzio, 7th November, 1637, and to Alfonzo Antonini, 20th February, 1638.

[2] Cf. Nat. Ed., vol. vi, *passim*; vol. xi, p. 41; vol. xii, pp. 466, 494; vol. xiii, pp. 43, 46, 80, 90, 98, 106, 116, 142; vol. xviii, p. 423.

[3] Brahé thought that comets were the result of sudden condensations of the ether of space, while Kepler accounted them to be exhalations of the planets. Santucci held that they were produced in the heavens by the sun. See his *Trattato nuovo delle Comete*, Fiorenza, 1611. In later years, Galileo changed his opinion as to their 'probably terrestrial' origin.

warns astronomers not to engage in a discussion on the distance of comets before they assure themselves to which of these two classes of phenomena they are to be referred. The remark is in itself perfectly just, although the opinion which occasioned it is now known to be erroneous; but it is questionable whether the few observations which up to that time had been made upon comets were sufficient to justify the censures which have been cast on Galileo on account of it.

In the course of Guiducci's essay, some opinions of the Jesuit Father, Orazio Grassi, were so indiscreetly handled as to raise the ire of the Jesuits' College at Rome.[1] Grassi, under the pseudonym of Lotario Sarsi, published an onslaught on Galileo's cometary theory in a book called *The Astronomical and Philosophical Balance* (1619)—a violent pamphlet full of abuse of Galileo and his school. Friends, like Prince Cesi and Mgri. Ciampoli and Cesarini, now advised that the master himself should take up the fight; but ill health and the troubled state of the religious and political horizons prevented the appearance of his reply for four years. At length, on October 19, 1622, he sent the manuscript of *Il Saggiatore* (The Assayer) to Mgr. Cesarini in Rome, and for five months it passed from hand to hand among the members of the Accademia dei Lincei, who examined it carefully and (with the author's consent) altered some passages which might possibly have given a handle to his enemies. The Papal *Imprimatur* was granted on February 2, 1623, and the book appeared at the end of October with a dedication to Pope Urban VIII, and under the auspices of the Accademia dei Lincei. This celebrated work is a masterpiece of dialectics, for the author not only dexterously avoids the snares laid for him by Father Grassi and his abettors, but brings defeat and ridicule upon them at every turn. He takes, in order, the mistakes of his adversary, his false citations and false deductions, his errors of logic, of geometry, of physics, of astronomy, and exposes them all; and this is done so courteously and in such sparkling style that *Il Saggiatore* deserves its reputation as a model of dialectic skill, and an ornament of classical Italian literature.

The book (Fig. 10) was a great success, but it intensified the

[1] Soon after the appearance of the comets, a discussion upon them took place in the Collegio Romano. It was published early in 1619 under the title *De Tribus Cometis anni 1618: Disputatio Astronomica*, &c. It is interesting to note that Mazarin, then a boy of sixteen, took part in this discussion.

bitterness of the Jesuitical party, and the General of the Order forbade the members to speak of it, even among themselves. It is important to note that the Pope was delighted with it, and had it read aloud to him at table. Early in 1625 the book was denounced anonymously to the Inquisition as a veiled defence of the Copernican doctrines, and a movement was begun to have it prohibited, or, at least, 'corrected'. This attempt, however, failed, and brought only further discredit upon the agitators.

IV. The Trial and Abjuration

1. *Galileo's Plea for Copernicanism*

On the election of Cardinal Maffeo Barberini to the Papacy as Urban VIII, August 8, 1623, Galileo conceived the idea of going to Rome to offer his congratulations in person, and to use his influence with the new Pope to obtain, at least, toleration for the Copernican doctrines, now no longer subject to the weighty opposition of Cardinal Bellarmine, who died two years before. Remembering the warmth of Barberini's regards for him while Cardinal, Galileo had much to hope from a Pontiff so enlightened. He was encouraged, moreover, by hopeful reports from friends in Rome. Prince Cesi, writing October 21, 1623, was able to tell him: 'Under the auspices of this most excellent, learned, and benignant pontiff science must flourish. . . . Your arrival will be welcome to his Holiness. He asked me if you were coming and when, and, in short, seems to love and esteem you more than ever.' Rinuccini, Cesarini, Ciampoli, and others wrote in the same strain. Ciampoli, the Pope's private secretary, wrote on March 16, 1624: 'It is certain that the longer your coming is deferred the more it is desired by all those gentlemen who esteem you, and keep you green in their memory. You will find in the Holy Father no ordinary affection towards your person.' Ill-health, however, and then the bad weather and worse roads, intervened, but, at length, Galileo set out for the Eternal City on April 1, 1624.

All Rome was aware of the favour in which the Pope held Galileo, and his letters express great satisfaction with his reception, but as regarded the object which was nearest his heart, he made no progress. Within six weeks he had had six long interviews with Urban VIII, who always received him most affably, and allowed him to bring forward all his arguments in support of the Copernican theory; but all to no purpose; for while the Pope listened to his arguments, he would not grant his entreaties for even a passive toleration of the new doctrines.

Finding that his efforts to get the decree of March 5, 1616, revoked were of no avail, Galileo resolved with a heavy heart to return home, though the Pope had loaded him with favours which must have seemed like mockeries. His Holiness promised him a pension for his son, and sent a picture for himself; then two medals—one of gold and one of silver, and quite a number of Agnus Dei! Not content with these marks of favour, he addressed an official letter to the Grand Duke, on June 8, in which, to the no small chagrin of Galileo's enemies, his Holiness not only did full justice to our philosopher's services to science, 'the fame of which will shine on earth so long as Jupiter and his satellites shine in heaven', but laid special stress on his religious sentiments. Yet Galileo departed with the object of his visit unsatisfied.

2. *Dialogue on the Two Chief Systems of the World, the Ptolemaic, and the Copernican* [1]

Nevertheless, from various indications in the ecclesiastical world in the next two years, 1624–6, Galileo was led to think that the advocates of Copernicanism had now little to fear, provided that the defence was so circumspectly handled as not to outrage the Inquisition's decree of March 5, 1616, which condemned the doctrine not as 'heretical', but only as 'rash'.[2] He resolved to push on to completion a work he had contemplated as far back as 1610. This was to be entitled *A Dialogue on the Flux and Reflux of the Tides.* From 1626 to 1630 he was almost entirely engaged on this work, but was interrupted by frequently recurring illnesses and family troubles.[3]

At length, by the beginning of 1630, he had practically completed this Dialogue, and in announcing the fact to his friend

[1] Cf. Nat. Ed., vol. vii; vol. xiii, pp. 104, 236, 260–4, 365; vol. xiv, pp: 49, 64–70, 79, 97, 120, 150–67, 278–85, 331; vol. xix, pp. 327–30.

[2] Indeed it was known that Urban VIII, as Cardinal, did not approve of that decree. Early in 1630, when discussing it with Campanella he said: It was never our intention and, if it depended upon us, that decree would not have been issued' (Nat. Ed., vol. xiv, p. 88).

[3] About this period he was often consulted, with others, on hydraulic questions connected with the flooding of Tuscan rivers (Nat. Ed., vol. vi, p. 613). Cf. Cambiagi, *Raccolta d'Autori che trattano del moto dell' Acqua*, Firenze, 1765–74; Napier, *Florentine History*, London, 1847.

Prince Cesi, he expressed his intention of going himself to see to the printing at Rome, where the state of affairs seemed still favourable for this enterprise. Galileo's disciple, Castelli, had been called from Pisa in March, 1626, to be mathematician to the Pope, and enjoyed great consideration with all the members of the Barberini family. This life-long friend, like Cesi, approved the design, and informed our philosopher (February 9) that Father Niccolo Riccardi, another old pupil and now chief censor of the press, had promised his assistance. Filled with hope and with his manuscript complete, Galileo at length set out on May 1 in a Court litter, and travelling fast arrived in Rome on the evening of the 3rd. He had a long audience of the Pope, and wrote on May 18 to Florence in high spirits: 'His Holiness has begun to treat my affairs in a way that permits me to hope for a favourable result'. But Galileo was far too sanguine, for toleration, to say nothing of the recognition, of the Copernican theory was as far off as ever. Urban VIII would not object to the publication, but certain conditions must be fulfilled. The title of the book, *Dialogue on the Flux and Reflux of the Tides*, was misleading and must be altered. The subject, being a discussion of the relative merits of the Copernican and Ptolemaic systems, should be indicated in the title. The subject, moreover, would have to be treated from a purely hypothetical standpoint, and this fact must be clearly set forth in a preface. Then, the book must conclude with an argument which the Pope communicated to Galileo in 1624, and which his Holiness considered unanswerable. That argument was as follows: 'God is all-powerful; all things are therefore possible to Him; *ergo* the tides cannot be adduced as a necessary proof of the double motion of the earth without limiting God's omnipotence.' Rather than forgo the publication of a work towards which he had laboured and thought for over thirty years, Galileo consented to these conditions.

Meanwhile the manuscript had been submitted to Father Riccardi, who, after certain revisions and alterations, granted permission for the printing in Rome. Thus, by the end of June, 1630, Galileo was back in Florence with his manuscript revised and corrected, and with the ecclesiastical *Imprimatur* for its publication in Rome, on the understanding that a preface and conclusion would be added in accordance with the Papal wish. Publication seemed imminent, yet for another twenty months our author was tormented with obstacles and delays on the part of the censors, now in Rome, now in Florence. It would be little profitable to set out in detail these quibbling complications; but the work was at last issued from the Florentine press in February, 1632. It was in Italian, was dedicated to Ferdinando II of Tuscany, and bore the title, 'Dialogue of Galileo Galilei, Lyncean, Mathematician Extraordinary of the University of Pisa, Philosopher and First Mathematician of the Most Serene Grand Duke of Tuscany; where in meetings of four Days are discussed the Two Chief Systems of the World, indeterminately proposing the Philosophical and Natural arguments, as well on one side as on the other'.

The dialogue is carried on by three interlocutors, of whom two adduce the scientific reasons for the double motion of the earth, while the third tries to defend the opinions of the Ptolemaic and Aristotelian schools. Galileo gave to the defenders of the Copernican doctrine the names of two of his warmest friends, both long dead—Filippo Salviati of Florence (d. 1614), and Giovanni Francesco Sagredo of Venice (d. 1620). Salviati is the special advocate of the Copernican doctrine; Sagredo is witty, impartial, and open to conviction. To him are allotted such objections as have real force, as well as lively illustrations and digressions, which would be inconsistent with the gravity of Salviati's character. Simplicio, a name borrowed from the noted Cilician commentator of Aristotle who wrote in the sixth century, is a confirmed Ptolemaist and Aristotelian, and produces, as occasion requires, all the arguments of the peripatetic school; and as these fail to convince he has recourse to all the arts of sophistry.

The condition that the Copernican doctrine is to be treated as an hypothesis is ostensibly complied with. If Salviati or Sagredo show the untenableness of some Ptolemaic axiom, or add a stone to the Copernican structure, a remark is interpolated by one or other to weaken the effect. When we remember its history we cannot be surprised that the preface or introduction has no logical agreement with the contents of the Dialogue. The conclusion agrees no better than the preface with the body of the work. At the end of the fourth Day, which is almost wholly taken up with the tides, comes naturally the Pope's 'unanswerable' argument of 1624. Salviati treats it accordingly: 'It is', he says, 'an admirable and truly angelic argument, and perfectly in accord with that which, coming from God Himself, permits us to discuss the constitution of the world—doubtless with the view of preventing by exercise the diminution and enfeeblement of our intellectual faculties, while withholding from us the power of fully comprehending the works of His hands.'

The contents of the Dialogue can be given only in outline. Salviati opens by defining its object, which is to examine all the physical arguments evoked for and against their opinions by the defenders of Aristotle and Ptolemy on the one hand, and of Copernicus on the other. In a few words, the Aristotelian doctrines amounted to a statement that, whereas things earthly are imperfect and full of change, things heavenly are eternal, unchangeable, and perfect. Salviati proves that this statement, in the spirit in which it was usual to accept it, is untenable. The telescope shows imperfections on the sun's surface, while the recently observed new stars (of 1572 and 1604) are instances of change in the heavens. He thus prepares the way for a still wider departure from Aristotelian theory; he insists that the time has come to consider the nature of the world *de novo*, suggesting that Aristotle, had he the opportunities which the telescope afforded, would himself have realized the inadequacy of his own teaching.

Salviati proceeds to point out certain resemblances between the earth and moon and the more distant heavenly bodies. It was admitted that the moon shines only in virtue of the sunlight falling on her. The idea that the earth might similarly appear luminous to an inhabitant of the moon is less familiar and less readily accepted. And yet the visibility of the moon during a total eclipse of the sun, and the appearance 'of the old moon in the arms of the new' (as we now speak of it), are due most probably to reflected earth-light. The phenomena of Venus's phases are shown to be similar to those of the moon, and may be explained as due to the same cause. Venus, then, like the moon, owes her brilliance to the sun's light falling on her. The same probably applies to Mercury and Mars. The obvious inference seems to be that all these heavenly bodies are not so unlike the earth as men had thought. Points of resemblance there certainly are, and there may be many more which the distance of the planets alone prevents us from discovering. Salviati then refers to the spherical form common to earth, sun, moon, and planets, and suggests the existence of a common cause for that shape.

Here follow some remarks which show the idea of universal

gravitation hovering in Galileo's mind. He perceived the analogy between the power which holds the moon in the neighbourhood of the earth and compels Jupiter's satellites to circulate round their primary, and that attractive power which the earth exercises on bodies at its surface ; but he failed to conceive the combination of central force with initial velocity, and was disposed to connect the revolutions of the planets with the axial rotation of the sun—a notion which tended more towards Descartes' theory of vortices than towards Newton's theory of gravitation.

Having laid stress on the resemblance of earth and planets, Salviati proposes for them all a similar motion round the sun—one of the two main points of the Copernican theory. He shows how the apparent paths of all the planets can be thus explained, and in a far simpler way than by the Ptolemaic formula. On the Copernican hypothesis all motions of revolution and rotation take place in the same direction from west to east, whereas the Ptolemaic system requires some to be in one direction and some in another. A glance through a telescope turned towards Jupiter shows a family of small bodies circling round a great planet ; here one could see on a small scale the very thing that Copernicus had described as going on in the case of planets and sun.

On the second Day the discussion passes to the other chief point in the Copernican hypothesis—that the apparent daily motion of the stars is really due to the daily rotation of the earth on its axis. Here he breaks entirely new ground in his treatment of motion. His great discovery, which threw a new light on the mechanics of the solar system, was substantially Newton's first law of motion—' Every body continues in its state of rest or of uniform motion in a straight line, except in so far as it is compelled by force applied to it to change that state '. Putting aside any discussion of this ' force ', a conception first defined by Newton, and only imperfectly grasped by Galileo, we may interpret the law as meaning that a body has no more inherent tendency to diminish its motion or to stop than it has to increase its motion or to start, and that any alteration in either speed or direction is to be explained by the action on it of some other force. As it is impossible to isolate a body from all others, we cannot experimentally realize the state of things in which it goes on moving indefinitely in the same direction and at the same rate ; it may, however, be shown that the more we remove it from external influence the less alteration there is in its motion. Galileo illustrates this idea by a ball on an inclined plane. If the ball is projected upwards its motion is retarded; if downwards it is continually accelerated. This is true if the plane be fairly smooth and the inclination not very small. If now we imagine the experiment performed on an ideal plane, which is perfectly smooth, we should expect the same results, however small the inclination. Consequently, if the plane be quite level, so that there is no distinction between up and down, we should expect the motion to be neither retarded nor accelerated, but to continue without alteration.

Other more familiar examples are given of the tendency of a body when once in motion. This principle of motion being once established, it becomes easy to deal with several common objections to the motion of the earth. The case of a stone dropped from the top of a tower, which, if the earth be moving rapidly from west to east, might be expected to fall to the west, is compared to that of a stone dropped from the masthead of a moving ship. It is, therefore, entirely in accord with theory that the stone should fall as it does at the foot of the tower.

No objections to the hypothesis of the earth's rotation being found tenable, it is shown by Salviati how much more simple is the real motion proposed than the supposition that the universe revolves daily round a fixed earth. 'To make the universe revolve', he says, 'in order to maintain the immobility of the earth is as little reasonable as to require, in order to see Venice from the top of the Campanile, that the whole panorama should move round the spectator instead of his simply moving his head.'

The primitive notion of the stars as fixed in a crystal sphere had been long overthrown. And, supposing that they were distinct and independent bodies, it was difficult to imagine laws controlling their motion about a fixed earth that should result in revolutions timed uniformly for all and at the same time of enormous rapidity. Salviati makes the improbable practically impossible by referring to the phenomenon now known as the 'precession of the equinoxes', in virtue of which the direction of the earth's axis in space moves slowly, completing a revolution in about 26,000 years. As a consequence of this change, some of those stars, which in Ptolemy's time were describing very small circles and, therefore, moving very slowly, must now be describing larger circles at a greater speed, and vice versa. The system of stellar motions that would be necessary to account for all this would be inconceivably complex, whereas, on the Copernican theory, they are adequately explained by the rotation of the earth and a simple displacement of its axis of rotation.

A great part of the third Day is devoted to the question of stellar parallax. In this lay one of the most serious objections to the Copernican theory. If it was true that the earth swept round the sun in an orbit some two hundred million miles across, then it must follow that at one time of the year we should get a different view of the arrangement of the stars from that obtained six months later, when the earth was at the opposite point of its orbit. The nearer stars should undergo displacements in their apparent positions relative to those more distant. The answer to this was that these displacements probably did take place, but were too minute to be detected. But this answer, though strictly true, implied that the distances of even the nearest stars were great beyond all comprehension; and this in turn implied that the visible size of the stars indicated a real size of inconceivable dimensions. The latter difficulty was reduced by Salviati's assertion that the visible size of a star was an optical illusion; the telescope showed the stars to be sharp points, in contrast to the planets which though small to the eye really did possess visible dimensions. But the former difficulty remained, and nearly two centuries passed before Bessel made the first rough measurement of a stellar parallax. His method was essentially that suggested in the Dialogue, though the results obtained indicated for the star 61 *Cygni* a distance which would have astonished even Galileo himself.

Towards the end of the third Day reference is made to an annual rotation of the earth about an axis perpendicular to the plane of its motion, as postulated by Copernicus. But this third rotation is an unnecessary complication introduced by confusion in geometrical thought. That the actual state of things is quite simple Salviati illustrates by a reference to the motion of a ball floating in a basin of water. If the basin be held in the hand, the ball floating at or near the centre, and the experimenter turn round steadily on his feet, holding the basin in front of him, the ball remains in a position which is unaltered with reference to the walls and furniture of the room, although with reference to the man supporting the basin it might be said to have spun once completely round. And so with regard to the annual rotation spoken of by Copernicus, 'What other is the earth than a globe librated in tenuous and yielding air?'

The fourth Day is devoted to an examination of the cause of the tides, and is a development of his letter on the same subject to Cardinal Orsini in 1616. It is a singular circumstance that the argument on which Galileo mainly relied, as furnishing a physical demonstration of the truth of the Copernican theory, rested on a misconception. The tides, he says, are of three kinds—daily, monthly, and yearly. The first and principal ones are a visible effect of the terrestrial double movement, since they are the combined result of (1) the earth's daily rotation, and (2) the inequality of the absolute velocities of the various parts of the earth's surface in its revolution round the sun. The monthly tides depend in a secondary way on the moon's motion, and the annual tides, also in a secondary way, on the sun's action. These bodies, according to their positions relatively to the earth and to each other, produce inequalities in the earth's movements, and these inequalities are the cause of the monthly and yearly tides. To such notions Galileo attached capital importance, and he was inclined to ridicule Kepler's suggestion that the attraction of the moon was in some way the cause of the phenomenon. The influence of the moon on the tides had been recognized from ancient times, but a scientific explanation was not to be expected until the law of universal gravitation had been fully realized. Indeed, even now, with all the resources of modern science, the problem cannot be said to be completely solved.

This last part of the Dialogue is therefore of little value, and may be passed over. The chief work was to establish the Copernican theory, which, first promulgated in the days when human vision was unaided, had been found by Galileo to be supported by all the evidence that could be gathered by means of his telescope. The problem—if it may be still said to exist—takes a different form at the present day. So far are we now from the pre-Copernican theory of a fixed earth, that we look upon no single object in the universe as fixed. The sun itself has its motion amongst the other visible stars, and the present direction and rate of that motion are roughly known. Accordingly, the alternative which offered itself to the controversialists of Galileo's day, that either the sun or the earth is stationary, does not concern us any more ; both of these bodies are moving.

3. *Galileo's Second Encounter with the Inquisition. His Trial and Abjuration*, 1632–3 [1]

The publication of the *Dialogue on the Two Chief Systems of the World* raised a tumult in the ecclesiastical world, and especially among the Jesuits, who now resolved to pursue the author with the utmost energy. They claimed the monopoly of instruction and the first rank in the learned world, and were jealous of all intruders. Galileo was in every way inconvenient to them, and the more obnoxious in that he had already measured swords with distinguished members of the Order, Fathers Scheiner and Grassi. And now appeared his Dialogue, in which some old sores are reopened. The book was, therefore, denounced as a defence of Copernicanism under the flimsiest of disguises, as a gross violation of the admonition and decree of 1616, as an insult to the Pope himself. He was Simplicio—the Simpleton! Was not his unanswerable argument of 1624 put into the mouth of a simpleton, dragged in at the end, and summarily dismissed with a pious ejaculation ? Galileo was charged with daring, after solemn warning, to interpret the Scriptures to his own ends. It was bad enough, they said, to upset the old beliefs and spoil the face of nature with his celestial novelties, but at least he must be taught to leave the Bible alone. In this he was rebellious against Mother Church, and, further, deceitful in that he obtained the *Imprimatur* by suppressing material facts in his dealings with the censors. Although the safety of the Church and the vindication of its decrees were the ostensible reasons for the subsequent proceedings, it would not be far from the truth to say that revenge for an assumed insult was the primary and determining factor. Urban VIII and many of the high dignitaries of the Church, if not Copernicans at heart, were indifferent and cared little one way or the other.[2] As regards the Pope himself, we have seen evidence of his affection for Galileo and of his liberal sentiments, of his relish for the works on Sun-spots, on Floating Bodies,

[1] Cf. Nat. Ed., vol. xiv, pp. 372, 402 ; vol. xvi, various ; vol. xix, pp. 272–421.

[2] Even members of the Collegio Romano, including his old antagonists Fathers Scheiner and Grassi, were Copernicans in disguise. Cf. Favaro's *Adversaria Galileiana*, Serie Seconda, Padova, 1917, p. 27.

Il Saggiatore, &c. Then, we have his statements (1) that the Copernican doctrine is not heretical but only rash, and (2) that if it rested with him the decree of 1616 would never have been issued. All this seems to show that, if the question of a personal insult had not arisen, the Dialogue would at the worst have been put on the Index, as was the book of Copernicus, in 1616, 'until corrected'.

Early in August, 1632, the sale of the book was prohibited and its contents submitted to a special commission, who reported, after a month's session:

'1. Galileo has transgressed orders in deviating from the hypothetical standpoint, by maintaining *decidedly* that the earth moves and that the sun is stationary. 2. He has erroneously ascribed the phenomena of the tides to the stability of the sun and the motion of the earth, which are not true. 3. He has been deceitfully silent about the command laid upon him in 1616, viz. to relinquish altogether the opinion that the sun is the centre of the world and immovable and that the earth moves, nor henceforth to hold, teach, or defend it in any way whatsoever, verbally or in writing.'

On September 23 Galileo was ordered to appear in the course of the following month before the Commissary-General of the Holy Office in Rome. His friends pleaded his age and infirmities and the inclemency of the season, and begged that he might therefore be interviewed in Florence, but to no purpose. 'In Rome he must appear, and as a prisoner in chains if he will not come willingly', came the answer. To avert the most extreme measures the Grand Duke caused him to be informed (January 11, 1633) that it was at last necessary to obey the orders of the supreme authorities at Rome, and, in order that he might perform the journey more comfortably, Grand-ducal litters and a trustworthy guide would be placed at his disposal, and he would be lodged in Rome in the house of the Grand Duke's Ambassador. The pitiful impotence of an Italian ruler of that day in face of the Roman Church is painfully obvious in this decision. The Sovereign does not dare to protect his subject—even his old and respected tutor, but gives him up to the Inquisition, as if he were an alien malefactor.[1]

[1] The Venetian Republic was the only State in Italy that would have asserted its independence, as it had often done, and would have refused to hand over one of its officials to the Roman power. Indeed, when these proceedings began, Francesco Morosini of Venice offered to reinstate him in his old chair at Padua on any conditions that he chose to make, and to print his Dialogue in Venice.

On January 20, 1633, he left Florence, halted twenty days in great discomfort at the frontier on account of quarantine, and arrived in Rome on February 13. For the next four months the proceedings dragged on. He himself was inclined to hold out, his friends besought him to submit, he was ruthlessly haled backwards and forwards by his judges, questioned here and threatened there—the threat being a 'rigorous examination', an euphemism for physical torture. At last, on June 22, sentence was pronounced, and Galileo made his pitiful abjuration. After setting out in detail his various delinquencies, the Inquisitors conclude:

'Therefore, having seen and maturely considered the merits of your case, with your confessions and excuses, and everything else which ought to be seen and considered, we pronounce, judge, and declare that you have rendered yourself vehemently suspected by this Holy Office of heresy, in that (1) you have believed and held the doctrine (which is false and contrary to the Holy and Divine Scriptures) that the sun is the centre of the world and that it does not move from east to west, and that the earth does move and is not the centre of the world; and (2) that an opinion can be held and defended as probable after it has been decreed contrary to the Holy Scriptures, and, consequently, that you have incurred all the censures and penalties enjoined in the sacred canons and other general and particular codes against delinquents of this description. From this it is Our pleasure that you be absolved provided that, with a sincere heart and unfeigned faith, in Our presence you abjure, curse, and detest the said errors and heresies, and every other error and heresy contrary to the Catholic and Apostolic Church of Rome, and in the form that shall be prescribed to you. But that your grievous and pernicious error may not go altogether unpunished, and that you may be more cautious in future, and as a warning to others to abstain from delinquencies of this sort, We decree that the book *Dialogue of Galileo Galilei* be prohibited by public edict, and We condemn you to the prison of this Holy Office for a period determinable at Our pleasure, and by way of salutary penance We order you during the next three years to recite, once a week, the seven penitential psalms, reserving to Ourselves the power of moderating, commuting, or taking off the whole or part of the said punishment or penance.'

In conformity with this sentence, Galileo was made to kneel before the Inquisition, and make the following abjuration:

'I, Galileo Galilei, son of the late Vincenzio Galilei of Florence, aged seventy years, being brought personally to judgement, and kneeling before you, Most Eminent and Most Reverend Lord Cardinals, General Inquisitors of the Universal Christian Republic against heretical depravity, and having before my eyes the Holy Gospels which I touch with my own hands, swear that I have always believed, and, with the help of God, will in future believe every article which the Holy Catholic and Apostolic Church of Rome holds, preaches, and teaches. But because I have been enjoined by this Holy Office altogether to abandon the false opinion that the sun is the centre and immovable, and been forbidden to hold, defend, or teach the said false doctrine in any manner; and because, after it had been signified to me that the said doctrine is repugnant to the Holy Scripture, I have written and printed a book, in which I treat of the same condemned doctrine, and adduce reasons with great force in support thereof without giving any solution, and therefore have been judged grievously suspected of heresy, that is to say, that I held and believed that the sun is the centre of the world and immovable, and that the earth is not the centre and is movable, I am willing to remove from the minds of your Eminences, and of every Catholic Christian, this vehement suspicion rightly entertained towards me. Therefore, with a sincere heart and unfeigned faith, I abjure, curse, and detest the said errors and heresies, and generally every other error and heresy contrary to the said Holy Church, and I swear that I will never more in future say, or assert anything, verbally or in writing, which may give rise to a similar suspicion of me; and that if I shall know any heretic, or any one suspected of heresy, I will denounce him to this Holy Office, or to the Inquisitor and Ordinary of the place in which I may be. I swear, moreover, and promise that I will fulfil and observe fully all the penances which have been or shall be laid on me by this Holy Office. But if it shall happen that I violate any of my said promises, oaths, and protestations (which God avert), I subject myself to all the pains and punishments which have been decreed and promulgated by the sacred canons and other general and particular constitutions against delinquents of this description. So, may God help me, and these His Holy Gospels which I touch with my own hands.

'I, the above-named Galileo Galilei, have abjured, sworn, promised, and bound myself as above, and, in witness thereof, with my own hand have subscribed this my abjuration, which I have recited word for word, in Rome, in the Convent of Minerva, this 22nd June, 1633. I, Galileo Galilei, have abjured as above with my own hand.'

While the older writers generally go to one extreme and say that Galileo was tortured, thrown into a dungeon for years, or for the rest of his life, and was in physical fact a martyr, some recent authors have gone to the other extreme, and aver that he had no claim to much sympathy, that he had brought his troubles on himself by want of tact and temper. Others, again, blame him for not 'seeing this thing through'. Brewster, for example, compares him to the Christian martyrs, and finds him sadly degenerate. 'Had Galileo', he says, 'but added the courage of the martyr to the wisdom of the sage; had he carried the glance of his indignant eye round the circle of his judges; had he lifted his hands to Heaven, and called on the living God to witness the truth and immutability of his opinions, the bigotry of his enemies would have been disarmed, and Science would have enjoyed a memorable triumph.' Perhaps; but perhaps, on the other hand, his judges, instead of being cowed by the glance of his eye, would have delivered him to the stake, as they did Giordano Bruno earlier in the century (1600), and Marc' Antonio de Dominis only eight years before. Revealed truth may require its martyrs, and perhaps the blood of the martyrs may be the seed of the Church, but scientific truth is not thus established.[1]

After his sorrowful drama had concluded, Galileo was led back to the buildings of the Holy Office. And now that he and the Copernican theory were condemned with all the terrifying forms of the Inquisition, the Pope's wounded vanity was soothed, and he gave the word for a little mercy. Galileo was not to be kept in the prison of the Holy Office, but was banished to the villa of the Grand Duke of Tuscany at Trinità dei Monti, which he was to consider as a prison. A week later he was allowed to retire to Siena to the palace of Archbishop Ascanio Piccolomini (a former pupil in Padua), where he was to remain under the orders of the Archbishop, and on no account to leave the house, except to hear mass, without permission from Rome.

He was informed of this decision on July 2, and early on the 6th he shook the dust of Rome from off his feet. He reached Siena on the 9th and was warmly received by Piccolomini and other friends; but kindness could not make him forget that he was a prisoner. As the weary months rolled on, the old man became a little resigned to the situation, and began to occupy

[1] Cf. Bruno and Galileo, *Quarterly Review*, 1878, No. 290, p. 362, a Plutarchian contrast by John Wilson.

IL SAGGIATORE
Nel quale
Con bilancia esquisita e giusta
si ponderano le cose contenute
nella
LIBRA ASTRONOMICA E FILOSOFICA
DI LOTARIO SARSI SIGENSANO
Scritto in forma di lettera
All'Ill.mo et Reuer.mo Mons.re D
VIRGINIO CESARINI
Acc.o Linceo M.o di Camera di N.S
Dal Sig.re
GALILEO GALILEI
Acc.o Linceo Nobile Fiorentino
Filosofo e Matematico Primario
del
Ser.mo Gran Duca di Toscana.

FILOSOFIA NATURALE

MATEMATICA

IN ROMA M.D.C.XXIII.
Appresso Giacomo Mascardi.

F. Villamoena Fecit

Fig. 10.

DISCORSI
E
DIMOSTRAZIONI
MATEMATICHE,
intorno a due nuoue scienze
Attenenti alla
MECANICA & i MOVIMENTI LOCALI,
del Signor
GALILEO GALILEI LINCEO,
Filosofo e Matematico primario del Serenissimo
Grand Duca di Toscana.
Con una Appendice del centro di grauità d'alcuni Solidi.

IN LEIDA,
Appresso gli Elsevirii. M. D. C. XXXVIII.

Fig. 11.

himself with another of his great works, *Dialoghi delle Nuove Scienze*, the writing of which he spoke of as far back as 1610, in his letter to Vinta.

In November, 1633, thinking the time favourable, the Tuscan Ambassador in Rome began to move for a free pardon, but the Pope was not disposed to go so far, and pretended there would be a difficulty in getting the consent of the Holy Office—a patent evasion, as the decision rested solely with himself. At length, on December 1, 1633, the question of pardon or rather release from personal restraint came before the Congregation, the Pope presiding. It was refused, but Galileo was allowed to retire to his villa at Arcetri, near Florence, where he was to remain till further orders.

V. Declining Years (1634–42)

1. *Dialoghi delle Nuove Scienze* [1]

The Dialogues on the New Sciences ((1) on coherence and resistance to fracture, and (2) on uniform, accelerated, and violent or projectile motions) were begun in Siena in 1633, and were completed by the end of 1634. After the condemnation in 1633, the Holy Office placed Galileo's name on the list of authors whose writings *edita et edenda* were strictly forbidden. Galileo tried to publish at Vienna, only to find that all books printed there must be approved by the Jesuits, amongst whom happened to be Father Scheiner. He tried Olmütz and Prague, but at each place licensing and other difficulties cropped up. Finally he opened direct negotiations with the Elzevirs, and the work was issued from their press at Leyden early in 1638.[2]

The *Dialoghi delle Nuove Scienze* is practically a compendium, with later additions, of his early mechanical work in Pisa and Padua, and rightly did he call it, in his letter to Vinta of May 7, 1610, a new science invented by himself from its very first principles.

'Dynamics', says Lagrange, 'is a science due entirely to the moderns, and Galileo is the one who laid its foundations. Before him philosophers considered the forces which act on bodies in

[1] Cf. Nat. Ed., vol. viii, pp. 12 et seq.; vol. xiv, p. 386; vol. xv, pp. 248, 257, 284 vol. xvi, pp. 59, 72.

[2] See Fig. 11. In order to obviate trouble with the Holy Office, he pretended that the book was pirated from a manuscript copy, which he had given to Comte de Noailles (lately French Ambassador in Rome), to whom it is dedicated.

a state of equilibrium only, and, although they attributed in a vague way the acceleration of falling bodies and the curvilinear movement of projectiles to the constant action of gravity, nobody had yet succeeded in determining the laws of these phenomena. Galileo made the first important steps, and thereby opened a way, new and immense, to the advancement of mechanics as a science."[1]

In this new dialogue the discussion is carried on by the same speakers as in the Dialogue of 1632. The first two Days are mainly concerned with the resistance of solids to fracture, and the cause of their coherence. Their scientific value lies in the incidental experiments and observations on motion through resisting media. The debate opens with an examination of the current belief that machines built on exactly similar designs, but on different scales, were of strength in proportion to their linear dimensions. It is shown that the larger machine will equal the smaller in all respects, except that it will not be so strong or so resistant to violent actions. After explaining the strength of ropes made of fibrous or filamentous materials, he comes to the cause of coherence of the parts of such things as stones and metals, which do not show a fibrous structure. Presently, Salviati asks what is the nature of the force that prevents a glass or metal rod, suspended from above, being broken by its own weight or by a pull at the free end ? No satisfactory explanation is forthcoming ; but that suggested depends upon nature's so-called repugnance to a vacuum, such as is momentarily produced by the sudden separation of two flat surfaces. This idea is extended, and a cause of coherence is found by considering every body as composed of very minute particles, between any two of which is exerted a similar resistance to separation.

This leads to an experiment for measuring what is called the force of a vacuum. The experiment occasions a remark from Sagredo that his lifting-pump would not work when the water had sunk to the depth of 35 feet below the valve. This story is sometimes told as if Galileo had said jokingly that nature's horror of a vacuum does not extend beyond 35 feet ; but it is plain that the remark was made seriously. He held the then current notion of suction, for he compares the column of water to a rod of metal suspended from its upper end, which may be lengthened till it breaks with its own weight. It is extraordinary that he failed to see how simply this phenomenon could

[1] *Méchanique analytique*, Paris, 1788.

be explained by the weight of the atmosphere, with which he was well acquainted.[1]

We come next to the violent effects of heat and light. It is suggested that, perhaps, heat dissolves bodies by insinuating itself between their minute particles. The effects of lightning are mentioned and experiments with burning-glasses referred to. Then as regards light, Sagredo asks whether its effect does or does not require time. Simplicio is ready with an answer—that the discharge of artillery proves the transmission of light to be instantaneous, to which Sagredo cautiously replies that nothing can be gathered from that observation except that light travels more swiftly than sound; nor can we draw any decisive conclusion from the rising of the sun. Who can assure us that he is not in the horizon before his rays reach our eyes?

We next come to Aristotle's ideas on motion *in pleno* and *in vacuo*, and especially his assertion that bodies fall with velocities proportional to their weights and inversely proportional to the densities of the media through which they are moving. This proposition is examined in a strict scientific method. Heavy bodies of different weights are dropped in air to test the truth of the first part of the statement; and afterwards the motion of bodies rising or falling in liquids is considered; the result being to substitute for Aristotle's assumption that law of the motion of falling bodies which is the foundation of the science of dynamics.

After having discussed the first point and shown that the rate of fall is not proportional to weight, Salviati proceeds to examine the motion of bodies sinking or rising in water and other liquids; and brings forward experimental facts which, viewed in the light of Aristotle's statement, form a mass of contradiction. Putting then this antiquated theory aside, Salviati inquires what is meant by the rising of some bodies in a medium, and shows that only those bodies rise which are lighter than the medium. The rising of an inflated bladder in the air suggests that the atmosphere must have weight. Simplicio's assertion that it is on the contrary the bladder in this case that has levity is trivial, and is immediately disproved. Continuing his line of argument, Salviati points out that the question of rising or falling depends on the gravity of the medium as compared with that of the moving body; further, that when the motion of the body, either upwards or downwards,

[1] Nat. Ed. xi. 12, and xii. 33.

has once commenced, the different media offer different resistances to the motion, the heavier media, such as quicksilver and water, interfering more than air. We are thus led to the bold deduction 'that if the resistance of the media be wholly taken away all matter would descend with the same velocity'.

In the concluding part of this first Day the theory of the pendulum is applied to musical concords and discords, which result from the concurrence or opposition of vibrations of the air striking upon the drum of the ear. These vibrations may be made manifest by rubbing the finger round a glass set in a large vessel of water; 'and, if by pressure the note is suddenly made to rise to the octave above, every one of the undulations, which will be seen regularly spreading round the glass, will suddenly split into two, proving that the vibrations that occasion the octave are double those belonging to the simple note'. Galileo then describes a method he discovered by accident of measuring the length of these waves more accurately than can be done in the agitated water. While scraping a brass plate with an iron chisel, he occasionally produced a hissing or whistling sound, and whenever this occurred he observed the dust on the plate to arrange itself in small parallel streaks equidistant from each other. In repeated experiments he produced different tones by scraping with greater or less briskness, and remarked that the streaks produced by high notes stood closer together than those from low. Among the sounds produced were two, which by comparison with a viol he ascertained to differ by an exact fifth; and measuring the spaces between the streaks he found thirty of one equal to forty-five of the other, which is exactly the proportion of the lengths of strings of the same material which sound a fifth to each other.[1]

Salviati also remarks that if the sounding materials be different, as for instance if it be required to sound an octave to a note on catgut on a wire of the same length, the weight of the wire must be made four times as great, and so on for other intervals.

Salviati: 'I will now show you an experiment from which the eye will derive a similar pleasure. Suspend three balls of lead by cords of different lengths, such that, while the longest makes two oscillations, the shortest makes four, and the medium three. This will occur when the longest string measures 16 of any assumed

[1] This beautiful experiment has been largely used in modern times by Chladni, Savart, and Wheatstone, with very interesting results.

unit (inches, feet, or yards), the medium 9, and the shortest 4. Now, pull these three pendulums aside and let them go at the same instant. You will observe a very curious interplay of the strings, passing each other in various ways, but such that, at the completion of every fourth oscillation of the longest pendulum, all three arrive simultaneously at the same terminus, whence they start afresh to perform the same cycle. This combination of oscillations is precisely that which in music yields the interval of the octave and the intermediate fifth. By changing the lengths of the cords in such ways that their oscillations correspond with agreeable musical intervals, we shall see quite different crossings, but always such that, after a definite time and after a definite number of oscillations all the pendulums will reach the same terminus at the same moment, and then begin again and so on. If, however, the strings are altered so that the oscillations are incommensurable and never complete a series together, or if commensurable complete the series only after a long interval and after a great number of oscillations, then the eye is offended by a disorderly succession of crossing threads, just as the ear is pained by an irregular sequence of air waves.'

The second Day is occupied entirely with an investigation of the strength of beams—an amplification of his researches on the same subject dating back to 1609. Beyond Aristotle's remark that long beams are weak, because they are at once the weight, the lever, and the fulcrum, nothing appears to have been written on the subject before Galileo took it up. The discussion opens with a consideration of the resistance of solid bodies to fracture. This resistance is very great in the case of a direct pull, but is much less in the case of a bending force. Thus, a rod of glass or iron will bear a longitudinal pull of, say, 1,000 lb., while 50 lb. will break it if it be fastened by one end into a wall or other upright support. It is this second kind of strain that is here discussed.

'In order to arrive at the resisting values of prisms and cylinders of the same material, whether alike or not in shape, length, and thickness, I shall take for granted the well-known mechanical principle of the lever, namely, that the force bears to the resistance the inverse ratio of the distances which separate the fulcrum from the force and resistance respectively.'

He assumed as the basis of his inquiry that the forces of cohesion with which a beam resists a cross fracture in any section may all be considered as acting at the centre of gravity of the section, and that it breaks always at the lowest point. An elegant

PLATE XLI HALL OF THE GALILEO MUSEUM IN FLORENCE

result deduced from this theory is that the form of a beam, to be equally strong in every part, should be that of a parabolic prism, the vertex of the parabola being the farthest removed from the point of support. As an easy way of drawing the curve for this purpose, he recommends tracing the line in which a heavy flexible string hangs supported from two nails.[1]

The curvature of a beam under any system of strains is a subject into which, before the days of Newton, it was not possible to inquire, and even in the simpler problem considered by Galileo he makes assumptions which require justifying. His theory of beams is erroneous in so far as it takes no account of the equilibrium which must exist between the forces of tension and compression over any cross-section.

In the third Day we find no new physical facts of importance, and much of the time is taken up with theorems and formulae deduced geometrically from the phenomena of uniform and accelerated motion as dealt with in the first Day. The further discussion of that subject, however, leads to a more detailed statement of the principle of inertia; but the definition of uniformly accelerated motion at once introduced a difficulty. Salviati gives the correct description of it as that of a body which moves in such a manner that in equal intervals of time it receives equal increments of velocity.

There follows an interesting application of the results obtained. He examines the times of descent down differently inclined planes, assuming as a postulate that the velocity acquired was the same for all planes of the same height. This fact he had verified by careful experiments, although he was unable at the time to prove it mathematically.[2]

The fourth Day plunges at once into the consideration of the ' properties which belong to a body whose motion is

[1] This curve is not strictly a parabola, it is now called a catenary; but it is plain, from the description of it further on (in the fourth Day), that Galileo was aware of this fact. The catenary resembles the parabola, and, as he justly remarks, the resemblance is all the more striking if the string is so taut that the depth of the lowest point is less than a quarter of the distance between the two extremities.

[2] Viviani relates that, soon after he joined Galileo in 1639, he drew his master's attention to this. The same night, as Galileo lay in bed, sleepless through indisposition, he discovered the necessary mathematical demonstration. It was introduced into the subsequent editions of the Dialogues, sixth Day.

compounded of two other motions, one uniform, and one naturally accelerated. This is the kind of motion seen in a projectile.'

After some preliminary instruction in the properties of the parabola, Salviati returns to the subject of projectiles, and lays down the law of the independence of the horizontal and the vertical motions. A body projected horizontally would—but for its weight and external impediments—continue to move in a straight line; and Salviati contends that, as the effects of gravity acting by itself would be entirely downwards, gravity acting on the projected body can neither increase nor diminish the rate at which it travels horizontally. Therefore, whatever be the shape of the path or the direction of motion at any moment, the distance travelled horizontally may be taken as a measure of the time that has elapsed since motion began. He proves that on this assumption the path described has geometrical properties which identify it with the curve known as the parabola. His demonstration is essentially that now given in works on elementary dynamics.

After demonstrating the parabolic nature of the path, he inquires into certain points of interest with regard to it, and gives proofs of many of the elementary propositions which in modern text-books are associated with parabolic motion. He also draws up a table giving the position and dimensions of the parabola described with any given direction of projection, finding by this means what he would have been unable to give a strict mathematical proof of—that the range on a horizontal plane is greatest when the angle of elevation is 45°.[1]

2. *The Laws of Motion*

No sooner was the manuscript of these dialogues out of his hands (summer of 1636) than Galileo occupied himself with new projects. 'If I live,' he wrote on July 15, 1636, to Bernegger of Strasburg, 'I intend to put in order a series of natural

[1] 'In solving the problems of falling bodies and of projectiles, Galileo was essentially applying the principles of the Differential or Fluxional or Indivisible Calculus. If pure mathematics had attracted him as strongly as its application to physics, he would have thought these problems out, and would have founded the Fluxional Calculus, which is the glory of Newton and of Leibnitz.' Professor Jack, in *Nature*, vol. xxi, p. 58.

and mathematical problems which I think will be as curious as they are novel.' These were left unfinished, and now form the fifth and sixth Days, which were added to later editions by Viviani after Galileo's death.[1] The fragment of the fifth Day is on the subject of Euclid's definition of ratio (Book V, props. 5 and 7) and was intended to follow the first proposition on equable motion in the third Day's debate. The sixth Day contains his investigations on the force of percussion, on which he was employed at the time of his death.

'In the last days of his life', says Viviani, 'and amid much physical suffering, his mind was constantly occupied with mechanical and mathematical problems. He had the idea of composing two other dialogues to be added to the four already published. In the first he intended to give many new demonstrations and reflections on various passages in the first four dialogues, and the solution of many problems in Aristotle's physics. In the second he proposed to discuss, treating it geometrically, an entirely new science, *viz.* the wondrous force of percussion, which he claimed to have discovered and which, he said, exceeded by a long way his speculations on the same subject formerly published' (Nat. Ed., vol. xix, Part ii).

In these admirable dialogues Galileo does not formulate in definite laws the interdependence of force and motion. This was done for the first time by Newton at the beginning of his *Principia* (1687), and hence they are rightly called 'Newton's Laws of Motion'; but in justice to Galileo it must be admitted that he not only prepared the way for Newton, but supplied him with much of his materials. Thus, the first law—that a body will continue in a state of rest, or of uniform motion in a straight line, until it is compelled to change its state by some force impressed upon it—is a generalization of Galileo's theory of uniform motion. Since all the motions that we see taking place on the surface of the earth soon come to an end, we are led to suppose that continuous movements, such, for instance, as those of the celestial bodies, can only be maintained by a perpetual consumption and a perpetual application of force, and hence it was inferred that rest is the natural condition of things. We make, then, a great advance when we comprehend that a body is equally indifferent to motion as to rest, and that it equally perseveres in either state until disturbing forces are applied.

[1] Cf. Nat. Ed., vol. viii, pp. 321–62.

The second law—that every change of motion is in proportion to the force that makes the change, and in the direction of that straight line in which the disturbing force is impressed—is involved in Galileo's theory of projectiles. Before his time it was a commonly received axiom that a body could not be affected by more than one force at a time, and it was therefore supposed that a cannon-ball, or other projectile, moves forward in a straight line until the force which impelled it is exhausted, when it falls vertically to the ground.

The establishment of this principle of the composition of forces supplied a conclusive answer to the most formidable of the arguments against the rotation of the earth, and, accordingly, we find it in the second Day of the Dialogue of 1632. The distinction between mass and weight was, however, not noticed, and, consequently, Galileo failed to grasp the fact that acceleration might be made a means of measuring the magnitude of the force producing the motion. How far he was from this discovery may be gathered from a remark by Salviati, incidental to the main argument, to the effect that when different bodies are falling freely towards the earth's centre, 'the difference of their gravities has nothing to do with their velocities.'

Of the third of the laws of motion—that action and reaction are always equal and opposite—we find traces in many of Galileo's researches, as in his theory of the inclined plane, and in his definition of momentum. It is also adumbrated in his 'Della Scienza Meccanica' (1594), and in his latest ideas on percussion.

His services were little less conspicuous in the statical than in the dynamical division of mechanics. He gave the first direct and entirely satisfactory demonstration of equilibrium on an inclined plane. In order to demonstrate this he imagined the weight and the resistance to be applied to the ends of a bent lever whose arms were equal to the vertical and slant sides of the plane; then reducing the lever to a straight one it was easy to prove that the forces *in equilibrio* on the plane were also *in equilibrio* on the lever, and were to one another as the length to the height of the plane. By establishing the theory of 'Virtual velocities', he laid down the fundamental principle which in the opinion of Lagrange contains the general expression of the laws of equilibrium; while as regards that still obscure subject, molecular cohesion, he brought it for the first time within the range of mechanical theory.

3. *On the Moon's Librations* [1]

Just before his sight began to fail,[2] Galileo made his last astronomical discovery, which is now known as the moon's librations. This discovery was announced in letters to Fulgenzio Micanzio (November 7, 1637) and Alfonso Antonini (February 20, 1638). To the former he writes:

'I have observed a marvellous appearance on the surface of the moon. Though she has been looked at such millions of times by such millions of men, I do not find that any have observed the slightest alteration in her surface; but that exactly the same side has always been supposed to be presented to our eyes. Now I find that such is not the case, but that she changes her aspect, as one who, having his full face turned towards us, should move it sideways, first to the right and then to the left; or should raise and then lower it; or, lastly, should incline it first to the right shoulder, then to the left. All these changes I see in the moon; and the large, anciently-known spots which are seen on her face will help to make evident the truth of what I say. Add to these a further marvel, which is that these three mutations have their several periods—the first daily, the second monthly, the third yearly.'

Galileo was not long in detecting the causes of the apparent libratory or rocking movement. The diurnal or parallactic libration he saw was occasioned by our distance as spectators from the centre of the earth, which is also the centre of the moon's revolution. In consequence of this, as the moon rises we get an additional view of the lower part and lose sight of that portion of the upper part which was visible while she was low in the horizon. The causes of the other motions are not so easily explained, nor is it certain that Galileo himself understood them; his conjecture of a connexion with the tides is certainly wide of the mark.

The moon in revolving round the earth spins once on her axis, so turning the same side always towards the earth's centre. But this familiar truth is only approximate. The speed of rotation is uniform; but the speed of revolution in her orbit is not so, because that orbit is not a circle but an ellipse, in which (as is always the case with elliptic motion) the moving body travels faster while

[1] Cf. Nat. Ed., vol. xvii, pp. 212, 291.

[2] Early in 1636 his sight began to fail. By the end of June, 1637, the sight of the right eye was gone, and early in December following he became totally blind.

near the centre of attraction than when farther away. The result is that we see alternately a little round the eastern edge, and a fortnight later a little round the western.

The two librations, due to independent causes, have approximately the same period—about one month. Their effects, however,

Fig. 12. FACSIMILE OF DESIGN FOR A PENDULUM CLOCK.
Drawn by Vincenzio Galilei from his father's dictation.

vary according to the changing position of the earth in its orbit; and any particular phase of the libration is more nearly reproduced after twelve months than after one. Galileo was, therefore, justified in suggesting an annual period, although it is not customary at the present day to associate the annual period with any very distinct librations.

4. *Application of the Pendulum to Clocks* [1]

A few months before his mortal illness Galileo once more gave proofs of his mechanical genius. It has been remarked in the progress of science and scientific invention that the steps, which on looking back seem the easiest to make, are often those which are the longest delayed. The application of the pendulum to clocks is an instance of this. We have seen that Galileo was early convinced of the value of the pendulum as a measurer of time, and that as far back as 1582–3 he used it in the Pulsilogia; yet fifty-five years later, although constantly using it meanwhile, he had not devised a more practicable application than that described in his letter of June 6, 1637, to Lorenzo Realio, and in his 'Astronomical Operations' of 1637–8 for finding the longitude at sea.

'I make use of a heavy pendulum of brass or copper, in the shape of a sector of twelve or fifteen degrees, the radius of which may be two or three palms (the greater it is the less trouble in attending it). This sector I make thickest in the middle radius, tapering gradually towards the edges, where I terminate it in a tolerably sharp line, to obviate as much as possible the resistance of the air, which is the main cause of its retardation. This sector is pierced at the centre, through which is passed an iron bar shaped like those on which steelyards hang, terminated below in an angle or wedge which rests on two bronze supports. If the sector (when accurately balanced) be removed several degrees from the perpendicular, it will continue a to-and-fro motion through a very great number of oscillations before coming to rest; and in order that it may continue its oscillations as long as it is wanted, the attendant must occasionally give it a push so as to carry it back to large oscillations.

'Now to save the fatigue of continually counting the oscillations, this is a convenient contrivance,—a small delicate needle extends from the middle of the sector which in passing strikes a rod hung at one end. The lower end of this rod rests on the teeth of a horizontal wheel as light as paper. The teeth are cut like those of a saw. The rod striking against the perpendicular side of a tooth moves it, but when returning it slips over the oblique side of the next tooth and falls at its foot, so that the motion of the wheel will be in one direction only. By counting the teeth you may see at will the number passed, and, consequently, the number of oscillations or periods of time which you wish to measure. You may also fit to the axis of the wheel a second, with

[1] Cf. Nat. Ed., vol. viii, p. 451; vol. xvii, pp. 96, 212; vol. xix, pp. 647–59.

a smaller number of teeth and in gear with a third wheel having a greater number of teeth, and so on.'

It was chiefly because of the inadequacy of this method that the negotiations with the States-General were finally broken off. Now, in the second half of 1641, it occurred to Galileo that the problem could be solved by adding the pendulum to the ordinary clock as a regulator of its movements. He explained his idea to his son, Vincenzio, who made a drawing (of which we reproduce a facsimile, Fig. 12) from his father's direction. Before the plan could be tried Galileo fell ill, and died January 8, 1642. The matter was laid aside, but seven years after his father's death Vincenzio resumed it, and was engaged in constructing what would have been the first pendulum clock, when he, too, fell ill and died, May 16, 1649.[1]

[1] A working model of this clock, inscribed 'Eustachio Porcellotti costruito a Firenze l'anno 1883', is in the Science Museum, South Kensington, and keeps very good time.

THE HISTORY OF ANATOMICAL INJECTIONS

BY F. J. COLE

'THE purpose of injections,' says Robin with admirable brevity, 'is to make known to us the absolute vascularity of the tissues'; and the practice of the method, to quote Lacauchie, 'founded an hundred reputations, and filled the museums with beautiful preparations'. The history of a scientific method such as this, established alike by its aims and results, may well be traced in some detail.

It will not surprise many to learn that the scope of this inquiry cannot be kept within the terms of Robin's definition. No method of scientific research long retains its original character. The further we trace it back the more it changes, until the question when, how, and with whom it originated may indeed be debated but not resolved. Thus the earlier injections were of a general or random character, and were used to fill the larger cavities of the body, such as the bladder from the penis, or to explore the solid inscrutable bulk of tumours. Afterwards specific injections into blood-vessels were attempted, but even then the endeavour was to test or illustrate the doctrine of the circulation, or to demonstrate the local distribution of certain vessels.

The reception of the injection method did not quite follow the traditional course. The results obtained by it were too obvious and striking to be either questioned or ignored. It passed rapidly into general use, and quickly reached its point of maximum utility. For a time it monopolized attention, and its importance was so grossly exaggerated as to countenance the belief that all problems of anatomy and physiology might be solved by its means.[1] In 1727 the French Academy of Science could find no more worthy

[1] The method is honoured by a reference in Gibbon's *Decline and Fall* (first edition, 1788, vol. v, p. 429). He says: 'A superstitious reverence for the dead confined both the Greeks and the Arabians to the dissection of Apes and Quadrupeds; the more solid and visible parts were known in the time of Galen, and the finer scrutiny of the human frame was reserved for the Microscope and the Injections of modern artists.'

successor to Sir Isaac Newton than Frederik Ruysch—the most famous injector of his time. How unsound contemporary judgement may be could not have a more significant confirmation than the verdict of posterity on the work of these two men—a verdict which exemplifies the practical wisdom of Boyle's remark that 'we are not near so competent judges of wisdom as we are of justice and veracity'. The frenzy for injections finally exhausted itself, and as a means of research the method may almost be said to have served its purpose, and to have passed into the shadow of history. In stating this, however, certain modern developments must not be ignored. In 1886 Camillo Golgi introduced the intravitam method of fixing animal tissues by a vascular injection of the fixative, and in 1900 Flint studied the circulation in the living embryo by throwing an injection into the blood-stream.

There are manifestly several important questions which the injection method is peculiarly adapted to elucidate. The circulation of the blood can be demonstrated beyond question by an injection experiment; the existence of foetal and uterine portions in the placenta is establishable only with the assistance of the syringe; the connexion of the vas deferens, epididymis, and testis is made beautifully clear by mercury injections; and the independence of the blood vascular and lymphatic systems rests upon a series of careful injections. Again, certain negative results may claim attention. Descartes' speculation that nerve fibres were valvular tubes transmitting fluid contents was finally disposed of by injection experiments, in spite of the fact that at first such support was in some cases claimed for it. On the other hand, erroneous views were undoubtedly suggested and maintained by the injections of the earlier anatomists. Of these lapses mercury was the most frequent cause, and on that account it is now rarely if ever used. Even when a mercury injection is successful, the least abrasion results in an extensive bleeding of the mercury, and preparations so delicate were justly considered to have but a doubtful value. With other injection masses extravasation effects frequently resulted in serious error and controversy among anatomists, and tended to bring the method into disrepute. The disputes between the Hunters and the Monros owed something to this cause, and extravasations into minute extra-vascular spaces of a tubular character explain the 'capillicules' of some of the later French injectors.

The frequent use of the word 'art' indicates an appreciation of the beauty of the injected specimen by the earlier anatomists, and it is easy to picture their wonder as smaller and yet smaller vessels were disclosed. What, therefore, must their delight have been when, the unaided eye being able to see no more, the microscope revealed the amazing spectacle of the blood capillaries in all their profusion of number and form. We can scarcely blame them if they abandoned Science for display, and injections were undertaken merely because the result was picturesque. Pole says: 'The veins in the kidney of a cat run very superficial, and branch out in a manner peculiarly beautiful, which is the only inducement to make this preparation.' And again, an injected and corroded kidney of the horse 'makes a noble and beautiful preparation'.

In this article it is not proposed to refer other than briefly to the injection of living animals, or to the transfusion experiments which greatly occupied the attention of the Royal Society and the French Academy of Science in the second half of the seventeenth century. The history and literature of this subject has already been exhaustively dealt with, but it does concern also the anatomical application of the injection method, and to that extent must receive attention. Although certain transfusion and intravitam injections had been attempted before the doctrine of the circulation was established in 1628, there can be no question that Harvey's work was the fundamental and determining influence in these early injection experiments. Before 1650, however, the new doctrine, though indeed accepted, had not succeeded in energizing contemporary investigation, and it was only from this latter date onwards that a further advance is to be noted. The period of injection lies between 1650 and 1750, after which time interest in injections gradually declined as it was found that the new method was unable to realize the great expectations of its professors. Ruysch's activities are comprised between the dates 1665 and 1728, and hence coincide almost exactly with the middle of the injection period. The first injections, however, were not anatomical, but were undertaken for physiological or medical reasons. The object was to transfer blood from one living animal to another, and to test the effect of injecting various liquids, drugs, or air into the vessels of the living body. Great hopes were based on these haphazard and ruthless efforts. In 1667, and again in

1670, the French Academy conducted numerous injection experiments in the belief that the reinvigoration and even the rejuvenescence of mankind would be thereby accomplished, but they were quick to acknowledge the vanity and futility of their hopes. The failure, however, was not absolute, for it was soon perceived that the humbler but still important requirements of anatomy could be promoted by the new method, and in a short time the success of the anatomical injection provided an ample compensation for the failure of the physiological.

Before 1650 there exist only a few references to injection experiments in the literature of Biology. The idea was in the background, but its exploitation was delayed. In this early work we should expect the seeker to precede the syringe, and thus we find Aretaeus the Cappadocian,[1] who lived in the second or third century of the Christian era, demonstrating that 'one may pass a plate of metal from the vena cava connected with the heart to that by the spine, and from the spine through the liver to the heart; for it is the same passage leading upwards'. Galen[2] goes a step further, and studied the distribution of the cerebral vessels by inflating them with air through a tube. Thus the blood channels were distended, and acquired the relief necessary to enable them to be followed the more easily. According to Michele Medici, Alessandra Giliani of Persiceto, who died in 1326, filled the blood-vessels with liquids of different colours, which thickened and hardened as soon as they were injected, but did not decompose. This statement, however, lacks confirmation. The first reference to injections after the invention of printing occurs in the commentary of Jacobus Berengarius published in 1521.[3] He employed a syringe and injected the renal veins with warm water—'per syringam, aqua calida plenam'. Massa (1536)[4] inflated the kidney by forcing air into the renal vein, and Stephanus

[1] Aretaeus Cappadox, *De causis et signis acutorum et diuturnorum morborum*, Venice, 1552, 4to. *The extant works of Aretaeus the Cappadocian*, F. Adams, London, 1856, p. 280, 8vo.

[2] Claudius Galenus (*c.* 130–200), *De Anatomicis Administrationibus*, lib. ix, cap. 2. Ed. C. G. Kühn, Leipzig, 1821–33, 8vo.

[3] Giacomo Berengario da Carpi (1470–1530), *Commentaria cum amplissimis additionibus super anatomiam Mundini*, Bologna, 1521, 4to.

[4] Nicolaus Massa (ob. 1569), *Anatomiae Liber Introductorius, seu dissectionis corporis humani*, Venice, 1536, 4to.

(1545) [1] devised a pump to inflate the vessels with air—thus making their distribution more conspicuous to the unaided eye. Sylvius,[2] in his last work published shortly after his death, 'the work of his old age', states that the blood-vessels are better seen if they are inflated with air, or when a coloured liquid is introduced into them by means of little tubes. He used variously coloured fluids such as saffron and wines, but rejects the injection method because the liquid escapes when the vessels are cut and the preparation is thereby spoilt. In 1556 the Portuguese physician Amatus Lusitanus [3] filled the vessels with a liquid by means of a siphon, or forced it in through a tube by air pressure supplied from the mouth.

Eustachius [4] is undoubtedly one of the pioneers of the modern injection method, and is credited by later authors, such as Portal and Milne-Edwards, with having practised more elaborate injections than his writings justify. In his work on the structure of the kidney he describes the following procedure. The kidney was kept warm by applying to it a sponge soaked in hot water. Tubes were inserted into the renal artery and vein, and spirit or water was injected into the vessels by the force of the breath. The fluids entered the kidney, extended through its substance, and passed ultimately into the pelvis and ureter. In the converse experiment of injecting the ureter it was found that the injection passed into the tissue of the kidney and reached the arteries and veins. Again, the renal artery was inflated, and air found its way into the ureter. The erroneous belief in a direct connexion between the arteries and the uriniferous tubules, based as it was on injection experiments, survived for a long time, and was taught in the anatomical schools throughout the eighteenth century. It was not until Bowman published his work on the kidney in 1842 that the correct relation of the renal arteries and veins to the glandular tubules was first completely demonstrated. Eustachius also followed the course of the vessels by the tedious process of

[1] Charles Estienne (1504–64), *De dissectione partium corporis humani*, Paris, 1545, fol.

[2] Jacques Dubois (1478–1555), *In Hippocratis et Galeni physiologiae partem anatomicam isagoge*, Paris, 1555, fol.

[3] Johannes Rodriguez da Castello Bianco (1511–68), *Curationum Medicinalium Centuriae* 7, Basle, 1556, Cent. 4 fol.

[4] Bartolommeo Eustacchi (1520–74), *De Renum Structura*, Venice, 1563, 4to; *Tabulae Anatomicae*, Rome, 1714, fol. Text by J. M. Lancisi.

excarnation, in which, unlike later anatomists, he does not appear to have used any injection. Boerhaave is of opinion that Eustachius must have known of a more perfect method of investigation than inflation with air, because of the great exactness with which he traced the vessels, and von Haller considers that plates 13 and 27 of the *Tabulae*, published long after the death of Eustachius, represent the vessels so skilfully that it is hardly credible they could have been prepared without the assistance of fluid injections, unless, as Boerhaave adds, Eustachius brought to bear upon the study of anatomy an 'application and exactness more than human'. In recommending the injection of liquids he was certainly in advance of a time which favoured the distension of the vessels with air rather than filling them with a more solid medium. Thus Laurentius, in 1593,[1] ascertained that there was a connexion between the umbilical vein and the portal vein and inferior vena cava of the foetus by inflating the first named with a 'hollow bugle', and Crooke, in 1615,[2] refers to the necessity of having 'reeds, quils, glasse-trunkes or hollow bugles to blow up the parts'.

At about this time the possibility of diverting blood from the vessels of one living animal into those of another was first conceived and practised. The earliest writer to mention transfusion experiments was Magnus Pegel in 1604, and others are Andreas Libavius (1546–1616) in 1615 and Johannes Colle (ob. 1631) in 1628. It is to be noted that these works were published before Harvey's treatise on the circulation of the blood.

It is appropriate that the most important of the early injection experiments should have been made by Harvey himself—'that ocular philosopher and singular discloser of truth'. His treatise on the circulation, however, is silent on the matter of injections. In a letter addressed to Paul Marquard Slegel, dated March 26, 1651, and published later by Sir George Ent,[3] he returns to the criticisms of Riolan on the doctrine of the circulation, and adds the following remarkable piece of evidence:

'It may be well here to relate an experiment which I lately tried in the presence of several of my colleagues and from the

[1] André du Laurens (1558–1609), *Historia Anatomica Humani Corporis*, Leyden, 1593, 8vo.

[2] Helkiah Crooke (1576–1635), *Microcosmographia*, London, 1615, fol.

[3] Sir George Ent (1604–89), *Opera omnia Medico-Physica*, Leyden, 1687, 8vo.

cogency of which there is no means of escape. . . . Having tied the pulmonary artery, the pulmonary veins and the aorta, in the body of a man who had been hanged, and then opened the left ventricle of the heart, we passed a tube through the vena cava into the right ventricle of the heart, and having, at the same time, attached an ox's bladder to the tube, in the same way as a clyster-bag is usually made, we filled it nearly full of warm water, and forcibly injected the fluid into the heart, so that the greater part of a pound of water was thrown into the right auricle and ventricle. The result was, that the right ventricle and auricle were enormously distended, but not a drop of water or of blood made its escape through the orifice in the left ventricle. The ligatures having been undone, the same tube was passed into the pulmonary artery, and a tight ligature having been put round it to prevent any reflux into the right ventricle, the water in the bladder was now pushed towards the lungs, upon which a torrent of the fluid mixed with a quantity of blood, immediately gushed forth from the perforation in the left ventricle ; so that a quantity of water, equal to that which was pressed from the bladder into the lungs at each effort, instantly escaped by the perforation mentioned. You may try this experiment as often as you please ; the result you will still find to be as I have stated it.'

Priority of publication, however, for this experiment belongs to others. Marchettis, for example, who supported Harvey as against Riolan, in the presence of Thomas Bartholin in 1652, demonstrated that liquid injected into the arteries emerged by the veins.[1]

Passing over Riolan,[2] who practised inflation of the vessels for demonstration purposes, and Lyser,[3] who published the first general treatise on anatomical methods, in which a blowpipe is mentioned for inflating with air or filling cavities with water, but in which injections are not considered, the next contribution to the subject is by Glisson.[4] He injects with a tube to which a bladder containing the medium is attached. The tube is pushed into the vessel and tied, the bladder filled with the injection and bound to the tube, and the fluid forced into the vessels by compressing the bladder at first gently and then more firmly. The

[1] Domenico de Marchettis (1626–88), *Anatomia, cui responsiones ad Riolanum . . . additae sunt*, Padua, 1652, 4to.

[2] Jean Riolan, *fil.* (1577–1657), *Opuscula anatomica varia et nova*, Paris, 1652, 12mo.

[3] Michael Lyser (1627–60), *Culter Anatomicus*, Copenhagen, 1653, 8vo.

[4] Francis Glisson (1597–1677), *Anatomia Hepatis*, London, 1654, 8vo.

media employed are hot water by itself or mixed with milk, water coloured with saffron, and ink. A figure is given of his apparatus, but it is of little assistance in following the description.

We now come to 'that miracle of a youth—Christopher Wren', whose experiments have been described by Boyle [1] and Oldenberg.[2] According to the Philosophical Transactions of 1665, Wren (1632–1723) first suggested to Boyle at Oxford not later than 1659 [3] to ligature the veins of a living animal, open them on the side of the ligature nearest the heart, and inject with 'slender syringes or quills fastened to bladders'. He recommended as the subject 'pretty big and lean dogs'. This operation appears to have been frequently practised in Oxford, and also in London before the Royal Society. 'And they hope likewise, that beside the *medical* uses, that may be made of this *invention*, it may also serve for *anatomical* purposes, by filling, after this way, the vessels of an animal as full, as they can hold, and by exceedingly distending them, discover *new* vessels.' 'To *Oxford*, and in it, to Dr. *Christopher Wren*, this invention is due.' Wren, then a youth of twenty-four, injected wine and ale into the blood of a living dog by one of the veins, and noted that the animal became extremely drunk. The experiment he takes 'to be of great concernment and what will give great light to the theory and practice of physic', and Sprat refers to it as that 'noble anatomical experiment of injecting liquors into the veins of animals'. We may now complete what remains to be said on the history of physiological injections, which we have already seen were known to Pegel and Libavius long before Wren's first experiment in 1656. According to J. M. Vèrdries, a large ox at once died when Wepfer injected air into its jugular vein, and Major [4] asserts that transfusion experiments on dogs were practised by Hans Jurge de Wahrensdorf in 1642. Elsholz [5] injected medicines into the veins of man and dogs, and

[1] Hon. Robert Boyle (1627–91), *Of the Usefulnesse of Naturall Philosophy*, Oxford, 1663, 4to, pp. 62–5.

[2] Henry Oldenberg (1615–77), 'Account of the Rise and Attempts of a way to convey liquors immediately into the mass of blood,' *Philosophical Transactions*, London, 1665, 4to, vol. i, p. 128.

[3] An error. Wren's first experiment was undertaken in 1656.

[4] Johann Daniel Major (1634–93), *Prodromus inventae a se Chirurgiae infusoriae*, Leipzig, 1664, 8vo.

[5] Johann Siegesmund Elsholz (1623–88), *Clysmatica Nova*, Berlin, 1665, 12mo.

PLATE XLII. REINIER DE GRAAF

Tuus humillimus atque
observantissimus famulus
R. De Graaf

From the first edition of the *De Usu Siphonis* (1668). Signature from a MS. in the possession of the Royal Society (Photograph supplied by Dr. Charles Singer).

also practised transfusion of the blood. Lower (1666)[1] was the first to undertake transfusion of blood from an artery of one animal into the vein of another, and in the following year Jean Denys performed the same operation on man—'a circumstance of great exultation to the French'. Boerhaave, in commenting on this result, says: 'the experiment was soon received with great applause both through France and England, and great things were expected from it in the cure of diseases and the recovery of youth . . . but in a little time all these expectations disappeared, and the experiment was prohibited to be made on men by the public law'. In a letter to N. Steno, dated 1667, but not published at the time, Francesco Redi (1626–98) mentions that he had instantly killed two dogs and a hare by injecting air into the veins, but a sheep and two foxes died more slowly. Finally Clarke (1668)[2] states that he had been engaged for many years in injecting various liquids into the blood of living animals, but he had grave doubts as to its utility except for anatomical purposes. The minutes of the Royal Society for May 28, 1660, record that 'Dr. Clarke was intreated to bring in the experiment of injections into the veins'—which doubtless refers to the physiological experiment.

From Glisson up to the time of de Graaf sufficient progress was made in the methods of injection to justify the statement that de Graaf simply collated and fixed the knowledge of his time, but added to it little that was new. Wepfer[3] investigated the vascular supply of the brain by means of injections of saffron water, and gives the first correct description of the course and branching of the carotid artery and of the vessels of the brain membranes. Important results were achieved by Malpighi.[4] He recommends examining the vessels of the lungs under the powerful illumination of the rays of the sun. If this should prove inadequate

[1] Richard Lower (1631–91), 'The success of the experiment of transfusing the blood of one animal into another', *Philosophical Transactions*, London, 1666, 4to, vol. i, p. 352.

[2] Timothy Clarke (ob. 1672), 'Some anatomical inventions and observations, particularly relative to the origin of the injection into veins, the transfusion of blood, and the parts of generation', *Philosophical Transactions*, London, 1668, 4to, vol. iii, p. 672.

[3] Johann Jakob Wepfer (1620–95). *Observationes Anatomicae*, Schaffhausen, 1658, 8vo.

[4] Marcello Malpighi (1628–94), *De Pulmonibus*, Bologna, 1661, fol.

air may be driven into the main trunk of the pulmonary artery, when the vessels will swell up and appear in all their ramifications as if they were carved with a chisel. Even the smallest branches are dilated and stretched out like the branches of a tree. Or if a still more refined method be necessary, mercury can be syringed into the same vessel, whereupon all its branches up to the finest twigs will assume a beautiful silver colour. But the doubt which still, he says, tortured his mind, which his injection of liquids of different colours did not resolve, was whether the arteries and veins were discontinuous, and only connected up indirectly by a spongy parenchyma, or whether the junction was effected directly by microscopic vessels. He repeats Harvey's experiment, and finds that water or a black liquid syringed into the pulmonary artery emerges by the pulmonary vein. Sometimes also it emerges by the trachea. He is not misled by this surprising result, but realizes that injections may extravasate and give rise to erroneous conclusions. Subsequently he saw the blood capillaries in the lung of the frog, but injections played no part in that memorable discovery. A few years later, in 1666,[1] Malpighi is investigating the structure of the kidney by injection methods. He used ink, urine coloured with ink, a black or other coloured liquid, and a black liquid mixed with spirits of wine. A coloured fluid injected with a syringe into the renal artery reaches the smallest branches of the artery and the 'internal glands' (since called Malpighian bodies), so as to produce the appearance of a 'beautiful tree loaded with apples'. He was not able to inject these bodies from the veins, nor could the uriniferous tubules be injected either from arteries or veins. He regards the uriniferous tubule as the excretory duct of the Malpighian body, but did not succeed in actually demonstrating the connexion between them.

Bellini, who was only nineteen years of age when he produced his remarkable work on the structure of the kidney,[2] is said to have used an injection medium which melted with heat. He mentions mercury, which he appears to have employed for injection purposes, and talks of artificially distinguishing the vessels by injecting some coloured liquid into the renal artery and vein, but does not state the nature of the medium or the colours selected.

[1] Marcello Malpighi, *Exercitatio Anatomica de Renibus*, Bologna, 1666, 4to.

[2] Lorenzo Bellini (1643–1704), *De Structura Renum*, Florence, 1662, 4to.

Robert Boyle [1] is responsible for a striking contribution to the technique of injection. He says:

'And perhaps there may be some way to keep the arteries and the veins too, when they are empty'd of blood, plump, and unapt to shrink overmuch, by filling them betimes with some such substance, as, though fluid enough when it is injected to run into the branches of the vessels, will afterwards quickly grow hard. Such may be the liquid plaister of burnt Alabaster, formerly mention'd, or ising-glass steeped two days in water, and then boild up, till a drop of it in the cold will readily turn into a still gelly. Or else Saccarum Saturni, which, if it be dissolv'd often enough in spirit of vinegar, and the liquor be each time drawn off again, we have observ'd to be apt to melt with the least heat, and afterwards to grow quickly into a somewhat brittle consistence again.'

This is the first unequivocal mention of solidifying injection media, and it is noteworthy that two of them, plaster and gelatin, are in use at the present day, and also that Boyle does not suggest the addition of colour—an improvement with which he can hardly have been unfamiliar. Whether he ever carried his suggestions into practice is doubtful. It is true that Grew in 1681 asserts that Boyle was the first to use wax as an injection mass, but Boyle himself says, 'but I must not insist on these fancies'; and again in 1659, when commenting on the anatomical methods of de Bills, 'most of the ways I proposed to myself were as yet little more than bare designs'. Saccarum Saturni, it may be mentioned, is acetate or sugar of lead, the preparation of which by the old method is given by James.[2] It has a waxy consistency.

In his work on the anatomy of the brain, Willis [3] brings the new method into useful practice. The vessels of the brain, he says, are 'seen better and more distinctly, if you first squirt into the carotidick artery some black liquor'. He injects the rete mirabile of ruminants from the carotid with ink, and finds that the injection reaches the carotid of the other side. In another passage his instruments are mentioned, as follows:

'Let the carotidick arteries be laid bare on either side of the *cervix* or the hinder part of the head, so that their little tubes or

[1] Hon. Robert Boyle (1627–91), *Of the Usefulnesse of Naturall Philosophy*, Oxford, 1663, 4to.

[2] Robert James (1705–76), *A Medicinal Dictionary*, London, 1745, fol., vol. iii, art. Plumbum.

[3] Thomas Willis (1621–75), *Cerebri Anatome*, London, 1664, 4to.

pipes, about half an inch long, may be exhibited together to the sight; then let a dyed liquor, and contained in a large squirt or pipe, be injected upwards in the trunk of one side, after once or twice injecting, you shall see the tincture or dyed liquor to descend from the other side by the trunk of the opposite artery; yea, if the same be more copiously injected towards the head, from thence returning through the artery of the opposite side, it will go thorow below the *Praecordia*, even to the lower region of the body; when in the meantime, little or nothing of the same tincture is carried thorow the outward and greater jugular veins. Then the head being opened, all the arteries, before the entrance of the head, and the veins of the same band with them, will be imbued with the colour of the same injected liquor.'

Some years later, in 1672,[1] Willis was the first to inject an Invertebrate. In an involved passage describing the heart and gills of the lobster, he mentions injecting a 'black liquor' into the heart, from which it passed to the gills (!). He also injected the same liquid into the afferent branchial artery of a fish, and noted that it circulated through the gill and finally reached the dorsal aorta. Notwithstanding this significant result his ideas on the circulation of the blood in fishes are confused and difficult to follow. In 1675 [2] Willis investigates the structure of the lungs by means of injections of the arteries, veins, and bronchi. He notes how these three factors are associated in the lung tissue, and points out how impossible it is to grasp their relations, or make a representation of them, without injections. He recommends 'quicksilver, hot and flowing gypsum, wax mingled and made liquid with oyl of turpentine, or some such matter'. No colours are mentioned, which is again extraordinary, seeing that he was writing after Swammerdam, with whose work he must have been acquainted.

On March 27, 1667, Pecquet [3] described a supposed connexion between the thoracic duct, which he had rediscovered in 1651, and the renal vein in Man by inflating the duct with air through a quill. In this work he was assisted by Louis Gayant, one of

[1] T. Willis, *De Anima Brutorum*, London, 1672, 8vo. Also editions at Oxford and Amsterdam bearing same date.

[2] T. Willis, *Pharmaceutice Rationalis*, 2, Oxford, 1675, 4to.

[3] Jean Pecquet (1622–74), 'A new Discovery of the communication of the ductus thoracicus with the emulgent vein', *Philosophical Transactions*, London, 1667, 4to, vol. ii, p. 461. Translation of a letter which appeared in the *Journal des Sçavans*, 1667, 8vo., and elsewhere.

the Parisian comparative anatomists. On February 8, 1672, Pecquet [1] professes to have discovered by injections a direct connexion between the thoracic duct and the postcaval vein. He forced into the thoracic duct with a syringe hot milk or a substance which was fluid when thrown in hot but solidified on cooling. He gives no indication as to what this substance was, but states that the method succeeded 'in part'. Before the operation, the body was warmed by injecting nearly boiling milk. As regards publication, Pecquet slightly anticipates Swammerdam's 'invention' of a solidifying injection medium, and was himself anticipated by Boyle. Walter Needham, who criticizes Pecquet's paper in the *Philosophical Transactions* for 1672, and points out his mistakes, refers to the 'coagulating injection', but does not appear to have been greatly impressed by it.

A year before the publication of de Graaf's tract, 'Theodorus Aldes' [2] injected the arteries of a foetal cow with a black liquid, and found correctly that the uterine cotyledons were uninjected. The reverse experiment of injecting the uterine arteries was responsible for the erroneous belief that the placentae could be injected from the cotyledons. Monro, however, considers that in the latter case Slade's expression '"carried into the substance of the placentae" may signify no more than effused on their unequal pappy substance'.

The small tract by de Graaf, published in 1668,[3] is important not as an original contribution to the subject, but because it brings contemporary knowledge to a focus, and determines the fate of the injection method. He is the first to figure an injecting syringe of the modern pattern, and is credited with having injected mercury into the spermatic vessels. He says his attention had been directed to the subject of injections five years previously, owing to the great difficulty of tracing the blood-vessels by the methods of dissection then in use. By means of his syringe, however, it was possible to demonstrate all the arteries and veins of the body in a single day. Again, it was possible

[1] J. Pecquet, 'Une nouvelle découverte de la communication du canal thorachique avec la veine cave inférieure', *Journal des Sçavans*, Paris, 1672, 8vo.

[2] Matthaeus Slade (1628–89), *Dissertatio epistolica contra Gul. Harveum interpolata*, Amsterdam, 1667, 8vo.

[3] Reinier de Graaf (1641–73), *De Usu Siphonis in Anatomia*, Leyden and Rotterdam, 1668, 8vo.

to establish by experiment the circulation of the blood. A liquid injected into the carotid artery, after circulating through the brain, returned by the jugular vein, and if injected into the

FIG. 1. Engraved title of the first edition of the *De Usu Siphonis* (1668).

pulmonary artery it returned by the pulmonary vein to the left side of the heart. Other experiments of a similar nature are described. Thus de Graaf, as well as Malpighi and others, anticipates the publication of Harvey's pioneer experiment made in 1651. De Graaf describes the injection liquids used. They are all watery preparations and have no permanency. A beautiful dark blue colour, he says, may be obtained by the action of sal ammoniac on copper. If this is found too difficult to prepare, use may be made of colours extracted from flowers, such as the violet, cornflower, and rose. All these colours may be changed by appropriate reagents, so that if only one extract is to hand other colours may be prepared from it. Further colours available are gamboge and indigo, and a green is obtainable by mixing the two. He recommends gamboge and indigo because they are easily procured, and, unlike sal ammoniac, do not affect the metal of the syringe.

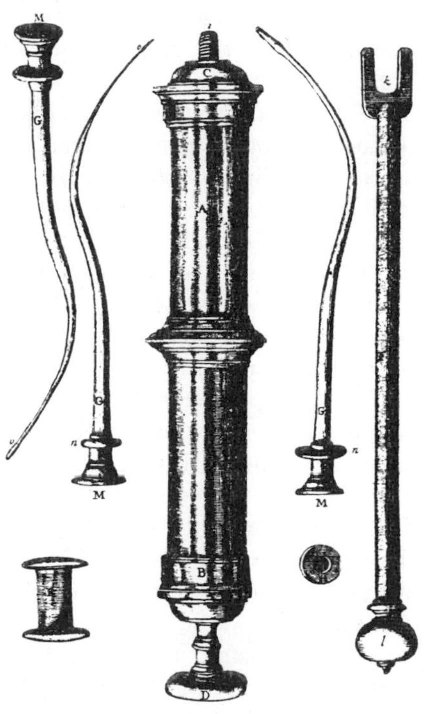

Fig. 2. De Graaf's injection syringe and accessories (1668)

A typical and important experiment is thus described by de Graaf. If the carotid artery is injected towards the head with the green medium one sees

'omnes partes, tam internas, quam externas, quae ab arteriis carotidibus sanguinem accipiunt, colore viridi tingi; quod iucundissimum spectaculum in cerebro exhibet, cuius superficies tam dextra quam sinistra, propter anastomosin, quam interse habent utriusque lateris arteriae ab iniecto colore viridi plurimas quasi arborum figuras repraesentare videntur, quae non sunt nisi arteriarum ramificationes per cerebrum excurrentes'.

Another example of de Graaf's ingenuity is to inject the coeliac artery with a green medium, and the mesenteric artery with a yellow. He was thus able to delimit in a very diagrammatic way the portions of the gut supplied respectively by these two vessels. Milk injected into the liver gives it a white colour, and similarly the gland becomes green if a green liquid is used, from which he concludes that the colour of the liver is dependent on the contents of its vessels. The thoracic duct is injected and distended with milk, and the uterine vessels are inflated with air (1672). He also investigates the connexion between the vas deferens, seminal vesicles, and urethra by injections, and in the same way shows that the penis can be erected after death by injecting the internal iliac artery. His injection experiments were largely responsible for the debateable view, held until comparatively recently, that the Mammalian seminal vesicle acts as a receptacle for the spermatozoa.

De Graaf's syringe is not dissimilar to the modern instrument. It is made of copper or silver. The canula is long and bent, and is screwed directly to the syringe, being tightened with a key and the joint made good by a leather washer. There is no stopcock. The piston is packed with thread, and the key is drilled out to contain brass wires for cleaning the canula. Straus-Durckheim in 1843 states that the injection syringe had hardly been modified since the time of de Graaf, and that Swammerdam's methods were still in use.

In a paper dated December 17, 1668, King [1] describes the structure of the testis. He holds that, apart from the arteries and veins, the testis consists of massed tubules—' is a mere scheme or congeries of vessels '. He proves that they are hollow by injecting them with coloured spirit. These tubules, according to King, can be demonstrated in the rat by dissecting off the tunica albuginea and shaking them out in water, as first shown by de Graaf.

The Parisian comparative anatomists of the seventeenth century [2] made little use of injection methods, but the few experiments they describe deserve attention. They study the structure of the kidney of man and the lion by injecting milk into the renal vein, and are thus able to dispose of the statement of Vesalius that the factors of the renal vein arise in the centre of the kidney. If an injection be thrown into the pulmonary artery of a dog, it traverses the lung and emerges by the pulmonary vein much more easily and quickly if the lung be kept inflated by a pair of bellows. In the tortoise they investigate the relations of the epididymis

[1] Sir Edmund King (1629–1709), 'Observations concerning the organs of generation ', *Philosophical Transactions*, London, 1669, 4to, vol. iv, p. 1043.

[2] Claude Perrault (1613–88), *Mémoires pour servir à l'histoire naturelle des animaux*, Paris, 1671, 1676, fol.

and testis, and conclude that the former is simply a convoluted tube. 'Having made an injection of a coloured liquor into this ductus, a great many other little ductus's were made to rise, which did not appear before, and which went from the testicle to the epididymis: these ductus's being enclosed in the membrane which retained the circumvolutions of the epididymis, and which fastened it to the testicle.' This result, combined with that described by King in 1668, may be said to inaugurate our modern knowledge of the structure of the testis and its related ducts, but it was not until 1745 that Haller put the whole matter in a convincing light.

Swammerdam [1] is usually regarded as the inventor of the solidifying injection mass. We have already seen that this claim cannot strictly be maintained, but it is also undeniable that Swammerdam stereotyped the method, and was responsible for its general adoption by anatomists after his time. Priority of publication may belong to Boyle and Pecquet, but it is to Swammerdam, one of the greatest practical anatomists of all time, that the credit rightly belongs. According to the life by Boerhaave, who is followed by Straus-Durckheim, the first wax injection was made in van Horne's house in Leyden on January 22, 1667, but this date is obviously an error, and is in fact contradicted not only by another passage in the same work, but also by Swammerdam himself, who states that in 1666, 1669, and 1670 he demonstrated his method to van Horne, Slade, Thévenot, and Steno. His results were to some extent incorporated in J. van Horne's Prodromus of 1668, but the facts seem to be that whilst van Horne suggested the wax method, it was Swammerdam who first reduced it to practice. Swammerdam says: 'Factum est, ut D. van Horne proponerem, structuram venarum et arteriarum beneficio iniectae rubrae vel virentis cerae detegi posse.' J. Hudde (1640–1704), the eminent Dutch mathematician, also proposed to Swammerdam the use of coloured injection media. Boerhaave adds that the three plates of six figures were sent to the Royal Society on May 1, 1672, accompanied by the actual specimen itself. The body of the published work is dated March 5, 1670 [1671], and the appendix May 1, 1672. The preparation of the uterus was in the collection

[1] Jan Swammerdam (1637–80), *Miraculum naturae, sive uteri muliebris fabrica*, Leyden, 1672, 4to.

of the Royal Society in 1681, when it was catalogued by Grew in the following words: 'The womb of a woman, blown up and dried. Together with the *spermatick vessels* annexed; and the *arteries* in the bottom of the *uterus*, undulated like the claspers of a vine; all filled up with soft wax. Also the membranous and round *ligaments* of the womb, the *ureters, bladder, clitoris, nymphae, hymen, Fallopian tube,* and the *ovarys,* commonly called the

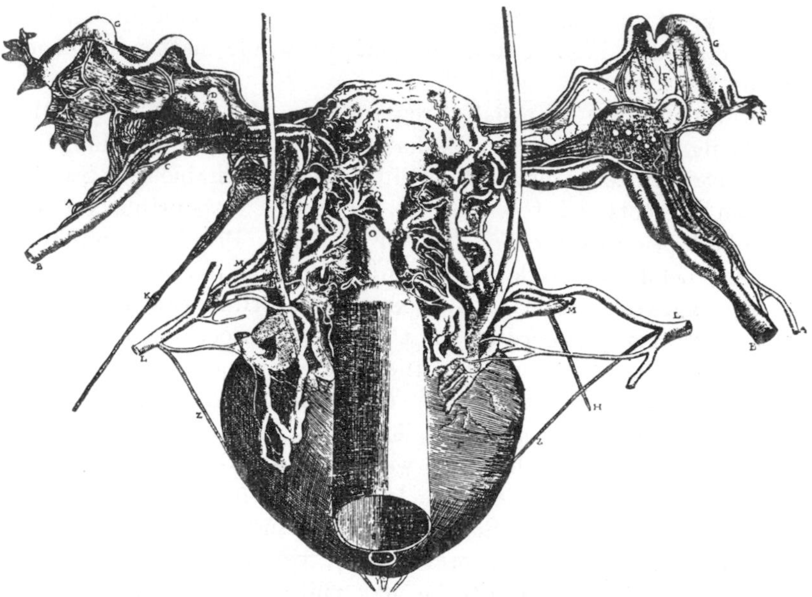

FIG. 3. Uterus injected with red wax by Jan Swammerdam (1671), who presented this specimen to the Royal Society.

testicles; all made most curiously visible, and given by Dr. *Swammerdam.* The descriptions and figures hereof may be seen in the same author's book, printed at *Leyden,* 1672, and presented to the *Royal Society.*' In 1781 the Royal Society's museum was handed over to the British Museum, and Swammerdam's injected specimens of the uterus, gall bladder, and spleen have since been lost sight of. Swammerdam gives a brief description of his method. He used pure white wax, to which, when liquid, the red, yellow, green, or other colour as required was added. The medium was injected as a hot liquid with a syringe, care being taken to avoid the admission of air into the vessels because this was found to impede the injection. Before injecting, the blood was pressed out of the larger vessels in order that they might quickly and easily fill. When cool the wax solidified in the vessels, and the result was a permanent preparation which could not bleed if it should be necessary to dissect it. The discovery was communicated to his friend Ruysch, who hailed Swammerdam in emphatic language as the inventor of the wax method. The uterus described in 1672 was injected with soft red wax, blown up, and dried. Swammerdam also practised colouring the wax differently for arteries and veins, and had previously in 1667 attempted physiological injections of acid liquids into the veins of living animals. In his history of bees, first published in 1738,[1] there is an interesting description of how he injected the blood-vessels of a Lepidopterous larva. He says the blood-vessels of insect larvae are so very delicate and transparent that they cannot ordinarily be discerned,

Fig. 4. Swammerdam's method of injecting the small vessels of insects (first published 1738).

'though there are inventions of art, by the assistance of which we may come to the knowledge of them. In Silkworms I succeeded by the following method. . . . I provide myself with a little glass tube, . . . which I take care to have made like a vial in the middle, at one end to be drawn out to the utmost smallness, and at the other end made thicker and broader, in order that the air blowing into it, may be conveniently forced in at this end: this done, I fill the little pipe with some thin liquor coloured, not, however, of a very penetrating kind, let in through the thicker end, and then with the greatest caution perforating the skin, I thrust the thinner end into the heart. This may be done easily enough. By these means, and then gently blowing into it, the heart, and many of the vessels shooting out from it, may be filled.'

It is generally admitted that Frederik Ruysch, the Professor of Anatomy at Amsterdam, is the apostle of the injection method,[2] which was in fact for some time referred to as the 'Ruyschian

[1] J. Swammerdam, *Biblia Naturae*, Leyden, 1737–8, fol.

[2] Frederik Ruysch (1638–1731), *Opera Omnia Anatomico-medico-chirurgica*, Amsterdam, 1721–5, 4to.

art'. Even his opponent Bidloo admits that he was a 'subtle butcher', who welcomed even a civil war which was to provide him with material for his studies. Ruysch was occupied with anatomy between 1665 and 1728, and from about 1675 onwards was recognized as the leading exponent of injection experiments. He was the intimate friend of de Graaf, Swammerdam, and Boerhaave. Ruysch's object was to produce finer and more complete and permanent injections than his predecessors. In the latter respect he was anticipated by Swammerdam, but he succeeded in being the first to obtain extensive injections of the blood capillaries. His preparations surprised 'even the most learned part of mankind', and, to quote Boerhaave, he was the first to demonstrate that there were not two arteries distributed alike throughout the whole body, for 'in the liver they appear like small pencil brushes, in the testicles they are wound up like a ball of thread, in the kidneys they are inflected into angles and arches, in the intestines they ramify like the branches of trees, in the uvea they form circles and radii, in the brain they are waved in and out in a serpentine course, in the omentum they are disposed something like the meshes of a net, and in almost every other part of the body they assume a different and peculiar structure'. Already in 1664, as he tells us himself, Ruysch was making injected preparations of the spleen of the calf which were publicly demonstrated with applause by J. van Horne.

Galen and the ancients believed that there were parts of the body, to which they gave the name 'spermatic', which had no vascular supply of any kind. This belief survived until the seventeenth century, when the fine wax injections of Ruysch demonstrated the occurrence of blood-vessels in bones, ligaments, tendons, and membranes—in fact in all living tissues almost without exception. Ruysch therefore may claim the credit of establishing the ubiquity of the vascular system.

Ruysch followed Swammerdam in injecting a substance which cooled in the vessels, and thus ensured the permanence of the preparation. His first attempts exceeded the achievements of Swammerdam, and vessels of extreme tenuity became visible to the naked eye. He was hence able to discover the thin fibrous periosteum of the auditory ossicles (1697), which had up to that time been regarded as naked, the vasa vasorum, the capillaries of the bronchial tubes, and the vessels of the choroid and other membranes. It is natural that he should exalt the significance of his own discoveries, and that he should regard this most subtle and intrusive vascular system as having an existence *per se.* In 1723, almost at the end of his life, his enthusiasm for injections is expressed in the following passage: 'Vah quantum est, quod nescivimus ante repletionis artem! quantum didicimus per illam! O mirus naturae in ultima vascula olim invisibilia appulsus, quo tam abscondita manifestantur! O grata repletio, numquam contemnenda deinceps, numquam laudanda satis!' He claims, for example, that by establishing the high vascularity of the cerebral cortex, he has disproved its glandular nature as asserted by Malpighi, and his substitution of the vascular for the glandular theory of the cortex tends to obscure the important advance which was actually made. He is unusually and unfortunately emphatic on this point. Of all the discoveries which he has been making for forty years, he says in 1705, the most important is the proof that the cerebral cortex is not a gland but only a mass of blood-vessels. His injections were so refined that the parenchymatous elements were overlaid and obscured by the multitude of vessels which sprang into view. This, and doubtless certain extravasation effects, induced him in 1696 definitely to formulate the doctrine that the tissues were only vascular networks variously arranged ('totum corpus ex vasculis'). Already in 1695 H. Ridley had stated this view, and as late as the eighteenth century William Hunter was still teaching that the glands consisted of a concourse of vascular elements and excretory ducts without any definite or apparent parenchymatous basis. Ruysch had injected the arteries and veins of certain glands, macerated them in water, and then unravelled them without finding anything but a tangle of injected vessels. Ruysch, however, did not hold that a gland was formed wholly of blood-vessels, but admitted that between the extremities of the vessels there existed a neutral pulpy substance. He believed, further, that the blood was poured directly into the factors of the excretory ducts of the glands, since his

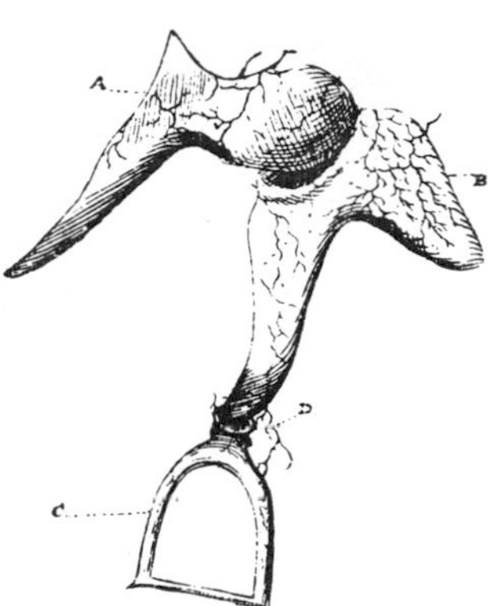

FIG. 5. Periosteum of the auditory ossicles injected by Ruysch (1697).

wax injection thrown into an artery emerged by the excretory duct without entering any intermediate non-vascular tissue or spaces on the way. 'The ceraceous injection of Ruysch being artfully impelled makes its way [via the uriniferous tubules] even into the pelvis of the kidney from the arteries' (Boerhaave). 'This was the principal ground for the famous, but now exploded theory of the existence of exhalant arteries with open mouths, which in the secretory glands opened directly into the excretory ducts' (Bowman). On this view there was no need for a parenchyma, but Ruysch was unable to deny the existence of a non-vascular tissue in the lingual glands, although he maintained that the lobules of the liver consisted of bunches of vessels without any parenchyma. His views on the ultimate structure of animal tissues were strongly supported by Cowper and Nuck.

As a result of Ruysch's experiments there arose a 'despotism of injections', which attributed the functions of the tissues to the specific disposition and peculiarities of their vascular supply. This disposition was believed to exist in infinite variations, and the diverse activities of the tissues were explained as the necessary consequence of such variations. Malpighi, who had correctly stated the relations between glandular tissue and its blood supply, was thus strongly opposed and beaten down by Ruysch and his followers, who denied the very existence of glandular tissue, and saw nothing in a gland but a subtle complex of blood capillaries. Two exceptions to this vascular autocracy were admitted—the ovary and the testis, the fabric of which Ruysch was never able to prepare by injection, and he concluded therefore that red blood did not penetrate into the essential parts of these glands. Nuck and Cowper (1697) also failed to inject the testis.

Another consequence of Ruysch's injections was an improved method of preserving bodies from putrefaction, for which his name became famous in all the schools of Europe. In 1666, by order of the Dutch Government, Ruysch, who was then in his twenty-eighth year, undertook to preserve by vascular injection the body of Vice-Admiral Sir William Berkeley, whose ship had been captured, and the admiral himself killed, when leading the van in the gallant but hopeless contest of June 1. The experiment, though made more difficult by the putrefaction of the body and the admiral's wounds, was wholly successful, and the Dutch chivalrously returned the body to England in condition as fresh, we are told, as that of an infant. Ruysch was encouraged to attempt other preparations of a similar nature for his own Museum, which has been described as a 'perfect necropolis, all the inhabitants of which were asleep and ready to speak as soon as they were awakened'. Ruysch himself says: 'Sunt mihi parvula cadavera, a viginti annis balsamo munita, quae tam nitide sunt conservata, ut potius dormire videantur, quam exanimata esse corpuscula.' Entire bodies of infants and adults were mummified by vascular injection of possibly some preparation of arsenic, with what success the following paraphrase of an eloquent passage in Eloy provides ample testimony: [1]

All the bodies which he injected preserved the tone, the lustre, and the freshness of youth. One would have taken them for living persons in profound repose—their limbs in the natural paralysis of sleep. It might almost be said that Ruysch had discovered the secret of resuscitating the dead. His mummies were a revelation of life, compared with which those of the Egyptians presented but the vision of death. Man seemed to continue to live in the one, and to continue to die in the other.

A quaint manifestation of Ruysch's interest in injections is to be found in his Museum. The skeletons are thrown into dolorous attitudes and provided with anatomical pocket handkerchiefs of injected omentum, their wrists are adorned with organic and injected frills, and even the bladder used to seal the mouths of the jars has been carefully injected. Of the methods by which his wonderful preparations were produced we have but scanty knowledge. Even his friend Boerhaave, who was occasionally present when Ruysch was injecting, is silent as to his methods. When Peter the Great, on his second visit to Amsterdam, acquired Ruysch's collections in 1717, having first seen and coveted them in 1698, it is said that he stipulated that the preparations should be accompanied by a description of the methods of the preparateur. This was accordingly drawn up by Ruysch, and ultimately found its way into the library of the University of St. Petersburg, founded in 1819. In 1742 an account of this manuscript, based on the copy in Ruysch's handwriting, was published by Joannes Christophorus Rieger. Rieger had been in the employ of Peter the Great, after whose death in 1725 he retired to Holland, where

[1] Obviously inspired, however, by a passage in Fontenelle's *Éloge de Ruysch*, published in 1731, which itself owes something to the rhetoric of Ruysch himself.

he lived over a bookseller's shop, and compiled the work which includes the description of Ruysch's methods. He states that he saw the manuscript by permission of the President of the Imperial Academy of St. Petersburg. Now, the Academy was only founded in 1725, so that he must have examined the manuscript before his retirement to Holland, and after the death of Peter the Great. It seems probable that Ruysch's collections and manuscript were for a time handed over to the care of the Academy, since a description of the preparations was issuéd by that body in 1741.[1]

From Rieger we learn the following details of Ruysch's procedure. Macerate the body to be injected in cold water for a day or two. Slit up the aorta and vena cava and press out the blood, and immerse the body in hot water from four to six hours. The injection used is in the winter suet or tallow, and in the summer a little white wax is added. Another mass may be prepared by mixing wax, turpentine, and resin. The colour employed is vermilion, or spirits of wine and vermilion, and the larger vessels are afterwards filled with wax to prevent the escape of the spirit. Two tubes are fixed in the aorta, one being directed forwards and the other backwards, and the vena cava is ligatured. The vessels are filled by means of two heated syringes, and the subject is to be moved about continually in cold water after the operation to prevent the gravitation of the heavy vermilion whilst the mass is setting. Having been injected the specimen is preserved in diluted alcohol, which Ruysch distilled himself from barley. Black pepper was added to assist its penetrative power. The strength of the alcohol was only about 67 per cent.—too weak for the purposes of permanent conservation, although no deterioration was observable during Ruysch's own lifetime. Hyrtl suggests that the secret was handed over to Rieger when it was seen that the preparations were becoming worthless, and we have the statement of Jesse Foot, published in 1794, that 'I saw the preparations, belonging to Ruysch, which are deposited in the Museum at Petersburg, going apace into decay', and before this, in 1748 Lieberkühn, who had examined examples of Ruysch's injection mass, considered it too fluid to last, whilst the preparations themselves did not stand microscopic examination. In order to display the

[1] Ruysch's second collection was catalogued by himself in 1724 and 1728, and by Abraham Vater, who had been one of his pupils, in 1736–40.

PLATE XLIII

FRIEDRICH RUYSCH

From the first edition of his collected works (1721)

ALEXANDER MONRO *PRIMUS*

From his collected works (1781)

injected microscopic vessels Ruysch cleared the tissues with oil of lavender or turpentine.

It is obvious in the Rieger document that Ruysch conceals more than he discloses. In the subsequent literature there are many conjectures by various anatomists as to the composition of Ruysch's media, but none of them carry the authority or conviction of the definite if meagre statements of Rieger. Besides injection, Ruysch was very successful in the use of the inflation method. The lymphatic vessels were inflated by means of glass tubes, dried, and dissected so as to expose the valves, of which Ruysch, though not the discoverer, was the earliest to publish a careful study in his first paper issued in 1665.

From Ruysch to Lieberkühn, who established the importance of microscopic injections, there is an extensive literature—not, however, of sufficient interest to be dealt with other than briefly. Blankaart [1] seems to have been the first to demonstrate *by injections* that the connexion between arteries and veins was not by a spongy parenchyma but by capillaries, a conclusion already reached by Malpighi, and confirmed a few years later by Lange [2] and Leeuwenhoek (1689). Caspar Bartholin,[3] the son of Thomas and the grandson of old Caspar, published two works dealing with injections when he was still very young, in the earlier of which he was accused by his contemporaries of subtle plagiarism. His advance on de Graaf is that he was the first to recommend systematically flushing out the vessels with water before throwing in the coloured injection, thus anticipating a modern refinement. He proceeds as follows: the part is steeped in tepid water in order to soften the clotted blood, which is then removed from the vessels by an injection of warm water. This preliminary operation is facilitated by ligaturing the return vein and suddenly releasing the ligature, when the sharp rush of the hot water carries the blood with it. He claims for this method a finer and more general injection, but this was questioned by later writers, who asserted that the water left the vessels, and produced misleading infiltra-

[1] Steven Blankaart (1650–1702), *Tractatus novus de circulatione sanguinis*, Amsterdam, 1676, 12mo.

[2] Christian Johannes Lange (1655–1701), *Disputatio de circulatione sanguinis*, Leipzig, 1680, 4to.

[3] Caspar Bartholin (1655–1738), *De diaphragmatis structura nova*, Paris, 1676, 8vo. *Administrationum anatomicarum specimen*, Frankfurt, 1679, 8vo.

tions in the surrounding tissues. Bartholin's apparatus was either a modified syringe or a specially constructed machine devised to facilitate his irrigation experiments. The principle of both, however, is the same, viz. that an unlimited supply of the injection fluid can be forced into the vessels without withdrawing the nozzle from the artery. This is effected by a T-piece and valves, but he admits that, owing to temperature difficulties, his contrivance is not suitable for wax injection, for which the ordinary syringe must still be used. Bartholin was the first to devise an injection apparatus with a continuous feed, but it found little favour until Straus-Durckheim in 1843, and Robin in 1849, developed the idea and recommended a more complex form of it. It is to be noted that de Graaf and Bartholin are the first to figure injection syringes. Bartholin injected air, water, and various coloured liquids, but does not favour wax. He produces a green by mixing gamboge and indigo, as already described by de Graaf.

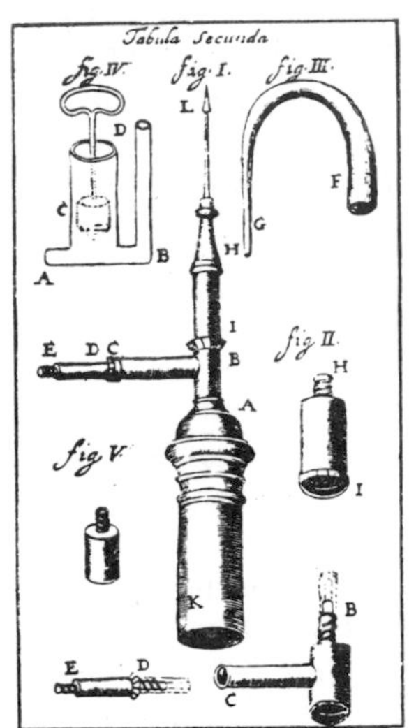

Fig. 6. Injection appliances of Caspar Bartholin (1679). The syringe can be recharged without disconnecting the apparatus.

Duncan is perhaps the earliest writer definitely to advocate the use of mercury,[1] and to contrast the arteries and the veins by injecting them with different colours. He employs usually melted wax thinned with turpentine and oil, and also a black liquid. His wax injection is 'according to the method of M. Swammerdam', except that he recommends that as the wax hardens quickly it is better to inject the animal when it is still alive! He refers to Swammerdam also as experimenting with mercury for injections. Duncan injected one colour first by the jugular vein, and then another colour by the vertebral artery. 'Thus the arteries and veins are easily distinguished by their different colour', and the communications between the arteries and veins are demonstrated. The experience which the great comparative anatomist Duverney derived from injections was strangely

[1] Daniel Duncan (1649–1735), *Explication nouvelle et mécanique des actions animales*, Paris, 1678, 12mo.

unfortunate.[1] In two papers written in 1679 and 1683, but not published until 1733, he anticipates the speculation of Ruysch by stating that the so-called solid tissues of the body are nothing but a miraculous concourse of different vessels, and that when all the liquids have been expressed from a tissue nothing is left but canals and vesicles. In the Stork his injections failed to reveal the lacteal vessels or the thoracic duct, the occurrence of which in Birds he therefore doubts. An injection thrown into the mesenteric vein passes into the cavity of the intestine, and conversely if a portion of the gut be filled with milk, ligatured at both ends, and then compressed, the milk is forced into the mesenteric vein. He holds that the chyle in the Bird passes via the mesenteric veins to the liver.

Dismissing Charleton,[2] who refers to 'various manual operations besides meer dissection', such as 'inflations, injections of divers liquors by syringes', Simon Lescot (*c.* 1615–90), the surgeon who is supposed to have introduced Swammerdam's wax method into France *c.* 1680, and de Heyde, who practised physiological injections on living dogs,[3] we come to Bidloo,[4] who inaugurated an injection experiment which, applied as it afterwards was to the blood system, became very popular in the eighteenth century. He filled the lungs with 'fused bismuth', afterwards removing the soft parts by corrosion, thus producing a permanent cast in metal of the cavities of the lung. By 'fused bismuth' must be understood not the pure metallic substance, which melts at too high a temperature for the purpose, but a complex mixture containing bismuth and mercury, some preparations of which had almost the consistency of wax.

The English comparative anatomist Samuel Collins [5] has never achieved the reputation which his work undoubtedly deserves. It is contemptuously regarded by Hutchinson as 'of less value than the head that is placed before it'—referring to the beautiful

[1] Joseph Guichard Duverney (1648–1730), *Histoire de l'Académie Royale des Sciences*, Paris, 1733, 4to, T. 1, pp. 278 and 363.

[2] Walter Charleton (1619–1707), *Enquiries into Human Nature*, London, 1680, 4to.

[3] Anton van der Heyde, *Centuria Observationum Medicarum*, Amsterdam, 1683, 8vo.

[4] Govard Bidloo (1649–1713), *Anatomia humani corporis*, Amsterdam, 1685, fol.

[5] Samuel Collins (1618–1710), *A System of Anatomy*, London, 1685, 2 vols., fol.

engraving of Collins by Faithorne which constitutes the frontispiece. Collins, however, made little use of injections, and his few experiments served but to mislead him. He employed white wax and vermilion, and as a result adopted views on the relations of

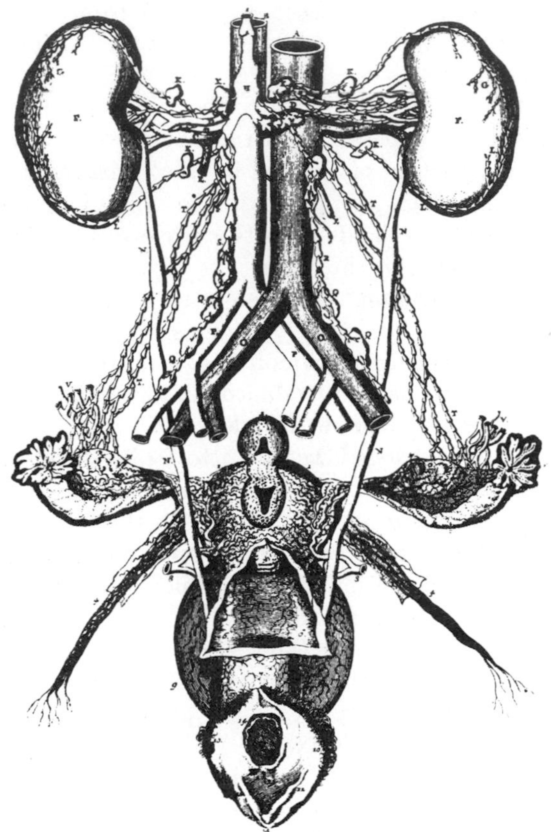

Fig. 7. The lymphatics of the urogenital organs injected with mercury by Anthony Nuck (1691).

the cephalic and genital arteries and veins contrary to the knowledge even of his own time. His contemporary Nuck,[1] on the other hand, wielded the syringe with admirable results. He injected red coloured wax and coloured fluids. The wax was mixed with oil to make it sufficiently fluid to reach the finest branches, but his chief medium was mercury, which he was the

[1] Anthony Nuck (1650–92), *De ductu salivali novo*, Leyden, 1685, 16mo. *Adenographia curiosa*, Leyden, 1691, 8vo.

first to put to extensive use. He also experimented with an amalgam of mercury and lead or tin, and a preparation called tinctura mercurialis, the composition of which is not given. Nuck's general description of the lymphatics is one of the most complete before Mascagni. He investigated the structure of the secretory glands by injecting their ducts, lymphatics, and arteries and veins, and also by inflation. He was the first to demonstrate, by means of mercury injection, that the lymph stream passes through and beyond the lymphatic glands and is not interrupted by them. He was also the first thoroughly to explore the constitution of the secretory and lymphatic glands and of the lymphatic system by mercury injections. But his experiments led him into numerous errors. He found that mercury passed from the lacteals and lymphatics into the arteries, and concluded from this that the lymphatics arose from the arteries. In the same way he deduces that the ramifications of the arteries are connected up with the factors of the salivary ducts. He inflated the arteries of the spleen and kidney and saw air pass into the lymphatics, and hence believed that the lymphatics were veins. It was not until 1757 that Monro secundus finally demonstrated that Nuck's errors were due to extravasation effects. A similar misconception was based on his injection with mercury of the glandular lobules of the mammary gland, when the mercury found its way into the blood-vessels—especially the arteries.

The botanist Camerarius [1] was one of the pioneers of mercury injection. He injected the testis from the vas deferens, and other organs, with milk, wax of diverse colours, and mercury, and also inflated them with air. Compare the preceding and later work on the injection of the testis of Perrault, King, and Haller. The inevitable mercury extravasation induced the belief that there was a connexion between the seminal tubules and the lymphatics. In 1686 A. van der Heyde injected the gastro-vascular canals of Aurelia with a black liquid. Contrary to expectation Leeuwenhoek [2] does not appear to have practised injections himself, but he mentions injections of hot wax, and gives the following interesting description of a mercury injection: 'A certain doctor of Physic,

[1] Rudolphus Jacobus Camerarius (1665–1721), 'De nova vasorum seminiferorum et lymphaticorum in testibus communicatione', *Ephem. Acad. Nat. Cur.*, Ann. 1686, 1688, Nürnberg, 1687, 1689. Dec. 2, Ann. 7, p. 432, 4to.

[2] Anthony van Leeuwenhoek (1632–1723), *Vervolg der Brieven geschreven aan de Koninglijke Societeit tot London*, Delff, 1689, p. 336, 4to,

to whom, among other persons, I had shown the circulation of the blood, told me that this circulation had also been exhibited to him by a chirurgical gentleman; and on my desiring to know how it was shown to him, he said by injecting quicksilver into an artery, which circulated back again through a vein; but when I asked him how they were assured that one of the vessels in which the experiment was made was an artery, and the other a vein, he answered that they were not certain as to that point. I also asked him what was the size of the vein in which the quicksilver, so injected, was circulated; to which he answered, that it was above a thousand times larger than those vessels in which he had seen the circulation of the blood at my house.' Leeuwenhoek was not convinced by this experiment, holding that it was possible the injection might return by an artery instead of a vein, and thus not necessarily demonstrate the circulation. On the other hand, Ray [1] believes that the structure of the body may easily be detected by blowing air into the vessels and drying the preparation, or by injecting melted wax or quicksilver with syringes. He accepts the doctrine of the tubular (i. e. non-parenchymatous) structure of the glands, and considers it wonderful that 'all the glands of the body should be congeries of various sorts of vessels cur'd, circumgyrated, and complicated together, whereby they give the blood time to stop and separate through the pores of the capillary vessels into the secretory ones, which afterwards all exonerate themselves into one common ductus'. In his work on the brain Ridley [2] found injection experiments very helpful. He says: 'Other bodies have been introduc'd by injection, as *tinged wax* and *mercury*, the first of which by its consistence chiefly, the other by its permanent nature and colour, contribute mightily towards bringing to view the most minute ramifications of vessels, and secretest recesses of Nature.' He prefers mercury to wax, the latter being too coarse for the finest vessels. 'By an injection with mercury I find scarce any nerves but what hath some such small ramifications of blood-vessels in them.' His 'chief hopes' were based on mercury, but he is almost the first to recognize that extravasations might make mercury injections 'altogether useless'.

[1] John Ray (1628–1705), *The Wisdom of God manifested in the works of the Creation*, London, 1691, 8vo.

[2] Humphrey Ridley (1653–1708), *The Anatomy of the Brain*, London, 1695, 8vo.

The Ruyschian art found an enthusiastic supporter in the arch-plagiarist Cowper.[1] Besides inflating the vessels with air and drying the preparation, he used as injection masses plaster of Paris, wax, and mercury. With the latter he injected the biliary ducts, kidney, lymphatic system, and carotid artery. He demonstrates by mercury injections that the maternal blood-vessels penetrate into the placenta,[2] but although he found, as related by Drake, that mercury injected into the uterine artery of a cow passed via the cotyledons into the foetus, he hesitates definitely to commit himself to the continuity of maternal and foetal blood. There was a strong tendency at the time to believe so natural an assumption, as witness the following passage from Drake (1707): 'On the other hand, if the arteries of the uterus were continued to the veins of the same part, and those of the foetus in like manner, without communicating with each other, their confluence in the placenta seems to be altogether impertinent and of no use, and the umbilical arteries and vein fram'd for no other service or purpose, than to give the blood room for an idle sally.' Cowper injected mercury into the lactiferous tubes of the mammary gland, and he confirms Nuck in finding it emerge by the blood-vessels. His injection experiments lead him into many other errors, as, for example, when he describes a mass thrown into the renal artery escaping by the ureter. Mercury is used, and also variously coloured wax, to inject other than vascular cavities, e. g. the Fallopian tubes. The arteries of a foetus are demonstrated by a wax injection. He uses wax of four different colours to inject the liver—hepatic artery, red; portal vein, 'dark' colour; post-caval, 'distinguishable' colour; and bile duct, yellow. The liver lobules were next roughly separated, slightly macerated, and the parenchyma removed with a stiff brush of hog's bristles. Wax is injected into the salivary and pancreatic ducts to ascertain how these ducts are formed within the gland. Cowper also filled the pulmonary passages with 'block tin', and macerated off the soft parts. These preparations of liver and lung are among the earliest corrosion experiments of which we have any record, but the idea is obviously borrowed from Bidloo, who in his turn owes something

[1] William Cowper (1666–1709), *The Anatomy of Humane Bodies*, Oxford, 1698, fol. Some copies are dated 1697, but the work was not published until May, 1698.

[2] Monro concludes that on this point Cowper is speaking *à priori*, and not from what he actually saw.

to the excarnation preparations of Spigelius and Glisson described in 1627 and 1654. Block tin in the seventeenth century was a more or less pure tin as now, and hence on account of its high melting-point could not have been employed as Cowper states. He must therefore have made use of an alloy.

A short but important paper by Homberg was written in 1699, but not published until 1702.[1] Homberg experimented with wax, mercury, and turpentine—not the modern oil of turpentine, but a resin of the consistency of honey. He is also the first, if not to introduce, at least to demonstrate the practicability of injecting metals of low fusibility. His metal injection consisted of equal parts of lead, tin, and bismuth, which remains liquid at a temperature below the scorching-point of paper. Subsequently Newton and d'Arcet found that by altering the proportions of these three ingredients, and principally by increasing the percentage of bismuth, the melting-point was reduced, and the addition of mercury carried the reduction still further. Homberg finds the consistency and melting-point of wax and turpentine unsuitable for injection, whilst mercury escapes from the smallest cut. He believes the principal difficulty to be air in the vessels, and he professes to have overcome this difficulty by means of a complicated piece of apparatus which inflated and dried the vessels, thus facilitating the escape of the air. This apparatus, however, was afterwards discarded as an unnecessary refinement, and he next tried the interesting experiment of exhausting the air in the vessels by means of an air pump before throwing in the injection—a method said to have been one of the secrets of Ruysch. Then he runs in the hot metal as above, and after it has cooled, he removes the soft parts by maceration or otherwise, so that he has finally a permanent metal cast of the cavities of the vessels. He is alive to the danger of attempting such an injection on material which has been in water, and he recommends that such material should be first dried for a day in the air pump. Homberg is undoubtedly the pioneer of the so-called corroded preparation—a method which lost interest in the nineteenth century but was never absolutely abandoned, whilst recently the popular use of the low fusible alloys has resulted in its re-introduction.

In the first half of the eighteenth century the sciences of

[1] Guillaume Homberg (1652–1715), 'Essais sur les injections anatomiques,' *Hist. de l'Académie Royale des Sciences*, Paris, 1702, Ann. 1699, 8vo.

Anatomy and Physiology were making great progress principally by means of injection methods. All the literature of the period refers to the still fascinating doctrine of the circulation, and the injection syringe was the most popular and trusted instrument of the time. Hovius [1] worked out the circulation in the iris, ciliary process, and lachrymal gland by injections of wax and mercury. He describes vessels in the cornea, although none are present in a state of health, and professes also to have discovered vessels in the lens and humours of the eye, and to have established a definite circulation in those parts. Méry's injections of air and water [2] represent a very definite advance towards an understanding of the mechanism of respiration, but the unsoundness of his methods, as pointed out by G. B. Bülffinger in 1732, was against a more important result.

Vieussens' injection experiments [3] can only be generally condemned, and it is doubtful whether he obtained a single valid result. In his earlier work on the brain he used spirits of wine or brandy tinged with a saffron colour, and also black and green liquors and ink, but he gives no account of his material or procedure. Only local injections are described. Thus coloured spirits of wine pumped into the carotid artery reaches the longitudinal sinus directly from the arteries of the dura mater—an error afterwards corrected by Ruysch and Rau. His later publications are more detailed as regards injections, and he is now using mercury. His injections of the latter substance, obviously conducted at too great a pressure, result in deplorable errors. Mercury thrown into the trachea passed into the blood-vessels of the lungs, if injected into the mesenteric artery it escaped into the cavity of the intestine, from the cystic artery it found its way into the gall-bladder, and from the uterine arteries it penetrated into the cavity of the vagina but not that of the uterus. As the result of his injections he deduces that there is a connexion between the arteries and the lymphatic system. The lymphatic duct is regarded as a lymphatic-

[1] Jacobus Hovius, *De circulari humorum ocularium motu*, Inaug. diss. Utrecht, 1702, 4to. Editio nova, Leyden, 1716, 8vo.

[2] Jean de Méry (1645–1722), *Histoire de l'Académie Royale des Sciences*, Paris, 1703, Ann. 1700. Ibid. 1708, Ann. 1707, 8vo.

[3] Raymond Vieussens (1641–1716), *Neurographia Universalis*, Lyons, 1684, fol.; *Novum vasorum corporis humani systema*, Amsterdam, 1705, 8vo.; *Dissertatio anatomica de structura uteri et placentae muliebris*, Cologne, 1712, 4to; *Expériences et Réflexions sur la structure et l'usage des viscères*, Paris, 1755, 8vo.

arterial-nervous apparatus, and he invents a new class of short-circuit vessels to explain the rapidity with which liquids taken into the stomach are removed by the kidneys, to account for which, he claimed, the Harveian circulation was too slow. Again, he injected the left coronary artery with coloured brandy, and describes the fluid as passing 'without violence' not only into the entire substance of the left auricle, but also into its cavity. Hence we have another of his theories that the coronary arteries and veins are not connected up by capillaries but by the cavities of the heart, with which both arteries and veins communicate by conspicuous apertures. An experiment of Vieussens which obtained wide currency and belief was the injection of the carotid artery of a living bitch with four pounds of mercury. He holds that the mercury 'without breaking any vessels, or the effusion of one drop of blood, passed through the placenta surrounding each whelp, and was pushed into the umbilical vessels themselves'. This experiment was twice repeated by Monro with quite different (but correct) results. The latter states that the mercury travelled into the very minute branches of the vessels of the maternal placenta, but none whatever reached the umbilical vessels or the foetus.

Noues,[1] who used rectified spirits with cinnabar ground in, and speaks of injections of coloured wax, has little of importance, but Schacher[2] carries Homberg's experiment a stage further, and is the first to *draw* the injection through the vessels with a vacuum pump. His mass, besides starch, is pig's fat, mutton suet, and spermaceti, each in combination with wax, to facilitate the filling of the smallest vessels, and the colours are vermilion, red oxide of lead, verdigris, Florence lake, the blue extract of corn-flower, and carmine. At this time Salzmann[3] describes injecting the thoracic duct from the lymph vessels in the neighbourhood of the renal vein, and Cheselden[4] injects the blood-vessels with wax and the lymphatics with mercury, but does not discuss his procedure. He

[1] Guillaume des Noues, *Lettres de Mr. de Noues à Mr. Guillielmini*, Rome, 1706, 8vo ; *Avertissement pour les anatomies toutes nouvelles de cire colorée*, Paris, 1717, 12mo.

[2] Polycarpus Gottlieb Schacher (1674–1737), *De anatomica praecipuarum partium administratione*, Leipzig, 1710, 4to.

[3] Johannes Salzmann (1672–1738), *Dissertatio encheiresis inveniendi ductum thoracicum*, Strassburg, 1711, 4to.

[4] William Cheselden (1688–1752), *The Anatomy of the Humane Body*, London, 1713, 8vo.

points out that the lymphatics of fish, first observed by T. Bartholin, may be seen in the mesentery without injection. Injection as a method of preservation was practised by Ravius,[1] many of whose preparations in the Anatomy Hall at Leyden were catalogued by his successor, B. S. Albinus, in 1725.

An early, if not the earliest, attempt at histological injection is described by Muys.[2] He first of all removed the blood from the vessels by an injection of warm water, and then threw in a coloured liquid. His object was to ascertain the vascular supply of muscle fibres, and he wrongly concluded that the muscle fibrillae were tubular, and that the capillary artery discharged 'a part of its liquor' into the cavities of the fibrillae. The latter, however, were too small to admit the coloured blood corpuscles, which therefore underwent fragmentation before entering the fibre. A figure supports this flight of the imagination. The short paper by Rouhault, written in 1718,[3] is important and exceptional, since no attempt is made to conceal laboratory methods. He says the mass usually employed is a mixture of hog's lard, white wax, mutton suet, and turpentine, or spirit of turpentine charged with a little wax. His colours are vermilion for arteries and verdigris or indigo for veins, and he claims to be the first to employ different colours for arteries and veins. In this, as we have already seen, he is mistaken. Before throwing in the mass, the blood is cleared out of the vessels, and the body warmed, by an injection of tepid water, after which the body is wrapped in warm linen. The above injection does not pass through the capillaries, and he only uses it to demonstrate the coarser circulation, admitting that it gives preparations inferior to Ruysch's. We have pointed out that Boyle was the first to suggest isinglass as an injection mass, but Rouhault may claim the credit of demonstrating its importance. He was not aware of Boyle's work, and the use of gelatine was suggested to him by Méry. He dissolved Ghent glue or isinglass in water, and in 1716 obtained a perfect injection of the finest vessels, the mass, for example, passing right through the placenta

[1] Johannes Jacob Rau (1668–1719), *De methodo anatomen docendi et discendi*, Leyden, 1713, 4to.

[2] Wyer Guillaume Muys (1682–1744), *Journal Littéraire*, La Haye, Jan., Feb., 1714, 8vo.

[3] Pierre Simon Rouhault (ob. 1740), 'Sur les injections anatomiques', *Histoire de l'Académie Royale des Sciences*, Paris, 1720, 8vo, Ann. 1718.

and emerging by the veins. He exhibited at the Academy the vessels of the placenta injected with different colours. Comparing his preparations with some believed to have been injected by Ruysch's method, he considered his own as good as Ruysch's, which latter he adds, had *not* been injected with wax. Rouhault also used spirits of wine coloured with orchanette or carmine, which penetrates equally well into the finest vessels, but, like mercury, the preparation cannot be dissected afterwards without bleeding, and hence this method is only suitable for whole preparations.

Boerhaave,[1] though less a practical anatomist than a commentator, brings to bear on the subject so intimate a knowledge of his contemporaries and their methods that his observations invite attention. He does not believe that the minutest vessels are demonstrated even by 'the most subtile coloured liquor'. For example, while the membranes of the testis are richly provided with blood, the 'pulp of the testicle is supplied with pellucid juices by arteriolae much smaller than the sanguiferous, and into which the Ruyschian injection will not enter'. It was commonly noted at the time, in the case of vermilion masses, that the waxy or other vehicle penetrated into vessels which would not admit the vermilion, and Boerhaave believed that the pressure exerted by the injector forced the injection into vessels which were not normally traversed by red blood, 'since no part of the cortex ever appears red without injection'. He thus supports Malpighi as against Ruysch in the controversy on the structure of the glands, holding that Ruysch's injections unnaturally dilated the blood capillaries and thus obscured or even destroyed the glandular tissue, nor will he assent to Ruysch's view that injection mass passes from the arteries into the excretory ducts of a gland. Boerhaave makes one statement the origin of which the writer has not been able to trace. He says that some English anatomists injected the carotid arteries with urine coloured with ink (a medium employed by Malpighi), and found that the nerves of the brain were tinged with the colour. When one of the nerves was sectioned it appeared full of black specks, which were supposed to represent the cavities of the nerve tubes. It will be remembered that at this time the theory of Descartes that nerve fibres were tubes transmitting fluid contents was generally accepted.

[1] Herman Boerhaave (1668–1738), *A method of studying Physic*, London, 1719, 8vo.

In 1720 Valentin[1] gives formulae for injecting the alveoli of the lungs and the blood-vessels with low fusible metals, the flesh being afterwards boiled away, leaving a cast of the injected cavities. He uses also a wax medium made of white wax mixed with mutton suet and coloured with verdigris or vermilion, which passes through capillaries and sets when cold. Albinus,[2] who claimed that the art of injecting coloured liquids into the vessels was a Belgian invention, employed a red liquid which stiffened when the vessels were distended with it. His injections were finer than Ruysch's, and he filled the minutest vessels in the skin, brain, and capsule of the lens. The latter, however, must have been in the foetus, as there are no vessels in the capsule of the adult. William Hunter, who examined his injections in 1748, was deeply impressed by them. In 1756 Albinus is combating vigorously Ruysch's doctrine that the human body is composed entirely of vessels.

Between 1721 and 1732, when Monro primus published his first paper on injections, several writers deal with the subject, but little is added to what was already known. Helvetius[3] supports Ruysch in holding that the whole body is almost nothing more than a prodigious assemblage of lymphatic and blood vessels, but considers that injections are deceptive in attracting too much attention to the blood-vessels. The latter belief finds an apt illustration in the work of Stukeley on the spleen,[4] whose injections of wax in two colours induced him to adopt the view that the spleen consists almost entirely of the ramifications of the splenic artery. He admits the presence of a few muscular fibres, but even the veins, according to this author, are only slightly concerned with the internal economy of the gland. According to William Hunter, who as a contemporary might be expected to know, Nathanael St. André (1680–1776) was one of the early English injectors, and Haller asserts that St. André claimed to have injected vessels in the epidermis with quicksilver. This

1 Michael Bernard Valentin (1657–1729), *Amphitheatrum Zootomicum*, Frankfurt, 1720, fol.

2 Bernard Siegfried Albinus (1697–1770), *Oratio, qua in veram viam, quae ad fabricae humani corporis cognitionem ducat*, Leyden, 1721, 4to. *Academicarum Annotationum*, Leyden, 1754, Lib. i, 4to.

3 John Claude Adrian Helvétius (1685–1755), *Idée générale de l'oeconomie animale*, Paris, 1722, 8vo.

4 William Stukeley (1687–1765), *Of the Spleen, its description and History*, London, 1723, fol.

sinister physician is also said to have possessed a collection of anatomical preparations, but, to quote Hutchinson, 'after his decease Mr. Christie's auction room bore abundant witness to the frivolity of his collections'. Nicolai [1] praises isinglass as an injection medium, but finds that it extravasates too readily, and is apt to fill cavities and leave the vessels empty. Isinglass is also recommended by Mauchart,[2] who again is the first to experiment with injections of plaster—previously recommended by Boyle. The desire to imitate Ruysch's preparations, and the difficulty of doing so owing to a lack of knowledge of his procedure, is referred to by Wagstaffe,[3] who reproaches Ruysch, then living, for concealing his methods. He refers to the futility of exhibiting injections as a mysterious curiosity, and not as a scientific invention available to all for the advancement of knowledge. 'Dr. Ruysch has given us several excellent and curious drawings of the *finest preparations* in the world; but we had certainly been more obliged to him, if he had communicated his observations on the manner of *preparing* them, and form'd from thence a noble, a just and a demonstrative *rationale* of the uses of the parts.' This lack of candour, however, continued to discredit the publications of anatomists, for much later, in 1784, Sheldon, when referring to the great improvement in the art of injection 'in late years', says that 'progress of the science has undoubtedly been much impeded by the mystery observed among anatomists, respecting the composition of their injections, and their method of dissecting, injecting, and preparing the different parts: a mystery which deserves the severest censure, and is unworthy of the character of a philosopher or a man'. He adds: 'I shall disclose, without reserve, whatever I am acquainted with on this head.' Jesse Foot, in 1794,[4] praises Sheldon for 'his unreserved discovery [disclosure] of the art of injecting', and endorses his disapproval of a proprietary anatomy. On the other hand, Drake, in a posthumous work,[5] explains and

1 Henricus Albertus Nicolai (1701–33), *De directione vasorum pro modificando sanguinis circulo*, Strassburg, 1725, 4to.

2 Burchard David Mauchart (1696–1752), *Programma Anatomicum de iniectionibus sic dictis anatomicis*, Tübingen, 1726, 4to.

3 William Wagstaffe (1685–1725), In Drake's *Anthropologia Nova*, London, 1727, 8vo, ed. 3, vol. i, p. xi.

4 Jesse Foot (1744–1826), *The Life of John Hunter*, London, 1794, 8vo.

5 James Drake (1667–1707), *Anthropologia Nova*, London, 1728, 8vo, Appendix, Plate 51.

figures a large injecting syringe holding twenty ounces. The intake is separate from the outflow, and both have cocks so that the syringe can be refilled without removing the nozzle from the vessel, as first practised by C. Bartholin. The outflow pipe is a flexible one of leather, and the nozzle is of brass or silver, and small enough to 'pass into the lacteal vessels and chyliferous ducts'. Lancisi,[1] by syringe injections of mercury, coloured water, and air, perpetuates the error of Vieussens, Verheyen and others that 'there are meanders, winding passages, and diverticula leading from the coronary veins into each of the four cavities of the heart, and that the blood makes use of these as outlets'. By means of these 'outlets' Lancisi believes that the blood 'alternately goes in and out with a kind of ebb and flow'. An attempt at histological injection was made in 1728 by Price,[2] who studied the structure of the villi of the ox and filled the vessels with wax, but no results of interest or importance accrued. Weiss,[3] improving on the practice of Ruysch, adopts a method which was almost at once developed and standardized by Monro and Cassebohm. He injects the human body by immersing it for a long time in warm water, and then throwing in first coloured turpentine to fill the smaller vessels, and afterwards a wax medium to distend and seal the larger ones. 'A gross mixture of wax, tallow, and vermilion' is used by Mortimer[4] to inject the arteries, and the same substance, but coloured with smalt, to fill the veins. This is one of the few references to a differential injection of arteries and veins since the method was first introduced by Duncan in 1678. A general description of wax injection is given by Thiesen,[5] and Trew[6] is an early experimenter with fluid plaster, which, however, he does not consider an efficient substitute for wax. He tried also the

[1] Giuseppe Maria Lancisi (1654–1720), *De motu cordis*, Rome, 1728, fol.

[2] Charles Price, 'Remarks on the Villi of the Stomack of Oxen,' *Phil. Trans.*, London, 1728, vol. xxxv, p. 532, 4to.

[3] Johannes Nicolaus Weiss (1702–1783), *Theses sistentes viscerum glandularum*, Altdorf, 1729, 4to; *De structura venarum*, ibid., 1733, 4to; *De aquae adminiculo in administratione anatomica*, ibid., 1733, 4to.

[4] Cromwell Mortimer (ob. 1752), 'Case of some uncommon anastomoses of the spermatic vessels in a Woman', *Philosophical Transactions*, London, 1730, vol. xxxvi, p. 373, 4to.

[5] Gottfried Thiesen (n. 1705), *De materia ceracea eiusque iniectione anatomica*, Königsberg, 1731, 4to.

[6] Christopher Jacob Trew (1695–1769), *Commercium Litterarium*, Nürnberg, Spec. 9, 1731, Hebd. 30, 1736, 4to.

vegetable resins, resina anime and sandarach, and isinglass, and shows that injected mercury passes through the capillaries and completes the circulation. Vater [1] was a pupil of Ruysch, and was trained in injection methods by him. He employed different coloured liquids and wax, and according to Haller was most skilled in the art of filling the vessels, producing results equal to those of his master. The short paper published by Monro primus in 1732 [2] is important, not on account of its originality, but because of the influence it exercised, and the method there recommended was very widely adopted, and is not even now entirely abandoned. Monro's article, for example, forms the basis of the chapter on injection by Daubenton in Buffon's *Histoire naturelle.* Monro says: 'Scarce any anatomical books describe with accuracy the method of injecting,' and again, 'few have hit on the art of injecting the very small capillary tubes'. He uses a brass syringe which includes a contrivance for re-filling without removing the nozzle from the vessel—as previously recommended by Bartholin and Drake. For 'subtile or fine injections' he proceeds as follows: Macerate in warm water to liquefy the blood. First inject coloured oil of turpentine to fill the very small vessels, and follow up at once with a coarser injection to distend the larger ones. The two media mingle in the vessels, so that it is impossible to tell from inspection that two have been used. The colours employed are vermilion for arteries and distilled verdigris for veins—mixed with clear oil of turpentine and filtered or decanted to get rid of the granules. His coarse injection is composed of tallow, one pound, bleached white wax, five ounces, and salad oil, three ounces. Melt and add two ounces of Venetian turpentine. Sprinkle in the colour and filter through a linen cloth. Some, if not most, of the Hunterian preparations in the Royal College of Surgeons must have been injected by this method, since when they are cut or dissected they bleed from the finer vessels only. Two years later Monro published a striking but neglected paper [3] in which the independence of the maternal and foetal bloods in the placenta

[1] Abraham Vater (1684–1751), *De iniectionis variorum colorum utilitate ad viscerum structuram detegendam*, Wittemberg, 1731, 4to.

[2] Alexander Monro, *primus* (1697–1767), 'An Essay on the Art of injecting the vessels of animals', *Medical Essays and Observations published by a Society in Edinburgh*, Edinburgh, 1732, vol. i, 12mo.

[3] 'An Essay on the Nutrition of Foetuses', *Medical Essays*, &c., Edinburgh, 1734, vol. ii, 12mo. Some copies are dated 1733.

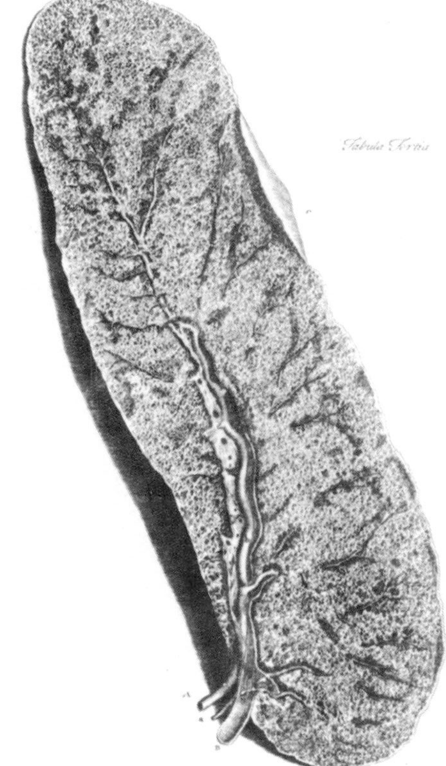

The coronary vessels injected by Ruysch (1704)

The spleen of the ox injected with wax by William Stukeley (1723)

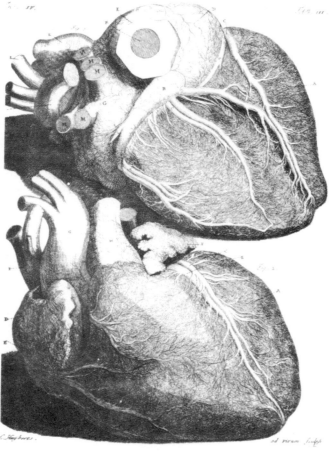

is for the first time placed beyond question by injection experiments. On this occasion he makes use of oil of turpentine coloured with vermilion, and mercury. He says:

'The liquors are not carried from the mother to the foetus, or from the foetus to the mother by continued canals, that is, the uterine arteries and veins do not anastomose with the veins and arteries of the secundines; but the extremities of the umbilical vein take up the liquors by absorption in the same way as the lacteal vessels do in the guts; and the umbilical arteries pour their liquors into the large cavities of the sinuses, or other cavities analogous to them.'

It should, however, be pointed out that in Monro's time, to quote his own words, absorption was a process 'whereby the small open orifices of vessels imbibe liquors lodged in the cavities of the body', and hence the above passage is not as strictly accurate as it appears. Monro proves his case by injections of the human subject, and also of cows, sheep, and dogs. When the uterine arteries are fully injected none of the medium passes into the umbilical vessels or the foetus, though it was searched for most carefully. Conversely, if the umbilical arteries are injected, not one drop of the mass could be found in the uterine vessels.

'I have tried injections of very different kinds so often into the vessels of the womb and secundines of cows, prepared in all the different ways I could contrive for making liquors pass from the one to the other, without having once made a drop to pass, that I cannot be more certain of anything, than that there is no anastomosis or continuity of these vessels in cows.'

It is somewhat difficult to understand how Monro reconciled the results of his experiments with his definition of absorption just quoted. In 1736 Monro describes an attempt to inject the testis from the vas deferens with mercury,[1] but he only succeeded in partially filling the epididymis, the convoluted tubular nature of which and its communication with the vas deferens being thereby demonstrated.

Hales[2] is responsible for an ingenious and characteristic innovation. In order that the force of the injection might be known and kept constant, which cannot be the case when a syringe is used, he provided for the necessary pressure by using a known

[1] 'Remarks on the Spermatic Vessels and Scrotum, with its contents' *Medical Essays*, &c., Edinburgh, 1736, vol. v, 12mo.

[2] Stephen Hales (1677–1761), *Haemastaticks*, London, 1733, 8vo.

column of the fluid to be injected, which column was maintained at such a level as to ensure a driving power equal to that of the arterial blood. Nevertheless, in spite of this precaution, extravasations occurred into the cavity of the gut. The apparatus consisted of a gun-barrel heated with boiling water, and having a brass cock at the bottom with which to regulate the height of the column. Sometimes two barrels were joined together, giving a height of ten feet. Hales confirms, what had previously been noticed, that vermilion is sometimes stopped by the capillaries, even when the vehicle of the injection passes into the veins. Before operating, the animal is heated with warm water, and occasionally Hales irrigates the vessels prior to throwing in the injection mass. The latter was either melted beeswax or a preparation which he attributes to 'Mr. Ranby' [John Ranby, the surgeon]. The latter medium consists of white resin and tallow, two parts of each, turpentine varnish, eight parts, and colouring matter (vermilion or indigo), three parts.

According to Cuvier, Nicholls [1] was celebrated for injections only inferior to those of Ruysch. In a manuscript copy of William Hunter's lecture notes in the possession of the writer, Nicholls is stated to favour coloured varnish as a fine injection. W. Hunter and Straus-Durckheim both assert that Nicholls was the first to make corroded preparations. We have seen that Bidloo, Cowper, and Homberg anticipated him by some forty years. The error is repeated in the *Dictionary of National Biography*, and Munk's Roll of the Royal College of Physicians. Kaau-Boerhaave [2] injected water into the cavity of the gut, and when the latter was compressed, the water was observed to pass into the veins and finally to wash the blood out of the portal vein. This experiment was held to prove that absorption was by the veins, and on the same grounds Duverney had previously concluded that the lacteal vessels were absent in birds. Kaau also studied how the lymph vessels might be made most conspicuous by injection.

Cassebohm,[3] who produced the first general treatise on

[1] Frank Nicholls (1699–1778), *Compendium Anatomico-Oeconomicum*, London, 1736, 4to.

[2] Abraham Kaau-Boerhaave (1715–58), *Perspiratio dicta Hippocrati per universum corpus anatomice illustrata*, Leyden, 1738, 12mo.

[3] Johannes Friedrich Cassebohm (ob. 1743), *Methodus secandi et contemplandi corporis humani musculos*, Halla, 1740, 12mo.

anatomical methods after Lyser, devotes an interesting chapter to injections. He adopts the sound practice of experimenting with but few and simple methods, but their possibilities are exhaustively canvassed, and very detailed directions are given. Two types of media are tested: (1) non-coagulating fluids injected cold, such as coloured spirit, varnish, and mercury; (2) coagulating liquids injected warm, such as wax, suet, and isinglass. An ordinary thick medium is prepared from yellow or white wax thinned with deer or goat suet, lard, or turpentine according to the time of the year and consistency required. For fine injections isinglass is recommended. The colours selected are vermilion for arteries and verdigris for veins. A successful method was found to be first to inject paper or copal varnish coloured with vermilion to fill the smallest vessels, and to follow it immediately with a waxy medium similarly coloured to demonstrate the larger trunks. This agrees almost exactly with the procedure of Monro. Cassebohm made many attempts to wash out the blood from the vessels with warm water before injecting the solidifying medium, but he is only partially successful, and does not recommend it. As a subject he prefers a newly born child, or an abortion of seven or eight months with the umbilical string intact. The body is warmed with water before injecting, and in the case of the foetus the umbilical vein is filled first with the non-coagulating and then with the coagulating medium. Thus both arteries and veins may be injected in a single operation. His methods of injecting the adult body are complicated, e.g. his general injections are made from at least ten different points in the arteries and veins. Before filling the veins he endeavours to rupture the valves by means of an iron rod.

Le Cat[1] appears to have been the first to investigate the natural relations of a part by filling the cavities with a solidifying medium, and then preparing optical sections. This is also the essential principle of the paraffin section. Le Cat made use of melted resin and wax or very thick glue. A combined alcoholic and waxy fine injection, in which the human element is literally represented, is recommended by Fabricius,[2] but not elsewhere referred to. Its

[1] Claude Nicolas Le Cat (1700–68), 'On the figure of the canal of the Urethra', *Phil. Trans.*, London, 1741, 4to, vol. xli, p. 681.

[2] Philippus Conradus Fabricius (1714–74), *Idea anatomiae practicae*, Wetzlar, 1741, 8vo.

composition is: Rectified spirits of wine, 22 ounces; resin (sandarach), 2 ounces; resin (elemi), 1 ounce. Dissolve slowly over a sand bath, and then add: yellow wax, 2 ounces; human fat, 6 ounces. Colour with vermilion or verdigris worked up in alcohol. At about this time Westphal published a short tract on anatomical injections.[1]

An important experiment was described by Haller in 1745.[2] He injects the testis with mercury from the vas deferens so as to make its structure plainly visible to the naked eye. He doubts if it is possible to demonstrate this in any other way. An excellent figure is given, which is, however, too delicate to be reproduced except as a metal engraving.[3] In 1749 Haller sent an account of his experiment to the Royal Society which was printed in the *Philosophical Transactions* for 1750. 'Let the epididymis', he says, 'be gently and carefully filled with quicksilver, by the ductus deferens, now and then pausing, or dipping the testicle in warm water, that the vessels, being gradually expanded, may give way; for a sudden repletion will be apt to burst the middle or upper part of the epididymis. By this method, it has often appeared to me, that the epididymis, through its whole length by which it adheres to the testicle except the head, is one subtile canal, which is capable of being unfolded, as was perceived by de Graaf.' Haller also demonstrated the vasa efferentia and coni vasculosi, the rete vasculosum, the vasa recta, and the seminiferous tubules. All these structures had been already described, but less completely and accurately, by de Graaf in 1668, and some of them by Aristotle. A portion of the mercury had even penetrated into the seminiferous tubules, thereby establishing their tubular nature, and enabling Haller to determine the exact course of the seminal fluid on leaving the testis. He also discovered the vas aberrans. By further injections of wax and mercury Haller worked out the relations and structure of the vesiculae seminales. For general injections, however, Haller preferred the turpentine and vermilion medium to all others. It is difficult to understand, after reading Haller's paper, what was left for the Monros and the Hunters to

[1] Andreas Westphal (1720–88), *De iniectionibus anatomicis*, Greifswald, 1744, 4to.

[2] Baron Albrecht von Haller (1708–77), *De viis seminis*, Göttingen, 1745, 4to.

[3] This figure appears in Quain's *Anatomy*, but the beauty of the original is somewhat lacking.

PLATE XLV

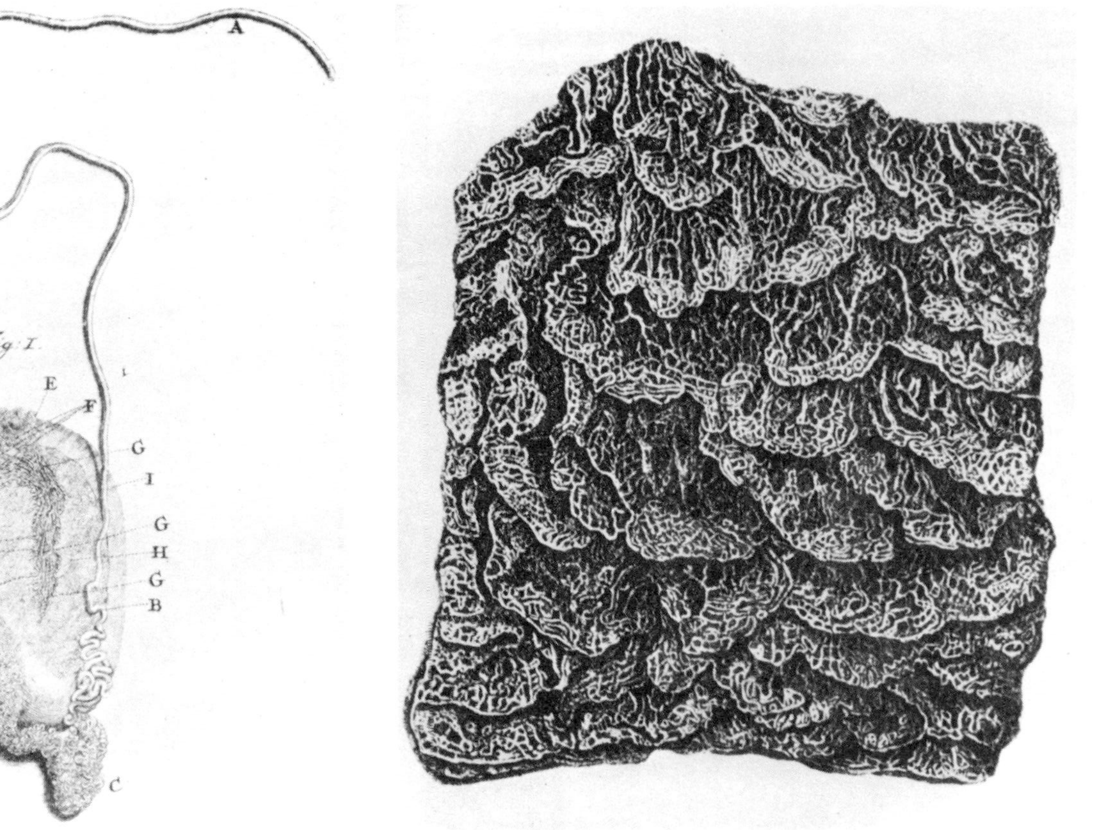

Testis injected with mercury from the vas deferens by Albrecht von Haller (1745)

Micro-injection of the mucous membrane by Johann Nathanael Lieberkühn (1745)

dispute about as regards priority of the injection of the testis with mercury. The only advance they made on Haller's results was the quite minor one of filling more completely the seminiferous tubules. The success of the experiment of injecting the testis with mercury from the vas deferens may be gathered from the statement of Bowman in 1842 that 'there are not ten specimens that can be pronounced at all full in the Museums of Europe; and there is no evidence that, even in the best of these, the injected material has reached the very extremity of the tubes'. On the other hand, in the abstracts of the *Philosophical Transactions* published in 1809, it is stated: 'Beautiful specimens of the serpentine vessels or seminiferous tubes of the testicles, filled with quicksilver, are to be seen in the private anatomical museums of this metropolis.'

Lieberkühn[1] shares with Ruysch the honour of having made the most important contributions to the practice of anatomical injection. He carried Ruysch's methods a stage further, and is one of the first successfully to inject the *microscopic* vessels, and to make corroded preparations. His material was drawn mostly from the human subject, and he uses the same injection mass for the large and small vessels, except that for the finest branches it is further thinned with turpentine and the quantity of colouring matter increased. Sixty of his microscopical injections are still in the Museum of Human Anatomy at Vienna University. These preparations include the gills of the pike, salivary glands injected by the artery, vein, and excretory duct, periosteum and perichondrium, lung of frog, tortoise, ox and man, the villi of the intestine, and the placenta. Other injections are at the University of Berlin, where they are available for inspection. According to Adams, there were a few preparations by Lieberkühn in the British Museum. Some are at St. Petersburg, and were described by C. F. Burdach in 1817. A number were purchased by G. C. Beireis on the death of Lieberkühn's son, and are stated to be very beautiful and manifestly superior to Ruysch's preparations in St. Petersburg. Lieberkühn's injections were in fact so good that, a century after they were made, Henle was using them for

[1] Johann Nathanaël Lieberkühn (1711–56), *De Fabrica et Actione villorum intestinorum tenuium hominis*, Leyden, 1745, 4to. Some copies are dated 1744. 'Sur les moyens propres à découvrir la construction des viscères', *Mémoires de l'Académie Royale des Sciences*, Berlin, Ann. 1748, 4to.

the purposes of original research. He is the first injector whose sections will bear the highest magnification, and have undergone no deterioration since they were made. His injection apparatus was described and figured by J. C. F. Bonegarde in 1741, i.e. before the publication of Lieberkühn's work on the villi. J. C. Bohl also refers to Lieberkühn's experiments before 1745, so that his reputation as an injector therefore must have been established before the publication of his most famous work. The plates in this tract on the mucous membrane were etched by Pierre Lyonet—perhaps the greatest of anatomical engravers. Lyonet's beautiful plates of the Goat Moth larva were published in 1760, and his first attempts at copper-plate engraving, as he tells us, were made in 1743. Hence the plates in Lieberkühn's memoir must have been among the first efforts of the master.

In his 1745 paper Lieberkühn describes how he succeeded in injecting the arteries and veins of the villi of the intestine separately with different coloured wax. He first attempted a simultaneous injection from the artery and vein with two syringes, having nozzles adapted to the relative capacity of artery and vein. This resulted, however, in a mixing of the colours where they met. He then substituted injecting first the mesenteric artery with red, until by naked eye inspection some, but not all, of the villi were injected. Then he threw a green injection of lower melting-point into the mesenteric vein. Hence the two injections might meet but could not mix. Some of the villi were green and others red, and at the places where the red and green patches were in contact the arteries of the villi were red and the veins green. Thus was Lieberkühn the first to demonstrate microscopically the vascular supply of the mucous membrane. This had been injected before by Ruysch and others, but a microscopic examination of their preparations by Lieberkühn himself showed that they had not succeeded in injecting the capillaries either completely or without rupture. But Lieberkühn's injection and inflation experiments led him to perpetuate one serious error. By pushing the injection too far, the mass escaped into the cavity of the gut. He thus concluded, and his views were accepted almost without question for some time after, that the blood-vessels of the villi communicated with the lacteal vessels, and that the latter opened directly into the lumen of the intestine, each villus having at its apex one or more orifices. This mistake was revived by Hewson, and even as late as 1849 Robin states that some anatomists still believed in the existence of these orifices. Before Lieberkühn's time the injections of Ruysch and his pupils had encouraged the belief that both the lacteal and the mesenteric blood-vessels opened directly into the cavity of the gut by large apertures. For example, a solution of indigo in urine forced into the intestine between two ligatures was found to pass into the lacteal vessels.

Lieberkühn's method of making corroded preparations, published in 1748, was as follows: Take white wax free from grease, add a fifth part of resin, a tenth part of Venetian turpentine, and of vermilion as much as is necessary. Inject the large vessels with care. Put the injected part into strong nitric or sulphuric acid diluted with water. Leave in the acid until the organic parts are destroyed, wash in water, and the result is a cast of the cavities of the large vessels. As, however, such preparations were very fragile, he devised a method of producing casts of the vessels in silver. Make a corroded preparation as above, and embed it in a paste made of two parts plaster and one part pulverized tiles in water. When set burn out the wax by graduated heat, and pour in the molten silver. Put the mould into vinegar, which disintegrates the plaster, and allows of its easy removal. To make a surface preparation of the vessels, the part is injected with the finer medium, and the vascular network is developed by directing on to it a strong stream of water, thus removing the organic matter. This gives very beautiful pictures in relief of the capillary network.

The eminent comparative anatomist Daubenton, who wrote the chapter on injections in Buffon,[1] occupies himself chiefly with a discussion of Monro's paper, and discloses no methods of his own. He describes in some detail fifty-five injected preparations in the Royal Museum at Paris, but does not state how they were prepared. Sue, in his general treatise on preserving and injecting methods,[2] gives practical directions for the preparation of injection media, and Quellmaltz [3] adds palm oil to the list of recommended

[1] Louis Jean Marie Daubenton (1716–99), 'Description du Cabinet du Roi'. In Buffon's *Histoire naturelle*, Paris, 1749, 4to, T. iii, p. 133.

[2] Jean Joseph Sue (1710–92), *L'Anthropotomie ou l'art d'injecter*, Paris, 1749, 8vo.

[3] Samuel Theodor Quellmaltz (1696–1758), *De oleo palmae, materie iniectionibus anatomicis aptissima*, Leipzig, 1750, 4to.

masses. Dr. Sandys is said by several contemporary writers to have been the first to make injected preparations transparent with oil of turpentine (*c.* 1750), but this useful and beautiful method was practised by Swammerdam, whose procedure was described by Schrader in 1674, and later by Ruysch.

We now reach the period of the second Monro and of the Hunters, whose teaching activities and Museums, if not exhibiting great originality in the matter of injections, did more perhaps than anything else to stereotype the anatomical injection in England. In January 1753 Monro secundus[1] injected the tubuli testis from the vas deferens with mercury, and hence confirmed the precise connexion between these two structures already established by Haller. A preliminary description with figure was published in 1754, and the complete work appeared in October 1755. Also in 1755, and again in 1757, he describes injecting the lymphatics with mercury, and concluded that they arose from the lacunar membranes and cavities of the body, and not from the arteries, as was at the time believed. In 1785 Monro investigated the structure of the gills of the Skate by injections of distilled oil of turpentine coloured with vermilion. Another, and quite novel, experiment was the injection of the lateral line canal of the Skate with water, air, milk, mercury, and oil of turpentine and vermilion. He discovered the tubules which arise from the main canals and open on to the surface of the skin, but committed the serious error of concluding that the lateral line system was a part of the lymphatic apparatus. He states that in 1765 he injected the mesenteric arteries with red wax, the corresponding veins with yellow wax, and the lymphatics with quicksilver, but the description of what must have been one of the earliest triple injections was not published until 1770. In the meantime, in 1769, Hewson had figured an injection of the mesentery of the turtle, in which the arteries had been filled with red wax, the veins with black wax, and the lacteals with mercury. Nearly all the early injectors confined their operations to Mammals, and Monro's work is therefore doubly interesting because of its strong comparative bias. Thus he was the first to inject an Echinoderm. Mercury was led into the tube feet of a common sea urchin, and he was able to

[1] Alexander Monro, *secundus* (1733–1817), *Essays and Observations Physical and Literary*, Edinburgh, 1754, 8vo, vol. i; *De Testibus et Semine in variis animalibus*, Edinburgh, 1755, 8vo; *De Venis lymphaticis valvulosis*, Berlin, 1757, 8vo; *The structure and physiology of Fishes explained*, Edinburgh, 1785, fol.

demonstrate the connexion between the tube feet, ampullae, and radial water-vessel. The subject was not an easy one and he made mistakes, but certain facts were correctly elucidated. Again in the same animal he injects the vessels of the intestine with mercury, and from them 'filled a beautiful network of vessels, not only on the intestines, but dispersed on fine membranes, which tie the intestine to the inner side of the shell'. The injections of Echinoderms to be seen in the Museum of Anatomy, Edinburgh University, were probably made by Monro.

In 1752, William and John Hunter[1] injected the epididymis and seminiferous tubules of the testis with mercury from the vas deferens, thus repeating the experiment first demonstrated by Haller. This injection was described by William Hunter in his lectures, but no account of it was published until 1757, and then only a very brief one. From 1746 William Hunter taught, but again did not publish before 1757, that the lymphatics were independent of the arteries and veins. This view was based on injections of both systems, it being found that an injection mass would not pass from the blood-vessels to the lymphatics or vice versa, and either system could be completely injected without affecting the other. Only by forcing the injection and rupturing the vessels could other results be obtained. The blood vascular apparatus was therefore a closed system of tubes without any direct connexion with the lymphatics, as was in Hunter's time, and indeed for long after, generally accepted. The presence of numerous valves in the lymphatics supported Hunter's view that the motion of the lymph was independent of the driving power of the blood-stream. The lymphatics were held to commence blindly from the lacunar surfaces and interstices of the body.

In the description of his plates of the gravid uterus, published posthumously in 1794, William Hunter states that he first injected the vessels of the foetal placenta from the navel string in 1743, but it was only when the plates were first issued in 1774 that this experiment was described. He says that he injected the placenta with wax of different colours—the uterine arteries red and the veins blue, but none of the injection mass passed into the vessels of the navel string. In the 1794 publication further details are added. The placenta of man and quadrupeds, he remarks, is

[1] William Hunter (1718–83), *Critical Review*, London, 1757, 8vo; *Anatomia uteri humani gravidi*, Birmingham, 1774, fol.; *An Anatomical Description of the Human Gravid Uterus*, London, 1794, 4to.

composed of two parts intimately blended—a foetal element, which is the continuation of the umbilical vessels of the foetus, and a maternal, which is an ' efflorescence of the internal part of the uterus '. Injection experiments carried out on man and quadrupeds demonstrated that the maternal and foetal bloods were absolutely distinct in the placenta, and both systems might be injected ' to an amazing degree of minuteness ' without the two masses mingling. Nevertheless, each mass completed the circulation of its respective part, and returned to the parental or foetal centre from which it originated. This hypothesis is, of course, an old one. It was originally suggested by G. C. Aranzio in 1564, denied by Dulaurens in 1598, Fabrici in 1600, and Noortwyk in 1743, reaffirmed by Harvey in 1651 and Needham in 1667, but was first clearly established by the injection experiments of Monro primus published in 1734, of which the subsequent work of William Hunter afforded ample confirmation. Harvey's reasoning on this point is remarkably sound, but it is apparent that he and the writers before him base their views not on observation but on probabilities. Indeed Fabrici, in supporting the view of the ancients, admits that he has no positive evidence to offer ' because the fleshy mass itself stands in the way of any accurate investigation '. When the second Monro and William Hunter were students, it was still believed that the maternal blood circulated through the foetus by the navel string, and returned to the parental vessels, in spite of the positive demonstration to the contrary by the first Monro. Hunter's belief in injection methods was deeply strengthened by his visit to Albinus in 1748, when the beautiful preparations of the Leyden Professor fired the imagination of the Scots anatomist.

William Hunter was teaching anatomy from 1746 to the year of his death in 1783. In an excellent manuscript transcript of his lectures in the writer's possession, unhappily without date, four lectures out of eighty-two are devoted to injection methods—a proportion large enough to emphasize the importance in which injections were held at the time. He states that nothing has contributed more to the promotion of anatomical discovery, and that ' there is no making a good practical anatomist without it '. His watery injections are made from glue, isinglass, or gum arabic, and for the finest injections he used turpentine thickened with a little resin. The directions given for preparing lead casts of the vascular and other cavities of the body are too similar to those published by Lieberkühn in 1748 to have been independently evolved by Hunter.

John Hunter,[1] in some notes written about 1770 but not published until 1861, contrary to modern practice, prefers stale material for injection purposes—even material which has been preserved in spirit. He soaks it in water until it has undergone a certain degree of putrefaction. This produces a relaxed condition of the vessels, and the blood is then cleared out by injections of warm water. His injection media are: resin and tallow; turpentine, hog's lard or tallow; hog's lard by itself or butter; glue or size; and isinglass. He notes that the finer the injection the more colour is wanted. The gravid uterus is injected simultaneously by the artery and the vein, so that the distribution of these vessels may be studied in the placenta. The colours selected are vermilion, King's yellow (a preparation of orpiment), blue verditer (a hydrated oxide of copper), and flake white. Green is made by combining yellow wax with the blue verditer. For corroded preparations, the material for which must be fresh and healthy, he recommends a vigorous injection of a mixture of wax, resin, turpentine varnish, and tallow.

After the Hunters, no important development in injection methods is to be recorded until the introduction of the 'soluble' form of Prussian blue, and carmine gelatine and other precipitates as colouring matters, in the nineteenth century. By the end of the eighteenth century the superiority of a solidifying gelatinous vehicle, such as isinglass or size, had, after many years of trial, at length asserted itself. Wax and varnish media were still employed for ordinary coarse and fine injections, but for injections of the capillaries intended to challenge the scrutiny of the microscope, gelatine was manifestly the most appropriate medium, and it is astonishing that recognition of this fact was so long delayed.

The remaining work of the second half of the eighteenth century may be briefly dismissed. Laghi[2] uses nut oil and thin glue or size. Lyonet,[3] the author of a great work on the anatomy of the

[1] John Hunter (1728–93), *Essays and Observations on Natural History*, London, 1861, 8vo.

[2] Thomas Laghi, 'De Iniectionibus', *De Bononiensi Scientiarum et Artium Instituto atque Academia Commentarii*, Bologna, 1757, 4to, T. iv.

[3] Pierre Lyonet (1707–89), *Traité anatomique de la Chenille*, La Haye, 1760, 4to.

Cossus larva, made little use of injection methods. He tried, however, injecting ink and coloured liquids into the heart of his larva. The results made him doubt whether the animal possessed blood-vessels at all, and he suspects that nutrition is effected by some means apart from the heart, for which another function must be sought. The connexion between arteries and veins by means of closed capillaries was confirmed by the injection experiments of Jancke,[1] but the work of the period is concerned rather with the lymphatic system, for the injection of which mercury was still the most popular medium. Hewson[2] states that the walls of the lymphatics, though thin, are strong, and will withstand a higher column of mercury than the blood-vessels. The lymphatics of fish are injected with mercury or thin coloured size from the ventral or abdominal lymphatic trunk, from which the medium passes into the entire lymphatic system. His experiments support the doctrine of the independence of the lymphatics, and, apart from extravasations, he holds that there is no connexion between them and the blood-vessels. Meckel's mercury injections[3] of the lymphatic and mammary glands and other organs resulted in the unfortunate conclusion that the veins are directly connected with the cavities of the tissues and thus absorb from them—a result at once questioned by Hewson. A new use of an old substance of unknown composition called cera punica or Punic wax was initiated by the injections of Walter.[4] His museum included a large number of beautifully injected preparations, but these apparently were injected with the usual red, yellow, and green wax. According to Hyrtl, Punic wax was soluble equally well in oil, spirit, and water, and combined readily with quicksilver. Wornum believes that it is prepared by boiling common yellow wax three times with sea water, to which a small quantity of potassium nitrate has been added. It would be difficult to

[1] Johannes Gottfried Jancke (1724–63), *De Ratione venas angustiores imprimis cutaneas ostendendi*, Leipzig, 1762, 4to.

[2] William Hewson (1739–74), 'On the Lymphatic System in Birds', 'On the Lymphatic System in Amphibious Animals', *Philosophical Transactions*, London, 1768–9, 4to, vols. lviii and lix, pp. 217 and 198; *A Description of the Lymphatic System in the Human Subject*, London, 1774, 8vo

[3] Johann Friedrich Meckel (1724–74), *Nova experimenta et observationes de finibus venarum*, Berlin, 1772, 8vo.

[4] Joannes Gottlieb Walter (1734–1818), *Observationes Anatomicae*, Berlin, 1775, fol.

PLATE XLVI

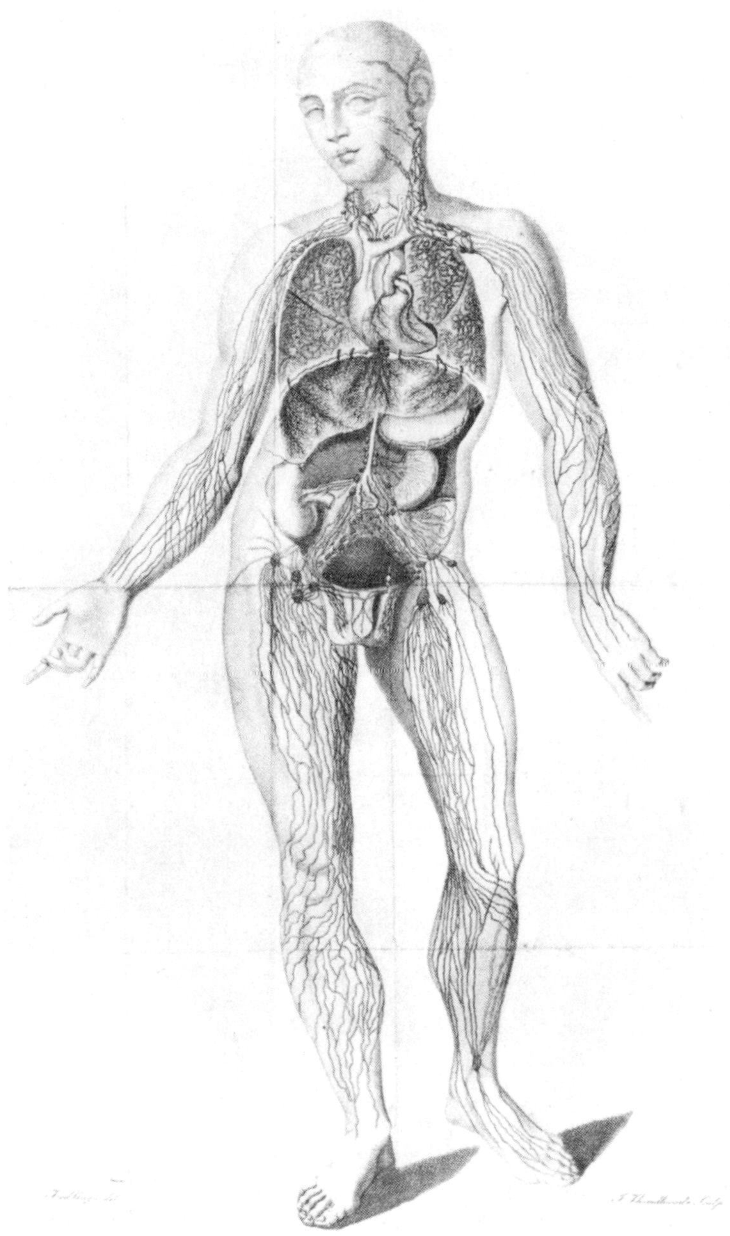

General scheme of the lymphatics of the human body based on mercury injections by William Cumberland Cruikshank and his pupils (1786)

decide which statement discloses a greater ignorance of the actual constitution of the substance in question. The first author to devise a method of injecting the uriniferous tubules of the kidney was Schumlansky.[1] This was done by forcing the injection into the arteries until the glomeruli were ruptured, thus permitting the injection to extravasate into the tubules, and to pass from thence into the pelvis of the kidney. The method was afterwards successfully practised by Bowman in 1842. Ruysch had accidentally produced the same result before, but he considered the passage of the mass from the arteries to the tubules to be effected by natural channels. Schumlansky was also the first to hold that the Malpighian body not only secreted the urine but constituted the origin of the uriniferous tubule, which therefore was the means by which the secretion was conveyed to the ureter. Further, with the assistance of a vacuum pump, he inflated the uriniferous tubules until the air reached and filled the tubules in the cortex of the kidney. Sheldon[2] considers that the lymphatics can seldom be injected by any other medium than quicksilver, except the thoracic duct, which he fills with a coloured 'coarse injection' compounded of yellow resin, pure mutton suet, and wax. He was the first to shape the end of the canula as in the modern hypodermic syringe in order to facilitate its entry into the vessels, and he recommends drying injected specimens and afterwards making them transparent with turpentine. Full directions are given for mercury injections. Faujas[3] prints an interesting

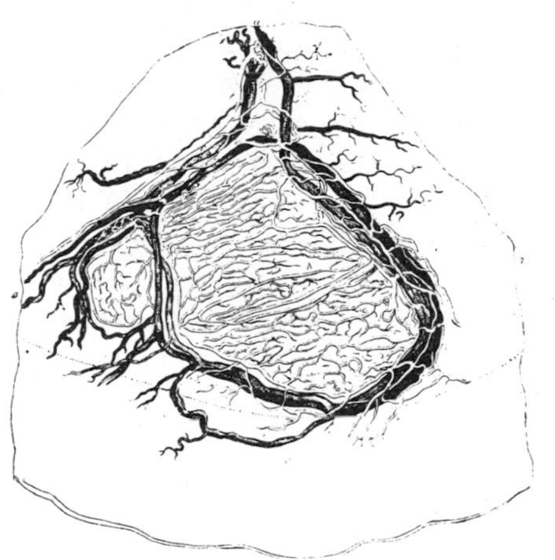

FIG. 8. Triple injection of the arteries, veins and lacteals of the mesentery of the turtle by William Hewson (1769).

[1] Alexander Schumlansky, *De Structura Renum*, Strasbourg, 1782, 4to.

[2] John Sheldon (1752–1808), *The History of the Absorbent System*, London, 1784, fol.

[3] Barthélemi Faujas de Saint-Fond (1741–1819), *Voyage en Angleterre, en Écosse, et aux Îles Hébrides*, Paris, 1797, 2 vols., 8vo.

and familiar account of an interview with Sheldon, who communicated to him his method of preparing mummies by injection. The body was injected at intervals with strong spirit saturated with camphor and diluted with a little turpentine. The skin was well rubbed with finely powdered alum, and a flesh tint imitated by a coloured preparation being thrown into the carotid artery. A varnish composed of powdered camphor and common resin was found to preserve excellently the soft parts of the body. A final injection of the alcoholic solution of camphor by the crural artery completed the process, and the body was then protected from the air by a double case of timber. The entire corpus of a young woman so preserved, after the lapse of many years showed no signs of decay, the arms remaining flexible and the flesh throughout almost as supple and elastic as in life. This specimen is now in the Museum of the Royal College of Surgeons.

There are injections preserved in the Museum of Human Anatomy at Vienna, prepared by Joseph Barth (1745–1818), said to be quite equal to those made by Lieberkühn. Barth used wax, mastic varnish, and a small quantity of fatty oil, to which a lavish proportion of vermilion was added. The English anatomist Cruikshank,[1] who enjoyed the patronage of Dr. Johnson and achieved the doubtful distinction of being referred to by De Quincey in his Murder Essays, has his own method of injecting the lymphatics. He first injects the arteries and veins of the part in question, and also the excretory duct if a gland—the object being to fill all the cavities which are not lymphatic. He then throws the preparation into water and allows it partly to putrefy. Gas collects in the only vessels uninjected, i.e. in the lymphatics, which therefore become visible. They can now be punctured, the air forced out, and filled with mercury. The work on the lymphatics by Mascagni,[2] published in 1787 but dating some years earlier, surpasses all other efforts of a similar nature, to whatever period they belong. He filled the lymphatics with mercury, and counter-injected the blood-vessels chiefly with glue and vermilion. He points out that the particles of the vermilion are only slightly larger than blood corpuscles. He tried also

[1] William Cumberland Cruikshank (1745–1800), *The Anatomy of the Absorbing Vessels of the Human Body*, London, 1786, 4to. Second edition, 1790, 4to.

[2] Paolo Mascagni (1752–1815), *Vasorum Lymphaticorum Corporis Humani Historia*, Senis, 1787, fol.

tallow, wax, and plaster. The mercury was introduced as usual by a gravity tube, in this case of glass, having two arms at right angles to each other. The vertical arm was the larger and contained the mercury, the horizontal arm being sufficiently tenuous to be introduced into the finest vessel. Poli [1] injected Lamellibranchs with mercury. The injection was pressed too hard and penetrated into the visceral ganglion, from thence forcing its way along the nerves radiating from the ganglion. Poli was hence led to conclude that the ganglion was a part of the lymphatic system and acted as a lacteal reservoir, the nerves corresponding with the lactiferous vessels. A novel and highly interesting paper was published in 1794 by Sir Anthony Carlisle.[2] He injected the excretory canals and reproductive organs of the cestode joint with coloured size, and was the first to work out the course of the former vessels, in which he correctly suspected the presence of valves. An excellent discussion of injection methods in use at the time, accompanied by practical directions, is given by Sir Charles Bell.[3] 'Injection', he says, 'has been the great instrument in the hands of modern anatomists.' His masses include preparations of varnish, wax, resin, tallow, size, oil, and turpentine. The colours are as usual, except that lamp-black is included. The general method is that of Monro—a thin non-coagulating medium followed by a stiff solidifying mass for the larger vessels. Bones are injected, decalcified, dried, and finally made transparent with turpentine. Pole's 'Instructor' has evidently been drawn upon.

The first general treatise in English on anatomical injection methods and technique was produced by Pole in 1790.[4] No original researches are embodied in the work, which is a comprehensive and thoroughly practical account of what was known at the end of the eighteenth century. The English Lyser of 1740 does not deal with injections. Four types of injection media are described—coarse (seven formulae), fine (six formulae), minute (six formulae), and mercurial. A cold injection, which sets after

[1] Giuseppe Saverio Poli (1746–1825), *Testacea Utriusque Siciliae*, Parma, 1791, fol.

[2] Sir Anthony Carlisle (1768–1840), *Transactions of the Linnean Society*, London, 1794, 4to, vol. ii.

[3] Sir Charles Bell (1774–1842), *A System of Dissections, explaining the Anatomy of the Human Body*, Edinburgh, 1798, fol.

[4] Thomas Pole (1753–1829), *The Anatomical Instructor*, London, 1790, 12mo.

some hours, is added on the authority of William Hunter. The coarse injections are preparations of wax; the fine, of varnish; and the minute, of gelatine. According to Pole, mercury was going out of fashion, and was seldom employed when other masses were available. It is significant of the expense of spirit and glass that most preparations are recommended to be dried and varnished.[1] Vermilion injections are directed to be preserved face downwards, so that when the colour settles in the vessels it will be on the side that is seen. There is no mention of fusible metal in the section on corroded preparations. Another text-book on practical anatomy was published by Hooper in 1798.[2] Its inspiration, however, is obviously derived from Pole, than which it is also less complete.

Between the opening of the nineteenth century and the inception of the modern period there is little of importance to record. In 1819 C. A. Rudolphi injected the digestive system of the Liver Fluke with mercury. Huschke[3] injected the uriniferous tubules of the kidney. The surface of the kidney was exposed to the action of a vacuum pump, and an injection of size and vermilion was thrown into the ureter. It penetrated along the tubules as far as the surface of the kidney, but the operation was not generally successful. Huschke saw the artery enter and leave the glomerulus in the Salamander, but denied that the glomerulus was continuous with the uriniferous tubule. Injections of Vertebrates other than mammals, and Invertebrates, were so much the exception that it is interesting to note that, as catalogued by Alessandrini,[4] such a collection was instituted in the University of Bologna in 1808. Fohmann[5] was the first to inject the lymphatics by random punctures in places where they form a network. This is the so-called 'ponction réticulaire', as distinguished from 'ponction directe', where the syringe is inserted into a specific and recognizable vessel. The method was revived by Rolleston in his

1 Cf. Cole, 'History of the Anatomical Museum', *Mackay Miscellany*, Liverpool, 1914, 8vo, p. 304.

2 Robert Hooper (1773–1835), *The Anatomist's Vade-Mecum*, London, 1798, 8vo.

3 Emil Huschke (1796–1883), 'Ueber die Textur der Nieren', *Oken, Isis*, Jena, 1828, Bd. xxi, col. 560–72.

4 Antonio Alessandrini (1786–1861), *Catalogo del Gabinetto d'Anatomia comparata della Pontificia Università di Bologna*, Bologna, 1854, 8vo.

5 Vincenz Fohmann (1794–1837), *Mémoire sur les vaisseaux lymphatiques de la Peau*, Liége, 1833, 4to.

PLATE XLVII

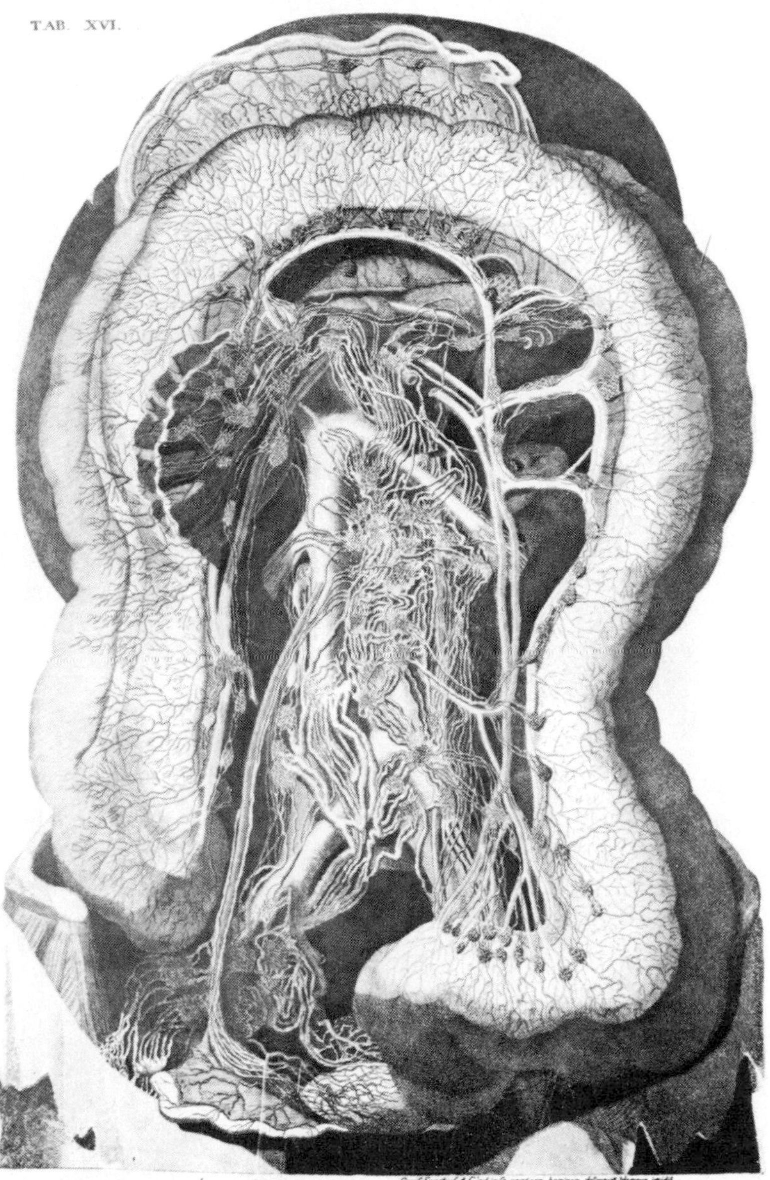

Mercury injection of the lymphatics of the human colon and abdomen by Paolo Mascagni (1787)

Harveian oration of 1873, and is still extensively practised. The expression 'double injection' was usually understood as referring to an injection first of a fine and then of a coarse medium, as practised by Monro; but it was sometimes used to distinguish the experiment of Doyère and Quatrefages.[1] These authors injected a dog successively with two different solutions which met in the vessels and there gave rise to a precipitate. Two of the substances employed were potassium chromate and acetate of lead, the result being a yellow precipitate of chromate of lead. In spite of the fact that this method was successfully exploited and spoken highly of by Bowman in 1842, it has never been developed, and indeed did not survive the criticisms directed against it by Robin in 1849.

A general and important review of injection methods, based largely on first-hand investigation, was published by the distinguished comparative anatomist Straus-Durckheim in 1843,[2] and the issue of this work may be considered to mark the termination of the historical period covered by this article. Straus-Durckheim, who speaks frequently of the expense of the materials employed, classifies injection masses into three groups: (1) coarse injections, of which fifteen are described; (2) fine injections, nine described; (3) corrosion injections, seven described. The substances tested are many and various. They include: yellow or white wax, tallow, lard, spermaceti, fatty oils, essential oils especially of turpentine and lavender, Venetian turpentine, Burgundy resin, hard resin, plaster, gelatine, white of egg, milk, water, alcohol, mercury, and fusible metal. He recommends spermaceti mixed with an essential oil because it penetrates the best, solidifies well, is transparent, takes a very brilliant colour, and any extravasations are easily removable. He has a cold solidifying injection mass made by the action of nitric acid on olive oil, but it does not appear to be very satisfactory. He finds that the setting of plaster masses may be delayed by the addition of gelatine. Gelatine he considers the best material for fine injections, and the variety of it he uses is isinglass. White of egg is diluted with water and coagulated with ferric sulphate—a procedure which has much to recommend it. Straus-Durckheim

[1] Louis Doyère (1811–63) and Armand de Quatrefages (1810–92), 'Sur les capillaires sanguins', *Extraits des Procès-Verbaux des Séances de la Société Philomatique*, Paris, 1841, 8vo, p. 17.

[2] Hercule Eugène Straus-Durckheim (1790–1865), *Traité pratique et théorique d'anatomie comparative*, Paris, 1843, 8vo.

regards mercury as one of the worst injections it is possible to employ—an opinion which had been gaining ground before his time, and became general soon after. His fusible metal for corroded preparations is made of bismuth, eight parts ; lead, five ; and tin, three parts. A small addition of mercury further reduces the melting-point. He experimented with the following colours : vermilion, carmine, indigo, Prussian blue, gamboge, verdigris,

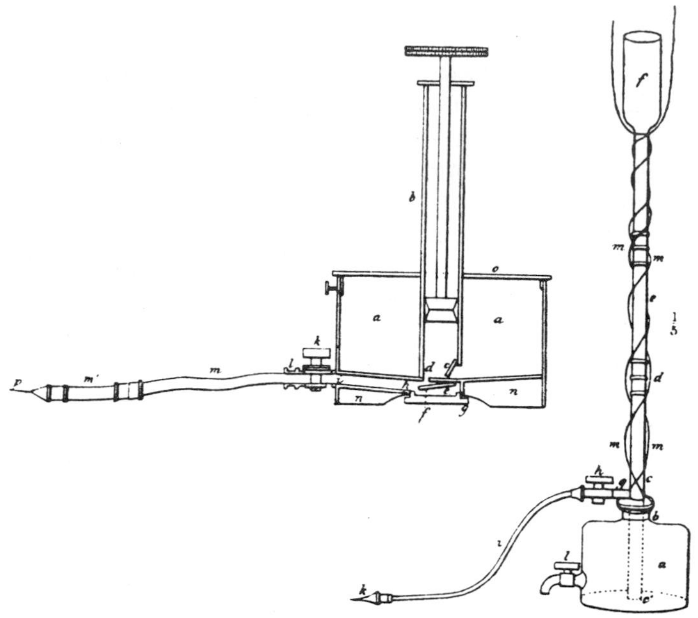

FIG. 9. APPARATUS DESIGNED BY HERCULE EUGÈNE STRAUS DURCKHEIM (1843).

In the top figure the syringe can be recharged without disconnecting any part of the apparatus. In the lower figure the necessary pressure is obtained by the gravitation of the mercury.

lamp-black, madder, 'orcanète' (a vegetable red dye extracted from the roots and bark of Anchusa), 'orseille' (a reddish purple extract of lichens), chrome yellow, Indian yellow, and neutral Potassium chromate. As regards instruments, for small injections the syringe of Swammerdam is satisfactory, but he prefers a more complicated injection pump which enables a considerable body of liquid to be thrown in without removing the canula from the blood-vessel. The figure (9) explains itself. Other simpler contrivances are described for injecting small animals such as Mollusca and Crustacea. For fine injections, and especially for the lymphatics, he devised an apparatus called the injector, which he himself had used for eighteen years (Fig. 9). It consists of the vessel *a* filled with the injection mass, which passes by the outlet *g* to the canula. The vertical tube *c* may be varied in height, and extends almost to the bottom of the vessel *a*. If now mercury be introduced by *c*, it accumulates at the bottom of *a* and forces the injection mass out by *g*, the actual pressure depending on the height of the column of mercury selected. This apparatus was afterwards modified by Robin. Straus-Durckheim also describes methods of inflation, of exhausting the vessels with a vacuum pump before injecting, and of filling the vessels of very large animals such as the rhinoceros and elephant.

SCIENCE AND THE UNITY OF MANKIND

By F. S. Marvin

This volume appears at a moment in world-history which will always be recalled for criticism or for admiration, probably for much of both. After a time of devastation and loss throughout the world unparalleled in history, a supreme attempt has just been made to combine mankind in a working union to prevent future conflicts and promote the peaceful solution of the international differences of the future. This is the greatest political fact of the moment. And in the sphere of knowledge science has reached the point at which its volume and complexity, its necessity for life and for education, are being at last fully realized in all civilized countries. There are thus two maxima before us, one in the political, the other in the intellectual world. We are in these studies interested in science, and especially in its history. But it would be a fatal mutilation of the history of science to attempt to treat it without regard to the accompanying social and political conditions of mankind. Whichever is root or trunk or flower in the tree of progress, the whole is vitally connected, and we are therefore bound as historians of science to ask what are the relations between our two maxima ; can we trace any direct relation between the growth of scientific knowledge and the unification of the world which has led to this new expression in the League of Nations.

The inquiry is large and difficult enough. The war seemed for the moment the heaviest blow which the cause of unity had ever borne. Yet it was the war which has given birth to the League of Nations. The war diverted, too, some of the best brains from the study of science, and has laid many of them to rest for ever. Yet it was to the war that we owe the recent triumphs of aviation, besides a multitude of minor inventions which may have their use in peace. Neither optimist indeed, nor cynic, can hold his own in this debate, and we shall be content with very wide and suggestive rather than demonstrated conclusions.

There is first the abstract question, the essentially social nature of science itself. It must be noted how science arises in the active mingling of the thought of many minds, and at times and places where men of various race and antecedents have met and interchanged ideas. Then comes the historical, or *a posteriori*, inquiry, how far the growth of science has been accompanied by a closer knitting up of the world as one community. This may be shown in a growing intimacy of all sorts of relations, in politics, commerce, culture, and religion. And the historical argument will lead finally to some consideration of the actual links which science has forged for the process of unification, and what it may do in the future to hasten it.

First, then, the abstract question, how far does the nature of science itself point to unity in mankind ? Science, i.e. organized or connected knowledge, is a social product. It made its appearance after communities of men on a fairly large scale had already been formed, and it was both the result and the means of a widening intercourse and communion between them. This fact, which is clear historically, is in no contradiction with the equally obvious fact that science in a rudimentary sense is present in the simplest inference by which the savage, or the animal for that matter, guides his action by the observation of certain sequences in the external world. The storm threatens, the plant poisons, the fire burns, and the most elementary of minds frames its conduct accordingly. 'All men', says Aristotle at the opening of the *Metaphysics*, 'desire to know', and it is by this differentia of conscious knowledge that the species is defined. But at the outset it was the necessities of life and not only or even mainly man's intellectual curiosity which prompted the forward process. By his quicker wits, his better memory, his greater readiness to fit means to ends, man was enabled to endure the trials of nature and to gain the mastery over the other animals, many of them far stronger physically than himself. Now all these desirable qualities are elements in the structure of science. Yet it is necessary for clearness of definition to limit our terms. There is a point at which man's growing knowledge and experience take on a distinctive character and make a fresh start, and this point was the birth of Greek philosophy in Ionia. It concurs—and the concurrence is significant—with the intense social activity which centred on the western sea-board of Asia Minor in the seventh and eighth centuries B.C. Here was the meeting-place of travellers and trade-routes both from the eastern empires of Assyria and

Babylon, and from the south and south-east, Egypt, Phoenicia, and Crete. Here the quick-witted Greek was in touch with all that could be reported by word of mouth from the centres of old priestly learning, here he could meet men who had seen the greatest wonders of man's activity on earth, and could set out himself to explore and to question the thinkers and the workers who could describe these things on the spot. Hence the figure of Thales and of those early Sages of Ionia must be regarded not only as characteristic types of the Hellenic race, but as the recipients and transmuters of wisdom and observations long anterior to themselves—the mouthpieces of a social product to which the toiling millions of the Nile valley and the Euphrates, the royal owners of wealth and authority, the priestly houses of record and study, had all contributed their part. And they again rest on the more primitive cultivators of the same soil in ages of stone. This social connexion is of the essence of science, and it spreads far, not only in space—as we see in this first appearance in the Middle East—but in time also. There is affiliation with the past as well as widespread intercourse in the present. Science was not born until these two conditions had been fulfilled, and to them it owes the special intensity of its social character, purer even than the social nature of language or religion or art. It is the sociality of reason itself, the 'connected experience' which, as Aristotle says in the same passage of the *Metaphysics*, differentiates the reason of man from that of the other animals. 'Of all the mental treasures of the race scientific truth alone compels general acquiescence.'[1] To it alone the differences of race or age or nationality are indifferent. In it alone we see the complete fusion of mind with mind which constitutes 'sociality'. And through the steady spread of this general acquiescence, or communion in truth, over all obstacles of ignorance or prejudices or remoteness on earth, the growth of humanity is best exhibited. Just as in Aristotle's view the 'desire to know' is the differentia of man from the other animals, so in the spread and acceptance of science we have the groundwork of humanity, the progressive Being *par excellence* known to us in the world.

It is interesting and significant that the perception of the universal nature, the sociality of reason, was, like science itself, only gradually acquired. It has kept pace through the ages with

[1] Sir Wm. Osler's Harveian Oration of 1906, Clarendon Press.

the growth of science, and is only in these latter days fully accepted, with the full acceptance of the dominating voice of science. To the classical Greek the light of truth appeared to be only granted to a small *élite* of mankind. Round the little island of right-thinking and enlightened men was a great ocean of barbarian outsiders, perverse in their mental processes and language, unintelligible to those who were on the way to truth. The Stoics, who were the universalists of the ancient world, made all men equal and all accessible to the dictates of reason. But they attained this position rather by evacuating reason of the scientific content which it had acquired during the Greek evolution than by seeking in all the humblest manifestations of reason the germs of the fuller growth. It was the universalism of Tolstoi rather than of Comte, though the Greek generalizing mind had made the premature attempt possible. Something of the same sort is true about the mediaeval theory, profoundly and impressively though Dante puts it. He tells us,[1] as Aristotle did, that the height of human power, the quality of man that makes him man, is thought, the power of understanding things. And he goes further than Aristotle towards the social source of knowledge. He is explicit that this endowment of reason is not an individual thing ; no one can thus think by himself. It belongs to man as a species, and only by the multitude of other men can any one man enjoy his faculty or increase it. On this side, then, Dante goes further and deeper than any one had gone before, for the universalism of Christianity was behind him and inspired his thought. But on the side of the nature of science he shared the arrested development inherent in the mediaeval position. These multitudes of men from whom the individual derived his reason and with whom he should enjoy it, were not co-operating agents in an infinite process of building up truth, but humble participants in a feast already provided by divine wisdom for those who had grace to hear the invitation and a pure heart to partake of the bounties offered. It is not till the re-birth of science in the sixteenth and seventeenth centuries that we can see both the social nature and the infinite progressiveness of knowledge first dawning together on the Western mind. ‘ Mind begets mind ’, said William Harvey, in a pregnant phrase, at the beginning of the seventeenth century, and it is in that century that we can first trace clearly the two

[1] *De Monarchia.*

connected, and hereafter dominating thoughts, of 'progress' and 'humanity', both based on the growth of a collective mind, exhibited in its most articulate form in the conclusions, constantly verified and constantly modified, of scientific truth. We cannot stay here to watch the gradual expansion of these ideas; but one feature in the nineteenth century demands some notice as perhaps the greatest single contribution to the conception of science as a social product. This is the contribution from archaeology. It was archaeology which established the notion of a firm, converging tendency in human thought towards common conclusions about ourselves and the world we live in. It has dispelled the barriers which seemed both to the Greeks and to the Catholic mediaevalists to divide mankind. It shows us the whole of our species struggling together from a common starting-point, by very similar steps to a common goal. The story thus unfolded is no more an idyll of brotherly love than any other part of history. It must indeed have been far less so, difficult though that may be to believe. Multitudes of races, as of individuals, fell by the way, more perhaps by the hands of fellow men than by other enemies. And yet the real progress which was achieved was by extended co-operation. The directing mind was social in its origin and social in its purpose, conscious or unavowed. And in the light of archaeology we can see that the same social processes of thought were at work wherever our first fathers set up a totem to protect their tribe, or sacrificed to the divinity of an ancestor or a nature-spirit, or devised some machine or word to make the common work more effective or the common emotion more intense.

This theory of the sociality of thought which Durkheim worked out fully, and perhaps too exclusively, in the sphere of religion, might, we think, be pressed further home than it has been at every stage. It would then appear that science, in the systematic sense, is the strongest of all links both between individuals and between groups of men. The friends of science must, on this subject, part company from emotionalists of the Tolstoi type. It is true that, either for association or disruption, emotion is for the moment a far more effective force. It is simpler, more obvious, more accessible to the masses of mankind. But the links of reason persist and spread, they go back to the distant past for those who can follow them, and while capable of indefinite reinforcement by emotion, they can maintain their hold and are a force in themselves, independent either of support or assault by passion. In this sense it is true that science is the fundamental bond of our race.

But we must turn from the difficulties of a summary appreciation of the most complicated of psychological questions to the clearer issues which are presented on the historical side. Here we may see—if we start on a backward journey from the present into the past—that the world is now more of a unity than it has ever been, and that science is at a maximum. These are firm starting-points, and as we look back, we may trace some three or four great moments or pre-eminent steps in the advance of knowledge linked up with the unification of the world. The earliest, the first occasion on which we can speak of science in the stricter sense, is the Greek construction of science and philosophy which extends from Thales and Pythagoras in the fifth and sixth century B. C. to Ptolemy, the continuer of Hipparchus, in the second century A. D. Connected with that is the formation of the Greco-Roman world at the head of which were men trained in Greek philosophy and attempting to apply the rules of reason to the congeries of human facts which had come together in their hands. The second would be the revival of science in the fifteenth and sixteenth centuries, based on the recovery of the work of the Greeks. Connected with that on the practical side is the discovery of the New World and the expansion of the West. Then, towards the end of the eighteenth century, begins the last and most rapid movement of all: it may be dated from the invention of the steam-engine. In this period, the last hundred and fifty years, the progress on both of our parallel lines has been unexampled. An increase in scientific knowledge, a new linking up of the world, have taken place, which completely eclipse in volume all that the earlier centuries have done of the same kind.

Our historical argument is to review very briefly these great stages and point out a few of the salient facts which support the theory that science and unity are essentially connected.

When we look first at the Greek evolution, we are struck rather by the marked incapacity of Greek cities and Greek statesmen to achieve any form of permanent co-operation than by their tendency to unity. If the argument for the unifying power of science rested here, its foundations would be bad. But it is of course impossible thus to isolate the prominent contesting Greek states from their environment, Athens from Alexander, or on the larger field, Greece from Rome. We should not be inclined to do so at all, were it not for the vicious tendency of the orthodox teaching of history which concentrates attention on the activities of Cleon and makes no serious estimate, probably does not even mention, Pythagoras, Hippocrates, Archimedes, or Hipparchus. And when we see how in the last two centuries B. C. the work and thought of the Greeks mingled finally with that of the Romans, it is clear that we must treat the two as complementary, parts of one whole, the organization of the Western world on the lines of the first scientific synthesis. Roman law is in fact a close analogue of Greek science, and in itself scientific. The two peoples, common in their racial origin, closely associated in the building up of city-state, were finally in the Roman Empire of the second century socially and politically merged; for the Empire rested intellectually as much on Greek thought as it did practically on Roman roads and legions and municipia.

It may be said that in this very vague and general way every form of orderly government might be called scientific. No doubt there is force in this criticism, and if we could follow it out, we should find that the impulse to seek orderly sequence in external nature, and to apply consistent rules to the management of men, are indeed near akin. But we may make some rather closer applications of Greek scientific notions to public order and unification in the Greco-Roman world. Think of the influence of astronomy. Pliny tells us in one of his letters that the Greek science of the heavenly bodies had done much to allay the superstitious fears of the populations of Asia and to produce calm and order in the public mind. And when one remembers the primitive explanations offered of eclipses and the panics which they have caused among untutored peoples, one can well understand the change which must have come over the general mental attitude in the eastern Mediterranean, as the Greek precise and natural explanation of such phenomena made its way. But there is another astronomical conclusion which seems of even greater moment for the unification of the world and has not been generally treated of in that connexion. That is the sphericity of the earth and the first perception of the truth that our planet is one of the multitude of spherical bodies, suspended or pursuing a certain path in space. This was a Greek achievement, and to the deeper vision of history which is now opening up to us, it will appear as one of the capital moments in human evolution. It seems to have been achieved by the time of Pythagoras. From that time on, but not before, it was possible for all the inhabitants of the earth to regard themselves as partners in one home with a common fate. Not till then could the ' humanum genus ' of the Romans, the ' Dear City of Man ' of the Stoics, begin to be envisaged with any reality. The subtle bias of such a thought towards the pacification and unity of the Roman world is obvious, though we cannot demonstrate its course of working. We may be as sure of it as we are of the world-wide desire at present for some effective confederation of mankind, though the conflicting policies of governments and newspapers will obliterate some of the approaches to the ideal for the future historian.

For us, reviewing now the approaches to the Greco-Roman system at its best, the main current seems to be the tradition of the Stoics and Cicero, down to men like the Elder and Younger Pliny, who were the backbone of the Empire. To Pliny the Elder Cicero ranked highest of all writers, save Homer alone, and Cicero was the man who, as quaestor in Sicily, showed his public spirit and his sympathy for Greek thought by rescuing the tomb of Archimedes from neglect. This type of Roman was no original thinker, not himself a contributor of any stones to the structure of science or philosophy, but being formed philosophically by Greece, and fully aware of the Greek superiority in abstract thought, he was able, in the Roman position of command, to spread ideas of intellectual as well as political order both East and West. East they spread more than West, for the area was already fertilized by Greek thought, and in that region the Greek tradition, in which a scientific habit of mind played its part, kept the Eastern Empire in being for a thousand years after the barbarians had broken up and resettled the Western. In the Catholic West, too, it was Greek thought which dominated and systematized the theology and philosophy of the Middle Ages, and thus formed again a link of unity.

But we must pass on to the re-birth of science in the stricter sense, when the men of the Renaissance brought their theories about nature into contact with the facts of observation and experiment. This was really the capital achievement of the Renaissance, outweighing the literary revival and the outburst of zeal for art, adventure, and enjoyment. The re-birth of science is often represented as the throwing off of bad Greek traditions, and, so far as it consisted in abandoning the idolatrous following of the mistakes of Aristotle and Galen, this is true enough. Yet all the leading thinkers who began the new scientific movement of the sixteenth and seventeenth centuries were in contact with the Greek masters who had worked in the same field before, and made use of their results. So far as the earlier thinkers had thought scientifically, so far as truth grows as a whole, just as mankind is a whole, this filiation is clearly essential unless the entire work is to be done over again. Descartes and Galileo were just as much subject to the condition in mathematics and physics as were the schools of medicine and philosophy. It was Arab thinkers who, in many cases, had kept the torch alight.

But there is one point of contact which is of special interest here, both as it concerns the unification of the world, and as a striking example of the successful union between science and practice, which was to become increasingly fruitful in later years. We spoke of the doctrine of the sphericity of the earth, and saw that it had been accepted some time before the meridian of Greek thought. Aristotle clearly stated it, and it was worked out in detail by Ptolemy. The maps of Ptolemy, with latitude and longitude fairly correct for the parts of the world known to him, were forgotten in the West for a thousand years, and replaced by imaginary constructions based on the supposed teachings of Holy Writ. The sphericity of the earth was, in fact, formally denied by the Church, and the mind of Western man, so far as it moved in this matter at all, moved back to the old confused notion of a modulated 'flatland', with the kingdoms of the world surrounding Jerusalem, the divinely chosen centre of the terrestrial disk. But at the beginning of the fifteenth century Ptolemy was recovered and translated into Latin, and by that time also intercourse with the Arabs and the journeys of Marco Polo and others to the East had prepared the way for a wider view. It was the re-establishment of the Greek doctrine of sphericity, the maps of Toscanelli based on Ptolemy, the currency which began again to be given to more scientific notions of geography, which inspired Columbus to seek the East by crossing the Atlantic. This is a salient instance of science stimulating the active powers of man to gain a completer knowledge of his earthly home, and through knowledge to bring it into one sphere of thought and action.

If we have an eye fixed on the ultimate goal of human federation, as we have in this essay, the expansion of the Western world which began at this time, the linking up of Europe with the Far West and the Far East by voyages of discovery, by commerce and by settlement, may well seem at least second if not first of all the achievements of the Renaissance, and ultimately far more important than the disruption of the Church and the rivalries of the new-born national states which absorb the larger part of the interest of the political historian of the period.

Another institution, also inspired by scientific thought, appeared at about this time, side by side with the Renaissance of physical science, even more potent perhaps in furthering the future unity of the world. This was Grotius's construction of international law, written in the first quarter of the seventeenth century, and contemporary with the work of Galileo and Harvey.

The analogy in the way of thinking between the first and greatest master of international law and the founders of physico-mathematical science is as close and striking as their coincidence in time. Both are clearly scientific. Just as in external nature thinkers from the time of the Greek Sages had been seeking laws, or forms, true for all geometrical figures of certain shapes, for all movement, celestial and terrestrial alike, so in Grotius a man arose who set out to ascertain what truths about man's nature in society might be assumed to be of universal validity and universal application. He was the first to look for these principles of universal right in man's own nature, apart from religious sanctions and superior to local fluctuations of time or race. On these universal principles he would build a universal code. Nothing obviously could be more consonant with, or more favourable to, our ideal of the unity of mankind. And this great effort appeared in the West at the same moment that the greatest of physical laws was being worked out by the labours of Galileo, and that Harvey was framing the first great mechanical law of physiology. Thus, if the statue of Grotius stands high, as it should stand, in the Courts of the League at Geneva, it should be recognized that the basis of his work, his own chief title to fame, rests in the application of scientific method to human affairs, and that this capital step was taken at the same time that modern science of the physical order was being launched on its triumphant career. Broadly considered, Grotius's work was part of the same movement, the constitution of truth by human reason acting on the observation of facts and reducing them to general laws by induction tested by experiment. And if we are told that in the case of international law, the experiments of history have proved disastrous to the generalizations, we shall reply that in the case of any law of life, and above all of human life, time for realization is of the essence of the problem. The infractions of international law no more destroy it than occasional murders destroy the general prevalence of the law—if we may call it a law—of conjugal and filial affection. Grotius stands firm, and for his justification we do not need a record of unbroken conformity to his laws, but of a progressive acceptance and enforcement of them by the nations. History, even the history of the war, has fully established this, and Grotius's ideal is brighter than ever as a beacon for a world travelling towards peace and international justice.

We have to pass now, much too hastily, from the time of these first great essays of modern science to our own, or recent times, in which the youth of the seventeenth century, with all his vigour of invention and enthusiasm of faith, may seem to have come of age, stained with many faults, but vastly stronger and more competent and well-grounded in a knowledge both of his errors and his power. We are still tracing the unifying process, assisted by science, and we claim to find it in spite of the wars of religion in the sixteenth century, the wars of trade in the eighteenth, the wars with France and Napoleon at the Revolution, or the Great War which has just closed. This last might perhaps be called a scientific war in its methods and engines, but it can no more be attributed to the growth of science as its cause than the wars of the Revolution could be ascribed to Lavoisier or Watt's engine.

The steam-engine may indeed be taken as the turning-point in the history of modern science, from which its growth, steadily progressive from the seventeenth century, became markedly quickened, especially in its applications to life and above all in those applications which promoted the unification of the world. Apart from its subsequent uses in linking up the world, Watt's invention is a classical instance of the linking up of science and practice, theoretical training and industrial profit. For the inventor was a man of general scientific attainments, especially on the mathematical side, and he was working at Glasgow in a department for study and not in an ordinary workshop, when

Newcomen's engine was brought to him for repair and led to his capital improvements. And the decisive thought was suggested to him by another scientist, Black, the professor of chemistry in the university. The same story, of the immediate linking of abstract and disinterested study with practical applications, may be told of the other principal agent in improved communications in the nineteenth century, the electric telegraph. It derives directly from the researches of Oersted and others into the nature of electricity, just as the waves of Hertz fifty years later gave us wireless telegraphy.

It is enlightening, from many points of view, to examine closely such cases as these because they exhibit clearly the active thought of the West in contact with facts, giving increased power in practice, and forming fresh associations between men. The contrast with Eastern, especially Indian, thought is palpable, the thought which turns constantly inwards and gives one the constructions of mysticism, the ideals of Nirvana. It becomes an obvious truth, as these contrasts and these connexions are realized, that the unification of the world, proceeding from the nucleus forged by Greco-Roman thought round the Mediterranean, is essentially the creation of that form of orderly, organized, and objective thinking which we have called science.

We have been dealing rather with some of the material and mechanical expressions of the unification of thought than with the spiritual unity itself, though the latter must be regarded as fundamental, and is as certain as the former. The similarities of life induced by the spread of great industries, the rise of great cities, the connexions of railway, steamship, and telegraph all over the globe, are obvious to every globe-trotter, and are by no means an unmixed good. Side by side with the revival of regional associations and nationalist feelings and self-government has gone a general flattening out of ancient peculiarities, a diminution of separate languages, a dying down of secluded cults and customs. It is hard to strike the balance between the two tendencies, but on the whole the world seems, to most of those who have seen much of it in recent times, a more uniform and, some would say, 'vulgar' place than it used to be. This, so far as it is true, is one side of the process of unification, the depressing side. No birth takes place without some loss and suffering, and if a true birth of a spiritual kind is taking place—which our whole argument tends to prove—then we may confidently hope that suitable forms and garments will follow to clothe the child of the future.

'It is no accident', Dr. A. N. Whitehead tells us in his recent book on the 'Organization of Thought', 'that an age of science has developed into an age of organization. Organized thought is the basis of organized action.' The words might have been taken as the text of this article. They sum up in the shortest and most unanswerable way the argument, that science being the 'organization of thought', the effort, and the result of the effort, to bring more and more facts of experience into an organic unity, must be, and has been, accompanied by a corresponding organization and unification of the beings who have produced that science. We have seen some of the external workings. It remains to look for a moment beneath the surface, and to consider how these spiritual forces may by cultivation, and with right direction from the general will, conduce to greater unity—of the desirable type—in future.

Social unity rests now, both nationally and internationally, in every unit large or small in which it exists, on the conscious co-operation of all the individuals who compose it. This is clearly and admittedly true in all civilized communities which have attained a national existence. It is true also of the large aggregates such as Russia, China, or India, which are struggling in various circumstances and difficulties to attain a conscious national existence. It was not the case in the great political aggregates of the past, the Roman Empire or the Holy Roman Empire of the Middle Ages. In these the unity was imposed from above. It is now to be attained from within, a task of incomparably greater difficulty and longer effort. But the new units, when organized, will have in them an enduring life of quite another order than the old. Within the family, the town, or the national state, besides the bonds of affection, there are the feelings of decency, of honour, of common interest, present more or less to the minds of all. Certain anti-social acts are never done except in moments of criminal excess; such acts, and others short of criminal, are reprobated and mostly checked by the social conscience of the community. Now, when we turn to the human community as a whole, we notice at once marked differences in the force and incidence of the various elements in the social conscience. Affection is far weaker: decency and honour are rather the reflections of the smaller social conscience than of the human conscience as a whole: the common interest is hardly felt, when we are thinking of all humanity face to face with the unplumbed perils of the universe. But considerations of the more strictly intellectual order, of human dignity, of the solidarity of mind, of our debt to the past and the possibilities of the future, all these feelings, which are social also, are stronger on the wider basis of humanity than on the narrower. But they require more cultivation: they are present consciously to the minds of only a minority even in the best educated countries; they seem to many of us to be unattainable by the mass. Yet they are essential to the unity of the future; and science probably provides the most accessible channel for their entry to the mind. In no other branch of human progress is the advance so clearly demonstrable from the past, and in no other is it so obviously the joint work of all civilized nations, and even of the uncivilized, all co-operating according to their gifts.

The teaching of history in a new spirit will be one of the means, perhaps the most widely applicable, of deepening the intellectual basis of unity. If, as we believe, the League of Nations becomes shortly the dominant political fact in the world, the attention of all the associated peoples will be directed perforce to the character and history of their fellow members and the basis of the union between them. As soon as this question is raised, we come to science in the various senses and manifestations which have been alluded to in this paper. Hence any teaching of history which may be favoured by the League—and there are good reasons why it should interest itself in the matter and give advice—must give a large place to the history of science as the field on which the nations have always worked most easily together, used one another's results and helped one another, except in cases such as dye-stuffs or munitions of war, where war-like or commercial rivalry has disturbed the natural harmony of truth. In this study the citizen, who will in an international system be more than the citizen of one state, may find the means of strengthening those social feelings of the more intellectual kind which are weaker and limited in the merely national sphere. The sense of human dignity cannot be better served than by observing the growth through the ages of that quality in mankind as a whole which Aristotle taught us to regard as the differentia of man as a species. Our sense of a debt to the past is most vivid in the case of that aspect of man's thought where congruence with the established results of earlier workers is most essential. And those earlier workers, even if hostile in the flesh, become, when dead, men *sans phrase*, all organs alike of one spirit whose nature is to struggle unceasingly for more strength, more clearness, and more comprehension.

Science here touches religion, as indeed any system of thought which involves the unity of mankind is bound to do. And the relation between the two suggests a contrast which has been often noticed before, but is of special interest to our argument and would lead us far if we had space to pursue it. The Middle Ages were noted for an internationalism in religion, which within its area—something like but something wider than the Roman Empire of the West—was the most intense and searching unity mankind has ever achieved or endured. The break came in the sixteenth century, and since that date it has been impossible, inconceivable even, to hold an oecumenical council for the whole West which would determine the right opinions to be held on any question of religion. But simultaneously with this dispersion on the religious side came a drawing together, a new internationalism, on the basis of science. The seventeenth century saw the establishment in the leading countries of the West of national societies and academies of science, which at once began to exchange visits, confer membership on one another, and advance by friendly rivalry. In the eighteenth century regular international co-operation began, and a multitude of permanent international associations followed, which are the most hopeful of all the prominent new organizations of the latest age. There are some hundreds of such, centred before the war chiefly in Brussels and The Hague; and their multiplicity is not due to difference of opinion but to variety of topic. They do not, as a Papal council would, declare the right opinion on a *filioque* clause either *ex cathedra* or through an official majority; but they welcome differences of view, supported by knowledge, on any vexed question within their range. They do not despair of agreement on any subject where evidence is available. But the agreement must be free, by conviction and not by force, by reason and not by authority. This is the unity of the future.

FOUR ARMENIAN TRACTS ON THE STRUCTURE OF THE HUMAN BODY

BY F. C. CONYBEARE

I. INTRODUCTION

ARMENIAN medicine was a closed subject until the appearance of Ernest Seidel's *Mechitar's Trost bei Fiebern* in 1908. That work, together with the texts here rendered, will enable the occidental reader to form some idea of the character and sources of mediaeval Armenian scientific ideas.

The British Museum Codex, Or. 6798 (Catalogue, no. 138),[1] is the source of our four treatises on the formation and Structure of the Human Body. It is a composite MS. formed of several distinct books and written by at least four different hands. The first section of this MS. contains our four treatises. This section consists of some 50 vellum leaves in double columns of 35 lines each. The writing is of the late cursive type called *notergir*, or notary's, small, but neat and clear. The titles of the sections or chapters are given in red. Folios 56–127 contain a second and separate book, written on paper in double columns of 32 lines, in a hand closely resembling the first. This second book gives our first two treatises over again in an identical text, with the same truncated colophon (see § 21 below). Either it was copied from the vellum book, or both were copied from a common source. As in the first book, so in this, the last tract is attributed to Gregory of Nyssa (§ 23 in our translation).

In addition to the two copies of our texts in the British Museum

[1] I would like here to correct an error in my catalogue of the Armenian MSS. of the British Museum. Misled by the numeral 3 affixed in the margin to the so-called treatise of Gregory of Nyssa on the folio of MS. 6798, I suggested that it was misplaced and should precede f. 4, in such a way that this treatise followed the Introduction and intervened between §§ 2 and 3. The folio was really in its right place, and the secondary text at f. 109b equally begins Gregory of Nyssa's tract after § 20 which describes the Seven Members, although § 21, on the Parts of the Body, is omitted.

volume, nine others are known to exist in a more or less complete state; of these, seven are in the Mechitarist Library at Vienna, one is at Munich, and one in the Bibliothèque nationale at Paris.[1] The four texts vary a good deal, but our version is practically identical with one of the Viennese (294). The great variety of readings indicates that the works had long been current and popular at the time when the variant MSS. were written.

The circumstance that the excerpt of Mechitar Heratzi (see § 1) occurs in at least two other MSS., and is followed by a passage which may be attributed to Asar (see next paragraph), suggests that the latter may himself have made the addition. As a physician Asar may well have had access to other works of Mechitar (who was a Cilician physician, *c.* 1150–1200) besides that on Fevers; that work the Mechitarists of Venice printed in Armenian in 1832,

[1] The Vienna MSS. 678 and 294 exactly agree with ours in texts and in the ordering of the contents. No. 658 of the same collection resembles it.

Another Vienna MS. no. 17 in Dashean's Catalogue, less allied in text to ours than no. 294, is a recent copy of an older MS. of unknown age; it gives the same contents in the same order.

The Vienna collection contains five other similar MSS. No. 436 is a late cursive. It begins with § 1 and ends with § 18, but Dashean's Catalogue gives no further information. No. 442 also contains it, and begins with § 1, but ends with § 17. The third MS., no. 466, begins with § 2, but what else it contains Father Dashean does not say. No. 254 is of the 17–18th century. The tract begins f. 7 r°, and contains §§ 1, 3, &c., but the codex is very defective through loss of leaves, and Dashean does not record its contents very fully. A fifth text is given in codex 540 of the same library, written in 1669. This contains the works of Galust, a physician of Amasia of that age. This text begins with § 1 and ends with § 20.

A Munich codex, no. 2, written in 1602, contains our text, beginning f. 71 r°, but I have only the barest details of its arrangement and contents, though Hunanean in his two volumes on ancient vulgar idioms of Armenia (Vienna, 1897), gives excerpts of—

A Paris MS., no. 108, ff. 29–45, contains the same text. Hunanean writes that he had examined ten MSS. of this work, and notes that two of them only, Vienna 294 and Paris 108, insert the excerpt from Mechithar Heratzi which I number §§ 3, 4. It ends with the words: 'So ends the description of the visual faculty by the will of God.' Our MS. continues without a break the rest of the discourse about the eye. Our MS. is thus a member of a close group consisting of three texts. Very slight differences divide them in respect of this excerpt; e. g. Vienna 294 and our MS. employ the word *Quawashs,* whereas Paris 108 substitutes *Vuslays,* and 294 and our MS. use the Middle Armenian plural *astarni,* whereas Paris 108 has the classical form *astarq*.

Hunanean cites § 1 according to the four codices, Vienna 294, Vienna 540 (Galust's codex), Munich 2 (written probably in 1602), and Vienna 17.

and a scholarly German version of it, with valuable commentary, was published in 1908 by Ernest Seidel. This treatise is much quoted in Asar's *Manual of Therapeutics* in the Mechitarist Library of Vienna (MS. 287), and is given twice over in a British Museum MS. (Codex f. 129 b and 41 a). It consists of 123 chapters, and cites many of the ancient Arabic and Armenian medical writers that are cited independently in the Great Tripartite manual of Medicine or Akhrapatin (i. e. Γραφίδιον) of Amirtovlath of Amasia, of which a magnificent vellum codex exists in the British Museum (Cat. Armen. MSS. 134, Or. 3712). Amirtovlath wrote *c.* 1466.

It is thus not impossible that § 1 of our tract is from the pen of this Asar, for in the preface to that writer's *Manual of Therapeutics* the comparison of the physician's art to that of the religious confessor (§ 1, paragraph 5) recurs in the same words. However, it is a commonplace often met with in Armenian medical treatises, so we must not attach too much importance to this. That one was copied from the other, or both from a common source, is, however, certain, since the language is the same. I translate herewith Asar's preface as it is found in our MS. (Brit. Mus., Or. 6798) and the Vienna MS. 287 :

'In the name of God the merciful and compassionate. A book of the healing art, as prescribed by wise philosophers and healing doctors for the understanding of man's nature and for ministration to the sick unto the uses of healing.

'For as by means of confession and true repentance they receive healing of soul, so likewise at the hands of healing doctors and with the aid of drugs they shall receive bodily health and be quit of their maladies. So now the humble in spirit, I, the servant of God's servants, the unlearned and much sinning Asar of Sebaste, have desired to collect the selection, and in brief to set forth according to our wants, a little out of much of the words of the philosophers, and to minister to sufferers, unto the uses of healing. And may the Creator vouchsafe health unto all according to his good and benevolent will, and to him be glory,' &c.

We may now devote a few lines to the discussion of the identity of the physician Abu Sayid, who is quoted in the first of our tractates (§ 1, paragraph 2). Two physicians of that name meet us in the history of Armenian medicine. The earlier was a contemporary of Gregory son of Vahram early in the eleventh century, and Amirtovlath cites his remedies more than once, e. g. in the following, which is given in Hunanean (ii. 415) :

'For liver disease due to fever we also copy out the remedy

used by Grigor son of Vahram much to his advantage at a time when he suffered in his liver through fever, and went to Mufarʌin (i. e. Nfkert) in the year 1037. And it was a prescription of Busayid and did him good. And his symptoms were these: pain in the back and right arm and heaviness of the hand, and internal stabbing pain in the back where the ribs fall away. When he lay down on his right side he felt acute pain and grew feverish, and wine and anything he ate hot gave him constipation (*or* ? aggravated it). So when this malady came on he went to Mufarʌin; as it was in winter time, the doctor gave him no medicine, saying: At this time of the year drugs will do you no good, for the man is frozen like the earth, and drugs are useless. But he gave him cool drinks, such as pomegranate liquor, &c., and let him eat what he liked. Then when spring came he prescribed him this treatment for forty days,' &c.

The Armenian prince evidently suffered from neuritis. Here is another mention of Abu Sayid from the same source:

'Another remedy which we have copied from Abu Sayid's manual of medicine, which Gregory son of Vahram used when his liver pained him and he went to Mufarʌin, as he did every spring, and derived great profit therefrom for several reasons: Take damask plums and twelve jujubes,' &c.

This Abu Sayid of 1037 was probably a Syrian or an Arab, but some of his writings were clearly preserved among the Armenians as late as the second half of the fifteenth century, either in the original or in Armenian translations.

Rather more than one hundred years later we have a notice of another Abu Sayid, a physician and savant, who was a friend and correspondent in turn of Nerses Shnorhali the Graceful, patriarch of Sis in Cilicia, who died A. D. 1173, and of Nerses of Lambron, bishop of Tarsus, who died 1198. Shnorhali, in his commentary on St. Matthew, states that he consulted this Abu Sayid about the reconciliation of the rival pedigrees of Jesus in the first and third gospels. He calls him a physician and savant, and wanted to know what solution was provided of the difficulty in Abu Sayid's Church. This proves that Abu Sayid was not an Armenian but a Syrian Christian. It was also at his request that Nerses of Lambron composed his tract on the *Names of City Builders*, published in the *Ztschr. f. Arm. Philologie*, 1903, I, p. 206. As Nerses wrote in Armenian, we infer that his friend Abu Sayid could at least read that tongue. He probably wrote in Syriac or Arabic.

It is impossible to say for certain to which of these personages the reference in § 1, the Prologue, refers. The circumstance that the tract is in the Middle Armenian idiom of Cilicia proves nothing, for if it was originally written in Syriac or Arabic, an Armenian might translate it as well later as sooner. There is, however, some evidence for Nerses of Lambron being the translator, in which case it is likely to be the work of the Abu Sayid who was his contemporary. This evidence consists of three notices to the effect that this Nerses composed such a treatise. The first is found in a short but anonymous life of him cited by Alishan, the modern historian of Armenia, in his volume of *Sisuan*, p. 91. The second is in a colophon printed, apparently from the MS., of Nerses's meditation and prayers in connexion with the *Dormitio Iohannis*. The third is a notice, printed in an edition of sundry works of Nerses printed at Cpl. in 1736, to the effect that he wrote the book on the *Formation of Man*. On the strength of these notices Hunanean inclines to believe that Nerses of Lambron was the translator. He was certainly familiar with Syriac, for we have Armenian versions of Syriac originals from his pen. Hunanean confesses himself unable definitely to fix the date of the tract from the language in which it is composed, but finds no difficulty, as we have seen, in attributing it to Nerses of Lambron, who died in 1198. I find myself a great affinity between its idiom and that of Mechitar Heratzi, the author of the work on Fevers. It would be out of place here to go into details, and I will mention only two striking facts. Both in it and in Mechitar we find *ukhtavoruthiun* for *akhtavoruthiun*, a sign, though one rarely encountered, of the phonetic decay of the vowel *a* in that age. Again, instead of writing *erkouorek* for 'testicles', both writers employ the form *ekavorek*. This is a rare form, so rare that Dr. Seidel, excellent scholar as he is, has not understood it. I confess that I can in general see no distinction between the Armenian style and idiom of Mechitar and that of the author, whoever he was, of our tract. It is possible that Mechitar, who was a friend of Nerses Shnorhali, and wrote his work on Fevers in the year 1184 (when Nerses was Patriarch of Sis), may himself have executed the translation of Abu Sayid at the wish of Nerses. It was a common thing for learned men to undertake such tasks at the behest of a prelate; and that may be the reason of Nerses of Lambron's name being attached to it.

As regards our fourth treatise, it is needless to say that Gregory of Nyssa, whose name is attached to it, had nothing to do with the work, and that the tract of which I have here (§ 23) translated the first few pages is falsely attributed to him. It awaits more complete treatment than I have been able to give it. It was the connexion in literary tradition of this Father of the Church with Nemesios which gave rise to such an extravagance. The work of the latter exists in Armenian, but has nothing in common with the work ascribed here to Gregory of Nyssa.

We have, then, in these four treatises a monument of the medical learning of the Armenians not later than the twelfth century. It would need a wider acquaintance with the many MSS. of these works than I have had the opportunity of making, to decide whether and how far the texts have been amplified by medical editors and scribes like Asar of Sebaste. We must not, for example, without further inquiry, attribute to the original form of the treatise the ascription to some planet or other of each organ of the body. These ascriptions in our work invariably come at the end of the section devoted to the particular organ, and may easily therefore be a later accretion from the pen of Asar, who in the colophon of § 21 admits that he in some way completed Abu Sayid's work, and who no doubt incorporated in it §§ 3, 4.

Like the later medical schools of Europe, the Armenian was dominated by Arabic learning. Most of the technical terms used are Arabic, much disguised in their Armenian dress. Equally so are the names of Greek medical writers that often came first through Syriac, and from Syriac through Arabic. Bagarat, for example, a common and distinguished name in Armenia and Georgia, disguises Hippocrates. In other Armenian medical treatises we have Archigenes disguised as Ardjidjanes or Ardjiasus, Paul (of Egina) as Flaus, Oribasius as Arpisaus, Rufus as Upufaus, Diogenes as Deudjanis, and so on.

II. Contents of British Museum Text

[Press-mark Or. 6798 ; Cat., no. 138, fo. 2–11]

The text that we here print contains four separate works, of somewhat different style and motive :

I. § 1–§ 18 is a complete and systematic treatise on the structure and functions of the organs of the body. It is perhaps the work of the physician Abu Sayid, who lived in the twelfth century, but his work has been amplified by one or more medical scribes such as Asar.

The British Museum MS. has a colophon at the end of § 18 attesting that 'This Book was written by Halathzaden', and we are asked to remember Astuadsatur the Elder, 'our father'.

Whether 'this book' refers only to the treatise which precedes is not clear. It is, however, an indication that §§ 19 and 20 formed once a work separate from the treatise.

II. § 19–§ 21 is a separate and more theoretic work, which deals chiefly with the numbers of the various organs and with their relation to the mental and spiritual qualities and with the causes of disease. The colophon expressly states that it is imperfect.

In the British Museum Codex Or. 6798 this colophon, § 21, is truncated. But in the Codices of the Mechitarists' Library at Vienna, 678 and 294, which otherwise presents a text identical with ours and which was copied from 678 in A. D. 1625, it runs thus:

'*Now, Brethren, our original was very imperfect and faulty*, but, by the help of God, Asar of Sebaste (Sivas), the scribe and true disciple of the book, having with excessive erudition given his leisure to foreign works, with much labour was *barely able to bring it to so much accuracy as this.* But *you that are aided by it, bear in mind the sinful* much toiling Asar the Scribe and myself, the sinful penman who has soiled the pages of the paper and am also called the penman. Mark, O beloved among sages, to accept from me the word of the Apostle Paul, that there is won of your goodwill the grace of the Lord' . . .

It is highly improbable that the author of this longer notice would have gone out of his way to incorporate in it the phrases italicized from the shorter notice. The MS. which thus enables us to restore the colophon was written in 1625 at Ispahan by one Paul the Monk.

III. § 22 is a short note on the relationship of the various organs to each other. It was, as its colophon tells us, transcribed by, if not the work of, one Halathidy.

IV. § 23 is a spurious work attributed to Gregory of Nyssa (*c.* 331–*c.* 396) on the formation of the foetus in the mother's womb. Only the first part is here rendered by way of giving an idea of its contents and character and of identifying it.

The tractates here translated consist of the following sections, the numbers affixed being my own and not those of the MSS. :

I. 1. Introduction.
2. Concerning the Head and Brain.
3. Concerning the Eyes.
4. Concerning the Muscles of the Eye.
5. Concerning Vision (in the text no title is given).
6. On the Ear.
7. On the Nose.
8. On the Mouth.
9. On the Heart.
10. On the Lungs.
11. On the Liver.
12. On the Spleen.
13. On the Kidneys.
14. On the Gall.
15. On the Bladder.
16. On the Testicles.
17. On the Stomach.
18. On the Guts.

II. 19. On Sinews (*or* Nerves), Ducts (*or* Veins), and Blood in General. Begins: 'The all wise God formed the joints of man . . .'
20. On the Seven Members (*or* Organs) whereby man hath Life. The first is the Brain.
21. Colophon.

III. 22. The Parts of the Body.

IV. 23. A work attributed to Gregory of Nyssa, beginning: 'Man is said to be of four constituents.'

I have to thank Dr. Singer for supplying me with photographs of the British Museum MS. as well as for many suggestions; and my gratitude is especially due to Father P. N. Akinian of the Mechitarist Convent of Vienna for the care with which he has revised my translation, correcting it in numerous passages and furnishing the right meanings of many obscure terms.

III. Translation of the Four Tracts

TRACT I

1. *Concerning the formation of man and the creation of all the members* (or *organs*) *of man, by the will of God.*

Of truly able select philosophers and healing doctors, Hellenes and Greeks, for the understanding of the nature of man's body, mouldings and members, bones and articulations, ducts (*or* veins) and sinews. How they were created and what are their respective natures or functions, whereby they supply with moisture all the members [1] of the body.

But also the exciting causes (lit. movements) of diseases and the remedial aiding of the same, and the operation, as understood by the great physicians Galen, Aristotle, and Bagarat (Hippocrates), by whom [the remedies] were disseminated among Greeks and Assyrians and Persians and Indians, among Hellenes and Arabs, and were disseminated unto all the corners of the world by the Giver of grace from above, and there were able men of all races. And at their behest and by their words many investigated and made themselves wise and able. Among whom was also one called Abu Sayid, who took from the books of the chief physicians sincerely and concisely, and bestowed on us this treatise, correct and succinct, unto the praise and glory of God, who fashioned creation and equipped all with utilities as he willed and made all, and what he commands comes to be.

And who is able to search out the deep things of God ? for whatsoever God made is exceeding good.

God made water, earth, sea, and dry land, beings of fire and beings of clay, beings spiritual and those that breathe, animals and birds, plants and vegetables, and all else. God made the body of Adam, and vouchsafed to him rational spirit and charged him to love God and keep his commandments. And the love and science of man, to discover this was the art of healing, by means of healing doctrine and co-operation of drugs to minister to the suffering unto the uses of health.

For by means of the confession of sins and acceptance of repentance a man shall receive healing of soul, and at the hands of physicians and with the aid of drugs he shall receive bodily health by the will of God, as saith the prophet : He that hath not bodily health, cannot serve God in spirit.

The wise [2] Galen says that God created man like a city, having twelve gates by which drugs and foodstuffs enter, while superfluities go out, whereby the system (*or* person) is constantly aided.

And of these twelve gates, of which we spake, two are eyes, two ears, two nostrils, one the mouth, two the breasts, one the navel, and two exits, one for discharging water and the other the posterior.

But there are two great channels on the two sides of the haunches (buttocks, *lumbus*, *ilium*, or *coxa*). And on each side of the haunch 180 channels open, which makes 360 ducts in movement, from all of which the members derive material and are strengthened.

Also as there are four winds which blow over the world, from East, West, South, and North, and as the year is divided into four seasons, Spring,

[1] *Member* is used both of internal organs, heart, liver, &c., and of external limbs.

[2] Perhaps the treatise of Abu Sayid begins here rather than with § 2. In any case § 1 up to this point must be a composition of Asar's, the editor or redactor of Abu Sayid's work as we have it here.

Summer, Autumn, and Winter, so the life of man is divided into four portions. For in childhood and the first age, man's nature is hot and moist, because it is dominated by blood and eastern air and follows the spring season, for the nature of spring is hot and moist. But in the second age, while youth lasts, nature is hot and dry, being dominated by bile and southern air, and follows the summer season, for the nature of summer is hot and dry.

And the third portion of life is in nature cool and dry, being dominated by bile (savta, χυλός) and western air, and it follows the autumn season, for the nature of autumn is cool and dry. And when man enters the fourth portion, it is old age, and his nature is cold and moist, for it is dominated by phlegm (palλam, *pituita*) and north wind, and follows the winter season, for its nature too is cold and moist.

The wise say that when God created Adam it was springtime, and night and day each consisted of twelve hours, and the sun was in Aries in the first degree. Therefore when spring comes, everything turns green and sprouts up out of the ground; and all animated beings are stirred, and humours (lit. minglings) of body are subtilized, and blood and bile ferment and are rarefied. Wherefore all physicians have bidden in spring days to bleed and imbibe purgatives, for all men's humours in this manner are made to ferment and in these days grow soft and rarefied. But there are formed in the person of man four kinds of liquid, salt and bitter, sweet and ill-smelling. Salt liquid is of the eyes, for were it not salt it would melt the fat of the eyes. And bitter liquid is of the ears, for were it not bitter, flies and creeping things would enter the ears, and a maggot would be there and do harm. And sweet liquid is of the mouth, which receives the savour of things eaten. And ill-smelling liquid is in the loins, whence comes seed, and offspring is generated therefrom.

So far so good.

2. *Concerning the structure of the Head and the Brain.*

The wise Galen says that God made the human head and set within it the brain, and of all wisdom and faculty of movement did he set the seat therein. God made the human head in three portions. The first he made the place of sensation and of light's filaments of the eyes. The second portion he made a vessel of the consciousness and of the intelligence (antidjeli, *djeldsch*). And the third portion he made the place of guarding, that whatever it sees and understands, therein it may guard and study it.

And God devised the head of seven layers (*tapaλa*) and seven membranes, in every layer one membrane. For all these are a protection of the brain that there may reach it suddenly no whit of mischief (? *zenuthiun*) nor any pain. The first layer is the hair; the second the skin; the third the flesh; the fourth the bone; the fifth is another skin enclosed within the bone; the sixth is a skin over the brain and interiorly another; the seventh is the brain. And God instituted all this protection of the brain that chill or heat should not be able to penetrate to it, and do harm (? *zên*) to it; for the brain is master of the house and in command of the heart. And the heart is sovereign of the whole person, and than the heart or brain there is no more excellent (*aλêk*) member in the body. For health and life reside within the heart, but intelligence (*or* consciousness) and initiative in the brain.

And the hair of man was made by God and devised out of consumed (or burned) blood, and in proportion as the consumed blood increases, the hair takes increase (? λ*alapa*) and waxes strong and long; and more and more this arises from the blood which is consumed, for while the man is alive and healthy it is made from pure blood. But when fleshiness (? *mis*) increases and phlegm (=*pituita*) in him, the hair does not sprout, and the man whose hair grows thin, his nature becomes phlegmatic; but if a man's head is grown bald, that is due to red bile. But man hath grace and shame in the brain, and when a blow (*or* pain) affects the brain, it causes a lack of two things in the man, and the man becomes unconscious (lit. silly) and insensible, so that he cannot know and understand what is good and what bad. And if the blow falls on the middle of the brain, which is the seat of domination and consciousness, or if it be excessive and concentrated, the mind fails and he swoons and can comprehend nothing.

But if the injury befalls the posterior cavity which is the retentive (? ἑκτικός), forgetfulness comes over him, so that he knows no more, either past or future. Medical men have said that they knew cases where men were so injured, with the result that they forgot the names of their own fathers and mothers who begat and bore them. And there were some who, when they were yawning, even forgot to shut their mouths, so oblivious did they become. And this affection visits men in grey-haired age more severely than at any other time, and interiorly is due to phlegm (*pituita*).

And there is, furthermore, a path leading from the brain to the heart, so that the brain incessantly has the heart in view; and when this path is open, a man is subject to the disease called swooning (*saqthay*=syncope); and it often happens that when this affection prevails, a man is bereft of consciousness and intelligence, and the colour of his face goes. And not a few physicians through ignorance of this affection imagine the patient to be dead and hand him over for burial, for one who is suffering from it is as it were dead and the colour leaves his face. For when the passage to the brain from the heart is blocked, the heart is unable to absorb water any more from the brain, and the life of the heart and its warmth are unable to reach the brain, with the result that the latter is congealed; and then this affection occurs and manifests itself in the man. For if life and illumination and substance were not in (*or* from) the heart, the brain would quickly be congealed and the man die.

But if there were not coolness and moisture in the brain to counteract the heart's heat, this heat would rapidly consume the entire person. And God made the brain cool and damp, and set the moon over it as its controller [1] (*or* arbiter), but made the heart hot and dry, and set the sun over it as its controller. And these two by their nature give strength and support to one another, and substantiate each other, and they fare well with God's help.

3. *Concerning the structure and formation of the Eyes.*

The great Mechithar has said that every physician who wishes to tend the eyes must study the structure of the eyes, treatment of which is [given] by philosophy; and in these recipes (*vuslays*) [2] the eye was not described, not even in the *alrapatin* (γραφίδια, *pharmakopoiia* or *therapeutica*). So then, I, Mechithar, was minded to describe the formation of the eye in brief, and relate how many tunics (*tapala*) there are which are linings (*astaṛni*) of the eyes, or how many *ṛutupat* which are humours of the eyes, or how many tendons *azaltunae* (Arab. *zala*) which are muscles of the eyes. And I describe the visual (*or* contemplative) spirit, where it belongs and how it proceeds and progresses along conjoined (*or* equal) fibres which they name the *Lusénion* (i. e. retina): it is necessary to know this, because ancient

[1] *tanuter*, an astronological term, lit. house-lord.

[2] *Nuslay* or *Nuskhay* in Paris MS. 108. Our MS. and Vienna 294 read *Qunnash*; ? an Arabic word.

philosophers have somewhat contradicted each other concerning the membranes of the eyes ; for some say there are six ; others five, others four, others three, and some two tunics. But I hold with Galen, for Galen and his following said there are seven [of] the eyes, and three humours and nine muscles, and I mention all, one by one, by the aid of God.

We now describe the tunics of the eyes.

Now the first tunic that is interior attached to the bone, which the Tadjiks (Arabs) call the *zulpie*, which is to be translated the hard body (i. e. sclerotic) ; it is more sinewy and firm and hard than the other membranes, which is why they so named it, and it attaches to the bone which separates the bony ruggedness (?) from the eyes, and protects the eyes and prevents mischief (*zahokn*, accus. of *ahok*) from getting into the eyes. For the first meninx (*mizλ*) is interior, and of the cranium there are two meninges, the one (attached to) the bone and the other attached to the brain. Now the one attached to the bone, hard is its body and sinewy, and it possesses many veins (*or* ducts) from the artery, and its use is to keep off the bone's ruggedness and weight from the brain, and prevent injury (*tchahok* ?) to the brain.

And the second meninx which attaches to the brain is more delicate and soft and pure than the first, so that it may not weigh on the brain, and its body likewise is subtle [1] of veins, and of the artery. And this membrane which is named sclerotic, is engendered of (*or* ? engenders) this meninx which attaches to the cranium.

And the second tunic is that which is named *shmima*, i.e. placenta ; this they call *sekin* (choroide), and they call *shmima* the skin which covers a child in its mother's womb, and it is born with this membrane ; by this metaphor they have illustrated it, and have called it by this name (viz. *mater*). And it belongs to this meninx which lies upon the brain, as we mentioned before.

The third tunic which they have named *lapaqia*, which is to be translated *ark* (i. e. retina), for it has the semblance of an ark, wherefore it was so called, and the thing owes its origin to the placental (? *shmima*) membrane.

But after this come three humours. There is a humour which they have named glass (*zudjadji*), which is to be translated *apikeni* [2] (i. e. vitreous), because it has the semblance of white glaze. Wherefore they have so named it.

Now this is followed by a humour which they have named *djaliti*, which is to be translated *sarneni* (or crystalline), pure and resplendent and circular, and you must know that the crystalline is a conspicuous (lit. glorious) and precious appurtenance of the eyes. . . . For through it arises the visual (faculty) and perception of colours and forms, and its roundness is to the end that it may not incur mischief (*tchahok*) and adverse shocks, impinging on it. For the reason that the arteries in those places remain at peace from the roundness, and the crystalline (ternic) is in the middle of the eyes like a ball, held in the midst, or like a (central) point in a circle, and it is surrounded and protected by all the membranes, and humours subserve the precious (thing), in order to ward off mischief (*tchahok*) and secure its welfare continually

And after this is the fourth tunic which they name *yanqaputhia*, which is translated *sardosteni* (arachnoid), because it resembles a spider's web, and is subtle and limpid and pure, wherefore it is so called. And it lies between the crystalline and the white of egg in the middle, that mischief (*tchahok*) may not from its humours happen to the crystalline.

1 Perhaps the sense is ' delicately veined and supplied with arteries '.

2 In Paris 108 *apakini*.

And besides this there is an humour which they name subtile white of egg (i.e. aqueous humour), because it resembles egg in whiteness, and therefore they have so named it.

Fifth is the tunic which they call *yanapia*, to be translated *khaλoλeni* (vine, uva), because it resembles that fruit of the vine, wherefore it is so named.

And the sixth membrane is named *kharnopia*, to be translated *eldschereni* (= *cornea*), which is why they have so named it. In itself it is exteriorly limpid and resplendent and smooth as if hard skinned, and this is the reason why when you open the eye you see the image.

But Gelianos (? Galenus) said these three tunics were one (or the first) meninx, and appealed in witness to the fact that when *karha* arises, which is a tumour (*or* ? blister, pimple, &c), and if it issues upwards into the *aponeurosis* (*mizλ* or ? = *amnios*), it is quickly inflamed (lit. boiled) and at once opens and the *spin* (cicatrix) due to it is soft and there is not a white facula, and a remedy is quickly ascertained.

But if the tumour (?) is in the second *mizλ* (*aponeurosis* or membrane), it inflames (*efi*) and opens late, and the *spin* (? cicatrix) which is due thereto (?) is thicker and white in colour, and a remedy is quickly ascertained.

But if the tumour penetrates the third *mizλ* (? membrane), it is late to inflame and the *spin* (? cicatrix), which comes in it, is denser and firm and the colour, a white facula (? *dschah*), and no remedy whatever is known.

And if the tumour is large, so that it bursts itself, or the matter (*Khltv* = ὕλη, χυλός) be acrid (*sur*, *or* sharp), so that it opens, and the uvea appears exposed, if it appears small, they call it a musca (fly in the head).

If it appear large, they call it a *bepp*[1] in the head, and name it *karhay* (i.e. wound), and also they call it *bath* and *kháλuart* (abscess).

And there is a seventh tunic called *multhahimay*, to be translated *Koshrads* (= *conjunctiva*), and it is inside like a mantle (or shelter) to protect exteriorly, and it spreads out upon the membranes, wherefore they gave it this name, for it is a cartilage (*khrdjtam*). Wherefore they make a wide perforation in it and let pass the water, named tube inside like a *kamsh* (i.e. *gamysh* = reed-tube), and open the eyes by God's will.

4. *Concerning knowledge of the Muscles of the Eye and their function.*

And be it known that the eye has four chief muscles : one, on the upper side (*dih*), which draws up the eye, towards the eyebrow, and one on the lower side, which draws the eye down, towards the nose and cheek, and one on the side of the source (or fountain), which pulls the eyes towards the eyebrow, and another on the side of the ear ; and there are four to the four sides, strong (? *brnen*), and if to any one of them humour penetrates, and they relax, the eye droops and drops and goggles, for it lengthens and is drawn back obliquely. But if one of them is affected by dryness, it is drawn back to the other in the same way, and that distorts the eyes, for whatever dries up the eyes drags them up and makes a squint, and this is the cause of squinting.

And there are two other muscles which they call *thevq* (wings). It is they that move the eye in a circular direction, up and down and from side to side, as a man desires.

And there are three other muscles destined to control the tube of the nerve fibre, at the end where the pupil is, so as to concentrate and keep the light in it. And if the mixture of these muscles be moistened and softened,

[1] *Bepr*, Persian for leopard.

or if they be lacerated by any outside shock, the light is poured out and dispersed all over the eyeball (lit. fruit), which they name *inthishar*, and which they translate *vathats*, i.e. outflow.

But if their composition dries up, and they contract, as for example a thong (or rope), falling into water, relaxes and is stretched ; and if it falls into fire conversely contracts and shortens (lit. comes together), so now the muscles, when they are wet, become *inthishar*, and, when they dry up, draw together the tube of the sinew at its origin, and the pupil is compressed. And the pupil appears the eye of a needle. And this is what they name compression or contraction.

And there are three sinews (muscle or nerves) which control the upper lids of the eyes, and their function is to draw the two lids down. And one draws up the lid and opens the eye, but the lower lid has no sinews, and for that reason is not moved, and if it moves does so unnerved (lit. by non-muscle or non-sinew).

And these are the nine muscles in the eye we wrote of, three supplying the upper lid.

Now I describe the tube of the muscle (*or* sinew *or* nerve) and say whence it is generated, and how it goes to the eye, and its use or function. And we must remark that the head has three cavities : one on the hinder side in the occiput, one on the front side in the forehead, and one in the middle. And the hinder cavity contains the faculty of memory,[1] the middle one the understanding, which is the brain, and the front one the senses. And from the front cavity springs an united (*or* conjoint) nerve, which they name the first pair or conjoint, and at middle distance (*or* in the middle of its position) it has as it were a tube, and proceeds straight along the right side to the right eye, and leftwise to the left eye. And when they reach the inner part of the bone of the forehead, they confront each other and mingle and become one, and separate afresh at the same spot, and pass along the right side to the left eye, and leftwise to the right. And the reason why God so arranged is that in case one eye be blinded, the light collects in the other and is gathered in it adventitiously. And we have evidence of this in the fact that (he that) wishes to see things clearly, that is, things dim or afar, covers one eye for the light to collect in the other eye. And this Nature has taught us and made clear, and this is the tube of the nerve, along which visual faculty passes and reaches the eye.

Now I have to mention the visual soul, whence it arises and how the visual (faculty *or* object) comes to be through it,[2] when the stomach dissolves and exhausts (or presses) the food, and sends it on to the liver, and the liver receives and concocts it, until it is converted into blood ; and in this concoction an exhalation rises as from all things in process of being cooked (*or* boiled), and this air, so far as it is in the liver, they name a spirit of Nature. Now what of it is limpid and pure (decent *or* normal *or* temperate) goes to the heart, and there is named vital spirit, and what is pure in it ascends by the ducts to the brain and enters the firm meninx, which is within the cranium, and there circulates along all the ducts and is further concocted and purified. And then it enters the second meninx, which is above the brain. In the same way it circulates these along all the ducts and is cooked and purified.

But nature which stood in need of this warm vapour knew well how to refine it, wherefore she made the road a long one and the passages narrow.

[1] Read *isholuthiun* for *ishol.*

[2] The transition is so abrupt that words may have dropped out.

And so it enters the front cavity of the head, and is there named the perceptive spirit, and when it is there sufficiently refined it enters first the conjoint nerve (*or* muscle), which contains it in its midst like a tube, and so it passes to the eyes, where it is named the visual spirit and accomplishes vision, through the moisture of the crystalline, and through the mediation of the tunics. So ends the description of the visual faculty by the will of God.

5. And you must know that the benevolent God set surely the faculty of vision in the middle of seven tunics, and made them a protection that no ill may reach it; by way of convenience also in order that the brain's moisture may do it no harm.

And as for the water of the eyes, he provided it that the heart's warmth and wind might not do harm to the eyes. And on the outside of the eyes God has provided two sets of hair, one on the lower lids and one on the upper lids.[1] The one he made for the sight and the other to carry away moisture from the eyes; for if there were not eyelashes, the water of the eyes would continually be leaking out. For see you not that when the lashes are kept away from the eyes, tears come regularly. If there were not upper lashes a man could not see anything from afar. The eyelashes enable him to see what is near, and the upper lids what is afar. And the latter it is that keep off the sun sufficiently for it to do no harm to the eyes. And a man whose eyelids are removed is like a channel into which the water enters, while there lack trees and grass along the banks; the channel is ruined by the soil filling it up. For the trees and grass along the banks kept it from being damaged. So with a man's eyes; when the hair is cut away from the lids he is liable to many maladies of the eye. God made man's eyes of light, and the sun takes light from light. When a child is separated from its mother and in that hour the moon is in star chamber of Zohal (Saturn), the child becomes *atchiku* (blind or ophthalmic), for when the moon is in a foul star chamber, the seed of disease enters the eye.

6. *On the formation of the Ears.*

God fashioned the ears for hearing, and the audition of man was arranged by him within the brain. And he appointed two ducts leading from the brain outwards to the ears. And God made the ears like a strain of music (*or* like a goblet) *speurô* (? spiral),[2] which when you strike it, gives a sound like a flute; and what is heard, it transmits to the brain which takes it in, and makes of it what is convenient.

And many men are deaf from their mothers' wombs, and that is owing to the fact that these two channels are blocked up. But those who being grown up go deaf, the cause is this: that from a superfluity of bilious matters a vapour ascends to the brain, and from the brain issues into the ear to the two channels and fills them up, and stops the hearing. For the passage by which we hear is then blocked. And often enough the cause resides in the phlegm (=*pituita*), for an excess of this affects these two channels, and the ears, and stops them up, and denies you hearing. Scabies (ψώρα), however, does not create phlegm. But of black bile and of red bile and of blood a superfluity creates severe scabies, and they interfere with and prevent hearing.

God made the ear like a cellar (?) and built its entrance with a twist, so

[1] The word *Irander* 'upper lids' usually means 'eyebrows', but to so render the passage would make nonsense of it.

[2] The Vienna MSS. read *spetrô*, which recalls *spectrum*, but that gives no good sense in the context.

that foul air might not enter, nor an injurious sound easily enter the ear, and reach the brain and do harm. And (God) put in a naturally bitter liquid, and did so to check any harmful insect that might like to enter the ear, for it encounters the bitter liquid and is prevented from entering ; for because of its smell and taste they do not venture within, but flee and go back.

And God made more precious than the whole person the eye and ear ; but some physicians have said that the ear is more excellent than the eye, because, when it is dark and night-time, the eye cannot see anything, whereas the ear, even if it be night or day, hears everything, and, what is more, hears better by night than by day. And some physicians have rated the eyes higher than the ears for the reason that, if you do not hear a thing with your ears, you cannot tell another about it ; whereas the eye sees everything, and describes it to another person. And by way of example they adduce the thunder and lightning, but the eye sees the lightning and only afterwards the ear hears the thunder.

But God made the ear because of the brain, so that whatever is heard, is sent on to the brain, as the gentleman's doorkeeper does, who does not allow everyone, especially the unsuitable, to enter his master's chamber (*or* court), until he receives an order to do so from him, and then lets him enter. Just so the ear, which does not allow the unsuitable to come in.

7. *Concerning the formation of the Nostril.*

God formed the nostrils in order to perceive all good or bad odours, and to expel and disperse all the brain's superfluities. And he fashioned the nose like a conduit to protect the brain. It gathers and keeps therefrom any superfluities. All this issues out by the nostrils.

And from the nostril a passage was devised by God to the windpipe, which when the mouth is shut lets the breath issue forth ; for were it not so, the mouth would have to remain open, and there would be a great risk of creeping things entering the mouth and doing harm, so that many would die of it.

And God made the human mouth like a box of which the padlock (*kuplaq*) is a rampart gate.[1] And the nose was created downwards, firm and solid out of cartilage and free from the food that is being chewed, so that when the mouth is full, the breath can issue forth through it. And the fumes of bilious matters issue forth by the nostrils, and the breathing is not hampered nor the man strangled by them. And these fumes and superfluities in this way cannot enter the brain and harm it. For there is then an ample passage kept open, and, as long as it is open, this injury cannot befall a man. And the discharge of the nostril was made salt in flavour so that foul irritants (? *havaj*)[2] may not enter, and the brain remain uninjured. And if a thousand times the nostrils were open still creeping things would not enter, for its liquid is salt, and its presiding star is Lucifer. But God made the countenance (*eres*) of man fiery and temperate, and lo, it is clear that winter does not freeze it with cold nor summer torment it with heat, nor is it burned up by the sun, by the command of God.

8. *Concerning the formation of the Mouth.*

God fashioned the mouth for eating, and made its liquid sweet, in order to receive the savour of all things ; and he fashioned the tongue in it in order by its continual movement to keep the mouth moist, and we receive the flavour on the roof of the mouth (i. e. through the palate) with the tip of the tongue ; for have you not remarked that when the tongue is cut off, the

[1] Probably the reference is to a battlement projecting above the main portal of a fortress and perforated beneath with apertures.

[2] The Vienna MSS. read *haut*, Arabic for *air*.

savour of things is lost and no longer perceived? And the liquid which is collected in the mouth is generated in the liver; and when it ascends from the liver, it supplies blood as far as the windpipe, and then by God's command it becomes white and turns into water. But if red bile is left in the liver, it causes the savour of the liver to be bitter; and the water which is in the mouth becomes bitter. But if black bile is left in the liver, the savour turns sour, and the mouth becomes sour. But if phlegm is left in the liver, the savour of the mouth becomes salt; but if there be only pure blood in the liver, and the liver be healthy, then the savour of the mouth is sweet and *mah thatil* (?).

And God made in the mouth a passage to carry to the whole person flavours. But the presiding star of the mouth is Jupiter, while as that of the tongue was appointed Mercury, and that of the teeth is the same Mercury. For have you not remarked that when a man is in Mercury, he is soft and twistful in his speech, and cannot talk trenchantly, and a man who lacks teeth cannot fully enunciate his words?

9. *Concerning the formation of the Heart.*

God fashioned the heart as sovereign of the entire person, and made its nature hot and dry, placing the entire heat of the person therein, and constituting it the abode of warmth, so that the whole person is heated therefrom. The heart itself is perpetually in movement and has no rest, like the heavens which know no repose at all. So too the heart. And God made the heart like a tree which is broad at one end and narrow at the other. And the entire heart consists of two chambers.[1] There is one chamber on the right which contains living blood, and one on the left which contains air. But from the heart proceed four ducts, of which each divides into 32, so that in all there are 128 ducts, which the Arab calls its arteries. And together with these ducts the heart is in continual movement; and they are evidence of life and death. For when skilled physicians lay their fingers on the duct which is called *madjas* (radial artery?) they diagnose from every side the ailments connected with death or life, and apply the requisite drugs. And of these four ducts we mentioned, two are distributed to the upper parts of the body, and two to the lower, and they give force to the entire body, so that it is strengthened by them and keeps healthy, and as its presiding star was appointed the Sun.

10. *Concerning the formation of the Lungs.*

God fashioned the lungs and constituted their nature cool and moist, and made it the chamber of phlegm which is *malas*. And God made the lung as the breath of life of the whole body, and placed it between the heart and stomach to be continually the refrigerator (?) of the heart, and allay (?) its ardour. And it also allays the heat of the liver, and prevents the heat and ardour of either of them from consuming the body. And God has constructed the lungs like an oven, that the air in it may constantly circulate within. If it were not so, a man could not at all keep his bowels (*or* middle) cool. And God made the lungs an air-chamber, to distribute the wind over the body, and wherever it is wanted it is made to go as required. And its presiding star he made to be the moon.

11. *Concerning the formation of the Liver.*

God made the liver to preside like a good or evil star over the whole body, and made it the fountain (*or* source) of blood. And he made its nature hot and moist, and constituted it the furnace of the stomach, so that any

[1] Or, ? 'And the circle of the heart is to be set firm upon two chambers'.

food which enters the stomach is cooked by the liver by its heat. And when the food is cooked, the liver draws its virtue (*or* strength) into itself, and in accordance with its nature converts it into blood, and distributes it over all the members, and feeds and strengthens the body.

At first it sends to the heart the simple (*or* clear) exhalations (*or* vapours) which issue from itself, and having reached the heart they are afresh distilled or rarefied and so rise to the brain. But the bitterness which like sulphur (?) escapes upwards, it sends to the gall to become yellow bile (*safra*) ; and the other crude (uncooked or undigested) blood it contains as a moisture and heavy liquid, it gives to the lung, to become phlegm (*pituita*) ; and the blood, which possesses fat and is crude and mixed with water, it gives to the kidneys, to be cooked and become semen ; but whatever is not accounted for goes to the *falabusht* or bladder, which discharges it without. But the rest of the burned blood is collected at the buttocks (*or* on the bottom) as sediment (τρύξ), and the *aqpuq* (?) gives it to the spleen to become bile.

And in this way the liver feeds all the members of the body.

But if it should happen that some indisposition affect the liver in the way of heat or cold, moisture or dryness, or if it be due to obstruction caused by constipation, the liver is oppressed and cannot get to their destinations its contributions to the various members, and these superfluities are left in the liver and cause it injury ; and enfeeble not it alone, but all the members, and the body is consumed and peaked and is in danger.

But if the liver is oppressed by cold, and cannot warm the stomach, so as to cook (*or* digest) the food, the latter is left undigested (uncooked) and becomes bile, and thin water ; and this water the liver cannot take up (? absorb, lit. bear weight of), but it disperses it through the body along the fibres and joints (*or* seams), and it causes paralysis and uselessness (*lakva*).

But if the food is left undissolved and passes down through the bowels, it causes *thuqmay* (? dyspepsia) and *lopindsch* (? *gulundj* = tranchées), and looseness above and below.

But if the liver is overcome with heat, and the stomach be thrown into disorder and prevented from disposing properly of the food and from sending its contribution to its master, and if the passage for bile (*or* of the gall) be blocked, then the liver is unable to tolerate this yellow bile (*safra*), but sends it abroad into the system (*or* person) and renders it *sralan* (? *deλnuthiun*, i. e. yellow.) And if the red bile penetrates to the lung, it stabs. And if to the spleen, it swells it up, and the colour turns yellow and the belly is enlarged and a morbid condition set up. And there are various other diseases akin to this, I mean in connexion with the blood, and the black bile, which, if it remain in the liver, causes divers maladies.

And there issue from the liver five huge ducts and spread over all the members, and its central duct is what they term the basilica (*pasilik*), and the lower one is the *catholice*, which they call *aqhal* ; and the upper one what they call *kifal* (cephalic). But two other ducts issue from the liver, from which it passes to the artery.

And all the ducts issuing from the liver are incessantly in motion, and the divisions of all the ducts, which come from (*or* are in) the liver, are 232. And God made Jupiter to preside over the liver and other ducts. But the liver is one of the great members, which feeds and nourishes all the members by the will of God.

12. *Concerning the formation of the Spleen.*

God made the spleen like a stable, which collects in itself all the excrement, and sundry burnt blood which there is in the liver and ducts is all gathered into it; for the liver contributes to it all the burnt blood. And there is a duct from the liver direct to the spleen, along which all burnt blood is sent straight to the spleen; and a duct from the spleen goes to the heart, which contributes cold to the heart, which counteracts the heart's heat lest it burn the person. And the duct from the spleen to the liver, when the spleen fails,—then the burnt blood therefrom issues to the liver. And the liver cannot sustain its weight, but disperses it to the members of the body, and renders a man *savtakot* (? melancholic) and carbunculous and pockmarked and *quthesh*, and otherwise diseased.[1]

And the cause of all this is burnt blood, which gathers in the spleen and damages the liver and excites these complaints. For the liver cannot carry the load and disperses it into the members. But if this burnt blood goes straight to the heart, it induces melancholy and epilepsy.[2] But melancholy is of several kinds, and one sometimes laughs, sometimes weeps, sometimes is calm, speechless, and insensible.

If this burnt blood collects and goes straight to the brain, it renders a man epileptic. And this ailment has three kinds like melancholy. One as it were besets a man and draws him from aloft downwards; another impels him as it were to stand erect and then fall down, and he foams at the mouth, and makes foul noises, and dashes foot and hand and head on the ground, and many pass urine. Another kind is that which drives a man to a spring or to water, and he will declare that good children beset him, and talk with him and beat him. And all this is caused by burned blood collecting in the spleen and overburdening it and swelling it up; and next it causes *isthisλayilahmi*,[3] which means a tumid condition of the human system and the face grows fat; and if red bile is mixed therewith, it produces pleuritis (lit. malady of the ribs).

God made the spleen cool and dry, and made it the chamber of black bile, and the taste is sour. And Zohal (Saturn) is appointed its presiding star.

13. *Kidneys.*

God made the kidney the seat of desire, and it was made more delicate and subtle than other seeds, and it was set firm in the middle by a sinew (*or* fibre), and from the sinew enclosed firmly in a pelt which protects [and] covers this pellicle of the kidneys so that the heat of the heart may not consume it.

And a duct of blood issues from the heart and goes to the brain, and from the brain descends along two ducts behind the ears and passes under the muscles through the one full of blood and through the other full of breath (*or* air), to the kidneys. And it descends and is cooked (*or* digested) and the water (*or* liquid) is afterwards separated and what is left becomes semen, and of it the child is engendered in the woman's womb by the command of God. The kidneys are so made that from the right and left hand one there is a passage for the water (*or* liquid). And they were made the abode of desire by God and of seed, and Lucifer appointed their presiding star.

14. *Concerning the formation of the Gall.*

God made the gall bladder the abode of bile which is red bile, because whatever of heat and dryness there would be in the person is all collected by

[1] Father Akinean renders his MSS.: 'And on the body warts are formed and eruptions are caused and pockmarks and erysipelas, and pigs (a kind of malady) and other forms of sickness.'

[2] *malas kath.* The first word elsewhere (Section 10) is interpreted phlegm. Akinean: pesanteur de tête or heaviness after drink. I believe it is μέλας, and that the words here used = μέλαινα σταγών, *atra gutta.*

[3] ? Elephantiasis. Akinean: Hydropisii.

the liver and sent straight into the gall bladder; and it has its entrance; but it has no way of issuing out. And God made its fibre (sinew) cool and moist, and its pelt (δέρμα, *corium*) hot and dry, that the cold and moisture of the fibre and the heat and dryness of the pelt may take away the bile of the liver and correct the same. For were it not so, there would be heat of the bile and it would consume the kidneys.

And its presiding star is *Marekh* (Mars).

15. *Concerning the Bladder.*

God made the bladder a vessel for the water of the entire person, so that whatever water is secreted from the stomach and liver may be collected in it.

But the bladder has two tunics. The interior tunic is the urinary receptacle (lit. water-pourer), and the outer tunic is for the semen, which descends from the kidneys along the outer tunic and proceeds to the right-hand testicle, where it is collected and then discharged by the urethra to fall into the woman's womb, where it engenders the child by divine command.

And stone is formed in the semen and is deposited in the bladder, and the urine and the stone pass out together. And there are some men who cannot pass their water; for if the vessel of the urine becomes hot and dry, the urine comes out yellow and mixed with bile, and the passage burns and blocks the mouth of the bladder.

But if the penis (?) turns hot and humid, and then it is filled, but the man cannot retain or keep the urine, and it often issues forth, and if it should happen at any time that it forcibly hinders (micturition), then the man's body suffers relaxation, and the receptacle of urine is distorted; and if for this reason (the urine) cannot any more enter (the urethra), and if no passage be found by which it can enter afresh, then [the urethra] often will burst and the man die.

God made the bladder cool and humid, and appointed the Moon to preside over it.

16. *Concerning the formation of the Testicles.*

When the testicles were constituted by the command of God, man's strength was placed in them, and they were made the place of semen and of desire. And the desire of seed goes out from them and is spread over the body and the eyesight, which desires and impels a man to gratify his passion.

And a man's beard absorbs stuff from the eggs (i. e. testicles). You can make the experiment yourself. For a man entering the bath, before he washes lays them in sesame oil; and when he quits the bath after washing, his beard becomes lustrous and soft and sleek.

And the eggs constitute the difference between man and woman; for if they are cut, a man becomes beardless, like a woman. Nor can he any more beget a child, and they are one of man's most important members. The wise Galen declares there to be four important members in a man's body, viz.: 1, the brain; 2, the heart; 3, the liver; 4, the eggs. Were they not one of the most important they would not be so allied with the heart, as you can see for yourself. If they are violently seized and insulted, a shivering comes into the heart and it faints, and unless they are quickly released, the man at once dies. And the presiding star is Lucifer or Zôhray.

17. *Concerning the Stomach.*

God made the stomach of sinew and flesh, and around the mouth of the stomach runs copious sinew, but on the inside on the floor of it copious flesh; for this reason that flesh is hot and sinew cold. And so the food is assembled in a hot place and there concocted; for in a cool place things cannot be cooked. And the sinew was for this reason set above, in order to draw the food and drink into it. And whether the food is at one time plentiful and at another scanty, makes no difference. For however much you drag the sinew it relaxes, and when you let it go it returns again to its position; whereas if you drag the flesh it at once is lacerated. And in the stomach are four virtues (*or* powers). The first they call *Djazip* (Arab. *pull*), because it is a dragging force, the second *masiq* (Arab. *retain*) which is retaining. The third they call *Hazim* (Arab. *digest*) or digestive, and the fourth *Tafiay* (Arab. *repel*) or expulsive, and sending out. But even if things are a thousand times light or small they require seven hours in order to be cooked (digested) and melted down. But if the stomach is hot and vigorous, and thoroughly dry or wet, food is rapidly digested before the fifth hour. But if the stomach be cold and wet, the digestion takes place otherwise. If the food be very light and defective, it is quickly digested and assimilated. And whatever one eats or drinks, the stomach assimilates it into itself and there it is digested; for on the right side is the liver and on the left the spleen. Above it again is the heart, and below it the gall; behind is the spine and in front the abdomen. And when the food and drink descend into the stomach then the inner mouth of the stomach shuts, so that the food and drink do not flow out until it is properly cooked and digested and its nutritive virtue appropriated, and sent on to the liver to make blood, and to be apportioned to all the members as suits them. And then it opens a passage to what is useless and conducts it out, and physicians call this the *pap* (i. e. *bab*), which is doorkeeper or pylorus.

18. *Concerning the Guts.*

God made the guts in seven parts, for that is the statement of Bagarat (Hippocrates). But Galen declares for six guts. For the first one which is contiguous with the stomach is not reckoned by him among the guts, as it is by Bagarat; where it quits the stomach and divides off, it has another rôle than the guts, and is named the pylorus, because it discharges a different function. Moreover the six guts form groups of three, because three of them are thick and three slender.

The first thick one they call *ithnayashara*, which is twelve (*duodenum*), and another *λulin*, which is vocal. And another *musthaλim*, which is foundation and sensation of the guts. But of the slender ones, the first is called *sayim*, which means dry gut, another *avar* or crooked, and another *taλiλ* or minute. In all, three thick and three thin, apart from the pylorus which adheres to the stomach. But all food and drink taken by a man is collected in the stomach, which receives the nutritive elements as required, while the useless are presently dispatched without by the wind, for the latter descends to the lung.

Colophon.

This book was written by Halathzaden, an unworthy sinner, dust and ashes. Remember Astuadsatur (i. e. Theodore) the Elder, our father, and Christ remember you.

TRACT II

19. *Concerning the formation and devising of Bones and Sinews and Ducts* (or *Veins*) *and all Blood.*

The all-wise God formed the joints of man of bones 248 in number, and articulated them together, and bound them fast with sinews, and cemented them into one body. And he set in them ducts for the vitalizing and irrigation (lit. inebriation) of the entire body.

Now in the head man has five bones, and in the entire *qnateλ* (i.e. temple) is one bone, and the teeth consist of 32 bones. The whole number of bones in the head is 39. In the neck and vertebra [1] (*or* ? tibia) are two bones.

But in the hand, i.e. upper limb, in the shoulder are two. In the *dsil* or elbow is one bone. In the upper (?) arm one bone, in the forearm two bones. In the palm of the hand 35, which makes all the articulations of the bones of the hands 41, with a corresponding number for the hand on the other side.

In the back bone and vertebrae [2] there are 18. In the right hand ribs there are nine bones, in the left hand ribs eight.[3]

And in the feet (?), in the *Djur* there are two, in the member (μελος) one, in the knee two, in the shank (*sṛunq*) two, in the vertebrae of the foot 35, which makes all the bones of the foot to number 42, with a corresponding number in the other one.

But a woman has 252 bones, four more than a man, and these four bones are in the generative space of the woman, and they interlock like fingers; and when a child is about to be born, by the command of God they divide back from one another and the child issues forth, and as soon as it is separated from the mother, they close again and join by the command of God.

But God devised the sinews (*or* nerves) of man 248 in number, first the great sinews which are bonds (*or* yokes) 25, and then are divided, and become lesser (*or* thin) ones, 228 in number. And the sinews were made by God to hold fast the bones. But the ducts of man were made by God to replenish with moisture all the members of the body.

But God also set apart in the human brain 14 ducts, and attached every duct to a member, so that all the members are sustained and nourished by them, and are invigorated by one another and fortify the person.

And the ducts send and stimulate the person to movement, and the tongue to speech, and run also to eye and ear and other senses. For four ducts lead to the eyes and move them. Two communicate light to the eyes and two move them. But these two ducts which give light are void of blood and dynamical, so that the eyes are *datzmov* (?) sustained from the brain. When these two ducts are cool and wet, a man sees well from afar and ill from near; but when they are hot and dry, one sees well from near and ill from far. And if they are violently harmed a man goes blind.

And two ducts run to the ear and are empty and bloodless. Their rôle is to guard the ear and transmit whatever it hears to the brain. And God created the ear a hard cavity and spiral, so that a loud sound cannot suddenly penetrate, nor water, nor insect get in and reach the brain and injure it. But if these two ducts are impeded by superfluities, then the ear is rendered deaf. And two empty ducts lead to the nose, to receive sweet smells and send them to the brain, and by night they keep ready.

And God made the nose a grotto (*or* fortress) to keep continually open by night and day, and to keep wet, in order to drain away all superfluities and preserve the brain clean and pure, so that the latter may remain untroubled by epilepsy and swoon, and from paralysis and from *lakua* (? helplessness) and lunacy (? melancholia), and from headache and all other diseases of the brain.

And there is a duct in the palate of the mouth which is sensitive to all savours, sweet or bitter, salt or saltless, sharp and acid, which pull and draw the brain to the controlling principle (?).

And another duct is in the tongue, which enables the tongue to talk.

[1] *lisern*, κνήμη, or σφόνδυλος, or vertebra. [2] *lisern* above so rendered.

[3] One rib having been abstracted to constitute Eve.

And as the tongue is the hand of the jaws, which collects the food and gives it to the teeth for them to pulverize it; in case anything escapes the teeth, the tongue collects it afresh and directs it to the teeth, like grain which is thrown with the hands between the millstones, the tongue being the hand of the jaws.

And two ducts lead to the hand, and each of these two ducts is divided into two. And one runs toward the right hand and one toward the left, and these regularly keep the hands in movement.

And two ducts go to the feet, which equally keep them in movement.

And one duct runs to the two testicles from the brain, to move them and hold them firm and enable them to fulfil its desire. It is the instrument of desire; and if it is hindered from acting, they use drugs and infusions to re-establish and clean the duct, to enable it to function afresh.

And a woman's ducts are eight more in number than a man's, and she has four ducts which allow her to fulfil her female functions, and four in the breast whence milk issues, two to one teat and two to the other, which makes up the number of all the female's ducts to 248, by the will of the ineffable and incomprehensible God.

20. *Concerning the formation of the Seven Members, whereby man hath life;* the first member is the brain, second heart, third liver, fourth lungs fifth reins, sixth gall bladder, seventh spleen, which makes up the seven organs in man that are not external, as are eyes, brows, nose, ear, tongue, hands, feet, testicles (eggs). For one of these may be cut off, or two or three of them, and the man survives; whereas if a single one of these internal members on which his life depends fails, he dies irreparably.

And we are to know that perception, sensation, motion, wisdom and vigilance reside in the brain, whose nature is cool and moist. Life and light and strength and redness in the heart, whose nature is dry and hot. Joy and laughter and gladness in the liver, whose nature is hot and moist. Breath (*or* spirit) and voice and *yejul* (? *hival*) in the lung, whose nature is cool and moist. Desire and power to copulate and seed are of the reins, whose nature is hot and moist. Irritability and anger and love of power in the gall, whose nature is hot and dry. Shame and rancour and impatience and suspense in the spleen, whose nature is cool and dry. And as long as these seven members which occupy man's interior are of even tenor and free from superfluities, a man is healthy; but if one of them chances to be injured, you may recover health with God's aid by means of wise physicians and drugs, for God has bestowed wisdom on experienced physicians so that they can make good by remedies the injury according to its nature.

But if the nature of the members gets astray and inverted, so that the brain turns dry and hot, or the heart cold and wet, or the liver cold and dry, or the lung hot and the gall cool, in such conditions of their natures, the possessor of them invariably falls sick,[1] and it is hard to heal him, nay, he can only die.

And God made in the windpipe a bit of flesh like a column, which is called *zankik* (? epiglottis or uvula), to protect the breath, so that no food or drink go astray and diverge from the passage to the stomach, and enter upon that by which the breath issues from the lungs. This going astray the epiglottis prevents. But if on a sudden food should pass along the passage by which the breath issues from the lungs and fall into the said pipe and obstruct the breath, then a man is oppressed and sick (lit. over-hot). In such a case the breath is constricted, and it is an effort to draw it; for if it cannot find its way out, the man will die. For there is no interior passage by which

[1] *uchtavor* for *achtavor*.

it can escape, since the lung reposes (?) obliquely (?) on the stomach, and therefore has no passage from within.

And in respect of the lung physicians have said that it continually ventilates the lung ; for if the lung wing did not waft the air in from outside through the lung and so enable the heart to drive its heat without, then its heat would wax so great as to consume and burn up the system, and the man would die. On the other hand, except for the heat of the heart, which the lung receives from it, life could not stir in the man. And there are these two airs, of which the one issues out and the other enters in. The one which enters is cold and is called *pisaṛ* (Pers. *bisear*, i.e. numerous) or *eṛat* or ample and copious ; but that which goes out is stifling hot and is called *khupat*, i.e. wanton or proud and potent (?).

And skilled physicians diagnose death or life by this criterion : when they lay their fingers on the duct and understand.

And they call the liver *aλsamelptan* (?), that is the purveyor of the body and feeder. For when the meat and drink are collected in the stomach and digested, the nutritive and useful elements are passed on by it to the liver which receives them and converts them into blood and sends it streaming to all the members, and so nourishes and sustains the strength of the person.

Thin and clear blood is given to the heart, but the sour (*or* hard *or* rough) element of it to the gall, while such as is coarse (*or* thick) and watery goes to the lung, and what has a moderate amount of fat in it to the reins, and the dregs of the blood goes to the spleen. But such part as has little moisture and red (*or* yellow) bile goes to the bowels. For ill-digested and unwholesome matters descend into the gut from the stomach, for the bile to rarefy them. They are discharged for convenience, and any red bile or food becomes a superfluity in the liver, and descends readily into the guts.

And as long as the liver is warm and wet, it is vigorous of function and quick to digest and quick to pass on ; but if it be enfeebled, then the liver at once cannot digest and pass the food on. And this unconcocted blood remains in the gut and becomes scurf (*or* scab) and induces colic. And should pale bile increase in the liver and engender blood that like a fibre exists there and moves, the man will take too little nutriment and nature will not assimilate the food. And pale bile cannot give blood to the gall, because there is (only) enough of it in the liver, and it would cause harm and disease in the man.

But when there is too much of phlegm (*pituita*) in the liver, so that it cannot send blood to the gall, as in the case of the pale bile we wrote of above, in this case it is badly liquefied and becomes water and sets up pains of *sokhay* (ill nutrition), so that the body swells. And when the black bile finds no passage to the spleen and the road is blocked, that gives rise to *Hillath* (Arab. languor), like pox, etcetera.

And of all the diseases which affect man, all the causes can be traced to the liver being feeble and unable to send on its superfluities (? secretions) to their places, and to distribute them to the members. For the health and life of the whole person is in the liver, by the care of God.

21. And now, brethren, our original was very imperfect and faulty, and we were barely able to bring to so much accuracy as this. And you that are aided by it, bear in mind the much-sinning servant, and may God have mercy on you.

TRACT III

22. *The Parts of the Body.*

The brain is a part of the body containing the bond (or conjunction) of life. Bone is the casing of the brain. Sinew is the (instrument) of the body in movement. The duct (or vein) is the (instrument) of the blood. Blood is the nutriment of the body. The breathing is the (instrument) of spirit. Spirit is the (instrument) of prime movement. The belly is the instrument of primal fecundity (or enjoyment). The liver is the instrument for the replenishment of the veins with blood. The heart is the origin of living (or psychic) life or origin of the life of the soul. The soul is of one breath. The hands are instruments in giving and taking. The feet are instruments subservient to going one's way. The mouth is an instrument of voice and of taking food. The tongue is an instrument of speech and of taste. The sense organs are instruments of sensation. The eye is an instrument of seeing; the ear of hearing; the nose of smell.

Body is the instrument of touch. Bile is a refinement (*or* subtilizing) of blood. *Malas* (? = phlegm, *pituita*, mucus) is a refinement of bile. *Mamatz* (chyle) is a superfluity from first food. Seed is moisture with breath in transmutation of blood suitable unto life or a superfluity of the last food agreeable, and resembles a creative agency that begets him, or a nutritive soul (?). Creation and birth, or the first food of a living being in transmutation of body.[1]

And glory be to Christ, God our hope.

It is to be remarked that there are three sources (or beginnings, fundamentals), to wit, the cerebellum (or meninx) whence spring the sinews (*or* nerves) which run together to the neck, pass to the throat and divide towards the shoulders, hands, and rest of the body.

This nerve (*or* sinew) binds together and secures all the joints of the bones, and gives play and movement to the feet and hands and all the animal members. . . .

But the heart is the source (*or* principle or foundation) of the arteries; for it possesses two chambers on the right side to collect the blood, which it takes from the liver, and on the left side it contains air collected, which it takes from the lung. And on this side it has four ducts without blood, full of warmth, and it divides into 32 several ducts, which makes up 108 ducts, by means of which is distributed warmth and life (over) the entire body. And these ducts are called *madjas* (? μέλος).

And they reveal disease and health in man. But the liver is the source of the blood of the ducts that issue from it, five in number full of blood, and three of them of great size. And we call the uppermost duct *Kifal*, and the middle one *Basilica*, and the lower one *Akhal* (?). From these branches off a duct from the forearm. And all the ducts from the liver separately number 232, and by them blood is transmitted all over the person.

But be it remarked that the body is attended by these three, to wit, sinew (or nerve), artery, and blood duct. For strength descends from the head by means of sinews (*or* nerves), and distributes it to all the body and so affords sensation; and the blood feeds and strengthens, and an artery affords warmth and life to the entire body.

And the much sinning Halathidj the Elder ye shall remember, and may God remember us in the day of judgement.

TRACT IV

23. *Concerning the Human Organism.*

Of the great Gregory of Nyssa.
An abridgment.

§ 3. Man is said to be of four constituents, for there are four causes of the genesis of all beings, fire and air, water and earth. And water is

[1] The last lines concerning seed are unintelligible.

a seminal nature, and generative after the semblance of woman. But fire is energizing after the semblance of nature; for two of the elements are creative or active, fire and air, and two are passive, water and earth. For fire and air are male, while water and earth are female; because the humidity of water without warmth of fire is barren, whereas warmth of fire and humidity of water, intermixed, are fruitful for generation.

The first synthesis (?) then of the four elements is fire; because more subtile and ever moving and nimbler is the nature of fire, like the nature of desire in us which they call an irrational movement, i.e. passionate (θυμικόν) and concupiscent, whereby the living (being) is released, for without these it is impossible for the living being to be released and constituted.

The first movement then is desire on the part of the fire in us, by which one approaches a woman, and the instruments of the seed are two ducts which descend from the head to the reins and lead to the groin, the one duct is full of blood, the other is full of air; and when the duct of blood reaches the groin, the blood is turned into seed, as in the breasts of women, where blood is turned into milk. And when one approaches a woman, the seed falls into the woman's womb, in the same way as seed of plants which falls into the earth. And in turn the seed of the woman, like milk in a vessel, is something warm, and the mingled seed becomes for her like the curds of milk, and in the union of seeds they are clotted with each other and form the foetus, and thereupon is fulfilled the word of Job which says: 'Like milk thou milkedst me and like cheese thou didst coagulate me.' And there is fulfilled the prophet's word which says: 'The two shall be made into one flesh', for out of the seed of man and woman is made the child, out of the twain one flesh made substance, like a river divided into two streams and again united in one. Now by the coagulation of the seed in the woman's womb it is turned back into blood during six days, and on the sixth day there is fashioned the worm, as the prophet says, 'I am a worm and am not a man.' For the first man was only completed on the sixth day and moulded by the hands of God, but now it is from corruption and from passions that he is engendered. And on the fortieth day members are completed for him, and then he is created the spirit of God, as to whom he alone knows who created him; as he said: 'The spirit shall go forth from me and all breaths I made' (? from myself). But a female child is completed in 70 days and then receives the spirit. For a male child is assigned to the right side of the mother and the liver and gall are on the right side, and are hot and easily make the child to grow, as a flat country with vines quickly matures the fruit. And a female child is assigned to the left side of the mother, and there is the spleen, which is cold. For that reason the child's growth is long, like fruit on a mountain which ripens late.

But in the moulding of a child, according to the nature of the parents it comes into being; according as they are diseased or healthy, so is constituted the alloy of flesh. Wherefore it is necessary for a man and woman to choose a time when they are untainted by sickness, for then their child will be healthy, like husbandmen who choose a time for sowing. For if the ground is dry, or again too muddy and wet, the seeds are damaged and give no healthy fruit. How much more he who sows rational seed for the engendering of children in rational soil? Wherefore it is necessary to make good choice, that the child may be healthy and free from all complaints. But when the child begins to sprout in its mother's womb it receives food through the umbilicus and grows like a melon or a gourd which receives water through its stalk, etc.

STEPS LEADING TO THE INVENTION OF THE FIRST OPTICAL APPARATUS

By Charles Singer

The story of the early stages in the evolution of our two most important optical appliances, the *microscope* and the *telescope*, is essentially one and can only be told as one. With the construction of working models of these two instruments their history begins to diverge and can be treated separately.

The microscopes and the microscopic work of the classical observers, Leeuwenhoek, Malpighi, Hooke, and Kircher, have frequently been described, and have been discussed by the present writer in two previous articles.[1] In one of these can be found also a description of some of the microscopical work of the first Accademia dei Lincei (1603–30) and of its early members, Federigo Cesi (1585–1629), Francesco Stelluti (1579–1651), Fabio Colonna (1567–1650), and Joannes Faber of Bamberg (1570–1650). These men all died before the work of the classical microscopists had begun.

Better known are the labours of the earliest scientific designers and constructors of the telescope, Galileo and Kepler, whose work, like that of the earlier microscopists, was completed in the first half of the seventeenth century. The most important documents relating to the discovery of the telescope are printed in the pages of the monumental *Edizione nazionale* of the works of Galileo edited by Professor Favaro. In English the story is told in the older *History of Physical Astronomy* by Robert Grant, in an article in the present volume, and in an essay by G. Moll in the first volume of the *Journal of the Royal Institution.* Special mention must also be made of the remarkable achievement of J. Hirschberg, whose *Geschichte der Augenheilkunde* is a permanent monument of its author's skill, learning, and industry, and is of far wider scope than its title suggests.

In the pages which follow we push the inquiry further back and set forth in chronological order the actual advances in the practical knowledge of optics which rendered possible the construction of the earliest compound optical instruments. These

[1] Charles Singer, 'Notes on the Early History of Microscopy', *Proceedings of the Royal Society of Medicine*, 1914, vol. vii (*Section of the History of Medicine*), pp. 247–79, and 'The Dawn of Microscopical Discovery', *Journal of the Royal Microscopical Society*, 1915, pp. 317–40. These papers contain bibliographical notes on the material in question.

advances were more closely linked, in certain stages, with progress in ophthalmic anatomy than has usually been recognized, and we dwell on this aspect in some detail. On the other hand, the account of the last stage in our story, the work of Galileo and Kepler, we pass over briefly as it is more readily accessible.

1. Among *writers of classical antiquity* there are a number of references to *burning-glasses.* It is not always clear whether a concave mirror or a crystal ball or a glass globe full of water is intended. The earliest of these references is that of Aristophanes in the *Clouds* (*c.* 424 B.C.), which runs as follows:

Strepsiades. I have found a very clever method of getting rid of my suit, so that you yourself would acknowledge it.

Socrates. Of what description?

St. Have you ever seen this stone in the chemists' shops, the beautiful and transparent one, from which they kindle fire?

Soc. Do you mean the burning-glass?

St. I do. Come, what would you say, pray, if I were to take this, when the clerk was entering the suit, and were to stand at a distance, in the direction of the sun, thus, and melt out the letters of my suit?

Soc. Cleverly done, by the Graces.

St. Oh! how delighted I am, that a suit of five talents has been cancelled.

Another unmistakable reference to the caustic power of such globes is that of Pliny the elder (A.D. 23–79), who says that 'if one holds glass globes containing water (*cum addita aqua vitreae pilae*) against the sun they give out such heat that they burn clothes'.[1] The most unequivocal statement that the magnifying power of these globes was known is perhaps that of Seneca, who wrote (*c.* A. D. 63): 'Letters, however small and dim, are comparatively large and distinct when seen through a glass globe filled with water'.[2] That these burning-glasses were sometimes made of solid crystal we again learn from Pliny, who states, 'I find among doctors that there is nothing thought better for cauterization of the human body than a crystal ball (*crystallina pila*) placed against the rays of the sun'.[3]

Among the best known of the passages in classical writings that are alleged to describe lenses is Pliny's famous reference to Nero's emerald. Some have thought that the stone was cut as a concave lens and thus enabled the Emperor, who had weak

[1] *Historia naturalis,* xxxvi. 26 [67].

[2] *Quaestiones naturales,* i. 6.

[3] *Historia naturalis,* xxxvii. 2 [10].

sight, to see more clearly, but for this there is no real evidence. It would rather seem that it was used as a mirror. Nothing can really be made of the passage, but for what it is worth we give it here: 'When an emerald is of flattened shape (*corpus extensum*) it will render the image of an object against which it is held as well as a mirror, and Nero the Emperor was wont to view the combats of gladiators by means of an emerald'.[1] It should be added that the evidence is that Nero was hypermetropic or astigmatic, not myopic, and that his difficulty was with near not distant objects, so that he would not be helped by a concave lens.

In works on optics reference is sometimes made to a so-called 'convex lens' of rock crystal found by Layard at the ruins of the palace of Nineveh. The object is now in the British Museum, and critical examination has shown that the theory of its lenticular nature is untenable. It is probably of the nature of a jewel or ornament, since its surface is cut in small facets and not ground smooth. If used as a lens its power would be about ten dioptres.

From a number of sites of classical antiquity crystal balls have been recovered and these may or may not have been used as burning-glasses. The point is doubtful, but it is certain that they are not lenses in the usual sense of the word. It may further be added that there is no unequivocal reference in all ancient literature to a lens in our sense, except perhaps in connexion with the anatomy of the eye. Even here the idea is absent from the Hippocratic collection (sixth to fourth century B.C.) and from Aristotle (fourth century B.C.), and it can scarcely be said to have emerged even in Celsus (first century B.C.) who, having described the parts of the anterior chamber of the eye, tells us that, 'under these is a drop of humour resembling the white of an egg, from which proceeds the faculty of vision. By the Greeks it is called *crystalloides*.'[2] In this term he includes the structure which we now call the *lens*.[3]

2. *Euclid* (third century B.C.), in his *Optics*, considered that light passed in straight lines. Like nearly all the ancients he considered an essential factor in vision was something proceeding *from* the eye. Thus he opens his work: 'We assume that the

[1] *Historia naturalis*, xxxvii. 5 [16].

[2] Celsus, *De Medicina*, vii. 7.

[3] Many works on Optics contain reckless statements on the optical knowledge and optical apparatus of the ancients. These are effectively dealt with by T. H. Martin, 'Sur des instruments d'optique faussement attribués aux anciens par quelques savants modernes', in the *Bullettino di bibliografia e di storia delle scienze matematiche e fisiche*, vol. iv, Rome, 1871.

visual rays proceed in straight lines, leaving a certain space between them. The figure formed by the visual rays is a cone having its apex at the eye and its base at the object seen. Only those things are seen on which the visual rays impinge, and those things are not seen on which they do not impinge.'[1] The Euclidian origin of this work was long disputed. By some the author was thought to be Theon of Alexandria, who lived in the fourth century A. D., and was perhaps the father of Hypatia, but the work is now believed to be a genuine product of Euclid.[2] In addition to the *Optics* of Euclid there are two other works, entitled *Catoptrica* or *De Speculis*, that were circulating in his name in a Latin form during the Middle Ages. These were certainly spurious. One was translated from Greek and the other from Arabic; the Greek work is almost certainly by Theon.[3]

3. *Cleomedes* (first half of first century of Christian Era), in his *Cyclical Theory of Meteors*, made a beginning of the study of refraction. He referred to the bent appearance of rods partially immersed in water, and he knew that an object lying in an opaque basin, and just obscured by the brim, could be rendered visible by pouring in water. He applied the same principle to the atmosphere and suggested that the sun, even where below the horizon, might, under certain circumstances, be visible. It is remarkable that he failed to give a practical application to this view of atmospheric refraction, for he rejects, as incredible, statements of his predecessors that in certain eclipses the sun seems to be still above the horizon while the eclipsed moon rises in the east.[4]

[1] The text of the ὅροι here translated runs as follows: Ὑποκείσθω τὰς ἀπὸ τοῦ ὄμματος ὄψεις κατ᾽ εὐθείας γραμμὰς φέρεσθαι διάστημά τι ποιούσας ἀπ᾽ ἀλλήλων. καὶ τὸ μὲν ὑπὸ τῶν ὄψεων περιεχόμενον σχῆμα εἶναι κῶνον τὴν κορυφὴν μὲν ἔχοντα πρὸς τῷ ὄμματι, τὴν δὲ βάσιν πρὸς τοῖς πέρασι τῶν ὁρωμένων. καὶ ὁρᾶσθαι μὲν ταῦτα, πρὸς ἃ ἂν αἱ ὄψεις προσπίπτωσιν, μὴ ὁρᾶσθαι δέ, πρὸς ἃ ἂν μὴ προσπίπτωσιν αἱ ὄψεις.

[2] The *Optics* of Euclid has recently been translated by G. Ovio, *L'Ottica di Euclide*, Milan, 1918. The definitive edition is that of J. L. Heiberg and H. Menge, *Euclidis Opera Omnia*, Leipzig, 1883–95, vol. vii, 1895, *Euclidis optica, opticorum recensio Theonis, Catoptrica cum Scholiis antiquis*. Interesting information concerning this work of Euclid may be found in J. L. Heiberg, *Literargeschichtliche Studien über Euclid*, Leipzig, 1882, and T. L. Heath, *The Thirteen Books of Euclid's Elements*, Cambridge, 1908.

[3] It is printed by Heiberg and by Ovio (see above). The Arabic work is discussed by S. Vogl, 'Über die (Pseudo-) Euclidische Schrift de Speculis,' in the *Archiv für Geschichte der Naturwissenschaften und der Technik*, i, p. 419, Leipzig, 1909.

[4] The work of Cleomedes has been edited by H. Ziegler, Leipzig, 1891. Some valuable notes on his work are to be found in T. L. Heath, *Aristarchus of Samos*, Oxford, 1913.

4. *Rufus of Ephesus* (flourished about A. D. 100) improved on the description of the eye by Celsus, and for the first time in literature, so far as we know, refers to the lens as 'lentil shaped' (φακοειδής). The passage occurs in the work *On the naming of the parts of the human body*, and runs as follows: 'The fourth tunic encloses the crystalline humour (τὸ κρυσταλλοειδὲς ὑγρόν); at first this had no

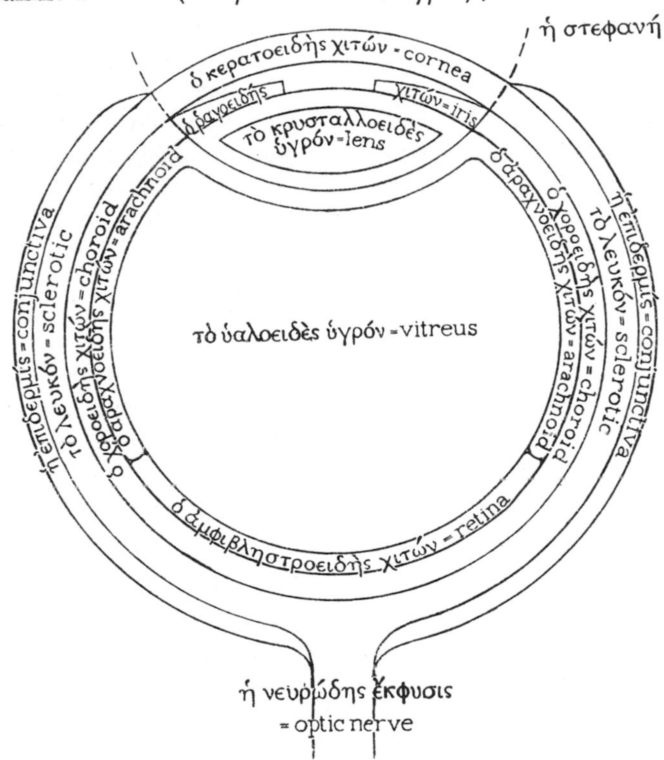

FIG. 1. THE STRUCTURE OF THE EYE RECONSTRUCTED FROM THE DESCRIPTIONS OF RUFUS OF EPHESUS (FIRST CENTURY C.E.)

special name, but later it was named *lentil-like* (φακοειδής) on account of its form, and *crystalline* on account of the character of its humour (κρυσταλλοειδὴς δὲ διὰ τὸ ὑγρόν).' Elsewhere he refers to it as 'disk shaped' (δισκοειδής).[1] The description of the eye given by Rufus is distinct enough for a diagram representing his views to be reconstructed (Fig. 1). The work of Rufus was, however, unknown to the Latins of the West in the Middle Ages who copied

[1] The works of Rufus of Ephesus are available in an edition by C. Daremberg and E. Ruelle, Paris, 1879.

and modified the less accurate description of Galen (131–201). That view was barren and we need follow it no further. But the works of Rufus became known to the Arabian writers—largely through the influence of Rhazes (860–932)—and among them (see sections 7 and 8) we shall trace a profitable line of optical advance.[1]

5. *Hero of Alexandria* (flourished second half of first century) advanced the knowledge of optics from the experimental side. He showed in his *Catoptrika* and his *Dioptra* that light is reflected from a surface at an angle equal to the angle of incidence.[2] His instrument, the Dioptra was, in effect, a theodolite and depended for its working on the principle of the equality of these angles. The *Dioptra* was unknown in the West until quite modern times,[3] but the Arabian optical writers were acquainted with it much earlier.[4]

6. *Ptolemy* (died about A.D. 155), in his *Optics*, continued the study of refraction begun by Cleomedes, and applied the experimental method to this subject. He showed that luminous rays, in passing from one medium to another, are deflected, and he attempted to measure the deflection. This work of Ptolemy was written in Greek, and has been lost. It was translated from Greek into Arabic and, in the twelfth century, from Arabic into Latin. Only the Latin version survives, and its attribution to Ptolemy is doubtful.[5] This work was exceedingly important for

[1] The theories of the structure of the eye among Greek writers may be examined in a series of diagrams prepared by H. Magnus, *Die Anatomie des Auges in ihrer geschichtlichen Entwickelung*, Breslau, 1906. Some of these figures need revision. The propagation of Galen's ophthalmic anatomy in the West and its progressive deterioration is traced in a group of learned contributions by K. Sudhoff, extending over a number of years, in the *Archiv für Geschichte der Medizin*.

[2] The works of Hero are edited in the Teubner series by Wilhelm Schmidt and Hermann Schöne, *Heronis Alexandrini opera quae supersunt omnia*, Leipzig, 1899–1903. There is an unusually able and exhaustive but anonymous account of Hero in Pauly-Wissowa's *Real-Encyclopädie der klassischen Altertumswissenschaft*.

[3] The text of the *Dioptra* of Hero was first made known in the West by a translation by L. Venturi in his *Commentari sopra la storia e la teoria dell'ottica*, Bologna, 1814. The earliest manuscript was discovered by Minoïdes Mynas on Mt. Athos in 1841, and is of the eleventh century (Paris, Bib. Nat. Supp. gr. 607).

[4] The knowledge of Hero displayed by the Arabian writers is discussed by E. Wiedemann, *Sitzungsberichte d. phys.-med. Soz. zu Erlangen*, xxxvii–ix.

[5] G. Govi, *L'Ottica di Claudio Tolomeo da Eugenio Ammiraglio di Sicilia . . . ridotta in Latino sovra la traduzione Araba di un testo Greco . . . conforme a un codice . . . Ambrosiano*, Turin, 1885. See also A. Favaro, 'L'Ottica di Tolomeo', in the *Bullettino di Bibliografia e di Storia delle Scienze matematiche e fisiche*, xix, Rome, 1886.

its influence on mediaeval optics both in the East and West, and Alhazen particularly was dependent upon it.[1]

Greek science in its creative aspect practically expired with the second century. Its last optical work is that of Ptolemy, unless we include the treatise that has been attributed to Theon (see 2, above), in which case we must consider it as prolonging a feeble life until the fourth century. We have seen, moreover, that among the ancients there is no definite evidence of knowledge of lenses apart perhaps from the lens of the eye. But even had lenses been known they could hardly have been clearly described, for the prevalent view, popularized by Plato, that regarded vision as something *emanating from* the eye, would have prevented appreciation of their nature.

7. The fall of Greek science was followed—after a long interval—by the rise of the Arabian system which was based on it. The earlier Arabian optical writers, Jacob al Kindi (750–850)[2] and Costa ben Luca[3] (864–923), for example, were concerned only with the translation and codification of Greek works of science. More important for our purposes are certain later Arabian philosophers who were led to abandon the Platonic theory of emanation and were thus in a better position to appreciate the nature of the eye. The most important of these was Alhazen.

Alhazen, or Al Haitham (Ḥasan ibn al Ḥasan ibn al Haitham, Abu Ali, 965–1038), was an Arab of Basra, who abstracted the work of the older Greek optical writers but exhibited some originality. He devoted much space and skill to the development of the optical effects of curved mirrors. He had a fairly clear notion of the nature of refraction, and improved the apparatus of Ptolemy for measuring the angle of refraction in different

1 See S. Vogl, *Roger Baco*, Erlangen, 1906, p. 26.

2 Al-Kindi = Jaqub ibn Ishaq ibn al Sabbah, al-Kindi, abu Jusuf wrote many astronomical and mathematical works including one, *De speculis ustivis* that has not been rendered from Arabic unless it is represented by the Bodleian MSS. Digby 168 (XXXVI), fol. 129, *Alkindi ex libro perspectiva* or Canon. Codd. Misc. 370 (II), fol. 240–end, *Alkindi liber de radiis stellatis*. See, however, Heinrich Suter, 'Die Mathematiker und Astronomen der Araber', in the *Abhandlungen zur Gesch. der math. Wissenschaften*, X, Leipzig, 1900. In the same journal (XXVI, 1912) is an article by A. A. Björnbo, 'Alkindi Tideus und Pseudo-Euclids drei optische Werke', including a work of Alkindi, *De causis diversitatum aspectus*.

3 Costa ben Luca also wrote a work, *De speculis ustivis*. This does not seem to have been translated; cf. Ferdinand Wüstenfeld, *Geschichte der arabischen Aerzte und Naturforscher*, Göttingen, 1840, p. 49, and Moritz Steinschneider, *Die hebräischen Uebersetzungen des Mittelalters*, Berlin, 1893, §§ 310, 312, 336, and 342.

media. As regards his knowledge of the structure of the eye, not only was his ophthalmic anatomy better than that of most of his predecessors (Fig. 2), but he considered that vision resulted from rays coming to the eye from the object, and opposed the view, current up to the eighteenth century, that explained vision as involving the emanation of something from the eye. He placed the lens or *humor crystallinus* in the centre of the globe of the eye, and considered that in it the external impulse became converted into the sensation of vision. He speaks also of the enlarging power of a segment of a sphere of glass. The optical work of Alhazen, which is of enormous proportions, was translated into Latin by an unknown writer of the later twelfth or earlier thirteenth century.[1]

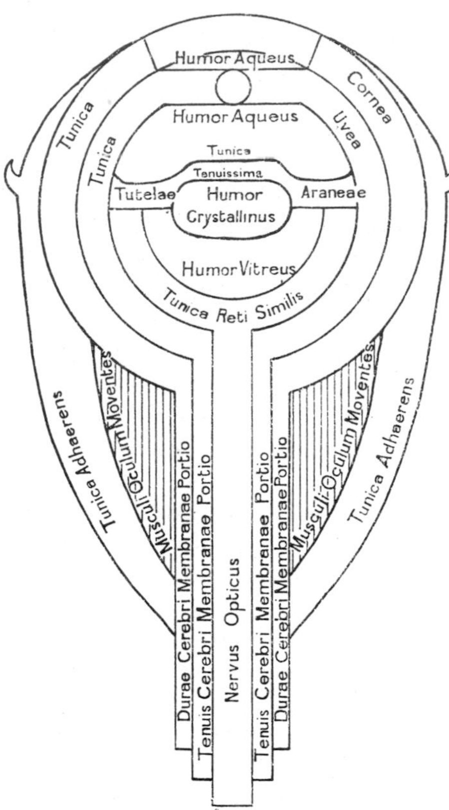

FIG. 2. THE STRUCTURE OF THE EYE, AFTER ALHAZEN

Redrawn from the printed edition, Bâle, 1572. The diagrams of the eye in MSS. are identical with this. The best MS. seen by the writer is of the thirteenth century in the Crawford Library of the Royal Observatory at Edinburgh. The diagram of the eye in it is very similar to that here presented.

8. *Avicenna* (980–1037). An important writer on Optics, contemporary with Alhazen but even more afflicted with the *verbositas arabica*, was the encyclopaedic Bokhariote, Avicenna. His views on optics are diffused through a perfect multitude of writings, but a relevant aspect of his teaching is succinctly expressed in a comparatively small work *On Physics*, which has not been translated in full from the Arabic. The passages on vision have, however, been rendered into German, from which we retranslate as follows :

[1] The work was printed by F. Risner at Bâle in 1572. A valuable discussion

'The organ of sight is the crystalline humour [lens] in the pupil. They are wrong who think that the act of seeing arises from something that goes out from the eye to the perceived object and collides with it. For if this something be material the difficulty would be that in the eye would have to be a body big enough to spread over half the vault of heaven.' He then goes on to say that the emanation theory has great difficulty in explaining the less apparent size of distant objects, 'for if impressions take place through contact [of the emanation], then the impression would be of a size exactly the same as the actual size. But if on the other hand the impression takes place by means of a flow *towards* the crystalline humour, we maintain that the more distant object would then appear the smaller. The proof is as follows (Fig. 3):

'Let the circle *E F* be the lens around the centre *G*, and let *A B* and *C D* be two objects of equal size. . . . Then the triangles *A G B*, *C G D* have an equal base but *C G D* is taller. Further, the angle *C G D* subtends the arc *H K*, and the angle *A G B* the arc *L M*, so that the arc *L M* is greater than the arc *H K*. The optical picture of *A B* imprints itself on *L M*, and that of *C D* on *H K*, and so it is that the optical picture of the further objects imprints itself as smaller.'[1]

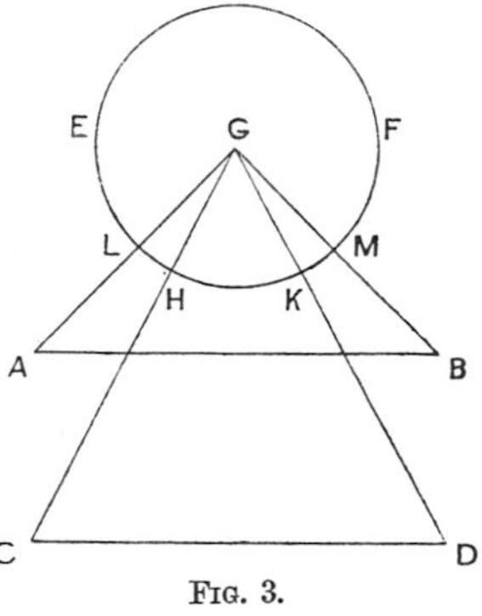

FIG. 3.
Avicenna's diagram to explain the less apparent size of distant object.

In the Qanûn of Avicenna there is a description of the Anatomy of the Eye. This only concerns us here in that the lens is regarded as flattened, instead of simply spherical as in most mediaeval

of Alhazen's work by one who is both an expert physicist and an Arabic scholar is contained in E. Wiedemann, 'Zu Ibn al Haitams Optik', in the *Archiv für die Geschichte der Naturwissenschaften und der Technik*, iii, p. 1, Leipzig, 1910.

There are a number of good MSS. of the Latin translation of Alhazen. Those known to the present writer are:

- Corpus Christi College, Oxford, CL, fo. 1–114 v, thirteenth century.
- Crawford Library, Royal Observatory, Edinburgh, thirteenth century.
- Trinity College, Cambridge, 1311, fo. 1–165, thirteenth century.
- Peterhouse, Cambridge, 209, fo. 1–111, fourteenth century.
- British Museum, Sloane 306, fo. 1–175, fourteenth century.
- British Museum, Royal 12 G17, fo. 102v–end, fourteenth century.
- Bodleian, Ashmole 424, fo. 3–355 fourteenth–fifteenth century.
- Bodleian, Digby 104, fo. 80–end, fifteenth century.

[1] E. Wiedemann, 'Ibn Sinas Anschauung vom Sehvorgang', *Archiv für die Geschichte der Naturwissenschaften und der Technik*, iv, p. 239, Leipzig, 1912.

descriptions, both Eastern and Western. The Anatomy of the Eye, as set forth by Avicenna, was conveyed to the West in the Latin translation of the Qanûn by Gerard of Cremona, executed towards the end of the twelfth century.[1]

9. *Witelo* (first half of the thirteenth century) was a Pole, who made an extensive study of the work of Alhazen in its Latin translation.[2] He was also familiar with the views of Avicenna. His own work grew out of these writers and is perhaps an improvement on them. Thus he drew up a table of refractions for three media—air, water, and glass—from which it could be seen that the angle of refraction did not vary directly according to the angle of incidence. It is doubtful, however, to what extent these tables were original or the result of direct observation. Witelo knew also, through Alhazen, of the magnifying power of sections of glass spheres, and expressed the wish that he possessed one of these in order to examine small objects.[3]

10. *Roger Bacon* (1214–94) accomplished real advances in the knowledge of optics. His work was based primarily on Latin translations of Arabian writers, and especially on Witelo's development of the views of Alhazen and Avicenna. He is distinguished from his predecessors, however, by his clear conception of the value of experiment, and by the evidence in his works that, having made a serious and continuous effort to discover the laws of refraction and reflection, he sought to apply his knowledge to the improvement of the power of vision. In this he is a real pioneer, and is in the truest sense the father of microscopy. Moreover, in the writings of Bacon there emerges a clearer though still inaccurate conception of the nature of a lens. He seems to think that there is only one refraction in the case of a lens but two in a spherical burning glass, and it is clear that the lenses he used were sections of spheres and were thus plano-convex.

[1] There are a large number of editions of this vast work, of which the first is probably that of Milan, 1473. The passage on the Anatomy of the Eye is in Book III, Fen III, chapter i, and it is translated direct from the Arabic in P. de Koning's *Trois traités d'anatomie arabes*, Leiden, 1903, p. 660.

[2] Not, as is sometimes asserted, from the Arabic. It is practically certain that Witelo knew no Arabic.

[3] The works of Alhazen and of Witelo were printed together by F. Risner at Bâle in 1572. An account of all that is known of Witelo, together with parts of his *Perspectiva* reprinted from the manuscripts, has been set forth by Clemens Bauemker in his *Beiträge zur Geschichte der Philosophie des Mittelalters*, Munich, 1908.

It is easy to exaggerate the claims of Bacon, and the wildest statements are often made about his discoveries. There is no evidence that he ever made a telescope nor any but a simple microscope, but he had an idea of the nature and property of lenses, and, groping with the instinct of genius, he did vaguely foresee both telescope and compound microscope.

The following passages will serve to indicate the stage he had reached in optical knowledge. It will be observed that in the first of the passages Bacon figures and refers to the *object* as though

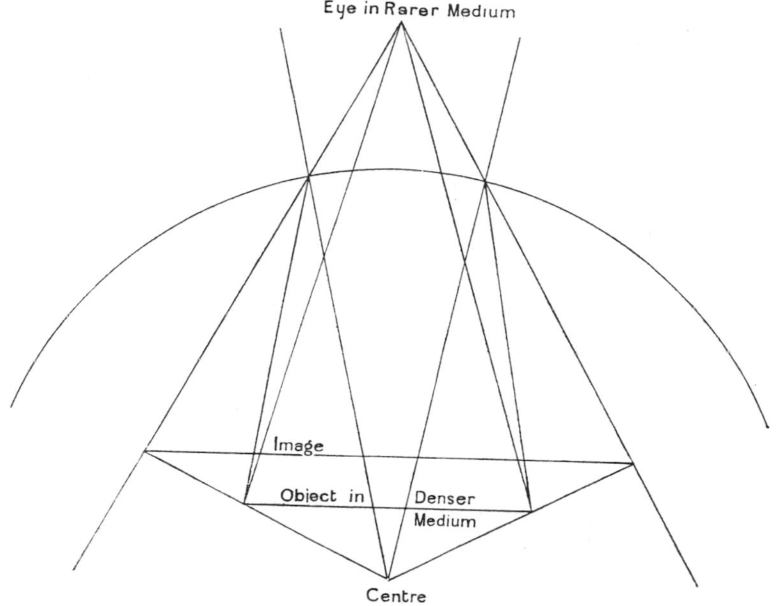

FIG. 4. DIAGRAM OF ROGER BACON TO ILLUSTRATE OPTICS OF LENS

it were itself in the denser medium of which the lens is composed. He thus confuses the optical action of the lens with that of a liquid in which an object is immersed. Perhaps he is led to do this by his knowledge that the optical results of immersion in a liquid had been investigated by his predecessors, or perhaps by their descriptions of the process of vision as taking place within the supposed central crystalline sphere of the eye. It is also apparent that he does not realize the refractive action of the *plane* surface of his plano-convex lenses.

'If any one examine letters or other minute objects through the medium of crystal or glass or other transparent substance, if it be shaped like the lesser segment of a sphere, with the convex side towards the eye, and the eye being in the air, he will see the letters far better, and they will seem larger to him. For, according to our Canon (Fig. 4) concerning a spherical medium beneath which the object is placed, the centre being beyond the object, the convexity being towards the eye, all causes agree to increase the size, for the angle in which it is seen is greater, the image is greater, and the position of the image is nearer, because the object is between the eye and the centre. *For this reason such an instrument is useful to old persons and to those with weak eyes, for they can see any letter, however small, if magnified enough.* But if a larger segment of a sphere be employed, then, according to our Canon (Fig. 5) the size of the angle is increased, and also the size of the image, but propinquity is lost because the position of the image is beyond the object, the reason being that the centre of the sphere is between the eye and the object seen. Therefore such an instrument is not of so much use as the smaller portion of a sphere.'

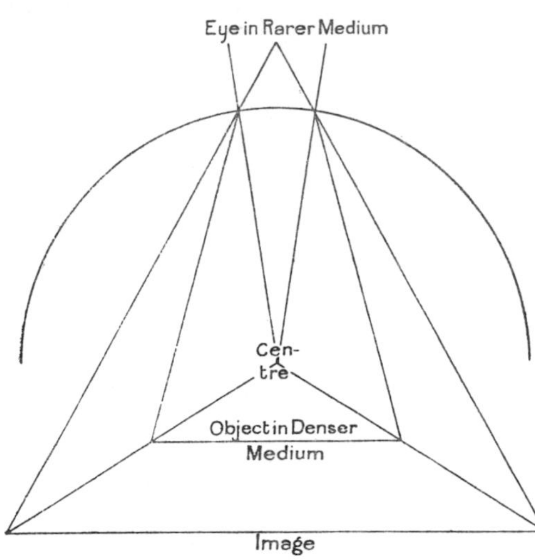

Fig. 5.

DIAGRAM OF ROGER BACON TO ILLUSTRATE OPTICS OF LENS

'Objects are greater when the vision is refracted; for it easily appears by the above-mentioned canons that very large objects may seem to be very small and conversely, and those at a great distance away may seem very near and conversely. For we can so form glasses and so arrange them with regard to our sight and to objects that the rays are refracted and deflected to any place we wish, so that we see the object near at hand or far away beneath whatever angle we desire. And so we can read the smallest letters or count grains of sand or dust from an incredible distance owing to the magnitude of the angle beneath which we see them, and again the largest objects close at hand might be scarcely visible owing to the smallness of the angle beneath which we view them; for *it is on the size of the angle on which this kind of vision depends, and it is independent of distance save per accidens.* So a boy can appear a giant, a man seem a mountain, and in any size of angle whatever, for we can see a man under as large an angle as though he were a mountain and make him appear as near as we desire. So a small army might seem very large, and though far away appear near, and conversely: *so, too, we could make sun, moon, and stars apparently descend here below,* and similarly appear above the heads of our enemies, and many other similar marvels could be brought to pass, that the ignorant mortal mind could not endure the truth.' (*Opus Majus*, Part V.)

'As to double refraction, what is causally manifest with regard to it we can verify in many ways by the results of experiment. For if any one hold a crystal ball or a round urinal flask filled with water in the strong rays of the sun, standing by a window in

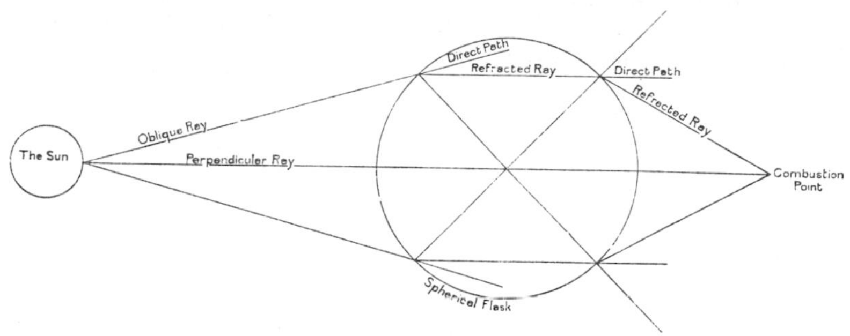

FIG. 6.
DIAGRAM OF ROGER BACON TO ILLUSTRATE OPTICS OF BURNING-GLASS

face of the rays, he will find a point in the air between himself and the flask at which point, if any easily combustible substance is placed, it will catch fire and burn, which would be impossible unless we suppose a double refraction. For a ray of the sun coming from a point in the sun through the centre of the flask is not refracted, because it falls perpendicularly on flask, water, and air, passing through the centre of each (Fig. 6). . . . But all the (other) rays given forth at the same point in the sun from which this perpendicular ray comes are necessarily refracted in the body of the flask, because they fall at oblique angles, and since the flask is denser than air, the refraction passes between the straight path and the perpendicular drawn from the point of refraction to the centre of the flask. And when it passes out again into the air, then, since it comes upon a less dense body, the straight path passes between the refraction and the perpendicular drawn from the point of refraction, so that the refracted ray may fall upon the first perpendicular which comes without refraction from the sun. *Now, since an infinite number of rays are given off from the same point of the sun, and one only falls perpendicularly on the flask, all the others are refracted and meet at one point on the perpendicular ray which is given off along with them from the sun, and this point is the point of combustion.*

On it are collected an infinite number of rays, and the concentration of light causes combustion. But this concentration would not take place except by double refraction, as shown in the diagram.' (*Opus Majus*, Part VII.)

'*Glasses (perspicua) can be so constructed that objects at a very great distance appear to be quite close at hand, and conversely.* Thus we read the smallest letters from an incredible distance, number objects, however small, and make the stars appear as near as we wish. . . . Also objects can be made to appear so that the greatest seems the least, and conversely; what are high appear low and short, and conversely; and what is hidden appears manifest. . . .

'But among the more subtle powers of construction is this of directing and concentrating rays by means of [instruments of] different forms and reflections at any distance we wish, where whatever is subjected to them is burned. . . . But greater than any such design or purpose is that *the heavens might be portrayed in all their length and breadth on a corporeal figure moving with their diurnal motion, and this would be worth a whole kingdom to a wise man.* Let this, then, be sufficient as an example, although an infinite number of other marvels could be set forth.' (*De Secretis Operibus Artis et Naturae.*)[1]

11. *John Peckham* (died 1292), Archbishop of Canterbury, was the author of a work on optics entitled *Perspectiva communis.* His views were mainly derived from Alhazen, though perhaps partly also from Bacon. He is important here as having drawn wide attention to optical principles. He refers to the possibility of concave lenses.[2]

12. *The Invention of Spectacles.* The names of Salvino d'Amarto degli Amarti of Florence and Alessandro della Spina of Pisa (both *c.* 1300) have become associated with the special

[1] The above passages are translated from J. H. Bridges, *The Opus Majus of Roger Bacon*, Oxford, 1897, and J. S. Brewer, *Fratris Rogeri Baconi opera quaedam hactenus inedita*, London, 1859. The best account of Roger Bacon's views on Optics has been given by E. Wiedemann, *Roger Bacon und seine Verdienste um die Optik*; S. Vogl, *Roger Bacon's Lehre von der sinnlichen Spezies und vom Sehvorgange*; and J. Würschmidt, *Roger Bacon's Art der wissenschaftlichen Arbeiten, dargestellt nach seiner Schrift De speculis.* All these three papers are contained in the volume edited by A. G. Little, *Roger Bacon, Essays contributed by various writers on the occasion of the commemoration of his birth*, Oxford, 1914.

[2] Peckham's optical work exists in a number of manuscripts, and has often been printed. The first edition is dated from Milan, 1482. It has been analysed by S. Wilde in his *Geschichte der Optik*, Berlin, 1838, Teil I, p. 83. The recent work on Peckham by P. H. Spettmann, *Die Psychologie des Johannes Pecham*, Münster, 1919, contains practically nothing on optics.

application of lenses for use as spectacles. Lenses, as we have seen, were known to Roger Bacon, who suggested also their use in aiding vision. D'Amarto and Spina are said to have applied the principle thus suggested. The evidence for this statement is not entirely satisfactory. It is as follows:

Salvino d'Amarto. An eighteenth-century Florentine antiquary named Leopoldo del Migliore found at the Church of Santa Maria Maggiore at Florence a gravestone with an inscription which may be thus translated: *Here lies Salvino d'Amarto degli Amarti of Florence, the inventor of spectacles. May God forgive his sins. He died anno Domini 1317.*[1] The gravestone no longer exists.

Alessandro della Spina. He was a monk of the Dominican convent of St. Catharine at Pisa. In a chronicle of that institution, prepared in the fourteenth century and not long after his death, the latter event was placed in the year 1313. It further recorded that *Brother Alessandro della Spina, a worthy and modest man, could fashion cunningly whatsoever was shown or described to him. Spectacles (Ocularia) were first made by him and given out with a cheerful and willing heart.*[2] This chronicle is lost.

Whatever the value of these statements there is collateral evidence that spectacles in fact came into use towards the end of the thirteenth century. (*a*) In the 1729 edition of the Dictionary of the Accademia della Crusca, under the heading *Occhiali*, we read that Giordano da Rivalto of Piacenza, a monk of Pisa, in a sermon delivered on February 23, 1305, said that 'it is not twenty years since there was discovered the art of making spectacles to see better, one of the best and most necessary of arts. . . . I have myself seen and spoken to the man who first discovered and made them'. (*b*) The naturalist Francesco Redi (1626–94) published in 1678 a manuscript which was dated 1289, in which was the following passage: 'I am so weighted with years that without the glasses called *occhiali* I could neither read nor write. These have been lately invented to the convenience of poor old people whose sight is enfeebled.'[3] (*c*) Bernard de Gordon, who was physician at Montpellier from 1285–1307, makes mention of

[1] Migliore's work, which we have not seen, was published in 1684. It is mentioned by J. J. Volkmann, *Historisch-kritische Nachrichten von Italien*, Leipzig, 1770–1, i, p. 542, but in Volkmann's time the gravestone no longer existed.

[2] *Allgemeines historisches Lexikon*, Leipzig, 1709.

[3] Francesco Redi, *Lettera sopra l'invenzione degli occhiali di naso*, Florence, 1678.

spectacles as *oculus berellinus* in his *Lilium medicinae* (written 1305, first printed Lyons 1474). This is probably the first medical work containing such a reference. It is amusing to observe that he mentions them only to recommend his own eyewash which will, he says, render them useless ! [1]

The actual application of convex lenses as spectacles can thus be placed as beginning about 1280. Roger Bacon's reference enables us to place the idea on which they are based somewhat further back, say to 1260 or to 1250. It is apparent, from what has been said, that the use of spectacles spread rapidly about the turn of the thirteenth century.[2]

13. *The introduction of concave lenses* should be treated as a stage separate and distinct from the invention of ordinary convex spectacles. Although there are many mentions of myopia in ancient and mediaeval literature, and although John of Peckham seems to have known of the possibility of concave lenses, they did not come into use until a much later date than that given for convex spectacles. Little is known of the introduction of concave glasses and the earliest references are to works of art. The first of these are probably Lucas Cranach the elder's 'Adulteress before Christ' in the Pinakothek at Munich, painted about 1500, and Rafael's portrait of Pope Leo X in the Pitti Palace, painted probably in 1517. Both of these portray spectacles, the reflex of which proves them to be concave. Leo X is known to have been a myope. These pictures are, however, very exceptional cases, and concave lenses do not appear to have become generally accessible to myopes until after the middle of the sixteenth century.[3] Maurolico (1554) understood them quite well. Jacques

[1] Bernard describes his eyewash as *tantae virtutis quod decrepitum faceret legere literas minutas sine ocularibus !* For the dates of Bernard see R. R. von Töply, 'Die chronologische Reihenfolge der Schriften des Bernard Gordon', in the *Mitteilungen zur Gesch. der Medizin und Naturwissenschaften*, vi, p. 94, Leipzig, 1907, corrected by K. Sudhoff, 'Zur Schriftstellerei Bernard von Gordon', in the *Archiv für Gesch. der Medizin*, X, p. 162, Leipzig, 1916.

[2] The question of the invention of spectacles has been frequently discussed. One of the latest writers who has traversed this field is V. Rocchi, 'Appunti di storia critica del microscopio', in the *Rivista di Storia critica delle Scienze Mediche e Naturali*, January, 1913. The best general account is that of Emil Bock, *Die Brille und ihre Geschichte*, Vienna, 1903. In English there is Carl Barck's *History of Spectacles*, Chicago, 1907, taken largely from Bock, and G. H. Oliver's *History of the Invention and Discovery of Spectacles*, London, 1913.

[3] See Jakob Stilling, *Untersuchungen über die Entstehung der Kurzsichtigheit*, Wiesbaden, 1887.

Houllier of Paris (1498–1562) was the first physician to order them, and was soon followed by Mercuriali (1530–1606) in Italy and Forestius (about 1590) in Holland.

14. *Leonardo da Vinci* (1452–1519) represents the turning-point between mediaeval and modern in the history of Optics. If his ever active mind was continually throwing out hints the value of which was not appreciated for centuries, yet no man reflected better than he the floating opinion of his time. Leonardo suggested the use of sections of glass spheres and spectacle lenses for magnification, and he made an attempt to explain the action of convex lenses—he probably knew nothing of concave lenses—in cases of presbyopia. This condition he regarded as due to loss of power to converge, and he writes thus of the helpful action of convex lenses :

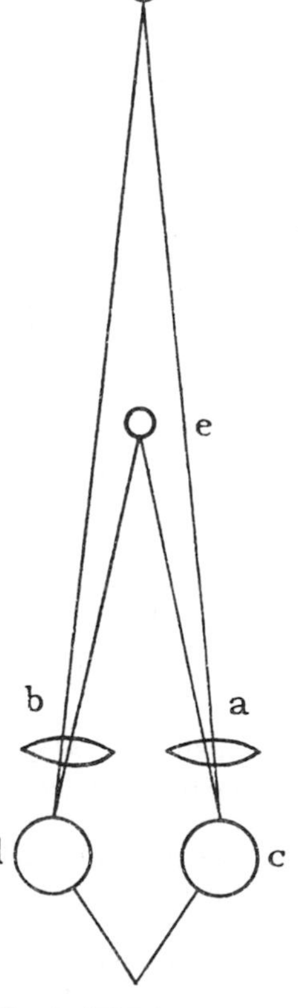

Fig. 7. THE ACTION OF SPECTACLE LENSES, according to Leonardo, redrawn from *Codice Atlantico*, fol. 244 r.

'A demonstration how lenses help vision. Let *a b* be the lenses and *c d* the eyes. When these become aged the object which they were accustomed to see easily at *e* with axes diverted from the straight line of the optic nerves [can no longer be seen so near]. For owing to the advance of age this power of turning becomes weakened so that it is not possible to converge without great pain to the eyes, and it is then necessary to move the object further off, that is from *e* to *f*, where it can be seen better but in less detail. Now, interposing the lens, the object can easily be seen at the proximity customary for youth, i.e. at *e*. This is so because the image of the object passes to the eye through a compound medium both rare and dense, *rare* as to the air between object and lens, and *dense* as to the thickness of the glass of the lens itself. Wherefore the direction turns in its course through the glass and diverts the line so that the object is seen at *e* as though it were at *f*, with the added advantage of not diverting the axis of the eye from its optic nerve, and as it is nearer it can be seen and recognized better at *e* than at *f*, especially if minute' (Fig. 7).[1]

[1] The passage here translated is in the *Codice Atlantico* at the Ambrosian

Leonardo's solution of the action and nature of lenses is erroneous, but in this and many other passages we see him groping after a better theory of vision than was current in his day.

His description of the anatomy of the eye is very faulty though perhaps better than that of his mediaeval predecessors. He dis-

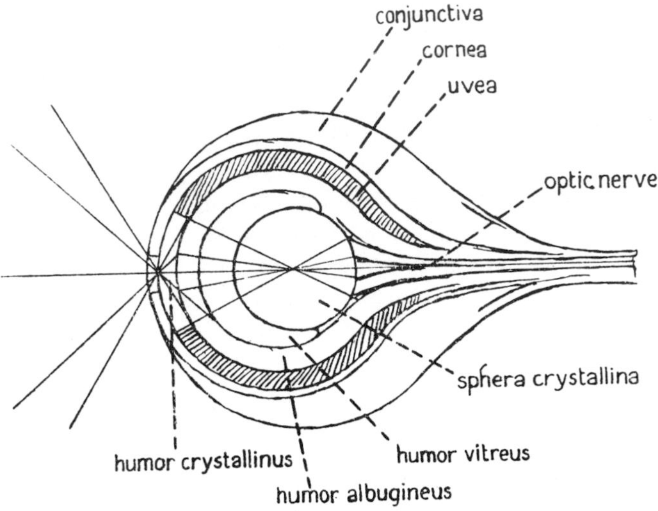

FIG. 8. THE STRUCTURE OF THE EYE, according to Leonardo. Redrawn from *Codice Atlantico*, fol. 337 r. The legends are supplied from other passages.

tinguished a *humor crystallinus*, which is really the 'lens' of modern notation distinct from the *sphera crystallina* or central

Library at Milan (published in facsimile by the Regia Accademia dei Lincei, 1899–1904). As the translation is difficult and in parts doubtful the original Italian is here given, *Codice Atlantico*, fo. 244 r. :

'Pruova come li occhiali aiutan a la vista. Siano li occhiali a b e li occhi c d quali per essere invecchiati, bisognia che l'obbietto, che soleano vedere in e con gran facilità e forte piegare il loro assis dalla rettitudine de' nerbi ottici, la qual cosa, per causa della vecchiezza, viene tal potenzia di piegare a essere indebolita, onde non si po torcere sanza gran doglia d'essi occhi, sì che per necessità son constretti a fare più remoto l'obbietto, cioè da e a f, e lì poi lo vede meglio, ma non alla minuta ; ora, interponendo l'occhiale, l'obbietto è ben conosciuto nella distanzia della gioventù, cioè in e, la qual cosa accade perchè l'obbietto e passan all'occhio per vario mezzo, cioè raro e denso, raro per l'aria ch[e è] tra l'occhiale e l'obbietto, e [den]so si è per la grossezza del vetro [delli] occhiali, onde la rettitudine [. . .] piega nella grossezza del vetro [e to]rce la linia a d in modo ch[e ve]dendo la cosa in e, esso la vede come in i f, per comodità di non piegare l'assis dell'occhio da' sua nervi ottici, e per vicinità la vede e conosce meglio in e che in f, e massime le cose minute.'

body of the eye (Fig. 9), but he still placed the visual function with this latter body, as did Vesalius a generation later.[1] He gives a clear figure in illustration of his view of the structure of the eye, showing also the path which he believed the rays to follow. This figure we here reproduce, adding the anatomical notation (Fig. 8). It shows the rays crossing each other twice within the globe of the eye before they reach the optic nerve. This double crossing was introduced to obviate the formation of an inverted image, the existence of which was a great stumbling-block to the Renaissance optical investigations. Leonardo stands alone in his time in showing the rays in the *humor crystallinus* pursuing a course as though through a pin-hole.

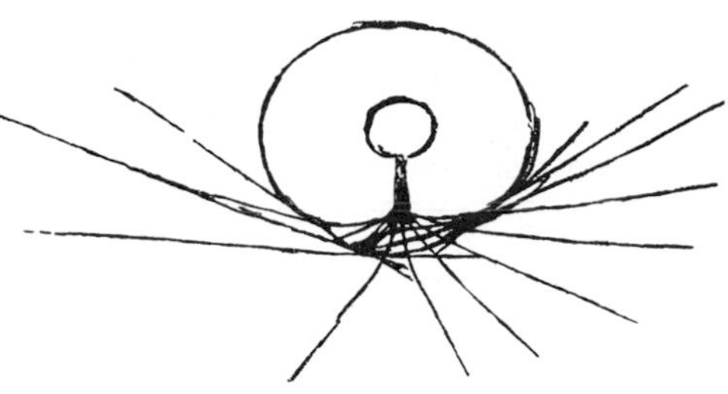

Fig. 9. DIAGRAM OF EYE, from Leonardo, showing the *sphera crystallina*, the supposed site of vision in the centre of the globe. (*Quaderni*, iv, fol. 12 v.)

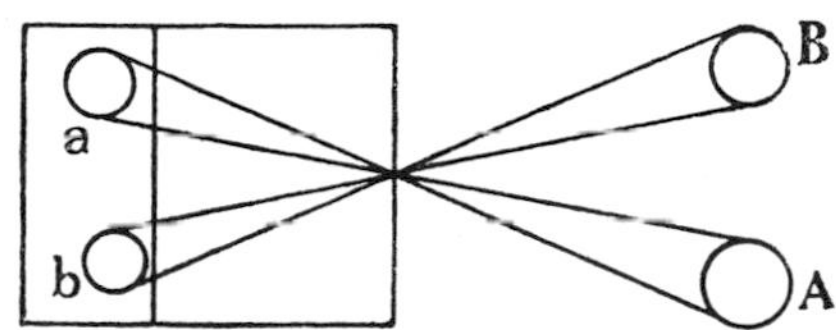

Fig. 10. THE ACTION OF A PIN-HOLE CAMERA, according to Leonardo. (*Codice Atlantico*, fol. 216 r.)

Now this conception of the action of a lens as equivalent to a pin-hole offered a fruitful line of advance. In Leonardo's own hands it led incidentally to the development of his idea of a camera obscura, the germ of which is contained in the following simple statement :

'As all lines can meet in a focus (*punto*) without interfering with each other—since they are incorporeal—so also can all *species* of objects. And as every point is within sight of the object opposite to it and every object is within sight of the point opposite to it, the diminished rays of such species can pass through this focus (*punto*) ; after passing they re-form and their size again increases but their impressions appear reversed ' (Fig. 10).[2]

[1] *Studies*, vol. i, p. 121.

[2] The Italian of this passage, quoted by J. P. Richter, *The Literary Works of Leonardo da Vinci*, London, 1883, i, § 81, is as follows : ' Siccome in un punto passan tutte le linie sanza occupatione l'una dell' altra per essere incorporee, cosi possono passarvi tutte le spetie delle superfitie, e siccome ogni dato punto rede ogni antiposto obbietto, e ogni obietto rede l'antiposto punto naturale, ancora

The word *species* here used by Leonardo is a technical term, which refers to an hypothesis of the nature of vision prevalent in the Middle Ages and derived from antiquity.[1] According to this view all visible objects are constantly giving off impalpable thin shells from their external visible surfaces. These meet with the visual spirit or emanation from the eye, and having so collided pass back together into the eye as images. Reference to this theory will be found in the description of the eye by Manfredi in the first volume of this series.[2]

What may be termed Leonardo's 'pin-hole theory', combined with his conception of the nature of the eye and its central *sphera crystallina,* led him to suggest a practical form of camera obscura (Fig. 11), and the diagram he constructed for this apparatus is important as the earliest figure of an optical apparatus other than a spectacle lens. In this apparatus a glass globe is placed just behind the opening, and the eye placed at the focus of the globe observes the distant object. The diagram shows a double crossing of the rays as in his figure of the anatomy of the eye (Fig. 8).

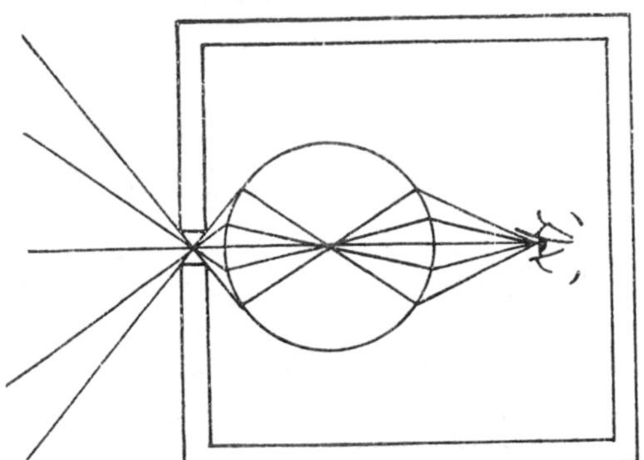

FIG. 11. A CAMERA OBSCURA, according to Leonardo. Redrawn from *Codice Atlantico,* fol. 337 r.

Leonardo's contributions, though original and valuable, remained inaccessible for nearly four centuries. It does not, however, necessarily follow, as has been too often assumed, that they were without influence, and it is at least possible that too much stress has been laid on Leonardo's intellectual isolation. Man of genius though he was, he was also a child of his age and he represents, though with the instinct of genius, the ideas of his age.[3] Thus we

per esso punto possono transire i diminute razzi di tali spetie, dopo il transita delle quali si riformeranno e ricresceranno le quantita di tali spetie.'

1 It is set forth, for instance, in the *Timaeus* of Plato, which was very influential in the Middle Ages through the early Latin translation of Chalcidius.

2 *Studies,* vol. i, p. 118.

3 Leonardo left his scientific remains in a state of confusion, and they have

find a generation or two later his views on the camera obscura developed by Digges, and his 'pin-hole theory' and view of the double inversion of images in the eye given a mathematical direction by Maurolico.

15. *Girolamo Fracastoro* (1478 ?–1553) was a suggestive writer, who devoted considerable space to a rather confused account of refraction. In the course of this discussion he has the following passage:

'[Not only the character but] also the position of the medium affects the appearance of the objects seen, as may be observed with spectacle lenses (*in specillis ocularibus*). For if the lens be placed midway between eye and object, it appears much larger than if the lens is made to approach the object or the eye.[1] . . . Glasses (*specilla ocularia*) may be arranged of such density that if any one looks through them at the moon or at any star they appear near and hardly higher than the steeples.[2]

It is possible that he was here contemplating a bilenticular apparatus.[3]

16. *Francesco Maurolico* (1494–1577) made perhaps the first published independent attempt, after Bacon and Peckham, at a mathematical analysis of the optics of the lens and of the eye. He is thus the predecessor of Kepler. There was another important service that Maurolico rendered to optical science. He dethroned the lens (*humor crystallinus*) from its pride of place, occupied ever since Greek times, as the seat of vision. There were, however, two great errors from which he could not free himself. He believed, firstly, that every point of an object seen must correspond to only a single visual ray, and secondly, that the image in the eye is not reversed. His view, however, that short sight was due to excessive and far sight to defective power of the crystalline lens, though erroneous, was yet a fruitful and in its day not untenable hypothesis.[4]

17. *Leonard Digges* (died 1571 ?) was the first to whom can be definitely attributed the construction of a bilenticular system.

suffered much by time, misuse, and neglect. It is impossible to give a bibliography here, but his optical results are summarized by E. Solmi, 'Leonardo da Vinci e il metodo sperimentale nelle ricerche fisiche', in the *Atti e memorie della R. Accademia Virgiliana di Mantova*, Mantua, 1905, and by O. Werner, *Zur Physik Leonardo da Vincis*, Berlin, 1911.

[1] *Homocentrica*, ii. 8.

[2] *Homocentrica*, iii. 23.

[3] The *Homocentrica* in which these passages occur was first printed at Venice in 1538. The scientific value of this work is discussed by the present writer in an article in the *Annals of Medical History*, vol. i, p. 1, New York, 1917.

[4] Maurolico's work, *Photismi de lumine et umbra*, was printed in Venice in 1575.

The evidence for this statement rests on the following passage in a work printed in 1571 by his son, Thomas Digges (died 1595):

'Marueylouse are the conclusions that may be perfourmed by glasses concaue and conuex of circulare and parabolicall fourmes, using for multiplication of beames sometime the ayde of glasses transparent, which by fraction should unite or dissipate the images or figures presented by the reflection of other. By these kinds of glasses or rather frames of them, placed in due angles, ye may not only set out the proportion of an whole region, you represent before your eye the lively image of euery towne, village, etc., and that in as little or great a space or place as ye will prescribe, but also augment and dilate any parcell thereof, so that whereas at the firste apparance an whole towne shall present it selfe so small and compacte together that ye shall not discerne any difference of streates, ye may by applycation of glasses in due proportion cause any peculiare house or roume thereof dilate, and shew it selfe in as ample fourme as the whole towne first appeared, so that ye shall discerne and trifle or reade any letter lying there open, especially if the sonne beames may come un to it, as playnly as if you wer corporally present, althouh it be distante from you as farre as eye can discrye. But of these conclusions I minde not here more to intreate, hauing at large in a volume by it selfe opened the miraculous effectes of perspective glasses.'

Digges's system appears to have been combined in some manner with a camera obscura. Unfortunately, his further description of it was never published.[1]

In a second edition of the same work, printed in 1591, Digges tells us that:

'My father, by his continuell painfull practices, assisted by demonstrations mathematical, was able, and sundrie times hath, by proportional glasses, duly situate in convenient angles, not only discovered things farre off, read letters, numbered pieces of money, with the very coyne and superscription thereof, cast by some of his freends of purpose upon downs in the open fields, but also seuen miles off declared what hath been done in priuate places.'

18. *Felix Plater* (1536–1614) made an important addition to the optical theory of the eye, and rid himself of at least one of Maurolico's fundamental errors. In a work printed in 1583 he clearly ascribes to the crystalline lens and to the retina the functions with which we are accustomed to credit them. 'The *humor*

[1] The work of Thomas Digges in which this passage occurs is entitled *A Geometrical Practise named Pantometria,* and was printed in London in 1571.

crystallinus ', he says, ' is the lens of the nerve of vision (*perspicillum nervi visorii*) and, placed in front of it and of the foramen of the pupil, it collects as rays the *species* impinging on the eye, and sends them forth magnified to the circuit of the retina so that it may the more easily appreciate them. . . . The primary organ (*pars primaria*) of vision, the optic nerve, on reaching the eye expands into a hollow retiform hemisphere. This receives and distinguishes the *species* and colours of external objects by the light which, passing into the eye through the foramen of the pupil, is distributed (*representata*) by its lens as has been said above.' [1]

19. *Giambattista della Porta* (1540–1615) is the first to whom can be attributed the actual combination of lenses in the form of a microscope. This statement rests on the evidence of the following passage in his *Magia naturalis*: ' Concave lenses enable one to see far off more clearly, while convex ones make near objects more discernible.' He was thus apparently myopic, and he goes on to say that ' with a concave lens things afar appear smaller but plainer, with a convex lens you see them larger but less distinct. If, however, you know how to combine the two sorts properly, you will see near and far both large and clear.' In later years, when the microscope became a recognized instrument, much larger claims were made by and for Porta, but there is no real evidence that he made any effective practical application of his idea.[2]

But Porta made a more valuable contribution to optics than his accidental suggestion of the possibility of a microscope and telescope which he did not follow up, for his is the first printed book containing a clear comparison of the action of a lens to that of a pinhole camera. By this suggestion he made possible the mathematical developments of Kepler. ' Just as objects illumined by the sun ', says Porta, ' are set forth on a sheet of paper intercepting the beam from a small hole in a shutter, so does the light passing through the foramen papilla form images of the objects seen by the action of the crystalline lens (*subiens in crystallinum*).'

20. *Hieronymo Fabrizio* of Aquapendente (1537–1619), the teacher of Harvey, is worthy of a brief mention here as following up the optical work of Maurolico in the field of anatomy. This admirable and accurate observer in his *Tractatus de oculo visusque organo* (Padua, 1601) was the first to figure an eye with a definite

[1] Felix Plater, *De partibus corporis humani structura et usu*, Basel, 1583, p. 187.

[2] Porta's *Magia naturalis* was first printed at Naples in 1558, but the passages in question do not occur in it, nor in any edition until that of 1588.

lens placed in the front part of the globe. His conservative attitude, however, prevented him from accepting the views of Plater and he still placed the seat of vision in the lens. It is a pity that a somewhat cumbrous learning and an undue respect for authority marred the value of this as of much of the other work of the great Paduan professor.

21. *The Empirical Discovery of the Telescope and Compound Microscope.* There are at least three claimants, all Dutchmen, that have to be seriously considered: Zacharias called Jansen, Jan Lippershey, and James Metius otherwise known as Jacob Adrianzoon. On whichever our choice may fall, however, the date of the discovery may be placed between 1591 and 1608, and probably nearer the later than the earlier date. The reader may be reminded that at this period lenses both concave and convex were well known and habitually used for spectacles. The profession of 'spectacle maker' had become quite recognized and lucrative although little scientific attention had as yet been directed to the nature and application of lenses. The three claimants may be considered separately.

(*a*) *Zacharias,* son of Jan, and known as Jansen (1580–16 ?) of Middelburg, is usually regarded as the first who actually constructed a microscope. His first attempt was the result of an accident. It appears that while still a lad and at work in the shop of his father, who was a spectacle maker, he happened to place two lenses in a tube and found that they acted as a microscope or telescope. Effective instruments were constructed by him in the first decade of the seventeenth century. The evidence that Jansen was really the first constructor of these bilenticular instruments rests on the testimony of Willem Boreel (1591–1668), the Dutch Ambassador to France. Boreel's evidence is given in a letter by him to Pierre Borel (1620–71), which runs as follows:

'I am a native of Middelburg, the capital of Zeeland, and close to the house where I was born . . . there lived in the year 1591 a certain spectacle maker, Hans by name. His wife, Maria, had a son Zacharias, whom I knew very well, because, as a neighbour and from a tender age I constantly went in and out playing with him. This Hans, or Johannes, with his son Zacharias, as I have often heard, were the first to invent microscopes, which they presented to Prince Maurice, the governor and supreme commander of the united Dutch forces, and were rewarded with some honorarium. Similarly, they afterwards offered a microscope to the Austrian Archduke Albert, supreme governor of Holland. When I was Ambassador to England in the year 1619, the Dutchman Cornelius Drebbel of Alkomar, a man familiar with many secrets of nature, who was serving there as mathematician to King James, and was well known to me, showed me that very instrument which the Archduke had presented as a gift to Drebbel, namely, the microscope of Zacharias himself. Nor was it (as they are now seen) with a short tube, but nearly two and a half feet long, and the tube was of gilded brass, two fingers breadth in diameter, and supported on three dolphins formed also of brass. At its base was an ebony disk, containing shreds or some minute objects which we inspected from above, and their forms were so magnified as to seem almost miraculous.'[1] His instruments were probably made of two convex lenses.

(*b*) *Jan Lippershey* of Wesel (flourished 1608). It is evident that at or about the period 1608–10 there were other Dutch mechanics on the track of the discovery of Zacharias, for Borel goes on to say that

'long after, in 1610, after much research, these individuals (i. e. Zacharias and his father) invented, in Middelburg, the long sidereal telescope with which we gaze at the moon and planets and presented one of them to Prince Maurice, who deemed it prudent to conceal the invention in order to make use of it in military operations. While rumours were abroad regarding the invention, a stranger came to Middelburg, and having asked for a spectacle maker, was shown by mistake into Lippershey's shop. From the questions put by the stranger, Lippershey, being a shrewd man, was able to detect the construction of the instrument, and having succeeded by this means in making telescopes, he was generally considered the real inventor of them. However, the mistake was soon discovered, for Drebbel, upon his return to Holland, proceeded with Adrian Metius to Jansen's shop, and purchased telescopes of him.'

This account treats Lippershey as a mere imitator of Zacharias. There is however evidence, though not of an entirely convincing character, that the discovery was really made independently by Lippershey. When Borel investigated the matter about 1645 he examined several witnesses, three of whom attributed the discovery of telescopes to Lippershey, one of them placing it before 1605 and the other two in 1609 or 1610. Moreover

[1] This passage is contained in a work by Pierre Borel, *De vero telescopii inventore cum brevi omnium conspiciliorum historia*, The Hague, 1655. Borel produces other witnesses for Zacharias; one of them, his son, gives the year as 1590; the other, his sister, brings it down ambiguously to 1611 or 1619. Neither is worthy of prolonged consideration.

Antoine-Marie Schyrle de Rheita (1597–1660), a competent mathematician and acquaintance of Kepler, writing in 1645, tells also that Lippershey invented the telescope in 1609. Lippershey, according to him, happened to place a concave before a convex lens and found that the weathercock of a neighbouring church appeared larger and nearer. He fitted the glasses in a tube and amused his customers with this instrument. The Marquis of Spinola, who was then at the Hague, bought the instrument and presented it, according to de Rheita, to the Archduke Albert of Austria by whom the invention was made known. The whole story is suspiciously like that of Zacharias Jansen, and the two are perhaps a confusion of the same original account.[1]

More satisfactory evidence in Lippershey's favour has been obtained from the Acts of the States-General preserved in the Government archives at the Hague.

'On the 2nd of October, 1608, the Assembly of the States took into consideration the petition of John Lippershey, spectacle maker, a native of Wesel, and an inhabitant of Middelburg, inventor of an instrument for seeing at a distance. A committee was appointed to confer with Lippershey, for the purpose of ascertaining whether it might not be possible to improve the instrument so as to enable one *to look through it with both eyes.* On the 4th of the same month it was resolved that certain of the members should test the instrument of Lippershey, by observing with it from the turret of Prince Maurice's mansion. It was further resolved, that if the perspective should be found useful, an engagement should be entered into with the inventor, to execute three such instruments of rock crystal, and that he should be enjoined not to divulge the invention to anybody. On the 6th of the same month, the Assembly agreed to give Lippershey 900 florins for such an instrument. On the 15th of December they examined the instrument invented by Lippershey *to see with both eyes,* and approved of it; but as many others had a knowledge of this new invention to see at a distance, they did not deem it expedient to grant him an exclusive privilege to execute such instruments. However, they gave him orders to execute, for the use of the Government, two other instruments to see with both eyes, allowing him the same remuneration for his services as in the first instance.'[2] The

[1] A.-M. Schyrle de Rheita, *Oculus Enoch et Eliae, sive Radius sidereo-mysticus,* Antwerp, 1645.

[2] The application and findings of the committee that sat on Lippershey's petition were published by G. Moll, 'On the first invention of telescopes collected from the notes and papers of the late Professor van Swinden,' *Journal of the Royal Institution,* vol. i, pp. 319 and 483, London, 1831; see also R. Grant, *History of Physical Astronomy,* p. 521, London, 1852.

instruments of Lippershey were almost certainly made of a convex and concave lens and therefore did not invert.

(*c*) The claim of *James Metius* (Jacob Adrianzoon, died *circa* 1629) may now be set forth, and a consideration of it enables us to understand why the States-General did not fully accede to the petition of Lippershey. There is documentary evidence that while the claim of Lippershey was still under consideration by the States-General, viz. on October 17, 1608, James Metius of Alkmaar put in a petition to obtain the exclusive right to sell an instrument of his invention for making distant bodies appear larger and more distinct, and that at the time of his petition he knew of the instrument 'that had lately been offered to the States *by a citizen and spectacle maker of Middelburg*', i. e. presumably by Lippershey. This petition survives in the correspondence of Christiaan Huygens (1629–95) and has recently been published.[1]

Of James Metius we can form a rather better picture than of the other two claimants. He was the son of Adrian Metius (Adrianzoon or Anthonisz, 1527–1607) who became Burgomaster of Alkmaar in 1573. Adrian, who was born in Metz and hence got his name *Metius*, has a place in the history of pure Mathematics for his solution of π as equal to $\frac{355}{113} = 3{\cdot}1415929\ldots$[2] This Adrian Metius had two sons, Adrian and James. Adrian Metius (1571–1635) was himself a mathematician and was doubtless the man of that name who is referred to by Borel as purchasing a telescope along with Drebbel at Zacharias Jansen's shop in Middelburg. The other son, James, was less gifted. Of him Descartes wrote:

> 'It is about 30 years since one named Jacques Metius, a man who had never studied (but who had a father and brother that followed mathematics as a profession), though he himself loved to make mirrors and burning-glasses, . . . having by him glasses of various shape, had the good luck to look through two of them, of which one was a little thicker in the centre than at the edge and the other was much thicker at the edges than at the middle; and he luckily put them in the ends of a tube, and thus the first telescopes were made. It is on this model only that all the others which have been seen since have been made, without any one yet,

[1] In the *Œuvres complètes de Christiaan Huygens*, vol. xiii, fasc. 1, p. 591, The Hague, 1916. Huygen's view of the history of the invention, with which our own conclusions are in accord, is given on p. 436 of the same volume. The passage is translated from the Dutch in the appendix on page 533.

[2] Cp. Moritz Cantor, *Vorlesungen über Geschichte der Mathematik*, 4 vols., vol. ii, 2nd edition, Leipzig, 1900, p. 599.

so far as I know, having adequately determined the shapes that the glasses ought to have.' [1]

There are conflicting elements in these accounts, but from them can be drawn the following deductions which seem to sum up the probabilities.

i. The invention of the microscope probably preceded that of the telescope.

ii. The invention of the microscope was the work of Zacharias Jansen after 1591 and before 1608. It was perhaps formed of two convex lenses.

iii. The invention of the microscope by Zacharias Jansen was followed by that of the telescope about 1608 by Lippershey and Metius. It was the military applicability of the apparatus that drew attention to it.

iv. This first telescope was of the Galilean type with a concave eye-piece and a convex objective.

22. *Galileo* (1564–1642) was the *effective* discoverer of both telescope and microscope. The event has often been described and may be referred to the early part of 1609. The story, in his own words, is briefly thus :

'About ten months ago', he says, 'a rumour reached me of an ocular instrument made by a certain Dutchman by means of which an object could be made to appear distinct and near to an eye that looked through it, although it was really far away. . . . And so I considered the desirability of investigating the method, and I reflected on the means by which I might come to the invention of a similar instrument. A little later, making use of the doctrine of refractions, I first prepared a leaden tube, at the ends of which were placed two lenses, each of them flat on one side, and as to the other side I fashioned one concave and the other convex. Then, moving the eye to the concave one, I saw the objects fairly large and nearer, for they appeared three times nearer and nine times larger than when they were observed by the naked eye. Soon after I made another more exactly, representing objects more than sixty times larger. At length, sparing no labour and no expense, I got to the point that I could construct an excellent instrument so that things seen through it appeared almost a thousand times greater and more than thirty-fold nearer than if observed by the naked eye.' [2]

In another work he says :

'Some would tell me that it is of no little help in the discovery

[1] The passage will be found in the *Dioptrique*, and can be referred to in the *Œuvres de Descartes* published by C. Adam and P. Tannery, vol. vi, Paris, 1902, p. 82, and note on p. 227.

[2] *Siderius Nuncius*, Venice, 1610.

and resolution of a problem to be first of all in some way aware of the true conclusion and certain of not being in search of the impossible, and that therefore the knowledge and the certainty that the microscope had indeed been invented had been of such help to me that perchance without that I should not have discovered it. To this I reply that the help rendered me by the knowledge did indeed stimulate me to apply myself to the notion, and it may be that without this I should never have thought of it. Beyond this I do not believe that knowledge to have facilitated the invention. But, after all, the solution of a problem, thought out and defined, is a work of some skill, and we are not certain that the Dutchman, the first inventor of the telescope, was not a simple maker of ordinary lenses who, casually arranging glasses of various sorts, happened to look through the combination of a convex and a concave one placed at various distances from the eye and in this way observed the effect that followed thereon. But I, moved by the knowledge given, discovered it by a process of reasoning.'[1]

23. The telescope used by Galileo had of course a concave eyepiece and a convex objective. Specimens of the telescopes actually used by him—though not of the earliest type—are still in existence and are on exhibition at the Galileo Museum in Florence. Galileo, in spite of his claim to have invented the telescope on a theoretical basis, did not really arrive at a successful solution of the optics of the lens, and his account of the path of light in the bilenticular system is unsatisfactory. It was improved by *Kepler* (1571–1630) in his *Dioptrice* (Cologne, 1611), and he at the same time suggested that form of telescope and of microscope, consisting of two convex lenses, which has developed as our modern instrument. Kepler had already some years earlier made a mathematical analysis of the properties of a simple lens in his *Ad Vitellionem paralipomena, quibus astronomiae pars optica traditur*, Frankfort, 1604, and his optical labours were now crowned by his description of the bilenticular system. It should be added that in the year of the appearance of Kepler's *Dioptrice* (1611) appeared the work of Marco Antonio de Dominis (1566–1624), *De Radiis Visus et Lucis in Vitris Perspectivis et Iride*, containing suggestions somewhat similar to those of Kepler. In view of the well-known dishonesty of the notorious Archbishop of Spalatro, it is difficult to determine to what extent these ideas were his own. In any event the priority of effective demonstration rests with Galileo, and of the publication of mathematical analysis with Kepler.

SEE APPENDIX, PAGE 533.

[1] *Il Saggiatore*, Rome, 1623.

HYPOTHESIS

By F. C. S. Schiller.

§ 1. Deep down in the heart of man lives a chronic *Fear* of the weird world in which he finds himself so inexplicably plunged and is so imperiously prompted, by hereditary instinct—and the fear of greater evils—to struggle for his existence to the last breath. This Fear is usually kept under by the fabricated order of the social life, and kept drugged by the traditions, conventions, creeds, and mechanisms of a social *routine.* But under abnormal conditions it breaks out, and renders him a prey to wild and senseless 'superstitions'; which are in truth survivals from the times when as yet no 'science' had been devised to reassure men by its claim to curb the chaos of happenings, superstitions which still betray our unavowed distrust of the exorcisms of the verbal formulas which we significantly call the *laws* of nature—in the *hope* that nature will *obey* them. This Fear moreover begets not only the superstitions but the gods: we may expect to find traces of its universal influence, *camouflaged* of course, disseminated also through the doctrines of the sciences, and revealing itself as an obstinate bias or persistent prejudice against such endeavours as run counter to it.

For it must not be supposed, of course, that Fear is the only attitude which the world inspires in man. Fear alone would be utterly inadequate to keep us alive. It must be matched with Courage—even if only that of desperation—so as to generate a prudent circumspection and to lead to successful action. Accordingly, in his dealings with nature man cannot always remain cowering on the defensive. He must *react,* must show initiative, must attack the problems of his life, must *experiment* boldly, and manipulate and alter the given conditions, 'speculating' and running risks. And, if he is defeated, as he always is at first, he must never say die, but must try again, more intelligently.

§ 2. The interplay between these antagonistic and yet complementary attitudes may be traced through all life, and we may expect to find it operative also in the science of Logic. We shall further find that these attitudes appeal differently to the different schools of logic. To an *intellectualistic* logic it will seem self-evident that sound thought must play for safety and endeavour to proceed by *valid* steps to an *assured* conclusion. It will eschew therefore or minimize, so far as may be, risks and hazardous guess-work; it will admire and revere formulas and forms, which, it will fondly believe, cannot go wrong, but assure success, like spells, to those who have mastered them. Hence the perennial charm of formulas like a 'syllogism' or a 'universal' and their prominence in logics of this kind. They function as 'words of power'; to pronounce them is deemed enough to flood with light the obscurest problem, to rehearse them, like magic charms, seems essential and sufficient for salvation.

To a *voluntaristic* logic this attitude seems merely superstitious. It demands not words but deeds. So it will value a formula strictly according to its achievements in actual use. It will not labour, vainly, to construct a fool-proof form, nor imagine that by clinging to it faithfully any one can guard himself against any error. It will perceive that the value, the truth, the meaning, the difficulty, and the danger, of every form lie in the application thereof, and that to risk the application of a principle, is to run a risk of error. For every principle may be misapplied to a case it does not really fit. But seeing that this risk in real reasoning is as unescapable as death, this bolder logic will not shrink from it. It will recognize and sanction the running of risks, provided that they are taken open-eyed and not blindly, and the desire for logical *value* is not frustrated by being exalted into a demand for absolute *validity*. It will approve of operations expressive of spontaneous, and still more of conscious, experimentation, of intelligent anticipation, of the mental attitude of unending inquiry, of an insatiate appetite for new truth and more knowledge.

In short, it will recognize that *knowing is an adventure*, in which to progress we must not be content to take our stand always on firm ground, whether of 'solid fact' or 'fundamental principle', but must learn to *swim* the breakers of Doubt, and win our way through the ocean of Becoming to the Happy Isles in that immeasurably distant West, where, if anywhere, those weary of activity can rest in eternal bliss, because every problem has been solved and every aim attained.

§ 3. In no department of logical doctrine do the contrast and clash of these two tendencies in logic come out more copiously and instructively than in the treatment of Hypothesis. For this is just where the need for cognitive enterprise must impress itself even upon the most reluctant minds. Accordingly Hypothesis receives a somewhat Cinderella-like treatment in intellectualist logics. It is disparaged and intellectualized as much as the facts will allow, though, since Bacon's egregious failure, no one has dared to prohibit it altogether. But a meagre chapter is the most they can endure to waste upon it. Even out of that they are careful to expunge the real essence and significance of the mental attitude which dares to tackle nature with hypotheses and endeavours to enmesh the monsters of the deep in the flimsy fabrics of man's imagining. The scope and value of hypotheses are studiously minimized, and their use is restricted by regulations which purport to curb their licence and to ensure their soundness. But the simple truth is that intellectualist logicians are *afraid* of them.

The matter can best be elucidated by a few illustrations which will show how valiantly, and yet how vainly, logicians struggle against conceding the right of experimenting with hypotheses, and refuse to regard it as chief among the *intellectual* Rights of Man.

§ 4. We may consider first the most modern of the large 'orthodox' *Logics*, that of Professor Bosanquet. When we look up its definition of Hypothesis[1] we find that it is called "any conception by which the mind establishes relations between data of testimony, of perception or of sense, so long as that conception is one among alternative possibilities, and is not referred to reality as a fact".

There are several points to note in this definition.

(1) The essence of Hypothesis is said to be a *conception*, i. e. a mental *content*, not a mental *attitude*—the attitude to wit of *inventing* a mental content, of *holding it in suspense*, of taking it as a possibility not as an assured fact, of assuming it experimentally for the argument's sake and in order to test its value by its consequences.[2]

[1] Vol. ii, ch. v, *init.*

[2] Although his definition had thus subsumed 'hypothesis' under the notion of 'conception', Professor Bosanquet nevertheless subsequently has occasion to reverse this relation, and to say (p. 160) that (certain) "conceptions may be regarded as hypotheses in course of development and proof". This would seem

(2) The stress plainly falls on the 'conception' and the 'relations' it establishes rather than on the mind's activity and intention in forming its conceptions and establishing the relations by applying them. Yet the mental attitude of hypothesis-building is evidently the motive force in the whole process: it is this which selects and uses the conceptions which seem likely to solve the problems which engross the mind. This is particularly plain in cases where the conceptions used are vague, inchoate, dubious and plastic, and in fact *most* 'hypothetical': for we can then feel how masterfully the hypothetical attitude *moulds* our conceptions as we employ them on our data and develop them to suit its purposes.

(3) Professor Bosanquet's definition retains the traditional, but unworkable, antithesis between 'hypothesis' and 'fact'. This however is nothing but an obsolete dogma which has come down from the naïve days when logicians believed that 'facts' were solid and immutable and the firm 'foundations' of the sciences. It has unfortunately survived unceasing demonstrations in the history of every science that its 'facts' are relative to its 'theories' and 'principles', that anything may be *feigned*, if it is not found, and taken as fact *hypothetically*, and that there is no hard and fast line to be drawn between hypothesis and fact, because the conclusions of hypothesis consolidate into 'facts' when they are sufficiently confirmed, and may then themselves form the basis for further hypotheses. The difference between fact and fiction is not that in the former the mind is somehow coming in contact with an extra-mental independent real, while in the latter the *intellectus sibi permissus* runs riot and is creating freely out of nothing; we are in *both* cases manipulating mental contents, not, that is, 'the real' facts, but facts-as-they-seem-to-us in the apprehension of the imperfect knowledge of a limited intelligence—nor yet *mere* fictions, but fictions-suggested-by-the-real we *suppose* ourselves to be encountering, and *believed* to be applicable and relevant to that. When such a scientific fiction [1]

to admit that *some* conceptions are essentially hypothetical instruments for exploring nature withal; and he even says (p. 159) that "in an ultimate sense there is no knowledge without Hypothesis". Why then not adopt this ultimate sense? Would not a little more daring declare *all* knowledge to be the great hypothesis by which man, alone of living beings, endeavours to control his world?

[1] I am not here using the term in its strict technical sense as an assumption, *known* to be false, but also methodologically *convenient*.

turns out to be *very* relevant, what more natural and proper than that it should be raised to the rank of ‘ fact ’ ? The history of scientific conceptions is full of illustrations, though none, perhaps as instructive and romantic in its vicissitudes as the history of the ‘ atom ’, a metaphysical invention in its initial stage, long rejected by the best scientific opinion as unprofitable, then revived and turned into a good working-hypothesis of chemistry, criticized anon as unverifiable and barely holding its ground, as a convenient fiction against its ‘ energetic ’ critics ; until finally by a dramatic revolution the hypothesis of atomic dissociation explained the facts of radio-activity, and incidentally, turned the atom into a fact in the very act of exploding it and depriving it of the ultimateness and indivisibility its very name had implied for ages. But the lesson of this history will not have been truly learnt until it has inculcated into men of science a proper indifference towards the ontological pretensions of the atom’s hypothetical successors, the ‘ electrons ’ and ‘ quanta ’, and they are content to accept them as the smallest coins in the reckonings of science, because, and so long as, they serve their purposes and perform this function.

§ 5. J. S. Mill’s account of Hypothesis [1] is not at any rate open to the charge of excluding from its definition the essential attitude of *supposing*. It declares that “ an hypothesis is any supposition which we make (either without actual evidence, or on evidence avowedly insufficient) in order to endeavour to deduce from it conclusions in accordance with facts which are known to be real ; under the idea that if the conclusions to which the hypothesis leads are known truths, the hypothesis itself either must be, or at least is likely to be true ”.

This however is not so much to define Hypothesis itself, as a mental attitude and product of mental activity, as to define its cognitive function, i. e. what he conceives to be the proper *use* of Hypothesis in scientific inquiry. This is why he equips his definition with a number of restrictive regulations. Had he merely wanted to describe the procedure itself, he would have *begun* by pointing out that “ an hypothesis being a mere supposition, there are no other limits to hypotheses than those of the human imagination ”, and then have gone on to investigate how the human

[1] *System of Logic*, Book iii, ch. 14, § 4.

imagination is actually limited in its guesses. This would probably have convinced him that its play was both limited by, and relative to, its experience of reality, and that it was not therefore urgently necessary to make the regulation of this activity too strict.

Interpreting in accordance with his general theory of logic, Mill goes on to remark that " hypotheses are invented to enable the Deductive Method to be earlier applied to phenomena ". This they do by taking the place of the first " indispensable " (*sic !*) step in the Method, viz. ' induction ', so that " the law is not acquired by direct induction from the facts but assumed as a hypothesis ", after which ' ratiocination ' and ' verification ' proceed as before. He contends, however, that in some cases the verification attains to the ' validity ' of an experiment according to the canon of the Method of Difference, and that then the hypothetical ' law ' is fully proved. But the assurance given by this Method " cannot be obtained when the cause assumed by the hypothesis is an unknown cause imagined solely to account for the observed effects." In the discovery of the law of gravitation, " if it had not been previously known that the planets were hindered from moving in straight lines by some force tending towards the interior of their orbit ; though the exact direction was doubtful ; or if it had not been known that the force increased in some proportion or other as the distance diminished, and diminished as the distance increased, Newton's argument would not have proved his conclusion." " The most genuinely scientific hypothesis ", therefore, must not be " destined always to remain an hypothesis, but be of such a nature as to be either proved or disproved by comparison with observed facts." In such cases alone " verification is proof " (*ib.*, p. 15), and this was the meaning of Newton's maxim that the agency alleged by a hypothesis must be a *vera causa*.

Now in this account there is, amid much sound sense, an admixture of error, which is traceable to the desire to force scientific procedure into conformity with the ideal of logical proof.

(1) If Hypothesis can take the place of Induction in the ' Deductive Method ' and a verified hypothesis can become a fact, with the highly desirable result of accelerating scientific procedure, is not the bottom knocked out of Mill's whole conception of inductive science ? His ' facts ' are no longer necessary, seeing that they may with advantage be replaced by hypotheses ; nor

are they absolute facts, but 'facts' accepted in virtue of their relation to verified hypotheses. It does not really matter whether the *data* from which we reason are wholly hypothetical, or believed to be 'facts', or something betwixt and between that is alleged but disputed; nor is there any need to quarrel about their precise status. It is more advantageous and expeditious to go on investigating. For if we are willing and able to verify our conclusions, we shall presently become aware of the value of our *data*, which does not depend on their initial claims, but on the results they lead to.

(2) The notion that a hypothesis can only be proved by disproving all alternatives and showing that no other can satisfy the *data*, must surely be dismissed as a chimera. The only alternatives that can be either proved or disproved are those which have suggested themselves to human ingenuity, and there is no reason to think that this has, at any time, exhausted the possibilities, or to presume that it will ever do so. Moreover, the test proposed by Mill, viz. the conformity with fact of the deductions from the true hypothesis and of these *alone*, is essentially illusory.

For the reason, *inter alia*, that such conformity can never be shown to be *exact*. No sense-organ and no instrument can ever measure exactly, no observation can wholly eliminate the errors of observation, no observation can be repeated absolutely, and no two observers will absolutely coincide in their estimates. Hence, however closely the observations approximate to the deductions from the hypothetical 'law' intended to explain them —and any such approximation is likely to be exaggerated by the influence of the hypothesis on the observer—it can never be argued that they conform with *this* law *and no other*. It is always possible that some other law, or combination of laws, would fit the observations as closely as, or more closely than, that accepted. The alternatives might often be less *simple*; but simplicity is an appeal to man's convenience, not nature's. The simplicity of the 'laws' first tried is methodological; it is right to use the simplest hypothesis, if it suggests itself. Moreover, the simplicity of our first approximations is often deceptive; our laws of nature are *statistical*; they formulate the *average* only of what may turn out to be on closer inspection seriously divergent activities, and it is well known that statistics smooth out individual differences. Moreover, curiously enough, the progress of science has, since

Newton's day, quite familiarized physicists with the idea that even the law of gravitation may not be as exactly true as was supposed.[1] It may only express, roughly, the motions of large masses, without applying to 'electrons'; and even astronomers are now led to speculate whether the observed fact of the much more rapid motions of the 'older' types of star should not be taken to mean that gravitation is not an initial endowment, but an *acquired habit*, of celestial bodies.

Lastly, it should be remembered that the alternative hypotheses, which seem to be equally capable of interpreting the facts, are often worlds apart in their character. It hardly matters to the calculations of physics whether bodies are taken to be solids moving through an 'ether' that cannot be shown to impede their motions, or areas of diminished density in an adamantine ether. It hardly matters to biology whether the adaptations of organisms be regarded as teleological or as pseudo-teleological. In ethics, determinism + a little ignorance and habit + a little indetermination will provide interpretations of human actions that seem equally complete. The political philosopher, though not the party man, will recognize that the conservative and the innovator will always be able to make out a case for the diametrically opposed views they take of social changes. And so forth and so on. Alternative hypotheses therefore must always be admitted to be possible in principle, even where there are none actually contending for acceptance.

It follows that Verification never can be proof, if 'proof' be taken to involve a claim to *absolute* truth. No verification will ever prove more than that the observations are *compatible* with the hypothesis they test. That does not prevent them from being equally compatible with other hypotheses, if these lead to very similar deductions, and if the observations are not exact (as we have seen they never are), or have been vitiated by the influence of a dominant hypothesis. Should a philosopher be disposed to be too sanguine about the capacity of 'facts' to impose themselves on prejudice and to resist misinterpretation, he should be invited to consider how overwhelming were the 'proofs' of witchcraft until about two hundred years ago.

[1] Of course had this been written in 1919 instead of 1918, one could not but have referred here to the triumphant verifications of Einstein's theory of relativity and the consequent superseding of Newton's law of gravitation.

(3) Newton's doctrine about a *vera causa* should no longer be allowed to impede the freedom of scientific movement. If it is intended as an *ex post facto* description of a hypothesis which *has* worked satisfactorily, it is the merest truism. For *after* we have solved our problem we can of course always say *which* of the alleged ' causes ' was a ' true cause '. If, on the other hand, it is meant as a rule of research, it is futile and false. For then *ex hypothesi* we have not yet found a true cause, but are still looking for one. If the maxim means that no causes are to be tried but such as are similar or analogous to causes already accepted, it does indeed chime in well with man's instinctive conservatism, and may claim perhaps a certain methodological merit as proclaiming that it is well to try familiar agencies before resorting to the unknown, somewhat superfluously, because this is what both imagination and indolence will prompt us to do; but how does this pledge *nature* to submit to analysis by ' causes ' which are familiar enough to seem ' true ' to man at any stage of his knowing? As a matter of scientific history this presumption has perpetually misled him. How much ingenuity and effort has he not wasted for aeons in attempts of this sort? How persistently has he not tried to construe physical happenings on the analogy of his own agency by the method of ' animism '? How many generations of scientific men have not insisted with Aristotle that the effective clue to nature must lie in the qualities which things have for our senses, and tried to build a system of physics out of the antitheses of hot and cold, dry and moist, heavy and light, &c.?

In § 5 Mill gets much nearer to a right estimate of Hypothesis. He disallows Newton's dictum *hypotheses non fingo*, and declares that "nearly everything which is now theory was once hypothesis". He points out that hypotheses are necessary *to guide experimentation*, for "even in purely experimental science some inducement is necessary for trying one experiment rather than another", that "the process of tracing regularity in any complicated, and at first sight confused, set of appearances is necessarily tentative: we begin by making *any* supposition, *even a false one, to see what consequences follow from it*; [1] and, by observing how these differ from the real phenomena, we learn what corrections to make in our assumption. The simplest supposition which accords with the

[1] Italics mine.

more obvious facts is the best to begin with; because its consequences are the most easily traced. This rude hypothesis is, then, rudely corrected, and the operation is repeated . . . until the deductive results are at last made to tally with the phenomena."

In the first of these remarkable passages Mill recognizes that it is not enough to conceive truth as awaiting discovery and the mind as arriving at it by the systematic exploration of the totality of reality. Reality is too vast to be exhausted thus in a finite time by minds which cannot afford to fritter away their brief span of life in aimless 'fool' experiments: hence the enormous importance of the sagacious hypothesis which takes them straight into the vitals of a problem. It is recognized, secondly, that the real guarantee of a good hypothesis does not consist in its conformity with abstract rules, formulated in advance of the investigation; the hypothesis is made good by being progressively knocked into shape in the process of its verification by the facts it has to account for. Hence it is not really requisite that a hypothesis should be correct, complete, or even probable, to start with; if it is plastic and corrigible, it will gradually assume a more and more valuable shape. The one essential it must have is that it must suggest a *convenient method* of exploring the subject it concerns; provided it does this, it may permit itself the use of the most patent fictions, and may leave aside, as 'difficulties' to be dealt with later, the incoherences which its incompleteness involves. Thus once the physicists had been led by the analogy between the propagation of light and that of waves in a liquid to assume a 'luminiferous ether', they did not allow the contradictions which arose out of conceiving this ether, either as continuous or as atomic, to deprive them of a valuable working hypothesis. It is simply not true that gaps and contradictions are fatal to the scientific status of a hypothesis, so long as it remains the best available and a hope remains of finding a remedy for its defects.

In § 6 Mill rightly points out that in the actual state of our knowledge more than one hypothesis may be competent to explain the facts, and that the correct anticipation and prediction of results subsequently found to occur in fact, though impressive, is not strictly a 'proof' of a hypothesis.

Both points are important, and have been too little regarded. The coexistence of alternative interpretations, about equally capable of explaining the facts, and consequently in dispute, is common, and is indeed the normal condition in the growing stratum of every science: usually indeed it does not last, as regards any particular scientific question, because new facts accrue which do not remain neutral, but reinforce one side until it wins a definite superiority. If Mill, however, had taken his illustrations from the hypotheses of philosophy, he would speedily have perceived that a conflict of hypotheses need not end in any decisive victory of either side. For philosophy is full of problems which persist from generation to generation, and are as far as ever from a solution by common consent, because no facts can be discovered which cannot be made to fit into either of the rival hypotheses. To cite only two clear cases, no facts have ever been found, and probably none can even be conceived, which would decide the dispute between determinism and indeterminism or between pessimism and optimism.

Similarly, Mill is essentially right when he says 'it is not a valid reason for accepting any given hypothesis that we are unable to imagine any other which will account for the facts'. For verification neither is, nor can it ever become, proof absolute. It is committed by its method to the formal flaw of Affirming the Consequent:[1] when it argues that a hypothesis is true because the facts that follow from it are observed, this lacks cogency, because these same facts, together with others still unknown, might follow still better from another hypothesis not yet formulated. And we cannot lay it down that a hypothesis shall only be accounted true when it alone can account for the facts. For even though it had vanquished all its rivals, which seems to be an impossibility in many fields of inquiry, it would still only hold its title as the *Rex Nemorensis* held his priesthood. It would still be liable to the challenge of new hypotheses to come, and sooner or later, it would become obsolete and go under. All that can be said, therefore, for a hypothesis which successfully holds the field is that, though it is not absolutely proved and cannot claim absolute truth, it can be accepted as true provisionally and *donec corrigatur*. And is not this, after all, the most sensible thing to do? It is to hold it *as good as true*, and to give *it* all the practical privileges of truth, and *us* all the assurance we practically need,

[1] Cf. my *Formal Logic*, ch. xxii, § 6.

without encouraging a groundless scepticism which appeals to an abstract possibility there is nothing to support. Only we reserve to ourselves the right of exchanging our hypothesis for a better, should one become available, and are, perhaps, a shade more likely to look out for it. But what harm is there in that ?

§ 6. Dr. Whewell's extensive knowledge of the sciences and their history could not but exercise a beneficial influence on his views about the logic of science. Accordingly, (1) he is distinctly right, as against Mill, in emphasizing the 'colligation of facts' by a *new* conception as the really creative act in scientific induction; and if Mill, instead of 'reflecting' on and elaborating established sciences, had attempted to organize a nascent science just emerging out of chaos, or to investigate a pseudo-science, he would soon have cured himself of the comfortable belief that the formation of a conception capable of apprehending the facts was any man's job, and quite a secondary achievement compared with the collection of the facts themselves.[1] Similarly, if our Hegelizing logicians had considered such problems of real knowing, they would not be so satisfied to regard the exposition of the powers and virtues of 'the Universal' as the whole duty of logic, but would perceive that the mere existence of universals was not of the slightest value to the sciences so long as nothing was said about the processes of contriving and selecting one that was relevant to, and operative on, the particular problem engaging scientific attention, and so long as the difference between the wrong and the right universal was held to be no concern of logic's. They might then perceive that, no matter how exalted was the metaphysical dignity ascribed to Universals, the universals in actual use are human conceptions, and that every conception *in its application* to the *data* is a *hypothesis*, which has to be *selected* from a number of competitors; also that the use of universals to analyse experience is itself a vast hypothesis, of which it is wiser neither to underestimate the risk nor to exaggerate the success.[2]

[1] For illustrations of this see below, § 14.

[2] The situation is only obscured, and not bettered, by talking about 'the *concrete* universal', which is to be found in nature, and contrasted with the 'abstract' universal, which is characteristic of human thought. For though it is true that we treat real things *as if* they were conglomerations of universals, and as represented by conceptions we have devised, and though, *when we have hit upon the right ones* which work satisfactorily, we are enabled to predict and control the course of events (for many purposes), we are not on this account

(2) Whewell concerns himself with the *discovery* of the right conception which colligates the facts. It demands 'sagacity' and "commonly succeeds by guessing; and this success seems to consist in framing several tentative hypotheses and selecting the right one".[1]

Here the mention of sagacity rules out the suggestion of any formal or mechanical or inevitable way of arriving at truth. Mental activity in the form of guessing is recognized, and the freedom to guess, and take the consequences, seems to be conceded. This is to rely on the consequences of hypotheses in the process of verification to sift them, and to establish their value *a posteriori*. The only point to which exception need be taken

entitled to assert that things really are such as we *feign* them to be. Now 'the concrete universal' is precisely a phrase for covering up this logical gap. It is intended to beg the question of the real validity of our thought, and to conceal the risk we take in our conceptual analysis of experience. It takes for granted what is really the hypothesis to be established, viz. that the relation of individual beings to the 'natural' kinds, in which we find it convenient to group them, is essentially the same as that of our conceptions and their exemplifications. That is, it tries to validate the claim of our thought to yield an accurate representation of the real by affixing the same label, 'universal' both to the real and to our method of manipulating it. Yet it is clear that there are great differences between the two cases. Thus the bond of unity which holds together a 'natural kind', if and in so far as we can really trace it *in rerum natura* and cannot be driven to confess that it is merely a convenient grouping for human purposes, is certainly very different from the connexion between the examples of the same conception. In a natural kind, like the 'species' of biology or the 'elements' of chemistry, the generic identity of its members appears to be constituted by a common descent and to be arrived at through a process of generation (or some analogue), in which an initial unity is progressively differentiated and dissipated by individual variations. In an 'abstract universal', the unity is arrived at by a *feat* of the mind which sets aside the recognized differences between cases as *irrelevant* (or non-significant, or unessential). In view of this and other differences it is not easy to justify the application of the same name 'universal' to both cases, especially by a logic which professes the ideal of 'the whole truth and nothing but the truth', and so debars itself from discarding *any* differences as '*irrelevant*'.

The original form of the hypothesis of real universals, as it appeared in Plato, was more modest or more honest. For Plato, while fully convinced of the validity of universals and of their superior reality, did not ignore the recalcitrance of phenomena. He refused indeed to regard it as something *more* than the universal, which alone had true being; it was condemned as a detraction from Being, as a taint of Not-Being, which corrupted the Being of the sensible world into Becoming, and delivered it over to change and decay. But he never attempted to slur over the discrepancy between the ideal and the sensible by a mere label.

[1] *Novum Organon Renovatum*, p. 59.

is the implication that there is *one* hypothesis which can be treated as *the* right one and that this is available for selection. Instead of ' the right one ' we should read ' the *best* one '.

(3) Creativeness in inventing hypotheses, and sagacity in selecting among them, are so essentially the secret of scientific discovery that this error is hardly serious, especially as Whewell is careful to note that "to try wrong guesses is with most persons the only way to hit upon the right ones".[1] He is emphatic (p. 80) that "the framing of hypotheses is for the inquirer after truth, not the end, but the beginning of his work", and that he must be "diligent and careful in comparing his hypotheses with the facts" (ibid.). "The discoverer has thus constantly to work his way onwards by means of hypotheses false and true" (p. 83), continually to test them. As such a test Whewell mentions the power to predict results, which shows that "the hypothesis was valuable, and, at least to a great extent, true" (p. 86). Where such predictions extend to cases of "a kind different from those which were contemplated in the formation of our hypothesis" and a "consilience of inductions" occurs, the conviction of the truth of a hypothesis becomes irresistible.

(4) Whewell duly emphasizes the necessity of taking risks. "Advances in knowledge are not commonly made without the previous exercise of some boldness and licence in guessing" (p. 79). He condemns the "barren caution which hopes for truth without daring to venture upon the quest of it" (p. 126), and does not even "gain its end, the escape from hypothesis" (p. 127).

§ 7. From Dr. J. Venn may be culled not a few valuable hints about the nature of Hypothesis.[2] (1) While distinguishing hypotheses from guesses by the seriousness and importance of their subject, he identifies it, or ' Supposition ', as an *attitude*, that of putting forward a conception or mental picture tentatively and doubtfully, in the hope that it may turn out to be true.

(2) He traces back "this important and far-reaching procedure" to "the necessities of daily and primitive life", to "that very familiar state of things in our practical life in which we are in doubt between two or more alternatives". It is this situation which generates "the use of the particle ' *if* ' : in other words, the use of the hypothetical proposition". But "starting from this

[1] *Ibid.*, p. 79. [2] *The Principles of Empirical Logic*, chap. xvi.

narrow practical application, the same form has developed and adapted itself to far wider uses of science and of fancy ".

(3) So it can be used to unravel a historical problem as well as to forecast an action, and so in its 'constructive' (or rather *reconstructive*) use becomes " a method of inquiry or investigation ".

(4) " Once started on the career of framing hypotheses . . . we find it difficult to stop." Hypothesis goes on to embrace both Analysis and Synthesis, because it employs itself in separating wholes into parts, and reconstituting them in new wholes.

(5) Even if Hypothesis be taken more narrowly, no answer can be given to the question—What hypotheses are allowable in science ?—without reference to the *purpose* of the hypothesis. "The cautions against rashness in framing hypotheses" are on a par with such " sanitary advice as the injunction not to over-tire oneself ", and though " the presumption must of course always be against the man who advances a really novel hypothesis ", it " should not take the form of denouncing him for making it ". " All that is wanted is not restriction of the hypotheses that are made, but only reticence in such as are published or declared."

All this is excellent, as is Professor Carveth Read's comprehensive regulation that no hypothesis "is of any use that does not admit of verification (proof or disproof) by comparing the results deduced from it with facts or laws. If so framed as to elude every attempt to test it by facts, it can never be proved by them, nor add anything to our understanding of them ", and his inferences that "to be verifiable an hypothesis must be definite", but that " except this condition of verifiability, and definiteness for the sake of verifiability, without which a proposition does not deserve the name of an hypothesis, it seems inadvisable to lay down rules for a 'legitimate' hypothesis ".[1] It is, of course, sound sense to demand that a hypothesis should have a definite meaning, but strange that it should be necessary to insist on this. However, no one who has noted how rarely logical doctrines are applicable to actual knowing will dispute the wisdom of this stipulation.

§ 8. Our critical survey of what is said about Hypothesis by the different sorts of logicians should not only have given us some idea of the problems involved, but also have paved the way

[1] *Logic*, pp. 269, 270.

for a more constructive treatment. This will naturally start from a consideration of Hypothesis in its most comprehensive sense, as a mental attitude or distinctive activity. For it is here that the primary difference lies between enunciating a fact and entertaining a hypothesis. In the former case alone is a (formally absolute) claim made to propound a truth that is a reflection of reality ; in the latter, no such direct relation to reality is alleged. This claim may, of course, be false, and indeed often is, but when it is made, not *pro forma*, but with full conviction, it seems to carry with it a feeling of assurance, highly valued by dogmatic minds, and a pretension to finality, which is calculated to close all discussion of the question. Or, more commonly, it means that nothing questionable has yet been discovered, and that critical reflection on the apparently given 'fact' has not yet begun. As a rule it is the attitude of the positive, prosaic mind, for which all is 'fact' or nothing at all, which does not discriminate between 'facts' and appearances and can literally *see* nothing but facts, and is incapable of illumining and extending them by the play of fancy and imagination, because it shies at questions and detests doubts. This is a very useful type of mind for many mundane purposes, though it is not really adequate to cope with a world full of deceptive appearances, in which everything genuine can be imitated, and everything is capable of *camouflage* and mimicry, for defence or aggression: for the advancement of knowledge, however, its use is very limited, and as a type of mind it is normally to be put on a *lower* intellectual level than that which is able to function hypothetically. It is also to be regarded as an *earlier* type, though, as Dr. Venn pointed out (§ 7), practical life must very soon have given men occasion to develop their imagination, to assume the attitudes of inquiry and doubt, to consider alternatives and to choose among them.

The mental attitude which entertains hypotheses on the other hand, and can take 'fact' as hypothetical and possibly unreal, means an intellectual revolt against mere givenness. It has become critical of appearances, and has partially freed itself from the oppression of brute fact. It meets reality with an active response, and does not merely submit to whatever comes along. It feels free to anticipate reality by its guesses, to question it, to experiment, to distrust and doubt appearances, to rearrange the world, at least in thought, to *play* with it, and with itself. For

Hypothesis is a sort of game with reality, akin to fancy, make-believe, fiction, and poetry. In the hypothetical attitude 'facts' have ceased to be accepted at face value, to be just fact, and become capable of being *symbols*, whose suggestions are more important than their bare existence. Whatever they may be in reality, they are no longer *fixed* in the mind, but afloat; not being fixed ideas, they can be moved about and played with. But, like games in general, this play has a serious function. By loosening the connexion between what the real is (or seems) and what we think about it, it enables us to think it other, and *better*, than it is; and so, guided by our hypotheses and 'ideals' (which are postulates), we can set to work to *make* it other, and better, than it was.

Moreover, it is by this hypothesis-building habit that science touches poetry on the one side, and action on the other; for it is akin to both. (1) The play of fancy and the constructive use of the imagination reveal the creativeness of human intelligence; by their use the scientist becomes a 'maker' like the poet, and surprises the secrets of nature long before he is in a position to prove his hypotheses. In both cases human power is limited, often wofully enough; but it suffices to infuse into the *routine* of nature a breath of *novelty* that keeps it fresh and capable of progress. (2) Yet on the other side, this hypothetical attitude mediates between thought and action, and helps to break down the superficial distinction between the theoretic and the practical. It drives the scientist out of the purely receptive attitude, and makes him a *doer*. For to entertain a hypothesis is to hold a mental content hypothetically, and this is to hold it experimentally, which, again, is to operate on it and to manipulate it. And this is surely to *act* on it, though the alterations we enact by our thought-experiments cannot be rendered as visible as the changes wrought in the contents of a test-tube by a chemical experiment. But their magnitude may be estimated by comparing the crude *data* from which a science starts with the scientific facts into which they are transformed. They differ usually far more than do the rainbow and the drops of water to which it is traced.

§ 9. It will be evident from the above sketch that logicians have good reason to fight shy of the hypothetical attitude. It is not a thing to be easily caught and tamed by rules and regulations. It is not a procedure that lends itself to formal description. Its reasoning cannot possibly be made to seem ' valid '. Nor can it be made ' safe '. It inevitably takes risks, and so must be condemned as ' dangerous '. Of course, it *is* dangerous, with all the danger inherent in life and action. But does this justify its condemnation ? Only, surely, if there exists a safer and a better way for logic to pursue. But logic can hardly any longer cling to two of its favourite delusions which prompted it to condemn Hypothesis. The first is that formally valid reasonings are *possible* ; the second that they are *safe*.

Now to the first it must be objected that *no* reasoning can be ' valid ', because none can be put in a form so valid that it will actually be valid in *all* its applications. No perfection of syllogistic form, no super-excellence of any universal ' essence ', will yield an absolute guarantee that the conclusions deducible from them will not on occasion *fail in fact.* The universal rationality of man is quite inadequate to guarantee actual rationality in the case of every lunatic, or even of any one whatsoever under stress of circumstance. The universality of quadrupedality as an attribute of the dog will not prevent a particular dog from having to hop about on *three* feet ; nor, to take graver instances, will the complete success of the law of gravitation in accounting for astronomic movements guarantee its application to the intramolecular motions of atoms and electrons. In short, every ' law ', formula, or ' universal ', runs a risk of failure when it is applied to a new case ; and is not every *further* case to some extent a *new* one ? In all ' deductive ' proof something essential is *added* to the conclusion it anticipates, by the observation that it also actually occurs. It is only empirically that we learn whether or not the laws and ' universals ' we are employing apply to the ' case ' we have selected.[1]

As, in this way, we can *never* be sure in advance that a formally valid reasoning will not in actual use develop a defect we may classify, alternatively, as an ' Ambiguous Middle ' or a ' Fallacy of Accident ',[2] it follows that *no* formally valid argument has a right to regard itself as (absolutely) safe. And it may further be pointed out that we cannot avoid running risks by sitting still and refusing to budge. For if we will not take any ' unsafe ' step, we shall assuredly miss the new truth that may be won by specula-

[1] *Formal Logic*, chap. xvi, § 10.

[2] Cp. *Formal Logic*, p. 200.

tion, and undoubtedly run the risk of remaining stuck in whatever error attaches to our actual 'knowledge'.

It is no use quarrelling, therefore, with the hypothetical attitude, and the risks it involves. It is an attitude life, logic, and science all require of us, and there is no way of avoiding risks in any of them. If our consciousness of this becomes oppressive, we can bethink ourselves of the consoling thought that life also requires us at times to *forget* our hypotheses, or to take them as *practical certainties*, when we have to act on them.

§ 10. When Hypothesis is taken in this very comprehensive sense it becomes clear why it is not to be confined to a single chapter of inductive logic, nor to be satisfied even by making much of the hypothetical judgement and interpreting universal propositions as 'hypothetical'. Allied with the volitional attitude of postulation, Hypothesis reveals the essential creativeness of human intelligence. It pervades all mental life and penetrates every logical notion. Accordingly its workings can be traced throughout every topic of logic. "If every 'fact' rests on selection and involves an experimental analysis of the given, and if every 'law' is provisional and in need of confirmation, it follows that there is something hypothetical about every act of thought, and that the distinctions between fact, interpretation, theory, hypothesis, and guess are plastic and fluid." [1]

Similarly, it may be shown (1) that the syllogistic (or any other) form of inference is nonsense unless the premisses are taken as hypotheses, and (2) that in their scientific use conceptions *are* hypotheses, and that in the absence of good hypotheses, *alias* 'suitable conceptions', a science makes no progress.

§ 11. Ever since Aristotle, logicians have been assuring the students of their science that scientific reasoning, in order to demonstrate its conclusions, must start from *true* premisses. Yet from the first this dogma has involved them in the most serious embarrassments. It became necessary to explain how the sciences came by their requisite supply of initially certain premisses; but the doctrine of self-evident and intuitively certain principles devised to supply this need, was philosophically untenable,[2] and flagrantly false to the facts of scientific history. The sciences do

[1] *Formal Logic*, chap. xxii, 3.

[2] *Formal Logic*, chap. xviii, § 3; *Scientific Discovery and Logical Proof* in these *Studies*, vol. i, p. 243 f.

not *start* with an assurance that their principles are true, but gradually acquire it as they progress, select their principles, develop their conceptions, and ' scrap ' their failures.

Moreover, even after this difficulty was supposed to be surmounted, it was still necessary to explain how a conclusion deduced from premisses known to be true could have any cognitive value. The form of ' demonstration ' soon incurred the charge that its conclusion was either already included in the truth of the premisses, or else begged the very point it claimed to ' prove '. The syllogism was either a tautology or a *petitio principii*. If in laying down the principle that ' all men are mortal ' we *have* included the case of ' Socrates ', the conclusion that ' he is mortal ' is already known and is nothing new, so that the ' demonstration ' is superfluous : if we have *not*, we are not entitled to assume that he is, and the demonstration begs the question. After much beating about the bush, the ' orthodox ' logics have of late agreed to say that the way out of this deadly dilemma is so to interpret the major premiss as to disclaim for it any assertion about individuals, and to take it as the enunciation of a law of nature or a ' universal ', which by the minor premiss is applied to the case in hand. But, as I have shown,[1] this interpretation is no way out of the difficulty. It merely requires the dilemma to be slightly re-worded. The interpretation in question does indeed avoid a *verbal* reference to particular cases in the major premiss. But it does not escape thereby. For it covertly assumes both that a principle or universal *can* apply to cases without going wrong, and that it *does* so apply to the case in point. When this assumption, however, is questioned, it collapses, precisely as did the other interpretations which are now abandoned. Thus, however strenuously we insist that *all men are mortal* really means *humanity entails mortality*, and *Socrates is a man* really means *humanity is predicable of Socrates*, we cannot declare him mortal on the strength of these premisses, unless it is true and certain that they apply to him for the purpose (whatever it is) of the particular argument. We cannot rely on the verbal identity of the middle term, but have to show that humanity is predicated of Socrates in the minor premiss in precisely the same sense as it entails mortality in the major. And about this a question can always be raised.

[1] *Formal Logic*, chap. xvi, § 9.

Hence, as was pointed out in § 9, we can never know with absolute certainty whether our syllogism is really sound and our conclusion will in fact follow. But even if we could know this, and know it in advance, we should only have proved that we *did* know this. If we knew then that human mortality covered the case of Socrates, the 'inference' about his mortality proved nothing new: whereas, if we knew it not, but nevertheless asserted a conclusion which implied it, we should simply have begged the question.

There is, however, an easy and reasonable way out of the whole puzzle. Instead of taking the premisses as *certain*, we may take them as *hypothetical*, and construe the whole 'demonstration' as an experiment with, and test of, their truth. The conclusion will then be conceived as an anticipation of experience, and its empirical verification will be a confirmation of the truth of its premisses. It then becomes easy to conceive the whole argument as what it always must be in an actual context, viz. as an incident in an inquiry relevant to some doubt or dispute. And then there will be no ground for any charge of *petitio* or lack of novelty. For we shall not have *assumed* that the conclusion is certain to come true; we shall merely have prepared ourselves for the *news* of its coming true, if it should do so, and have enabled ourselves to *explain* it, by deducing it from premisses of which it helps to augment the credit.

This is by far the most natural and intelligible interpretation of 'demonstration', and the only one that does not discredit its very form. There should be all the less difficulty about accepting it, seeing that the alleged need for true premisses is an entire delusion. There is no difficulty whatsoever in reasoning from any sort of premiss, false, hypothetically true, or more or less probable. The reasoning proceeds just as well whether the premisses *are* true or merely *supposed* to be. And it is quite possible to start an inquiry with doubtful, and to conclude it with assured, premisses. We have merely to admit that the empirical truth of conclusions reacts on the premisses from which they were drawn, and strengthens their claim to truth.

That hypothetical reasoning is quite feasible was admitted by the first philosopher to recognize hypotheses *eo nomine*. Plato, in the *Republic*, Book VI, has a brilliant sketch of the relations of science and metaphysics, in which he declares that scientific principles are essentially 'hypotheses' in need of proof. This proof they can only get if they can be derived from an 'unhypothetical principle' which 'takes away the hypotheses', by attaching them to the self-evident supreme truth of the 'Idea of the Good'. Until they have been thus derived, they are neither certain nor intelligible, and so it is the function of the philosopher thus to derive them. He proceeds, however, from these same 'hypotheses' in his ascent to the First Principle. He treats them literally as 'hypotheses' (=things *laid down*, supports to stand on), i.e. as 'stepping-stones and starting-points' he traverses on his way to the point at which the all-embracing, all-explaining, self-proving, and self-evident Good reveals itself to his intelligence, and he, when he has grasped it, is guided back by it to the lowest limits of the intelligible world, deducing and assuring every step, and so the original 'hypotheses', by its means.

Now, up to a point, this is an excellent account of the procedure by hypotheses. They are essentially 'starting-points and stepping-stones'. It is true also that scientific principles are initially hypotheses, but do not remain so. Where alone Plato errs is in assuming that they can be converted into full truths only by a metaphysical deduction. The whole cogency of his argument depends on this assumption that there is no other way of establishing a principle. But this is false in fact. Principles can also be established *empirically*, by the success of their working. Hence the sciences do *not* depend upon metaphysics, but upon their own achievements in modifying, improving, and correcting their initial assumptions, until their hypotheses grow into theories, consolidate into 'facts', and become adequate to the problems of each science. A 'hypothesis', then, can *change* its logical character in the natural progress of a science.

§ 12. We have already suggested (in § 4, note) that the analysis of the flow of events by means of conceptions may be regarded as a great cognitive hypothesis, and have seen (in § 7) that to suggest a 'suitable conception' for the apprehension of a subject, which is a really creative act of thought, must be at first hypothetical. Both points, however, need some elaboration and illustration.

The first is a statement of the assumption that reality is knowable, and can be known by us. This assumption is of course indispensable, and is made in every inquiry. But it is a mistake to represent it on this account as an '*a priori* necessary truth'.

For it is only made *methodologically*, and as a working hypothesis. Not, that is, as a method which is *certain*, but as one which is *worth trying*. If the course of events persistently conspired to defeat it, and we were in consequence unable to use it to *advantage*, we should be forced either to devise another assumption or to despair of 'knowing'. Actually it is successful enough to be deemed a legitimate *postulate* and a *methodological assumption* so convenient that we are resolute to use it. But in spite of its merits and attractions, we can still perceive that it has not wholly lost the hypothetical character it started with.

§ 13. The second point demands fuller treatment. To show that conceptions are, in scientific use, hypotheses, two courses are open to us. We may either argue from the general character of their use or may illustrate concretely from the specific difficulties of various sciences. The former method will prove convincing, if we stop to consider the problem of how among the mass of conceptions, good, bad, and indifferent, appropriate and futile, which all appear to be, in a general way, relevant to a particular problem in a science, a selection is made of those which are actually tried. Clearly, this problem cannot be passed over—as it has been. The logician ought to warn the scientist that he must make a selection, and that if he selects wrongly, he will fail. Also, that it is his duty to select the *best* conceptions, if he wishes to advance his subject. Hence it is clear that since any conception used is selected, and may be selected badly, or not so well as it might be, the scientist is always running a risk of error and failure, and acting on the *hypothesis* that he *has* picked the right, or rather the *best*, conceptions for his purpose. As this hypothesis is never fully proved—and indeed is always in the long run disproved by the progress of the science—it follows that the conceptions of all the sciences should be entertained hypothetically.

§ 14. The illustrations of the second method are potentially infinite. For every science has been held up, at some stage or other of its career, by the lack of conceptions that would give it a real grip of its subject-matter and enable it to analyse it and to deduce consequences which could be verified in fact. In some sciences this condition has been chronic, but in others it has only proved a temporary obstacle.

At the moment such an obstacle appears to exist in what has been, for the last three hundred years, a very progressive science, physics. The following account of its trouble is taken from a lecture on 'Radiation and the Electron' by Professor R. A. Millikan of Chicago University, which appeared in *Nature*, Nos. 2534 and 2535 (May 1918). He begins by saying that "recent developments in the domain of radiation are of extraordinary interest and suggestiveness, but they lead into regions in which the physicist sees as yet but dimly—indeed more dimly than he thought he saw twenty years ago"—for the reason that "experiment has outrun theory, or better, guided by erroneous theory, it has discovered relationships which seem to be of the greatest interest and importance, but the reasons for them are as yet not at all understood". Hence, "one of the great unsolved problems of modern physics". It concerns the theory of light, and arises out of a phenomenon, by no means rare in science, and common in philosophy, viz. that a hypothesis, which at one time seemed to have been defeated and destroyed, is revived by subsequent developments, and resumes the contest.

In this case the corpuscular theory had been progressively worsted by the ether-wave theory, because, unlike the latter, it could explain neither the facts of interference, nor the wireless waves, nor the greater speed of light in air than in water, and its speed's independence of the source of the light: so by the end of the nineteenth century it appeared to be dead, while its competitor had not only explained all the known facts, but repeatedly predicted new ones.

Then its troubles began. It was found that the X-rays detached negative electrons (or 'corpuscles') from about one in a thousand billion of atoms in the space they traversed—why not from the rest? Similarly, ultra-violet light falling on a metal was found to expel negative electrons. Further, the energy with which these electrons were projected was found to be the same whatever the intensity of the light that liberated it, and the same was found to hold also in the case of X- and γ-rays. These new facts did not fit in with any wave-theory, but were suggestive of a corpuscular theory. "For if the energy of an escaping electron comes from the absorption of a light corpuscle, then the energy of emission of the ejected electron ought to be independent of the distance of the source, as it is found to be, and furthermore, corpuscular rays would hit only a very minute fraction of the atoms contained in the space traversed by them. This would explain, then, both the independence of the energy of emission upon intensity and the smallness of the number of atoms ionised."

Sir J. J. Thomson first tried to mediate between the two conceptions by "assuming a fibrous structure in the ether and picturing all electro-magnetic energy as travelling along Faraday tubes of force conceived as actual strings extending through all space". This hypothesis, mythical as it sounds, and unable as it was to explain the facts of interference, was carried further by Professor Einstein in 1905. He "assumed not only that the energy emitted by any radiator kept together in bunches or *quanta* as it travelled through space, but also that a given source could emit and absorb radiant energy only in units which are all exactly equal to $h\nu$, ν being the natural frequency of the emitter, and h a constant which is the same for all emitters". It followed from this assumption that "the energy of emission of corpuscles under the influence of light would be governed by the equation $\frac{1}{2} mv^2 = v\,e = h\,\nu - p$, in which $h\,\nu$ is the energy absorbed by the electron from the light-wave or light *quantum* . . . p is the work necessary to get the electron out of the metal, and $\frac{1}{2} mv^2$ is the energy with which it leaves the surface". At the time when this formula was constructed there was no experimental evidence to support it, though it had been suggested by some experiments of Professor Planck's; but it has now resulted from ten years of laboratory work that "in the discharge of electrons by light this equation of Einstein's seems to predict accurately all the facts which have been observed". Hence "it must certainly be regarded as one of the most fundamental and far-reaching of the equations of physics . . . for it must govern the transformation of all short-wave-length electro-magnetic energy into heat energy".

In spite of this success, however, "the semi-corpuscular theory, out of which Einstein got his equation, seems to be wholly untenable, and has, in fact, been pretty generally abandoned, though Sir J. J. Thomson and a few others seem still to adhere to some form of ether-string theory". Two objections are fatal to it: it cannot account for the facts of interference, and there is positive evidence against any fibrous structure in the ether. The 'oil-drop' experiments have shown the granular structure of electricity by discontinuous changes in the velocity of an electron when the charge in it is varied while the electric field remains constant; but the converse experiment shows the *lack* of a discontinuous change, when the charge remains constant and the field is varied, and has disproved the fibrous structure of the field. "Despite, then, the apparently complete success of the Einstein equation, the physical theory of which it was designed to be the symbolic expression is found so untenable that Einstein himself, I believe, no longer holds to it, and we are in the position of having built a very perfect structure, and then knocked out entirely the underpinning, without causing the building to fall. It stands complete and apparently well-tested, but without any visible means of support."

Professor Millikan then proceeds to consider whether the explosive emission of energy by atoms in definite *quanta* cannot be ascribed to a gradual absorption of energy inside the atom. "It is necessary to assume, if the Thomson-Einstein theory is rejected, that within the atom there exists some mechanism which will permit a corpuscle continually to absorb and load itself up with energy of a given frequency, until a value at least as large as $h\nu$ is reached. What sort of a mechanism this is we have at present no idea." It entails "a type of absorption which is not due either to resonance or to free electrons. But these are the only types of absorption which are recognized in the structure of modern optics. *We have as yet no way of conceiving this new type of absorption*[1] in terms of a mechanical model". Still, this type of explanation, "though as yet very incomplete, seems to me to be the only possible one. . . . Yet the theory is at present wofully incomplete and hazy. About all we can say now is that we seem to be driven by newly discovered relations in the field of radiation, either to the Thomson-Einstein semi-corpuscular theory, or else to a theory which is equally subversive of the established order of things in physics". Manifestly, then, the path of physics is here blocked, until some *new* conception, suggested, it may be, to some creative mind by some further experiment, pronounces its 'Open, Sesame!'

§ 15. As a good example of a science hung up for ages, in a manner strongly suggestive of a lack of appropriate conceptions, we may consider the sad case of Psychology. Here we have a science of apparently enormous potentialities and pretensions, of universal interest, of great antiquity, upon which many

[1] Italics mine.

generations of thinkers have lavished much time, ingenuity, and enthusiasm. Yet disappointingly little has been made of it. After more than two thousand years of strenuous cultivation, it still has no laws but only technical terminologies, no *consensus* about methods and principles but a swarm of discordant 'schools', no definite limits and no assured territory but far-reaching claims and perpetual border-wars with all its scientific neighbours. It has 'descriptions', but none adequate to the subtleties and shades of the processes they describe: nor has it any real control of the mind and the power to predict its operations. Indeed, it does not seem quite sure even that it has got a mind, any more than that the 'soul' exists, from which it draws its name. For one of its latest fashions is seriously endeavouring, under the flag of 'Behaviourism', to construct a psychology from which the conception of consciousness has been eliminated altogether.

The obvious explanation, to which this condition prompts, is that Psychology has not so far succeeded in getting hold of appropriate conceptions. A glance at those it has tried to use would seem to confirm this suspicion. They seem to be a very 'scratch' lot, borrowed from all sorts of quarters, ill-defined and ill-adapted to any sort of scientific testing, and never reducible to systematic order and logical coherence, except at the cost of suppressing half the facts they are supposed to account for.

For example, it would seem that Psychology has never been able to make up its mind even on the fundamental issue whether the mind is one or many. From the first (Plato), eloquent asseverations of the indiscerptible unity and simplicity of the mind or 'soul' or 'self' have been jostled by elaborate classifications of its division into 'faculties'. Psychology has always been fond of faculties, with a foolish affection, and after every official disclaimer of their value may always be caught reinstating them more or less covertly, though it has never been shown how any mechanism of faculties, whatever its complexity, could explain even the simplest mental operations,[1] or even how any 'faculty'

[1] The advocates of 'faculties' always have finally to admit that the analyses effected by their aid are illusory, because actual psychic life always actuates them all. Instead, however, of regarding this fact as a confutation of their method, they only feel prompted by it to descant on the superior unity of the mind. But if the mind is so perfect a unity that none of its faculties will ever operate singly, what was the use of feigning a classification which divided it into a plurality of radically different faculties?

could be more than a futile *ex post facto* replica of the function it professed to explain. No doubt the source of these embarrassments is that, for *various purposes,* it *is* imperative to treat the soul, now as one, and now as many. But Psychology has never been able to make clear what is the scientific case for this variety of purposes, or how they are to be brought into accord with each other. A sympathetic critic must conclude that no conception suitable for expressing simultaneously both the unity and the plurality of psychic life has yet been devised.

Again, he will not be able to resist the conviction that the analogies from the external world, with which, as a nascent science, Psychology was forced to commence, have served it ill. The analogies of the 'spirit' and 'psyche' (soul) with a *breath* (*spiro,* ψυχᾶν), and of the mind or soul (*animus, anima*) with a *wind* (ἄνεμος) were not really suited to express the uniqueness of its nature. All the comparisons of the soul with an external '*thing*', which prompted primitive man (logically enough) to tell fairy-tales about the wicked giant who kept his soul for safety's sake in the guise of a bird in a cage, or locked up in a box, in the strong-room of a castle in an inaccessible island in a remote lake well out of harm's way, and scientific psychologists to fabricate, less picturesquely, pseudo-sciences about the 'combinations', 'complexes', 'complications', 'associations', &c., of fictitious 'elements' that (unlike the giant's 'soul') always eluded capture, have proved fallacious and unworkable. They failed because soul-structures are not permanent, like rocks and houses, or even trees.

But even if we learn, from these failures, more wisely to express our psychological hypotheses in *functional* terms, and to speak of 'processes' and 'attitudes' and 'activities', instead of 'soul-substances', 'sensations', and 'ideas', we are still only groping towards an adequate expression of the subject's singularity. The same objection holds more or less against all the conceptions, schemes, and technicalities of psychology—*they do not work.* But this is no reason why the psychologist should despair. He should recall rather how long and fruitlessly sciences like physics have had to struggle before they arrived at their present scientific rank, and should redouble his efforts to bethink himself of something new and suitable, opening his mind to suggestions from every *imaginable* quarter, and trying the hypotheses which occur to him, not by his antecedent prejudices, but by the success of their applications. Only there is *one* condition he must not overlook, in his satisfaction with the comprehensiveness, symmetry, and beauty of any system of psychology he may devise: whenever he has discovered or invented a set of conceptions which seem to him adequate to his subject, he must not shrink from proving the correctness of his *analysis* of the mind by a synthesis, and must show himself capable of *reconstructing his own*. For no scientific doctrine must be allowed to evade the test of application.

§ 16. A just appreciation of the functions of Hypothesis in the widest sense, and of the value of the attitude of supposal, will naturally conduce to greater sympathy with the non-scientific forms thereof, as well as to leniency in regulating those which are employed in scientific enterprise. For we shall then see that it is quite right and proper for the products of this creative attitude to differ according to the purpose for which they are employed and the nature of the subject they concern.

As for the narrower sense of Hypothesis, its function in the service of science is to *think the new*, whether it appears as an unprecedented fact, or demands the formation or reformation of conceptions. So long as it serves its purpose, much may be forgiven it. It need not be 'safe' or 'valid'; it need not accord with old analogies and time-honoured prejudices, nor be complete, and free from 'difficulties', or even 'contradictions'.[1] It need not have confuted all its rivals, in order to be accepted as the *best*; nor need it last for ever, and defy change.

The sole essential of a scientific hypothesis is that it should *work*—relevantly of course to the problems of the science. This postulate, however, naturally articulates itself into three.

(1) A scientific hypothesis must have a *definite meaning*, as was said in § 7. This does not mean that its terms should be conceived as rigid and unchangeable, but that its meaning must at any rate permit of deductions being drawn from it, and put to the test of experiment. In other words, a hypothesis which cannot be applied means nothing—scientifically. Such hypotheses are rare in the stricter sciences, but common enough in ethics and politics, and rampant in philosophy, where theories have a strong tendency to assume an inapplicable form, as a protection against criticism. What, for instance, is the meaning of a purely formal

[1] For apparent 'contradiction' as an incident in the growth of meanings, cf. these *Studies*, vol. i, p. 241.

law of duty, like Kant's 'Moral Law', which cries out against the degradation of every sort of application? What is the meaning of nearly all the catchwords of politics, of 'democracy', 'autocracy', 'liberty', 'equality', 'social justice'? What is the function of realities and 'essences' that are unknowable, of 'substances' that cannot appear, of qualities that are 'occult', of designs that are 'inscrutable', of truths that are 'unverifiable'?

(2) Inapplicable hypotheses, then, are meaningless just because they evade testing, and thus defy the *second* requirement of a scientific hypothesis, that *it must be such as to admit of definite tests,* which determine by their issue whether it is true or false. This proviso condemns, not only hypotheses which decline to be tested altogether, but also those which are *too* accommodating. A hypothesis which professes to explain everything, but is compatible with anything, is not scientific: for it, too, refuses to take the risk of refutation. Such hypotheses were once prevalent—in mediaeval science any event might be ascribed to 'witchcraft' or the Devil's agency—and they still linger on in philosophy, where 'the Absolute' is a palmary example of such illusory explanation. The *testing* of hypotheses is their *verification,* and is, in principle, an *unending* process. But it is needless to regulate it further by logic, because both the nature of the tests, and their value, are relative to the problems to be solved, and the actual investigators of a science alone are the competent authorities to decide these matters.

(3) The only condition the logician is entitled to suggest is that the hypothesis accepted should be the one that *works best,* i.e. better than any alternative within the purview of the science. He can urge the true scientist to be ever on the look out for the best hypothesis, and not to be obstinate in clinging to old hypotheses, to which he has become attached, when the new are better. He should, therefore, cultivate the qualities of mind which will enable him to act thus, and, in the interests of truth, try to become as tolerant and open-minded as his idiosyncrasy permits. Nor should he try to suppress the alternative hypotheses by mere authority and to repress discussion.

§ 17. For inquiry demands an abundance of hypotheses. In the beginnings of a subject, especially, it is most important to keep a long and varied list of alternatives before the mind. For only so shall we have our attention called to the variety and intricacy of the facts, whose significance does not usually lie on the surface, and can to some extent guard ourselves against overlooking any relevant feature of our problem.

In the 'facts' alone we cannot put our trust. For (1) the facts are precisely what we have, and hope, to ascertain. The 'facts' we start with are what we have to examine, and need not be more than apparent; the 'real facts' have to be extracted from them by critical experiments and the hypotheses which guide them.

(2) It should be recognized also that the 'facts' are always *more or less relative to the hypotheses* which apprehend them, and in terms of which they are described. They take their colour from the hypotheses they serve. Violently antagonistic hypotheses will appeal to unrecognizably different 'facts', and to get an adequate supply of facts we may need a plurality of hypotheses.

(3) Yet the facts themselves, in cases where we can get them apart from hypotheses, are often amiable enough. They will display a charming *ambiguity,* and fit into several hypotheses with (approximately) the same facility. We have to labour therefore to overcome their flabbiness, and to find facts sturdy enough to withstand assimilation by hypotheses, and to lend themselves to 'crucial' experiments. It is precisely because we cannot find such facts that so many of the ultimate issues of philosophy cannot be decided.

For all these reasons, then, science should habitually reckon with a plurality of hypotheses, and eschew the sharp antithesis between 'the right' hypothesis and the many 'wrong' ones. Its concern is really with the *relative values* of the hypotheses in active service.

§ 18. We may conceive ourselves to have acquired in the foregoing discussions a clear conception of the function of the hypothetical attitude in scientific research, and to have liberated the latter from the severe restriction of the former in the name of logic. But it remains to reconcile the scientific attitude towards Hypothesis with the practical, and to show that they are not so discrepant as they are usually supposed to be.

At first sight it would seem that, however valuable the attitudes of doubt and inquiry, the methods of systematic questioning, the exploration and weighing of alternatives, the unending remodelling and reconstruction of beliefs, might be for the purpose of advancing knowledge, they were at any rate entirely unsuited to the life of action. For action demands finality and decision, a choice between alternatives, a firm stand on stable principles, a suppression of doubts, a fixity of purpose, a going on resolutely on a chosen path without looking back or looking round. It must fix its eye upon its aim undeviatingly, and resist the distraction of hypotheses.

All this is true, but it is not the whole truth. For if it were, the action would be *blind*. The action, however, which our life requires is *circumspect*. It always presupposes more or less consideration of alternatives, before a judgement is passed and a decision is taken. And this process occurs equally in action and in science. There is no radical antagonism between theory and practice.

That it is not *easy* to reach equal proficiency in both stages of successful action, or of successful inquiry, may cheerfully be admitted. The good theorizer is not necessarily the most ingenious experimenter or the most careful observer. So the narrow-minded fanatic often has more driving-power than the mind that looks before and after. But it not infrequently drives him too fast, on the road to destruction. As we cannot often get a genius 'four-square without a flaw', we have everywhere to allow for the defects of men's qualities. And the difficulty of combining these two conflicting attitudes is precisely one which life involves for us all.

It may also be admitted that the attitude of doubt, the entertaining of hypotheses, the holding of judgement in suspense, is irksome, and even intensely repugnant, to many minds. They demand *certainties*, and are eager for *assurances*. They think doubt torture. They shrink from risks, but, if they needs must take them, prefer to 'go it blind'. The traditional theories of theology, philosophy, and logic have always catered for these minds.[1] They have even pandered to their weaknesses. They have represented knowledge as proceeding from truth to truth, and not from error. They have claimed an *absolute* certainty, where all we have (or need) is certainty enough to live by. They have confounded doubt with sin and scepticism, and discouraged the attitudes of inquiry, experiment, and faith.

But these theories are profoundly inadequate, and the minds they flatter are neither the noblest, nor the best adapted to the

[1] Cf. The *Proceedings* of the Aristotelian Society, 1918, p. 267 f.

nature of the world. The opposite bias, which actually *enjoys* danger and courts adventures in the realm of thought, and has no desire to rest content with the established order, but rejoices in activity and plunges into movements, finding

> Life's treasure in an endless quest,
> And peace of mind in infinite unrest,

is no doubt too rare to extort recognition, and involves a renunciation of ultimate ideals which perhaps ought not to be lightly sacrificed. But it exists, and though *intellectual* courage is much rarer than physical, it, too, has its place in the development of the human spirit. And a spirit of *fortitude*, which, though it does not *revel* in risks, sees the necessity of running them, and is willing to accept them, should not be beyond the compass of a reasonable logic.

SCIENCE AND METAPHYSICS

By the late J. W. Jenkinson

This paper was read by my husband some years ago to a Philosophical Society in Oxford. He had at the time no thought of publishing it; but, acting upon the advice of his friends, I have now decided to do so. I am much indebted to Dr. R. R. Marett for very kindly reading over the manuscript and preparing it for the press. This is the paper to which reference is made by Dr. Marett in his Biographical Note prefixed to my husband's posthumous work, *Three Lectures in Experimental Embryology* (Oxford, Clarendon Press, 1917).

C. Jenkinson.

Πάντες ἄνθρωποι, says Aristotle at the opening of the *Metaphysics*, τοῦ εἰδέναι ὀρέγονται φύσει. A striving after knowledge is implanted in the nature of all mankind; and all philosophy, all love of wisdom, begins in a wonder which the child feels equally with the savage—a wonder which expresses itself first in astonishment and awe at the visible mysteries of nature that surround him, and afterwards in perplexity before the invisible secrets of his own soul. As the child grows into a man this vague bewilderment is gradually replaced by a definite method of inquiry controlled by the reason; and the same process is observable in the evolution of the knowledge of the human race.

The earliest philosophers of ancient times, such as Thales, turned their attention first of all to the material universe. But, owing partly to the entire want of a scientific method, and partly to the absorbing interest of the problems of practical life, the pursuit of the physical sciences soon fell into disrepute. Meanwhile, the need felt by the more earnest thinkers of supplying some ultimate reason for the maxims and conventions that formed the moral repertory of the ordinary citizen was instrumental in evolving a stupendous system of metaphysics; and the ablest philosophies of later days have found it difficult to go beyond it.

At the same time, the study of purely mathematical sciences led to the elaboration of the syllogism. This, however, was not an instrument for the prediction of fresh instances from general laws ascertained by rigorous induction. Hence it became later in scholastic and ecclesiastical hands an engine of tyrannous oppression, which sought to enslave the human intellect to dogmas derived from patristic premisses.

But in more modern times the old order of thought has reasserted itself. Under a kindlier fortune freed from the bondage of tradition, the sciences have been enabled to develop. By the use of induction, man proceeded to inquire into the operations of nature, into the meaning of those mysteries around him that were the first to attract the attention of the primitive philosopher. Such inquiries have proved so successful, while their pursuit has been so absorbing, that those who have prosecuted them have been often tempted to suppose that solely in the knowledge obtained by these methods lies the sum total of possible human wisdom; that in a connected view of all their results consists the true philosophy, the only attainable system of the universe.

Whether this claim be justifiable or not we shall presently have to inquire. For the moment let us confine our attention to attempting to answer the questions: What is meant by a science? What is it in modern science that has led to such brilliant results? and how far is there a single method common to all the sciences?

With regard to the purely mathematical sciences of geometry and arithmetic, it is obvious that in them the deductive method alone is employed in order to develop all that is contained in the universally acknowledged, axiomatic laws of space and number. These sciences, therefore, do not, of themselves, lead to any knowledge of such events as occur in space and time, or in time only—forming the subject-matter of the sciences called inductive, with which we are chiefly here concerned.

It may very possibly be objected that it is improper to speak of an object, such as a stone, as an event. For the stone seems at first sight the permanent possessor of unalterable properties. A moment's reflection, however, will show that even such a relatively simple object as a stone is, as truly as more complex objects, but a passing assemblage of atoms. As the mineralogist and geologist would tell us, these atoms have certainly not always been so combined, are even now imperceptibly separating from one another, and will in the future be scattered to all the ends of the earth, perhaps of the universe.

An object, then, may properly be styled an event, or better a concourse of events. Each event is determined by certain others that have preceded it, and determines in like manner those about to follow; being thus related to the past on the one side, and to the future on the other. Now it is the business of science to take those concourses of events which the ordinary man calls natural objects, and unravel the combination. It must separate as it were the constituent threads of the complicated plexus, and discover what events have determined those constituting the given phenomenon, that is to say, have always been found invariably and necessarily to precede them; and, if possible, also what events will in turn be determined by them. In the case of the stone, for instance, its position and shape, the minerals that compose it, and the chemical constitution of these have all been determined by certain definite events in the past; while all in their turn will help in determining other events such as sea-bottoms, mountains, or animals and plants in the future.

But, in order to give an intelligible explanation of an object, science must also, apart from this task of discovering what made it, and what it does, attempt an answer to the further question, What is it made of? There is, therefore, in every science an ultimate something, mass for instance, or the molecule, or the atom, or the living cell, but in all cases something, to which it is sought to reduce natural objects, and from which they are regarded as having been derived in accordance with the particular point of view of the science concerned. At the same time, every science has an ultimate conception, such as molecular motion, or chemical affinity, in terms of which it endeavours to express those particular changes in natural objects to which it pays attention.

It will thus be seen that science, in attempting to determine, in Aristotelian language, the material cause of an object, also proceeds to the discovery of the efficient cause, and vice versa; and that in both cases it is seeking for the antecedent necessary conditions of the existence of that object. When, on the other hand, we look for an answer to the question, What does the thing do? we are inquiring not into the causes but into the effects of the object—in other words, into those phenomena which invariably and necessarily follow the given phenomenon, and are, in short, its properties or functions. When we say that an object has such and such properties, we mean simply that under certain conditions it will do certain things, will produce such and such effects. Matter, for instance, has the property of weight, that is, it will fall to the ground; copper sulphate has a blue colour, that is, affects the retina in a particular way; Canada balsam has a certain refractive index, that is, alters the direction of light passing through it at a certain angle; and so on.

All these examples have been taken from the class of inanimate objects; and, as a matter of fact, we do not commonly speak of the properties, but of the functions, of living things. In both cases alike, however, property or function means a tendency to produce under the appropriate conditions certain definite effects. At the same time we must never forget that, while we study the effects of one event, we are equally engaged in studying the causes of others. Hence, stated generally, the essential aim of a science may be said to consist in discovering the processes by which the natural objects that are immediately at its disposal have come to be derived from those simpler, and for it, ultimate elements whereto it endeavours to reduce them.

Now, unfortunately, in the method by which a science proceeds on this path of discovery, considerable confusion is, at the very outset, frequently caused by our not always referring our observations to a trustworthy standard. Sometimes we do indeed speak accurately, that is to say, quantitatively. Thus, when we say that weight is a property of matter, we mean that matter falls to the ground with a certain acceleration expressed in the easily ascertainable units of the metre and the second. In other cases, however, a property of a body is taken to mean its relation to the human nervous system, which as a measuring instrument is very delicate, easily disturbed, and highly variable. Hence the property—any colour or taste, for instance—has the great disadvantage of being merely qualitative and not quantitative, and is therefore not capable of that accurate observation and comparison which is the first essential in all sciences whatever. And, in the second place, as the arrangement in which the objects are found and observed is more or less chaotic, so also will be the table of observations, however carefully it be drawn up, and however accurately measured. This is bound to happen unless the objects are constantly compared with one another and classified—unless, that is, an attempt is made to proceed from the particular instances, so as to arrive at general conclusions. Now, when we consider what it is that has made modern science so fruitful in results, the attention is at once drawn to a principle of paramount importance. *Vere scire*, said Bacon, *est per causas scire*. The application of this maxim has abolished, to a large extent at any rate, the old vicious *inductio per enumerationem simplicem*. It led to the supplementing of mere observation by experiment, and so substituted a rational method of inquiry into the facts and operations of nature for the blind gropings and fantastic hypotheses formerly in vogue.

In order, then, to be able to arrive at such a generalization as shall be not a mere summing-up of facts already observed, but shall include hitherto unobserved instances—in short, in order to make an induction—the test of causation must be applied. Observation, the first stage in the process constituting scientific method, when it has been supplemented by experiment, and guided, in the search for new facts, by hypothesis or incompletely justified generalization, will lead to the second step, classification. The latter is equivalent to the establishment of general laws, that is to say, of statements as to what kind of events invariably and necessarily precede certain other kinds of events; whence we ascend through successively higher and higher stages to those laws of nature which in every particular science covers all the facts that come within its cognisance.

Whether there is discoverable any one law of nature such as would embrace all the ultimate laws of the particular sciences, and therefore all particular facts whatsoever, is a speculation of which the discussion may be for the present postponed.

The third process involved in scientific method is deduction—prediction, that is, that new facts, as yet unobserved, will fall under the laws already ascertained. While the laws are thus verified, the facts in their turn are said to be explained, when it is shown that they really are particular instances of those laws. In precisely the same way, in pure mathematics, theorems are said to be proved.

Let us now see how far these conceptions are being, or can be, carried out in the particular sciences.

Taking a simple example from dynamics, let us suppose we are investigating the phenomenon of falling bodies. From mere observation of such bodies, which are seen to pass through equal spaces in very different times, we could hardly arrive at a satisfactory generalization. The Peripatetics, indeed, had formularized the law that bodies fall through equal spaces in times inversely proportional to their weights; but this proposition Galileo easily disproved by letting stones of unequal weights fall from the tower of Pisa.

If, however, experiment be introduced, say, after the manner of Galileo, or by observing the fall of bodies *in vacuo*, we find, on the one hand, that the weight, or as we should now say the mass, of the body makes no difference to the time it takes to fall, and that the observed differences in time are due to the media through which it passes. On the other hand, we arrive at the general truth that the time depends on the force of gravitation in that particular place, and on the original distance of the body from the earth. This law can be verified as often as we please by fresh observations; which are themselves explained when it is shown that they are particular instances of the law.

The causes which, in this case, experiment has succeeded in eliciting from the crude observations are all cases of the action of one body on another, either in the form of attraction, or of direct impact—forces that can be easily measured, and ascertained from the molar motions of the bodies themselves.

Again, from mere observations made on boiling liquids it might be thought that the cause of boiling was merely the direct application of heat. But as soon as it was shown, by placing liquids in an air pump and withdrawing the atmosphere, that is, decreasing the pressure, that they were thrown into a state of ebullition, it became evident that some wider explanation must be sought for the facts. Now in all cases of boiling the liquid was found to decrease in volume by the escape of vapour at its surface. The attention of observers was thus drawn to a study of the effects of heat and pressure on these vapours, and it was found that the vapour tension was constant for a given temperature, increased as this increased, and at the temperature of the boiling-point of the liquid was equal to the atmospheric pressure. Boiling therefore in all cases consisted of the equalization of the two; and could be produced either by increasing the former, or by diminishing the latter: the boiling-point of a liquid not being constant, but varying with the atmospheric pressure. Moreover, as was seen in the previous example, if a new liquid were to be discovered, it could both be predicted and be verified that these relations hold good.

The same method is adopted in chemistry. Early observers were led, simply through the neglect of experiment, to put forward a theory that regarded combustion as an escape of the so-called phlogiston from the burning substance. The simple experiment of weighing the substance before and after its combustion of course soon upset this. Again, a very large number of substances under certain conditions emit light and heat, and these phenomena might be supposed to be identical with those of combustion. But the experiment of passing an electric current through a platinum wire *in vacuo* proved that, while both light and heat were emitted, the alteration of the substance found in all cases of combustion had not taken place. The generalization has thus been arrived at that combustion is a chemical union between two substances taking place with efficient energy to develop light and heat; and that these are therefore effects of it, while its causes are chemical affinity and certain conditions of temperature and pressure. Here, as before, a new phenomenon would be explained if it could be shown to be a particular instance of combustion, which itself is a case of the higher laws of general chemical reaction. The necessary antecedent phenomena, however, cannot in this case, so far as we yet know, be identified with merely mechanical forces acting between bodies whose molar motions can be easily ascertained. Chemical affinity is a name for certain links in the chain of cause and effect that are imperceptible to us, but are believed in because they form our only means of explaining the phenomena. The atomic theory provides the ultimate conception of chemistry, and it is justified by the verification of predictions made from it.

What is true of the imperceptibility of the effects of heat and light in purely physical phenomena is additionally true of them in chemical ones; for here the motions that we suppose heat, light, and electricity to induce are no longer merely inter-molecular, but intra-molecular as well.

Thus far the conception of scientific method originally laid down has been justified. In the three sciences from which examples have been taken, the knowledge attained is a knowledge of the general laws of causation—of the relation to one another of events in time, that is, of the determination of those which follow by those which precede. When, however, we come to the consideration of biology, it is by no means so easy to say that these conceptions enter into the logic of the science, or even that, in its present condition, it has any method at all. Indeed, there are some who would have us believe that the ordinary laws of cause and effect do not operate in this sphere; and that if we wish to make certain progress we must desert this path and seek another. Now, leaving quite out of the discussion the question what life is, that is to say, what have been its determining causes, if indeed any such exist, we may, as does the chemist with the atom, take as the ultimate conception of biological science the activity of the living cell. The living cell, even in the most simple of its forms, is not a mere undifferentiated mass of protoplasm with certain chemical and physical properties. It is an organism, a body possessing a definite structure, of which the various parts, or organs, such as, say, cilia, the nucleus, or a contractile vacuole, perform in relation to the environment certain functions such as motion, excretion, or reproduction. The collective performance of the functions constitute the phenomenon known as Life, these having as their apparent object the maintenance of the individual and the species. This is as true of the higher multicellular animals and plants as it is of the lower ones. In their case, however, the organ is not a part of the structure of a single cell, but is composed of one or more multicellular tissues that have taken upon themselves special functions. But in both cases it is the physiological division of labour that makes the organism what it is; organisms in all their variety forming the subject-matter of biological science.

In definite accordance with the function it performs, every organ has a form and structure of its own. Hence, in zoology at any rate, the mass of material is so great that the science has become divided into the two branches of morphology and physiology. The latter seeks to determine by accurate observation and experiment the exact functions of the various organs. On the other hand, it is the aim of the morphologist carefully to describe the very large number of varieties of structure found in organisms and organs, not only in their adult condition, but in all stages of their development; this latter study being separately classed under the special head of embryology. Further, the morphologist attempts to classify his observations by the aid of the conception of homology, those organs being spoken of as homologous which have similar spatial, though not necessarily temporal, relations, at least from a developmental standpoint, to the other organs and the whole organism, quite regardless of the function that they perform.

The existence of such orderly groupings can find an explanation only in the belief that the common possession of these homologous characters indicates descent from a common ancestry, a belief which finds strong support in the testimony of palaeontology. A natural classification would be one which coincided with this descent; and, if it were possible to determine the latter by other than purely anatomical means, we might perhaps know why the characters are grouped as we find them.

The scientific method of morphology, then, is to observe large numbers of facts, and to classify these in co-ordinate and subordinate groups on the supposition, unfortunately almost entirely untested at present by experiment, that this classification represents a natural descent going ultimately back to the simplest living elements. Indeed, if we ask ourselves what causes have led to all the varieties in structure and function which we see, we are bound to admit that neither morphology nor physiology has given any answer, at any rate any complete answer, to the question. Physiology, as we have seen, only determines the functions of organs, that is, the effects they produce; and, although this method, tested as it is by experiment, allows true generalizations to be made from which it would be safe to predict, still it gives no answer at all to the question which is being asked. The reason for this is that it regards the organism as the permanent seat of the same unalterable functions eternally performed for the sake of its own perpetuation, and not as something whose structure and correlated functions have had a definite history, and out of which organs of a totally different kind are probably going to develop.

In morphology the generalizations that are made are for the most part empirical. It has been found by experience that characters common to a large group of animals are not generally variable as regards their occurrence within the limits of subordinate groups, such as genera or species. We are as certain, for instance, that the next frog we open will possess an anterior abdominal vein, a character which it shares with a very great many other animals, in other respects often widely different from it, as we are that all men are mortal. But we are quite ignorant of the causes that have led to the possession of this character. Were the question propounded, it is perhaps not improbable that a morphologist would reply, 'because the frog is an amphibian'. But this, strictly speaking, is no explanation at all, as it is hardly necessary to point out. In the bare statement as it stands, there is no more involved than that an object possesses a certain property or character, because it belongs to a class of things, of which all have been found to possess the same character. Implied, however, therein must be the belief that some common cause has led to the existence of the common character, that in fact, the frog in this case is a special instance of some general law or laws. If we knew what causes had led to the existence of an anterior abdominal vein in the ancestral amphibian; if we knew how far heredity was able to preserve an old character in opposition to the new ones constantly coming into existence; and if we knew the laws of the mutual correlation of the different organs: then not only should we know why every frog that we open has this vein, but should likewise be able safely to predict whether or no any newly discovered creature, which for other reasons must be included in the class, would possess it also.

In order, therefore, that morphology may become the science which at present it is not, it seems to be of the highest importance that organs and organisms should be regarded as events determined by definite and necessary antecedent conditions, such as are to be found in the environment, in hereditary sources, and in the mutual interactions of the various organs. If we do not accept the influence of the first as a factor in phylogeny, we shall be reduced to the '*innere Ursache*' of latter-day German biologists. If, on the other hand, we do, we have to choose between Lamarckianism and natural selection. In the latter case experimental evidence must be obtained of variation, as well as of the selection of the 'fittest'. The direct influence of the environment upon the individual has lately been investigated by experimental embryology. Moreover, all three factors of organic evolution have in recent years been submitted, with considerable success, to statistical treatment. By this means it has been possible to give an experimental demonstration of variation, of natural selection, and of the correlation between different organs in the same individual, or between examples of the same character in successive generations.

It is necessary here to advert to a method of explaining biological phenomena which has been, and is, persistently used by a certain school of writers and thinkers, and, unfortunately, gains credit from the loose language into which biologists too often allow themselves to slip.

The doctrine of Teleology asserts that, in some way, the form which a rudimentary organ will eventually assume is determined by the function that it will ultimately perform. Now as a matter of fact there is a great deal of truth in this statement, although not if it be taken in the manner in which it is ordinarily understood. As far as can be gathered from writings of the teleologists, what is meant is that *at the time* at which a particular organ began to develop into the form in which we now find it—at the time, for instance, when the complicated auditory apparatus of the vertebrata was merely a shallow pit on the surface of the head—the preceding necessary condition of such development was the fact that long ages after it was to perform a particular function which it did not exercise originally.

As that which follows cannot in any sense be said to determine *in time* that which precedes, the theory further involves the belief that the predetermining condition was an act of volition on the part of a conscious being in whose mind there was *at the time* an idea of the function that the organ was some day to perform: an act of volition that, presumably, has been and is repeated for every separate organ, and for every stage in structure and function through which every organ has passed or will pass. Now, even if such a preceding and necessary condition existed, it would not be ascertainable by any scientific method, there being no reason to believe in the existence, say, in Jurassic times, of a sentient being capable of such volition. But those who uphold the doctrine seem to have fallen into a confusion between two different things, the formal and the final cause.

The material, efficient, and formal causes, if we mean by the last the idea of the effect in the mind of a sentient being, all precede in time the occurrence of that effect; and this kind of teleology is not, as it is asserted to be, a doctrine of final, but one of formal, causes. The final cause stands for the use to which an object is to be put, the effects it will produce, the function it will perform, which obviously succeed in time the existence of the object itself. The final cause, then, cannot be taken as ever determining in time the existence of the object, and is therefore a conception which belongs not to science at all, but to metaphysics. At the same time it must be admitted that misleading language is very frequently employed in this connexion by biologists themselves. How often, for instance, is it said, that the form of an organ depends on its function, or that an animal has such and such a structure in order that it may perform a certain function. This statement may be perfectly true in a metaphysical sense, but it tells us nothing of the conditions which determined the origin of the animal or organ. It may be a very convenient shorthand expression for the facts. Yet, speaking scientifically, it would be more accurate, even if more circuitous, to say: the conditions in the past have been such that this particular animal or organ does now, as a matter of fact, perform this particular function, organ and function being co-ordinate effects of a common cause.

The only necessary conditions of a phenomenon ascertainable by science are those material and efficient causes which precede it. At the base of biology, then, there certainly lies the belief and hope that it may be found possible in some way to determine what these are. Even so we may never get to know what the immediately preceding conditions are, not at least till we know what life is. No more can the chemist get at the immediately preceding conditions of a chemical reaction, or the physicist at those of the transformations of energy, until the former knows what an atom is, or the latter understands the laws of molecular motion.

Lastly, in the so-called mental and moral sciences it cannot, unfortunately, be said that any but the most empirical generalizations have been made, owing to the difficulty of applying the test of experiment. Yet there is, indeed, some hope that experimental psychology, as a branch of the physiology of the brain, and the comparative study of psychical facts may prove fruitful in scientific results. But in spite of this disadvantage, it is none the less true that the states or acts of consciousness known as feeling, thought, and will are, like other phenomena, in all cases predetermined accurately by the mutual reactions of character and environment; character being built up on the one hand of hereditary tendencies, and on the other of habits contracted by the repetition of similar actions. The subject-matter of psychology, therefore, in admitting of treatment along strictly deterministic lines, resembles that of all the other sciences. For it is as true of it as it is true of them that similar events are always predetermined by similar causes—a conception without which the whole fabric of human knowledge would crumble into chaos, and all intellectual effort be a profitless waste of time.

But, although the various sciences may justly hope to be able to explain their own proper facts in terms of their own ultimate conceptions, it is not so easy to say whether those who have set about the attempt to effect a transition from one science to another have much reason to be sanguine of success.

It is true that during the last half-century we have become more or less accustomed to look upon the universe as essentially a continuous process in time, during which events have become successively more and more complex. Such a view is summed up in the well-known definition of evolution as a change from an incoherent, indefinite homogeneity to a coherent, definite heterogeneity; and there is certainly some evidence for the truth of this conception. We know, for instance, that the more specialized forms of life, the higher animals and plants as we call them, are developments of a comparatively recent date; that these were preceded by simpler and more primitive types; and that earlier still was a period in which Life itself had not yet come into existence. Further, there may be, perhaps, some reason for believing that, in the remotely early history of this planet, inorganic phenomena were less complex than we now find them to be, having in fact since then undergone a physical and chemical differentiation from the primaeval terrestrial substance. But whether this be so or not, life at least had once a causal beginning, and the gap existing between the sciences of biology and of chemistry and physics is a breach not in the continuity of nature, but in that of human knowledge.

It may appear to some thinkers quite pertinent to urge that, for the practical purpose of making immediate progress, it is better, at least for the present, to keep the sciences apart. But, so long as any region of nature remains unexplored, it is illegitimate, and will be found impossible, to forbid it to any competent inquirer. Indeed, one highly trustworthy experimental chemist has already offered some explanation of one of the functions of the living organism, as observed in the simplest forms ; while the assiduity with which the sciences of thermo- and electro-chemistry are being pursued tends perhaps to show that the hope has not been abandoned of establishing a connexion between other branches of knowledge. But whether the ultimate conceptions of the several sciences are reducible or not to a common conception ; whether or no life, chemical affinity, and molecular motions can be expressed in the same terms as are observed molar motions, while all phenomena become deducible from the laws of these : there yet remains for the complete philosopher the necessity of inquiring into what is involved in his ultimate conceptions of time, space, matter, motion, and the rest. Or, in the event of it being found impossible to bring these ultimate conceptions together under one, he must show clearly how they are all but different ways of looking at those abstractions which every man makes for his own particular purposes from the concrete facts of ordinary experience. Such an inquiry might fitly be styled ontology, the knowledge of being in its various forms. Thereupon for the materialist nothing more remains to be investigated. Ontology is identical with metaphysics, and the desired system of the universe is attained. But there is one fact which the materialist forgets to analyse, and materialism absolutely fails to explain, a fact which is indeed the hardest to understand of all, and that is knowledge ; and, with knowledge, those other facts of self-consciousness, feeling, and will. The philosopher who ignores this question has no claim to put forward his system as complete. Even though no metaphysical system is final but only an approximation towards, or, better, the expression of, an absolute system, still it is incumbent on us to attempt to form some, if it be but a very imperfect, theory of the universe. Otherwise we must be willing to acquiesce permanently in a position of philosophic doubt, and to make that which should be but the beginning the end of philosophy, namely, a scepticism, which indeed cannot be argued with, but will inevitably refute itself by doubting its own existence.

Assuming, what very few would be found to deny, the existence of knowledge, what is meant by it ? Knowledge is a relation between a subject, or knower, and an object, or known. But, says the physical realist, I am my nervous system : and knowledge means simply the sum total of the relations between my nervous system and the external world. All that is meant, therefore, to take a concrete instance, when I say I know that trees are green, is that the external objects known as trees act upon, and are related to, my nervous system, in so far as regards that quality known as colour, in a certain definite manner. The nervous system retains these impressions in exactly the same way as the wax retains the impression of the seal, and thus memory, imagination, and so forth, are easily explained. According to this supposition the human being is nothing more than an exceedingly complicated machine, reacting variously to its environment—such a machine as one might perhaps hope to make by combining, say, a photographic camera with a phonograph and a thermometer. Moreover, they who uphold it must, if consistent, believe that each of these instruments possesses knowledge, and therewith, presumably, an illusion as to its own self-consciousness. For, on this view, self-consciousness, and the belief in a subjective as of a nature distinct from that of the objective, in a noumenal as opposed to the phenomenal, is the merest illusion. The universe is reducible to a number of series of objects coexisting in space and following one upon another in time; the illusory ego being simply the merely objective fact that certain parts of the series are gathered together into what we commonly term the experiences of individual personalities.

Is such a theory tenable? Can knowledge be explained in this way as the result of impressions falling, through a series of relationless atoms, upon the *tabula rasa* of a nervous system? Is there no difference between human consciousness and the negative used to take a combination photograph, or the wax cylinder that registers the successive sounds in a concert-room?

On the contrary, the nature of knowledge necessitates the belief that there must be something which unifies these atoms of sensation into an experience, beyond the mere objective fact that they are so unified. Again, this something, this knowing subject, must be itself outside the objects which it relates, and therefore is in no sense a product of, or co-ordinate with, them. For, if it were, it would thereby immediately lose its subjectivity and become merely objective. Further, knowledge being essentially a relation between a subject and an object, if it be said that it is impossible to arrive at any certainty with regard to the existence of the subject, it becomes thereby equally impossible to be certain of the relation. Hence the materialist is forced to confess that, though he has a knowledge of objective phenomena, he does not, and never can, know that he has that knowledge. Materialistic writings, of course, are full of such phrases as idea, conception, law, and the like. These imply a certainty in regard to the existence of knowledge, and by their use the self-contradiction involved in the materialistic attitude stands self-condemned.

Knowledge, then, instead of being a relation between purely objective phenomena, is a relation between me, a subject, and certain objects; this relation being variously termed sensation, perception, or conception, and its seat or necessary condition being my nervous system. And, let us be careful to remember, we never can, in our knowledge of phenomena, get beyond this relation. 'I know such and such a thing' means, therefore, that I am related to, in unifying, certain objects of sensation, perception, or conception. These unified data of experience I term objective phenomena. Myself who unify them I term subjective, noumenon. These phenomena, including psychical facts, which are the successive acts of unification performed by the knowing subject, appear to me under the forms of space and time; and they bear objective relations to one another, which relations it is the function of the sciences to investigate.

My own and other bodies, together with their nervous systems, are among these phenomena; and among these relations are those between what is commonly called the external world and the nervous systems of other bodies, and, by inference, of my own. 'The baby new to earth and sky' is serenely unconscious that he has a body, much more a nervous system. And we might, conceivably, remain permanently in this condition; though it would not, therefore, follow that we should be devoid of knowledge.

My body only means a particular set of phenomena of which I am nearly always conscious, and my nervous system a set of phenomena which, as scientific investigation has taught me, conditions, that is, is the invariable accompaniment of, my having any knowledge at all.

The appearance of these phenomena under the form of space leads us to confuse 'I' with 'my body'. Again, it leads us to imagine that 'I' am spatially located in my brain. Impressions come from external objects to the brain, and then get in some mysterious way to 'me', who retain them in my memory as ideas. It is this futile tendency, even when the purely materialistic hypothesis has been discarded, to persist in endeavouring to find a medium between consciousness and the external world, still regarded as endued with a reality of its own, that makes the path easy for those who confound the objective relation of one phenomenon to another, of my body to other objects, with that subject-object relation which is the very essence of knowledge.

And now let us inquire a little more deeply into the nature of this relation. It was the great merit of Kant to have insisted as against the passive-substance theories of Locke and Berkeley, which Hume drove to their logical conclusion in a sensational atomism, that alike in sensation—for sensation is a critical faculty —in perception, and in conception the subject is active; unifying the unrelated data of experience; imposing relations on these according to certain laws of its operation, that is to say, in space and time; and so attaining to a certainty of its own existence, though never being the final term in the series which it unifies.

What, then, of the object of knowledge—that which, having been received in experience, the mind informs?

We have already seen the fallacy of supposing that external objects, having an existence of their own in space, produce impressions on the nervous system which in some inconceivable way become transformed into non-spatial sensations and ideas. And it is a cognate error which holds that there remains over, after we have stripped the object of knowledge of its admittedly ideal attributes, a residuum, an ultimate cause of sensation, unknown and unknowable, but clothed with a separate reality of its own—a thing-in-itself—which impinges upon us in experience and causes in us a knowledge of its attributes, so that there is between mind and matter an intermediate world of unreal ideas.

For it is impossible, upon this dualistic hypothesis, to understand how that which is so far out of relation as to be unknowable can so far come into relation as to give a manifestation of itself, and have its matter subjected to the informing operations of mind. On the contrary, the object of knowledge is nothing but the sum of its own attributes and relations. It is not unknowable things-in-themselves, producing in us unreal ideas, that constitute objective reality, but the knowable and nothing else. The real is that which can be synthesized, an objective relation which can and must be itself related to some subject. Only existing therefore as the knowable, the real is also necessarily ideal. Thus the distinction between perception and conception, between impressions and ideas—erroneously regarded as respectively real and unreal—comes simply to this. The individual object of perception is only the meeting-point of universal conceptions, *plus* the here and the now; which latter resolve themselves into relations of coexistence with and succession to other objects. Perception, in perceiving the particular, to adopt Aristotle's famous simile, arrests one universal, round which the rest rally like fleeing troops on a battlefield; while in conception the mind grasps some only of these universals. The objects of both are alike ideal, and they are real in the same sense of the word. The real is the ideal, and the ideal is the real. The material, objective, phenomenal universe is to be expressed only in terms of the subjective and noumenal, that is, of self-consciousness. Not that it becomes therefore any the less objective. But it is impossible to explain the subjective in terms of the objective; and, if, as Hegel put it, dualism is philosophic death, no other alternative is left but to say that the phenomena which constitute the object of knowledge are the creation of the mind itself.

It still remains for us to inquire *what* mind? Are we to say with the solipsist, *Alles ist ich, ich bin alles*: and to call all that comes to us in experience equally real; to ignore all other criteria of reality but this; and, in fact, to obliterate the distinction between reality and unreality, and almost that between self and not-self? Or, rather, must we not remember that the data of experience are independent of us; and accept the inference, drawn from the existence of other bodies similarly constituted, and having similar relations in time and space to our own, that personalities other than ourselves but knowing the same universe as our own exist—an assumption, indeed, without which we could never give a consistent account of our own knowledge at all? Hence, must we not believe in a difference between reality and unreality, such as lies at the bottom of what is, after all, common belief? For, to the ordinary man, objective reality means that which is coherent with either his own or other people's experience. To the scientific man, the real means that which is intelligible to himself and others—that which falls into its place in an organized system. To the moralist, the real is the rational and the rational the real, a sense which includes the two others. Ultimately, then, the question, Is a thing real ? must mean, Is it related absolutely as it seems to me to be related ? And the further question, What is that whereto that which is an absolute relation is related ? compels us to acknowledge an absolute mind. The object of such a mind is, not a world of ideas midway between subject and thing-in-itself, the phenomena of Kant's noumena, the knowable manifestations of Herbert Spencer's unknowable. Reality, facts, a world of ideal relations, not self-existent entities, or unrelated atoms, but an organically connected whole of thought completely thought out—herein consists the object of knowledge to an absolute mind. Whether such a mind could be spoken of as personal is an open question, which it is hardly necessary here to discuss. But it may be observed, firstly, that the idea of personality, which gains in force from the mischievous confusion between body and person, seems to be inseparably bound up with the moral and intellectual imperfections of human individuals ; and, secondly, that this mind must be regarded not so much as something outside individual minds, differing from them in capacity, yet related to them as they are to each other, but rather as an inner harmonizing activity between them, of which they are, so to speak, the expressions, and in which they are included.

To such an idealism it is often objected that it is as true to say that nature makes the understanding, as that the understanding makes nature—that the world makes God, as that God makes the world. This, indeed, is the dualistic position in which Kant leaves us, the subject which gives the form not being more necessary for knowledge than the object which supplies the matter. But, quite apart from the inconceivability of the existence, outside thought, of that metaphysical phantom the thing-in-itself, in knowledge the subject, not being derivable from the object, is the more important. For, logically, that is, so far as validity is concerned, the active is prior to the passive, as form or εἶδος to ὕλη or matter. Hence the ὕστερον πρότερον of materialism, which tries to explain mind, the subject, by matter and motion, the object. It starts with the most elementary categories of being and not-being in order to evolve the universe therefrom, instead of beginning, as Hegel does, with self-consciousness, the highest category of all, and, let it be added, the most fully knowable. For, since knowledge is essentially a relation, where this knowledge is most intimate, it is most transparent ; where subject and object are one, there and there only is knowledge complete.

Now it appears at first sight that there still remain obstacles which the doctrine of evolution offers to the acceptance of such a system as this ; and the nature of these obstacles is twofold. Firstly, there is the metaphysical difficulty : evolution presents to us the objective universe as essentially a continuous process in time. All its events are strictly and absolutely determined by preceding events. Its culminating point is the genesis of consciousness, which, therefore, it would appear, having undergone a development in time, is the product of the non-conscious. It is the doctrine of becoming, not of being. According to idealism, on the other hand, that ultimate reality which is found in self-consciousness is essentially timeless. Idealism can explain being in terms of knowing, but it seems as if it could not explain becoming. Secondly, we have what may be called the moral difficulty. Evolution puts before us 'Nature red in tooth and claw with ravin'—nothing but a selfish struggle for existence, with many imperfections, many degradations, and all its small advance at a tremendous cost of individual life. This of course is the problem of evil ; and its solution, although very largely a question of temperament—for a pessimist will be a pessimist to the end of his days—also depends on the meaning attached to the word 'perfection', which can itself only be satisfactorily thought out by a consideration of what is involved in the idea of time.

In the first place it must be insisted that, with regard to the development of consciousness, no intuitionist theories are allowable which invoke cataclysms to explain what is explicable by known laws—which talk of spiritual influxes coming no one knows whence or how, and arbitrarily and inexplicably descending upon the brain of the greater ape, and there developing into thought, feeling, and will. We must acknowledge the development of consciousness in time, of mind with body, if we accept the fact of evolution at all. The stages that we imagine the race to have passed through are paralleled in miniature in the embryogeny of each individual. If, therefore, what is true of the individual is also true of the race, and if in the growth of the embryo from the germ—and this resembles exactly in all essentials of structure so lowly organized a creature as an amoeba, which forms the starting-point for the evolution of the whole animal kingdom—

we can fix no point at all, draw no hard and fast line, for the first appearance of consciousness; then it is impossible to believe otherwise than that the rudiments of consciousness have been present, even as they are undoubtedly present in their now living allies, in those lower forms from which the human race has been derived.

And yet mind does not evolve from what may be called, for the sake of convenience, matter; nor is mind a product of physical forces, nor a higher form of life. The psychical is not the physical, but an aspect of it. Just as new properties come into being when oxygen and hydrogen unite to form water—properties which could not have been foretold from the known peculiarities of the elements, though the molecular motions might doubtless have been predicted from the movements of the atoms; or just as blue colour appears when particles vibrate with a particular periodicity, though the vibrations are not blue: so also does the psychical arise when the physical has attained a certain degree of complexity, and with the further development of the one series the other marches side by side.

Now, if the view that has been taken above be true—if it be the understanding that makes nature—then time must be looked upon, not as the self-existent succession of things-in-themselves, but as a law of the intelligence—a form, to use Kant's expression, under which the mind, itself out of time, sees the phenomena which it makes. Mind is not here and now, or then and there, but simply is. If it appear to evolve in time, it is because time is the form under which it sees the phenomena on the progressive knowledge of which this development depends. To object, therefore, that in this case, say, the palaeozoic fauna can never have existed, because there was then no mind for which they could exist, or, in the same way, that only those parts of space exist which are being actually perceived, is entirely beside the point. For all that is meant by saying that certain things existed at a certain time, or in a certain place, is that they appear to the mind in a certain order of precedence and succession, or in one of coexistence, with regard to other things. Since this order is not, any more than the things, self-existent, it makes no difference whether the mind, for which alone it and they exist, has reached any particular stage of development or not. They do not exist for us who are now and here, but they existed or will exist then and there for us. Mind, in short, is not in time, but time is in mind. And, lastly, if this were not so, if time were ultimately self-existent, then—and the same reasoning applies to space—either there would necessarily be an infinite regress and progress of events in time; or there would be a period of actual finite time in which events happen, preceded, and to be succeeded by, infinite periods of possible time in which events might but do not happen: both of which suppositions seem unthinkable. On all grounds, therefore, we are compelled to call time a form of our perception. The ultimate explanation of existence is thus to be found not in a temporal *prius* out of which all things emerge, but in a logical *prius* which they presuppose, and towards which temporally they seem to move.

What is to us a time-process from the less perfect to the more perfect, what is to us the development of the notion from pure being to the absolute idea, is really not construction but reconstruction. That which, to use Aristotle's language, is χρόνῳ πρότερον, or earlier in time, is λόγῳ ὕστερον, or later in importance. Metaphysics must explain existence, not aetiologically, as science does and *must* explain its phenomena, but teleologically—in terms, that is, not of its origin but of its validity, or, in other words, of its final cause; because everything is not merely that which it sprang from, nor even that which it seems to be, but also that which it may become, though we cannot say of it 'it is' till it has become. Following Hegel, then, we must rationalize the universe, not by supposing that which is the presupposition of the lower categories to have developed in time out of them, but by explaining it in terms of the highest category, self-consciousness; which, in seeking the objective as that through which alone it can realize itself, compels itself to look upon the universe of its own creation under the forms of time and space.

Let us turn to the second difficulty. It is impossible to attempt the solution of the problem of evil—as impossible as we have found it so to explain knowledge—by a dualistic reference to the unmanageable ὕλη which the δημιοῦργος or divine artificer never quite succeeds in reducing to shape.

Evil and good are essentially relative terms. Evil, according to the conception of the evolutionist, means failure on the part of the organism to adapt itself to a new environment, provided that in so doing it also fails to progress. For it is not true, as has been sometimes asserted, that adaptation to environment, the mere persistence of an organism, is the only test of excellence; and that, therefore, to quote a well-known example, the tadpole which becomes a degenerate ascidian is as 'good' as that which has gone on developing till it has reached the man. On the contrary, it is possible to speak in a perfectly scientific, that is, aetiological, way of the persistence, or non-persistence, of the development or degeneration, of the specialization or reversion, of an organism as facts due to adaptation, variability, the struggle for existence, and so on; and yet also to speak of a higher, or better, organism as one that has attained, by adapting its own variability to an ever-changing environment, to an increased and increasing division of physiological labour. At the same time, it must be freely admitted that the word 'good' or 'higher', as used by the evolutionist in this sense, is one borrowed in the first instance from the terminology of ethics, and used there to describe the development of the moral character.

Now this differentiation must involve a greater limitation by one another of the different organs in the organism; and so it is that any one organ has in a higher stage far less scope for individual exercise than in a lower. Hence, what once contributed to survival may now, if indulged in to the same degree, cause pain, failure, sin, or whatever we choose to call it. 'For man when he has climbed to the top of the ladder kicks over that by which he rose, and brands with the name of sin those qualities once good, now bad, which enabled him to be what he is.' It is therefore impossible to speak of any action, or quality, as being in itself either good or bad; and quite futile therefore to suppose that perfection is 'some far off, divine event to which the whole creation moves'. If perfection, or unconditional good, exist at all, it is to be found here and now, if we could only see it. And, it is just in the process itself that it consists, namely, in the effort, at the expense of necessary error and pain, towards self-realization; which, be it added, is the only ultimate criterion, as it is the only ultimate end, of conduct.

Now, in that particular form of evolution which it is the function of ethics to study, we find, as a matter of fact, that all advance, in the sense above indicated, is accompanied by a continually increasing control over the environment—that is to say, by the development of the will. Will is thought intensified, the momentary determination by the organic consciousness of its own environment. Looked at as part of the time-series, these acts are conditioned by one another, and by the environment, in the same way as knowledge of an object is conditioned by previous knowledge and by the object itself; and, if time were ultimately self-existent, the will would not, and never could, be completely free. But, just as thought is logically prior to, and the presupposition of, that out of which it seems to develop, so will—though conditioned in time by its own previous acts, as a function of that self-consciousness which is realizing itself in the irrational order of time and space which it makes for itself—is free. For the development of the will accompanies the advance of knowledge. Where the organism is a self-conscious subject, survival, or moral equilibrium, implies knowledge; while failure, or the loss of moral equilibrium, means ignorance. Thus it is that true virtue is always *μετὰ λόγου*, because it implies control over any possible environment. Perfect freedom would be perfect understanding.

If these considerations be just, then the gigantic time-process in which science has found the objective, phenomenal universe to consist, in which all events are rigorously predetermined by others, and during which the evolution of conscious beings has taken place, can only receive its full and ultimate meaning thus: namely, when it is shown, by an inquiry into the credentials of science—in other words, into what is involved in the fact of knowledge—that, outside phenomena, there is something else to which they owe their none the less real existence, namely, a noumenal or subjective. Such an inquiry may fitly be termed philosophy *par excellence*, or metaphysics. Now where does the system of metaphysics that has here been outlined, whether true or false, seek to find the ultimate essence of existence? Not in a crudely anthropomorphic God, as conceived, unfortunately, by too many religions. Nor again in that objective force, the far away first cause of all things that are, which others have vainly sought to deify. It seeks to find it in that subjective consciousness, that union of thought, feeling, and will, the elements of which each apparently separate individual may find hidden in his own soul. Different though the aims and methods of the man of science and the philosopher appear, still each has much in common with the other. For a faith in the ultimate rationality of the universe is the presupposition of all scientific endeavour and of all speculative effort ; and the philosopher and the man of science alike are content to

> Strive and hold cheap the strain,
> Learn nor account the pang,
> Dare, never grudge the throe.

Herein they find that *εὐδαιμονία* or ' blessedness ' which, as one of the greatest of philosophers if not also of scientists would have told us, is the *τέλος τέλειον*, the end in and for itself, of all human aspiration whatsoever.

A SKETCH OF THE HISTORY OF PALAEOBOTANY

WITH SPECIAL REFERENCE TO THE FOSSIL FLORA OF THE BRITISH COAL MEASURES

By E. A. Newell Arber [1]

Introduction

The history of a particular Science depends in no small degree on certain events, which may be more or less peculiar to that branch of knowledge, or common only to a series of related studies. In the case of Palaeobotany, as of Palaeozoology, the factor which more than any other has controlled and still continues to dominate its progress is, literally, the unearthing of documents which are still legible. These ' sources ' (i.e. the fossils themselves) have first to be found and then removed from Mother Earth. In this respect Palaeobotany presents a great contrast to the Experimental Sciences. We are absolutely dependent on what we have dug out, and since, speaking generally, it is very far from easy actually to locate the evidence in a series of otherwise barren rocks, our progress depends largely on a succession of happy chances occurring often only at long intervals. The cherished belief of to-day in regard to some particular point in fossil botany may have to be shamefacedly abandoned to-morrow, simply because something more complete which demands a revision of the received opinion, has been literally turned up.

No doubt in the history of this, as of other sciences, there is a hinterland in the form of a prehistoric period, so-called. Among the curios known and commented on from quite early times down to the mediaeval period, there were some which happen to be fossil plants. Perhaps the most celebrated of these cases is that

[1] Owing to the death of the author before this paper was revised, the responsibility of editing it has fallen to his wife.

PLATE XLVIII

Fig. 1. *Cycadeoidea etrusca*, Cap. and Solms, Pl. I, Fig. 2 of Capellini, G., and Solms-Laubach, H., *I Tronchi di Bennettee dei Musei Italiani*, Mem. R. Accad. d. Scienze di Bologna, Ser. 5, T. II., 1892.

Fig. 2. Edward Llhuyd, *Lithophylacii Britannici Ichnographia*, 1699, Pl. IV (reduced). 186, *Lithopteris*; 188, *Lithosmunda*; 189, *Lithosmunda minor*; 191 and 197, *Trichomanes*.

of the silicified stem of a species of *Bennettites* (plants dimly related to the modern Cycads), which was no doubt unearthed from Jurassic rocks in Italy something like three thousand years ago (Plate XLVIII, Fig. 1). At any rate, this particular fossil was known to the Etruscans who about 1000 B. C. inhabited the district where Marzabotto now stands. Very probably it was the only example any of them had ever seen; at any rate they appear to have regarded it as an object of superstitious reverence and great rarity, since on the occasion of the death of some important member of their community, it was placed in his tomb in addition to other offerings indicating respect. However, to re-bury a fossil is but to preserve it anew for posterity, for some years ago this particular specimen was dug up again and formed the subject of brilliant botanical researches by a celebrated German professor. Further, upon it is founded our present knowledge of the structure of the cones of this particular plant, constituting one of the most astonishing advances ever made in fossil botany. The present resting-place of this specimen is a museum at Bologna.

Again, every one is aware that the ancient Phoenicians knew of Amber, which is part of a fossil plant. Jet, another vegetable fossil, was also familiar to the ancients. Many other instances might be given, the written history of which begins in classical or mediaeval times. But in this brief sketch it is not proposed to explore this hinterland further. The real history of the science, as a science, dates only from the Renaissance, and to reckon even from that period to the present day involves the inclusion of a lengthy early stage which, strictly speaking, was Pre-scientific.

The Pre-scientific Period

In the earliest stage in the printed history of our science, such fossil plants as found their way into a collector's 'cabinet' were mere curiosities. It was not admitted—in fact it was universally denied—that they were relics of plants which had actually lived on the surface of the earth in a bygone age. Geology as a science did not exist then, nor indeed until after the Renaissance. Fossil plants were mere sports of nature (*lusus naturae*). Curiously enough such things actually occur not infrequently, cases where purely mineral concretions simulate plant remains in shape and form in an extraordinary degree being not uncommon in the older rocks. It was therefore not unnatural to regard all such cases as obvious frauds, and the interest in them lay chiefly in the curiosity aroused by the completeness of the delusion. At the same time, despite this fact, the science was advancing under the influence of the 'cabinet' and its curios.

The birth of the Renaissance in Italy shed new light on the origin of true fossils, and that light travelled quickly. The Italians, among them Leonardo da Vinci, realized that many of the contents of the cabinets were really of organic origin, and, once this was admitted, an added interest naturally attached to them. They were no longer mere curios. The first intellectual stimulus had arrived. Attention became focussed on them, speculation became rife, and as the result geology through palaeontology was born. At the same time it must not be imagined that palaeobotany had by now become a science, using that term in the modern sense. It had not. It had been born and was on its way to become a science. So far that was all, and for many years the study remained pre-scientific, rather than scientific.

And here we may pause to note the change induced in the terminology of the contents of the cabinet by the more exact knowledge which was slowly becoming evolved. The term 'fossils' originally meant anything dug up out of the earth. Sands, clays, marbles, &c., were as much fossils as anything else delved out of the earth's crust. With the coming of the scientific renaissance, a fresh distinction in terms was inevitable. *Real* or *natural* fossils were distinguished from what were called *extraneous* fossils. The former were what we should now describe as lithological, petrological, and mineralogical specimens. 'Extraneous fossils', to-day universally known simply as 'fossils', were the relics of what originally had been organized bodies, either animal or vegetable. The growth of collections and the dawning realization of these and other distinctions, seem to have led to the preparation of manuscript lists, enumerating the contents of the cabinets.

The next great step was obviously the printing and publication of such catalogues, including not only the names which the owner gave to the objects, but descriptions of their characters. These catalogues—which of course at first included every mineral object regarded as possessing any interest at all—in so far as they related to fossil plants, represent the sum total of the palaeobotany of the period. The earliest of these catalogues which contains any reference to British fossil plants was published in 1699 by Edward

PLATE XLIX

Fig. 3. One of the original cabinets belonging to John Woodward (1665–1728), whose geological collections form the nucleus of the Sedgwick Museum, Cambridge. On the open front a volume of Woodward's MS. Catalogue is shown, and four of his fossil plants.

Llhuyd (= Lhuyd = Luidius) (1660–1709), who was keeper of the Ashmolean at Oxford. This work is termed *Lithophylacii Britannici Ichnographia,* and a second edition was published in 1760 after Llhuyd's death. This catalogue was illustrated, and several Coal Measure plants are represented by fair figures. They were, however, like all the others catalogued during this pre-scientific period, nameless, by which is implied that they were not referred to genera and species, a much later device dating from the time of Linnaeus. For instance, Fig. 188 on Plate 4 of the first edition (Plate XLVIII, Fig. 2), which is probably intended to represent *Alethopteris Serli* (Brongn.), is described as 'Filix florida Mineralis sive *LITHOSMUNDA* Cambrobritannica. E fodinis Glamorganensibus superiùs dictis'. Again, Fig. 202 on Plate 5, probably *Annularia sphenophylloides,* is described as '*RUBEOLA* mineralis. E fodinis Actonensibus'.

EDWARD LLHUYD (1660–1709)
From an initial vignette in the Register of Benefactors, Ashmolean Museum, 1708.

The specimens figured by Llhuyd were mostly derived from coal-mines in Glamorganshire and Somerset and the Forest of Dean, and a few from Denbighshire and Flintshire.

After the publication of Llhuyd's work, the collection of fossil plants received an immense additional mental stimulus from the initiation of the great controversy on the *Flood Theory,* towards the end of the seventeenth century. It was then held that there had been one great deluge, the flood of Noah, and that all fossils were the productions of this catastrophe. Clearly, then, the explanation of the origin of all fossil plants was to be sought for in the early chapters of Genesis. They were all of precisely the same geological age, and all had been entombed in water-borne sediments during a comparatively brief period. These contentions are fully set forth in the next great handlist—*A Catalogue of English Fossils* (1728–9), by Dr. John Woodward.

Woodward (1665–1728) was a medical man, and a Professor at Gresham College, London. By his will, he left his large collections, to which the catalogue in question relates, to the University of Cambridge, and at the same time founded the Woodwardian

Chair of Geology in that University. This collection was contained in several cabinets, and these are carefully preserved to this day in the Sedgwick Museum of the University in exactly the condition in which they were left by Woodward (Plate XLIX, Fig. 3). The Woodwardian collections are undoubtedly the oldest still in existence, and in them one can actually see 'Cabinets' as they stood at the end of the eighteenth century. Further, it is these very cabinets which have formed the foci round which the present enormous and valuable geological collections belonging to the University have crystallized out, so to speak, in the last two centuries.

Woodward, in common with the celebrated Swiss geologist, Scheuchzer, was one of the great Apostles of the Flood Theory. In the preface to an earlier work on this hypothesis (*Essay toward a Natural History of the Earth*, 1695) he says:

'It will perhaps at first sight seem very strange, and almost shock an ordinary *Reader* to find me asserting, as I do, that the whole Terrestrial Globe was taken all to pieces and dissolved at the Deluge, the Particles of Stone, Marble, and all other solid Fossils dissevered, taken up into the Water, and there sustained together with Sea-shells and other Animal and Vegetable Bodies; and that the present Earth consists, and was formed out of that promiscuous Mass of Sand, Earth, Shells, and the rest, falling down again, and subsiding from the Water. . . . The other Instance I make choice of shall be of the Universality of the Deluge.'

In Woodward's catalogue of English Fossils (*An Attempt Towards a Natural History of the Fossils of England*, &c.) the theory of the Deluge was pushed to extremes, and efforts were made to fix the season during which it took place. In the second volume (1728) of this work (p. 59) we read:

'The Hazle Nuts, digg'd up in *England*, are rarely such as appear to be ripen'd. The Pine Cone are in their vernal State; as are all the Vegetables, and the young Shells. The Deluge came on, and a stop was put to their further Growth, at the End of *May*.'

Again, in the first volume (p. 21), which appeared a year later, we find the following remarks:

'Of all the Fossil-Nuts I have ever seen, either in the *North*, the *Isle of Wight*, or any other Part of *England*, tho' some few, perhaps by reason of some particular Advantage of Situation and

PLATE L

Fig. 4. David Ure, *The History of Rutherglen and East-Kilbride*, Pl. X (reduced).

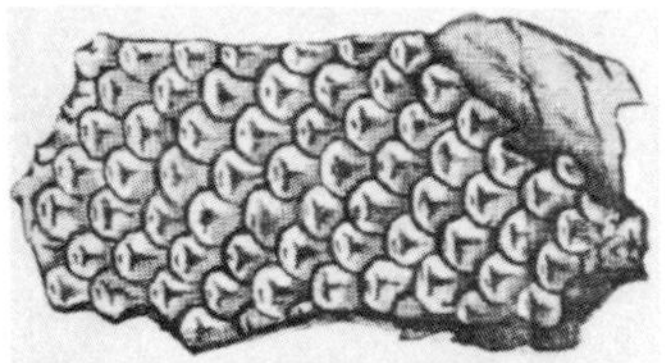

Fig. 5. E. M. da Costa, Phil. Trans. 1757, Vol. 50, an impression 'from a coal-pit in Yorkshire'.

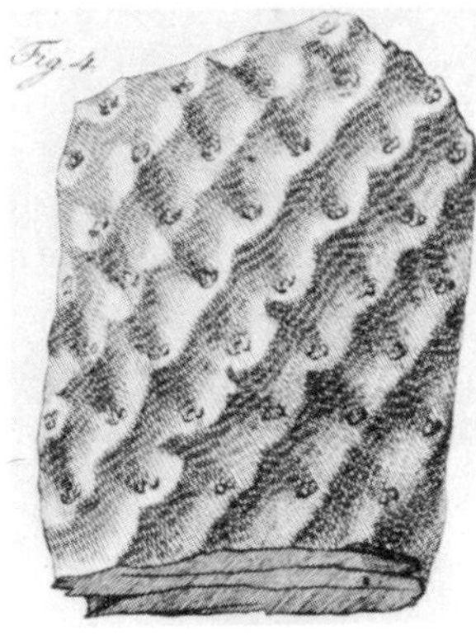

Fig. 6. E. F. Schlotheim, *Die Petrefactenkunde*, 1820, Pl. XV, Fig. 4, *Palmacites.*

Fig. 7. W. Martin, *Petrificata Derbiensia*, 1809, Pl. X (reduced), *Phytolithus Filicites* (*striatus*) = *Alethopteris lonchitica* (Schl.)

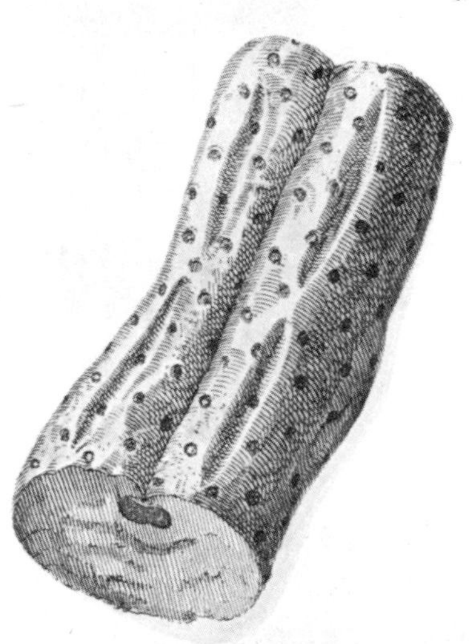

Fig. 8. W. Martin, *Petrificata Derbiensia*, 1809, Pl. XI (reduced). *Phytolithus Plantites* (*verrucosus*) = *Stigmaria ficoides*, Brongn.

Sun, are somewhat larger ; yet the generality of them appear to be of about the Growth and Condition that Hazel-Nuts usually are at the end of *May* or the beginning of *June* ; and that the Deluge began at that time of the Year.'

The dogmas of the Diluvialists long continued to attract attention to fossil plants. In 1758, for instance, James Parsons (1705–70), in a paper published by the Royal Society (*Philosophical Transactions,* vol. 50. pt. 1 for 1757, p. 396), on Eocene fossil fruits from the Isle of Sheppey, disputed Woodward's conclusion that the Noachian deluge took place in May, since, according to Parsons' specimens, ' there are the stones of fruits, found fossil, so perfect as to make one imagine they were very ripe, . . . which would induce one to think that the deluge happened nearer Autumn '. (p. 402.)

Apart from catalogues of cabinets, other records of fossil plants began to appear in the eighteenth century in the accounts which were published of the history, antiquities, and especially the curiosities, of particular regions, districts, or places.

For instance, in Robert Plot's (1640–96) *Natural History of Oxford-shire,* which appeared in 1677, a Stigmarian rhizophore (Plate III, fig. 11) is figured, which Plot concluded (p. 98) ' seems to represent a *Carp* or *Barbel,* the best of any Fish which I have yet compared it with '. Another work, nearly a century later, David Ure's *History of Rutherglen and East-Kilbride,* published in Glasgow in 1793, contains the earliest illustrations of fossil plants from Scotland (Plate L, Fig. 4). No less than four plates (X–XIII) of very fair figures, considering the period, are devoted to them, and they are also closely compared with living genera. The earliest British *memoir* (as opposed to catalogue or guide book) solely concerned with Carboniferous fossil plants was published in 1758 in the *Philosophical Transactions* of the Royal Society of London (vol. 50, pt. i for 1757, p. 228, Pl. v). This is the prototype of a vast series of subsequent papers, many of them having appeared in the same publication. The author of this paper, Emanuel Mendes da Costa (1717–91), entitled it ' An Account of the Impressions of Plants on the Slates of Coals '. It describes briefly specimens of *Sigillaria, Stigmaria* and other genera from various coalfields (Plate L, Fig. 5). The occurrence of such fossils in coal-bearing rocks had previously been put on record by Woodward and several foreign observers, but in Da Costa's paper an attempt is made to examine and discuss these fossils themselves with more thoroughness than is found in earlier writings.

However, about the beginning of the nineteenth century new ideas began slowly to prevail. It was seen that if any real progress was to be made with the study of fossil plants it was necessary not only to describe them, but to indicate them binomially (i.e. to refer them to definite genera and species) on the system which Linnaeus had applied some years previously to living plants and animals. Further, a somewhat later period witnessed the final overthrow of the Flood Theory and the birth of stratigraphical geology. It came to be recognized that plant-bearing rocks were not all of the same geological age, but belonged to different periods in the history of the earth's crust, as is witnessed by the diversity of fossil floras which they contain. It is, of course, impossible here to trace the rise and progress of this great biological conception, but it may be noticed that, once its truth was admitted, the necessity for the binomial usage was felt to be even more pressing than before. Thus the systematic side of the subject received a great impetus.

The binomial system of nomenclature, as applied to fossils, did not originate in this country. It must regretfully be admitted that British workers played no part at its birth. It was only in quite modern times that British contributions began to exert any profound influence on the study of fossil plants.

The Transition to the Scientific Period

The Scientific Period begins with the adoption of the binomial system and the foundation of systematic stratigraphy. The transition period is beautifully seen in the works of a single writer, of whose memoirs only two contain any reference to fossil plants. This was a German, Ernst Friedrich, Baron von Schlotheim (1764–1830), of Gotha. Schlotheim in 1804 published his *Beschreibung merkwürdiger Kräuter-Abdrücke und Pflanzen-Versteinerungen.* In this book he gives good figures of a number of Coal Measure plants, but no names were applied to them. This book belongs to the pre-scientific period. But in 1820, sixteen years later, he published his *Petrefactenkunde auf ihrem jetzigen Standpunkte durch die Beschreibung seiner Sammlung versteinerter und fossiler Überreste des Thier- und Pflanzenreichs der Vorwelt erläutert,* in which the plants figured in the first memoir were named binomially. Plate L,

Fig. 9. James Parkinson, *Organic Remains of a Former World*, 1804, Vol. I, Pl. V (reduced).

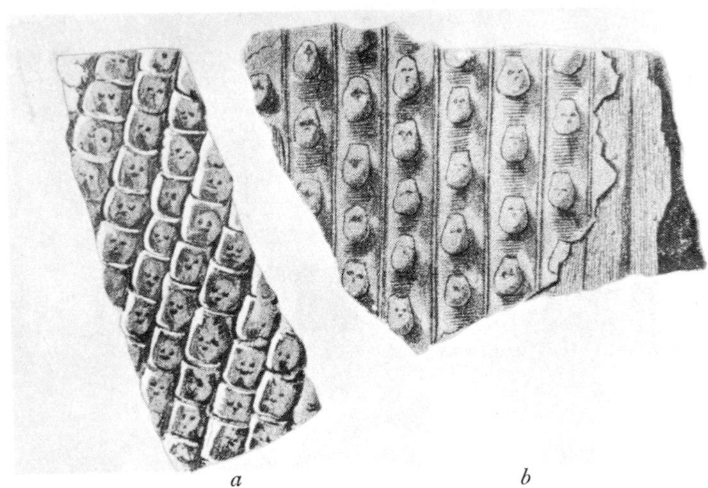

a *b*

Fig. 10. H. Steinhauer, Trans. Amer. Phil. Soc. 1818. *a*, *Phytolithus tessellatus* = *Sigillaria tessellata* (Steinh.); *b*, *Phytolithus notatus* = *Sigillaria notata* (Steinh.)

Fig. 6, shows one of the illustrations from this work. Since no one in the interval between the appearance of these two volumes had applied other names to the same fossils, the names in question still stand, but the interval of sixteen years cost Schlotheim the honour of being actually the first to apply the binomial system to fossil plants. The credit for this deed was gained by Steinhauer in 1818 (see p. 481). Before, however, we pass to consider the earliest stage in the Scientific History of Palaeobotany, there are other works belonging to the transition period to which a passing reference should be made.

In 1809 William Martin (1767–1810) published at Wigan his completed *Petrificata Derbiensia.* The earlier parts of this work had in reality been obtainable some years previously. The first four numbers were published in 1793, and were called *Figures and Descriptions of Petrifactions collected in Derbyshire.* The object of the work was to furnish ' coloured figures of extraneous fossils '. The illustrations, which included a few fossil plants, give very fair, though not really accurate, representations of the specimens themselves. Let us see, however, what the descriptions of the fossils were like. We will take two examples. What we now call *Alethopteris lonchitica* (Schl.), Plate 10 (reproduced in Plate L, Fig. 7), is described as '*PHYTOLITHUS FILICITES* (*striatus*) frondc bipinnatâ : pinnis ovato-oblongis ' &c. ; and *Stigmaria ficoides,* Plate 11 (reproduced in Plate L, Fig. 8), as '*PHYTOLITHUS PLANTITES* (*verrucosus*) trunco subcylindrico subramosa ' &c. Here we see some primitive attempt at a generic name, but the distinction between the name and the diagnosis hardly exists.

In 1804 the first of three volumes of James Parkinson's *Organic Remains of a Former World* appeared, a most remarkable work considering the state of knowledge of that time. Parkinson (?–1824) was a medical man in practice at Hoxton, then *near* London. The sub-title of this work was ' an examination of the mineralized remains of the Vegetables and Animals of the Antediluvian World ; generally termed Extraneous Fossils '. Here we see two fundamental ideas of the period clearly exhibited—firstly, that extraneous fossils were replicas of plants and animals which *had* once lived, and secondly, that all such fossils were nearly contemporaneous and dated from just before the Flood, which itself was the cause of the *present* existence of fossils. These conceptions marked a great advance on those of Llhuyd, who was ignorant of the real nature of the objects which he described. Parkinson, however, fully understood how fossil plants had come to be preserved, and he carefully distinguished between impressions, casts, and petrifactions. The variety of petrified woods known to Parkinson was remarkable. Among others he describes or mentions the Tertiary wood of Loch Neagh in Ireland, the Cretaceous Palm and other woods of Egypt, the Forest layers in Peat bogs, the Oligocene lignites of Bovey Tracey in Devonshire, and the Permian Star stones (Psaronius) of Chemnitz, Saxony. He also discusses jet, leaves buried in amber, and plant-bearing clay-ironstone nodules, as well as shale impressions. He comments on the 'great difficulty of ascertaining even the genera of the plants which are thus preserved . . . dorsiferous plants and cacti most common', and on 'fossil flowers—their existence doubtful'; 'the tender and almost succulent substance of the petals, stamina, and pistilla, will furnish very little reason for supposing that they should resist a destructive resolution, sufficiently long to allow them to pass through those chemical changes, by which such a length of duration would be given to their original forms, as would secure their passing unchanged, in their figure, from the vegetable to the mineral kingdom'. Parkinson figured certain Coal Measure plants both from shale and clay-ironstone impressions, as well as several types of fossil wood, some of them from polished slabs, but not, it is important to notice, from microscopic sections. An example of his illustrations is shown in Plate LI, Fig. 9.

The Scientific Period

The Scientific period of fossil botany dates from the year 1818, when Steinhauer first described binomially certain British Coal Measure plants in a memoir published in America. In the century which has since elapsed we can distinguish two fairly well-marked phases in the scientific study of fossil plants, one, which we may term the Pioneer period, extending from about 1818 to 1870, and a later Modern period from about 1870 to the present day.

(i) *The Pioneer Stage*

The work of the Pioneer stage was essentially the collection and collation of evidence. The fossil-bearing rocks of all geological ages and in all civilized lands were studied. This work was originally and essentially palaeontological, and was chiefly carried out by those whose inclinations were on the whole geological rather than botanical. Some were, it is true, professed botanists, and a few even great botanists, but the majority of the earlier workers on fossil plants, while by no means ignorant of the knowledge of the day as regards living plants, were professedly palaeontologists or connected with geological institutions. It is only in the Modern Period that the study has attracted a majority of strictly botanical workers, though even now some of the best-known names are those of palaeontologists. Similarly, with few excèptions, the fossils collected during the last century have found a home with other fossils, i.e. in the geological rather than the botanical museums of the world.

But from whichever side the subject may have been approached the result was the same. A flora from one particular set of rocks in one particular place was described and illustrated in comparison with what was already known of similar floras elsewhere. The types new to science were specially distinguished and compared with what appeared to be their nearest relations among known genera or species. The occurrence of other forms already well known to science was duly noticed, with special reference to the occasion on which they had been previously described and figured. All new and rare specimens were so far as possible illustrated. This work was thus purely systematic.

Among the many authorities of the pioneer period it will only be possible here to indicate a few of those who laid the foundations of this work especially in this country. Some of these wrote perhaps only a single short memoir on a fossil flora of one particular age from a certain district. Several of these contributions are, however, just as much landmarks in the history of the science as the larger works of others, which roamed indiscriminately over the whole range of fossil plants of every geological age and from every region where their occurrence was known to the authors. The earliest memoir, in which fossil plants of any age and from any country were binomially described, was, as we have already mentioned, published in America in 1818, by the Rev. Henry Steinhauer in the *Transactions* of the American Philosophical Society of Philadelphia (N.S., vol. i, No. 18, p. 265, 1818). The plants studied were a few Lycopods and Calamites, chiefly from the Yorkshire and Somerset coalfields, and all were referred to a single genus *Phytolithus*. Excellent figures were given of the specimens. One of them (Plate LI, Fig. 10) survives as a type, *Sigillaria tessellata* (Steinh.), in addition to *S. notata* (Steinh.), which by some authorities has been regarded as a distinct species. But for the very limited number of fossils described, this memoir would no doubt rank as of much greater importance than it does. It should, however, be noticed that it antedated all the palaeobotanical writings of both Brongniart and Sternberg—in fact, all scientific works on fossil plants. So far as I am aware it is the earliest printed memoir in any country in which binomial terms were applied to fossil plants. The example set by Schlotheim and Steinhauer was swiftly followed by the publication by a Bohemian, Kaspar Maria, Graf von Sternberg (1761–1838), of the *Versuch einer geognostisch-botanischen Darstellung der Flora der Vorwelt*, Hefte I–VIII (1820–38), the first part of which appeared in the same year as Schlotheim's second work. In 1828 there also appeared the first instalment of the great *Histoire des Végétaux fossiles* (1828–38) by Adolphe Théodore Brongniart (1801–76). In an earlier paper published in 1822 Brongniart had already applied generic and specific terms to certain fossil plants ('Sur la classification et la distribution des végétaux fossiles', *Mém. Mus. Hist. Nat.*, vol. viii), and a complete 'Prodrome' of his *Histoire* had appeared in 1828. The works of Schlotheim, Sternberg, and Brongniart, which appeared in whole or in part between the years 1804 and 1838, may be regarded as the foundations of systematic Palaeobotany. The work of Schlotheim as regards fossil plants, if limited, was good in quality, but of the three writers Brongniart shows forth as the greatest scientist. His *Histoire*, now a rare and valuable work but still in every-day use, is the most cherished possession of the palaeobotanical researcher, and the one of the older works to which he turns most frequently. One of his plates is shown on a reduced scale in Plate LII, Fig. 11. It is no exaggeration to say that Brongniart first described a larger number of the Coal Measure species which occur in Britain than any other author before or since, and many of his determinations were founded on British specimens. By 1825 appeared Artis's *Antediluvian Phytology*, the first scientific work published in Britain on British fossils by a British subject. Edmund Tyrell Artis (1789–1847) was an archaeologist of some note in Northamptonshire. His incursion into fossil Botany was a brief one, and he only published this one small volume, containing twenty-four

PLATE LII

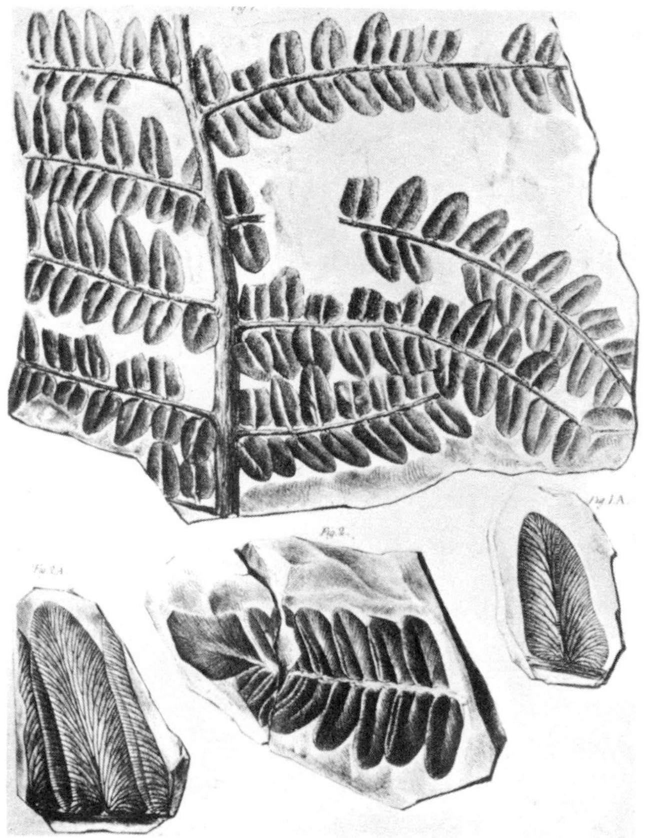

Fig. 11. A. Brongniart, *Histoire des Végétaux fossiles*, Livr. V, 1830, Pl. 68 (reduced). 1, *Neuropteris grangeri*, Brongn., 2, *Neuropteris flexuosa*, Sternb.

Fig. 12. E. T. Artis, *Antediluvian Phytology*, Pl. VII, *Filicites Osmundae* = *Neuropteris Osmundae* (Artis)

plates of Coal Measure plants from Yorkshire. But the work was thoroughly up to date, and the binominal system as applied to fossil plants, then only seven years old, was here fully expressed. The scheme of genera is also much more varied than in any work preceding it. Most of Artis's fossils are types, the specific names of which are still in use, e.g.,

Pecopteris Miltoni (Art.).
Pecopteris (*Dactylotheca*) *plumosa* (Art.).
Alethopteris decurrens (Art.).
Neuropteris osmundae (Art.) (Plate LII, Fig. 12).
Megaphyton frondosum (Art.).

Artis's book was followed in 1831 by the first instalment of Lindley and Hutton's *Fossil Flora of Great Britain* (1831–7), the three volumes of which contained 230 plates, with explanatory text. Plate LIII, Fig. 13, is a reproduction of a typical plate. This work was the product of a combination of scientist and collector. John Lindley (1799–1865) was undoubtedly one of the greatest botanists of the Early Victorian period, not only in England but in the world. He occupied for many years the Chair of Botany at University College, London, and is particularly famous for his work on the systematic classification of plants and on the Orchid family. William Hutton (1798–1860) was a Newcastle merchant, who was an ardent amateur collector of fossil plants, but there is no evidence that he possessed any scientific knowledge of them. The arrangement between these authors was that Hutton in Newcastle got together as many examples of the flora as he could; then he had drawings made and sent the illustrations, and occasionally the specimens also, from time to time to Lindley, the locality records often getting lost or mixed on the way. Lindley described the plants, and the results were published in parts conjointly. The plants figured in the *Fossil Flora* were of every geological age, from Lower Carboniferous to Tertiary, though a majority were derived from the Coal Measures of Britain. The figures on the whole were exceedingly inaccurate, and have caused endless trouble to later workers. Dr. Kidston in 1891, however, went into the whole matter by a comparison of the plates with the original specimens so far as they could be traced, with a view to deciding what was of value and what was useless among these records. Either the artists employed had no skill in the accurate representation of such objects, or, as is more probable, neither authors nor artists had any conception of the absolute necessity for strict faithfulness as regards detail in the illustrations. This is a fault common to much of the older literature, though Brongniart's *Histoire* is a conspicuous exception in this respect. It is only when in actual practice we come to compare a specimen with a published figure with a view to deciding whether they are or are not identical, that the need for absolute accuracy in the drawing comes home to us. The descriptions of the plates, chiefly by Lindley, are not brilliant; and, for the reasons already indicated, the records of ' locality ' are sometimes unreliable. But in one respect these volumes were good—the nomenclature applied to these fossils. The majority of the specimens figured in Lindley and Hutton are preserved in the Hancock Museum, Newcastle-on-Tyne.

I have entered somewhat fully into this work because it unfortunately remains to this day the one illustrated British book containing a general account of our Coal Measure plants. It is, however, far superior to its only British contemporary, Edward Mammatt's *A Collection of Geological Facts and Practical Observations intended to elucidate the formation of the Ashby Coal-field*, which appeared in 1834. This is a large quarto, containing no fewer than 102 plates devoted to the fossil flora of the particular coalfield. The text relating to the plates occupies only five pages, and is restricted to the names of the fossils and previously published figures of the same. Both the fossils themselves, the quality of the illustrations, and the absence of descriptions, combine to render this handsome work mere waste paper so far as a knowledge of Coal Measure plants is in question. With the completed publication of Lindley and Hutton's *Fossil Flora* in 1838, the pioneer period of Scientific Palaeobotany in this country was fairly begun.

(ii) *The Early Scientific Study of Petrifactions*

We now have to turn to other matters affecting certain fossil plants preserved in a particular manner. It had long been known that in various regions petrified woods occasionally occurred, in which the anatomical structure was more or less preserved by replacement by a mineral salt such as calcite or silica. The invention of the microscope led naturally to the idea of investigating

PLATE LIII

Fig. 13. J. Lindley and W. Hutton, *Fossil Flora of Great Britain*, Vol. III, 1837, Pl 222, Figs. 1 and 3, *Trigonocarpum olivaeforme*; Figs. 2 and 4, *T. nöggerathi.*

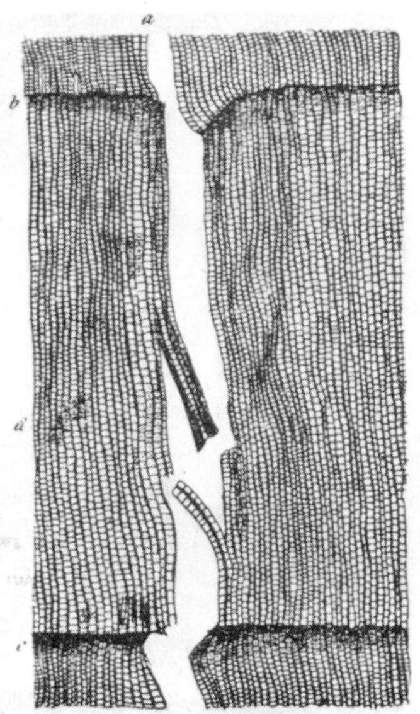

Fig. 14. W. Nicol, *On Fossil Woods from Newcastle, New South Wales.* Edin. New Phil. Journ., Vol. 14, 1833, Pl. III, 1, section of 'petrified conifera'.

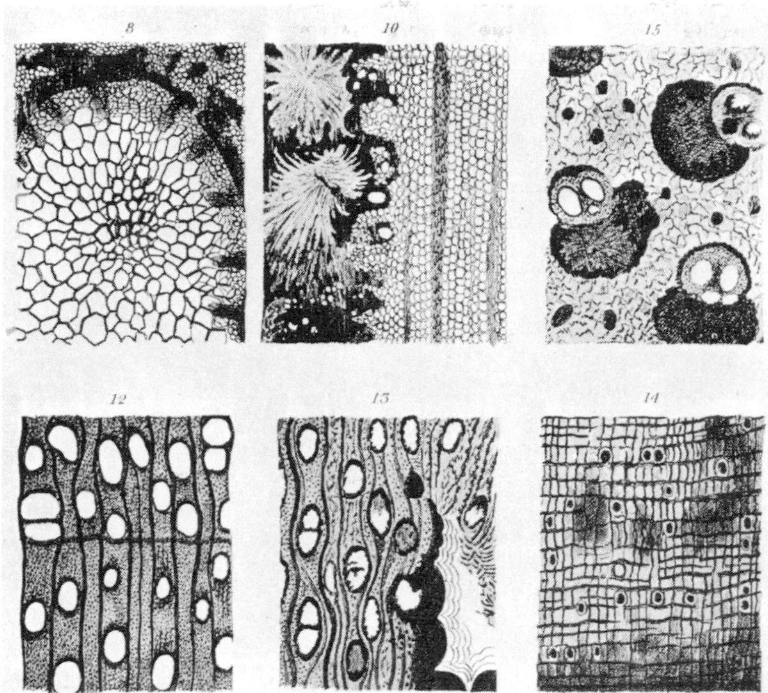

Fig. 15. H. T. M. Witham, *Observations on Fossil Vegetables*, 1831, sections of petrified tissues.

the anatomy of these fossils. Who it was who first overcame the initial difficulties of preparing sections of these woods sufficiently transparent to transmitted light and thus rendered them capable of being examined under the microscope, is a somewhat disputed point. The credit has generally been given, and I believe rightly, to William Nicol (? 1768–1851) of Edinburgh, who is remembered as the inventor of the Nicol prism. Nicol himself gave the credit to his lapidary, Sanderson, who in his turn had no doubt been inspired by Nicol himself. At any rate, Nicol did actually form a collection of thin slices of such woods, most of which are now in the British Museum, though a few are at Cambridge and in Stockholm. He published four papers on these between 1831 and 1835 in the *Edinburgh New Philosophical Journal* (vols. x, xiv, xvi, and xviii), but none of the specimens which he described were from the British Coal Measures. Plate LIII, Fig. 14, shows one of his drawings. After the completion of the publication of the third volume of Lindley and Hutton's *Fossil Flora* in 1837, the record of original work in Britain on fossil plants is almost a complete blank until about 1870, when the modern period commences. During this interval of more than thirty years, several famous botanists made brief excursions into the domain of fossil plants, notably Robert Brown (1773–1858) in 1851, and Sir Joseph Hooker (1817–1911) in 1848. John Morris (1810–86), a geologist in London, appears, however, to have been regarded as the chief authority on Carboniferous plants during this period, but he did little original work. His *Catalogue of British Fossils*, which went through two editions in 1843 and 1854, is, however, remarkable not only for the completeness with which all British fossil plants previously recorded were indexed, but as being the first and so far the last work of its kind published in this country. There is, however, one name of importance which in Britain served as a link between the days of Lindley and Hutton's flora and the modern school—Edward William Binney (1812–1881). Binney, a Manchester coal and paraffin merchant, was an amateur geologist and palaeobotanist who produced over a period of many years a large number of very modest contributions on all sorts of Palaeozoic fossil plants. Of these, those which relate to Carboniferous petrifactions are the most important, and form a connecting link between the works of Witham and Williamson in the history of the anatomical study of British fossil plants. Binney's memoirs on these matters, which especially related to Carboniferous Lycopods and Equisetales, and the seed *Trigonocarpus*, were published between 1855 and 1872.

The earliest results from Coal Measure materials to be published were from the pen of another writer, who repeatedly acknowledged his indebtedness to Nicol not only for material but for the method of obtaining the thin sections. This was Henry Thomas Maire Witham of Lartington in Yorkshire (1779–1844). In 1831 Witham published the first of two works, of similar contents, on the structure of fossil woods from various parts of the world. This was called *Observations on Fossil Vegetables, accompanied by representations of their internal structure as seen through the microscope*, and it was dedicated to William Nicol. The latter part of the title is particularly significant in the history of the study of fossil plants. Plate LIII, Fig. 15, shows some of Witham's anatomical drawings. In the same year Witham also published a memoir on the internal structure of a very important Coal Measure plant, *Lepidodendron Harcourtii*, With. These are the earliest memoirs illustrating microscopic sections published in any country. The second and augmented edition of Witham's book appeared in 1833, only two years later than the first, under the title *The Internal Structure of Fossil Vegetables found in the Carboniferous and Oolitic Deposits of Great Britain.* In the first edition the plants were not named binomially. They consisted of sections of the Lower Carboniferous genus *Pitys* from the neighbourhood of the Tweed and Edinburgh, Coal Measure Dadoxylons from the Newcastle coalfield, fossil woods from the Lias of Whitby, ? Permo-Carboniferous wood from Australia, and Tertiary wood from the Island of Eig or Egg (*Pinites Eggensis*, L. & H.), with Dicotyledons and Palms from Antigua, also of Tertiary age. In the second edition the chief additions were *Lepidodendron Harcourtii* and *Anabathra pulcherrima*, the latter a Lower Carboniferous type. In this edition some of the plants were named binomially.

There was little in these early works of Witham and Nicol which related to the flora of the Coal Measures, though Witham did, in fact, describe a Dadoxylon of that age from the Newcastle coalfield. These researches, however, are very important as being the forerunners of the great anatomical school of Williamson in later years, and of the modern period. With this later period the work of Witham and Nicol is linked by a few papers on the petrifactions of the Coal Measures described by Lindley and Hutton

PLATE LIV

W. C. Williamson

Fig. 16. William Crawford Williamson 1816–1895

in their *Fossil Flora* (1831–7); memoirs on this subject were published by Sir Joseph Hooker in 1848, Robert Brown in 1851, Hooker and Binney in 1855, and William Carruthers in 1867–70.

The earliest work on Carboniferous petrifactions on the Continent was Brongniart's classic on *Sigillaria elegans*, published in 1839, so in this path of progress the English school were distinctly first.

This period ends with the scientific discovery of 'coal balls' in 1868, first foreshadowed by Binney between 1855 and 1862, on which he himself, and in a still greater degree Williamson (Plate LIV) working simultaneously, but independently, established a few years later the anatomical side of palaeobotanical study on a sure footing (Plate LV, Fig. 17, is an example of Williamson's drawing).

(iii) *The Intermediate Period in the Study of Incrustations*

With regard to 'impressions', the modern period in Britain starts about the year 1881, when Dr. Kidston published his first scientific paper, and on the Continent some five years earlier.

We have now to consider what was being done in this country at Coal Measure impressions between 1837 (when Lindley and Hutton's third volume was completed) and 1882. This period of nearly half a century undoubtedly represents a great gap in the history of progress so far as the Coal Measure plants of this country are concerned. John Morris (1810–86), Professor of Geology at University College, London, was the chief authority on such fossils in the fifties, but he did little that was of an original nature. For the rest we have very brief but important contributions by Binney and by Carruthers, and that is all. Comparatively speaking, this period of forty-five years is a blank in so far as Coal Measure plants are concerned, though some progress was made with other floras. The object of the workers of the earlier or pioneer stage of the scientific period was, as a rule, only to record and figure. An illustration was given of a particular fossil which was named binomially and the record of its place of origin was also added. Little more was attempted. The outlook was wholly or chiefly taxonomic. If comparisons were embarked upon, it was usually merely in order to discuss the fossil in connexion with some living plant, often with a view to showing its identity with, or close relationship to, the recent type. Perhaps the outcome of these discussions appears to-day to be more negligible than it really was, owing to the constant error of all the older workers, not one of whom appears to have ever even dreamed of how remote the Carboniferous flora is from the British flora of to-day, both in time and in affinity. This mistaken outlook curiously enough lingers most persistently even among some modern workers; this is all the more remarkable, seeing that it is the achievements of this very school itself which have placed the facts beyond doubt. Speaking generally, it is still not sufficiently recognized how very remote this flora is in affinity from the vegetation of to-day of any part of the globe. It is true that we now know better than to attempt to include types of Carboniferous Pteridosperms in recent genera of Ferns, but we are still too anxious to assign fossils of that age to living families or groups. For the reasons just indicated, it is thus the descriptive rather than the systematic results of the pioneer period which are of value. This type of work naturally is not confined to the period in question. A very large proportion of modern publications on fossil plants are purely systematic and descriptive. And such will always continue to be the case. These are the necessary foundations on which everything palaeobotanical is built. Until a fossil is named, figured, and described it does not exist scientifically.

(iv) *The Modern Period*

The peculiarity of the researches of the modern period, which dates from about 1870, is that while purely descriptive records have been continued unabated, yet a large number of investigations of a special nature have also been undertaken. These may perhaps be best described as intensive cultivations of fossils, in many cases already known to science. A set or group of fossils has been reinvestigated more minutely and on broader lines than would be suggested by a purely taxonomic outlook. The results of these reconsiderations have often been most remarkable, e.g. the establishment of the group Pteridospermeae in 1903. These investigations are to a very large extent purely botanical in outlook and have as much a bearing on recent Botany as on all branches of Palaeobotany. At the same time the geological applications of other studies relating to the occurrence and distribution of Coal Measure plants have proved very important, not only on the pure, but even on the applied side of Geology, and some of these again affect botanical notions of evolution and kindred questions. It is

PLATE LV

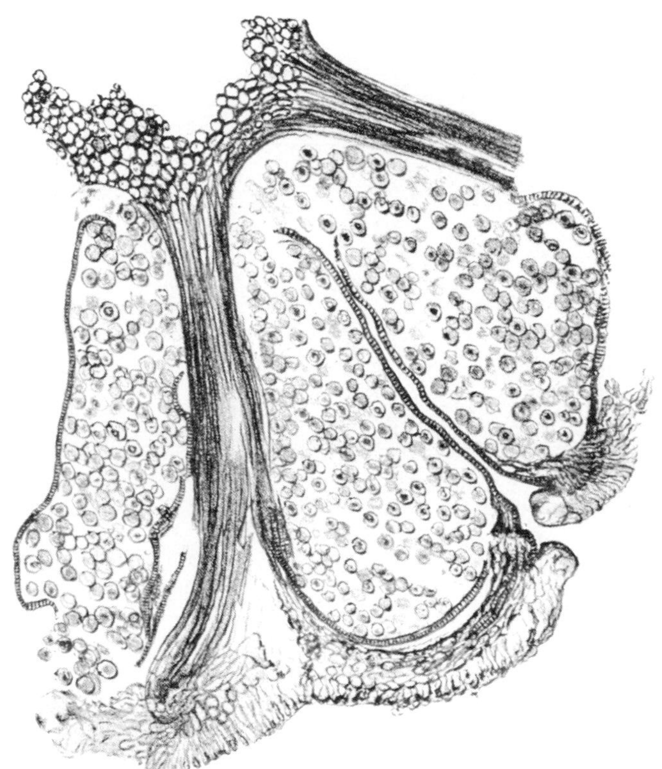

Fig 17. *Calamostachys Binneyana*

From a drawing by W. C. Williamson of the transverse section of the cone

MS. Sloane 4014 fo. 82 r

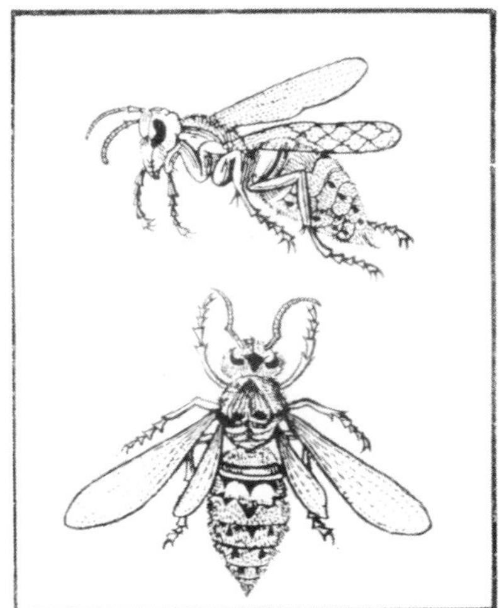

Hornets drawn by Thomas Mouffet before 1589

quite impossible to enter here into these advances in detail; to do so would be to set forth the present position of the subject in all its aspects—a long and laborious undertaking. Suffice it to say that these results, considering that barely half a century has elapsed since the beginning of the period, are certainly very remarkable in many directions and offer great inducements to future workers. If at first they hardly attracted the attention which they deserved, they have, since about the year 1895, received full recognition by the kindred sciences, especially Botany. This modern renaissance arose independently both in Britain and in France at almost the same period. Both schools have been very closely in touch with one another, and each has helped and learnt from the other: while naturally much good work has been done in many other countries, there has been no such body of continuous and linked research on fossil plants as has existed during the last fifty years in France and England. If anything, the French began a few years earlier (Renault 1868, Grand'Eury 1877, Zeiller 1878), but the dates of Williamson's and Kidston's first special researches are so nearly simultaneous, and they arose so independently, that the difference is negligible. Most of the initiators of the great French Modern School have now passed from us, including the two outstanding names of Charles René Zeiller (1847–1915), that Prince of Palaeobotanists, and Bernard Renault (1836–1904). In this country we have lost William Crawford Williamson (1816–95), the Father of Palaeobotanical Anatomy (Plate LIV, Fig. 16), but most of the other workers are happily still with us, among whom the names of Dr. D. H. Scott and Dr. R. Kidston are household words to all interested in Coal Measure plants.

ARCHIMEDES' PRINCIPLE OF THE BALANCE AND SOME CRITICISMS UPON IT

By J. M. Child

No modern student of the history of the science of mechanics can start without confessing indebtedness to the brilliant work of Ernst Mach, and especially to his *Science of Mechanics*. There are certain points of criticism in it, however, with which one cannot agree; and some, indeed, which the average mathematician fails to grasp, through lack of analytical power. The possession of this, however, seems prone to obsess its owner, and to overpower his imaginative qualities. Now, it is just these imaginative qualities that must be drawn on to a large extent, when discussing the work of the ancient Greeks. Thus, it is Mach's criticism of the work of Archimedes that has given rise to this Essay.

In his *Science of Mechanics*, p. 14,[1] Mach writes:

'From the mere assumption of the equilibrium of equal weights at equal distances is derived the inverse proportionality of weight and lever-arm! How is that possible? If we were unable philosophically and *a priori* to excogitate the simple fact of the dependence of equilibrium on weight and distance, but were obliged to go for *that* result to experience, in how much less a degree shall we be able, by speculative methods, to discover the *form* of this dependence, the proportionality!

'As a matter of fact, the assumption that the equilibrium-disturbing effect of a weight P at the distance L from the axis of rotation is measured by the product $P.L$ (the so-called statical moment), is more or less covertly or tacitly introduced by Archimedes and all his successors. For when Archimedes substitutes for a large weight a series of symmetrically arranged pairs of small weights, which weights *extend beyond the point of support*, he exploys in this very act the doctrine of the centre of gravity in its more general form, which is itself nothing else than the doctrine of the lever in its more general form.

[1] By this title is meant the English edition published by the Open Court Pub. Co., 1907; trans. McCormack. The supplementary volume, edited by P. E. B. Jourdain, 1915, will be referred to as Mach's *Supplement*.

'Without the assumption above mentioned of the import of the product *P.L*, no one can prove that a bar, placed *in any way* on the fulcrum *S*, is supported, with the help of a string attached to its centre of gravity and carried over a pulley, by a weight equal to its own weight (Fig. 1). But this is contained in the deductions of Archimedes, Stevinus, Galileo, and also in that of Lagrange.'

That the view taken by Mach is not beyond criticism is shown by the fact that, in two notes on pp. 513, 514 of the *Science of Mechanics*, Mach himself mentions an essay by O. Hölder *Denken und Anschauung in der Geometrie*, in which the author upholds the correctness of the Archimedean deductions against the criticism of Mach. Further, in the *Supplement*, Mach calls attention to a 'beautiful paper by G. Vailati', *La dimostrazione del principio della leva data da Archimede*, which appeared in the *Bollettino di bibliografia e storia delle scienze matematiche* for May and June 1914. In this paper, Vailati takes, as Mach says,

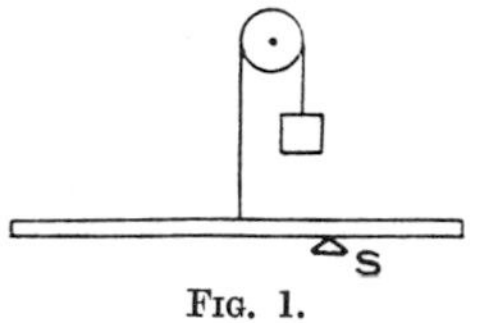

FIG. 1.

'the side of Hölder against my criticism of Archimedes' deduction of the law of the lever, but partly too Hölder is criticized. I believe that every one may read Vailati's exposition with profit and, by comparison with what I have said on pp. 17–20 of the third edition of my *Mechanics*, will be in a position to form a judgment upon the points at issue.'

Thus the aim of Mach, as he says later in the note, has been to strive to disturb the impression that the general law of the lever could be deduced from the equilibrium of equal weights on equal arms; and thereafter to show where the experience that already contains the general law of the lever is introduced. Hölder argues against the correctness of Mach's criticism, while Vailati shifts the ground of the debate, by suggesting that Archimedes 'derives the law of the lever on the basis of general experiences about the centre of gravity'. Mach concludes his note by remarking that 'if the reader has derived some usefulness out of this discussion, I am not very particular about maintaining every word I have used'.

The third paragraph of the quotation from the *Science of Mechanics*, as given above, can be dismissed. In the fourth German edition, it is omitted in this connexion; and, on p. 514 of the *Science of Mechanics*, there is a direction that, instead of this paragraph, there should be inserted on p. 20, at the end of the first paragraph, the passage that is given below ; it is to be noted that there is no direction given, as there should be, for the deletion of the paragraph from p. 14. The diagram is the same (Fig. 1), and the new context reads :

'A knife-edge may be introduced at any point under a prism suspended from its centre without disturbing the equilibrium, and several such arrangements may be rigidly combined together so as to form apparently new cases of equilibrium. The conversion and disintegration of the case of equilibrium into several other cases (Galileo) is possible only by taking into account the value of *PL*.'

This cannot possibly be contained in the deductions of Archimedes, for according to Archimedes it is wrong, since to a balanced load there cannot be added an unbalanced load without disturbing the equilibrium. Using modern phraseology, if the knife-edge S were placed *in any manner, at any point of the bar except at its middle point, so as to actually touch the bar even in the slightest degree*, the equilibrium would be disturbed ; in fact such an arrangement might be used as an indicator of the first moment of contact with extreme delicacy. The revised version of the paragraph certainly does not state that the equipoise is equal to the weight of the bar, at least not directly ; but this is the only conclusion we can come to, if we consider that the bar is supposed to be suspended from its centre before the knife-edge is introduced, and secondly that the two figures are identical.

The second half of the second paragraph ought also to have been noted for deletion. Its place is taken by the following passage (p. 513) :

'It is very obvious that if the arrangement is absolutely symmetrical in every respect, equilibrium obtains on the assumption of any form of dependence whatever on the disturbing factor *L*, or, generally, on the assumption $P.f(L)$; and that consequently the particular form of dependence *PL cannot possibly* [1] be inferred from the equilibrium. The fallacy of the deduction must accordingly be sought in the transformation to which the arrangement is subjected. Archimedes makes the action of two equal weights to be the same under all circumstances as that of the combined weights acting at the middle point of their line of junction. But,

[1] I have here correctly given the words to be in italics ; McCormack has wrongly italicized the word 'particular'. The original is : 'demnach kann aus diesem Gleichgewicht unmöglich die bestimmte Form *PL* abgeleitet werden.'

seeing that he both knows and assumes that the distance from the fulcrum is determinative, this procedure is by the premise unpermissible, if the two weights are situated at unequal distances from the fulcrum. If a weight situated at a distance from the fulcrum is divided into two equal parts, and these parts are moved in contrary directions symmetrically to their original point of support; one of the equal weights will be carried as *much towards* [1] the fulcrum as the other weight is carried *from* [1] it. If it is assumed that the action remains *constant* [1] during such procedure, then the particular form of dependence of the moment on L is implicitly determined by what has been done, inasmuch as the result is only possible provided the form be PL, or be *proportional* to L. But in such an event all further deduction is superfluous. The entire deduction contains the proposition to be demonstrated, by assumption if not explicitly.'

Leaving the criticism of these passages for a while, and coming to the historical side of the question, we find that Mach gives prominence only to the work of Huygens, Lagrange, Leonardo da Vinci and Galileo, with a mere passing reference to Guido Ubaldi del Monte. There is no mention of Guldinus's work, *Centrobaryca*, or *De Centro Gravitatis*, in which that author starts with the Postulates of Archimedes (of which he quotes nine), three axioms of Guido Ubaldi, the hypothesis that there is but one centre of gravity, followed by eight Propositions, with corollaries, taken from the first book of Archimedes 'on Equiponderants'. From this work it would seem that Vailati's view was not the view taken by Guldinus. For we cannot assume that Guldinus would found the whole of his work on Archimedes' Principle of the Balance, if he had thought that this principle had been obtained from the consideration of general experiences about the centre of gravity. Guldinus quotes freely from a commentary on the works of Archimedes written by Guido Ubaldi; this commentary is not mentioned by Mach.

Then, again, there is no mention made of the work of Cavalieri, given in the fifth section of his *Exercitationes Sex*, wherein is to be found the definition of the term 'moment', and that too in phraseology which seems to suggest that it was first used by Cavalieri as a definite nomenclature; and this, again, suggests that he was the first man to *definitely* see its importance as

[1] The equivalents in the German were set for italics; I have also given a more correct rendering of 'nähert', the words 'near to' having an ambiguous meaning.

a *principle*, apart from haphazard use of it in practice. He notes the fact that a body may have different moments according to its position. Coupled with this work is the work of Pascal on the cycloid; the first section of 'The Letters of Amos Dettonville', which contained the 'Method of the Centre of Gravity', may be assigned with almost certainty to a study of the above-mentioned exercitation of Cavalieri; the whole of Pascal's method depends on the principle of the 'balance', and yet Pascal is not mentioned in this connexion by Mach. These two men, Cavalieri and Pascal, gave Leibniz the idea for his centrobaryc analysis, and are, if for that reason alone, of extreme importance. For Leibniz inverts the usual process, and obtains a fundamental theorem in the calculus from the theorem that the moment of an area is equal to the sum of the moments of its parts.

Fig. 2.

Now the work of Cavalieri and Pascal was, and is, considered as an extension of the work of Archimedes; with Cavalieri and Pascal, it is the equilibrium-effect and not the turning-effect, or any disturbing-effect, that is always considered; even though the former defines and uses the 'moment', after he has derived the idea from Archimedes, while Pascal expressly uses the word 'balance'. In fact, it is not the law of the *lever*, but of the *balance*, or even more correctly the law of *balanced things*, that is primarily considered by Archimedes, and applied geometrically for geometrical purposes: thus the Latin title of 'Equiponderants' is a most suitable rendering. What Guldinus thought of the principle is evident from the vignette on the title-page of the first volume of his *De Centro Gravitatis*, a freehand sketch of which is here given (Fig. 2).

Of course it is not suggested that Archimedes was unaware at the start of the proportionality of the arms of the balance. Gow, in his *History of Greek Mathematics*, states that Aristotle remarks that cheating tradesmen would shift the centre of their balances towards the scale in which the weight lay (*Mech. Prob.* 1, fin.); and remarks that weights in equilibrium on a lever are inversely proportional to the arms inversely. Even if, as considered by Wohlwill (see *Supplement*, p. 6), the 'Mechanical problems' are not by Aristotle, there is every possibility, if not probability, that the law was known to Archimedes. My contention is that Archimedes, 'in his Grecian mania for demonstration, strives to get round this', as Mach himself says, on p. 28 of the *Science of Mechanics*; and that he succeeds without *the introduction of the idea of moments*; which is in opposition to Mach, who finishes his sentence with the words, 'his deduction is defective'.

I propose, therefore, to endeavour to show how the deductions of Archimedes might have been obtained without even a knowledge of the principle of moments. Since these deductions lead to the inverse proportionality of the arms of the balance, my first step must be to show that *the principle itself is actually contained in the postulates*, and therefore that there is no need for it to have been introduced by Archimedes at any intermediate step of his reasoning.

Of the nine postulates, the only ones that concern us at present are:

1. Equal weights at equal distances balance.
2. If weights balance at any distances, and to one of them a weight is added, then they will not balance; but there will be a downward motion of the weight to which something has been added.
3. Similarly, if something is taken away from one of the weights, they will not balance; but there will be a downward motion of the weight from which nothing has been taken away.
8. If magnitudes balance one another at any distances, things that are equal to these magnitudes will also balance at the same distances.

Now, as Mach states (p. 513, see quotation given above), these postulates contain the principle that the distance is determinative; in other words, that for a given weight set at a given distance, there is (i) *but one weight at a given distance*, and (ii) *but one distance for a given weight*, with which a balance can be effected.

For, it is not too much to assume the 'common notions', stated by Guido Ubaldi, in addition to these postulates. The first two of these axioms are respectively:

1. If from equiponderants, equiponderants are taken away, the remainders are equiponderant.

2. If to equiponderants, equiponderants are added, the wholes are equiponderant.

These together form a statement of what is called 'the principle of the superposition of loads'; and this principle is one of the, if not *the*, most important fundamental principles of mechanics; apparently Mach does not consider it, perhaps on account of its obvious nature. The third axiom is:

3. Weights that balance the same thing are equal.

This, in other words, is the statement that, if there are two weights, for any given distance, that will balance a given weight at a given distance, then these two weights must be equal; that is, there is but the one weight for the given distance. Further, if there are supposed to be two distances for one given weight, that will effect a balance with a given weight at a given distance, then two equal weights will balance one another at these unequal distances; which is obviously contrary to the postulates.

Thus, in modern phraseology, there is established a one-to-one correspondence between P (the weight) and L (the distance at which it is set), in order that a balance should be established with a given weight at a given distance. If we now make the assumption, made by Mach without any statement as to the grounds of the assumption, that P is some function of L, we can make, to start with, a more general assumption as to the form of the function. The assumption of a function is justified on the hypothesis of the simplicity and regularity of nature; Mach says that 'equilibrium obtains on the assumption of *any* form of dependence whatever on the disturbing factor L, or, generally, on the assumption $P.f(L)$'; but this is hardly justifiable. If we consider ourselves justified in assuming functionality, we must, at any rate to start with, make a more general assumption. The equilibrium-effect (note that it is not necessary to suppose a turning-effect, or any disturbing-effect of any sort, or to consider such in any way), must be the general function $f(P, L)$. It is not until we consider the one-to-one correspondence that we can say that this function must be of the form

$$a.P.L + b.P + c.L + d.$$

But, having obtained this form of the function, it is evident that it must be zero when (1) $P=O$, (2) $L=O$, (3) $P=O$ and $L=O$; hence b, c, d, are all equal to zero, and thus the effect is *proportional* to *P.L.* In other words, *the inverse proportionality of the arms is contained in the postulates*; and it only remains for Archimedes to deduce it by accurate reasoning with the means at his disposal; and this he can do *without introducing the idea of the statical moment at any step in the course of his reasoning.*

The next step we have to make is to give Archimedes' reasoning, and Mach's version of it; and to criticize the latter. To do this, it is necessary to give a summary of the first six propositions of Archimedes.

Prop. I states that 'weights that balance at equal distances are equal'; Prop. II states that 'unequal weights at equal distances do not balance, but give a preponderance on the side of the greater weight'; Prop. III states that, 'for unequal weights at unequal distances, the greater weight is at the less distance'. These are merely more or less obvious converses to the postulates, or rather corollaries to them. In Prop. IV there is a transition from weights to magnitudes (from βάρεα to μεγέθεα in the Greek), and we have the introduction of the term that is usually translated 'centre of gravity' (κέντρον τοῦ βάρεος). Archimedes proves, by a *reductio ad absurdum* depending on the second postulate, that 'the centre of gravity of a combination of two equal magnitudes is at the middle point of the line joining the centres of gravity of the two magnitudes'. In the proof Archimedes states that the fact that it lies in the straight line has been already proved, and the remark is merely repeated by Eutocius in his commentary; we are left in doubt as to whether this proof has been given elsewhere by Archimedes or by others. The term κέντρον βάρεος had been invented before the time of Archimedes, according to Gow, who seems to have considered the fact that Archimedes uses it without defining it; just as I use the fact, that Cavalieri carefully defines the term *momentum*, to reason that Cavalieri was the first to consider the term as signifying something equivalent to what we now call a moment, or at least to use it in this definite sense. That this was not the general sense of the word *momentum* is clearly shown by the preliminary note made by Eutocius.

'Both Aristotle, and Ptolemy after him, teach us that the tendency downwards (ῥοπή) is a certain sort of gravity and levity common to all things. Further, in the works of Plato, Timaeus says that all this downward tendency comes from gravitation; indeed, he thinks that levity is but a lack of it. . . . Now, in this book, Archimedes considers that the centre of this downward tendency [this ῥοπή, which Torelli translates by 'momentum *quod dicitur*'] is a point, such that if the plane, of which it is the centre, is suspended by this point, it will still remain parallel to the horizon; and for two, or more, planes it is the point such that, if the balance is hung from this point, it will remain parallel to the horizon.'

Thus, with Archimedes, there is no idea of a centre of parallel forces; the centre is merely the *balance-point*. The modern equivalent is *centroid*. The whole nomenclature is merely for *a pictorial purpose*; it is just as if Archimedes had said: You know that equal weights will balance at equal distances, and so on; well, the same idea (*and it will be useful to employ the same language*) can be adapted to geometrical reasoning about planes and things, *which have no weight*. We will consider that they have a 'weight-centre', at which we can consider that the whole plane is condensed; and that a balance can be established between two planes, and suchlike, just in the same way as with weights hung on a balance. Then, Archimedes, in Prop. IV, starts his work on planes; this start being marked by the transition in the original from 'weights' to 'magnitudes'. I think there is corroboration of this in the fact that, in later work, Archimedes uses an axis of balance, and not merely a point of suspension, without making any reference to the change. With this view, also, it would have been impossible for him to have given a geometrical proof that, when condensing his planes at their 'weight-centre', the magnitude at the centre is the sum of the magnitudes of the planes that are condensed; which is the equivalent of the fact that, with weights on a balance, the force on the fulcrum is the sum of the suspended weights. This assumption is made by Mach without any remark; and it is a greater assumption than any that he accuses Archimedes of making, and *itself includes the whole 'law of the lever'*. For, if two equal weights are supported by a double weight at the centre, we have, by taking moments round either end of the balance, the fact that a given weight will be in equilibrium with one twice as large, when the arms are in the ratio of 1 to 2. The proof for balanced planes depends on the first postulate, the superposition of loads, and the postulate that two loads that balance a given load are equivalent to one another. Thus suppose that M and N (Fig. 3) are two equal magnitudes having A and B as their centres; and let C be any other point not on the line AB. Join AC, BC; let D, E be their middle points; join DE, and draw any line CFG to cut DE in F, and AB in G. Then it is clear that CF is equal to FG. Now the magnitude M at A can be balanced about DE, i.e. about the middle point of AC, which is on DE, by a magnitude equal to M at C; and similarly, N at B can be balanced about DE by a magnitude equal to N at C. Therefore M and N together can be balanced by a magnitude equal to the sum of M and N, situated at C. But the latter can be balanced by a magnitude equal to the sum of M and N, situated at G, since $GF = FC$. Therefore, M at A and N at B are equivalent to the sum of M and N at G, some point on the straight line AB; *if, and only if, it is known that the 'weight-centre' of* M *at* A *and* N *at* B *is on the line* AB.

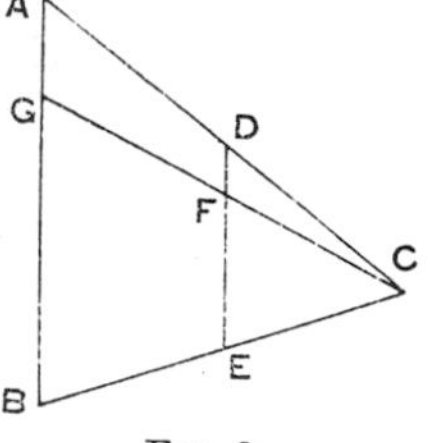

Fig. 3.

Thus the fundamental postulate is that the centre is on the straight line joining the respective centres of the two magnitudes; Archimedes says this has been proved, and Eutocius merely repeats the assertion, without giving any information as to how, when, where, and by whom the demonstration has been given; this is the reason why I have called it a postulate. It is to be noticed that the above proof only shows that the sum of the magnitudes can be taken at a 'weight-centre', somewhere on the line AB; the position of the point on that line has to be determined by other means, since, in the demonstration given, we can only consider an axis of balance parallel to AB.

Once this Prop. IV is demonstrated, I therefore contend that the language of Archimedes is merely figurative and pictorial. This view is corroborated by the concluding sentence of the proof of Prop. VI; the Greek is: *τοῦ μὲν ἄρα* A *κειμένου κατὰ τὸ* E, *τοῦ δὲ* B *κατὰ τὸ* Δ, *ἰσορροπησοῦντι κατὰ τὸ* Γ. In this it is to be noted that the preposition *κατὰ* is used in all three cases; and, although this preposition is capable of a variety of meanings, it is highly probable that it would retain the same meaning in the same sentence. Again, although Torelli translates the prepositions by . . . *ad* . . . *ad* . . . *in*, making a change in the third case, it is to be noted that he uses *in* (with the accusative = upon) and not *circa* (= about). I shall, therefore, in what follows, speak of 'a balance at the point P, or at the line AB'; the latter phrase being short for 'a balance at any point on AB'. I shall use the term 'centroid' or 'centre' also, instead of 'centre of gravity', to emphasize the point that it is merely a balance-point, or weight-centre at which the whole plane can be assumed to be condensed, *figuratively* speaking; in this way, I believe I shall give the true interpretation of Archimedes' work.

Prop. V extends Prop. IV to the cases of any number of magnitudes; in it, according to the figures given by Torelli, it is not necessary that the magnitudes should all be equidistant, so long as they are symmetrical. Nor does the demonstration demand this; all that is required is that, for any *pair* at equal

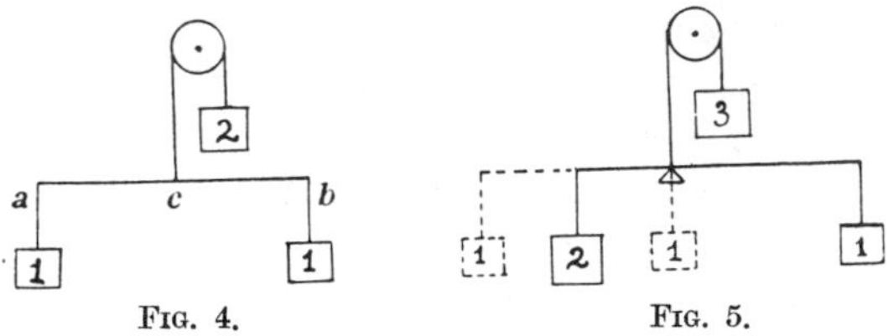

Fig. 4. Fig. 5.

distances on each side of the middle, the 'gravities' should be equal. We are here forcibly reminded of the definition of the 'centre of gravity' given by Boscovich in his *Philosophiae Naturalis Theoria*, a definition that is purely geometrical in principle. Boscovich considers all bodies to be made up of points, having equal gravities; and the gravity of a body is therefore proportional to the number of points in it. With this idea, he defines the position of the centre of gravity (retaining, as he remarks, the name, but not using the idea of gravity at all), and proves the existence of the centre for all bodies. The gist of what Boscovich says is as follows. A plane can be found parallel to any given plane, such that the sum of the distances from it of all the points to the right of it is equal to the sum of the distances of all the points to the left of it. Three such planes, mutually at right angles, determine a point; this point is the centre of gravity. Any plane through the centre of gravity has the equal-distances property. If several points are situated at the same place, as in a particle, then the corresponding distance must be taken as many times as there are points in the body; which comes to the same thing as multiplying each such distance by the mass of the particle. The idea of Boscovich I take to be an almost exact equivalent of that of Archimedes.

Before giving Archimedes' proof of the famous Prop. VI, I will quote, from pp. 10, 11 of the *Science of Mechanics*, Mach's attempt to 'reproduce in general outlines the train of thought by which Archimedes endeavours to reduce the general proposition of the lever to the particular and apparently self-evident case'. The two equal weights 1 suspended at *a* and *b* are, if the bar *ab* be free to rotate about its middle point *c*, in equilibrium. If the whole be suspended by a cord at *c*, the cord, leaving out of account the weight of the bar, will have to support the weight 2. The equal weights at the extremities of the bar supply accordingly the place of the double weight at the centre (Fig. 4).

'On a lever (Fig. 5), the arms of which are in the proportion of 1 to 2, weights are suspended in the proportion of 2 to 1. The

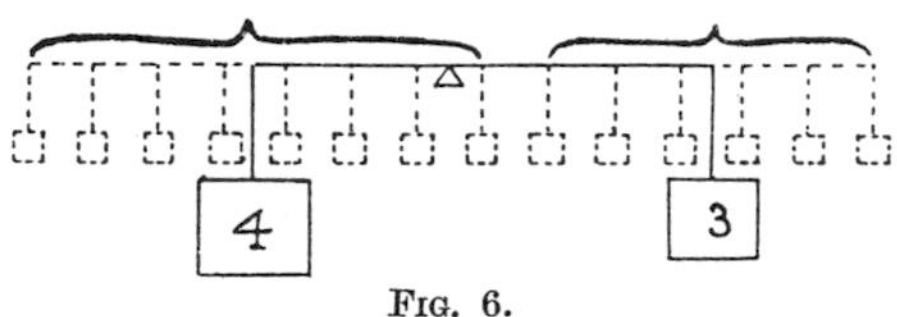

FIG. 6.

weight 2 we imagine replaced by two weights 1, attached on either side at a distance 1 from the point of suspension. Now again we have complete symmetry about the point of suspension, and consequently equilibrium.

'On the lever arms 3 and 4 (Fig. 6) are suspended the weights 4 and 3. The lever arm 3 is prolonged the distance 4, the arm 4 is prolonged the distance 3, and the weights 3 and 4 are replaced respectively by 4 and 3 pairs of symmetrically attached weights $\frac{1}{2}$, in the manner indicated in the figure. Now again we have perfect symmetry. The preceding reasoning, which we have here developed with specific figures, is easily generalized.'

It is upon this 'reproduction of Archimedes' train of thought' that Mach founds his objections; and I contend that all the difficulties are of Mach's own making. First of all Archimedes does not use a pulley, nor apparently does he view the balance in elevation, but in plan. He does not load his 'lever' with weights suspended from the bar; but his bits of planes are attached to the line of centres, anyhow as far as the wording of the demonstration goes, though the figures show them to be laid balanced across the line. He does not produce the bar *after* it has been once loaded, nor does he *replace* a weight 2 by two weights, each equal to 1. In fact Mach misrepresents Archimedes in every possible way, as will be seen by comparison with the proof of Prop. VI, as given by Archimedes, reproduced below.

Proposition VI

Commensurable magnitudes will balance at distances inversely proportional to their gravities.

Let A and B be commensurable magnitudes, and let their centres be A and B; also let ED[1] be any straight line; and suppose that the magnitude A is to the magnitude B as the length DG is to the length GE. It is required to show that G is the weight-centre of the magnitude which is made up of A and B taken together (Fig. 7).

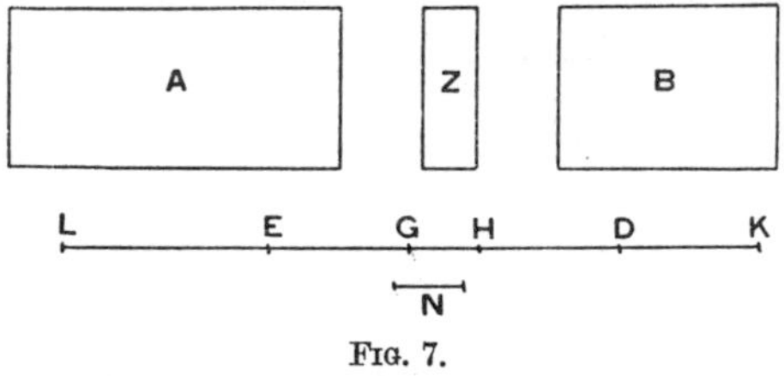

Fig. 7.

For, since $A:B=DG:GE$, and also A and B are commensurable; hence, DG and GE are commensurable, straight line to straight line; therefore they have a common measure; let this be the straight line N. Take DH, DK, each equal to EG, and EL equal to DG. Then, because $DH=EG$, therefore $DG=EH$; hence also $LE=EH$. Now, LH is double of DG, and HK of GE; therefore N will measure both LH and HK, since it measures their halves. Also, since $A:B=DG:GE$, and $DG:GE=LH:HK$, each being double of each; therefore also $A:B=LH:HK$.

Now, whatever multiple LH may be of N, suppose that A is the same multiple of a magnitude Z; so that $LH:N=A:Z$. But we have $KH:LH=B:A$; hence, $KH:N=B:Z$; that is to say, KH and B are equimultiples of N and Z respectively. But A is also a multiple of the same magnitude Z; therefore Z is a common measure of A and B.

Hence, if LH is divided into parts that are each equal to N,

[1] Thus worded, the exact meaning of the last sentence of the paragraph is obscure.

and *A* also into parts each equal to *Z*; then the number of parts in *LH*, that are equal in magnitude to *N*, will be equal to the number of parts in *A*, that are equal to *Z*. Hence, if to each of the parts of *LH* there is applied one magnitude equal to Z, so that its 'weight-centre' is situated at the middle point of the part; all the magnitudes taken together will be equal to *A*; and the 'weight-centre' of the magnitude formed from all these magnitudes taken together will be the point *E*. For the whole set are even in number, on account of the fact that $LE=EH$.

Moreover, it can be shown in a similar manner that, if to each part of *KH* there is applied a magnitude equal to *Z*, so that its 'weight-centre' is situated at the middle point of the part, all these magnitudes taken together will be equal to *B*, and the 'weight-centre' of the magnitude formed from all these magnitudes will be the point *D*.

Consequently, *A* is, in effect, applied at the point *E*, and *B* at the point *D*.

But, with this arrangement, we have a set of magnitudes all equal to one another, situated in a straight line, and their 'weight-centres' are equidistant from one another; also their number is even. It is plain, therefore, that the 'weight-centre' of the magnitude formed from all these magnitudes is the middle point of the straight line joining the centres of the two middle magnitudes. But, since $LE=GD$, and $EG=DK$, therefore the whole $LG=$ the whole *GK*. Hence, the 'weight-centre' of the magnitude formed from all the magnitudes taken together is the point *G*.

Therefore, *A*, applied at *E*, and *B*, applied at *D*, will balance one another at *G*.

Let us review the account given by Mach, in the light of the foregoing demonstration. First of all, even supposing that Archimedes had wished to give a pictorial illustration of Prop. VI, introduced maybe with some such words as, 'Just as in the case of a lever, we see, &c.', it is difficult to imagine him using a string and pulley. He would, I think, have been far more likely to have used a more delicate test, one that was free from all difficulties introduced by the friction of the pulley and the rigidity of the cord, and so on; namely, the method of suspending his first balance from one extremity of another balance, as in Fig. 8. Secondly, his figures give no hint of suspension to either a material or a non-material lever; they are simply supposed to be placed *at* certain points that are all in a straight line; and although pictorially the effect of the string is to localize the point of application, yet the hanging weight introduces an idea that is quite foreign, and conveys a wrong impression. Hence, I consider that if Archimedes had wished to illustrate the condensation of the 'weight' at the balance-point, by means of the force on the fulcrum of a lever, he would have shown it in plan, as in Fig. 9.

We next come to the actual facts, as given in the words of Archimedes, which are wrongly represented by Mach. We note that Archimedes takes any straight line, divided in the required proportion, performs certain geometrical operations upon it, and proves certain facts about it and the two magnitudes under

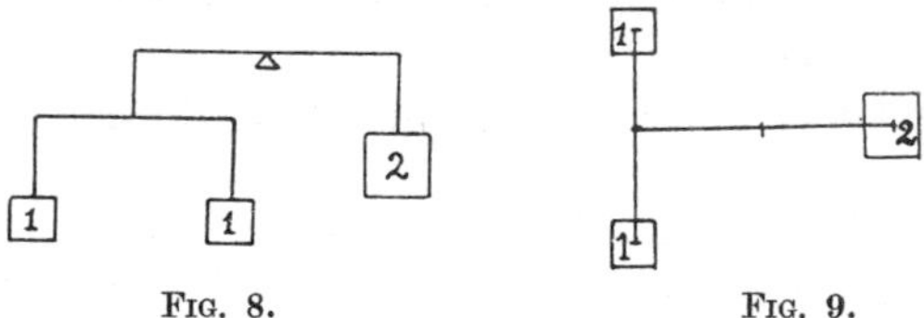

FIG. 8. FIG. 9.

consideration, before ever there is any talk of loading the line. There is no hint as to any material difficulty in loading the line, such as would occur if the line were short and the magnitudes large. Thus there is no distinction made, such as can be comprehended from the figures given in Fig. 10, where for the same length of line we have different areas of rectangles as our magnitudes; in the bottom figure, the parts cannot be 'laid' horizontally on the line. His idea evidently is that they are, so to speak, strung on the line, so that their centres coincide with certain points on it; and there is no notion of any special disposition with regard to those centres. He *first of all loads his line uniformly,* the number of loads being determined by the particular common measure that he takes as that of the two segments of his original straight line. He then finds two equivalent loads to the uniform load; and these are then stated to be equivalent to one another; there is no replacing of the one principal load by the other, in the manner suggested by Mach.

Having thus criticized the explanation of Mach, I am bound to give a better explanation, if I can, for others to criticize. This I will endeavour to do, trying to make it match the reasoning of

Archimedes step by step. The foundation of the mathematical demonstration of a science must be certain perceptions, or fundamental principles, at least for the earliest investigator. Although these in some cases apparently are flashes of inspiration, in general they are not so; what is termed the 'instinctive cognition', in the case of such principles as those of 'symmetry' or 'superposition of loads', is merely the result of automatic or involuntary classification of all experiences that have impressed the brain, though maybe unnoticed at the time, nor indeed until their cumulative effect, like that of a cumulative poison, has become great enough. If we wish to explain this cognition to others, we

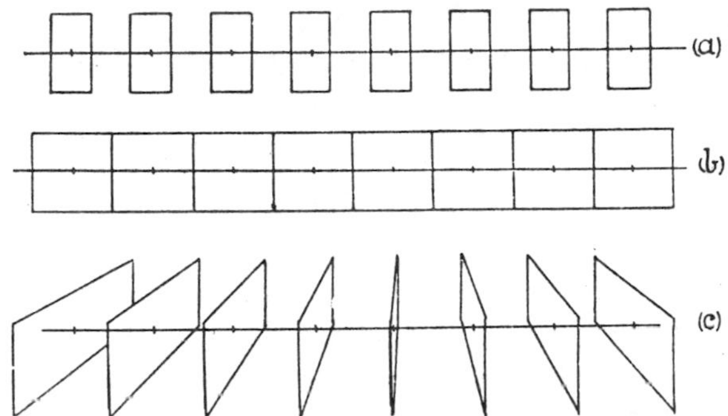

FIG. 10. Diagram (*c*) is drawn in perspective; the other two can be considered as being in either plan or elevation.

must appeal to the experiences from which it has been derived. Thus, Archimedes' postulates and first three propositions are an appeal of this nature. It is just as if he said: You know what happens in the case of a balance, equal weights balancing at equal distances, and so on. Well, the same idea can be used in the case of straight lines and planes. Although these have no weight, we can imagine that, when a straight line is compared with a straight line, we can consider that it has, so to speak, a weight proportional to its length, and that this weight can be supposed to be condensed at its middle point, and it balances about this point. Similarly, in the case of planes, we can speak of such things as rectangles, and circles, when compared with things of the same nature as themselves, as having weights proportional to their areas, that these weights may be considered as being condensed at their balance-points, which are their geometrical centres. With this idea, it is fairly clear that the 'weight-centre' of two equal planes or straight lines is the point about which they balance, that is the point about which they are symmetrically placed, *as far as their respective 'weight-centres' alone are concerned*; and so on. A modern teacher might illustrate Prop. V with a pack of cards, pinning each card by a pin through its centre to the appropriate point on the blackboard; and thus give his class a good idea of the principle of symmetrical disposition. In a similar manner, by an appeal to experience the principle of superposition of loads could be introduced, and applied to things having in reality no weight. Starting, then, with these preliminary notions, I suggest the following interpretation of the steps of the reasoning of Prop. VI, giving at the same time the steps from which Archimedes derived his general enunciation.

Suppose we start with an even number of equidistant planes or lines, say six, arranged with their centres all in a straight line, A, B, C, D, E, F. Then their weight-centre is their balance-point G, half-way between C and D. Hence, if H is any other point in the straight line, the whole six will balance, with respect to the point H, a plane, of six times the area, with its weight-centre at a point K, such that $HK=HG$; for the six planes can be supposed condensed at their weight-centre G, and would then balance six at K (Fig. 11 *a*).

Again, the four magnitudes A, B, C, D, will balance four at a point M_1, where HM_1 is equal to the distance of H from the middle point of BC; and the two at E, F will balance two at L_1, where HL_1 is equal to the distance of H from the middle point of EF. Hence, superposing the two loads, the whole six will balance, with respect to H, four at M_1 and two at L_1 (Fig. 11 *b*).

Similarly, the six will balance two at M_2, where HM_2 is equal to the distance of H from the middle point of AB, and four at L_2, where HL_2 is equal to the distance of H from the middle point of DE (Fig. 11 *c*).

Now reverse the first balance, left to right, and superpose it upon one of the others, say on that given in (*b*); then we have the balance shown in (*d*). But we can remove the six pairs at equal distances; and the balance shown in (*e*) is left. But 6 at G will be balanced by 6 at K; hence K is the weight-centre, balance-point or centre of gravity of 2 at L_1 and 4 at M_1. The reasoning is general for any even number of equidistant equal magnitudes. This is essentially the demonstration given by Archimedes, except that he shortens the proof, as soon as he has considered the equivalence in figures (*a*) and (*b*), by immediately stating that the two right-hand sides are equivalent to one another.

I submit that this is a more correct interpretation of the reasoning of Archimedes; and also that this gives a more feasible explanation of the manner in which he arrived at his general proposition. That is to say, he did not reduce the harder cases

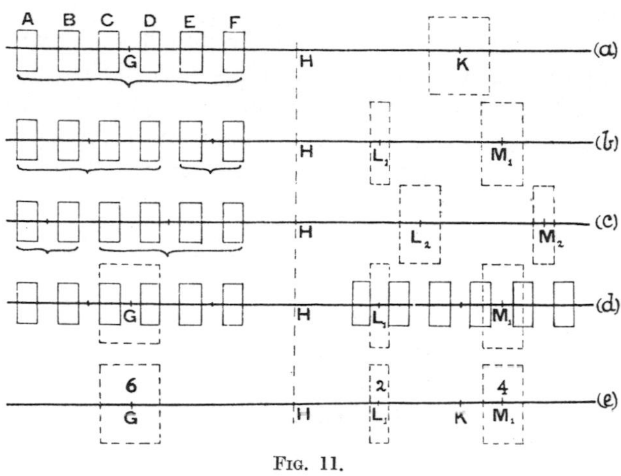

Fig. 11.

to the simple case; but, from a general consideration of the general case of symmetrically placed magnitudes, *he was led* to the case of non-symmetry. I am convinced that, as a *mechanical idea*, the notion of 'moment' does not enter into his reasoning. I will go even further; I do not believe that, at the time of composition of the first book on Equiponderants, Archimedes would have admitted such a notion as the foundation of a rigorous mathematical demonstration. Further still, I do not believe that he *ever* attained to the idea embodied in the formula $\Sigma mx/\Sigma m$, as applied to finding the centroid.

For, in the first book of Equiponderants, after extending Prop. VI to incommensurables, he proceeds to find the centroid of a triangle. Here, he is forced to give a complicated indirect proof, because mathematical accuracy will not permit him to omit the triangles, by which a strip of a triangle cut off by two lines parallel to a side differs from a parallelogram, *no matter how narrow that strip may have been taken.* Thus, instead of saying that the centres for all such strips lie on the median that bisects the side, and therefore the centre of the triangle lies on this median, he gives a proof of which the following is a brief account (see Heath's *Archimedes* for the full proof). That is to say, he does not feel justified in applying Prop. VI and VII except to magnitudes which have a *geometrical centre of point-symmetry,* such as a line, a rectangle, and a parallelogram (and it should be noted that for the latter there is not line-symmetry, which corroborates my contention that the essential idea of Archimedes was condensation of the magnitudes at their centres, irrespective of their disposition). Neither, at this stage at any rate, does he appreciate the idea of the infinitely narrow strip; or, at least, he refuses to admit it into accurate reasoning, although he uses it as a method of *discovery.*

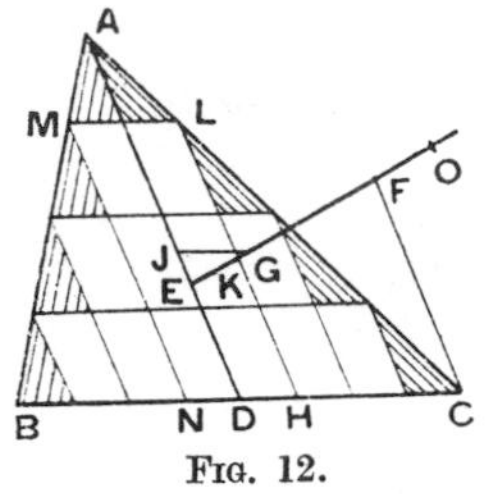

Fig. 12.

Proposition XIII

If possible, let the centre of ABC be at G, to the right of the median AD; draw GJ parallel to BC to meet AD in J; divide BD, DC into parts equal to DH, where DH is less than GJ; complete the figure by drawing parallels.

Then the centre of the unshaded figure is on AD, at E say. Join EG, cutting LH in K, and a parallel to AD through C in F; this point must then lie completely to the right of the triangle ABC.

Now the ratio of all the shaded triangles to the whole triangle ABC is as AL to AC, i.e. as EK to EF. But, since G is the centre of the triangle, and E that of the unshaded portion, then the centre of the shaded portion is at O, where

$$EG : EO = \text{shaded portion} : \text{whole triangle}$$
$$= EK : EF.$$

Now, since $EG > EK$, it follows that $EO > EF$; that is to say, O, the centre of the shaded triangles is still more to the right, than F, of the triangle; which is impossible.

Now the only reason, that I can see, for the employment of an indirect proof, is the rigorous manner in which Archimedes proceeds with his chain of propositions. For, if he allowed the mechanical idea to enter the reasoning, he has at his command a proof by the method of exhaustions. Thus suppose we consider one of the strips of a triangle cut off by parallels to the base. Suppose *PQRS* (Fig. 13) is such a strip, and that *T*, *U* are the middle points of the parallel sides. Let the base *QR* be divided into an even number of small parts, *QX*, *XY*, . . . *LU*, *UM*, . . . *ZW*, *WR*; complete the figure as in the diagram, where, as in Archimedes' figure for Prop. XIII, *PX*, *KL*, *NM*, *SW*, are all parallel to *TU*.

Then *PQZS* is a parallelogram, *T* and *L* are middle points of opposite sides; hence its centre is in *TL*, i.e. to the right of *KL*; *hence considering* LT *as the knife-edge of a lever*, the whole strip *PQRS* will preponderate on the right, and its centre will be to the right of *TL*, i.e. still more to the right of *KL*. Similarly it is to the left of *NM*. Hence, the narrower the strip, i.e. the smaller the parts into which the base is divided, the narrower is the parallelogram which includes the centre of the strip.

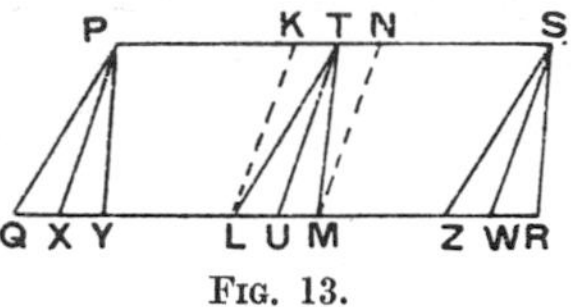

Fig. 13.

It follows by a continued application of the theorem that the centre of two magnitudes combined into one lies on the line joining the respective centres of the two magnitudes, that the centre of the triangle *ABC* in Prop. XIII lies to the left of *LH*, and to the right of *MN* in Fig. 12, *no matter how small the parts* ND, DH *may be*. Hence, it follows, by the method of exhaustions, that the centre lies on *AD*.

Now, this is distinctly the same train of reasoning that runs through the postulates and the first three propositions, before the transition is made to magnitudes; so that I am convinced that my suggestion that these are merely illustrative is the correct interpretation of them. That Archimedes, later at any rate, used the method of infinitely small strips, or, what comes to the same thing, the idea that an area is made up of parallel lines, and a solid of parallel planes, is proved by the palimpsest MS., discovered by Heiberg, of which a translation has been published by Heath, as a supplement to his *Archimedes*. A translation has also been given, in the *Monist* for April, 1909, by Miss Lydia G. Robinson, and this has been reprinted in pamphlet form and published by *The Open Court* as 'The Method of Archimedes'. It is to this pamphlet that I shall make reference in what follows, to avoid reprinting the original propositions, as given by Archimedes. In the introductory remarks of this letter to Eratosthenes, Archimedes states (see p. 8, l. 11): 'much that was made evident to me through the medium of mechanics was later proved by means of geometry, because the treatment by the former method had not yet been established by way of a demonstration'. I will leave the consideration of the suggestions I have so far made to my readers, in the light of this pregnant passage: and I will pass on to my last point, that Archimedes does not use anything that can truly be said to be equivalent to a general moment theorem based on the formula $\Sigma mx/\Sigma m$, even in his 'Method' of discovery. To prove this most effectively, I will give modernized versions of the theorems in Miss Robinson's pamphlet (referred to as *R.M.A.*), and make remarks upon them as I proceed.

THEOREM I (compare with *R.M.A.*, p. 10).

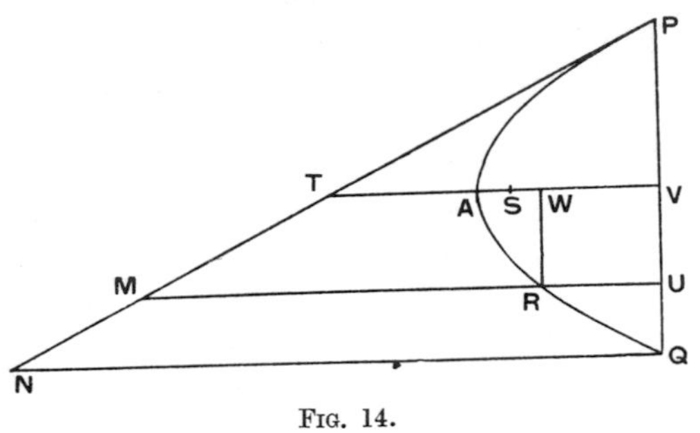

FIG. 14.

We have $4\,AS.RU = 4\,AS.(AV - AW) = PV^2 - RW^2 = PU.UQ$;

$4\,AS.MU = 4\,AS.TV.PU/PV = 2\,PU.PV$:

Hence $MU : RU = 2\,PV : UQ$, or $MU.UQ = RU.PQ$.

Summing, we have $\Sigma MU.UQ = \Sigma RU.PQ = \Sigma PQ.RU$;

i.e. (segment APQ) . PQ = moment of PQN about QN

$= 4.\Delta APQ \,.\, PQ/3$

$\therefore$ 3 (segment APQ) = 4 (triangle APQ).

Observe that the proof is valid for a non-symmetrical segment, if PQ is taken parallel to the tangent at the vertex of a diameter TV.

THEOREM II (compare *R.M.A.*, p. 11).

We have (Fig. 15) $AE^2 + EG^2 = AG^2$, i.e. $EF^2 + EG^2 = EH^2$.

Hence, multiplying by π, and summing, we have all the circles of the cone formed by the rotation of ACD together with all the circles of the hemisphere formed by the rotation of ABC are equal to all the circles of the cylinder formed by the rotation of $ABDC$.

But the cone is one-third of the cylinder; therefore the hemisphere is two-thirds of the cylinder.

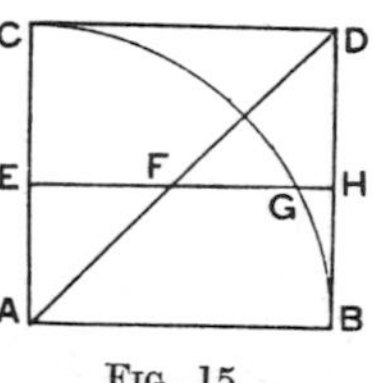

Fig. 15.

The first of these proofs is a mere version of the original; but the second is different altogether in form from the original. The method is so obvious, if the *summation* is admissible, that it would have been impossible for Archimedes to have missed it; but it does not lend itself to the method of balancing. Hence I suppose that it was not used owing to the infinite summation involved. The second point to be observed in the originals, as given by Archimedes, is that the lines in the first theorem and the circles in the second, are balanced in pairs; that is, Archimedes at this date uses nothing but the simple case of *two* balanced magnitudes, and then superposes loads; he makes no use of a general moment theorem. Finally, when he was investigating a quadrature, as in these theorems, his effort was to find a fourth proportional, if we use the mechanical idea for a moment, to (1) the distance, (2) a constant distance, (3) the weight, as represented by his line or circle; if the fourth proportional happened to work out as an elementary line or figure, of which he knew the dimensions, his success was assured. So that, here it is the *quotient* of the weight and the distance, and not the *product* that is in question.

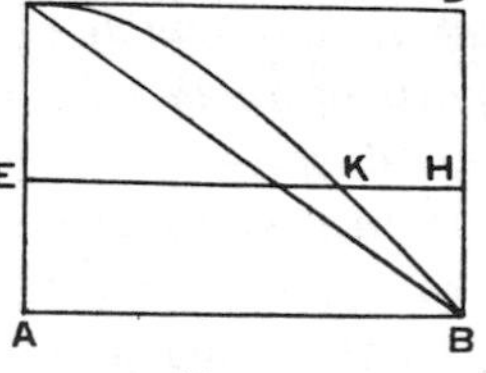

Fig. 16.

Theorem IV (compare with *R.M.A.*, p. 15).

Since CKB is a parabola, we have

$$EK^2 : EH^2 = CE : CA;$$

or

$$EK^2 . CA = EH^2 . CE:$$

hence, multiplying by π, and summing, we have

$$\begin{aligned}(\text{vol. of conoid}) . CA &= \text{sum of moments of the circles of the cylinder about } CD\\ &= \text{moment of the cylinder about } CD\\ &= (\text{vol. of cylinder}) . \tfrac{1}{2}CA;\end{aligned}$$

hence vol. of conoid is half that of the cylinder.

But the vol. of cone is one-third that of the cylinder ; hence the vol. of the conoid is one and a half times that of the cone.

There is but one point worth noting in this theorem, namely, that Archimedes quotes the volume of the cone in terms of the cylinder ; just as in Theorem I he quotes the position of the centre of a triangle. The former of these reduces to the latter. For, in Fig. 17, if $EF' : EF = \pi . AB : AC$, we have

$$\pi . EF^2 = CE.EF' ;$$

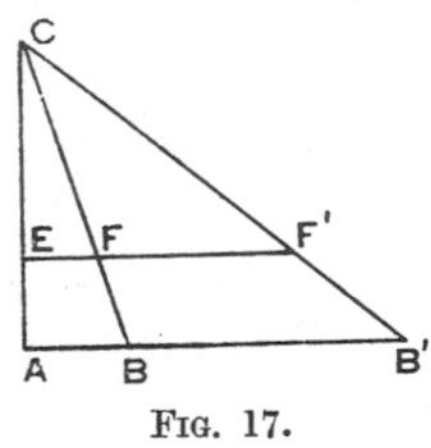

FIG. 17.

hence, summing, all the circles forming the cone are together equal to the moment of the triangle ACB' about a parallel to AB' through C, i.e. to the area ACB' multiplied by the distance of its centre from C ; from which we have

$$\text{vol. of cone} = \tfrac{1}{2}\, CA.AB' . \tfrac{2}{3}\, AC = \tfrac{1}{3}\, \pi . AB^2 . AC.$$

The question therefore at once arises : why, since Archimedes has demonstrated the position of the centre of a triangle, does he not give the volume of a cone as a corollary from it, as a sample of his method ? Of course the answer may be that, as Archimedes says, the volume of the cone in terms of the cylinder had been enunciated by Democritos and proved by Eudoxus. I suggest, however, that the reason is far different ; and that it is closely connected with my contention. Archimedes, as we see in Theorem I, is quite prepared to say that a triangle, ' where it is ', is equivalent to its area condensed at its centre ; *but he is not prepared to say that the triangle, where it is, is equivalent to the sum of the moments of its parts, or to the area condensed at the centre. when he is unable to ' balance ' each of these parts separately against parts of another figure.* Apparently, there is no way in which this could have been done ; and he is not prepared to adopt the infinite summation that is inherent in the general formula $\Sigma mx / \Sigma m$, when one side of his equation is a *pure summation* and the other a *summation of moments.*

THEOREM V (compare *R.M.A.*, p. 17).

Since, in Fig. 18, APQ is a parabola, we have

$$PN^2 : QM^2 = AN : AM = LN : QM,$$

$$\text{i.e. } PN^2 : LN^2 = AM : AN ;$$

$$\text{or } PN^2 . AN = LN^2 . AM.$$

Hence, multiplying by π, and summing, we have

$$\Sigma\,(\pi\,.\,PN^2.\,AN)=\Sigma\,(\pi\,.\,LN^2)\,.\,AM.$$

Therefore, if K is the centre of the conoid, we have

$$(\text{vol of conoid})\,.\,AK=(\text{vol. of cone})\,.\,AM\,;$$

$$\text{i.e. } AK:AM=2:3,$$

for the conoid is one and a half times the cone.

In this theorem, the method is the converse to that used for Theorem I; here a centre has to be found by means of a known cubature, and the product of 'weight' and 'distance' is involved, since the circles of the given figure have to be kept 'where they are'. Still Archimedes balances the pairs of circles, *one pair at a time,* and does not use a general moment theorem.

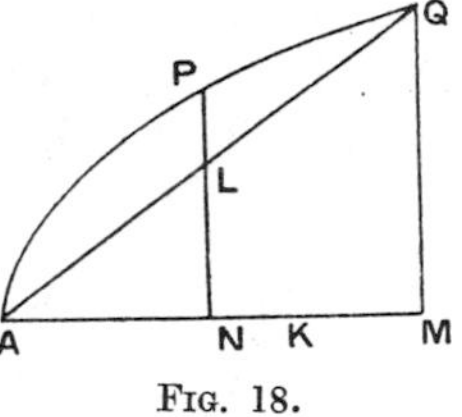

FIG. 18.

Archimedes might have determined the volume and the centre of the conoid at one and the same time, with a single figure, if he had so chosen; and, if he had done so, it would have been a most elegant example of his method. I cannot imagine why he did not

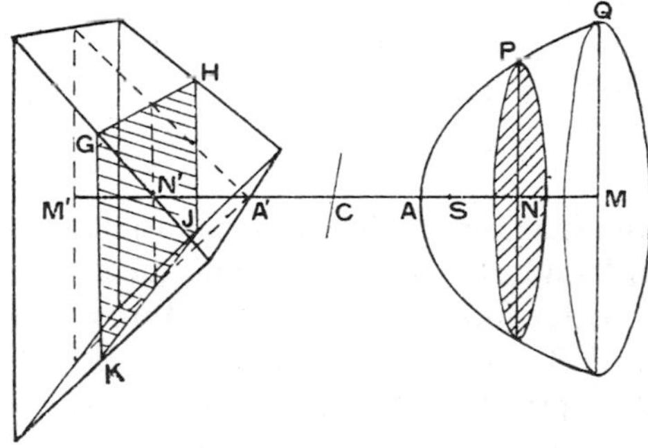

FIG. 19.

do so, except that, for the sake of clarity, the figure must be given in perspective.

Thus, in Fig. 19, if the breadth of the right-angled isosceles prism is π times the semi-latus-rectum of the generating parabola, it is easily seen that there is a 'balance', *involving nothing but equal magnitudes at equal distances,* between the circles of the paraboloid and the rectangles of the prism.

For, the circle of radius $PN = \pi\,.\,PN^2 = 4\,\pi\,.\,AS\,.\,AN$; and the rectangle $GKJH = GK \,.\, GH = 2A'N' \,.\, 2\,\pi \,.\, AS$; and these are equal if $AN = A'N'$.

Hence, the vol. of conoid = vol. of prism

$$= (\text{base of prism}) \,.\, \tfrac{1}{2} A'M'$$
$$= (\text{base of conoid}) \,.\, \tfrac{1}{2}\, AM$$
$$= \tfrac{1}{2}\ (\text{circum-cylinder}).$$

Also, since the two solids will balance at the middle point, C, of AA', NN', MM', &c., and their magnitudes are equal to one another, their centres must be at equal distances from C. That is to say, the distance of the centre of the conoid from A is two-thirds of AM.

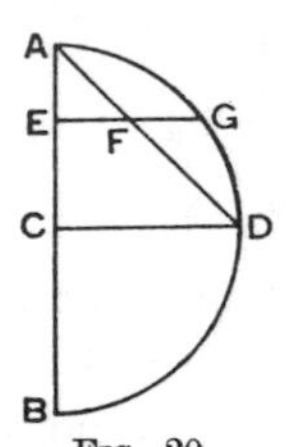

FIG. 20.

THEOREM VI (compare *R.M.A.*, p. 19).

In Fig. 20, we have $EG^2 = AE.EB$,

hence
$$EG^2 + EF^2 = AE.EB + AE^2 (= EF^2)$$
$$= EF.AB.$$

Multiplying by $\pi \,.\, AE$, and summing, we have

$$\Sigma\, (\pi \,.\, EG^2 .\, AE) + \Sigma\, (\pi \,.\, EF^2 .\, AE) = \Sigma\, (\pi \,.\, EF^2).AB\,;$$

i.e. (hemisphere).AX + (cone).AY = (cone).AB,

where X and Y are the centres of the hemisphere and the cone respectively.

Now, $AY = 3\ AC/4$, and the hemisphere is double the cone; therefore we have

$$2\ AX + 3\ AC/4 = 2\ AC$$
$$\text{i.e. } AX = 5.AC/8.$$

In this theorem again, Archimedes has to consider the product of the 'weight' and the distance; but he evades the moment summation by balancing in pairs. He states that, if the circle whose radius is EF is transferred to a point K on CA produced, such that $AK = AB$, it will balance the two circles together, whose radii are EF and EG, where they are. There is another point about this theorem to be noticed, namely, that Archimedes does not complete the numerical part of the proof of his enunciation; he stops short at the determination of the balance between the solids. As the given figure contains two letters, designating the generating rectangles of two cylinders which are not referred to in the proof, it seems probable that there is here a hiatus in the manuscript, and the rest of the proof goes on as in Prop. VIII, which has the same cylinders and the same letters to designate their generating rectangles. It will be useful to see how Archimedes treats the matter, *evading the infinite summation, and the reasoning of an algebraical character*; so I will now give a simplified version of Archimedes' own demonstration, merely eliminating to some extent the prolixity of the argument.

THEOREM VIII (compare *R.M.A.*, p. 21).

If, in Fig. 21, EAZ is half a right angle, we have

$$PL^2+AP^2=AP.PC+AP^2$$
$$=AP.AC,$$

or $PL^2+PO^2=AP.AT$, where AT is taken equal to AC.

Hence, a circle with radius PL together with a circle of radius PO is to a circle with radius PO as AT is to PO (or AP). Therefore

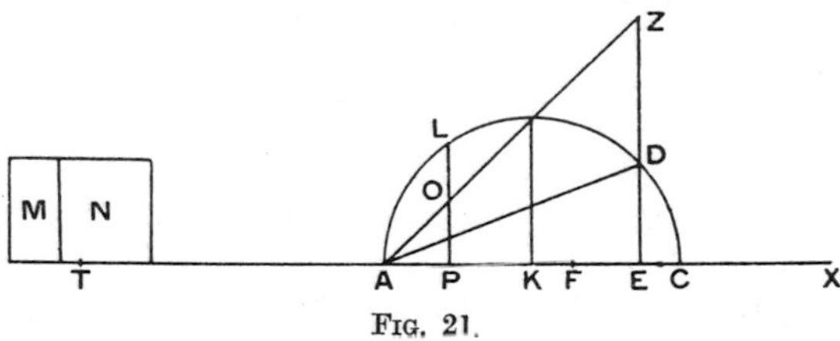

FIG. 21.

the spherical segment generated by $ALDE$ together with the cone generated by AEZ, where they are, will balance the cone generated by AEZ, set with its centre at T.[1]

Let F be a point such that $AE=4\ EF$, then F is the centre of the cone generated by AEZ. Let the cylinder generated by M and N together be equal in volume to this cone; also let the part of it generated by M, at T, balance the cone, where it is. Then the cylinder generated by N, at T, will balance the spherical segment, where it is. Hence (here I invert the argument of Archimedes for the purpose given in the footnote [2]), we have

$$\frac{\text{segment generated by } ALDE}{\text{cylinder generated by } N}=\frac{AT}{AK},$$

where K is the centre of the spherical segment generated by $ALDE$.

[1] This corresponds to the point at which the demonstration of Prop. VI breaks off.

[2] This, I think, corresponds more with an *attempt to find* the ratio in question, than what is actually given by Archimedes, who *starts* the last part of the argument with the ratio, and then proves that K thus taken is the centre.

Now, if CX is equal to the radius of the sphere, it has been proved (*Sph. and Cyl.*, ii. 2, Cor.) that

$$\frac{\text{segment generated by } ALDE}{\text{cone generated by } AED}=\frac{XE}{EC};$$

but

$$\frac{\text{cone generated by } AED}{\text{cone generated by } AEZ}=\frac{ED^2}{EZ^2}=\frac{AE.EC}{AE^2}=\frac{EC}{AE};$$

also, since cone generated by AEZ balances the cylinder generated by M, and is equal in volume to the two cylinders generated by M and N together, we have

$$\frac{\text{cone generated by } AEZ}{\text{cylinder generated by } M}=\frac{AT}{AF}, \text{ and } \frac{\text{cone generated by } AEZ}{\text{cylinder generated by } N}=\frac{AT}{CF}.$$

Hence, since $AT=AC$, by combining the ratios given, we have

$$\frac{AC}{AK}=\frac{XE}{EC}\cdot\frac{EC}{AE}\cdot\frac{AC}{CF}, \text{ or } \frac{AE}{AK}=\frac{XE}{CF}.$$

Now

$4\ FC=AE+4\ EC$, and $4\ XE=2\ AC+2\ EC=2\ AE+6\ EC$;

therefore $AE : AK=2\ AE+6\ EC : AE+4\ EC$,

and $AK : KE=AE+4\ EC : AE+2\ EC$.

Now compare the above demonstration with what follows, a proof which might have been given by Archimedes, if he had used the general moment theorem. Starting with the equality $PL^2+PO^2=AP.AC$, and multiplying by $\pi.AP$, and summing, we have, since $AP=PO$,

$$\Sigma(\pi\,.\,PL^2.\,AP)+\Sigma(\pi\,.\,PO^2.\,AP)=\Sigma(\pi\,.\,PO^2).AC,$$

or (seg. gen. by $ALDE$) . $AK+$(cone gen. by AEZ) . AF
$=$ (cone gen. by AEZ) . AC;

i.e. seg. gen. by $ALDE$: cone gen. by $AEZ = CF : AK$. But,

$$\frac{\text{seg. gen. by } ALDE}{\text{cone gen. by } AED}=\frac{XE}{EC}, \quad \frac{\text{cone gen. by } AED}{\text{cone gen. by } AEZ}=\frac{EC}{AE}$$

and $CF : AK = XE : AE$; and the rest follows as before.

Now, it is seen that the introduction of the two cylinders generated by M and N have materially complicated the argument; and it stands to reason that Archimedes would not have introduced them except for a very good purpose. This purpose, it seems to me, is nothing else but the means of avoiding the summation, which I have shown leads so easily to the required ratio $CF : AK$. The same thing is apparent in the next three theorems; and I should not give these, except on account of their extreme ingenuity, and because I think that the perspective figures I have used may enable my readers to follow Miss Robinson's pamphlet more easily, than is the case with Archimedes' figures which are in plan only. There is also another reason; the figures for the first two theorems in one case have the elementary planes across the line taken as the balance-beam, and in the other case along it; showing, as I have stated, that Archimedes is concerned with the centre only of these planes, and its position, and not at all with the disposition of his planes with regard to this centre. It may be, too, that this has something to do with the fact that the figures are given in plan only, for in the perspective view I have had to make them stand on edge on the line of balance; but I have done this only to avoid complicating the figure by introducing a parallel through the centres, and it does not militate against my argument. The reader must imagine that the letters of the proof refer to the parallel through the centre, which would be similarly divided to the base line; Archimedes, by giving only the plan, effects this automatically. The mention of the word 'balance-beam' above reminds me that I have not yet made a note of the fact that in these theorems Archimedes frequently makes use of the phrase 'Think of so-and-so as a scale-beam', according to Miss Robinson's translation; even here the language is figurative and employed for producing a better mental picture. In Prop. VI of the *Quadrature of the Parabola* we find a more or less distinct statement by Archimedes, in the following words:[1] '*Imagine* that which is proposed to be considered to be viewed in a plane upright to the horizon and the letters AB (?); then *deem* points on the same side of AB as D to be *below*. . . .'[2] He *thereafter* goes on to speak of AB as part of '*the* scale-beam' (ζυγόν), a term which is not used in the first book of Equiponderants. This shows very clearly that, at least in his own estimation, Archimedes, led to it by a consideration of the known properties of the lever, proves geometrically the principle of balance for planes and other magnitudes, and then uses the mechanical idea afterwards, and

[1] Νοείθω δὲ τὸ, ὅτε ἐστὶ τὸ ἐν τᾷ θεωρίᾳ προκείμενον, ὁρώμενον ἐπὶ ὀρθοῦ ποτὶ τὸν ὁρίζοντα καὶ τὰς AB γραμμάς· ἔπειτα τὰ μὲν ἐπὶ τὰ αὐτὰ τῷ Δ κάτω νοείθω.

[2] The italics in this quotation are, of course, mine.

language derived therefrom, simply as a pictorial means of fixing the ideas.

THEOREM XI (compare *R.M.A.*, p. 23).

In Fig. 22, we have

$$\begin{aligned} \text{parallelogram } KE : \text{parallelogram } HE &= FK : FH \\ &= DO : ML \\ &= CD : CM \\ &= CN : CM, \end{aligned}$$

where CN is equal to the radius of the cylinder.

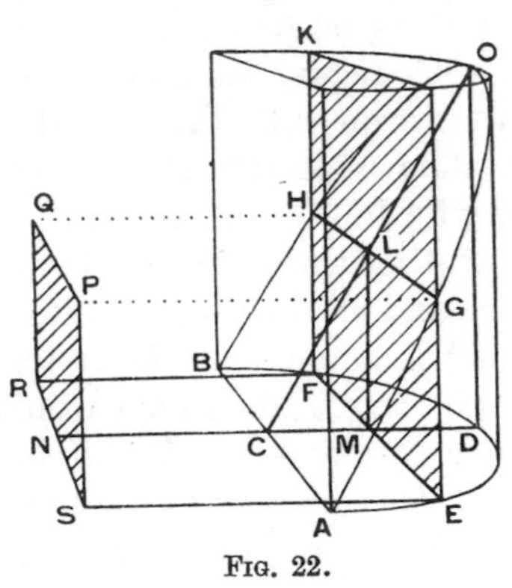

FIG. 22.

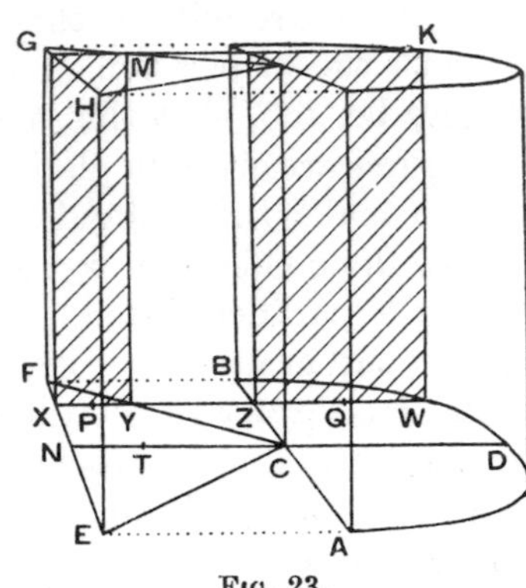

FIG. 23.

Hence, the parallelogram HE, if placed in the position $SRQP$, will balance at C the parallelogram KE where it is. Therefore the section $ABDO$, condensed at N, will balance the semi-cylinder where it is, at the same point C.

THEOREM XII (compare *R.M.A.*, p. 21).

The semi-cylinder on ABD as base and the prism on FEC as base can be cut by planes in corresponding parallelograms ZK, XM (Fig. 23). Also there will be another pair on the other side of NCD at an equal distance from it. . . .

(Here there is a hiatus in the manuscript ; the reasoning can, however, be easily supplied as follows.)

Suppose that P and Q are the middle points of XY and ZW respectively. Then we have

$$2\,PZ = 2\,XZ - 2\,XP = AB - XY = AB - FX = AB - BZ = AZ\,;$$

hence, $$2\,PZ.XY = AZ.ZB = ZW^2 = 2\,ZW.ZQ,$$

and $$XY : ZW = ZQ : ZP.$$

Hence, the parallelograms standing on XY and ZW are in the ratio of ZQ to ZP also; and, when combined with a similar pair on the other side of the diameter NCD, and equidistant from it, for the sake of symmetry, it is readily seen that there is a balance at C; and therefore also for the semi-cylinder and the prism.

Hence, if T is the point on CN immediately below the centre of the prism, i.e. the centre of the triangle ECF, we have

$$\text{vol. of section} : \text{vol. of this prism} = CT : CN = 2 : 3;$$

for the section, condensed at N, has been shown to balance the semi-cylinder, where it is; and it is therefore equivalent to the prism where it is.

Hence the volume of the section is one-sixth of the prism standing upon the square in which the base of the cylinder is inscribed.

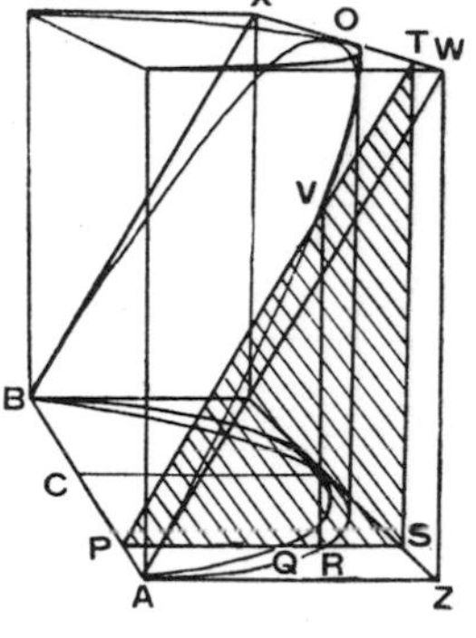

Fig. 24.

Theorem XIII (compare *R.M.A.*, p. 26).

If, in Fig. 24, BQA is a parabola, PR is a mean proportional between PS and PQ.

Hence, $\Delta PST : \Delta PRV$
$= PS^2 : PR^2$
$= PS : PQ.$

Now, for all sections, the antecedents of the equal ratios are constant; hence the sum of the antecedent triangles is to the sum of the consequent triangles as the sum of the antecedent lines is to the sum of the consequent lines; that is, the cylinder-section is to the prism whose base is AZW as the parabola AQB is to the rectangle ZB, i.e. as 2 is to 3. Therefore the cylinder-section is one-sixth of the whole prism circumscribing the cylinder.

It is especially to be noticed, about the first two of these last three theorems, that in the demonstrations we have an exact equivalent to a change in the order of integration; the first summation being that of planes parallel to AB, and the second that of planes parallel to CD. When the form of the result has been obtained, the ratio 2 to 3 gives the clue to the parabola, and its ratio to the circumscribing parallelogram; and the geometrical proof is then fairly evident. In precisely the same way, the mechanical ideas of the postulates and the first three propositions have given him the clue to the *idea of geometrical balance of magnitudes having no weight*; the proofs of the propositions on Equiponderants thereafter, starting with Prop. IV, are rigorously geometrical, although couched in terms reminiscent of their mechanical suggestion.

Now, what is effected in the balance proved in Prop. XII (p. 518) is the determination of the centroid of a semi-cylinder, or rather that of a semicircle; where the semi-cylinder and the prism are regarded as summations of equal and similarly situated planes perpendicular to each of the other sets given. For the proof really is that the triangle FEC balances at C the semicircle ABD; hence, if V is the centroid of the semicircle, we have

$$\text{triangle} : \text{semicircle} = CV : CT,$$
$$\text{i.e. } CV : CD = 4 : 3\pi.$$

Lastly, we here see that Archimedes has the idea of a solid as the sum of planes, and a plane as the sum of lines, fully developed as a working notion, even if not as a means of rigorous proof. I take it that, in a similar way, he considers the line as a sum of *points*, and not as a succession of short straight lines; or else, surely, we should find in this connexion the determination of the centroid of a circular arc, which would give a most excellent example of the 'Method'.

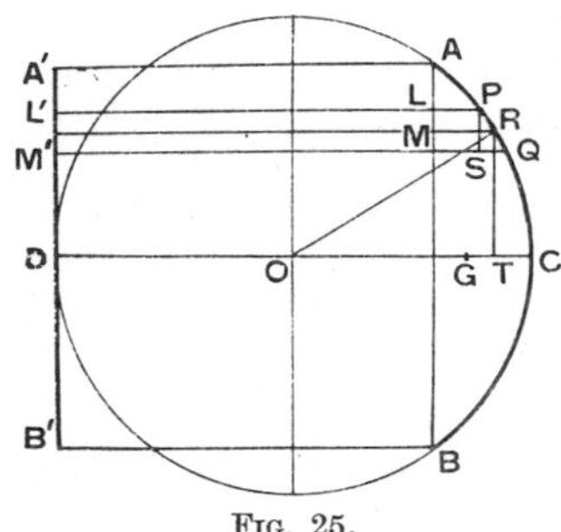

Fig. 25.

Thus, in Fig. 25, if the arc AB is divided into a large number of parts, so that a part, PQ, can be considered as a straight line having its centroid at its middle point R, then, on account of the similarity of the triangles PQS, ORT we have $PQ : OR = PS : OT$; and, therefore, the arc PQ, where it is, will balance the chord LM, transferred to $L'M'$. Hence, the arc AB, where it is, will balance the chord AB, transferred to $A'B'$; and, if G is the centroid of the arc, we have

$$(\text{arc}).OG = (\text{chord}).(\text{radius}).$$

A neat modern version can be given by 'taking moments' about the diameter perpendicular to OC.

ARISTOTLE ON THE HEART

By Arthur Platt

1. *Introduction.* It appears to me that there is room for a fresh examination of the Aristotelian description of the heart, that some of the problems and explanations of those problems previously offered may be set forth more coherently and clearly than hitherto, that a little common sense is all that is needed to clear away some of the difficulties, and, finally, that some portions of Aristotle's account may and should be interpreted in quite a new manner.[1]

In extracting his meaning it is often necessary to go behind the fragmentary and obscure wording of his statements and see what was really in his mind. The treatises of Aristotle are often of the nature of note-books for lectures; he puts down sentences intelligible to himself which he can amplify and make clear, but which by themselves are bewildering fragments. Besides this they have suffered terribly in the process of transmission to us, and are full of grievous blunders committed by scribes; whole passages have often fallen out and we can only guess what was in them; other bits have been added by people too ignorant to avoid supplying nonsense for sense. In fact when we come across a really thorny passage we can only deal with it if we have undergone a long and special training in Aristotelian scholarship which is an art by itself. And then most Aristotelian students, like the present writer, can lay no claim to be called anatomists. But perhaps it is an advantage in dealing with Aristotle not to know too much modern anatomy; one is more nearly in the position of an author who knew little enough himself from a modern standpoint.[2]

[1] I cannot help thinking that Professor D'Arcy Thompson despaired too soon in attacking this question in his translation of the *Historia Animalium* (iii. 3). I am truly sorry to find myself so often at variance with his notes upon the passage.

[2] Some years ago I submitted most of this paper to Sir George Thane; he disagreed with my actual novelties, and I put it aside and thought no more of it. Lately I disinterred it and on looking at it again was audacious enough to think

2. *Aristotle's Earlier Statement.* In the third chapter of *De Somno* Aristotle tells us that the finer and purer blood goes from the heart to the head, the thicker and more turbid to the lower part of the body. He tells us that the heart has three chambers, that the right is connected with the *vena cava*, the left with the *aorta*, while the middle chamber is 'common to' the other two, i.e. communicates with both, and separation of the blood into finer and thicker is effected in this middle chamber. But he adds that details must be added later, for 'to distinguish about these things is more fitting for other discussions'.

When we look at the 'other discussions' we find a totally different theory: it is much more elaborate and shows that its author had investigated the heart carefully. I think, therefore, we must assume that the statement in *De Somno* is the earlier, and that Aristotle completely changed his views on the question later on, for it is not credible that he should have first developed so minutely the account given in the *Historia Animalium* and have then contradicted it by the short and cursory account of the *De Somno* with the added observation that you must look elsewhere for details, knowing all the time that the details were quite different. At the same time it is not a necessary conclusion that the *De Somno* as a whole is earlier than the *Historia Animalium* and the *De Partibus* as wholes. It is more probable that Aristotle kept on working at all these treatises alongside of one another, and it is not possible to arrange the completed results in order of date. Moreover, when he had changed his views on any particular point, he did not always cancel the discarded earlier statement. Hence there are many contradictions in his writings.

3. *Aristotle's Later Statement.* In the *Historia Animalium*[1] he says: 'All hearts contain chambers, but in very small animals even the largest is barely visible, in animals of intermediate size the second also may be seen, in the biggest all three. . . . The largest chamber is on the right and highest up, the smallest on

that I was in the right in my conception of what Aristotle meant. I showed it therefore to Professor Elliot Smith and Dr. Singer, who encouraged me to publish it. Sir George's criticism has caused me to improve my nomenclature in several places, and I am grateful to him for that and for his patience and kindness in discussing the whole question with me. I know that this note will look sufficiently ridiculous, but it is only in interpreting desperately difficult Greek that I venture to dispute his verdict.

[1] iii. 3, 9, 10.

the left, that which is intermediate in size is between the two; both are much smaller than the largest.' He also says that the aorta springs from the middle chamber, a very important difference from his earlier statement which we shall presently discuss. And he says not a word about the middle chamber communicating with the other two, nor about the blood being differentiated in it. Instead of that he substitutes a new theory, that all three chambers communicate with the lungs. I shall presently translate and comment on the whole passage, but these are the most important points for our present purpose.

At the same time in the *De Partibus Animalium*,[1] arguing that large animals have more cavities than small ones, he says: 'And it is still better that there should be three chambers, in order that there may be one common principle or origin; now the middle and odd is a principle or origin.' In the same chapter Aristotle definitely contradicts the earlier statement about the blood being differentiated in the central chamber; it is better, he now says, that it should be kept distinct in the right and left chambers.

4. *Galen's Explanation.* According to Galen,[2] Aristotle's 'third cavity is that at the broad end of the heart, being part of the right chamber, not a distinct third chamber, and there runs from it a vein (φλέψ) into the lung, whereof the tunic is the same as that of the artery'. That is to say that Aristotle, in Galen's opinion, made the mistake of supposing the right ventricle to be two distinct chambers instead of one, though it is truly hard to see how he could have fallen into so gross an error. Aristotle's 'third cavity', then, in Galen's view, is the upper part of the right ventricle, his middle cavity is the lower part thereof, and the remaining one is the left ventricle.

It is palpable that this explanation does not suit the later statement of the *Historia Animalium*, but if we apply it to the earlier statement of the *De Somno* it is easy to see why Galen proposed it. If, as Aristotle asserted in *De Somno*, the right and left chambers were attached to vena cava and aorta respectively while another chamber lay between them, then what Galen calls the 'third cavity' could hardly be anything else than what Galen supposes, however gross the error imputed to Aristotle may be. The mention of the 'vein whereof the tunic

[1] iii. 4, 24.

[2] *De venarum arteriarumque dissectione*, cap. ix. Kühn, vol. ii, p. 817.

is the same as that of the artery' (φλὲψ ἧς ὁ χιτὼν ὁ αὐτός ἐστι τῷ τῆς ἀρτηρίας) leaves surely no doubt about Galen's meaning the upper part of the right ventricle, and in this explanation he omits the auricles altogether. The 'vein' in question is of course the pulmonary artery, which was regarded as a vein by the early anatomists (*vena arterialis*) because it is derived from the right side of the heart and appears from its position to belong to the venous system.

But it altogether passes comprehension to understand why Galen should ignore the vastly superior theory of the *Historia Animalium.* If, however, you suppose that Galen really thought he was explaining *that,* you must admit that he made a complete failure of it, and that it is an extraordinary accident that his attempt to explain one statement accounts for another. At the same time Galen is not entirely to be trusted in his comments upon his predecessors; he sometimes refused to look facts in the face when they seem derogatory to any ancient whom he respects. Thus he endeavours to save Plato's credit on the absurd theory (which Plato took from the Hippocratic school) that what we drink goes into the lungs, and his defence of this is disingenuous. So it is of course possible that he was talking somewhat at random about the third Aristotelian chamber.

5. *Vesalius's Explanation.* Vesalius thought that Aristotle was misled by 'the membrane of the left ventricle', i.e. the flap of the mitral valve, into thinking that this ventricle consists of two separate chambers.[1] He thus agrees with Galen in accusing Aristotle of splitting up one chamber into two, but he differs from Galen in saying that he splits the left ventricle instead of the right and he gives a plausible reason to account for the mistake. But above all he is trying to explain the later and developed account of the *Historia,* and escapes the strange error into which Galen fell when he treated the earlier account as if it were a fair representation of Aristotle's opinion. According to Vesalius, then, what is now called the aortic vestibule is Aristotle's middle chamber, and the rest of the ventricle is the left chamber. Therefore this theory will account so far for the later statement. But it certainly is hard to see how the flap of the valve could have so misled Aristotle, and there is a very great difficulty in the way besides. For Aristotle distinctly says that the *left* chamber is the

[1] *De fabrica corporis humani,* Lib. VI, cap. 12.

smallest, and on the Vesalian explanation the *middle* is much the smaller of the two. Consequently this theory must be rejected; it does not suit either the earlier or the later statement.

6. *The Auricular Theory of Huxley and Ogle.* To explain the later statement we are thrown back upon the theory advanced independently by Huxley[1] and by Ogle.[2] The left chamber, according to them, is the left auricle, the middle chamber is the left ventricle. This accounts for the details of the later statement perfectly so far as we are concerned with them for the present, and I hope to show that it will account for those we shall meet with later on. If the left chamber is so small a thing as the auricle, that is why Aristotle could not make it out in 'animals of intermediate size'. If the left ventricle is the middle chamber, that is why Aristotle says the aorta issues from the middle chamber. And should it be objected that he says his middle chamber is much smaller than the right one, the answer to this is that it really would appear to be so in an animal which had been suffocated. And the right auricle is entirely ignored by him because in a suffocated animal it would appear to be part of the vena cava.

7. *The Text of the later Statement reconsidered.* We may now turn again to the passage in the *Historia Animalium.*[3] 'The investigation is difficult, as has been said; it is only possible to acquire adequate information, if any one takes an interest in such things, in the case of animals suffocated after previous starvation.'

Aristotle made his dissections therefore on the bodies of strangled animals. We know what he would find. The right auricle and ventricle and the pulmonary artery would be gorged with dark blood; the left auricle and ventricle, the pulmonary veins, and aorta, would be empty. Therefore the right part would be easily made out; the left would have collapsed and be difficult to make out.

'The following is the nature of the blood-vessels. There are two vessels in the thorax, in the neighbourhood of the spine but lying within this[4] [i.e. in front of it], the greater in front and the lesser behind the greater. And the greater is more on the right, the less on the left; this latter some call aorta or suspender because the sinewy part of it has been observed even in the dead.'

[1] T. H. Huxley, *Nature*, November 6, 1869, republished in *Science and Nature*, p. 180.

[2] W. Ogle, *Aristotle on the parts of animals*, London, 1882, pp. 197–9.

[3] iii. 3. 6–14.

[4] I read ῥάχιν μὲν ἐντὸς δὲ κείμεναι ταύτης.

The greater vessel is of course the vena cava, both superior and inferior, the two (with the intervening right auricle) being treated as one by Aristotle as they were later by Vesalius in his *Tabulae sex.* The aorta is so called because it is sinewy like a strap or ' suspender ' of a wallet or the like,[1] and the heart seems to hang from it. This sinewy part has been observed ' even in the dead ', that is to say, even when the vessel has been exposed in a man killed by a sword-stroke or something of the kind, without the previous starvation and the suffocation which Aristotle considers rightly to make investigation easier.

' These vessels ', Aristotle continues, ' have their origin in the heart. For whereas they preserve their character unimpaired and remain vessels when they pass through the other organs, wherever they actually do so run, the heart on the contrary is as it were a part of them. This is more especially so with the vena cava, but applies also to the aorta, both vessels taking both an upward and downward course and the heart being in the middle of them.'

Aristotle had, of course, no idea that the arteries and veins anastomosed at their terminations. When he says that they ' pass through the other organs ' he has been misled into thinking that they do so, because they would be hidden by fat, &c., and would look as if they did. He adds, ' Where they actually do so run ', or ' Where they happen to do so ', which shows that he knows it to be exceptional. The main point here is that the heart is like an enormous dilation of the vessels and of the same sort of substance, whereas the other organs are bodies through which a vessel passes (if at all) as a foreigner through a strange land. His medical predecessors had said the vessels had their origin in the head !

' All hearts ', he proceeds, ' contain chambers, but in very small animals even the largest is barely visible, in animals of

[1] This was correctly explained by Vesalius : ' Magna igitur arteria (quam Aristoteles eo quod nervosam ipsius partem vel in mortuis conspici posse diceret, instar vaginae fortassis Macedonibus ἀορτῆς nuncupatae, ἀορτήν appellavit) . . .' *Corp. Hum. Fab.*, Lib. III, cap. 12. Only he is wrong in supposing that Aristotle invented the name. *Aorta* has no connexion whatever with *air* ; Aristotle had never even heard of the notion that arteries contain air, which notion was hatched in the next century by Erasistratus.

intermediate size the second also may be seen, in the biggest all three. . . . The apex of the heart points forward, as already observed, and the largest chamber is on the right and highest up, the smallest on the left, that which is intermediate in size is between the two; both are much smaller than the largest.

'All these are connected with the lung by perforations, but this is not plain except in the case of one of the chambers, because of the smallness of the passages. [The connexion of the largest chamber with the lung is this]: the vena cava is attached to the largest chamber, that which is uppermost and on the right, then through the middle of the cavity extends again a blood-vessel [to the lung], as if the chamber were a part of the vessel in which part the blood forms a lake. The aorta [springs] from the middle chamber [and is also connected with the lung] only not in the same way but communicates [with it] by a narrower air-passage (syrinx).'

This last paragraph has been, I believe, hitherto entirely misunderstood, owing to the exceeding obscurity of connexion which is unluckily characteristic of the writer; I also suspect that some words have been lost. Those in brackets have been here added to make the connexion plain. The last sentence without that addition is utterly unconstruable, or at any rate utter nonsense.

Aristotle held that the lung existed for the purpose of bringing the air and the blood into touch with one another. He also held that blood was formed in the heart; I cannot quote any passage to show positively that he thought this was done in *all* his three chambers independently of one another, but there can be no reasonable doubt, I suppose, that he did think so. He then is anxious to insist that the lung is connected with all three chambers and so with the blood all over the body. Thus he lays stress first on the connexion of vena cava, heart, pulmonary artery and lung, and that much is to me plain. Next comes the left ventricle, from which springs the aorta, and the aorta too is connected with the lung by a narrower *syrinx*. This word in Aristotle is always used of air-passages; Hippocrates uses it for a fistula, but I do not find that it is ever used for a blood-vessel. In this particular sentence the idea that it means a blood-passage of some kind has led to mistranslation and confusion; it was the consideration of this word more than anything else which led me to my view of the meaning of the whole sentence, and no explanation can be considered reasonable which takes *syrinx* here to mean a blood-vessel.

But if the *syrinx* is not a blood-vessel and if it connects the aorta in Aristotle's opinion with the lung, what is it ? Aristotle was anxious to show that the aorta connected with the lung. Now there *is* a passage which joins the aorta not far from its starting-point, the ductus arteriosus; in the foetus this is a blood-vessel connected with the pulmonary artery, but in the adult it is closed up, and forms the ligamentum arteriosum. An early anatomist exploring the thorax without the light of previous experience or older records would not be likely to make out the real connexion, but he would be not unlikely to come upon this ligament in attempting to disengage the aorta. Suppose that Aristotle did this: he was looking for his connexion with the lung, he found this ligament broken off and jumped to the conclusion that it was the connexion he wanted. This is no unworthy accusation against him; he did sometimes find what he looked for when it was not there, as many have done since. But there was something peculiar about the passage, and not being obviously a blood-vessel he called it a *syrinx*; had it been more manifestly a blood-vessel he would have called it *phleps*. That it is not a hollow tube is true, but this is no serious objection, for he also calls the optic nerves 'passages' (πόροι), doubtless thinking that incapacity to see a real 'passage' in such things with the naked eye did not prove their non-existence.

He cannot have meant one of the bronchial arteries, because they would have been called by him φλέβες.

The word for word translation of the passage runs thus: 'But the aorta from the middle only not so but by a much narrower syrinx communicates.'[1] Consider the words *not so*, *syrinx*, *communicate*. If we suppose the meaning to be that the aorta 'communicates' with the heart, why should he use such a verb at all ? Elsewhere he speaks of the aorta as 'having its origin in the heart' or 'running from the heart' or some such phrase. As a matter of Greek, *κατὰ σύριγγα κοινωνεῖ* cannot mean 'joins (the heart) by an aperture', which it is commonly supposed to mean; the syrinx cannot possibly be identified with the aorta itself or the aperture thereof; it must mean some other 'pipe', and that an air-pipe. The aorta therefore springs from

[1] *Historia Animalium*, iii. 3, 513^{b} 6.

the middle chamber and communicates with something else by some air-passage which is distinct from the aorta. Aristotle is plainly thinking, as it seems to me, of the connexion of heart and lung. And ‘ not so ’ is intelligible in my translation, but not in any other.[1]

Thus the two last sentences in the second paragraph on p. 527 contrast the connexion of vena cava and aorta with the lungs; the former communicates by the pulmonary artery which is a prolongation of the vena cava *through* the right ventricle, the latter not by sending off a similar branch through the left ventricle (*not so*) but by the ligament which is *much narrower* than the pulmonary artery.

There still remains one sentence of this paragraph, in which I accept the reading and interpretation of the great majority of commentators: ‘ But whereas the vena cava runs *through* the heart, the aorta runs *away from* the heart.’ [2] This is an insistence again upon the same thing and needs no further comment. But we should expect now to hear how the third chamber, the left auricle, is connected with the lungs. Of course it would be by the pulmonary veins. But instead of going on to this, Aristotle proceeds to point out other differences between vena cava and aorta, and to explain the distribution of the venous and arterial systems over the body. Though he more than once asserts that all his three chambers are connected with the lungs he nowhere explains all three connexions together. But a passage elsewhere in the *Historia Animalium* [3] shows us clearly what he meant. ‘ The passages from the heart (blood-vessels) lie above (the air-passages in the lungs); there is no common passage (i.e. no opening by which air and blood can mix), but the vessels receive the air

[1] The only instance of σῦριγξ in the sense of a blood-vessel of any kind is, I believe, in a line of Apollonius Rhodius (iv. 1647), wherein a fabulous bronze monster is said to have had a σῦριγξ αἱματόεσσα. It would take a bold man to argue from a solitary poetic use against the established use in the medical and scientific writers; an ornithologist cannot ascribe a ‘ fiery heart ’ to a bird because a poet chooses to do so.

Empedocles spoke of some sort of σύριγγες λίφαιμοι, but this very epithet means ‘ bloodless ’, and nobody can tell what Empedocles meant. It cannot be denied that Aristotle supposed him to refer to φλέβες, but neither can it be denied that Aristotle must have misinterpreted him; he writes a horrible poetical jargon which is often as obscure as *Sordello*.

[2] *Historia Animalium*, iii. 3, 513^{b} 7.

[3] *Historia Animalium*, i. 17, 496^{a} 30–3.

on account of their apposition (i.e. through the walls of the air-passages and vessels) and transmit it to the heart. For one of the passages (pulmonary artery) carries it into the right cavity, *and the other* (*pulmonary veins*) *into the left.*' This second passage must be the pulmonary veins, not the aorta, because according to the *Historia Animalium* the aorta belongs to the middle cavity. And this *left* cavity here, which thus is connected with the lung by a passage corresponding to the pulmonary artery can surely be nothing whatever except the left auricle. This indeed seems to me the strongest proof to be had of the correctness of the auricular theory.

In the chaotic state in which Aristotle's works have come down to us, for whatever reason, it is not to be wondered at that this explanation should be here omitted. There are plenty of other like omissions in him. The passage continues:

§ 13. 'And the great vessel is membranous and skin-like, but the aorta is narrower than it and very sinewy; as it stretches away further towards the head and the lower parts it becomes altogether narrow and sinewy.

§ 14. 'There runs first upward from the heart a part of the great vessel towards the lung and the junction of the trachea (reading with Karsch ἀρτηρίας for ἀορτῆς), a great and undivided vessel (i.e. the main channel of it does not bifurcate into equal halves but as a whole it may be called undivided), but still there do split off from it two branches, the one to the lung (i.e. pulmonary artery, for as we have seen this is regarded as a prolongation of the vena cava through the heart), and the other to the spine and the last cervical vertebra (i.e. the azygous vein).

§ 15. 'As the lung is double, so the vessel which runs to it first bifurcates and then runs along each air-channel and each perforation.'[1]

When we look at §§ 14 and 15 together we can hardly doubt that Ogle is right in his interpretation of these two branches as pulmonary artery and azygous vein. But the way in which Aristotle puts his statement is very confused. For he first says that the vena cava runs up from the heart *towards* the lung, and then says it begins by sending off a branch to the lung. Now the branch which it sends off to the lung, if it is the artery, goes off before the main vena cava starts upwards from the heart on its

[1] *Historia Animalium*, iii. 3, 513^b 7–9.

principal course, and Aristotle knew this quite well, for he has just explained it all. We must fall back upon the fact that his works are of the nature of an 'open lecture-book', in which are jotted down additional notes, often leaving earlier views not entirely obliterated but showing through a later stratum imposed upon them. We have here a jumble of this kind, but it is possible to see what he really meant.

With regard to the words 'towards the lung and the junction of the trachea', I feel sure that this is the right reading. The texts say 'the junction of the aorta',[1] but this means nothing. Junction with what? How can the vena cava superior, the 'undivided vessel' which is *not* the pulmonary artery, have any connexion with the aorta? Substitute therefore *arteria* for *aorta*, (and *arteria* in Aristotle means the trachea). The junction of this, taken in connexion with the context, obviously means the place where the trachea splits up into the lungs. He is thinking of the course of the vena cava superior as an undivided vessel up to the point where it splits into the innominate veins, and this point is in the region where is what he calls 'the junction of the trachea'.

There is still one passage on which I have something to say. After going on to describe the vessels of the lungs and the azygous vein, Aristotle continues in § 17: 'This is the way in which these branches are split off from the great vessel. But above these ἀπὸ τῆς ἐκ τῆς καρδίας τεταμένης the whole vessel again divides into two regions', i.e. the vein does not merely give off a vein this time but divides as a whole into two equal parts running to two distinct regions. The words I have left in the Greek are unconstruable. Read instead ἀπὸ τῆς (or ἐκ τῆς) καρδίας τεταμένη. Then the meaning is 'the whole vessel as it lies stretched from the heart', and this clause is added to show that, after the digression about pulmonary artery and azygous vein, we are going back to the vena cava itself, which as already said runs upward from the heart.

[1] It was this unhappy corruption of the text that drove Ogle into supposing that Aristotle was talking in this place of the ductus arteriosus, an impossible notion, because (1) he examined adult animals as he has told us himself; (2) it is the 'great undivided vessel' of the vena cava which is said here to run up to the 'junction' and that 'undivided vessel' is carefully distinguished here from the pulmonary artery.

With the whole description it is interesting to compare the following account from a work attributed to Hippocrates, but of uncertain date and authorship. I have emended the Greek, which is unintelligible in the standard editions.[1] 'The original vessel which, running along the spine and through the region behind the throat and the bronchus is fixed into the heart, sends off from itself a good-sized vessel with many mouths at its point of leaving the heart (!) and thence makes a *syrinx* through the lung into the mouth, which is called *arteria* (i.e. the trachea).' Such is angiology before Aristotle! Like other Hippocratics the author starts his blood-vessel from the head, whence it comes down to the heart and sends off the pulmonary artery: here again that artery is regarded as a branch of the vena cava. Then this runs through the lung and issues from it as the trachea, which is thus itself a prolongation of the vena cava. But again we find that the word *syrinx* turns up as soon as ever the author gets plainly into touch with an air-passage.

[1] *De ossium natura*, cap. 5; Littré, ix. 171; Kühn, i. 514.

APPENDIX TO ARTICLE X

[*For reference in text see pages 410 and 411*]

Translation from the Dutch by Miss M. Kuenen of a passage in the *Œuvres complètes de Christiaan Huygens*, vol. xiii, fasc. 1, p. 591, The Hague, 1916.

'Copy of the copy exhibited by Adrian Van der Wal, 1682, whence it appears that Jacob Metius was not the inventor of the Telescope but rather Lippershey of Middelburg.

To their Honours, the States-General of the United Netherlands.

Jacob Adriaenssoon, son of Mr. Adriaen Anthonissoon, ex-burgomaster of the Town of Alkmaer, deposes that for a period of two years, during the time remaining over from his trade, he has been occupied in searching out such hidden arts as had been attained by predecessors in the use and application of glass. He has come to the conclusion, by experiments with a certain instrument which he was using for another purpose, that the sight could be extended to such a degree that things invisible or indistinct by reason of distance, could be clearly seen by its aid. When he had observed this he experimented to improve it further, and he progressed so far with his instrument that a thing can be seen as far off and recognized as clearly as with the instrument recently shown your Honours by a citizen and optician of Middelburg, according to the judgement of His Excellence [*i. e. Prince Maurice*] and of others who have experimented with the respective instruments one against the other.

Although the Suppliant's instrument is made of very poor material and for one experiment only, yet he does not doubt that it would greatly improve with better materials. Moreover, he also expects and hopes so to improve the invention itself so that more useful results may be expected. But he fears that meanwhile some one may anticipate him by copying or imitating his instruments, building on the foundations which he (by God's grace) has laid with his own skill, industry, and thought and thus deprive him of the fruits that he may justly expect therefrom. He therefore humbly begs your Honours to grant him a patent, whereby all who have not previously had or used this invention be prohibited from copying it entirely or in part, or from buying or selling such instruments made by unlicensed persons without the special consent of the Suppliant, at risk of confiscation of such instruments plus the sum of 100 guilders for each.

Finally, for twenty years, or as otherwise agreed upon, to render the Suppliant in respect of the use and service of the instrument for the Commonwealth, such honours as your Honours, with your usual goodwill and discretion, shall decide.

In margin was written that the Suppliant was admonished to investigate further to bring his invention to the greatest perfection, and suitable steps shall be taken about his request for a patent.

Enacted Aersens, October 17th, 1608.

[*Aersens was Secretary to the States.*]

After comparison, this and the original request signed by his own hand, were found to coincide word for word.

In Alkmaer the 8th Nov. 1677. Quod attestatur Joh. H. Metius 1677.'

[*This Joh. H. Metius of 1677 is, naturally, not the same as the Jacob Metius of 1608 whose invention is in question.*]

INDEX

PRINTED IN ENGLAND
AT THE OXFORD UNIVERSITY PRESS

HISTORY, PHILOSOPHY AND SOCIOLOGY OF SCIENCE

Classics, Staples and Precursors

An Arno Press Collection

Aliotta, [Antonio]. **The Idealistic Reaction Against Science.** 1914

Arago, [Dominique François Jean]. **Historical Eloge of James Watt.** 1839

Bavink, Bernhard. **The Natural Sciences.** 1932

Benjamin, Park. **A History of Electricity.** 1898

Bennett, Jesse Lee. **The Diffusion of Science.** 1942

[Bronfenbrenner], Ornstein, Martha. **The Role of Scientific Societies in the Seventeenth Century.** 1928

Bush, Vannevar. **Endless Horizons.** 1946

Campanella, Thomas. **The Defense of Galileo.** 1937

Carmichael, R. D. **The Logic of Discovery.** 1930

Caullery, Maurice. **French Science and its Principal Discoveries Since the Seventeenth Century.** [1934]

Caullery, Maurice. **Universities and Scientific Life in the United States.** 1922

Debates on the Decline of Science. 1975

de Beer, G. R. **Sir Hans Sloane and the British Museum.** 1953

Dissertations on the Progress of Knowledge. [1824]. 2 vols. in one

Euler, [Leonard]. **Letters of Euler.** 1833. 2 vols. in one

Flint, Robert. **Philosophy as Scientia Scientiarum and a History of Classifications of the Sciences.** 1904

Forke, Alfred. **The World-Conception of the Chinese.** 1925

Frank, Philipp. **Modern Science and its Philosophy.** 1949

The Freedom of Science. 1975

George, William H. **The Scientist in Action.** 1936

Goodfield, G. J. **The Growth of Scientific Physiology.** 1960

Graves, Robert Perceval. **Life of Sir William Rowan Hamilton.** 3 vols. 1882

Haldane, J. B. S. **Science and Everyday Life.** 1940

Hall, Daniel, et al. **The Frustration of Science.** 1935

Halley, Edmond. **Correspondence and Papers of Edmond Halley.** 1932

Jones, Bence. **The Royal Institution.** 1871

Kaplan, Norman. **Science and Society.** 1965

Levy, H. **The Universe of Science.** 1933

Marchant, James. **Alfred Russel Wallace.** 1916

McKie, Douglas and Niels H. de V. Heathcote. **The Discovery of Specific and Latent Heats.** 1935

Montagu, M. F. Ashley. **Studies and Essays in the History of Science and Learning.** [1944]

Morgan, John. **A Discourse Upon the Institution of Medical Schools in America.** 1765

Mottelay, Paul Fleury. **Bibliographical History of Electricity and Magnetism Chronologically Arranged.** 1922

Muir, M. M. Pattison. **A History of Chemical Theories and Laws.** 1907

National Council of American-Soviet Friendship. **Science in Soviet Russia: Papers Presented at Congress of American-Soviet Friendship.** 1944

Needham, Joseph. **A History of Embryology.** 1959

Needham, Joseph and Walter Pagel. **Background to Modern Science.** 1940

Osborn, Henry Fairfield. **From the Greeks to Darwin.** 1929

Partington, J[ames] R[iddick]. **Origins and Development of Applied Chemistry.** 1935

Polanyi, M[ichael]. **The Contempt of Freedom.** 1940

Priestley, Joseph. **Disquisitions Relating to Matter and Spirit.** 1777

Ray, John. **The Correspondence of John Ray.** 1848

Richet, Charles. **The Natural History of a Savant.** 1927

Schuster, Arthur. **The Progress of Physics During 33 Years (1875-1908).** 1911

Science, Internationalism and War. 1975

Selye, Hans. **From Dream to Discovery: On Being a Scientist.** 1964

Singer, Charles. **Studies in the History and Method of Science.** 1917/1921. 2 vols. in one

Smith, Edward. **The Life of Sir Joseph Banks.** 1911

Snow, A. J. **Matter and Gravity in Newton's Physical Philosophy.** 1926

Somerville, Mary. **On the Connexion of the Physical Sciences.** 1846

Thomson, J. J. **Recollections and Reflections.** 1936

Thomson, Thomas. **The History of Chemistry.** 1830/31

Underwood, E. Ashworth. **Science, Medicine and History.** 2 vols. 1953

Visher, Stephen Sargent. **Scientists Starred 1903-1943 in American Men of Science.** 1947

Von Humboldt, Alexander. **Views of Nature: Or Contemplations on the Sublime Phenomena of Creation.** 1850

Von Meyer, Ernst. **A History of Chemistry from Earliest Times to the Present Day.** 1891

Walker, Helen M. **Studies in the History of Statistical Method.** 1929

Watson, David Lindsay. **Scientists Are Human.** 1938

Weld, Charles Richard. **A History of the Royal Society.** 1848. 2 vols. in one

Wilson, George. **The Life of the Honorable Henry Cavendish.** 1851